W9-AFB-937

TREES, SHRUBS, AND WOODY VINES
OF NORTHERN FLORIDA
AND ADJACENT GEORGIA AND ALABAMA

Frontispiece. Sabal palmetto, the State Tree of Florida. (Illustration by R. P. Elliott, reproduced with the artist's permission.)

TREES, SHRUBS, AND WOODY VINES OF NORTHERN FLORIDA AND ADJACENT GEORGIA AND ALABAMA

by

Robert K. Godfrey

WITH THE MAJORITY OF ILLUSTRATIONS BY

Melanie Darst

CARL A. RUDISILL LIBRARY
LENOIR-RHYNE COLLEGE

The University of Georgia Press

Athens and London

TO
ANGUS KEMP GHOLSON, JR.
"Not to pay a debt but to acknowledge it"

Q K
154
. G 63
1988
1 5 5/0 4

May 1992

© 1988 by the University of Georgia Press
Athens, Georgia 30602

Illustrations by Melanie Darst © 1988 by Robert K. Godfrey

All rights reserved

Set in 9/11 point Aster

The paper in this book meets the guidelines for
permanence and durability of the Committee on
Production Guidelines for Book Longevity of the
Council on Library Resources.

Printed in the United States of America
92 91   5   4   3   2

Library of Congress Cataloging in Publication Data

Godfrey, Robert K.
Trees, shrubs, and woody vines of northern
Florida and adjacent Georgia and Alabama.

Bibliography: p.
Includes indexes.
1. Woody plants—Florida—Identification. 2.
Woody plants—Georgia—Identification. 3. Woody
plants—Alabama—Identification. I. Title.
QK154.G63 1988   582.1'5'09759   87-35840
ISBN 0-8203-1035-2 (alk. paper)

British Library Cataloging in Publication Data available

The research for this book was supported by funds
provided by Tall Timbers Research, Inc.

Contents

Preface

The woody plants, trees, shrubs, and woody vines included in this book are those known to be native to northern Florida and adjacent southern Georgia and southern Alabama, and those that have been introduced into cultivation and have since become naturalized. The geographical coverage for Florida includes all of northeastern Florida north of a line from Flagler Beach, Flagler County, on the Atlantic Coast to Horseshoe Beach on the Gulf Coast and all of the Florida Panhandle. The arbitrary boundary approximates the northern boundary for the geographic coverage of *Trees of Central Florida* by Lakela and Wunderlin (1980). The geographical coverage for Georgia and Alabama is intentionally indefinite, because I did not know how to designate an arbitrary northern boundary line that would be meaningful.

In dealing with identification and description of woody plants (apart from woody vines), the question of the distinction between shrubs and trees inevitably arises. The usual distinction involves form and size: a shrub has more than one stem from the base, attaining only low stature, and not having a single well-defined crown; a tree, on the other hand, has a single main stem, a single, more or less definite crown, and a height of at least 3–5 m. Given the variability of form and stature occurring in woody plants, this distinction is very arbitrary. Indeed it is scarcely useful in a great number of cases, especially since the beginning student, at least, will often be looking at a given individual plant at a single moment in time, also perhaps for the first time. I do use the terms *tree* and *shrub*, and my sense of the difference is one of "Gestalt," that is, the sense of the general, integrated potential of form and patterns that I perceive from long experience. That, I think, is how the terms came into being in the first place; the definitions came from those who felt a need for exactness ("scientificness") and lost sight of the art in it.

The genesis of this floristic work entailed much reference to and dependence upon more comprehensive manuals and floras as well as monographs and revisions of various groups and local floristic treatments. My own taxonomic concepts and judgments inevitably derived in substantial part from a blend of those of predecessors and contemporaries, and to them I cannot repay the debt owed apart from that implied by the citation of their works in the list of references at the end of the book. Others have given aid in a great variety of more tangible ways.

The Tall Timbers Research Station, Tallahassee, Florida, has sponsored research for this book by awarding me a Henry L. Beadel Fellowship grant and by providing staff and financial support, which made possible most of the line drawings by Melanie Darst. I am appreciative of the stimulating interest of Dr. E. V. Komarek, Executive Secretary, Tall Timbers Research, Inc., and that of the former and present directors of the station, D. Bruce Means and J. Larry Landers, respectively. Cavell L. Kyser adeptly and graciously placed the manuscript on a word processor; to her I am very greatly indebted. She endured a great deal of travail, always pleasantly and with good humor, as bits and pieces of manuscript were handed to her, and as many corrections and changes came her way.

The Department of Biological Science at Florida State University kindly provided work space in its herbarium as well as full use of herbarium facilities. The herbarium curator, Loran C. Anderson, has always been hospitable and helpful during the twelve years of my "tenure" as professor emeritus.

To Melanie Darst I convey my respect and admiration for her artistic productions, and my warm personal regard for her as well. Her work was, for her, much more a labor of love than a source of income. R. M. Cook, Jr. (Wilderness Graphics, Tallahassee) contributed the endpaper drawing, and I thank him very much for it.

To Angus K. Gholson, Jr., I express my sincere appreciation for his enthusiastic support of this project, for his companionship and help on the field trips that we made almost weekly, sometimes more than once weekly, during the course of this research, and for his perceptiveness and insights regarding many taxa. He, more than anyone, helped me procure materials for illustration, in particular by informing me when and where specimens of a given taxon were at a stage in their life history when they exhibited characteristics that could be illustrated. As parts of the manuscript were completed in draft form, Angus very carefully scrutinized them, pointed out typographical and other errors, and offered critical suggestions. I am much indebted to him. To him and to Mrs. Gholson I tender my thanks for the continual hospitality extended by them during the course of my work.

W. Wilson Baker and Steve W. Leonard, who carefully and systematically explored ravines and steepheads to the east of the Apalachicola River in Florida, discovered several taxa of woody plants not known by me to occur there, as well as several that had not been seen there for about a hundred years. In addition, Angus Gholson and I sometimes explored in their company—a pleasurable experience indeed.

Suzanne Cooper rendered inestimable help by reading many of the first drafts of manuscript (for families as I finished them). I very much appreciate her labors in this tedious task, and, moreover, her interest in the project and enthusiastic encouragement for it.

John Popenoe, Director, Fairchild Tropical Garden, very kindly permitted me to use several delineations of superior quality by Priscilla Fawcett that are owned by the garden. A few illustrations by Vivian Fraser are from Correll and Correll (1972); it is a privilege to be able to use them. The illustration of *Toxicodendron toxicarium*, by Regina O. Hughes, is reproduced from Reed (1970) and is gratefully acknowledged. I am also grateful to the University of Florida Presses for permission to reproduce a few illustrations from Kurz and Godfrey (1962).

I am indebted to several persons who provided assistance, each with a particular plant group, some of them by permitting me to see parts of their own manuscripts: George W. Argus (*Salix*); James R. Burkhalter (*Sesbania*); George P. Johnson (*Castanea*); Robert W. Simons (*Bumelia*); Walter S. Judd (Ericaceae: *Lyonia* and *Rhododendron*). Michael O. Moore's contribution of the treatment for the genus *Vitis* is especially acknowledged, and for it I am very much indebted to him.

Others who rendered assistance in a variety of ways and to whom I am grateful are Wilbur S. Duncan, David W. Hall, Robert Kral, Kent D. Perkins, Daniel B. Ward, and David H. Webb.

I visited the herbarium at the University of Florida, Gainesville, on numerous occasions to check specimens, distributions of taxa, and the like. The members of the herbarium staff were always cordial and helpful, and I remain in their debt.

I wish to thank all my predecessors upon whose published work I have relied. I should hope that this work will add something of value to the work of those who have gone before. I also acknowledge the value to me of the various kinds of aid provided by my contemporaries, and I am greatly in their debt for support and encouragement. However, I must accept responsibility for the contents of this book, and any shortcomings that may be found are solely mine.

Introduction

This manual is intended, via keys, descriptions, and illustrations, to aid in identification of the kinds of woody plants, native and naturalized, occurring in the geographic area of coverage described in the Preface.

The process of learning to know plants is cumulative. Initially one may, for example, learn the identities of two or three oaks in one's locality. Then one may be stimulated to learn several more, then others. All of this requires real effort. With each plant learned, however, the next becomes easier.

It is hoped that as a person's knowledge increases, he will gain a fuller and keener appreciation for the fundamental importance of plants in the world in which we live. The human animal, like all others, is totally dependent upon green plants, since they alone can convert inorganic chemicals into organic foods and in the process help to maintain certain essential atmospheric gases in proper balance.

There is a widely held but fallacious view that one cannot know a plant (or an animal) until one has learned its name. On the contrary, a good observer, curious and inquisitive and sensitive to beauty, may learn a great deal about a plant without knowing its name; but learning the name, the scientific name in particular, opens avenues for verbal communication with others and makes it possible, for example, to search for information of varied kinds from pertinent popular and scientific literature (such as the references cited at the end of this book). This inquiry, in turn, will probably lead to questions about the plant's habitat, its usual associates (the generalized community of which it is a component), and its possible uses, not just to humans but to other organisms as well.

I have sought, both in the construction of the keys and in formulating the descriptions, to minimize to some extent the use of technical jargon by substituting words or phrases more familiar to the nonspecialist. Because specialized terminology is a kind of shorthand that allows more information to be supplied in less space, I have used it here freely. Most of the technical terms are defined in the glossary that follows this introduction.

Insofar as I have found it possible to do so, I have used vegetative characteristics in the keys, especially the major set of keys, because, in most instances, reproductive structures are present on a given plant relatively briefly during a given season. Despite the effort to rely upon vegetative characteristics, there were a few places in the major set of keys where, as a last resort, reproductive characters were used. In most of the other keys, those within the taxonomic treatment, reproductive characteristics were used more liberally.

A special effort was made to provide original descriptions. When one is seeking to verify an identification made by using the keys, the descriptions will permit one to check more detailed features of a plant. On the other hand, use of the descriptions may serve to sharpen one's perceptions of, or enhance interest in, constellations of characteristics which make a given kind unique. Descriptions of genera (in cases where more than one species for the genus was treated) and descriptions of individual species were made in sufficient detail to instruct the critical user and to stimulate further interest.

Most of the drawings were made by Melanie Darst especially for this book over a period of approximately five years. They are uneven in various respects,

1

particularly in the amount of detail. The constraining factors were time and (not least) money. Most of the drawings were done from living material that I procured (in some instances with the help of other persons) across the seasons during the course of drafting the book.

Considerable effort was expended in search of places where rarities or relative rarities might occur, in determining how various kinds grew in combination to form generalized vegetational patterns in relation to environment, in scrutinizing for morphological characteristics especially useful in construction of keys or in formulation of descriptions, and in making voucher specimens. In the aggregate, these endeavors entailed a very great many miles of travel by automobile and much tramping about on foot.

I have been engaged in field work over much of the area of coverage for nearly thirty years. Collections made during this period were deposited in the Florida State University Herbarium and represent a significant proportion of the data base for this book. I have explored and collected a great deal more across northern Florida than I have in adjacent southern Georgia and southern Alabama. For presence and distribution of plants in southern Georgia, I have relied principally upon the work of Thorne (1954) and Faircloth (1971); for southern Alabama upon that of Clark (1972).

The sequence of presentation of major groups is as follows: Gymnosperms, with families in alphabetical order; then Angiosperms, first Monocotyledons and then Dicotyledons, with families of each in alphabetical order. Within families, the genera are in alphabetical order; within genera, however, sometimes the species (if more than one), are in alphabetical order and sometimes they are not, depending upon whether I felt it would be helpful to have descriptions of certain species next to or near certain others for easier comparison.

Genera are described if more than one species of a genus is included; if only one species of a genus is included, then the description is of that species (without separate generic description).

Following the description of a given species, there is a very generalized indication of the habitat or habitats in which, in our area of coverage, the species is known to occur and a general indication of where in the area it does occur. This is followed, in most instances, by a notation (enclosed in parentheses) of the overall geographic range of the species. Following that, for most of the trees, the National Champion Big Tree for the species is cited (taken from the National Register of Big Trees, American Forestry Association, Washington, D.C., 1980).

PLANT NAMES

The scientific name (binomial: generic name and specific epithet with the authority or authorities for the name usually abbreviated) is given for each taxon treated. That name is the one which, insofar as I can determine, is the correct one. Synonymous scientific names, if any, for a given taxon are in italics at the end of the description for that taxon. In most cases, the synonymous names are those used by Small (1933).

For most taxa, one or more common or vernacular names follow the scientific name. Common names (in English or any other modern language) have a very limited, often very local, usefulness. The same name may be applied to several very different kinds of plants, or as many as a half dozen names may be applied locally to a single kind; thus common names may be bewildering. Some common names are now familiar to very few lay persons because for a long time our

2

societal focus has been mainly on subjects other than natural history. In addition, there are those "uncommon" common names that are fabricated to resemble vernacular names; they are merely translations into English of the specific (adjectival) epithet, combined with an appropriate noun. Common names used in this book were mainly taken from other similar works, and their selection, I admit, was casual. In the case of woody plants, trees especially, vernacular names are perhaps used rather more consistently than for herbaceous plants.

COUNSEL TO THE USER OF THIS BOOK

Technical botanical terminology is commonly used in botanical literature for the sake of brevity and accuracy. Command of this technical terminology, especially the more commonly used terms, will add much pleasure to the use of this and other similar botanical treatments.

The user of this book will probably soon come to realize that numerous morphological characteristics used in the keys and descriptions cannot be perceived without the use of a magnifying device. Simple, small magnifiers such as a 10-power lens carried in the pocket or suspended from the neck will enable one to observe these characteristics. For the sincere and enthusiastic student, a stereoscopic microscope at his desk will make it much easier to see many described characteristics to better advantage. Individuals, it would seem, are reluctant to consider purchasing such a microscope because of its cost. A stereoscopic microscope, however, is no more expensive than the equipment that amateur photographers, for example, buy as a matter of routine.

Have at hand, both in the field and at the desk, a simple 15-cm ruler.

Carry a small notebook in which to record information easily observed but often quickly forgotten—for notes about stature of a plant; its general habit; the kind of place in which it is seen growing; variation on different individuals of the same kind growing near each other, or even on a single individual; bark characteristics; whether leaves are simple or compound; flower and fruit color, and so on.

The leaves of many kinds of trees, shrubs, and vines are remarkably uniform for the species. It is true, however, that for numerous species leaf characters vary a great deal from one plant to another, or even on different parts of the same plant. Attention is called to this in particular cases, and one must learn to expect it and to deal with it. In general, it is wise, especially for one just beginning to study trees, to make observations from leaves of higher branches, those mature enough to bear flowers and fruit. Leaves of tree saplings, of especially vigorous shoots, or of sucker shoots, very often differ very much in size and form from those of fertile branches. Cognizance of such variation is necessary to an understanding of a species in which it occurs; such variation, poorly perceived or understood, may be confusing or misleading, particularly to a novice.

The senses of smell, taste, and touch, in combination with other relatively simple characters, can assist in identifying certain woody plants. (Caution should be exercised when handling plant parts exuding milky or viscid sap, for sometimes these have poisonous properties).

ON SEEKING THE AID OF A SPECIALIST

Persons engaged in identification of plants or plant materials sometimes wish to seek aid from someone presumed to have more expertise than themselves. Such aid may be sought by addressing an inquiry to the curator of the herbarium at a university. The courteous procedure is to ask in advance by telephone or letter (preferably the latter) whether material may be sent for identification or for checking identification. If permission is granted, the following suggestions may be helpful:

1. Send only dried plant material. Press the material in a plant press. If a plant press is not available, carefully spread the material between several thicknesses of newspaper; place over the newspapers an object heavy enough to apply considerable pressure; change the papers daily until all the moisture is absorbed from the specimen. If you are sending specimens representing more than one individual plant, identify each with a number for purposes of communication, or if there is more than one sheet bearing parts from a single individual, identify each sheet with the same number.

2. Be sure that the parts sent are representative of the plant to be identified. Include flowering or fruiting branches if available, together with parts of stem from a previous year's growth that will show leaf scars and bark.

3. When packing for shipment, reinforce the package with stiff cardboard to prevent breakage, then box or wrap securely.

4. Supply supplemental information not disclosed by the specimen itself: the stature of the plant, any particularly significant features of the bark of lower stems; flower color or color of ripe fruit if known to you; whether parts are glaucous when fresh, since glaucescence is often obscured or significantly diminished in the drying process; for vines, the means of climbing if that is not evident on the specimen itself; habitat; specific locality from which the specimen was taken; collector's name; and date collection was made.

SOME ENVIRONMENTAL PERSPECTIVES

Our area of coverage, being in the southern coastal plain, harbors a considerable number of evergreen or winter-green trees and shrubs, some of which significantly accent the winter landscape. These include at least the following: all the gymnosperms except *Taxodium* spp., the several species of pine being notably conspicuous; the several species of palm, *Yucca* spp., and switch-cane (*Arundinaria gigantea*); *Smilax* spp., several species of oak, live oaks (*Quercus virginiana*, *Q. geminata*, and *Q. minima*); those oaks which are usually at least partially winter-green, laurel oak, water oak, and diamond-leaf oak (*Q. hemisphaerica*, *Q. nigra*, and *Q. laurifolia*); southern magnolia and sweet-bay (*Magnolia grandiflora* and *M. virginiana*); American holly, myrtle-leaved holly, yaupon, gallberries, and dahoon (*Ilex opaca*, *I. myrtifolia*, *I. vomitoria*, *I. glabra*, *I. coriacea*, and *I. cassine*); buckwheat-tree (*Cliftonia monophylla*); sweetleaf (*Symplocos tinctoria*); loblolly bay (*Gordonia lasianthus*); laurel cherry (*Prunus caroliniana*); red bays (*Persea borbonia*, *P. humilis*, and *P. palustris*); Florida anise (*Illicium floridanum*); wax-myrtles (*Myrica cerifera*, *M. heterophylla*, and *M. inodora*); staggerbushes (*Lyonia ferruginea* and *L. fruticosa*); fetterbush (*L. lucida*); dog-hobble (*Leucothoe axillaris*); *Agarista populifolia*; sparkleberry (*Vaccinium arboreum*); shiny and glaucous blueberries (*V. myrsinites* and *V. darrowi*); rosemary (*Ceratiola ericoides*); minty rosemarys (*Conradina canescens* and *C. glabra*); jessamines (*Gelsemium sempervirens* and *G. rankinii*); honeysuckles (*Lonicera sempervirens* and *L. japonica*); cross-vine (*Bignonia capreolata*).

The casual observer traversing northern Florida and southern Georgia and

4

southern Alabama probably gains the impression of a landscape dominated by pine trees. There is, however, under our relatively equable, warm, temperate climate, a relatively great diversity of kinds of plants indigenous to the area. In response to naturally recurring complexes of physical and biotic environmental factors, fairly definite habitat types exist, and in each of them, recurring vegetational assemblages—that is, communities—occur. That is nature's way.

The perturbations attendant upon human occupancy and use of the land, however, are many and diverse, and as a consequence our "forestscape" has been disordered to a very considerable degree. Some habitat or community types remain in a quasi-natural state and can be referred to without great risk of misunderstanding. On the other hand, the impingements upon the natural order of things have been occurring for a long time and are currently accelerating, some of them becoming much more drastically impairing. Thus habitat and community distinctiveness becomes ever more blurred.

Not least among the disturbances are contemporary silvicultural practices, particularly "tree farming," which involves widespread clear-cutting, "site preparing" with heavy mechanical equipment, and then planting to pine. It may be noted in passing that pine plantations in general are not closely analogous to pine forests. "Site preparation" by mechanical means very greatly alters the topsoil and drastically changes—in some places virtually obliterates—the native flora and fauna, both of which are integral parts of what constitutes a forest.

Glossary

Abaxial. Pertaining to the side of an organ away from the axis, such as the lower surface of a leaf.

Abortion. Imperfect development or nondevelopment of an organ.

Abortive. Defective or barren.

Achene (Akene). A generally small, hard, dry, indehiscent, 1-seeded fruit, 1-locular; the seed free from the pericarp (ovary wall).

Acicular. Slender and pointed, resembling a needle.

Actinomorphic. Descriptive of a flower or set of flower parts which can be cut through the center into equal and similar parts along 2 or more planes; radially symmetrical.

Acuminate. Tapering to a pointed apex in such a way that the sides of the taper are more or less concave.

Acute. Sharply angled, the sides of the angle essentially straight.

Adaxial. Pertaining to the side of an organ toward the axis, such as the upper surface of a leaf.

Adnate. Referring to fusion of unlike structures or parts, such as staminal filaments at least partially attached to petals or the corolla tube.

Adventitious. Said of buds, roots, etc., which grow in irregular or unusual places.

Aggregate. Crowded into a cluster; a number of separate fruits from a single flower aggregated together, as in the "fruiting cone" of *Magnolia*.

Alternate. Said of leaves occurring one at a node; said also of members of adjacent whorls in the flower when any member of one whorl is in front of or behind the junction of two adjacent members of the succeeding whorl.

Alveolate. Pitted, honeycombed, as are the surfaces of some seeds or achenes.

Alveoli. Pits or depressions suggesting a honeycomb.

Anastomosing. Netted; especially applied to veins so connected by cross veins as to form a network.

Androecium. A collective term applied to all structures of the stamen whorl or whorls.

Androgynous. Having staminate flowers above the pistillate in the same inflorescence.

Androphore. A support or column, formed by fusion of filaments, on which the stamens are borne.

Anther. The pollen-bearing part of the stamen.

Anthesis. The time at which a flower is open; the time during which a plant is in bloom.

Antrorse. Directed forward or upward.

Apetalous. Without petals.

Apical. At the tip or summit.

Apiculate. Terminated abruptly by a small, distinct point, an apiculus or apicule.

Apomictic. In general, reproducing without sexual reproduction; often used to denote seed production that does not involve a sexual process.

Appendage. Any attached supplemental or secondary part.

Arachnoid. Covered with long hairs so entangled as to give a cobwebby appearance.

Arborescent. Approaching the size and habit of a tree.

Arcuate. Moderately curved-arching.

Aril. An appendage growing at or about the hilum of a seed.

Aristate. Awned; provided with a bristle at the apex.

Articulate. Jointed; provided with places where separation may take place.

Ascending. Directed or rising upward obliquely.

Asepalous. Without sepals.

Atomiferous-glandular. Having a very tiny, beadlike glandular exudate on a surface.

Attenuate. Gradually narrowed to a long point at apex or base.

Auricle. An earlike appendage.

Awn. A stiff, bristlelike appendage, usually at the end of a structure.

Axil. The angle found between any two organs or structures.

Axillary. Occurring in an axil, as a bud in a leaf axil.

Banner. The upper, broad, more or less erect petal of a papilionaceous flower of the pea family; also called standard.

Barbellate. Provided, usually laterally, with fine, short points or barbs.

Bifid. With 2 lobes or segments, as at the apices of some petals or leaves.

Bilaterally symmetrical. Said of corolla or calyx (or flower) when divisible into equal halves in 1 plane only; zygomorphic.

Binomial. The combination of a generic name with a specific epithet, which gives the scientific name of an organism.

Bisexual. Pertaining to an individual flower having both male and female parts.

Biternate. Said of compound leaves with three main divisions which are themselves divided into three parts.

Blade. The expanded portion of a leaf, petal, or other structure.

Bloom. The white powder, dust, or waxy covering sometimes on stems, leaves, flowers, or fruits.

Bract. A reduced leaf or small leaflike structure, particularly one subtending a flower or an inflorescence branch.

Bractlet. A small bract in a secondary position.

Branchlet. Ultimate portion of a branch.

Bristle. Stiff, strong, but slender hair or trichome.

Caducous. Falling off quickly or early, as is the case with stipules in many kinds of plants, for example.

Callus. A hard protuberance or callosity; new tissue covering a wound.

Calyx. The outer series of perianth parts of a flower, commonly green in color, frequently enclosing the rest of the flower in the bud stage, occasionally colored or petallike (petaloid), in some groups greatly reduced or lacking.

Calyx tube. Tube formed by wholly or partially fused sepals. Not the floral tube of perigynous or epigynous flowers.

Campanulate. Bell-shaped, usually applied to calyx or corolla.

Cancellate. Latticed, or resembling a latticed construction, usually said of a surface such as that of an achene or seed.

Capillary. Very slender, threadlike, hairlike.

Capitate. Arranged in a head or gathered into a very close cluster; also refers to hairs (trichomes) which are gland-tipped, i.e., capitate hairs.

Capsule. A dry, dehiscent fruit originating from 2 or more fused carpels, such as the fruit of *Hypericum*.

Carpel. The unit of structure of the female portion of a flower.

Cartilaginous. Firm and tough but flexible.

Castaneous. Chestnut-brown.

Catkin. An erect, lax, or pendent, scaly-bracted spike bearing unisexual, apetalous flowers such as the inflorescence of willows or the staminate inflorescence of hickories and oaks.

Caudate. Having a long taillike terminus.

Channeled. Having a deep longitudinal groove.

Chartaceous. Having the texture of thin but stiff paper.

Choripetalous. Corolla with petals distinct from one another.

Cilia. Marginal hairs.

Ciliate. With marginal hairs that form a fringe.

Circumscissile. Opening or dehiscing by a line around the middle, the top coming off as a lid.

Clavate. Club-shaped, gradually broadened upward.

Claw. The narrowed, stalklike base of some sepals or petals.

Cleft. Cut about half way to the midvein.

Clonal. Usually used to describe cloning by vegetative reproduction, the seemingly separate plants having arisen from rhizomes, stolons, or roots of a single or of neighboring "parent" plants.

Clone. A group of genetically identical individuals, resulting from asexual, vegetative multiplication, in woody plants often resulting from stems arising from subterranean runners.

Coherent. Having like parts united.

Colony. A stand, group, or population of neighboring plants of one species, the origin having been clonal, from seeds, or both.

Coma. A tuft of soft hairs, as at the apices or bases of seeds.

Comose. With a tuft of soft hairs, a coma.

Compressed. Flattened, especially flattened laterally. *See* Obcompressed.

Connate. United; applied to like parts or organs, such as sepals or petals, or opposite, sessile leaves whose bases are united around the stem, then usually referred to as connate-perfoliate.

Connivent. Approximate but not organically united.

Convoluted. Said of parts rolled or twisted together when in an undeveloped stage, as in some corollas in the bud stage.

Cordate. With a sinus and rounded lobes at the base, the overall outline usually ovate; often restricted to the base rather than to the outline of the entire organ; heart-shaped.

Coriaceous. Leathery.

Corolla. The inner, usually colored or otherwise differentiated, whorl or whorls of the perianth.

Corymb. A racemose type of inflorescence in which the lower pedicels are successively elongate forming a more or less flat-topped inflorescence, the outer flowers opening before the inner.

Crenate. Shallowly rounded-toothed; scalloped.

Cuneate. Wedge-shaped; narrowly triangular with the acute angle downward.

Cusp. A strong or sharp point.

Cuspidate. Tipped with a short point.

Cyme. A type of inflorescence in which each flower is terminal either to the main axis or to a branch.

Cymose. Arranged in a cyme.

Cystoliths. Intracellular concretions, usually of calcium carbonate.

Deciduous. Falling off at a certain season or stage of growth.

Decompound. More than once compound or divided.

Decurrent. Extending downward, applied usually to leaves in which the blade is apparently prolonged downward as two wings along the petiole or along the stem.

Decussate. Opposite leaves in 4 rows up and down the stem, alternating in pairs at right angles to each other.

Dehiscent. Opening and shedding contents; said of stamens and fruits.

Deltoid. Equilaterally triangular.

Dentate. Having marginal teeth pointing outward, not forward.

Determinate inflorescence. Said of inflorescences in which the terminal or central flower of the axis or branch axes opens first, thus arresting further elongation of a particular axis.

Diadelphous. In two sets, as applied to stamens when in two, usually unequal, sets.

Dichotomous. Forking in pairs.

Digitate. Fingered, with several members arising from one point.

Dimorphic. Having two forms.

Dioecious. Said of a kind of plant having unisexual flowers, the male and female flowers on different individual plants.

Disc flowers. The radially symmetrical flowers of the head in Compositae, as distinguished from ligulate ray flowers.

Distal. Farthest away from the center, the point of attachment, or origin.

Distichous. Two-ranked, on opposite sides of a stem and in the same plane.

Divaricate. Widely divergent.

Divided. Referring to the blade of a leaf, etc., when it is cut into distinct divisions to, or almost to, the midvein.

Dorsal. Pertaining to the back; the surface away from the axis.

Downy. Covered with short, weak hairs.

Drupe. A fleshy or pulpy fruit with the inner portion of the pericarp hard or stony and enclosing the seed or seeds; usually 1-locular and 1-seeded, sometimes more than 1–locular and more than 1-seeded.

Echinate. Bearing prickles.

Elliptic. An outline that is oval, narrowed to rounded at the ends and widest at about the middle (as the outline of a football); ellipsoid, a solid with an elliptical outline.

Emarginate. Said of leaves, sepals, or petals, and other structures that are notched at the apex.

Enation. An outgrowth of the surface of an organ.

Endocarp. The inner layer of the wall of a matured ovary.

Entire. Margins without teeth, lobes, or divisions.

Epigynous. Pertaining to a flower having sepals, petals, and stamens borne at the summit of a floral tube, the ovary being within and fused with the floral tube, the ovary thus said to be inferior.

Epipetalous. Said of stamens when they are inserted on the corolla.

Erose. Uneven; said of margins that give the appearance of having been torn, or of margins with very small teeth of irregular shape and size.

Even-pinnate. Said of compound leaves having an even number of leaflets, this usually easily determined because there is a pair terminally.

Evergreen. As used in this book, indicating retention of at least some leaves for at least two seasons. See winter-green.

Excrescence. An outgrowth; a disfiguring addition.

Excurrent. Projecting beyond the edge or margin, as the midrib of a leaf or bract.

Exfoliating. To peel off in thin layers, shreds, or plates, as the bark of some trees, such as that of sycamore.

Exocarp. The outer layer of the wall of a matured ovary.

Exserted. Extending beyond some enclosing part.

Eye. With reference to a flower, the marked center, the marking often being a color difference.

Falcate. Curved like a sickle.

Fascicle. A bundle or close cluster.

Filament. The stalk bearing the anther.

Filiform. Threadlike, long and very slender.

Fimbriate. Cut into regular segments and appearing fringed.

Flabellate. Fan-shaped.

Flaccid. Weak, limp, soft, or flabby.

Flexible. Pliant, not stiff.

Flexuous. More or less zigzagging.

Floccose. With flocs, tufts, or patches of soft, cottony or woolly hairs.

Floret. One of the small flowers of a close or compacted inflorescence (a relative term).

Floricane. The stem at flowering and fruiting stage (of a bramble, *Rubus*).

Foliaceous. Leaflike.

Foliolate. Pertaining to leaflets of a compound leaf, as 3-foliolate, 8–foliolate, etc.

Follicle. A fruit developing from a single carpel and dehiscing along one suture, such as the individual fruit of the aggregate in *Magnolia*.

Fusiform. Tapering from approximately the middle to both extremities.

Gamosepalous. Having a calyx with sepals to some extent united. Same as synsepalous.

Geniculate. Bent abruptly at the nodes, zigzagging.

Gibbous. A distended, rounded swelling on one side, as on a calyx or corolla tube or segment.

Glabrate. Becoming glabrous with age.

Glabrous. Without pubescence.

Gland. A secreting part or appendage.

Glandular. Bearing secretory glands or having any glandular secretion.

Glaucous. Having a frosted or whitish waxy bloom or powdery coating (glaucescence).

Glomerate. Disposed in compact groups.

Glutinous. With a gluey or sticky exudation.

Gregarious. Growing in large colonies, individuals associated but not united.

Gynoecium. The pistil, or pistils of a flower taken collectively.

Gynophore. Stalk or stipe of a pistil.

Habit. Term used for the growth form of a plant.

Hammock. A relatively locally applied term for a fertile area, usually higher than its surroundings, characterized by having chiefly hardwood vegetation; also an elevated "island" of woodland surrounded by marsh or swamp.

Hastate. Arrow-shaped, with basal lobes that spread nearly or quite at right angles.

Herbage. Aboveground, nonwoody parts of a plant.

Hilum. The scar at the point of attachment of an ovule or a seed.

Hispid. Rough with stiff or bristly hairs.

Hirsute. Rough with coarse or shaggy hairs.

Hyaline. Of thin membranous texture, usually transparent or translucent.

Hypanthium. An expansion of the receptacle of a flower that forms a saucer-shaped, cup-shaped, or tubular structure (often simulating a calyx tube) bearing the perianth and stamens at or near its rim; it may be free from or united to the ovary; in this book usually referred to as the floral tube.

Hypogynous. Pertaining to a flower having sepals, petals, and stamens, with their point of origin below the ovary, the ovary thus said to be superior.

Imbricated. Overlapping like shingles on a roof.

Imperfect. Having either stamens or a pistil (or pistils) but not both; unisexual.

Incised. Having deeply cleft margins.

Included. Not projecting beyond an enclosing part.

Indehiscent. Said of fruits that remain closed and do not shed their seeds.

Indeterminate inflorescence. Said of those kinds of inflorescences in which the lower or lateral flowers of the axis or of branches of the axis open first, the potentially terminal flower of the axis or branches of the axis open last, thus the growth or elongation of the main or lateral axes is not arrested by the opening of the first flowers.

Inferior. Usually referring to the position of the ovary with reference to the insertion of the other flower parts, the other flower parts being inserted on the rim of a floral tube adnate to the ovary, thus appearing as though at the summit of the ovary.

Inflorescence. Mode of flower bearing.

Infrastipular. Below the stipules.

Infructescence. The inflorescence in a fruiting stage; collective fruits.

Internode. The portion of a stem between nodes.

Involucre. A group of closely placed bracts that subtend or enclose an inflorescence or a single flower.

Involute. Said of margins that are rolled inward (toward the adaxial side).

Irregular. Said of a flower or parts of a flower members of which are unlike, commonly used for bilaterally symmetrical or zygomorphic.

Keel. A prominent longitudinal ridge; the two lower, united petals of a papilionaceous flower.

Labiate. Lipped, as in a calyx or corolla, such as the calyx or corolla of flowers of members of the mint family.

Lacerate. Said of a margin torn irregularly.

Laciniate. Cut into narrow lobes or segments.

Lacunate. With air spaces or chambers in the midst of tissue.

Lamellate. Made up of thin plates.

Laminate. Composed of thin layers.

Lanate. Woolly, with long, intertwined, curled hairs.

Lanceolate. In outline, broadest toward the base and narrowed to the apex, several times longer than wide.

Lateral. Belonging to or borne on the sides.

Leaf Scar. Scar left after the falling of a leaf.

Legume. A 1–locular fruit, usually dehiscent along two sutures, bearing seeds along the ventral suture; a leguminous plant.

Lenticel. Corky spot on young bark, arising in relation to epidermal stomata.

Lenticular. Lens-shaped, biconvex.

Ligule. In Compositae, the flattened, straplike portion of the ray flower.

Limb. With reference to a significantly tubular calyx or corolla, the expanded portion above the throat.

Linear. Long and slender with parallel or nearly parallel sides.

Lip. The upper or lower part of a bilabiate calyx or corolla.

Locule. A compartment of an anther or an ovary.

Loculicidal. Said of capsules that dehisce along the back of the locule.

Lunate. Crescent-shaped.

Marsh. A tract of wet land principally inhabited by emergent herbaceous vegetation.

Membranous. Having a thin, soft, pliable texture.

-merous. A suffix that means having a specified number of parts, usually used with a numerical prefix, as 3-merous or 5-merous.

Mesic. Pertaining to conditions of medium moisture supply.

Midrib. The central or main vein of a leaf or similar structure.

Monadelphous. Stamens united in one group by union of at least part of their filaments.

Monoecious. Said of a kind of plant having unisexual flowers, the male and female flowers on the same individual plant.

Mucro. A short and small abrupt tip.

Mucronate. With a short and small abrupt tip.

Naturalized. Of foreign origin, but established and reproducing as though native.

Neutral flower. Said of a sterile flower composed of a perianth without sexual organs.

Node. The place upon a stem which bears 1 or more leaves or bracts.

Nut. A hard, indehiscent, 1-seeded fruit, for example, the fruit of the beech, oak, hickory, etc.

Nutlet. Diminutive of nut; loosely applied to any small, dry, nutlike fruit.

Obcompressed. Flattened at right angles to the primary plane or axis, as, for example, achenes of some composites that are flattened at right angles to the radius of the receptacle. *See* Compressed.

Oblanceolate. In outline, broadest above the middle, shortly tapered to the apex, long-tapered to the base, much longer than broad.

Oblate. Flattened at the poles (as is a tangerine).

Oblique. Slanting; unequal-sided.

Oblong. In outline, longer than broad, the sides nearly parallel for much of their length.

Obovate. Inverted ovate.

Obsolete. Not evident or apparent; rudimentary, vestigial.

Obtuse. Bluntly angled, the angle more than 90 degrees, less than 180.

Ocrea (plural, ocreae). A tubular sheath around the stem (derived from stipules), used chiefly in the Polygonaceae.

Ocreola (plural, ocreolae). The smaller or secondary sheaths, as in the inflorescence of Polygonaceae.

Odd-pinnate. Said of compound leaves having an odd number of leaflets, this usually easily determined because there is a single terminal leaflet.

Opposite. Said of leaves or bracts occurring two at a node on opposite sides of the stem. Said of flower parts when one part occurs in front of another.

Orbicular. Approximately circular in outline.

Ovary. The basal part of the pistil having ovules within.

Ovate. In outline, broadest at or near the base, not greatly longer than broad (roughly the outline of an egg).

Ovoid. A three-dimensional object with an ovate outline.

Ovule. Reproductive organ within the ovary in which the female gametophyte is produced and which after full development becomes the seed.

Palate. In a two-lipped corolla, a projecting part of the lower lip which closes the throat, as in a snapdragon.

Pale. In Compositae, a chaffy receptacular scale or bract subtending the individual flower of a head; in general, a small, thin scale or bract.

Palmate. With three or more lobes or veins or leaflets arising from one point.

Pandurate. Fiddle-shaped.

Panicle. A loose, irregularly compound inflorescence with stalked flowers, i.e., the branches racemose.

Paniculate. Arranged in a panicle.

Papilionaceous. Butterflylike; said of the flowers of those members of the pea family that have the corolla composed of an upright banner or standard and two lateral wings, each representing a petal, and a keel composed of two petals variously united.

Papillose. Descriptive of a surface beset with short, blunt, rounded, or cylindric projections.

Pappus. In Compositae, designation for the collective scales, bristles, crown, etc., at the summit of the ovary (or achene) and outside the corolla.

Pectinate. Said of a structure that is cleft into divisions in such a way as to resemble a comb.

Pedicel. The stalk of a flower in an inflorescence.

Peduncle. The stalk of a flower borne singly or the stalk of an inflorescence.

Peltate. Said of a plane structure that is attached at a point on its surface, not attached marginally, for example, as in the leaf of *Menispermum*.

Pendent or pendulous. Hanging.

Perfect. Having both functional stamens and pistil or pistils; bisexual.

Perfoliate. Said of opposite or whorled leaves or bracts that are united into a collarlike structure around the stem that bears them.

Perianth. The calyx and corolla taken together, or either one if one is absent.

Perigynous. Pertaining to a flower having the sepals, petals, and stamens borne at the summit of a floral tube, the ovary (ovaries) being within but free from the tube.

Persistent. Remaining attached.

Petal. The unit of structure of the corolla.

Petaloid. Said of a floral part (apart from a petal), or a bract, colored and resembling a petal.

Petiole. The stalk of a leaf.

Petiolule. The stalk of a leaflet of a compound leaf.

Pilose. With pubescence composed of scattered long, slender, soft hairs.

Pinna. A leaflet or a primary division of a compound leaf.

Pinnate. Having a common elongate axis, with branches, lobes, veins, or leaflets arranged divergently on either side.

Pistil. The unit of female function of a flower, may be composed of a single carpel or two or more carpels united.

Pistillate. Said of a flower bearing a pistil or pistils but not stamens; may refer also to a plant having only pistillate flowers.

Placenta. The structure or tissue within the ovary bearing the ovules.

Plait. A flattened, usually lengthwise, fold, as a piece of cloth doubled back on itself.

Plicate. Repeatedly plaited, as the leaves of some palms, for example, cabbage palm. *See* Plait.

Plumose. With hairlike branches, feathery.

Polygamo-dioecious. Polygamous but chiefly dioecious.

Polygamo-monoecious. Polygamous but chiefly monoecious.

Polygamous. Having unisexual and bisexual flowers on the same plant.

Polypetalous. With distinct or separate petals.

Pome. A fruit in which the ovarial portion is relatively small and the fleshy portion is derived largely from the floral tube, as in an apple or pear.

Prickle. A sharp, spinelike outgrowth of the bark or epidermis, as in roses and brambles.

Primocane. The first-year, nonflowering or nonfruiting shoot (of a bramble, *Rubus*).

Prostrate. Prone, said of stems that lie on the ground.

Procumbent. Trailing or lying flat but not rooting.

Puberulent. Minutely pubescent, the hairs soft, straight, erect, scarcely visible to the unaided eye.

Pubescent. A general term for hairiness.

Pulvinus (plural, pulvini). Swelling or dilation at the base of a petiole or stalk of leaflet (or at the summit of either, as well).

Punctate. With depressed dots scattered over the surface.

Pungent. Said of an odor or taste that is sharply penetrating; acrid.

Pustulate hair. Hair with an enlarged base.

Pyriform. Pear-shaped.

Raceme. An indeterminate inflorescence with a single axis, the flowers stalked.

Racemose. Arranged in a raceme.

Rachis. The axis of an inflorescence, or of a compound leaf.

Radially symmetrical. Said of a flower or set of flower parts which can be cut through the center into equal and similar parts along 2 or more planes; actinomorphic; regular.

Radiate head. In Compositae, a head having marginal, ligulate flowers.

Ray flowers. In Compositae, the flowers of the head having ligulate or straplike corollas borne marginally, or having all ligulate or straplike corollas.

Receptacle. The more or less expanded apex of a floral axis which bears the floral parts.

Regular. Said of a flower or of parts of a flower members of which are alike; commonly used for radially symmetrical or actinomorphic.

Remote. Considerably separated from one another.

Reniform. Kidney-shaped.

Repent. Said of a stem that is prostrate and rooting at the nodes.

Reticulate. Netted.

Retrorse. Having hairs or other processes turned toward the base.

Retuse. With a shallow, rounded notch at the apex.

Revolute. Said of margins that are rolled backward (toward the abaxial side).

Rhombic. An outline like a rhomboid, a parallelogram with equal sides, having two oblique angles and two acute angles.

Roseate. Rose-colored.

Rotate. Radially spreading in one plane.

15

Rugose. With a wrinkled surface.

Saccate. Having a saclike swelling.

Sagittate. Shaped like an arrowhead with the basal lobes pointing downward.

Salverform. Said of a corolla in which the tube is essentially cylindrical, the lobes abruptly spreading.

Samara. An indehiscent, winged fruit, as in *Ulmus, Fraxinus, Acer.*

Scabrous, scabrid. Rough and harsh to the touch.

Scandent. Climbing.

Scarious. Thin and dry, usually more or less opaque.

Scurfy. Having flaky or scaly surfaces.

Secund. Disposed on one side of a stem or axis, usually by torsion.

Sepal. The unit of structure of the calyx.

Septate. Partitioned by cross-walls.

Septicidal. Said of capsules that dehisce along the junction of their carpels opposite the septa.

Serrate. Having marginal teeth pointing forward.

Sessile. Not stalked.

Setaceous. With a bristle form.

Simple. Not compound; unbranched.

Slough. A wet place of deep mud or mire; a sluggish channel.

Smooth. Not rough to the touch.

Spatulate. Shaped like a spatula, i.e., gradually widening distally and with a rounded tip.

Spicate. Arranged in a spike.

Spicule. A small, slender, sharp-pointed piece, usually on a surface.

Spike. An inflorescence with a single axis, the flowers sessile.

Spine. A strong, sharply pointed, woody outgrowth from the wood of the stem.

Spinescent. Terminated by a sharp tip.

Spur. In woody plants referring to a lateral short-shoot with closely spaced leaves.

Stamen. The male organ of a flower; usually consisting of a stalk or filament and an anther producing the pollen.

Staminate. Said of a flower bearing a stamen or stamens but not pistils; may refer also to a plant having only staminate flowers.

Staminode, staminodium. A sterile stamen, or a structure without anther, borne in the staminal part of the flower.

Standard. The upper, broad, more or less erect petal of a papilionaceous flower; also called banner.

Stellate. Starlike; said of hairs that branch in such a manner as to radiate from a central point.

Stigma. The part of the pistil which is pollen-receptive.

Stipe. The stalklike basal part of an ovary, or of a fruit such as an achene.

Stipel. Stipule of a leaflet of a compound leaf.

Stipule. An appendage at the base of a petiole or a leaf or on each side of its insertion.

Stramineous. Straw-colored.

Striate. Marked with fine parallel lines.

Strigose. Surface clothed with stiff, often appressed hairs, these usually pointing in one direction.

Style. The usually attenuated portion of the pistil which connects the ovary and stigma. Sometimes there is more than one style per flower, or one and branched.

Subulate. Relatively narrow, with a long taper from base to apex; awl-shaped.

Suffrutescent. Obscurely shrubby, only in part becoming woody, not necessarily low.

Suffruticose. Diminutively shrubby; low and woody.

Sulcate. Grooved.

Superior. Usually referring to the position of the ovary with reference to the insertion of the other flower parts, the other flower parts being inserted below or around the base of the ovary.

Suture. A seam or line or groove; usually applied to the line along which a fruit dehisces; any lengthwise groove that forms a junction between two parts.

Swale. An open hollow or depression, usually wet at least seasonally.

Swamp. A wooded area having surface water much of the time.

Syncarp. A multiple or aggregate fruit derived from numerous separate ovaries of a single flower.

Taxon. A grouping of organisms given a proper name, or, a grouping that could be given a proper name but is not named as a matter of convention; or, more simply, any taxonomic group. Thus a "family" or a "genus" or a "species" is a category (a rank in the taxonomic hierarchy), while each of the following is a taxon: Aceraceae, *Acer*, *Acer rubrum*.

Tendril. A slender twining or clasping process, modified stem, leaf, or part of a leaf, by which some plants climb.

Tepal. Denoting a unit of the perianth when the sepals and petals are essentially alike and not readily differentiated.

Terete. Circular in cross-section.

Ternate. In threes.

Throat. The orifice of a significantly tubular calyx or corolla.

Thorn. A stiff, sharply pointed, specialized branch, often eventually becoming a short- or spur-shoot, as in *Bumelia* and *Crataegus*, for example.

Thyrse. Compact, more or less compound panicle, as often used; more correctly, a paniclelike cluster with the main axis indeterminate and the lateral axes determinate.

Tomentose. Densely covered with short, matted hairs.

Torus. *See* Receptacle.

Trichome. Any hairlike outgrowth of the epidermis.

Truncate. Cut squarely across, either at the base or apex of an organ.

Turbinate. Top-shaped, inversely conical.

Umbel. A flat or convex flower cluster in which the flower stalks arise from a common point (like the rays of an umbrella).

Umbellate. Arranged in an umbel.

Uncinate. Hooked at the tip.

Unisexual. Pertaining to an individual flower having either staminate or pistillate parts, not both; used to denote a plant having either staminate or pistillate flowers.

Valvate. Opening by valves or pertaining to valves; meeting by the edges and not overlapping, as the scales of some buds, the bud of *Carya cordiformis*, for example.

Valve. One of the segments into which a capsule splits; the partially detached lid of a certain kind of anther.

Vascular bundle scar. Scar left on the leaf scar, marking a leaf trace, after a leaf has fallen.

Verrucose. Warty.

Versatile. Turning freely on its support, as an anther attached near the middle and capable of swinging freely on the filament.

Verticil. A ring of organs, e.g., leaves, flowers, flower parts, at a given position on an axis; a whorl.

Vestiture. That which covers a surface, as hairs, scales, etc.

Villous. Covered with fine long hairs, the hairs not matted.

Viscid. Sticky.

Wet woodland. A wooded area having markedly fluctuating conditions of moisture, varyingly with and without surface water (as opposed to a swamp wherein surface water stands much of the time).

Wing. Any membranous or thin expansion bordering or surrounding or extending from an organ, for example, blade tissue winging a petiole, petiolar or blade tissue extending internodally and winging a stem, or the wing of a fruit or seed; the lateral pair of petals of a papilionaceous flower.

Winter-green. Condition in which leaves of a given season over-winter, fall in spring immediately prior to or coincidental with commencement of new shoot growth (as in liveoak, *Quercus virginiana*).

Zygomorphic. Said of the corolla or calyx when divisible into equal halves in one plane only; bilaterally symmetrical; irregular.

Artificial Keys
to the Higher Taxa of
Trees, Shrubs, and Woody Vines
of Northern Florida and Adjacent Georgia
and Alabama

GYMNOSPERMS

Coniferous, seed-bearing trees or shrubs having naked seeds (not enclosed by an ovary) borne on upper surfaces of scales of a cone, *or* in a fleshy structure with the basic structure of a cone and berrylike (juniper "berries"), *or* singly and partially or wholly surrounded by an aril.

1. Leaves pinnately compound, circinate in the bud, somewhat palmlike or fernlike, evergreen (save where planted northward of their natural range), disposed very closely spiraled in a crown on a stout, upright, subterranean stem apex of which is woolly, most of it (below the living leaves) covered by old, hairy, petiole bases. **Zamiaceae,** p. 66
1. Leaves needlelike *and* in fascicles of 2–4 (ours) sheathed at the base, *or* scalelike or awl-shaped and borne oppositely, *or* essentially linear or lance-linear and borne alternately.
  2. The leaves needlelike *and* in fascicles of 2–4 sheathed at the base. **Pinaceae,** p. 44
  2. The leaves scalelike or awl-shaped and borne oppositely *or* essentially linear or lance-linear, and borne alternately.
    3. Leaves scalelike or awl-shaped and borne oppositely. **Cupressaceae,** p. 41
    3. Leaves essentially linear, or lance-linear and borne alternately.
      4. The leaves all alike and all borne on extension shoots, evergreen, eventually falling singly from those shoots; plants dioecious; ovulate "cones" with a single, plump seed partially or wholly surrounded by an aril. **Taxaceae,** p. 56
      4. The leaves borne on an essentially dimorphic shoot system: (1) shoots that are indeterminate long-shoots (extension shoots) bearing scalelike or awllike leaves having buds in their axils, such leaves withering and falling separately, and (2) determinate short-shoots bearing numerous alternately arranged and spirally disposed leaves having no buds in their axils, the short-shoots and their leaves falling together in autumn; plants monoecious; scales of ovulate cones separating at maturity. **Taxodiaceae,** p. 61

ANGIOSPERMS

True flowering, seed-bearing plants, the seeds produced within the closed ovary or ovaries of the pistil or pistils of a flower.

Monocotyledons

Stems without a central pith or annular layers, but having woody fibers distributed throughout soft tissue. Embryo with a single cotyledon, the germinated seedling with a single seed-leaf. In general, parts of the flowers in twos, threes, fours, or sixes, not in fives.

1. Plants climbing by means of tendrils terminating stipules adnate to petioles; stems of lower portions of leader-shoots (excepting in *Smilax pumila*) bearing scattered, hard, sharp prickles. **Smilacaceae,** p. 84

1. Plants not climbing, without tendrils, stem without prickles.

  2. Leaf blades (of plants beyond the seedling state) large and fanlike, longitudinally segmented, segments united proximally and longitudinally corrugated, free distally and generally flat (when fresh).   **Palmae**, p. 75

  2. Leaf blades not fanlike.

    3. Plants with freely branching, slender, extensive rhizomes from which canelike aerial stems arise, forming extensive "canebrakes"; nodes well separated from each other, internodes hollow; bases of leaves sheathing the stem, blades relatively membranous.   **Gramineae**, p. 74

    3. Plant with thick, solid stems, in some kinds short and essentially subterranean, in others becoming elevated to several meters; leaves very closely spiraled, their bases thinly inserted on the stem, otherwise flared, overlapping and tightly appressed, above the flared bases narrowed into leathery, swordlike or daggerlike blades whose tips are very sharply spinose.   **Agavaceae**, p. 69

Dicotyledons

Stems commonly formed of bark, wood, and pith; the wood forming a zone between the other two and increasing, as the stem continues from year to year, by addition of a new layer to the outside, next to the bark. Leaves mostly net-veined. Embryo with a pair of opposite cotyledons, the germinated seedling with a pair of seed-leaves. Cycles of flower parts mostly in twos, threes, fours or fives, or their multiples.

KEYS TO SUBSIDIARY KEYS
(DICOTYLEDONS)

A. Plants woody vines or stems trailing along the ground; climbing achieved by means of tendrils, aerial adventitious roots on the stems, by bending or twining of petioles, stalks of leaflets, or axes of compound leaves, by growing upwardly beneath the bark of other plants, or by scrambling through vegetation, this effected by divaricate, rigid branches which serve to anchor the stems as they elongate, or by vigorous, fast-growing stems armed with hooked prickles.   Key 1, p. 20

A. Plant a shrub or tree, habit various but not climbing or trailing.

  B. Leaves compound.   Key 2, p. 23

  B. Leaves simple.

    C. Leaves or leaf scars opposite or whorled.   Key 3, p. 25

    C. Leaves or leaf scars alternate.   Key 4, p. 28

Key 1

Plants woody vines, climbing by various means (some kinds, if no objects close by upon which to climb, having surface, often elongate, runners from which erect branches may grow), or plants with stems trailing, never climbing.

1. Leaves compound.

  2. The leaves opposite.

    3. Leaflets 2 per leaf, the leaf axis bearing a branched tendril terminally (*Bignonia*).   **Bignoniaceae**, p. 171

    3. Leaflets (of most leaves on a given plant) 3 or more per leaf, tendrils none.

4. Climbing achieved by aerial, adventitious roots on the stem (*Campsis*).

Bignoniaceae, p. 171

4. Climbing achieved by bending or twining of petioles, stalks of leaflets, or leaf rachises (*Clematis*).   **Ranunculaceae,** p. 526

  2. The leaves alternate.

    5. Stems armed with sharp prickles (*Rosa, Rubus*).   **Rosaceae,** p. 544

    5. Stems unarmed.

      6. Leaflets 3 per leaf.

        7. Stems climbing by means of aerial, adventitious roots (or if nothing upon which to climb, stems trailing and giving rise to low, weak, aerial stems and such may be abundant as ground-cover, especially in woodlands) [POISON-IVY] (*Toxicodendron radicans*).   **Anacardiaceae,** p. 108

        7. Stems climbing by means of twining (or if nothing upon which to twine, stems trailing), commonly forming great masses of herbage covering banks, fields, or into and over trees (*Pueraria*).   **Leguminosae,** p. 419

      6. Leaflets mostly 7 or more per leaf; climbing by twining (*Wisteria*).

Leguminosae, p. 419

1. Leaves simple.

  8. The leaves opposite.

    9. Plants of maritime situations; leaves succulent; (not climbing but main stems commonly trailing).   **Bataceae,** p. 159

    9. Plants not of maritime situations; leaves not succulent.

      10. Plant not climbing but with stems decumbent, arching, trailing, clambering, or prostrate-trailing and rooting at the nodes.

        11. Leaves without stipules, their blades regularly toothed marginally; flowers borne in long-stalked, headlike spikes; corolla lavender (*Lantana montevidensis*).

Verbenaceae, p. 689

        11. Leaves stipulate, their blades entire marginally; flowers not as above.

          12. Plants with main stems prostrate-trailing and rooting at the nodes; stipules minute, deltoid; leaf blades not exceeding 2.5 cm long, many or most of them shorter than that, some of them as broad as long; flowers in axillary, singly stalked pairs (*Mitchella*).   **Rubiaceae,** p. 598

          12. Plants with some stems reclining or clambering; stipules broad-based and abruptly narrowed to subulate tips; leaf blades ovate, oblong, lanceolate, or obovate, to 6 cm long and 4 cm broad; flowers in axillary, stalked, lax racemes or panicles (*Chiococca*).   **Rubiaceae,** p. 598

      10. Plants climbing.

        13. Plants climbing by twining.

          14. Sap milky.   **Apocynaceae,** p. 133

          14. Sap not milky.

            15. Ovary inferior, the floral tube surmounted by minute teeth; fruit a several-seeded berry (*Lonicera*).   **Caprifoliaceae,** p. 180

            15. Ovary superior; calyx united below and with distinct and evident lobes; fruit a flattened, 2-valved capsule [corolla yellow].   **Loganiaceae,** p. 451

        13. Plant climbing by aerial, adventitious roots, or mainly by scrambling.

          16. Climbing by means of aerial, adventitious roots (*Decumaria*).

Saxifragaceae, p. 641

          16. Climbing mainly by scrambling or fast-growing leader-shoots with rigid, often thorn-tipped, divaricate branches that serve to anchor them as they grow among other plants (*Sageretia*).   **Rhamnaceae,** p. 533

  8. The leaves alternate.

    17. Leaf blades, many or most of them on a given plant, about as broad as or broader than long.

      18. Plant climbing by means of tendrils opposite at least some of the leaves.

Vitaceae, p. 694

      18. Plants climbing by means of twining stems.

21

19. Petioles attached to leaf blades a very short distance above the basal margins, i.e., peltate (*Menispermum*).   **Menispermaceae**, p. 473

19. Petioles attached to the leaf blades at their basal margins, not peltate.

  20. The leaf blades variably cordate-ovate, heart-shaped, or kidney-shaped, *or* broadly ovate and more or less truncate basally, unlobed, or if lobed, then the lobes merely shouldered or short and blunt, no large sinuses between them, lower surfaces relatively abundantly short-pubescent and soft-velvety to the touch.

    21. The leaf blades ovate-cordate, the larger ones on a given plant 10–15 cm long, unlobed save for the usually conspicuous basal cordations; flowers borne singly or in pairs from leaf axils, bilaterally symmetrical, bisexual, corolla none, calyx tube about 5 cm long, S-shaped, densely woolly, fruit a fluted-cylindric, 6-valved capsule (*Aristolochia*).   **Aristolochiaceae**, p. 156

    21. The leaf blades variably cordate-ovate, kidney-shaped, or ovate with bases truncate or nearly so, unlobed, merely shouldered-lobed, or with few short, blunt lobes and no conspicuous sinuses between the lobes, larger blades on a given plant 4–10 cm long; flowers very small, borne in supraaxillary, sometimes terminal, panicles, radially symmetrical, unisexual (plants dioecious); fruit a red drupe (*Cocculus*).   **Menispermaceae**, p. 473

  20. Leaf blades, many of them on a given plant, prominently 3–5-lobed and with deep sinuses between the lobes, lower surfaces with few, scattered, spiculelike hairs, not soft-velvety to the touch (*Calycocarpum*).   **Menispermaceae**, p. 473

17. Leaf blades, all or most of them, definitely longer than broad.

  22. Plant climbing by means of tendrils terminating at least some, short, lateral branches of the season, and terminating at least some inflorescences or infructescences (*Brunnichia*).   **Polygonaceae**, p. 521

  22. Plants climbing by other means.

    23. Mature leaf blades abundantly silvery brown-scaly beneath; [vigorous-growing leader shoots to some extent twining, otherwise scrambling effectuated mostly by rigid, thorn-tipped, divaricate branches that serve to anchor them as they grow into and through contiguous vegetation].   **Elaeagnaceae**, p. 229

    23. Mature leaf blades not scaly beneath.

      24. Plant climbing by means of leader shoots within the bark of other kinds of plants, chiefly pond-cypress and white-cedar, such shoots flat or flattish, having only minute scale-leaves but at intervals producing lateral buds from which lateral shoots break through the bark and become ordinary leafy shoots which in turn bear flowers and fruits (*Pieris*).   **Ericaceae**, p. 234

      24. Plants not climbing as indicated in first no. 24.

        25. Venation of leaf blades pinnate *and* the prominent major lateral veins angling-ascending conspicuously parallel to each other; leaf margins crenulate; [climbing achieved to some extent by twining but to a greater extent by scrambling of notably supple stems] (*Berchemia*).   **Rhamnaceae**, p. 533

        25. Venation of leaf blades pinnate, but the major lateral veins not or scarcely evident, or, if evident, then not angling-ascending conspicuously parallel to each other; leaf margins entire or serrate.

          26. Lateral veins of leaf blades not or scarcely evident; stems scrambling, many or most of the lateral branches of leader-shoots rigidly divaricate and serving to anchor the stems as they grow among themselves or through other vegetation; leaf margins entire; plants of wet places; flowering and fruiting in autumn, flowers numerous, in involucrate heads having both disc and ray flowers (*Aster*).   **Compositae**, p. 200

          26. Lateral veins of leaf blades evident but not angling-ascending parallel to each other; stems climbing by twining although they may run along the ground if nothing nearby upon which to twine; leaf margins serrate; plants of mesic woodlands; flowering in late spring, flowers small and inconspicuous, borne singly from the leaf axils on dangling stalks, unisexual, plants monoecious.   **Schisandraceae**, p. 650

Key 2
Plant a shrub or tree and the leaves compound.

1. Leaves opposite.
  2. The leaves palmately compound.
    3. Leaflets mostly 5 per leaf but varying from 3 to 9, lanceolate or lance-elliptic, margins generally entire, lower surfaces densely and compactly gray-tomentose; leaf scars raised, each subtending a small, densely brown-hairy bud (*Vitex*). **Verbenaceae,** p. 689
    3. Leaflets mostly 5 per leaf, infrequently 3, 4, or 7, elliptic, elliptic-oblong or -obovate, margins irregularly serrate or doubly serrate, lower surfaces, if pubescent, then not densely and compactly gray-tomentose; leaf scars shield-shaped or obcordate, not raised, subtending an obviously scaly bud (*Aesculus*). **Hippocastanaceae,** p. 377
  2. The leaves pinnately compound or 3-foliolate.
    4. Lateral buds enclosed by the bases of the petioles; leaf scars joining each other and encircling the stem (*Acer negundo*). **Aceraceae,** p. 98
    4. Lateral buds in the leaf axils; leaf scars not joining each other and encircling the stem.
      5. Trees; flowering before new shoot growth of the season commences in spring; flowers unisexual (rarely one bisexual), staminate and pistillate on separate plants; fruit a single-winged samara; margins of leaflets entire, or if toothed then the teeth few and relatively obscure (*Fraxinus*). **Oleaceae,** p. 496
      5. Shrubs; flowering occurring as shoot growth of the season advances or later; flowers bisexual; fruit a berrylike drupe or an inflated capsule; margins of leaflets finely serrate, teeth many.
        6. Leaves trifoliolate (rarely with 5 leaflets); flowers few, borne in loose panicles on developing shoots of the season; corolla erect-campanulate; fruit 3–5 cm long, with a thin, papery, inflated, bladderlike wall (*Staphylea*). **Staphyleaceae,** p. 656
        6. Leaves 1-pinnately compound or partially to wholly bipinnately compound; leaflet number usually 7 or more on most of the larger leaves of a given plant having 1-pinnately compound leaves and ultimate leaflets more numerous on more complex leaves; flowers small, borne terminally on branches in erect, more or less flat-topped, many-flowered cymes 1–4 dm broad; corolla rotate; fruit a berrylike, purplish black drupe 4–6 mm long and containing 3–5 stone-covered seeds (*Sambucus*). **Caprifoliaceae,** p. 180
1. Leaves alternate.
  7. The leaves palmately compound (stem and leaves bearing prickles) (*Rubus*). **Rosaceae,** p. 544
  7. The leaves trifoliolate, 1-pinnately compound, or decompound.
    8. The leaves trifoliolate.
      9. Stems and leaves bearing prickles (*Rosa, Rubus*). **Rosaceae,** p. 544
      9. Stems and leaves not bearing prickles.
        10. Stems bearing hard, sharp, simple or compound thorns; petioles winged (*Poncirus*). **Rutaceae,** p. 606
        10. Stems not bearing thorns; petioles not winged.
          11. (Three entries numbered "11".) Shrub, commonly multistemmed, woody below but distal branches often winter-killed; fruit a small, 1-seeded, indehiscent, nearly flat pod 5–8 mm long; [Corolla papilionaceous] (*Lespedeza*). **Leguminosae,** p. 419
          11. Shrub or small tree [surfaces of leaflets glandular-punctate]; fruit a samara, the wing conspicuous, completely surrounding the seed-body, notably punctate (*Ptelea*). **Rutaceae,** p. 606
          11. Plant a low shrub or woody vine; fruit a small, dryish drupe (*Rhus aromatica, Toxicodendron radicans* and *T. toxicarium*, the latter two POISON IVY and POISON OAK). **Anacardiaceae,** p. 108
    8. Leaves 1-pinnately compound or pinnately decompound.

12. The leaves, all of them on a given plant, 1-pinnately compound.

13. Leaflets, most of them on a given leaf, with a blunt tooth on either side near the base, each tooth bearing a gland on its lower surface [occasional leaflets with more than 2 teeth] (*Ailanthus*). **Simaroubaceae,** p. 652

13. Leaflets not as above.

14. Plant a low colonial shrub to about 8 dm tall, stems slenderly wandlike, the wood of both stems and roots yellow; flowers and fruits in flexuous, pendulous racemes or narrow panicles; leaflets very strongly serrate, or lobulate and the lobes strongly serrate (*Xanthorhiza*). **Ranunculaceae,** p. 526

14. Plant not having the above combination of characteristics.

15. Margins of leaflets toothed.

16. The margins of leaflets having indistinct, rather long crenations with sessile, small, round glands in the low sinuses between the crenations (*Zanthoxylum*). **Rutaceae,** p. 606

16. The margins of the leaves not as above.

17. Leaves having persistent stipules at least partly attached to the petioles (*Rosa*). **Rosaceae,** p. 544

17. Leaves without stipules.

18. Lower surfaces of blades or leaflets grayish or whitish glaucous; axillary buds enclosed by the petiole bases; crushed, fresh foliage not aromatic (*Rhus glabra*). **Anacardiaceae,** p. 108

18. Lower surfaces of leaflets not grayish or whitish glaucous; axillary buds in leaf axils; crushed, fresh foliage with a pungent aroma. **Juglandaceae,** p. 38

15. Margins of leaflets entire.

19. Leaves, all or most of them on a given tree, even-pinnately compound (terminal leaflets paired).

20. Leaflets relatively small, 1–3 cm long, oblong or lance-oblong, rounded or nearly so and apiculate apically (*Sesbania*). **Leguminosae,** p. 419

20. Leaflets relatively large, 5–15 cm long, lanceolate to lance-ovate, apices usually acuminate and the tips of the acuminations blunt. **Sapindaceae,** p. 628

19. Leaves 1-odd-pinnately compound (terminal leaflet solitary).

21. Stems, petioles, and proximal portions of leaf axes abundantly armed with purplish, gland-tipped vesture of very unequal lengths, the vesture in part becoming prickly (*Robinia hispida*) *or* stems (although not necessarily those of the current season or even all stems of previous years' growth) bearing paired hard, sharp, stipular spines at the nodes (*Robinia pseudoacacia*). **Leguminosae,** p. 419

21. Stems not armed.

22. Leaf rachises winged between the leaflets (*Rhus copallina*). **Anacardiaceae,** p. 108

22. Leaf rachises not winged.

23. Leaflets mostly oblongish or obovate, usually not over 5 cm long and 3 cm broad, their apices rounded although sometimes minutely apiculate; fruit a legume, not whitish.

24. Leaflets punctate beneath; pubescence of unbranched, unappressed hairs (*Amorpha*). **Leguminosae,** p. 419

24. Leaflets not punctate beneath; pubescence of hairs attached medially and with 2 strongly divaricate, appressed branches (*Indigofera*). **Leguminosae,** p. 419

23. [POISON SUMAC] Leaflets oblong, elliptic, or ovate, 5–10 cm long and 2–5 cm broad, their apices tapered to obtuse to short-acuminate tips; fruit a whitish, dry drupe (*Toxicodendron vernix*). **Anacardiaceae,** p. 108

12. The leaves (at least some of them on a given plant) decompound.

25. Leaves 2-, 3-, or 4-odd-pinnately compound.

26. Plant an evergreen shrub (some leaves present on stem of a previous season);

leaves 3–4-compound with 3 primary divisions from a joint at the end of a basally clasping petiole; ultimate leaflets having entire margins (*Nandina*).
Berberidaceae, p. 161

26. Plant a deciduous shrub or tree; leaves both twice- and thrice-compound on a given plant; ultimate leaflets, for the most part, with their margins toothed or toothed-lobed.

  27. Plant a shrub or small tree, the main stem canelike and armed with stiff, sharp, straight or curved prickles; leaf scars narrow and obliquely nearly encircling the stem (*Aralia*). **Araliaceae**, p. 153

  27. Plant potentially a medium-sized tree (often flowering/fruiting when of low, shrublike stature), the stem unarmed; leaf scars conspicuously 3-lobed (*Melia*).
Meliaceae, p. 471

25. Leaves even-pinnately decompound (or in the case of *Gleditsia* some of them on a given tree once-compound).

  28. Ultimate leaflets 3–12 mm long and 2–4 mm broad; stems unarmed or armed with short, paired, stipular spines.

  29. The ultimate leaflets conspicuously inequilateral, the midrib close to one margin of the leaflet; overall length of leaves 10–30 cm or a little more, few of them as little as 10 cm, and to 15 cm broad; twigs unarmed; flowers small, many of them borne in pink, powderpuff-like heads on branchlets of the season (*Albizia*). **Leguminosae**, p. 419

  29. The ultimate leaflets slightly inequilateral, the midrib nearly central; overall length of leaves 3–5 cm; twigs armed with paired stipular spines; flowers small, many of them borne in globular, yellow heads from buds on usually leafless woody twigs of the previous season (*Acacia*). **Leguminosae**, p. 419

  28. Ultimate leaflets 15–30 mm long and 5–12 mm broad; stems of most (but not all) individuals armed with relatively large unbranched or branched thorns (*Gleditsia*). **Leguminosae**, p. 419

Key 3

Plant a shrub or a tree.
Leaves simple, all opposite, *or* in some sometimes both opposite and whorled on the same individual, *or* leaves of lower stems opposite, upper ones alternate, *or* chiefly on seedlings, saplings, or sprouts (of *Broussonetia*), sometimes opposite, in whorls of three, and alternate on the same individual.

1. Leaves evergreen, some present on twigs of a previous season (not including plants some or most of whose leaves overwinter and drop in spring).

  2. Plant a shrub parasitic *and* epiphytic on the limbs and branches of various dicotyledonous trees. **Loranthaceae**, p. 454

  2. Plant not parasitic *and* epiphytic.

    3. Leaves stipulate or, if stipules shed, then stipular ring-scars evident.

      4. Leaf blades notably leathery, elliptic, elliptic-obovate, or oblong, 4–12 (15) cm long and 1.5–5 cm broad, apically broadly obtuse to rounded; fruit cone-shaped, containing a single seed which germinates while the fruit is still on the plant, the first root becoming 2.5–3 dm long, fleshy and clublike before the fruit drops; for the most part a plant of maritime situations (*Rhizophora*). **Rhizophoraceae**, p. 541

      4. Leaf blades of firm texture but not leathery, not exceeding about 6 cm long and 4 cm broad, ovate to oblong, lanceolate, or slightly obovate, apices obtuse, short-acuminate, or acute; fruit a white drupe 4–5 mm in diameter; seeds not germinating while fruit is on the plant (*Chiococca*). **Rubiaceae**, p. 598

    3. Leaves without stipules.

5. Herbage with a strong, distinctly minty aroma; one or both surfaces of leaf blades punctate and the punctae, for the most part, atomiferous-glandular (in some, the punctae and glandularness more or less obscured by dense pubescence); flowers bilaterally symmetrical, the corolla 2-lipped. **Labiatae,** p. 399

5. Herbage without a minty aroma; leaf surfaces, if punctate, not atomiferous-glandular; flowers radially symmetrical.

  6. Leaves needlelike (their surfaces not punctate), 8–12 mm long and 1 mm broad or slightly more, their adaxial sides very strongly revolute, rolled around so that their edges meet beneath, thus the abaxial surfaces completely or very nearly obscured (*Ceratiola*). **Empetraceae,** p. 231

  6. Leaves not needlelike or, if needlelike, then one or both surfaces punctate.

    7. Leaf blades very densely and compactly gray-pubescent beneath; plant generally in maritime situations (*Avicennia*). **Avicenniaceae,** p. 156

    7. Leaves not densely gray-pubescent beneath; plants not in maritime situations.

      8. The leaf blades punctate at least on one surface; corolla with 4 or 5 distinct, inequilateral and oblique, yellow petals; stamens numerous (in those kinds treated here); fruit a capsule (*Hypericum*). **Guttiferae,** p. 346

      8. The leaf blades not punctate; corollas white or cream-colored, united below, with 4 small lobes; stamens 2; fruit a bluish black, berrylike drupe (*Ligustrum, Osmanthus*). **Oleaceae,** p. 496

1. Leaves deciduous, none on twigs of a previous season.

  9. Herbage with a strong, distinctly minty aroma. **Labiatae,** p. 399

  9. Herbage not having a minty aroma.

    10. Leaf blades, all or some of them on a given plant, lobed.

    11. Venation of leaf blades definitely pinnate; bark of older, larger stems peeling into papery patches yielding a varicolored mottling effect; young stems of the season with a covering of copious, rust-colored, cottony pubescence; inflorescences large, showy, pyramidal panicles borne terminally on branches of the season (*Hydrangea quercifolia*). **Saxifragaceae,** p. 641

    11. Venation of leaf blades palmate or subpalmate (that is, having 2 lateral veins and the midrib subequal and arising together from the base of the blade); bark of larger, older branches not peeling as described above; inflorescences various, but not as described above.

      12. Leaves on branches of crowns of mature trees for the most part unlobed, some of them alternate, some opposite or subopposite, their blades seldom lobed; leaves on seedlings, saplings, or sprouts variably unlobed, with 1 small lobe on one side, with a large sinus on one side and somewhat mittenlike, or prominently and symmetrically 3–5-lobed, usually some of the leaves alternate, some opposite or subopposite, sometimes some in whorls of three; venation subpalmate (*Broussonetia*). **Moraceae,** p. 478

      12. Leaves all opposite, their blades, for the most part, uniformly, symmetrically, and palmately 3–5-lobed.

        13. Plant a shrub to about 2 m tall; margins of the leaf blades (including the lobes) dentate-serrate, the teeth short and blunt; inflorescences long-stalked, more or less flat-topped, compound cymes borne terminally on branchlets of the season (*Viburnum acerifolium*). **Caprifoliaceae,** p. 180

        13. Plant potentially a tree; margins of the leaf blades (including the margins of the lobes) entire, *or*, if toothed, then the blades generally 5–lobed, the terminal lobe and usually the upper two lateral ones prominently and sharply toothed; inflorescences racemes, panicles, or fascicles on short, few-leaved branchlets *or* from buds on woody stems before new shoot emergence (*Acer* spp.). **Aceraceae,** p. 98

    10. Leaf blades all unlobed.

    14. Leaves, the lower ones on the stem, opposite, becoming alternate upwardly; flowers small and in involucrate heads (*Iva*). **Compositae,** p. 200

    14. Leaves all opposite, or some opposite and some whorled.

15. The leaves with stipules that connect between the petiole bases and leaving, when they fall, a scar connecting between the leaf scars or quite around the stem. **Rubiaceae,** p. 598

15. The leaves without stipules, or, if stipules or stipule scars present, then not as described above.

16. Leaf blades broadly ovate, cordate to truncate basally, acuminate apically, 1–3 dm long, to 1.8 dm broad, longest ones about twice as long as broad, shortest ones about as broad as long (*Catalpa*). **Bignoniaceae,** p. 171

16. Leaf blades not ovate, or if ovate, then none of them as large as indicated above.

17. Plant a low, commonly colonial shrub of swampy or marshy places; stems usually arching, some of them with the tips reaching the substrate, rooting, and forming new stems; submersed portions of stems very soft-corky, emersed portions with bark exfoliating in long, thin, cinnamon-colored strips; inflorescences shortly stalked, short cymes, the opposite pairs giving a verticillate appearance (*Decodon*). **Lythraceae,** p. 454

17. Plant not having the above combination of characteristics.

18. Plants of maritime situations; leaf blades subsucculent; flowers borne in radiate, involucrate heads borne singly and terminally at tips of branches (*Borrichia*). **Compositae,** p. 200

18. Plants not of maritime situations; leaf blades not subsucculent; flowers not in involucrate heads.

19. Lateral veins of leaf blades relatively prominent (especially on their lower surfaces) and angling-ascending straight and approximately parallel to each other, or if some of them branched, then the branches straight and angling-ascending in a V-like manner (*Viburnum dentatum*). **Caprifoliaceae,** p. 180

19. Lateral veins not as above.

20. Buds, young stems and petioles, lower surfaces of leaf blades (especially the midribs), and axes of cymes (at flowering stage) more or less rusty-scaly or -scurfy (*Viburnum* spp.). **Caprifoliaceae,** p. 180

20. Buds, etc., not rusty-scaly or -scurfy.

21. (Three entries numbered "21.") Plant a shrub or small tree, usually with main stems arching or leaning, vigorous leader shoots with opposite, stoutish, stiff, hard, divaricate, recurved, or slightly ascending branches, such branches often thorn-tipped; flowers unisexual and staminate or bisexual and functionally pistillate (*Forestiera*). **Oleaceae,** p. 496

21. Plant an arching or weakly erect shrub or vinelike and scrambling among other plants; branches of vigorous leader shoots usually alternate (owing to one of the opposite buds remaining dormant), divaricate or nearly divaricate, slender, often spine-tipped; flowers bisexual, borne in axillary, and sometimes terminal, interrupted spikes (*Sageretia*). **Rhamnaceae,** p. 533

21. Plant with woody, erect stems, *or* plant suffrutescent, the upper stems usually herbaceous and (in our range) often becoming winter-killed; vigorous leader shoots not having hard, stiff, short, divaricate or recurved branches.

22. Leaf blades, some of them or all of them on a given plant, marginally toothed although the teeth may be inconspicuous.

23. Marginal teeth of the blades minute, inconspicuous, and gland-tipped (*Euonymus*). **Celastraceae,** p. 194

23. Marginal teeth of the blades not gland-tipped.

24. Margins of leaf blades irregularly disposed, with a few small and inconspicuous dentate teeth on a side; flowers borne singly or in 3-flowered cymes terminally on branchlets of the season (*Philadelphus*). **Saxifragaceae,** p. 641

27

24. Margins of leaf blades with numerous serrate or crenate teeth on a side; flowers not borne as above.

    25. Leaf blades with margins serrate, the teeth abruptly narrowed to pointed tips; flowers borne in stalked, many-flowered, compound cymes borne terminally on branches of the season (*Hydrangea arborescens*).   **Saxifragaceae,** p. 641

    25. Leaf blades with margins crenate-serrate or crenate, the teeth blunt or rounded; flowers borne in sessile or nearly sessile axillary cymes *or* in axillary, longish-stalked, headlike spikes.

        26. Flowers/fruits borne in sessile or nearly sessile, short, compound, axillary cymes; fruits 4–stoned, lavender-pink, magenta, or violet, globose drupes (*Callicarpa*).   **Verbenaceae,** p. 689

        26. Flowers/fruits borne on longish stalked, axillary, headlike spikes; fruits 2–stoned, metallic-blue drupes (*Lantana*).
        Verbenaceae, p. 689

22. Leaf blades with entire margins.

    27. The leaf blades, most of them on a given plant, not over 5 cm long and not over 3 cm broad, short-ovate, elliptic, oval, or subrotund.

        28. Woody stems with light brown exfoliating bark; inflorescences inconspicuous, axillary, short, bracteate spikes (*Symphoricarpos*).
        Caprifoliaceae, p. 180

        28. Woody stems with buffish, tight bark; inflorescences conspicuous, pyramidal, many-flowered panicles borne terminally on branchlets of the season (*Ligustrum sinense*).   **Oleaceae,** p. 496

    27. The leaf blades, most of them on a given plant, with considerably greater dimensions than given in no. 27 above, ovate, lance-ovate, varying to obovate.

        29. Stems usually arching; bark of year-old and older stems buffish and exfoliating; flowers borne singly or in 3-flowered cymes terminating branchlets of the season; petals 4, white and showy, open corolla 3.5–5.5 cm across; fruit a capsule (*Philadelphus*).
        Saxifragaceae, p. 641

        29. Stems not arching; bark not exfoliating; flowers not as described above; fruit not a capsule.

            30. Fresh leaves, when crushed, with a spicy aroma; flowers borne singly and terminally on short, lateral branchlets of the season; perianth parts (tepals) numerous, strap-shaped, greenish brown or purplish brown; fruits small nutlets enclosed in a fleshy, indehiscent floral tube; plant a shrub (*Calycanthus*). **Calycanthaceae,** p. 177

            30. Fresh leaves, when crushed, without a spicy aroma; flowers small, 4-merous, in one (of our) species borne in congested headlike cymes surrounded by four large, white, petaloid bracts, in other species borne in open, more or less flat-topped, compound cymes; petals white, greenish, or yellowish green; fruit a berrylike drupe; plants shrubs or small to medium-sized trees (*Cornus*).
        Cornaceae, p. 215

Key 4
Plant a shrub or tree. Leaves simple and alternately arranged.

(Note that on some kinds of plants the alternate leaves may be very closely approximate at the ends of branches or lateral spur shoots and detectable as alternate only after careful scrutiny)

1. Leaves sessile, scalelike, broadest basally and more or less clasping the stem, their abaxial surfaces punctate. **Tamaricaceae,** p. 668

1. Leaves not scalelike.

  2. Leaves succulent, sessile, linear-oblanceolate to spatulate (commonly with very short, leafy branches in the axils giving a fascicled appearance), variable in length to about 2.5 cm; lateral branches, many of them, spine-tipped (*Lycium*). **Solanaceae,** p. 654

  2. Leaves not succulent.

    3. Stems with ocreae, that is, tubular sheaths surrounding all or part of the internodes, the leaves articulated at the summits of the ocreae (*Polygonella*). **Polygonaceae,** p. 521

    3. Stems without ocreae.

      4. Flowers borne in involucrate heads. **Compositae,** p. 200

      4. Flowers not borne in involucrate heads.

        5. Leaf blades palmately or subpalmately veined (lowermost lateral veins and the midrib equal or subequal and arising together at the base of the blade or very slightly above it).

          6. Petioles with a pair of conspicuous red glands on the upper side at the summit and immediately adjacent to the bases of the blades (*Aleurites*). **Euphorbiaceae,** p. 278

          6. Petioles not having glands as described above.

            7. Upper surfaces of leaf blades with yellowish, calluslike growths in the axils formed by the midrib and lowermost, relatively prominent lateral veins (these lateral veins on some leaves on a given plant arising with the midrib at the blade base, in other leaves arising from the midrib somewhat above the base of the blade); fresh crushed leaves with a distinctly camphorous odor (*Cinnamomum*). **Lauraceae,** p. 409

            7. Upper surfaces of leaf blades without callosities as described above; fresh crushed leaves without a camphorous odor.

              8. Leaf blades with a very scant amount of blade tissue basally (no more than 1 cm in longitudinal extent), otherwise very conspicuously divided into 5–11 elongate, fingerlike or leaflike lobes or segments, the smaller leaves with fewest segments (*Manihot*). **Euphorbiaceae,** p. 278

              8. Leaf blades unlobed or, if lobed, then with a much greater proportion of undivided blade than as described above.

                9. The leaf blades, at least some of them on a given plant, lobed. (However, see first no. 14 below regarding leaves of mature crowns of *Morus*.)

                  10. The leaf blades predominantly and conspicuously with 5 more or less triangular-acute lobes giving much the effect of a half-star (*Liquidambar*). **Hamamelidaceace,** p. 372

                  10. The leaf blades not having lobing giving the effect described above.

                    11. Plant a multistemmed shrub to 1–3 m tall.

                      12. The plant unarmed; blades of leaves of vigorous leader shoots mostly with 3 major, blunt-tipped lobes, the terminal lobe much the largest, blades of leaves of secondary, tertiary shoots, etc., variably unlobed or irregularly weakly lobed; pubescence of parts stellate; flowers borne in short, erect, umbellike racemes terminating short, lateral shoots of the season (*Physocarpus*). **Rosaceae,** p. 544

                      12. The plant armed (at most of the nodes) with shortly needlelike, unbranched or basally 2–3-forked, sharply pointed spines; blades of all leaves with 3 equal major lobes; pubescence of parts not stellate; flowers borne singly and pendulous from one of the axils of the 2–several leaves of lateral short-shoots (*Ribes*). **Saxifragaceae,** p. 641

                    11. Plant potentially a tree.

                      13. The plant with leaves having foliaceous stipules encircling the twig, leaving when they fall a stipular line-scar around the twig; petioles with dilated bases within which lateral buds are formed; bark of older branches (and often of the upper trunk) peeling off in irregular, thin plates exposing tan, green, or chalky white inner bark; leaf blades generally as broad as or broader than long, their edges with irregular, conspicuous, salient teeth, or both lobed and toothed (*Platanus*).

Platanaceae, p. 519

13. The plant not having the above combination of characteristics.
  14. Sap milky or viscid; woody twigs tan; leaves of seedlings, saplings, or sprouts ovate in overall outline, variably unlobed, with one small lobe on one side, with a large sinus on one side and somewhat mittenlike, or prominently and symmetrically 3–5-lobed; blades of leaves of crowns of mature plants generally unlobed; margins of blades other than the lobing, if any, with many teeth (*Morus*).   **Moraceae**, p. 478
  14. Sap neither milky nor viscid; woody twigs green [buds brown-velvety]; leaf blades varying very much in size and form, 1–3 dm long, mostly as broad as or broader than long, on a given tree few of them broadly ovate and unlobed, usually 3–5-lobed, lobes of largest ones sometimes somewhat fingerlike, sometimes very broad; bases of blades varying from subtruncate to markedly cordate; margins of blades, apart from the lobing, if any, entire (*Firmiana*).

Sterculiaceae, p. 656

9. Leaf blades all unlobed.
  15. Petioles swollen or dilated (with pulvini) at their bases and summits; flowers appearing in spring before new shoot growth commences, borne from winter buds in sessile clusters, bilaterally symmetrical, corollas light to dark pink or magenta; fruit a flattened, linear-oblong legume (*Cercis*).

Leguminosae, p. 419

  15. Petioles without pulvini; flowers and fruits not as above.
   16. The petioles one-fourth to one-half as long as the blades; lowermost lateral veins of blades extending ± at right angles at the blade base; inflorescences axillary, with few to numerous flowers in long-stalked cymes, the stalk of each cyme united part of the way to a distinctive and conspicuous, straplike, oblong to spatulate bract, the bracteate inflorescences or infructescences evident during much of the growing season (*Tilia*).   **Tiliaceae**, p. 675
   16. The petioles less than one-fourth as long as the blades; lowermost lateral veins of the blades strongly angling-ascending; no straplike bracts, as described above, present.
    17. Leaves 2-ranked. (*Celtis* and *Ziziphus* in).

Ulmaceae and **Rhamnaceae**, pp. 677, 533

    17. Leaves more than 2-ranked. (*Ceanothus*).   **Rhamnaceae**, p. 533
5. Leaf blades pinnately veined (in some, however, only the midrib clearly evident).
  18. Leaves and leaf scars 3-ranked, i.e., alternating in such a way that they are in 3 vertical ranks on the stem, *or* 2-ranked, i.e., alternating on either side and thus in 2 vertical ranks.
   19. The leaves or leaf scars 3-ranked (*Alnus*).   **Betulaceae**, p. 161
   19. The leaves or leaf scars 2-ranked.
    20. Margins of leaf blades entire; pith of twigs diaphragmed, eventually chambered between the diaphragms.
     20a. Lowermost lateral veins paired, arising with midvein at base of the blade and appearing subequal with the midvein (*Celtis*).   **Ulmaceae**, p. 677
     20a. Lowermost lateral veins not as above (*Asimina*).   **Annonaceae**, p. 118
    20. Margins of leaf blades (at least some on a given plant) variously toothed, or undulated; pith of twigs homogeneous (except in *Celtis*).
     21. Principal lateral veins not equally prominent and not angling-ascending parallel to each other.
      21a. Lowermost lateral veins paired, arising with midvein at base of blade and appearing subequal with midvein (*Celtis*).   **Ulmaceae**, p. 677

30

21a. Lowermost lateral veins not paired, arising above the base of blade (*Stewartia*). **Theaceae**, p. 670

21. Principal lateral veins of the leaf blades equally prominent and angling-ascending straight and parallel to each other toward or to the leaf margins.

  22. Margins of leaf blades undulate (at least some of them on a given plant); apices of blades rounded (*Fothergilla, Hamamelis*). **Hamamelidaceae**, p. 372

  22. Margins of leaf blades not undulate; apices of blades acute or acuminate.

    23. The margins of leaf blades dentate or serrate but not doubly so.

      24. Principal lateral veins ending in the marginal teeth (*Castanea, Fagus*). **Fagaceae**, p. 288

      24. Principal lateral veins branching before reaching the blade margins; [bark scaly or flaky, grayish brown, exposing reddish brown inner bark as it sloughs] (*Planera*). **Ulmaceae**, p. 677

    23. The margins of leaf blades, for the most part, doubly serrate, sometimes unequally and irregularly so

      25. Leaf blades mostly triangular-ovate or subrhombic; bark of trunks exfoliating in papery, curled scales (*Betula*). **Betulaceae**, p. 161

      25. Leaf blades mostly with shape other than triangular-ovate or subrhombic; bark of trunks smooth to ridged and grooved, or exfoliating in irregular shreds or scales but not in papery, curled scales.

        26. The leaf blades mostly symmetrical and equilateral; infructescences usually present on a given tree throughout the season, the individual fruits subtended either by foliaceous bracts or enclosed in saclike bracts (*Carpinus* and *Ostrya*). **Betulaceae**, p. 161

        26. The leaf blades, at least some of them on a given branch, inequilateral; fruits shed by the time the leaves are fully developed *or* flowering and fruiting in autumn (*Ulmus*). **Ulmaceae**, p. 677

18. Leaves and leaf scars more than 3-ranked.

27. Stipular line scars present as a ring all the way around the twig at the nodes of younger branches.

  28. Leaf blades 4–6-lobed, the margins otherwise entire, the leaf apices truncate to broadly V-notched; fruits terminally winged samaras borne in erect "cones" (*Liriodendron*). **Magnoliaceae**, p. 458

  28. Leaf blades unlobed (save for basal auricles in some), margins entire, apices obtuse to acuminate; fruits folliculate, borne in erect "cones" (*Magnolia*). **Magnoliaceae**, p. 458

27. Stipule line-scars, if present, not as a line or ring all the way around the stem.

  29. Buds covered by a single, caplike scale; [flowers or fruits in catkins early in the growing season] (*Salix*). **Salicaceae**, p. 615

  29. Buds with more than one scale, or naked (without scales).

    30. Petioles narrowly to broadly winged *and* stems bearing sharp thorns (*Citrus aurantium*). **Rutaceae**, p. 606

    30. Plant not having the combined two features described above.

      31. Plant very often profusely clonal by subterranean runners; aerial stems low, not over 4 dm tall, usually less; flowers small, borne in terminal, panicled cymes (*Licania*). **Chrysobalanaceae**, p. 197

      31. Plant not as above.

        32. Plant a shrub with distinctive, leathery, tough, and very pliable bark; distal portion of annual increments of growth becoming dilated-funnelform and capped by short-conical, cottony-cushionlike, pseudoterminal bud, the lateral buds similar and borne above angular-diverging leaf scars; flowers from buds on woody twigs, in short-stalked clusters of 2–4, 3–4 hairy bud scales subtending the flower cluster (*Dirca*). **Thymelaeaceae**, p. 673

32. Plant not having combination of features described above.

33. Fruit a nut (acorn) seated in a cuplike or bowllike involucre of many imbricated scales (*Quercus*).   **Fagaceae,** p. 288

33. Fruit not as above.

34. Fresh foliage and/or twigs notably aromatic when crushed or bruised.

35. Leaves evergreen.

36. Surfaces of leaf blades (one or both) punctate-dotted (suitable magnification usually required in order to perceive the punctae).

37. Midribs of leaf blades evident but lateral veins not at all or only faintly evident; flowers solitary from the leaf axils, bisexual; pistils several to numerous in a circle; fruits folliculate (*Illicium*).   **Illiciaceae,** p. 381

37. Midribs and major lateral veins of leaf blades clearly evident; flowers unisexual (plants dioecious), borne in scaly-bracted, erect catkins axillary to leaf scars or leaves; pistil 1 in the pistillate flowers; fruit a waxy, nutlike drupe (*Myrica*).   **Myricaceae,** p. 483

36. Surfaces of leaf blades not punctate.

38. Margins of all leaf blades entire; stipules none (*Persea*).

Lauraceae, p. 409

38. Margins of some leaf blades on a given plant entire but some blades with spiculelike teeth on each side (such toothed leaves occurring especially on seedlings, saplings, and sprouts); stipules present at first, soon deciduous and leaving short, stipular line-scars to either side of the petioles or leaf scars (*Prunus caroliniana*).   **Rosaceae,** p. 544

35. Leaves deciduous, none on twigs of the previous season.

39. Leaf blades with serrate margins; stipules or short, stipule line-scars present (*Prunus alabamensis* and *P. serotina*). **Rosaceae,** p. 544

39. Leaf blades with entire margins, or if lobed, the margins entire other than the lobing; stipules none (*Lindera, Litsea, Sassafras*).

Lauraceae, p. 409

34. Fresh foliage and/or twigs not notably aromatic when crushed or bruised.

40. Plant a "woody herb," stems to about 3 m tall, the lower stem woody, wood very hard, bark brown, exfoliating in longish, thin strips, upper stems herbaceous (in our area, stems usually winter-killed to or nearly to the ground, new stems formed during the following growing season); leaf blades mostly lanceolate, varying to lance-elliptic or lance-ovate; uppermost herbaceous branches bearing flowers singly from the axils of their smallish leaves; flowers with 4, rarely 5, bright yellow, obovate petals about 2.5 cm long and as broad distally (*Ludwigia peruviana*).   **Onagraceae,** p. 518

40. Plant not having the above combination of characteristics.

41. Wood of stems relatively soft, not dense, extremely light in weight.

42. Petioles 0.1–1.4 cm long; leaf blades 3–8 cm long, lanceolate to linear-lanceolate, narrowly elliptic, or rhombic, their margins very finely appressed-crenate, the teeth tipped by tiny callosities; other than the midvein (lower surface), venation not or only faintly evident (*Stillingia*).   **Euphorbiaceae,** p. 278

42. Petioles 2.0–4.0 cm long; leaf blades up to 17 cm long, their margins entire; lateral veins evident on both surfaces, especially the lower surfaces (*Leitneria*).   **Leitneriaceae,** p. 448

41. Wood of stems relatively hard and dense, not notably light in weight.

43. Vestiture of young stems, petioles, at least the lower leaf surfaces, inflorescence axes, flower stalks, perianth, and fruits composed of rust-colored, gray or silvery scales.

44. Plant a shrub or vinelike scrambler; flowers with a floral tube base of which surrounds the ovary, above the ovary it is flaring-obconic and surmounted by 4 small, petaloid calyx segments; corolla none (*Elaeagnus*). **Elaeagnaceae, p. 229**

44. Plant a shrub or small tree; flowers with no floral tube (ovary superior); calyx small, (usually) 5-lobed, corolla urceolate and with (usually) 5 very small lobes (*Lyonia ferruginea* and *L. fruticosa*). **Ericaceae, p. 234**

43. Vestiture of parts, if any, not composed of scales.

45. Leaf blades having small, dark reddish purple glands along the midribs of their upper surfaces (*Aronia*). **Rosaceae, p. 544**

45. Leaf blades not having glands as described above.

46. Plants usually having thorns on some of their branches, or having lateral short-shoots that are thorn-tipped, or both. (Note, however, that an occasional individual in a local population of the kinds in this grouping may not bear thorns.)

47. Stipules or stipule scars none.

48. Sap not milky or viscid; flowers borne in shortly stalked, few-flowered clusters axillary to leaves; sepals 4, minute, not persistent below the fruit; petals 4, linear-oblong, 6–10 mm long, recurved distally, each bearing adaxially 2 conspicuous bands of dense, longish, pointed, stiff hairs the uppermost of which are extruded from the open corolla (*Ximenia*). **Olacaceae, p. 496**

48. Sap milky or viscid; flowers borne in umbels or fascicles axillary to leaves, each umbel usually having several to numerous flowers (infrequently borne singly or few per umbel), the umbels usually evident for a considerable period of time before any flowers reach anthesis; calyx small, united basally, 5-lobed; corolla 5-lobed, each lobe with a lateral appendage to either side near its base; adaxial surfaces of petals without hairs as described above (*Bumelia*). **Sapotaceae, p. 630**

47. Stipules or stipule scars present.

49. Sap of leaves and twigs milky; flowers unisexual (plants dioecious); staminate flowers in loose, oblong to globose, stalked heads axillary to leaves on spur-shoots, calyx 4–lobed, stamens 4; pistillate flowers many, compacted in axillary, stalked, spherical heads, each flower with a 4–lobed calyx and a pistil with a nearly spherical ovary and a greatly elongated style, the styles collectively imparting to the inflorescence a peripherally long-hairy aspect; collective "fruit" orangelike in size and form (*Maclura*). **Moraceae, p. 478**

49. Sap not milky; flowers bisexual, borne singly, in racemes, corymbs, umbels, cymes, or clusters, each with a floral tube (either adnate to the ovary or free from the ovary (or ovaries, if more than 1); floral tube bearing about its rim 5 calyx segments, 5 petals, and stamens in multiples of 5, seldom fewer than 10; fruit various but not as above (*Crataegus*, *Malus*, certain *Prunus* spp., *Pyracantha*). **Rosaceae, p. 544**

46. Plants not having thorns or thorn-tipped short-shoots.

50. Pith of twigs with transverse diaphragms.

51. Pubescence of various parts composed of stellate hairs; [plant flowering from buds on wood of the previous season before and as new shoot growth of the season commences; flowers pendent, bisexual, ovary inferior, floral tube bearing on its rim 4 small, triangular calyx segments, a 4–cleft or -lobed, white corolla; fruit indehiscent, 2– or 4–winged] (*Halesia*). **Styracaceae,** p. 660

51. Pubescence of various parts, if any, not stellate.

52. Pith of twigs transversely diaphragmed *and* continuous (homogeneous) between the diaphragms; flowers unisexual or bisexual; [plants androdioecious, that is, some individuals having only staminate flowers, others having bisexual ones; fruit a drupe] (*Nyssa*). **Nyssaceae,** p. 490

52. Pith of twigs transversely diaphragmed *and* chambered between the diaphragms.

52a. Plant evergreen, clearly some leaves of the previous season on twigs of the previous season throughout the current year (*Agarista*). **Ericaceae,** p. 234

52a. Plant deciduous (tardily so in *Symplocos* but no leaves of the previous season on twigs of the previous season after flowering period in spring, deciduous in autumn in *Diospyros*).

53. Twigs with a terminal bud (uppermost leaf or leaf scar with a lateral bud in its axil, terminal bud above it); [leaves, for the most part, overwintering but all of them falling usually just before or as flowering occurs; flowers bisexual, closely set in sessile, yellow, ball-like clusters on twigs of the previous season just before or as new growth commences; fruit an oblong, green drupe about 1 cm long having small, deltoid calyx segments at its summit] (*Symplocos*). **Symplocaceae,** p. 666

53. Twigs with no terminal bud (uppermost bud axillary to a leaf or leaf scar); [leaves deciduous in autumn; flowering occurring on shoots of the season after the leaves are approximately fully developed; flowers axillary to leaves, greenish and inconspicuous, functionally unisexual (plants functionally dioecious), the staminate usually in few-flowered clusters, the pistillate borne singly; fruit a several-seeded, globose or depressed-globose berry to 4–5 cm in diameter, the enlarged, persistent calyx at its base]. (*Diospyros*). **Ebenaceae,** p. 227

50. Pith of twigs homogenous and continuous (without diaphragms).

54. Plant having evergreen leaves (some present on twigs of the previous season), their blades very finely punctate on both surfaces, sometimes showing tiny resinous dots of exudate; flowers unisexual (plants dioecious), borne in erect catkins, perianth none; fruit a waxy drupe with a tuberculate surface (*Myrica*). **Myricaceae,** p. 483

54. Plant not having the above combination of characteristics.

55. Stipules or stipule scars present (both sometimes minute).

56. Plant a tree; leaf blades broadly ovate or deltoid-ovate, many of them as broad as long, larger ones to 10–15 cm long, at maturity thickish and stiff, petioles long and slender (blades making a clacking sound when blown against each other by the wind); flowers unisexual (plants dioecious), borne in pendulous catkins before new shoots emerge in spring (*Populus*).

Salicaceae, p. 615

56. Plant a shrub or tree not having the combination of characteristics described above.

57. Sap milky (observed by severing a branchlet of the season or a petiole); inflorescences narrow, spikelike racemes terminating branchlets of the season; flowers small, unisexual, for the most part the pistillate ones near the bases of the racemes, the staminate above.

58. Plant a fast-growing, small to medium-sized tree (flowering/fruiting when of small stature); leaf blades broadly ovate or rhombic-ovate, mostly as broad as or broader than long, their bases broadly rounded, truncate, or very broadly and shortly tapered (*Sapium*). **Euphorbiaceae, p. 278**

58. Plant a shrub, infrequently to 3–4 m tall, usually 1–2 m; leaf blades narrowly elliptic, oval, or lanceolate, their bases cuneate (*Sebastiana*).

Euphorbiaceae, p. 278

57. Sap not milky.

59. Buds naked (without covering scales), usually only the uppermost (pseudoterminal) one clearly evident, other lateral buds minute, if visible at all, and each appearing just as a tiny mound of hairs; leaf blades mostly oblongish, many of them slightly broadest just above their middles *and* the lateral veins prominently ascending straight and parallel with each other to near the leaf edges where they become much fainter, strongly curve and run close to the leaf edges (*Rhamnus*). **Rhamnaceae, p. 533**

59. Buds with covering scales; leaf blades variously shaped but not with venation as described above.

60. Vascular bundle scar 1 in each leaf scar; flowers usually unisexual (plants dioecious) or less frequently a few flowers bisexual (plants polygamodioecious); fruit a berrylike drupe with the calyx persistent at its base (*Ilex*).

Aquifoliaceae, p. 133

60. Vascular bundle scars (2) 3 in each leaf scar; flowers bisexual; ovary inferior and the other floral parts borne on the rim of a floral tube *or* ovary seated within the base of and free from the floral tube, the perianth and stamens borne on the rim of the floral tube; fruits various but not having a persistent calyx at their bases, the calyx if persistent borne at the summit of the fruit (*Amelanchier, Crataegus, Malus, Pyracantha, Prunus, Pyrus*).

Rosaceae, p. 544

55. Stipules or stipule scars none.

61. Pubescence of parts, at least some of it, composed of hairs 2–several-forked (mostly 2–5) from the base, *or* of stellate hairs, *or* usually a mixture of varying proportions of hairs 2–forked from the base and with a V-form, a U-form, a Y-form, or attached at the middle and with 2 divaricate, usually appressed branches.

62. (Three entries numbered 62.) Plant a tree to about 25 m tall (sometimes flowering/fruiting when of low stature); leaves leathery, evergreen (some present on twigs of the previous season); young stems and proximal portions of petioles pubescent, the hairs mostly 2–5-branched from the base, the lower leaf surfaces with similar but smaller hairs; flowers borne singly from leaf axils, large and showy, about 8 cm across the open, spreading corolla composed of 5 obovoid, white petals (*Gordonia*).   **Theaceae,** p. 670

62. Plant a shrub (or individuals of *Styrax grandifolium* rarely of low, treelike stature); pubescence of parts, some of it or all of it, stellate; flowers very much smaller than in no. 62 above, borne in racemes *or* singly and in few-flowered clusters.

63. Margins of most leaf blades distinctly and regularly serrate upwardly from slightly below their middles; inflorescences longish, narrowly somewhat spirelike, numerous-flowered racemes borne at ends of branchlets of the season; ovary superior; petals 5, separate; stamens 10, 5 in each of 2 whorls, anthers opening by terminal pores (*Clethra*).

Clethraceae, p. 197

63. Margins of leaf blades varyingly entire, slightly and irregularly wavy, or obscurely and irregularly dentate or dentate-serrate; flowers borne singly, in few-flowered clusters on lateral small-leaved short shoots, *or* in axillary, loosely flowered, short racemes; corolla united basally, 5-lobed; stamens 10, filaments adnate to the corolla tube and united in a ring basally, free above, anthers dehiscing longitudinally (*Styrax*).   **Styracaceae,** p. 660

62. Plant a small, deciduous, understory tree, the crown branches more or less spraylike and layered; pubescence of parts usually a mixture of varying proportions of unbranched hairs, 2 branched from the base and with a V-form, with a U-form, with a Y-form, or attached at the middle and with 2 divaricate, usually appressed branches; inflorescences stalked, more or less flat-topped, compound cymes; ovary inferior; calyx segments and petals 4 each; stamens 4 (*Cornus alternifolia*).   **Cornaceae,** p. 215

61. Pubescence of parts, if any, not as described in the first no. 61 above.

64. Stems when young (before becoming woody) having a pair of definite, but narrow, wings extending downward from a given node, generally one of the pair to the second node below, the other extending to the third node below (*Lagerstroemia*).   **Lythraceae,** p. 454

36

64. Stems not winged.

65. Fruit a septicidally or loculicidally dehiscing capsule developed from a superior ovary, *or* a fleshy berry or drupe developed from an inferior ovary; anthers attached basally, inverted at anthesis or during development, dehiscing at the base (the apparent apex) by slits, clefts, or pores and often having awns or spurs; petals at least partially united (excepting in *Befaria*). **Ericaceae,** p. 234

65. Fruit a dry, surficially winged or unwinged drupe; anthers erect, attached near their middles, dehiscing longitudinally; petals separate. **Cyrillaceae,** p. 222

Descriptive Flora

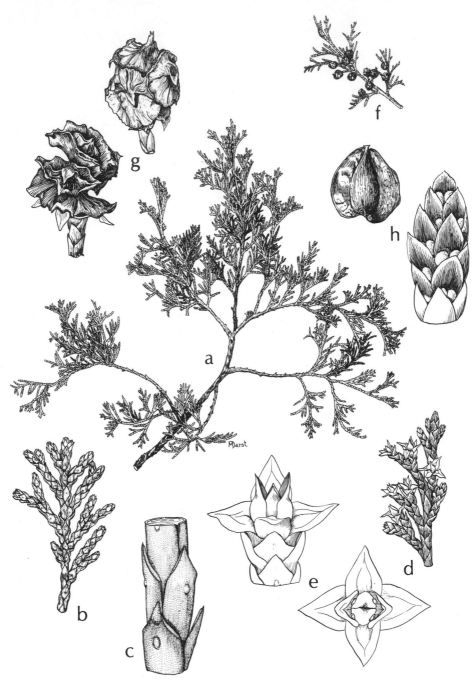

Fig. 1. **Chamaecyparis thyoides:** a, leafy twig; b, branchlet, somewhat enlarged; c, small portion of main axis of branchlet showing abaxial glands on leaves; d, branchlet with young ovulate cones at some tips; e, a tip with young ovulate cone, side view at upper left, face view at lower right; f, branchlet with maturing ovulate cones; g, mature ovulate cone before dehiscence, upper right, after dehiscence, lower left; h, male cone, at right, seed, at left.

Gymnosperms

Cupressaceae (CYPRESS FAMILY)

1. Branchlets disposed in 1 plane and spraylike; plants monoecious; ovulate cones with
very irregular surfaces, opening at maturity.   1. *Chamaecyparis*
1. Branchlets neither disposed in 1 plane nor spraylike; plants dioecious; ovulate cones
with smooth surfaces, berrylike, indehiscent.   2. *Juniperus*

1. Chamaecyparis

Chamaecyparis thyoides (L.) B.S.P. ATLANTIC WHITE CEDAR, JUNIPER.   Fig. 1

Tree to 20–28 m tall, trunks to 1 m d.b.h. Bark of leafless twigs brown, smoothish, that of large trunks dark reddish brown, narrowly fissured and with long,
flat, platelike ridges, on some specimens twisting or spiraling around the
trunks. Leafy branches and their branchlets oriented in approximately 1 plane
and appearing spraylike.

Leaves scalelike, opposite, successive pairs at right angles to each other. Lateral leaves of short shoots keeled on their backs, bases of a given pair with their
edges confluent or nearly so, their free tips angling outward or extending along
the sides of the next pair above, the latter referred to as facial pairs; some or
most leaves with an evident resinous gland more or less medially on the abaxial
surface; lateral branches of short shoots arising from the axil of one of a given
pair of lateral leaves, not from an axil of either of the facial leaves, thus account-
ing for the 1-plane architecture of the branch system; main elongation axes of
the branch system having scale leaves very much longer than those of short
shoots, succeeding pairs essentially alike, their confluent bases 7–8 mm long,
their tips appressed or slightly spreading, triangular and relatively sharply
pointed, resinous glands on the backs of the tips.

Male or ovulate cones borne singly at tips of short branchlets, the staminate
usually on branchlets at the extremity of a branch system, the ovulate farther
back. Male cones very small, with few opposite peltate scales, succeeding pairs
at right angles to each other, each scale bearing 3–4 anther cells marginally
beneath. Ovulate cones of peltate scales, arranged as in the staminate, usually
3–6 pairs, pointed or bossed centrally above, each with 1–2 ovules attached
basally; maturing cones, leathery, edges of scales tightly closed against each
other, more or less globular in outline but more or less angular because of the
way the edges of the scales meet, surfaces glaucous; mature cones mostly 5–8
mm in diameter, becoming woody and opening. Seeds 3–4 mm long, nearly as
broad, brown, winged laterally.

Chiefly inhabiting woodlands along clear or brown cool-water streams, sometimes in depressions in pine flatwoods, n.e. Fla.; Panhandle of Fla. from Liberty
and Gadsden Cos. westward, w. Ala., s.e. Miss. (Coastal plain, Maine to S.C.; cen.
w. Ga., n.e. Fla; cen. Panhandle of Fla. westward, s. Ala., s.e. Miss.)

Plants of n.e. Fla. and of the Panhandle of Fla. from Liberty and Gadsden Cos.
w. to about the Choctawhatchee River (and in s.e. Miss.) may be designated var.
thyoides; those in the w. Panhandle of Fla. and in Ala. may be designated var.

henryae (Li) Little (*C. henryae* Li). A key character of var. *thyoides* is that all or nearly all of the leaves have resinous glands on their backs; on the other hand in var. *henryae* the glands are restricted to leaves on the principal axes of a branch system. Bark of trunks of var. *henryae* is alleged to be *consistently* twisting on the trunks (Li, 1962), but I do not find this to be true. It is conspicuously twisting or spiraling on trunks of some specimens, not on others, this being the case for var. *thyoides* as well.

2. Juniperus (JUNIPERS)

Juniperus virginiana L. EASTERN RED-CEDAR. Fig. 2

Evergreen, aromatic, coniferous tree to about 30 m tall, the crown broadly pyramidal in outline (ours); if growing in the open, lower crown branches commonly persisting and reaching nearly to the ground. Bark of trunks thin, light reddish brown to grayish, sloughing in thin shreds.

Leaves of small young specimens, and sometimes of sprouts after injury, awllike or needlelike, 5–10 mm long, opposite or perhaps more commonly in whorls of 3, appressed to branchlets only basally then loosely spreading, not overlapping, adaxial surface flat or somewhat channeled, abaxial surface slightly rounded, tips very sharply pointed, glandless. As specimens reach a little age, leaves of branchlets gradually change to the form characteristic for mature specimens: short scalelike, 2–3 mm long, mostly opposite and 4–ranked, appressed and partially overlapping, triangular in outline, tips bluntly to sharply pointed, usually with an evident longitudinal, opaque depression on the back at first, this producing a raised gland as the season advances.

Male and ovulate cones borne on separate trees. Staminate cones usually evident in autumn on the tips of branchlets, subsequently gradually enlarging to 3–5 (6) cm long at maturity, then yellowish brown; on a great many trees present in profusion and somewhat prior to and at the time of discharging pollen in mid-winter, giving the trees a suffused green and gold aspect when seen from a little distance; scales of cones short, peltately attached, bearing 3–6 anther cells beneath. Ovulate cones borne at the tips of branchlets, inconspicuous at first, soon becoming irregularly knobby and bluish-glaucous, at maturity in autumn ovoid or ellipsoid, about 5–9 mm long, indehiscent and berrylike (juniper berries), mostly remaining bluish-glaucous; some trees bear relatively few ovulate cones, many trees bear them in great abundance, the latter having a suffused green–blue-green color when seen from a little distance; scales of cones fleshy, coherent, several in number, the lower sterile, the median and upper fertile, 1–ovulate; "berries" persisting into the winter until most of them consumed by birds. [Incl. *J. silicicola* (Small) Bailey (*Sabina silicicola* Small)]

Inhabiting a wide variety of kinds of sites, old fields, upland woodlands, especially in calcareous areas, moist to wet hammocks where limestone is beneath shallow soils, barrier beaches and shell mounds, banks of marshes or hummocks in marshes where limestone is subsurface, swampy areas with *Cladium jamaicese* (saw-grass) as principal ground cover and where limestone is subsurface, sometimes on natural levees or in alluvial woods where inundation is of short duration; local and often locally abundant throughout our area. (S. Que. and Maine to N.D., generally southward to cen. pen. Fla. and Tex.)

Most authors dealing with the flora of the southern U.S. distinguish *J. silicicola* from *J. virginiana*, generally considering it to occur coastally or near-

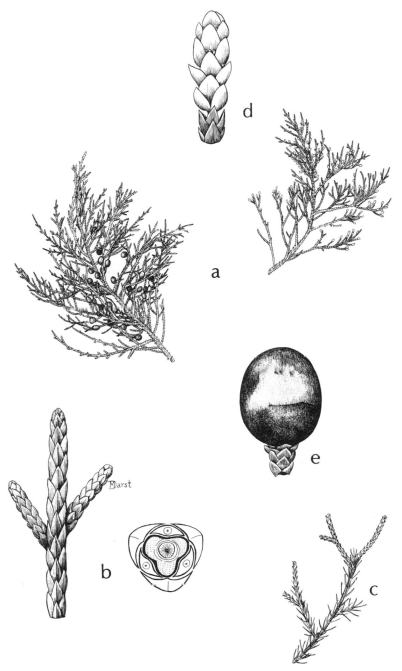

Fig. 2. **Juniperus virginiana:** a, branchlets, ovulate to left, staminate to right; b, enlargement of ultimate branchlet and at right diagrammatic cross-section of it; c, branchlet from very young plant; d, male cone; e, berrylike ovulate cone.

coastally from s.e. N.C. to Fla., thence westward to s.e. Tex. After much agonizing, I cannot perceive the differences that enable one to differentiate two species. (See Adams, 1986.)

Pinaceae (PINE FAMILY)

Pinus

Evergreen, resinous, cone-bearing trees. Bark of older trunks breaking up into laminated scales and plates, or in one of ours appearing ridged and furrowed. Plants have a central, leading shoot with strong apical dominance and form more or less conical crowns until about middle age, the branches quite secondary, more or less horizontal or ascending. In later age, the leader shoot loses its identity, the tree ceases to increase in height, the crown branches grow crookedly, the crown becoming more or less lobed or domed. Leader or long-shoots of the main stem or branches bear spirally arranged, scalelike primary leaves in the axils of which are produced determinate short-shoots (fascicles) composed of a basal sheath of scaly bracts surrounding the bases of needlelike foliage leaves (2, 3, or 4 to a fascicle in ours), the sheath and its needles eventually falling together. The range of lengths of needles, as described for the separate species below, may occur more or less intermixed on single branchlets although, usually, the longer ones predominate. Long- or extension-shoot buds form in the axils of scaleleaves at and near the tips of the main stem or at the tips of branchlets in late season and remain dormant over winter. Plants monoecious. Reproductive cones produced by a given plant, for the most part, only after several years of growth from the seedling, many more staminate than ovulate, each kind usually on separate branchlets, relatively infrequently both on the same one. Male cones occur closely aggregated and are extruded from scales of winter buds before the buds commence to enlarge (excepting in *P. serotina*), each cone having several imbricated scaly sterile bracts basally and many more spirally arranged, overlapping, fertile scales above, each of the latter with an upturned flange having a pair of pollen sacs on its lower surface, pollen being shed in late winter or in spring, after which the cones quickly shrivel and fall. Ovulate cones are borne on an enlarging and elongating long-shoot of the season, one or more per shoot; below the cone proper, the axis has several short, appressed, imbricated, sterile scales and distally longer and narrower sterile scales whose tips spread or recurve forming more or less a crown embracing the base of the fertile portion; the cone proper has some reduced, sterile scales basally above which are "scale-complexes" consisting of "bract-scales" each of which subtends an ovule-bearing scale having a pair of ovules adaxially. At this stage, in most of ours, the bract-scale is the more prominent, the tip of the ovuliferous scale shortly exserted above it. As the cone matures, the ovuliferous scale develops more rapidly and obscures the bract-scale, ultimately the two fuse, and the conspicuous woody scale of the mature cone is the product of the originally separate structures. After pollination, the cone (in ours) enlarges somewhat but does not attain conspicuous size during the first season; during the second season it grows rapidly to the adult size. In some of the kinds of our pines, the mature ovulate cones open promptly shedding their seeds while in some species some or all of them may remain closed for long periods, in the latter case becoming embedded as stems bearing them grow in diameter; in some species in which mature cones open promptly at maturity, the cones mostly fall over the winter or in early spring, in other species they may remain on the trees for several years in which case the trees appear conspicuously heavily laden with old cones. Seeds (in ours) each with a laterally embracing membranous structure that beyond the seed body is extended as a thin-membranous, papery wing.

Note: The male and ovulate cones of *P. elliottii* and *P. clausa* are first to reach anthesis in late winter or early spring, followed by those of *P. palustris*, then by *P. taeda* and *P. glabra*, finally by *P. echinata* and *P. serotina*, the time span being about six weeks.

1. Twigs from which needle-fascicles have been shed and smaller branches, smoothish, without scaliness or flakiness.

> **2.** Bark of stems as they grow larger gradually becoming scaly or flaky, ultimately bark of trunks breaking into large, laminated plates outer layers of which gradually slough; needles predominantly 2 per fascicle, 4–8 cm long, straight or moderately twisted; inner (adaxial) surface of scale of mature ovulate cone darker brown at the tip than the remainder of the scale.   1. *P. clausa*

> **2.** Bark of stems as they grow larger gradually becoming ridged and furrowed, ultimately on larger trunks prominently ridged and furrowed, the ridges, although laminated, very hard, tight, and not sloughing in conspicuous layers; needles 2 per fascicle, 5–10 cm long, usually all of them notably twisted; inner (adaxial) surface of scale of mature ovulate cone uniform brown throughout.   4. *P. glabra*

1. Twigs from which needle-fascicles have been shed notably scaly, flaky, or roughened by persistent scale leaves or their bases.

> **3.** Needles predominantly 3 per fascicle on a given plant.

>> **4.** Terminal buds about 15 mm in diameter, bud scales conspicuously silvery, their margins markedly fimbriate and their tips loosely spreading; longer needles 30–40 cm long, the fascicle-sheaths 1.5–2 cm long or a little more.   5. *P. palustris*

>> **4.** Terminal buds 10 mm in diameter or less, brown, or if grayish, the color a function of dried resin on the surface; scales of the buds appressed; needles 10–22 cm long, the fascicle-sheaths 0.8–1.2 cm long.

>>> **5.** Mature ovulate cones subglobose to broadly short-ovoid, 5–8 cm long, sometimes broader than long, stipitate.   6. *P. serotina*

>>> **5.** Mature ovulate cones elongated-conic, 10–13 cm long, considerably longer than broad, sessile.   7. *P. taeda*

> **3.** Needles predominantly 2 per fascicle, or a mixture of 2 and 3.

>> **6.** Longer needles on a given tree mostly 17–25 cm long (shorter on seedlings and saplings not over 1.5 m tall, varying from about 7–15 cm long); fascicle-sheaths mostly about 1 cm long; mature ovulate cones 8–15 cm long, for the most part falling from the trees by spring or early summer of the year after maturation.   3. *P. elliottii*

>> **6.** Longer needles on a given tree mostly 7–12 cm long; fascicle-sheaths 0.5–0.6 cm long; mature ovulate cones 3–6, mostly 4–5 cm long, usually persisting on the trees for several years so that the trees are constantly laden with old cones.   2. *P. echinata*

1. Pinus clausa (Chapm. ex Engelm.) Vasey ex Sarg. SAND PINE.   Fig. 3

A small to medium-sized tree, to about 25 m tall and to 4 dm d.b.h. or a little more. Bark of twigs from which needle-fascicles have been shed smooth, tan or grayish tan, that of trunks eventually thick, breaking into laminated gray plates from which surface layers slough exposing brown surfaces beneath. Winter buds 2 mm in diameter, scales grayish brown.

Leafy branchlets slender, needles 2 per fascicle, rarely a few with 3, mostly 4–

45

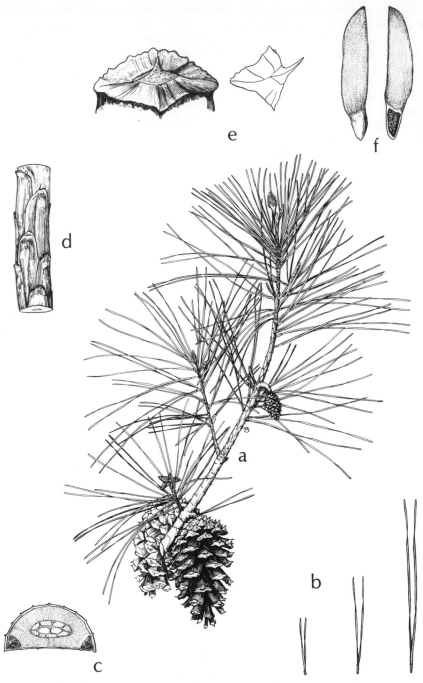

Fig. 3. **Pinus clausa:** a, branch (taken from tree in early spring) with (at base)
two ovulate cones having matured the previous season, (medially) an ovulate
cone which would mature during the current season, and (at tip) an ovulate
cone at anthesis; b, needle fascicles; c, diagrammatic cross-section of needle;
d, small portion of twig (from just above medial cone); e, apex of mature
ovulate cone scale; f, seed, two views.

8 cm long, slender, straight or moderately spirally twisted, medium green, often slightly yellowish green so that, when seen from a little distance, crown may have a dullish, relatively pale hue; sheaths about 5 mm long.

Male cones at anthesis pinkish-purplish, numerous from the slender winter buds, mostly about 10 mm long and 3 mm in diameter; pollen sacs yellow, sometimes with an orangish tint, pollen pale yellow. Ovulate cones yellowish green at anthesis; mature ovulate cones conic-ovoid when closed, ovoid-oblongish when open, 5–8 cm long, scales hard and inflexible, form of the umbo of the scale often such that the prickle is directed downward; inner (adaxial) surface of cone scale darker brown at the tip than the remainder of the surface. There is a great deal of variation from tree to tree, in some populations, at least, as to how long cones may remain closed (serotinous) after maturation. Seed, including wing, 1–2 cm long, wing about 5 mm broad. [Incl. var. *immuginata* Ward]

Inhabiting deep, well-drained, relatively infertile sands of ridges and hills, stabilized coastal dunes, near-coastal ridges, in some places in dense, even-aged stands, commonly in association with winter-green scrub oaks or with winter-green and deciduous scrub oaks intermixed. In much of pen. Fla., Panhandle of Fla., s.w. Ala.

In recent years, sand pine has been planted as a pulp crop more frequently than formerly on sites so infertile as to have been demonstrated to be unfavorable for growing slash pine.

National champion sand pine (1980), 6'6" circumf., 103' in height, 46' spread, Wekiva Springs State Park, Fla.

2. Pinus echinata Mill. SHORTLEAF PINE. Fig. 4

Medium-sized to large tree, to 25–35 m tall and to 1 m d.b.h. or a little more. Bark of twigs from which needle-fascicles have been shed notably scaly-roughened, brown; bark of medium-large to large trunks with distinctive flat, broad, heavy, reddish brown, fairly uniform, more or less rectangular, laminated plates from which outer layers slough gradually. Winter buds brown, about 3 mm in diameter.

Leafy branches relatively slender, needles predominantly 2 per fascicle, some fascicles with 3, 4–11 cm long, dark green, sheaths mostly 4–5 mm long.

Male cones few to numerous from the winter buds, greenish yellow at anthesis, 2–3 cm long, about 4 mm in diameter; pollen sacs and pollen pale yellow. Ovulate cones at anthesis with scales of fertile portion broader than long, more or less truncate apically, at least the margins roseate; mature unopened ovulate cones 4–6 cm long, ovoid, ovoid-oblong, or subcylindric, open cones ovoid to ovoid-oblong, sessile; cone scales hard and rigid, their exposed faces dull brown, prickle of the umbo short and straight, directed downward owing to the configuration of the face; cones opening promptly at maturity, most of them persisting on the tree for several years so that the trees generally appear heavily laden with old cones. Seed, including the wing, 15–20 mm long, seed body very dark brown, wing tan and slightly purplish.

A colonizer of open, upland areas (often together with *P. taeda*), especially old fields where even-aged stands are eventually produced. These are ultimately invaded and overtaken by hardwoods (and commonly a few individuals of *P. glabra*) so that after a relatively long time a mixed pine-hardwood forest prevails. At the present time, in a great many places in our range, such stands of mixed pine-hardwood have a relative profusion of uneven-aged individuals of

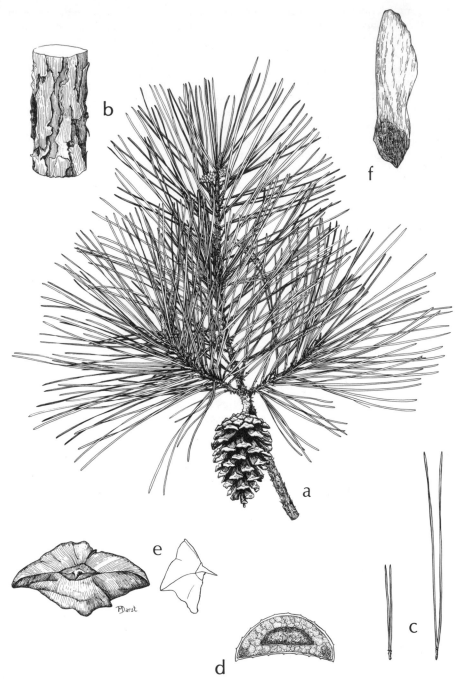

Fig. 4. **Pinus echinata:** a, branch with ovulate cone (at base) having matured the previous season, and (near tip of central branch) a cone of previous season that would mature the current season; b, piece of stem from well back of needle-bearing portion showing its roughness; c, relative lengths of shortest and longest needles; d, diagrammatic cross-section of needle; e, apex of mature ovulate cone scale; f, seed.

the shade-tolerant *Magnolia grandiflora* as a developing understory. In our range, *P. echinata* occurs in s.w. Ga., s. Ala., and adjacent Panhandle of Fla. (Cen. N.J., s.e. and s. cen. Pa., W.Va., s.e. Ohio, e. Ky., s. Mo., generally southward to Fla. Panhandle, e. Okla., e. Tex. (absent from the Mississippi River valley).

National champion shortleaf pine (1980), 10'10.5" circumf., 138' in height, 75' spread, Myrtle, Miss.

3. Pinus elliottii Englem. var. elliottii. SLASH PINE. Fig. 5

Medium to tall tree, to 40 m tall, to about 1 m d.b.h. Bark of twigs from which needle-fascicles have been shed much roughened by persisting scale leaves, that of trunks eventually breaking up into irregularly shaped, broad, flat, laminated plates from which outer layers slough more or less continuously. Winter buds brown, about 1.5 cm in diameter, their scales with fimbriate-pectinate margins.

Leafy branches stout, needles stoutish and rigid, slightly twisted, on different trees variably, 2 per fascicle, predominantly 2 with a few 3 per fascicle, or 2 and 3 per fascicle well intermixed, 10–22 (28) cm long; sheaths mostly 1.5 cm long.

Male cones few to numerous from the winter buds, uniformly deep purple, for the most part 3–6 (8) cm long and about 1 cm in diameter; at anthesis, sterile bases of ovulate cones conspicuous, turbinate, 1.5 cm long, lower scales brown with paler fimbriate margins, upper scales paler, margins hyaline, fertile part of cone pinkish-purplish, scales more or less glaucous; mature ovulate cones conical before opening, broadly conical to oblongish-conical when opened, 9–15 cm long and 8 or 9 cm across near the base, short-stipitate; cone scales hard and rigid, their exposed faces brown and sublustrous, prickle of the umbo short, straight although directed downward somewhat because of the configuration of the face; cones opening promptly at maturity, most of them falling overwinter and into spring. Seed, including wing, 2–3 cm long, seed body dark brown, wing usually purplish brown. [*P. palustris* sensu Small (1933)]

In low, commonly seasonally wet flatwoods, interdune hollows and near-coastal sands where often accompanied by much sand live oak; throughout our area. (Coastal plain, s.e. S.C. to s. cen. pen. Fla., westward to s.e. Miss. and s.e. La.)

Endemic to Florida, in southernmost Fla. and the lower Fla. Keys, and chiefly near the coast to Volusia Co. on the Atlantic Coast and to Levy Co. on the Gulf Coast, *P. elliottii* var. *densa* Little & Dorman (*P. palustris* sensu Small, 1933) occurs. It differs notably from var. *elliottii* by its seedlings, those of var. *elliottii* having stems thin and pencillike, elongating "normally" and with needle-fascicles relatively sparse; those of var. *densa* thick, somewhat carrotlike (Ward, 1963), not elongating appreciably for several years and bearing a relative abundance of needle-fascicles.

The pine most commonly planted in pine plantations in our area is slash pine.

National champion slash pine [var. *elliottii*] (1969), 8'7" circumf., 150' in height, 60' spread, Colleton Co., S.C.

National champion slash pine [var. *densa*] (1977), 11'5" circumf., 55' in height, 63' spread, Sarasota, Fla.

4. Pinus glabra Walt. SPRUCE PINE. Fig. 6

Medium- to large-sized tree, to about 40 m tall and to 8 dm d.b.h. Bark of twigs from which needle-fascicles have been shed, and of smaller limbs, brown to gray, smooth, that of older stems gradually becoming ridged and grooved, on large trunks the ridges dark gray, laminated, but the layers hard and tight, not

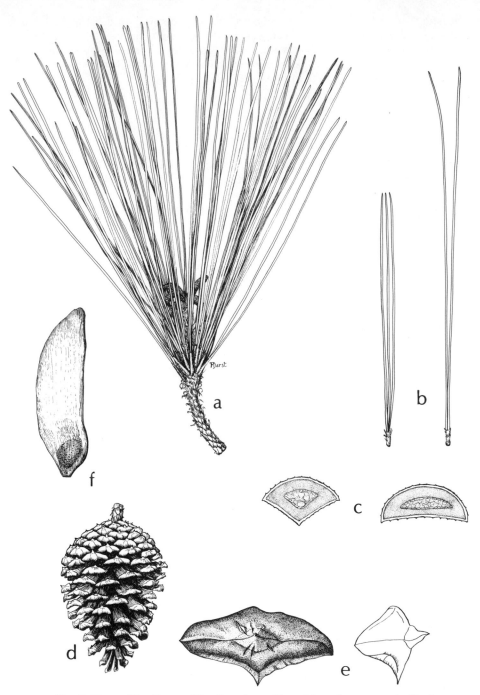

Fig. 5. **Pinus elliottii** var. **elliottii:** a, branchlet with male cones apically; b, fascicles with 2 and 3 needles; c, diagrammatic cross-sections of needles, from needle with 2 in a fascicle at right and 3 in a fascicle at left; d, mature ovulate cone; e, apex of mature ovulate cone scale; f, seed.

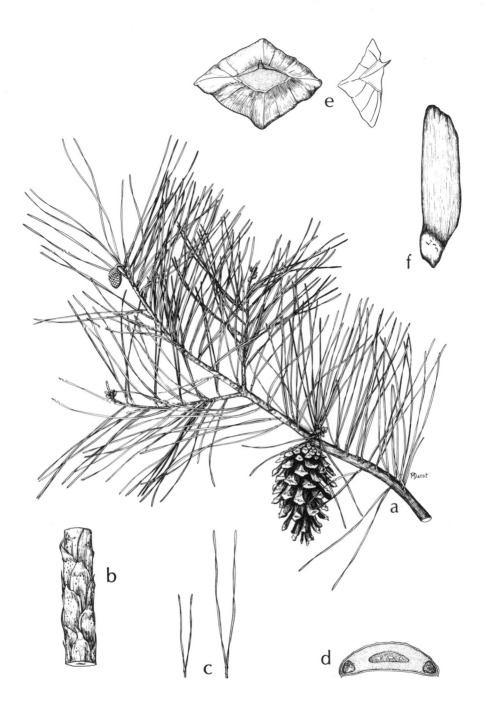

Fig. 6. **Pinus glabra:** a, branch, with (near its base) an ovulate cone having matured the previous season, and at tip a cone of the previous season that would mature the current season; b, piece of stem from which needles have been shed; c, needle fascicles; d, diagrammatic cross-section of needle; e, apex of mature ovulate cone scale; f, seed.

breaking up into large plates as is the case with our other pines. Winter buds grayish brown, 2 mm in diameter.

Leafy branchlets slender, needles 2 per fascicle, 5–10 cm long, spirally twisted, dark green; sheaths 5–7 mm long; crowns, when seen from a little distance, appearing dark green.

Male cones purplish yellow, few to numerous from the winter buds, mostly 5–8 mm long and about 4 mm in diameter; pollen sacs yellow, pollen yellow. Ovulate cones at anthesis with scales of the fertile portion broader than long, rounded at the summit, greenish and at least the distal ones with pink margins; mature ovulate cones opening promptly, ovoid before opening, oblongish-ellipsoid after opening; scales flexible, umbo with a small, straight prickle that tends to be directed upward, usually breaking off rather soon; inner (adaxial) surface of scale uniformly brown throughout. Cones brown at maturity, most of them remaining on the tree for 3–4 years, at least, and becoming gray, trees thus appear heavily laden with old cones. Seeds, including wing, 1.5–2 cm long, wing about 4–5 mm broad.

Generally growing scattered and intermixed with hardwoods in mesic or submesic forests of uplands, bluffs, slopes of ravines, in floodplain forests where elevated enough so that flooding is shallow and of brief duration; in much of our area of coverage although spotty in northeasternmost Fla. and perhaps absent in southeasternmost Ga. (Coastal plain, S.C. to n. Fla., westward to s. Miss. and s.e. La.)

National champion spruce pine (1980), 13′2″ circumf., 123′ in height, 45′ spread, St. Helena Parish, La.

5. Pinus palustris Mill. LONGLEAF PINE. Fig. 7

Mature trees varying considerably in height attained, larger ones to 30–40 m tall and rarely, if ever, as much as 1 m d.b.h. Twigs stout, 1–1.5 cm in diameter, those from which needles have been shed very much roughened by persisting scale-leaves; bark of trunks eventually thickish, breaking into laminated plates from which surface layers slough more or less continuously. Winter buds mostly 1.5 cm in diameter, the scales silvery and with conspicuous silvery-fringed margins.

For the first several years, the stem of the seedling elongates relatively very little, its short stem bearing many curvate-spreading needle-fascicles, the aspect being somewhat like a dense tuft of grass and sometimes referred to as the grass or broom stage; once stem elongation commences growth in height is relatively rapid.

Leafy branches stout, needles 3 per fascicle (or on occasional plants a few fascicles with 2 needles); needles mostly 15–30 cm long (occasionally to 45 cm long), sheaths 1.5–2.5 cm long, proximal scales brown, the distal ones silvery, thin-membranous, often gathered in such a way as to form concentric horizontal rings above which becoming frayed.

Male cones with the sterile basal bracts broad, brown-chartaceous and with transparent, thin-membranous margins; fertile portion of cones uniformly deep purple prior to anthesis, becoming yellowish, stout, about 1 cm in diameter, 2–6 cm long; pollen sacs and pollen very pale yellow; at anthesis, sterile bases of ovulate cones very conspicuous, the lower scales chestnut-brown with deeply lacerate, hyaline margins, uppermost scales wholly hyaline, fertile part of cone pinkish-purplish, scales somewhat glaucous. Mature ovulate cones opening promptly, long-conic before opening, ovoid-oblongish after opening, 15–20 (25)

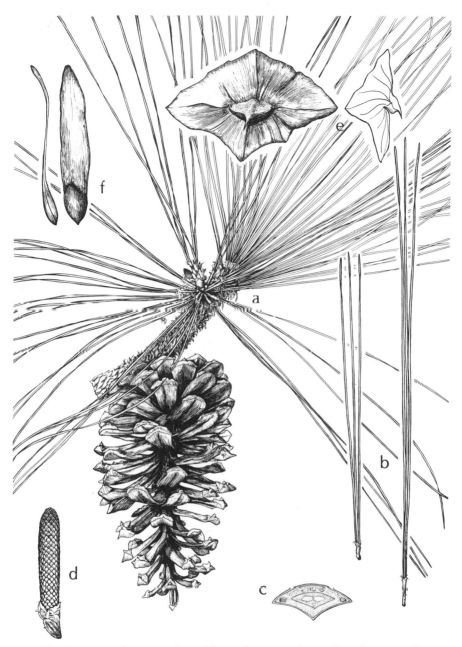

Fig. 7. **Pinus palustris:** a, branchlet with mature (opened) ovulate cone; b, needle fascicles; c, diagrammatic cross-section of needle; d, male cone, prior to anthesis; e, apex of mature ovulate cone scale; f, seed, two views.

cm long, about 12 cm broad near the base, brown; scales large, hard, 2–2.5 cm broad, umbo with a small, sharp prickle directed downward; cones generally gradually shed from the trees over the winter following maturation, some sometimes persisting a little longer. Seeds, including the wing, 3.5–4 cm long, the wings mostly about 1 cm broad, both seed body and wings brown. [*P. australis* Michx. f.]

Occurring on sandy or clayey-sandy ridges and hills with wiregrass (*Aristida stricta*) and deciduous scrub oaks, also in seasonally wet flatwoods or savannas where the oaks are absent, but with associated shrubs such as wax-myrtle, gallberries, saw palmetto, blueberry, staggerbush, etc.; throughout our area. (Coastal plain, s.e. Va. to s. cen. pen. Fla., westward to e. Tex.)

National champion longleaf pine (1977), 10'5" circumf., 125' in height, 63' spread, Angelina National Forest, Tex.

The longleaf pine is the State Tree of Alabama.

6. Pinus serotina Michx. POND PINE. Fig. 8

Small to medium-sized tree, to about 25 m tall and 6 dm d.b.h. Twigs moderately stout, their bark from which needle-fascicles have been shed roughened by apices of the hard, appressed scale-leaves, bark of trunks breaking into relatively narrow, laminated plates from which the outer layers slough gradually. Winter buds 5–6 mm in diameter, usually more or less coated with hardened resin which may become white. A feature unique to the pond pine (relative to other pines in our range) is that it has the capacity to form sucker-sprouts from the trunks or principal branches, particularly after injury from fire.

Needles predominantly in fascicles of 3, but sometimes with a few fascicles of 2 or 4 intermixed, 6–20 cm long, sheaths mostly about 1 cm long.

Male cones, at anthesis, mostly 3–4 cm long and 6–7 mm in diameter, exposed flanges of the scales greenish yellow or some or all of them with flushes of rose color; cones few to numerous and (in contrast to the condition in our other kinds of pines) borne on an already elongated and radially expanded winter bud, this at this stage up to 12 cm long and 8–10 mm in diameter; pollen sacs and pollen pale yellow. Sterile bracteate axis below the ovulate cone proper 1.5–2 cm long, all the bracts except the uppermost closely imbricated, chartaceous, chestnut-brown, the uppermost spreading about the base of the cone proper, their tips membranous-hyaline; fertile part of the cone with the lower scale of the scale complex very short, its apex truncate, the ovule-bearing scale with its expanded rose-tinted base exceeding the lower scale, its attenuated tip much exceeding it. Mature ovulate cones short-ovoid to subglobose when unopened, conic-ovoid when opened, 5–8 cm long, with short, thick stalks. Cones tend to remain on the trees for a long time, unopened on some individuals, opened on others; stems as they enlarge, first enveloping the stalks, ultimately at least the bases of the cones. Umbo of cone scale dark brown, terminated by a straight or incurved, sharp prickle. Seeds, including the wing, 2.5–3 cm long, wing 6–8 mm broad; seed body very dark brown, the wing tan.

Inhabiting poorly drained flatwoods, often intermixed with pond-cypress in wet flatwoods or flatwoods depressions and ponds, evergreen shrub-tree bogs or bays; in much of our area. (Chiefly coastal plain, s. N.J. and Del., s. to cen. pen. Fla. and Fla. Panhandle, s.e. and cen. Ala.)

National champion pond pine (1977), 9'7" circumf., 94' in height, 46' spread, Scotland Co., N.C.

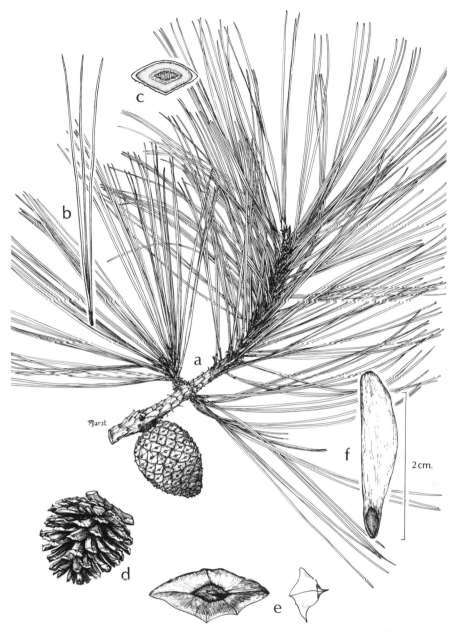

Fig. 8. **Pinus serotina:** a, branchlet with mature ovulate cone; b, needle fascicle; c, diagrammatic cross-section of needle; d, opened mature ovulate cone; e, apex of mature ovulate cone scale; f, seed.

7. Pinus taeda L. LOBLOLLY PINE.   Fig. 9

Large tree, to 25–40 m tall, and to 1 m d.b.h. or a little more. Bark of twigs from which needle fascicles have been shed roughened by persisting scale-leaves or their broken bases, that on trunks eventually breaking into relatively large laminated plates from which outer layers slough more or less continuously. Winter buds brown, 8–10 mm in diameter, margins of scales fimbriate-pectinate, the bud often partially covered with flecks of dried, whitish resin.

Leafy branches moderately stout, needles straight or slightly twisted, predominantly 3 per fascicle, an occasional fascicle with 2, 8–22 cm long, sheaths 1–1.5 cm long.

Male cones few to numerous from the winter buds, greenish-yellow to purplish, 1.5–3 cm long and about 8 mm in diameter; pollen cream-colored. Lower, empty scales of ovulate cones more or less chestnut-brown, only the uppermost spreading, fertile portion of cone dull green to roseate, about 1 cm long. Mature ovulate cones conic, 6–10 (12) cm long, sessile, opening promptly, many or most usually remaining on the tree for 3–4 years, thus trees usually appear heavily laden with cones; scales dull brown at first, usually gray by the time they fall; umbo of the scale with a short, sharp prickle usually directed downward owing to the configuration of the face of the scale. Seed, including the wing, 2.5–3 cm long, seed body blackish brown, wing buff-colored.

Inhabiting lowland woodlands, colonizing lowland and upland old fields (where eventually a component of mixed pine-hardwood forests—see comments under *P. echinata*); also colonizing borders of mixed woodlands and open adjacent areas; throughout much of our area. (Coastal plain and piedmont, southernmost N.J., Del., e. Md., southward to n. cen. pen. Fla., westward throughout Ala., much of Miss., La., e. Tex., s. Ark., s. Tenn.)

National champion loblolly pine (1980), 14'8" circumf., 163' in height, 91' spread, Urania, La.

Taxaceae (YEW FAMILY)

1. Needlelike leaves flexible and relatively soft, their tips pointed but not stiff and piercing to the touch; seed partially surrounded by a red, soft-fleshy cup, the seed and cup about 2 cm long and 1 cm in diameter.   1. *Taxus*
1. Needlelike leaves stiff, their tips sharply stiff-pointed and piercing to the touch; seed with its wholly adherent fleshy tissue olivelike, its surface dark green, purplish striped, about 2.5–3 cm long and 2 cm in diameter.   2. *Torreya*

1. Taxus

Taxus floridana Nutt. ex Chapm. FLORIDA YEW.   Fig. 10

An evergreen, glabrous shrub or small tree, to about 8 m tall, the branches disposed in an irregular fashion. Bark purplish brown, sloughing in plates, the twigs green and flexible.

Leaves needlelike, flat (the edges rolled downward in drying), linear, somewhat falcate, flexible, upper surfaces dark green, the lower with bright green midrib and edges, pale green otherwise, pointed at their tips but not sharp to the touch, mostly 2–2.5 cm long (lowermost on a branchlet shorter) and 1.5–2 mm broad, spirally arranged but the very short petioles twisted in such fashion

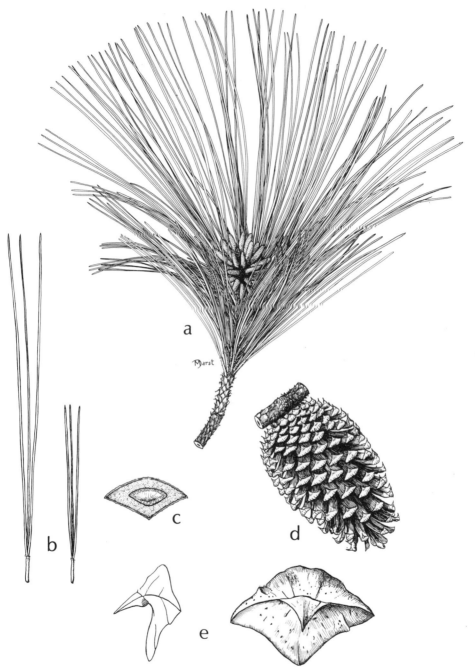

Fig. 9. **Pinus taeda:** a, branchlet with cluster of male cones; b, needle fascicles, representing the extremes of relative length; c, diagrammatic cross-section of needle; d, mature ovulate cone; e, apex of mature ovulate cone scale.

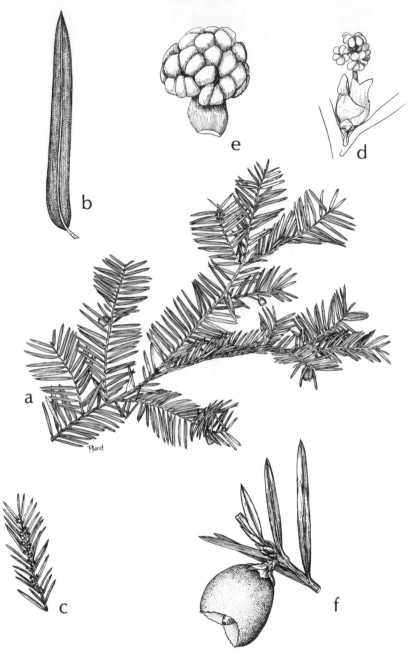

Fig. 10. **Taxus floridana:** a, branch with ovulate cones; b, leaf, enlarged (lower surface); c, branchlet with male cones; d, male cone at anthesis at which time fertile portion is elevated above the empty basal scales by elongation of the cone axis; e, fertile portion of male cone at early anthesis; f, mature ovulate cone.

that they appear in, or nearly in, one plane at either side of a twig. Twigs and foliage faintly aromatic when crushed.

Male and ovulate cones on separate plants in late winter or very early spring. Male cones small, occurring singly in leaf axils on branchlets of the previous season, sometimes in all leaf axils of a given branchlet, sometimes only in the distal portion of the branchlet, mostly disposed beneath the branchlet and easily seen if one is looking at branchlets from below; cones, at anthesis, with several broad, scarious, empty scales proximally beyond which, by elongation of the axis, the fertile scales are projected; fertile scales 6–8, peltate, 6–8-lobed, pollen sacs beneath the lobes. Ovulate cones occurring singly in only a few leaf axils, disposed beneath the branchlets, very inconspicuous at first and easily overlooked, consisting of an axis with a few sterile scales proximally, a single fertile 1–ovulate scale distally, this maturing in early autumn, composed of a single seed nearly or quite surrounded by a soft-fleshy, translucent, red, cuplike aril, the aril and seed about 2 cm long and 1 cm in diameter.

Birds consume most or all of the ripe ovulate cones, the aril is digested, the seeds pass in the birds' droppings.

Chiefly on lower slopes of wooded ravines, Apalachicola River Hills, e. of the river, from a little n. of Bristol, Fla. to a little n. of Torreya State Park; also in a white cedar swamp about 8 miles s. of Bristol.

National champion Florida yew (1971), 1'3" circumf., 18' in height, 28' spread, Torreya State Park, Fla.

2. Torreya

Torreya taxifolia Arn. FLORIDA TORREYA, GOPHERWOOD TREE, STINKING CEDAR.

Fig. 11

Relatively small, glabrous, evergreen tree, branches whorled, crown conical. Bark thin, shallowly fissured, sloughing in shreds, twigs green and stiff.

Leaves needlelike, linear-acicular, flat (edges rolled downward in drying), stiff, their tips sharp and piercing to the touch, bright glossy green above, light green beneath and with longitudinal grayish green stripes paralleling the scarcely evident midribs, spirally arranged but the very short petioles twisted in such fashion that the leaves spread in one plane to either side of the twigs. Twigs and leaves with a strongly pungent, rather disagreeable odor when crushed, the wood similarly aromatic.

Staminate and pistillate cones borne on separate plants. Staminate cones borne singly in the axils of proximal leaves or most leaves of branchlets of the previous season, more evident seen from below the branchlet than from above, each cone small, oval, with several scales each subtending 4 stamens. Pistillate cones small and at first not easily discerned, borne singly in few of the leaf axils on growth of the current season, each with several scales, usually with 2 ovules each surrounded by an adherent sac, the whole maturing in the second season into an olivelike seed wholly enclosed by and adherent to a leathery sac, about 3 cm long and 1.5–2 cm in diameter, its surface dark green, purplish striped, somewhat glaucous. [*Tumion taxifolium* (Arn.) Greene]

Chiefly inhabiting wooded ravine slopes and bluffs e. of the Apalachicola River from about Bristol, Fla., northward just into Ga.; a single isolated station in Jackson Co., Fla., to the w. of the river.

For the past 30–35 years or thereabouts, the Florida torreya has been afflicted with a blight, and of those plants still extant in its natural area perhaps none are

Fig. 11. **Torreya taxifolia:** a, branch with mature ovulate cones; b, leaf enlarged (lower surface); c, branchlet with immature male cones; d, male cone at anthesis.

of more than sapling size, most of them low sprouts, most if not all sprouts probably from the bases of top-killed specimens or from root sprouts. To be given an idea of the tree's potential size, there is a report (*Twin City News*, Chattahoochee, Fla., March 16, 1978) of a healthy planted specimen in Warren Co., N.C., having a near-basal circumference of 9 ft, a spread of 52 ft, and a height of 60 ft. It is estimated that it may have been planted there about 1830.

Taxodiaceae (BALD-CYPRESS OR REDWOOD FAMILY)

Taxodium

Small to large, monoecious, resinous trees having, when relatively young, a single main stem with branches forming a conical or more or less columnar crown; eventually, as lower limbs are self-pruned, the branches grow in an irregular spreading fashion so that the crowns of mature trees are more or less flat-topped. Trees growing under conditions of periodic flooding generally develop a basal butt-swell, the butt-swell usually about as high as the average annual depth of water during maximum flooding. Butt-swells usually develop surface irregularities called buttressing, the buttresses having the form of longitudinal, rounded flutes, or of more narrow and more or less planklike form. When subjected to periodic flooding, a distinctive feature is the production of "knees" projecting from the substrate, the knee arising from differential growth of the vascular cambium on the upper side of a shallow root.

Stems, subsequent to the seedling year, exhibit a basically dimorphic shoot system (but with an intergrading type explained below). There are shoots that are permanent, indeterminant long-shoots bearing scalelike or awllike leaves with buds in their axils; these shoots serve to increase the length of the main stem or of branches. Other shoots are annual, determinate, mostly unbranched, short-shoots having numerous, alternately arranged and spirally disposed leaves, the short-shoots and their leaves falling together in autumn. In addition, some shoots, commonly, but not always, increments of long-shoot growth late in the season, may have "ordinary" unbranched short-shoots proximally, and distally what appear to be short-shoots that bear, irregularly, short-shoot branches of their own and thus seem to be "compound" short-shoots. On such increments of growth, the unbranched short-shoots abcise in the usual way in autumn, and, for the most part, the "compound" short-shoots abcise in their entirety as well, but some of the latter may abcise their lateral short-shoots with the axis persisting as a long-shoot.

Some buds in the axils of scale or awllike leaves on currently growing long-shoots produce short-shoots right away, other may remain dormant for irregularly extended periods, becoming more or less buried by increase in diameter of the wood, later producing short-shoots whose origins are not readily apparent well back on the main stems of young plants or on increments of growth of branches several seasons old. In addition, when short-shoots abcise they generally do so slightly above their morphological bases where buds are formed that sooner or later produce short-shoots on increments of growth one to several years old. In spring, all of the new "green growth" is composed of short-shoots on wood of previous seasons; extension shoot growth commences somewhat later.

Leaves decurrent basally, alternate and spirally disposed. Those leaves borne proximally on growing long-shoots sometimes have the form of those on short-shoots and they are not appressed, all other leaves distal to these are scalelike or awllike and are closely appressed. Leaves on short-shoots of "young" plants, although spirally disposed, have their bases twisted so that they diverge from the axis in one horizontal plane (featherlike), thus pseudodistichous. In the case of pond-cypress, *Taxodium ascendens*, the featherlike or pseudodistichous nature of the short-shoots ceases to prevail on new increments of growth while the plants are still small (approximately 1–2 m tall). After this the leaves on short-shoots are obviously spirally disposed and angling-ascending or appressed and many or most of the short-shoots themselves, by twisting of bases, appear erect or nearly so on the twigs giving the branch system a very distinctive silhouette compared to that of bald-cypress, *Taxodium distichum*.

Male and ovulate cones occur on the same tree (plants monoecious). Buds of male cones are produced in relative profusion on rather long, more or less panicled tassels terminating late-season extension shoot growth (August and September in our area); they are dormant but conspicuous through the winter after short-shoots have abcised, rapidly expand and discharge pollen in late winter or very early spring prior to any new "green growth" appearing. In the process of expansion, the outer, lower, infertile scales loosen, the fertile portion of the cone is slightly extruded from the lower, infertile scales by elongation of the cone axis. The fertile portion comprises several scales spirally arranged, each bearing several anther sacks in two rows on the lower (abaxial) side. One or more buds of ovulate cones are formed near the twig tips in autumn and these expand into green-scaly, minute cones at the time of pollen release; scales spirally arranged, ovate-attenuate, each fertile one bearing two or more minute, bottlelike ovules on a cushion of tissue on its adaxial side. Ovulate cone matures into a fleshy, more or less globose "fruiting cone" 2–3 cm in diameter, its parts tightly compacted, exteriorly green-glaucous during maturation, the scales (composed of bract and ovulate scales fused) having become distally dilated and eccentrically peltate, woody at maturation and then separating, each bearing seeds basally. Seeds 2–3-angled or angled-winged.

1. Determinate short-shoots (of plants beyond the "juvenile stage"), most of them, ascending from the twigs in a vertical plane, thus secundly erect, a few of them variously spreading, often curvate; leaves rigidly awllike or acicular, alternating around the axis (not spreading in one horizontal plane), angling-ascending or appressed, apices gradually tapered to a point, 3–8 (10) mm long; bark of moderate to large-sized trunks usually dark brown, breaking into long, flat, vertical plates, thick, up to 2–2.5 cm from the vascular cambium to the outer edge of plates; *old* knees relatively short, commonly not over 3 dm high, nearly columnar, broadly short-conical, or moundlike, with thick bark over their summits.   1. *T. ascendens*

1. Determinate short-shoots spreading from the twigs, their leaves spirally arranged but by twisting of their free basal portions spreading laterally and featherlike (pseudodistichous); relatively young trees (still growing upwardly by extension of the central, main stem) having their short-shoots continuing to be strongly featherlike, their leaves pliable, flattish, linear or very narrowly linear-lanceolate, apices rather abruptly tapered to a short point, 8–20 mm long. As the tree ages and after the crown has attained considerable size, the leaves of short-shoots on succeeding increments of new growth are shorter, still shorter on the eventual flat-topped crowns of old trees; they are more acicular, often, not always, gradually reduced in length from the base of the short-shoot to its tip, the larger ones, fully developed, 4–5 mm long, at a glance sometimes appearing to be pseudodistichous, but careful close examination shows them to be spiraling about the axis; bark of trunks thin, usually not more than 6–7 (10) mm thick even on large trunks, sloughing in flat, thin strips; old knees tending to be irregularly conical in outline, often to 1 m high, occasionally to 2 m or a little more, their summits with thin bark.   2. *T. distichum*

1. Taxodium ascendens Brongn. POND-CYPRESS.   Fig. 12

Small to medium-sized tree. Bark of trunks eventually breaking into thick, long, flat, vertical plates, to about 2 (2.5) cm from the vascular cambium to the surfaces of the plates, bark of older trunks of some trees in some populations conspicuously spiraling. Knees generally not growing very high, old ones commonly not over 3 dm, more or less columnar or moundlike, sometimes very broad-based and shortly tapering to rounded summits, their bark thick, even over the summits. Twigs generally stoutish and stiff, straight or arching, but on some individual trees, the younger ones drooping or pendulous.

Short-shoots of young specimens (and of sprouts from cut-off stumps or bases of trunks killed by fire) by twisting of their bases spreading laterally and having flattish, longish, nearly linear leaves 1–1.5 cm long that by twisting of their bases spread laterally in one plane (indistinguishable from short-shoots of any small to medium-sized tree of *T. distichum*). Short-shoots of branches of specimens above 2–4 m in height much different: leaves shorter, 3–6 mm long, spirally arranged, strongly ascending close to the axis or upwardly appressed, the short-shoots themselves, many of them, although spirally disposed, twisting basally and held secundly erect on the twigs, commonly some of them spreading outward or downward and these often curvate, the general effect, however, is significantly that of erect or ascending short-shoots giving a very distinctive silhouette to the branches.

In our area, the male cones shedding pollen 3–4 (5) weeks later than in *T. distichum*. [*T. distichum* (L.) L. C. Rich. var. *imbricarium* (Nutt.) Croom]

Inhabiting wet sandy depressions or ponds in pine flatwoods, water-courses in pine flatwoods, on shores or near shores of ponds and lakes or shores of islands in lakes, sometimes in sandy-bottomed ponds where water stands 2–3 m deep for extended intervals, in near-coastal extensive flats with shallow soils over limestone (or marl prairies in s. pen. Fla.); throughout our area. (Coastal plain, s.e. Va. to s. pen. Fla., westward to s.e. La.)

National champion pond-cypress (1972), 23'8" circumf., 135' in height, 79' spread, near Newton, Ga.

2. Taxodium distichum (L.) L. C. Rich. BALD-CYPRESS.   Fig. 13

Potentially a large tree, to 40 m tall or more and to 4 m d.b.h. or more (although few trees of very large dimensions still extant). Bark thin (relative to that of *T ascendens*), even on old, large trunks, for the most part breaking into narrow, thin strips, usually not exceeding 1 cm from the vascular cambium to the surfaces of the strips. Knees potentially to 2 m high, mostly irregularly conical and narrow at their summits, their bark usually very thin at the tips.

Short-shoots of both saplings and small trees (perhaps to 10 m tall or more) laxly spreading, not at all secundly erect, the leaves, although spirally arranged, by twisting of their bases spreading laterally in one plane and distinctly feather-like, individually pliable, flattish, linear or very narrowly lanceolate, 8–20 mm long, apices abruptly tapered to a short point. As the tree ages and after the crown has attained a considerable size, the leaves of short-shoots on upwardly succeeding increments of extension growth are shorter, and still shorter on the

63

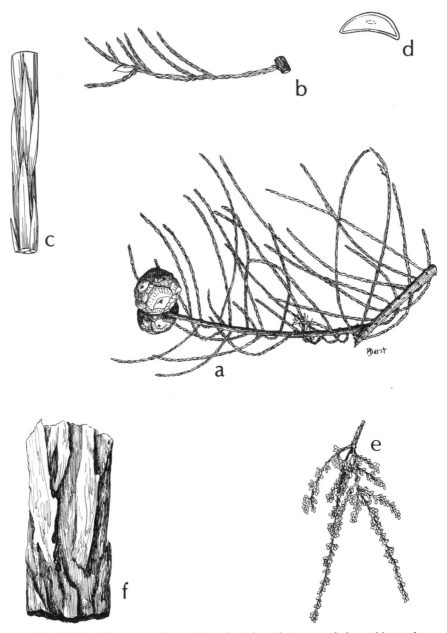

Fig. 12. **Taxodium ascendens:** a, branch with ovulate cones; b, branchlet with several short-shoots; c, portion of short-shoot with leaves; d, diagrammatic cross-section of leaf; e, tassels of male cones (in dormant winter condition); f, piece of bark from trunk 2 dm in diameter.

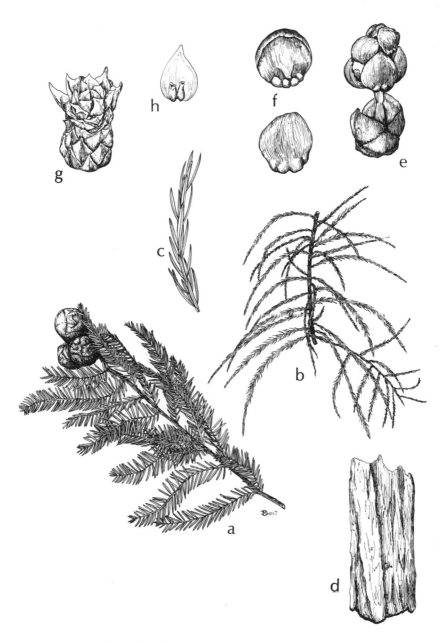

Fig. 13. **Taxodium distichum:** a, branch with ovulate cones (taken from a small, relatively young tree); b, portion of a branchlet (fully developed) from a branch on the high, spreading crown of a large, old tree; c, short-shoot from "b," somewhat enlarged; d, piece of bark from trunk 2 dm in diameter; e, male cone at anthesis; f, scale from fertile portion of male cone, 2 views, adaxial above, abaxial below; g, ovulate cone, at anthesis, from tip of a branchlet; h, scale from fertile portion of ovulate cone, adaxial view, with ovules basally.

eventual, more or less flat-topped high crowns of old trees; they are spirally disposed, more acicular, with longer tapering tips, the larger ones, fully developed, 4–5 mm long; these narrow short-shoots are generally loosely and laxly disposed on supple, spreading to pendulous branchlets, thus the silhouette of the branchlets having an aspect quite unlike that of *T. ascendens*. (Collections of *T. distichum* in at least some herbaria have a preponderance of specimens with featherlike short-shoots. This may reflect a bias on the part of collectors for selecting specimens with "typical" featherlike short-shoots; it is more likely a function of what a collector can reach, even with a long-handled tree-pruner.)

In our area, male cones shedding pollen 3–4 (5) weeks prior to its being shed in *T. ascendens*, and the spring flush of short-shoot growth from previous seasons' twigs or branches commencing earlier than in *T. ascendens*.

Inhabiting river banks and floodplains, lower reaches of spring runs or small streams from slopes behind floodplains, in and along major sloughs draining pine flatwoods, along sloughs and backwaters of near-coastal marshes, shores of impoundments where sometimes intermixed with *T. ascendens*, and in natural ponds and lakes associated with stream systems; throughout our area. (Coastal plain, Del. to s. pen. Fla., westward to e. Tex., northward in the interior to w. Ky., s. Ind., s. Ill., s.e. Mo., e. and s. Ark., s.e. Okla.)

National champion bald-cypress (1981), 53'8" circumf., 83' in height, 85' spread, Cat Island, La.

Watson (1983) in his study of "pondcypress" and "baldcypress" concluded, in essence, that morphological extremes are sufficiently distinctive as to be recognizable, but that there is a great deal of intermediacy. This led him to treat the bald-cypress as *Taxodium distichum* (L.) Richard var. *distichum* and the pondcypress as *T. distichum* var. *imbricarium* (Nuttall.) Croom.

For now, I prefer to recognize two species as named and characterized above because it is my perception that the vast majority of trees (populations) are thus distinguishable.

Zamiaceae (Cycadaceae) (CYCAD FAMILY)

Zamia pumila L. COONTIE, CONTIS, COMPTIES, COMFORTROOT.   Fig. 14

Low, palmlike or fernlike plant, circinate in the bud, the stem subterranean, upright, stout, smooth to wrinkled on older parts, unbranched or compactly branched, having little wood, but with thick, starchy pith and cortex; apex of stem woolly, most of it usually covered by old, hairy petiole bases and scales.

Leaves produced in annual flushes (or irregularly after fire), in a compact spiral crown at the summit of the stem or its branches, relatively few from any given crown in a single flush, evergreen for 2 or 3 years, however, so that they may be relatively numerous at any given time. Leaves petiolate, blades pinnately compound, at first with rusty pubescence, this quickly sloughing save at the base of the petiole; petiole length variable in relation to the overall leaf length which varies from 3–7 (16) dm long, mostly 10–30 leaflets on a side, disposed oppositely or suboppositely; leaflets long-linear, long-lanceolate, long-oblong, or oblanceolate, mostly 8–20 cm long, although some may be shorter or longer, 4–15 mm broad (although not so great a range of width on a single plant), closely parallel-veined, the number of veins varying in proportion to leaflet width, margins revolute, texture leathery and stiffish, apices blunt.

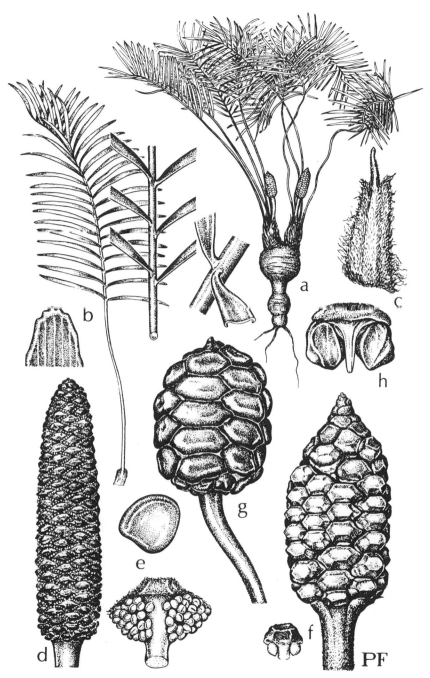

Fig. 14. **Zamia pumila:** a, male plant; b, at left, tip of leaf segment and, at right, leaf, next right, enlargement showing portion of rachis with basal portions of leaf segments, and next right, further enlargement of small portion of rachis and bases of leaf segments to show reflexed margins; c, cataphyll; d, microspore cone; e, below, microsporophyll, and above, microspore sac; f, on right, megaspore cone, and at left, megasporophyll; g, seed cone; h, megasporophyll with seeds. (Courtesy of Fairchild Tropical Garden.)

Leaves for the most part arching, disposition of the leaflets from the rachis varying from secundly erect in two rows to spreading laterally in one plane.

Male and ovulate cones produced on separate plants, sometimes in clusters, from the center of the crown of the stem, erect on stout, fuzzy-hairy stalks. Male cones at anthesis cylindrical or lance-ovoid, 5–8 cm long, the cone scales attached peltately in vertical, closely set ranks, the exposed outer faces of the scales hexagonal, cinnamon-brown, minutely fuzzy-hairy, each bearing beneath few to numerous sporangia. Female (ovulate) cones solitary, larger, stouter, their stalks stouter, mostly short-cylindric, the scales much like the male in structure, pubescence, and disposition (3–5 scales in 5 or 6 ranks), each bearing a pair of ovules beneath. Male cones not persisting long after pollen is shed. Ovulate cones maturing gradually over the season, the ripening seeds pushing the scales apart. Mature seed orange, 1.5–2 cm long, somewhat prismatic. [Incl. *Z. angustifolia* Jacq., *Z. floridana* A. DC., *Z. integrifolia* Ait., *Z. silvicola* Small, *Z. umbrosa* Small]

Inhabiting well-drained sands or sandy loams, often where soil is shallow over limestone, in Florida scrub, hardwood–cabbage palm hammocks, coastal shell mounds, slash pine–saw palmetto flatwoods, longleaf pine–deciduous scrub oak woodlands; n.e. Fla., mostly near-coastal, westward to Taylor Co., pen. Fla., and lower Fla. Keys; W.I. (See Eckenwalder, 1980.)

Angiosperms

MONOCOTYLEDONS

Agavaceae

Yucca

Evergreen perennial plants with stout main roots, stout stems with naked buds terminating the main stem or its branches which bear in a tight spiral long-lived, swordlike, daggerlike, or bayonetlike leaves with flared, partially overlapping bases. In some, the stem little, if any, elevated above the substrate, branches, if any, produced as offsets, in others, stem eventually elevated and branching sparingly. In general, the leaves narrowed above their flared bases, becoming broadest more or less medially, then narrowing to their apices which (in ours) are terminated by hard, spinose tips. Inflorescences showy terminal panicles. Flowers bisexual, radially symmetrical, ovary superior. Perianth of 6 creamy white to white, thickish parts (tepals) in two series of 3 each, the outer usually a little narrower than the inner. Stamens 6, in two series of 3 each, filaments relatively long, clavate, a little shorter than to about equaling the pistil, anthers very small. Pistil 1, ovary 3-locular, locules long and with many ovules "stacked" in each, style short and stout, terminated by a 3-lobed stigma. Fruit fleshy, berrylike and indehiscent *or* capsular and loculicidally dehiscent by 3 valves.

1. Leaves lacking filiferous fibers marginally; stems eventually well elevated above the substrate; stalks of the panicle seated within the ascending, uppermost leaves *or* stalks of the panicle long enough to elevate the base of the panicle 4–5 dm above the ascending, uppermost leaves; fruit indehiscent, berrylike, its stalk arched-recurved, thus the fruit pendent.

  2. The leaves dark green, not glaucous or glaucous only proximally, rigid-ascending at first, later spreading laterally, finally bending near their bases and declined; margins of leaves bearing irregular, opaque, short, hard, truncate or irregularly truncate-notched enations with sharp, tearing edges; stalk of the panicle very short so that one-quarter to one-third of the panicle is seated within the encircling erect-ascending leaves.

1. Y. aloifolia

  2. The leaves bluish green–glaucous when young, later grayish green, rigid-ascending at first, later the older ones relatively limber, bending more or less at their middles and disposed in a more or less "floppy" manner; edges of leaves with a narrow, opaque or brown-opaque margin that on some plants is smooth, on others bears minute, flat enations at least on distal margins, these rough to the touch or very sharp and cutting; stalk of the panicle usually long enough to elevate the base of the panicle even with or 3–5 dm above the tips of the encircling leaves.

2. Y. gloriosa

1. Leaves bearing loose, filiferous fibers marginally (most or all of the fibers may become detached from the oldest of the long-lived leaves on a given plant); stem little or only slightly elevated above the substrate; stalk of the panicle very long so that the panicle itself is elevated very far above the leaves; fruit a dry capsule, the stalk and capsule erect.

3. Y. flaccida

1. Yucca aloifolia L. SPANISH DAGGER.   Fig. 15

Plant with an erect, conspicuously leafy stem which at variable heights eventually produces a terminal inflorescence. The inflorescence terminates the

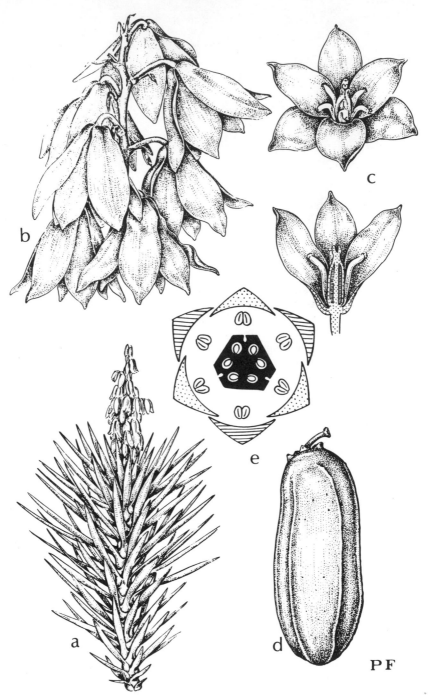

Fig. 15. **Yucca aloifolia:** a, upper part of flowering plant; b, distal portion of inflorescence; c, above, complete flower, below, flower longitudinally dissected; d, capsule; e, floral diagram. (Courtesy of Fairchild Tropical Garden.)

growth of the axis on which it is borne, but very soon one or more lateral branches are formed above which the leafy part of the axis bearing the inflorescence gradually deteriorates; if only one lateral bud is produced, it, by its subsequent growth, takes over as the erect leader shoot. Often a given stem may successively produce only one bud after flowering so that the stem after successive episodes of flowering gradually attains considerable height; owing to the longevity of the heavy leaves, the stem usually becomes top-heavy and topples whereupon its bud turns upwardly and the process may repeat itself. If more than one bud is produced after flowering, the uppermost generally becomes the erect leader-shoot and the subsequent growth from the other buds become lateral branches. This process frequently yields an openly branched crown which, again owing to top-heaviness, may cause all or much of the whole to topple. Besides the foregoing growth habit, buds often form near the base of a relatively elongate stem and these by the same process can form a thicket.

Leaves stiff and rigid, dark green excepting on their flaring bases and narrowed proximal portions above the bases where they are gray-glaucous; leaves erect when young, older ones becoming spreading, the oldest ones commonly bending near their bases and declined, the latter eventually dying and for a time the dead ones "skirting" the stem. Bases of leaves very broadly flaring and overlapping, abruptly narrowed above the bases, increasing gradually in width distally to midleaf or a little above, then gradually tapering to stiff, spinose tips. From plant to plant, or clone to clone, length and width of leaves varying considerably; on some plants, fully grown ones relatively short, about 2 dm long and about 2 cm broad at their broadest places, varying from those dimensions to 5–6 dm long and 4–5 cm broad on other plants or clones; margins bearing irregularly thin, stiff, truncate or notched-truncate enations with sharp, tearing edges.

Inflorescence with a short stalk so that one-quarter to one-third of the panicle is seated within the encircling young leaves; panicle columnar to narrowly ellipsoid, 5–6 dm long and about 2 dm broad. Flower stalks mostly about 2 cm long, arching so that the flowers themselves are pendent. Tepals variable in shape and length from flower to flower, even on a single plant, the outer three more or less oblongish, 4–5 cm long and 1–1.5 cm broad, the inner three more or less elliptic or oblanceolate, 4–5 cm long and 1.5–2 cm broad; filaments granular- or powdery-pubescent. Fruit an indehiscent, leathery berry, 8–10 cm long if well developed (which often it is not), about 3 cm in diameter. Seed black, sublustrous, obovate in outline, faces flattened, 5–7 mm long and about as broad distally. (Not infrequently fruit-set does not occur.)

Commonly cultivated or persisting from cultivation throughout our area; originally native on sand dunes, banks of brackish or salt marshes, on shores, and on shell mounds, Atlantic Coast from N.C. southward, and along the Gulf Coast to La.

2. Yucca gloriosa L. SPANISH BAYONET, MOUND-LILY YUCCA.

Plant having the same basic growth pattern as described for *Y. aloifolia*; the heavily leafy branches, however, when several of them are produced near the summits of older plants, are short, leaves of the several more or less interlacing, thus yielding a heavy moundlike appearance. This is in contrast to the very much more openly branched habit characteristic of old plants of *Y. aloifolia*.

Leaves having the same form as in *Y. aloifolia*, but when fully grown consistently longer and broader, mostly 6.5–10 dm long and about 5 cm broad medially, bluish green–glaucous throughout when young, grayish-glaucous when older; young leaves erect, older ones becoming limber and tending to arch or bend more or less medially, their distal portions somewhat "floppy"; margins with a narrow, longitudinal, opaque or brown band which, on some plants, is smooth to the touch, on others becomes somewhat frayed and rough to the touch, or on still others, at least the proximal margins may have tiny, sharp enations that tear one's skin.

Stalk of the inflorescence usually long enough to elevate the base of the panicle 3–5 dm above the encircling leaves; panicle 7.5–12 dm long, in outline somewhat pyramidal to elliptic, to about 4.5 dm broad medially. Tepals about equal in length, the outer three usually somewhat narrower than the inner three, on flowers of some plants all of them oval in outline, on flowers of other plants the inner ones oval, the outer ones narrowly elliptic, mostly about 4 cm long and the inner ones 2.5 cm broad and the outer about 2 cm broad; outer filaments narrower than the inner, their edges with short, flattish cilia, the surfaces otherwise more or less powdery-pubescent. Fruit an indehiscent, leathery, pendent berry 5–6 cm long. Seed black, lustrous, 6–7 mm long.

Infrequently (relative to *Y. aloifolia*) cultivated in our area; native on coastal dunes, s.e. N.C. to n.e. Fla.

Note: I have not observed fruit-set to occur on any plant of this species in the central Panhandle of Fla., where I see plants flowering each year.

3. Yucca flaccida Haw. BEARGRASS, SILKGRASS, ADAM'S NEEDLE.   Fig. 16

Plant with a short vegetative stem little elevated above the substrate, the leaves disposed rosettelike, eventually producing a stout, terminal inflorescence 2–4 m tall whose stalk (below the lowermost panicle branches) usually exceeds in length that of the panicle itself. The inflorescence terminates the growth of a given "rosette" which, after flowering and fruiting, generally perennates by a lateral bud or lateral buds from the short-stem, the original "rosette" very gradually deteriorating and very slowly dying. The new "rosette(s)" repeat the process thus becoming clumped, each one, after flowering, slowly deteriorating and dying so that eventually "plants" may be seen to be irregularly spread about. This process, when habitat conditions are favorable, may result in a large number of "plants" occupying a small area.

Young leaves of some plants conspicuously gray-glaucous, later becoming grayish green, leaves of other individuals not or inconspicuously glaucous. Fully grown leaves variable in length and breadth on a given plant, or with variable average dimensions from plant to plant, to about 8 dm long and 4 cm broad at their broadest places; leaves, above their somewhat flared bases, abruptly narrowed, then gradually broadening to somewhat above midleaf, above that gradually long-tapered to spinose tips; leaves rather rigid and erect at first, as they age often becoming limber and spreading in a "floppy" manner, often many of them arching or bending at about midleaf; leaf margins, as leaves are first emerging, with a narrow, opaque-brown longitudinal band which is somewhat frayed-fibrous at the moderately concave leaf tip; from this tip, one by one, filiferous fibers detach and peel back along the margin, the first to detach peeling approximately to the leaf base, this therefore, the longest, successive filiferous fibers detach and peel back shorter and shorter distances, the longer fibers, especially, curl and spiral, all of them soon bleaching to almost white; the leaves being long-lived, the fibers on the oldest ones often will have been broken off.

Stalks of the inflorescence glabrous, bearing stiffly erect bracteal leaves, the lowermost sometimes to 2.5 dm long, upwardly gradually becoming reduced in length, the uppermost usually about 2–3 cm long; axis of panicle and of

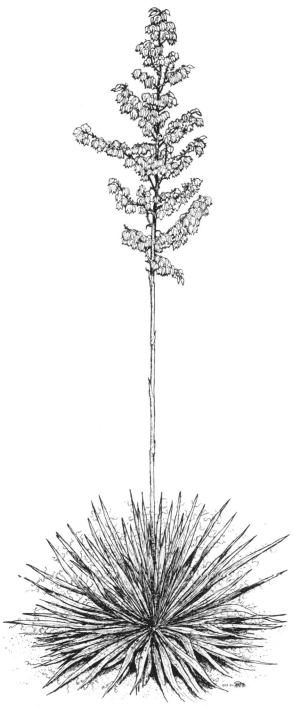

Fig. 16. **Yucca flaccida.** (Illustration by R. P. Elliott, reproduced with the artist's permission.)

branches and flower stalks glabrous and smooth on some, perhaps most, plants, roughly pubescent on others; flower stalks 1.5–2 cm long, rather lax at anthesis (stiffly erect in fruit). Outer three tepals sometimes slightly longer than the inner three, sometimes the inner slightly the longer, thus variable on a single panicle; outer ones lanceolate or narrowly elliptic, usually tapered to a distal point, 4–5 cm long and 1–2 cm broad, the inner ones broadly elliptic to ovate-elliptic, their tips obtuse, 4–5 cm long and 2.5–3 cm broad. Filaments equaling the ovary in length, their margins and surfaces bearing short, flat, membranous trichomes. Capsule erect, usually oblong or conical if fully developed, furrowed at the valves, abruptly narrowed apically to an apicule 3–5 mm long, mostly about 4 cm long and 1.5–2 cm broad. Seeds black, sublustrous, most of them obliquely half-round, faces flat, edges turned upwardly on both sides, 6–8 mm long. [Incl. *Y. smalliana* Fern. (*Y. filamentosa* L. var. *smalliana* (Fern.) Ahles)]

In our area, inhabiting a wide variety of sandy, open or semiopen sites, growing most vigorously in infertile sands; longleaf pine, scrub oak ridges and hills and semiopen pine plantations on such sites, open stands of sand pine–scrub oak, old fields, calcareous glades, open mixed pine-hardwood secondary woodlands, coastal sands, throughout our area and beyond.

Plants of *Yucca*, as described above, appear to me to be, in our range, the most widespread of those having filiferous fibers on leaf margins. The use of the name *Y. flaccida* as the correct one for plants as described is nothing more than an "educated guess" on my part, for I have achieved no appreciable understanding of the *Y. filamentosa* alliance, to which this belongs, from relatively recent treatments in botanical manuals or guides.

In addition to what is described above as *Y. flaccida*, there is a "form" present in parts of Liberty and Gadsden Cos., Fla., which baffles me. It occurs on former longleaf pine, scrub oak sandy ridges, which about thirty years ago were clearcut, then bulldozed and planted to slash pine. The slash pine grew very poorly, forming thin stands (most of it recently harvested). It is probable that the bulldozing operations widely distributed vegetative propagules of the *Yucca* plants present at that time and that these grew and flourished in the interim. The plants have the same general habit, including very tall inflorescences, as the *Y. flaccida* described above. The leaves are, in the main, angling-ascending, thickish and stiff, many of them irregularly, longitudinally corrugated, some of them with 1–2 twists, the average leaf is relatively short and broad, the larger ones to 4–6 dm long and 4–5 cm broad at their broadest places, their distal tapers broader and shorter than in *Y. flaccida* and the tips more concave adaxially. Whether or not plants like these occur more widely in our range, I do not know. In any event, I cannot attach a name to them, assuming they are sufficiently discrete to deserve one. There is, in the old literature, a plethora of names (at the species level, the varietal level, and the forma level) associated with variants of the *Y. filamentosa* complex.

Gramineae (Poaceae) (GRASS FAMILY)

Arundinaria gigantea (Walt.) Muhl. CANE, SWITCHCANE.

"Woody" perennial with hard, tough, subterranean rhizomes, aerial stems (canes) forming open or dense brakes. Aerial stems 1–8 m tall and to 3.5 cm in diameter, internodes hollow; stems at first unbranched, later freely branching, forming fanlike fascicles, the basal parts of shoots and primary branches with loose papery sheaths bearing short, narrow, pliable, lanceolate blades; toward the ends having several leaves with bases abruptly narrowed and petiolelike, the expanded portions 1–3 dm long, their sheaths overlapping; sheaths bearing several flat, scabrous bristles at the summit, these shed in age.

Flowering occurs at infrequent intervals (of unknown length), the flowering stems dying after maturation of the seeds. Flowering branches borne on the main stem, on primary branches, or on flowering shoots arising from the rhizomes, the latter mostly with bladeless sheaths. Spikelets solitary or usually in racemes or panicles, about 8–12-flowered, usually on long, slender stalks. Glumes 5–9-nerved, unequal, apically acute or obtuse, usually pubescent apically. Lemmas 11–17-nerved, papery, acute, acuminate, mucronate, or awn-tipped. [Incl. *A. tecta* (Walt.) Muhl., *A. macrosperma* Michx.]

Commonly, though by no means exclusively, inhabiting low-lying, moist to wet places, low woodlands of various mixtures and ranges of moisture conditions, river and stream banks, shrub-tree bogs and bays, sloughs and bayous; in addition, in woodlands on mesic and submesic slopes and uplands. My own observation is that brakes composed of large canes (to 10 m tall and 3.5 cm in diameter) occur in fertile, alluvial, river bottoms where elevated sufficiently that flooding is of short duration; elsewhere the canes tend to be very much shorter and of smaller diameter; throughout our area. (S. Md., Va. to s. Ohio, Ind., Ill., Mo., generally southward to cen. pen. Fla. and e. and s.e. Tex.)

Note: An exotic bamboo, *Phyllostachys aurea* Carr. ex A. & C. Rev., native to China, is frequently cultivated in our area. It naturalizes only to the extent that it spreads freely locally by stout subterranean rhizomes. In established clones, the aerial stems are straight, to 10 m tall or a little more and to 5 cm in diameter, often becoming yellow with age. Insofar as I know, it does not flower here.

Palmae (Arecaceae) (PALM FAMILY)

The species of palms represented in our area of coverage are five, three in the genus *Sabal*, one each in the genera *Rhapidophyllum* and *Serenoa*. (There are not, insofar as I know, any cultivated, exotic kinds which have become naturalized here.) Leaves very closely spirally arranged. In all of ours, the leaf blades more or less fanlike, the segments of the blade as it emerges from the bud very tightly folded in the manner of a closed Chinese fan. By the time the entire blade has emerged from the bud, the segments will have unfolded and fanned out (from the top downward), all the segments separating a part of the distance toward the base of the blade, *the undivided part of the blade being referred to as the palman*. In *Serenoa* and *Rhapidophyllum*, the segments all arise from the end of the petiole, the blade termed palmate; in *Sabal*, there is an extension of the petiole, a costa or "midrib," on the abaxial side of the blade and the segments fan out from it all along its length and the blades are then referred to as *costapalmate*. Adaxially, in *Rhapidophyllum* and *Sabal*, there is a relatively short to longish protuberance of the petiole at the base of the blade known as a *hastula* or "ligule"; this occurs on both sides in *Serenoa*. Inflorescences axillary to leaves. Flowers minute, radially symmetrical, ovary superior.

The leaves of seedlings and juvenile stages of growth differ greatly from those of plants having attained enough age to produce leaves with a form characteristic of the mature stage. The first leaf of the seedling has but a single segment, later ones 2, then 3, and so on until the mature form is attained. There is a great deal of variation in size of leafy crowns, number of leaves per crown, in length of petioles and size of blades. Some of the variance may relate to the relative age of different plants or even to the age of the plant when various of its relatively long-lived leaves were formed. Some of the variance doubtless reflects prevailing moisture regimes or relative fertility of the sites. Thus, in respect to these plants, it is more than ordinarily difficult to ascribe measurements, or particular attributes of form, other than in "ball-park" dimensions.

1. Petioles of leaves having on their edges short, hard, sharp, recurved prickles. 1. *Serenoa*
1. Petioles of leaves having the edges smooth.
  2. Sheath attached to either side of the base of the petiole large, fibrous, and weblike and with numerous, long, slender, very sharply pointed, longitudinally disposed "needles" extruding from it; leaf blades palmate; petioles and lower surfaces of segments of leaf blades with dull, grayish and somewhat silvery, fine scaliness.   2. *Rhapidophyllum*
  2. Sheath attached to either side of the base of the petiole without sharp "needles"; leaf blades costapalmate; petioles and lower surfaces of segments of leaf blades sometimes glaucous but without grayish or silvery, fine scaliness.   3. *Sabal*

1. Serenoa

Serenoa repens (Bartr.) Small. SAW PALMETTO.   Fig. 17

Stem usually more or less horizontal, eventually branching beneath the substrate or running more or less along its surface, the leafy extremity or extremities of the branches being at or just beneath the surface, commonly forming low thickets, or in some places an intricate, gregarious system of these forming more or less continuous, extensive colonies; less frequently with stems elevated above the substrate, obliquely to erect, to several meters long.

Petioles at base with an extensive, fibrous, brown, tubular web or fabric, above which slender, at first thinly glaucous (in most inland populations), 5–10 dm long, 0.8–1.5 cm broad, about the same breadth throughout, adaxially flat or nearly so, convex abaxially, its edges bearing short, hard, sharp, recurved prickles; at the distal extremity of the petiole on each side and slightly overlapping the base of the blade on each side, there is a brown, thin-membranous, hastula ("ligule") about 10 mm long, that on the adaxial side usually deltoid at first, often quickly frayed, that on the abaxial side usually broadly rounded and frayed. Leaf blades palmate, without a costa, green or yellowish green (in much of the range); palman of the blade longitudinally corrugated, very short at the position of the most lateral of the segments, gradually increasing in length toward the middle of the blade where it is about one-half the length of the free portions of the segments beyond, the latter essentially flat to either side of a slightly raised, narrow midrib abaxially and a slightly impressed midrib adaxially; palman, aside from its being corrugated, has the segments disposed in one plane in some leaves, in others the segments tend to be longitudinally gathered or puckered making the palman uneven and this causing the free portions of the segments to be disposed very unevenly; lateral segments narrowest and shortest, gradually increasing in length and width toward the middle; extremities of the segments attenuated or bifurcate, bifurcations with attenuated tips, the sinuses between the furcations varyingly 1.5–9 cm deep; larger blades on vigorous specimens to about 6 dm long and 9 dm broad, with up to 30 segments or a few more; blades less than half as large on specimens in many places and with correspondingly fewer segments. In Florida, at or near the Atlantic Coast, the

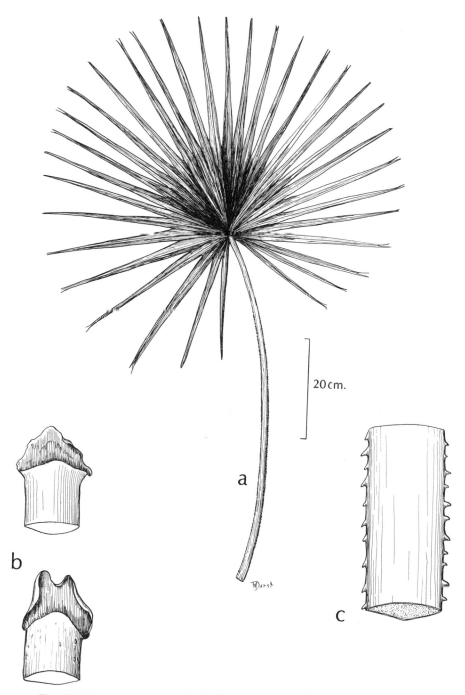

Fig. 17. **Serenoa repens:** a, leaf (drawn very much reduced); b, hastula ("ligule"), adaxial below, abaxial above; c, small portion of petiole showing prickly margins.

petioles and leaf blades (some populations) often averaging larger and notably glaucous and grayish green.

Inflorescences usually not exceeding the leaves, the axis bearing relatively coarse, sheathing bracts, those proximally on the axis empty, those more distal bearing more or less openly, paniculately branched spikes of small flowers. Flowers 5–6 mm long, bisexual. Calyx tubular below, with 3 minute lobes distally. Corolla exceeding the calyx, petals 3, nearly separate, reflexed at anthesis. Stamens 6, filaments connate basally; those opposite the petals with basal part of free portions broader than those opposite the sepals. Pistil 1, ovary superior, 3–ovulate, only one ovule maturing into a seed. Fruit drupelike, variable in size and shape, usually oblong-ellipsoid or somewhat pyriform, 15–25 mm long and 12–15 mm in diameter, black or bluish black when ripe.

Occurring in many kinds of habitats, often extremely abundant in both well-drained and poorly drained pine flatwoods, on deep sands of longleaf pine–scrub oak ridges and hills, in sand pine–oak scrub, coastal dunes, cabbage-palm "islands" in marshes, cabbage palm–hardwood hammocks; throughout our area. (Coastal plain, s.e. S.C. to the Florida Keys, westward to s.e. La.)

2. Rhapidophyllum

Rhapidophyllum hystrix (Pursh) Wendl. & Drude. NEEDLE PALM.   Fig. 18

Plant very slow-growing, stem not showing above ground for a long time, eventually erect or obliquely erect and eventually, on what I assume to be very old specimens, to 1 m high or a little more. The trunks of such specimens usually retain the fibrous leaf bases with their needles for many years and measured across these the stems may appear to be 5–6 dm in diameter; however, the stem proper, when finally exposed, is seen to be only 8–10 cm in diameter. Vigorous, healthy, old specimens may have 55 or more living leaves. Although I do not know, on the average, how many leaves may be produced each year, it would seem to be few and it would follow that individual leaves may live for a long time.

Petioles at base with an extensive, dense, fibrous, dark-brown, tubular web or fabric through which extend and from which are well extruded numerous long, slender, very sharply pointed, longitudinally disposed needles. Petioles, beyond their fibrous bases, elongate, much longer than the blades, somewhat triangular in cross-section, adaxial side flat, abaxial sides sloping to a rounded or obtuse angle; surfaces at first having a conspicuously dull grayish silvery, very thin covering that gradually breaks up into thin scales which tend to slough in patches leaving the surfaces more or less mottled; distally and adaxially the apex of the petiole protrudes over the leaf base as a hastula (ligule) of firm texture, broadly rounded, 3–10 mm long. Leaf blades palmate, without a costa, the palman very short relative to the length of the free portions of the segments, the palman varying in length from about 1.5 to 9 cm long, relatively obscurely corrugated; segments of blades on well-established plants varying from about 14 to 24, these often variable in width on a given leaf from 1 to 5 cm, usually broadest medially, the broadest ones and narrower ones generally interspersed; tips of segments varyingly truncate and shortly 2–4-notched, to weakly to strongly oblique and with 3–6 irregular, shallow notches; adaxial surface of segments, when they first unfurl, grayish silvery-scaly throughout but by the time fully expanded the scaliness sloughed save for that along the 1 or 2 raised nerves of the free segments and along the ridges of the corrugations of the palman, the

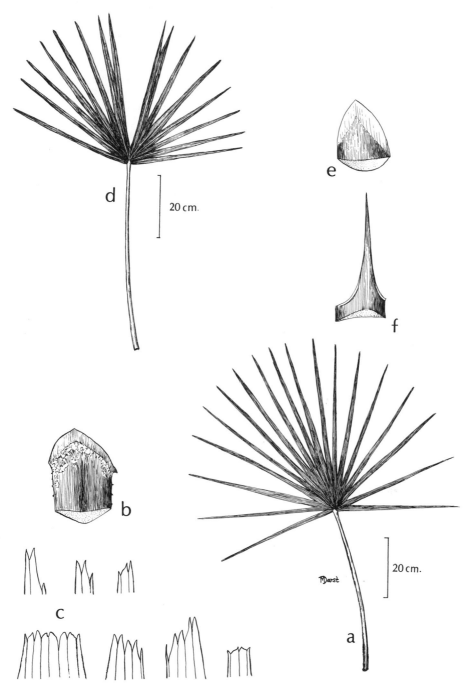

Fig. 18. a–c, **Rhapidophyllum hystrix:** a, leaf (drawn very much reduced); b, hastula ("ligule"); c, tips of leaf segments to show variation in notching; d–f, **Sabal minor:** d, leaf (drawn very much reduced); e, hastula; f, costa.

79

surfaces otherwise now dark green and glabrous; abaxial surfaces similarly silvery-scaly as they unfurl, the scaliness only moderately sloughing as the segments reach full size, the surfaces then having a subtle mottling of green silveriness.

Inflorescence axis proximally with several large, overlapping, empty scales sheathed at their bases, the distal axis inconspicuously scaly and with congested and very abundantly and densely floriferous branches, the whole inflorescence generally little extruded from the fibers, needles, and collected debris about the leaf bases. Flowers small, unisexual (plants usually dioecious, sometimes polygamodioecious). Staminate or predominantly staminate inflorescences more conspicuous than the pistillate and somewhat more extruded. Calyx united only basally, its 3 lobes ovate or triangular-ovate. Petals 3, longer than the sepals. Stamens, in the staminate flowers, 6, filaments slender, anthers about as long as the filaments, exserted. Pistillate flower with 3 pistils connivent basally, sometimes 1 or 2 of them aborting. Fruit drupelike, ovoid or globular, or squarish to pyriform if crowded during maturation, 15–20 mm in diameter, the surface brown to reddish- to purplish-brown to red, pubescent with a loose mat of tawny hairs that may be rubbed off very easily.

Fruits, many or most of them, often remain caught up among the fibers, needles, and debris around the leaf bases, some of them germinating there but not surviving very long; sometimes it appears that fruits may tumble out from small plants, or perhaps are scattered by predators, and germinate so that a great many individual seedlings or young plants may occupy a space as much as a meter across; evidently few survive.

Inhabiting woodlands of river bluffs, ravine slopes, wooded, spring-fed bottoms, wooded floodplains where flooding is of short duration, on exposed limerock of hammocks; local in much of our area. (S.e. S.C. to cen. pen. Fla., Panhandle of Fla. from Taylor and Jefferson Cos. westward to Walton Co., perhaps beyond, s. half of Ga., Ala., and s.e. Miss.)

3. Sabal

1. Leaves lacking filiferous fibers from the sinuses between the free portions of the segments and from the sinuses of bifurcations at the ends of segments; costa of leaf straight.
1. S. minor

1. Leaves (of plants beyond the seedling stage) bearing a filiferous fiber from each sinus between the free portions of the segments (eventually some such fibers may become detached); costa of leaf strongly curvate.

    2. Plant potentially a tree with unbranched stem to 25 (30) m tall; leaves (of plants beyond the seedling stage) with petioles 7–15 dm long and 2–4 cm broad, blades to about 12 dm long and 18 dm broad; plants occurring in moist to wet habitats.
2. S. palmetto

    2. Plant almost always with a subterranean stem (rarely with an emergent stem to 2 m high); leaves (of plants beyond the seedling stage) with petioles mostly 2.5–4 dm long and 0.6–2 cm broad, blades to about 6.5 dm long and somewhat broader than long; plants occurring in sand pine–oak scrub.
3. S. etonia

1. Sabal minor (Jacq.) Pers. **DWARF OR BUSH PALMETTO, BLUE-STEM.**   Fig. 18

Stem usually subterranean, the leafy crown just beneath the substrate, rarely (perhaps never in our area) emerging to 2–3 m high. Leaves very variable in size depending upon age of a plant or upon environmental conditions obtaining at a given site. Petioles smooth, thinly glaucous at first, often much longer than the blades, to 15 dm long, 1–2 cm broad, those of smaller leaves flat adaxially, those of larger ones to some degree concave, abaxial side rather strongly convex; hastula (ligule) adaxially at end of petiole and base of blade 7–10 mm long, varyingly triangular or its tip rounded, the texture hard. Blade costapalmate, costa straight, subulate, short (relative to those in our other two species of *Sabal*), 2–12 cm long (longest on largest blades); blade usually divided medially to the end of the costa, the two halves of the blade diverging so that there is a long V-shaped gap between them; variable in size, up to 7 (10) dm long and 12 (15) dm broad, number of segments varying with leaf size, larger leaves with up to 30 (40) segments; segments of the palman longitudinally corrugated, apart from the corrugations, spread in one plane, very short below the lateral segments, increasing in length to one-half to two-thirds the length of the entire blade at either side of the medial split; in size, the palman varying a great deal depending upon the size of the blade; free portions of the segments broadest basally (1.5–3 cm), tapering gradually distally, apically most of them equally or unequally merely notched to strongly bifurcate, the bifurcations stiff-textured, to 3 cm long or a little more; midvein of segments slightly impressed on adaxial surfaces and usually yellowish, narrowly elevated on the abaxial surfaces and green, numerous thin veins to either side of the midvein, these connected by cross-hatches on the abaxial surface; edges of segments not having filiferous threads detaching from the apices downward and hanging from the sinuses between them (as in our other two species of *Sabal*). Petioles and leaf blades with a very subtle bluish green hue when direct light shines upon them; abaxial surfaces of segments of leaves of seedlings with an evident bluish hue in any light.

Inflorescences usually exceeding the leaves, sometimes much exceeding them, as long as 10 dm from the base of the axis to the first floriferous branch, bracts of this part of the axis with long, light brown, tubular bases surrounding the axis, divided apical parts 3–6 cm long, dark brown; floriferous branches of the distal part of the axis subtended by similar but much smaller bracts, the branches well separated from each other, 6–12 cm long, each with a central axis from which spikes of flowers 3–9 cm long diverge alternately. Flowers minute, about 3 mm long, each subtended by 2 small, unequal bracts. Calyx tubular below, with 3 short, triangular, erect lobes above. Petals 3, oval, whitish, much exceeding the calyx. Stamens 6, filaments much longer than the anthers, the latter just exserted above the petals. Pistil 1, ovary 3-locular, 1 ovule in each locule (only 1 of which usually matures into a seed), style shorter than the petals. Fruit drupe-like, black, usually lustrous, orbicular to oblate, mostly 8–10 mm in diameter.

Chiefly occurring in lowland woodlands, sometimes on bluffs or slopes of ravines, sometimes in bottomland or flatland pastures; local throughout our area. (Chiefly but not exclusively coastal plain, n.e. N.C. to cen. pen. Fla., westward to Edwards Plateau, Tex.)

2. Sabal palmetto Lodd. ex J. S. Shult. and J. H. Shult. CABBAGE PALMETTO, SWAMP CABBAGE. Frontispiece. Fig. 19

Tree to about 25 (30) m tall, trunk diameter 3–4 (6) dm, often flowering/fruiting when of low stature; numerous dead leaves usually hang skirting the trunk beneath the crown of living leaves, these breaking off gradually somewhere along the petioles, bases usually persisting for a number of years before disintegrating to the extent that they crumble and fall away.

Leaves relatively numerous per crown, up to about 40 living ones on well-developed crowns. Petioles long and very stout, 7–15 dm long and 2–4 cm

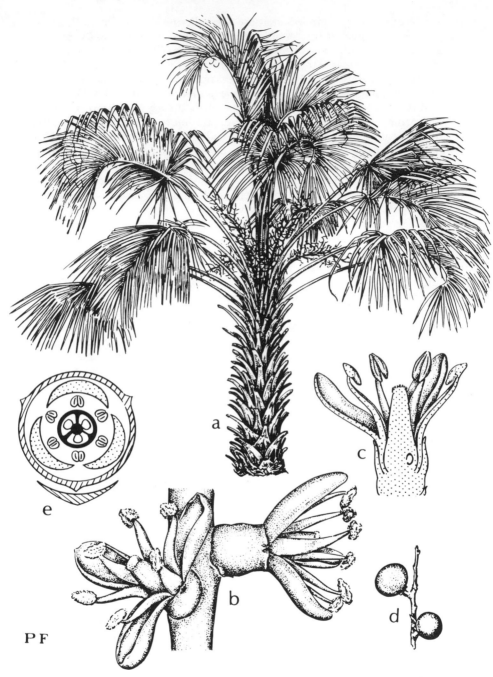

PF

Fig. 19. **Sabal palmetto:** a, habit; b, flowers; c, flower longitudinally dissected; d, fruits; e, floral diagram. (Courtesy of Fairchild Tropical Garden.)

broad, usually longer than the blades, plano-convex or concave-convex in cross-section, bases splitting as they expand, the splits eventually broadly flaring forming an inverted V-shaped gap between; usually the gap formed by one split petiole base is above a similar V-shaped gap formed by one side of each of two adjacent split petiole bases below, these together thus forming a rhomboid gap, and these collectively sometimes giving the appearance of a lattice. Adaxially a triangular-attentuate hastula prominently extends over the base of the blade 6–8 cm, its texture thickish and very hard and rigid, its edges usually upturned; on some blades the hastula is symmetric (in which case the base of the costa on the abaxial side is symmetric and the base of the blade is equilateral); on other leaves, the hastula is oblique (in which case the base of the costa on the abaxial side is correspondingly oblique and the base of the blade is inequilateral and oblique). Blades, the larger ones, about 12 dm long and 18 dm broad, the segments (40) 50–60 (90). Costa broadest basally and tapering gradually upwardly, eventually becoming so thin as to be imperceptible where the uppermost segments diverge from it. Palman longitudinally corrugated, very short at the position of the outermost leaf segments, gradually becoming longer inwardly to about three-fourths of the way to the costa, then gradually shorter; free portions of the segments mostly deeply bifurcate, bifurcations narrowly long-attentuated; as a leaf emerging from the bud unfolds, a single, filiferous fiber detaches from one edge of the free portion of each segment (the edge toward the costa on either half of the leaf) from its tip downward to the sinus between segments; a similar filiferous fiber detaches similarly from the edge of one of each pair of bifurcations of the segment; fibers generally remain attached to the sinuses, the longer ones between segments curl about conspicuously, the shorter ones between the bifurcations less so; leaf segments mostly 40–90 per leaf; costa of the blade strongly arching, and the segments disposed differently along various parts of the arch giving the blade a very distinctive form that appears different depending upon whether viewed from above, from the side, or from beneath.

Inflorescence large, as long as the leaves or longer, the lateral branches 4–5 dm long, each bearing alternately to about 15 spikes of flowers, the spikes floriferous throughout their lengths. Flowers minute, bisexual. Calyx cuplike, 1–1.4 mm long, with 3 short lobes on its rim. Petals 3, 2.5–3.3 mm long, ovate. Stamens 6, about equaling the petals, filaments united only basally, inserted on the bases of petals. Fruit drupelike, globose, 9–12 (14) mm in diameter, shiny-black, usually only 1 seed in each; seeds dark brown, nearly globose, mostly 8 mm in diameter.

In our area, occurring principally but not exclusively near the coasts, coastal strands and tidal flats, islands or elevated places in marshes, beach swales, moist to wet hammocks, s.e. Ga., n.e. Fla., Panhandle of Fla. westward to Gulf Co. (S.e. N.C. to s. pen. Fla., Fla. Panhandle; Bah.Is.)

The cabbage palmetto is the State Tree of Florida.

3. Sabal etonia Swingle ex Nash. SCRUB PALMETTO.

Plant almost always with an S-shaped or contorted subterranean stem, the crown bud held well below the soil surface.

Leaves few per crown, for the most part 5–10 living leaves at a given time. Petioles mostly 2.5–4 dm long and 0.6–2 cm broad, flat adaxially, strongly convex abaxially, glaucous at first, a hastula extended over the leaf base adaxially, varyingly 1.2–4 cm long, its texture hard and rigid, margins usually turned upward, varying from triangular and equilateral to notably oblique, the base of the blade correspondingly equilateral or oblique as in *S. palmetto*. Blades costapalmate, the costa arching as in *S. palmetto* but the blades average much smaller than in it; palman longitudinally corrugated, very short at the position of the lateral segments, gradually increasing in length toward the costa, free portions of all but the outermost segments deeply bifurcate, the bifurcations narrowly long-attenuated; filiferous fibers as in *S. palmetto*; leaf segments mostly 20–46 (56) per leaf.

Inflorescences varying from about half as long as to as long as the leaves, flowers essentially spicate throughout the length of the ultimate branches. Calyx cuplike, 1.0–1.4 mm long, 3 short lobes on its rim. Petals 3, ovate, 3.0–3.2 mm long. Stamens 6, equaling the petals, filaments slightly fused basally and inserted on the bases of the petals. Fruit drupelike, globose, 12–15 mm in diameter, shiny black, usually only 1 seed in each; seeds dark brown, nearly globose (the base somewhat flattened), mostly 10 mm in diameter.

Inhabiting sand pine–oak scrub and dune scrub, Florida, from Highlands to Clay and Columbia Cos., isolated stands of scrub in Citrus, DeSoto, Hernando, Manatee, Okeechobee, and Seminole Cos., coastal scrub from Dade to Volusia Co.

Smilacaceae

Smilax (GREENBRIERS, CATBRIERS)

Woody or herbaceous vines, climbing by means of tendrils which terminate stipules, the latter adnate to the petioles; petioles short, decurrent-winged, sulcate adaxially. Main stem and larger branches of most woody species variously armed with stiff, hard prickles. Woody species with tough rhizomes, some wholly or partially very large and tuberous. Flowers in umbels terminating axillary stalks, small, greenish, yellowish, or brownish, radially symmetrical, the perianth of six similar, spreading separate parts. Flowers unisexual, the staminate and pistillate on different individual plants (dioecious), staminate with 6 stamens, pistillate with 1 pistil, ovary 3–locular, style short, stigma of 3 relatively long stigmatic lobes. Fruit a berry.

Leader shoots of the season from rhizomes, especially those from thick-tuberous rhizomes, relatively stout, at first soft and succulent, grow extremely rapidly and attain considerable length before branching appreciably. Such shoots bear scale leaves and tendrils, the bladed leaves eventually produced distally on the smaller branches and branchlets. Tender young shoots may be used as a vegetable.

In some woody species, the leaves may vary a great deal from plant to plant, or even on the same plant, and identification may be somewhat less than easily achieved. It is very helpful to have carefully selected adequate material, including portions of the leader-shoots, especially when one is just commencing to learn identities.

1. Stems and petioles shaggy -pubescent; lower leaf surfaces densely and compactly pubescent.   5. *S. pumila*
1. Lower stems of leader-shoots and prickles abundantly clothed with brownish, stellate excrescences which render their surfaces scurfy to the unaided eye, *or* stems and prickles glabrous.
   2. Lower stems of leader-shoots and prickles scurfy.   2. *S. bona-nox*

2. Lower stems and prickles glabrous.

    3. The lower surfaces of leaf blades very heavily gray-glaucous.   3. *S. glauca*

    3. The lower surfaces of leaf blades not strongly gray-glaucous, if glaucous at all, the glaucescence very thin.

        4. Prickles on stems of leader shoots slenderly needlelike, lustrous dark brown or blackish.   8. *S. tamnoides*

        4. Prickles not needlelike, never lustrous dark brown or blackish throughout.

            5. Midvein on lower leaf surface much more pronounced than the principal lateral veins, the latter not or scarcely evident (raised).   4. *S. laurifolia*

            5. Midvein on lower leaf surface little if any more pronounced than the principal lateral veins, the latter evident.

                6. Leaf blades prevailingly lanceolate.   7. *S. smallii*

                6. Leaf blades prevailingly oblongish or ovate, or sometimes hastate.

                    7. Margins of mature leaf blade narrowly and tightly revolute by a thickened, cartilagenous band (vein) and with another similar vein close to it or the two juxtaposed so closely that they appear as one with a narrow groove; petioles not reddish; bark of lower (leader) stems pinkish-purplish.   1. *S. auriculata*

                    7. Margins of mature leaf blades thin, not banded; petioles usually reddish; bark of lower (leader) stems green.

                        8. Plants of well-drained places or where flooding occurs briefly; berries blue-black; leaves evergreen or semievergreen; edges of mature leaf blades for the most part not revolute, they and the edges of wings of petioles, sometimes proximally portions of major veins beneath, often but not always, having minute, flat, soft, toothlike enations.   6. *S. rotundifolia*

                        8. Plants of swamps or other places where surface water stands much of the time; berries bright red; edges of mature leaf blades usually revolute and usually not visible as viewed from above, they and wings of petioles without toothlike enations; leaves deciduous.   9. *S. walteri*

1. Smilax auriculata Walt. CATBRIER.   Fig. 20

Glabrous, evergreen, with tuberous thickened rhizome, this usually like a knobby chain or in knobby masses bearing brownish red scales when young. Stems clamber in intertwined tangles over low bushes or just in intertwined tangles exclusively their own, perhaps less frequently well into trees. In especially fertile sites, plants of this species grow much more luxuriantly and vigorously, for the most part having larger leaves, than they do on well-drained, infertile sands where they are more common. Leader shoots and major branches bear irregularly scattered prickles both at nodes and on internodes, their bases relatively broad then gradually tapered to a sharp point, tips of the prickles often blackish purple; bark of lower portions of fully mature leader shoots generally pinkish, orangish pink, or pinkish-purplish.

Leaves short-petiolate, petioles more or less twisted. Blades variable in size and shape, prevailingly oblongish, without lobes at the base or with varying degrees of basal lobing to definitely hastate, sometimes ovate, lanceolate, or suborbicular; bases rounded, obtuse, or cuneate, apices mostly rounded but generally apiculate or mucronate; most blades 3–nerved, two laterals arising with the midrib and arching-ascending so as to form as ellipse with the midrib central within it, reticulately veined between the nerves and between the lateral ones on the leaf edge, the latter usually having two thickened, juxtaposed veins along it, the two sometimes so close as to appear as one with a narrow groove, both surfaces dull green, the upper darker than the lower. Stalks of the umbels not exceeding the short petioles of subtending leaves. Berry glaucous, ripening through reddish or purplish to purple, finally black, usually remaining glaucous, sometimes shiny.

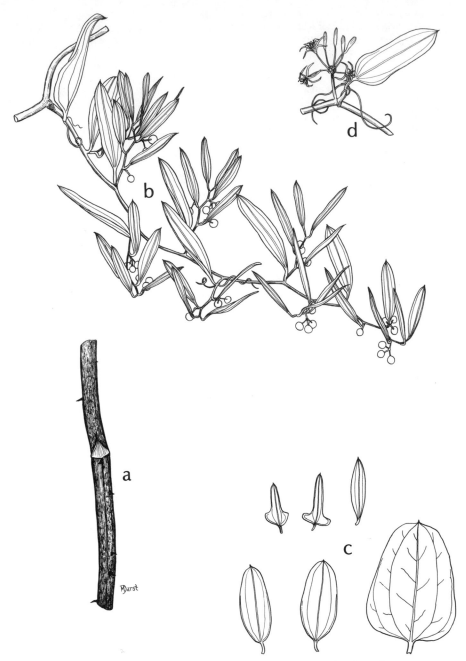

Fig. 20. **Smilax auriculata:** a, small portion of main leader shoot from near its base; b, a fruiting branch from relatively high on the stem; c, variable leaves; d, node with inflorescence.

Most commonly on deep sands of pine–scrub oak ridges and hills, coastal dunes and interdune hollows, less frequently in seasonally wet pine savannas and flatwoods, on stream banks and adjacent moist woodlands; throughout our area. (Coastal plain, N.C. to s. Fla., westward to La. and Ark.; Bah.Is.)

2. Smilax bona-nox L. CATBRIER, GREENBRIER. Fig. 21

Semievergreen with both subterranean runners and tuberous rhizomes, the latter abundantly clothed with short prickles having dark resinous tips. Lower stems, sometimes stems of main branches as well, and prickles, bearing abundant brownish, stellate excrescences (trichomes?) that render the surfaces scurfy to the unaided eye. Prickles scattered, both at nodes and on internodes, broadest at the somewhat flattened base, then gradually tapering and becoming terete, the tips sharp.

Leaves short-petiolate. Blades very variable in size and shape, usually thickish and stiffish at maturity, shiny green above, paler green beneath, often with pale green mottles or blotches; broadly ovate to lance-ovate, truncate to subcordate basally, rounded apically and with an abrupt short taper, or obtuse; in addition, leaves often hastate or fiddle-shaped; margin of leaf with a thickened cartilaginous band, this sometimes bearing prickles (the band usually more evident as viewed on the lower surface).

Stalks of the umbels 1.5 or somewhat more times as long as the subtending petioles. Berries shiny or dull black.

In a wide variety of habitats, upland and lowland woodlands, clearings, old fields, thickets, fence and hedge rows; throughout our area. (Md. and Va. to s. Ind., and Mo., generally southward to cen. pen. Fla. and Tex.; Mex.)

Identifiable in the field by the notably scurfy surfaces of stems and prickles, particularly on stems of leader shoots.

3. Smilax glauca Walt. WILD SARSAPARILLA. Fig. 22

Semievergreen, with long subterranean runners with knotty, tuberous thickenings, the runners bearing small prickles. Stem glaucous at first, in places often becoming mottled with black eventually, lower portions usually beset with slender prickles internodally and stouter, stronger prickles on nodes.

Leaves short-petiolate. Blades strikingly gray-glaucous beneath, the glaucescence composed of very tiny pillars of wax (as viewed with suitable magnification), prevailingly broadly ovate, truncate to subcordate basally, gradually tapered distally from the broadest part near the base, abruptly very short acuminate from the broadest part near the base, abruptly very short acuminate or apiculate at their tips, some leaves sometimes lanceolate and tapered basally; margins entire.

Stalks of the umbels slender, 1.5–3 times as long as the subtending petioles. Berries glaucous but tending to become shiny-black at full maturity.

Inhabiting sites of various kinds as for *S. bona-nox* and often growing in close proximity to or even intermixed and intertangled with *S. bona-nox* and/or *S. tamnoides*; throughout our area. (S. N.J., Pa., s. Ohio, Ind., and Ill., s.e. Mo. and s.e. Kan., generally southward to cen. pen. Fla. and e. Tex.; Mex.)

4. Smilax laurifolia L. BAMBOO-VINE, BLASPHEME-VINE. Fig. 23

Evergreen, with large, thick, heavy, tuberous rhizomes with reddish surfaces, its younger portions bearing large papery scales; cut sections at first fawn-colored,

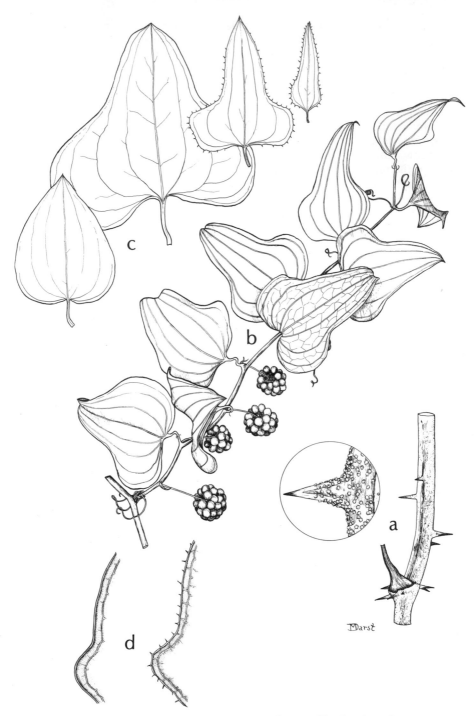

Fig. 21. **Smilax bona-nox:** a, small portion of stem of leader shoot from near its base, at right, and a bit of stem with prickle enlarged to show vestiture, at left; b, a fruiting branch from relatively high on the stem; c, variable leaves; d, edges of leaves enlarged to show variation.

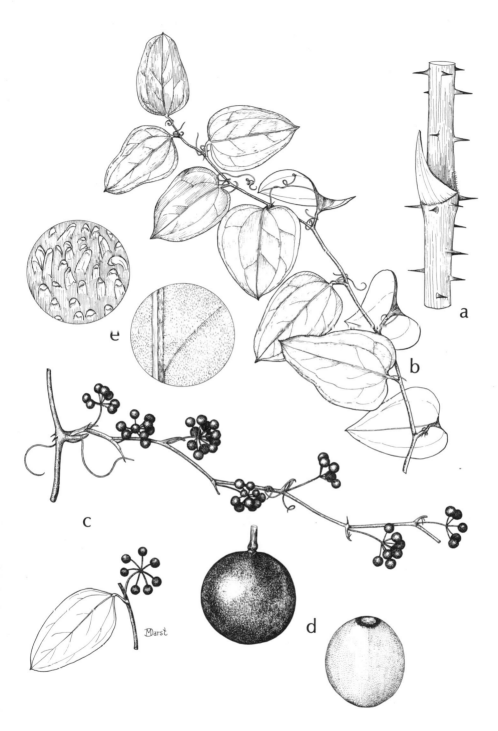

Fig. 22. **Smilax glauca:** a, small portion of leader shoot from near its base; b, a branchlet from high on the stem; c, fruiting branchlet, above, and node with leaf and an umbel of fruits, below; d, fruit, at left, and one of three stones of fruit, at right; e, portion of lower surface of leaf (drawn × 25), at right, and a tiny portion of same drawn from SEM photograph (× 250), at left.

Fig. 23. **Smilax laurifolia:** a, portion of leader shoot from near its base; b and c, fruiting branches, drawings from branches collected from the same stem and the same time (in February): b, growth of the previous season (fruits immature) and c, growth of two seasons prior (fruits mature); d, piece of stem with two leaves (drawn to same scale as leaves above) from plant growing in full shade.

turning pinkish red. Stems from older rhizomes very vigorous, up to 1–1.5 cm thick, bearing abundant stout prickles to 1 cm long on lower parts, these irregularly distributed and usually absent from nodes.

Leaves short-petiolate, petioles often twisted on leafy branches such that blades are held more or less erect-ascending. Blades mostly oblong, thick and leathery at maturity, rounded to broadly obtuse basally, their summits rounded and apiculate, occasionally blades of some leaves lanceolate or lance-elliptic and tapering at both extremities; midvein beneath much more prominent and raised than the principal laterals, the latter usually not raised or raised very slightly and scarcely evident.

Stalks of the umbels about equaling the petioles. Berries glaucous, becoming shiny-black at full maturity, ripening the second season after fruit-set and frequently persisting into or through the second winter.

Inhabiting shrub-tree bogs and bays, cypress-gum depressions, marshy shores of stream banks, swamps, places with prolonged inundation; commonly forming dense, impenetrable tangles, especially where no tall vegetation is present upon which to climb; throughout our area. (Chiefly coastal plain, cen. N.J. to s. Fla., westward to e. Tex., Ark., w. Tenn.; Bah.Is., Cuba.)

5. Smilax pumila Walt. SARSAPARILLA-VINE. Fig. 24

Evergreen, with slender, tan, subterranean, horizontal runners bearing brown, triangular-attenuate scale leaves, stems from the runners relatively very slender, without prickles, shaggy-pubescent, often more or less trailing, rarely over 5 dm tall, tendrils attached to stems of other small plants or objects, if any present, very often not climbing at all. (Growth in length of stems apparently restricted by abortion of terminal stem segments.)

Leaves short-petiolate, petioles shaggy-pubescent. Blades ovate to oblong-ovate or oblong, 4–10 cm long and 3–5 cm broad, 3-nerved, the two lateral nerves arching and forming an ellipse around the midvein; bases mostly cordate, apices rounded or broadly obtuse, the tips often apiculate; upper surfaces dark green, sparsely pubescent, lower surfaces grayish green, densely and compactly pubescent; margins entire.

Flowering late in the season; stalks of the umbels about equaling the petioles. Berries ripening over winter, bright red at maturity.

Mesic to xeric woodlands of various mixtures, sand pine–winter-green oak scrub; throughout our area. (Coastal plain, S.C. to cen. pen. Fla., westward to e. Tex.)

6. Smilax rotundifolia L. BULLBRIER, HORSEBRIER. Fig. 25

Semievergreen, glabrous, with long, slender, subterranean, nontuberous, seldom prickly rhizomes. Lower stems terete, those of branchlets more or less quadrangular; prickles numerous but not crowded, all on internodes, their bases moderately broad, tapering to sharp, usually dark brown tips.

Leaves short-petiolate, petioles usually reddish, their edges commonly having flat, soft, toothlike enations. Blades of leaves on leader shoots much larger than those of branches, varyingly reniform and broader than long (to about 12–14 cm broad and 8 cm long), or broadly ovate and longer than broad (to about 15 cm long and 10 cm broad), often markedly cordate basally, sometimes truncate or subcordate, tips prevailingly abruptly short-pointed. Leaves of branches relatively small, prevailingly ovate, less frequently lanceolate or elliptic, varying

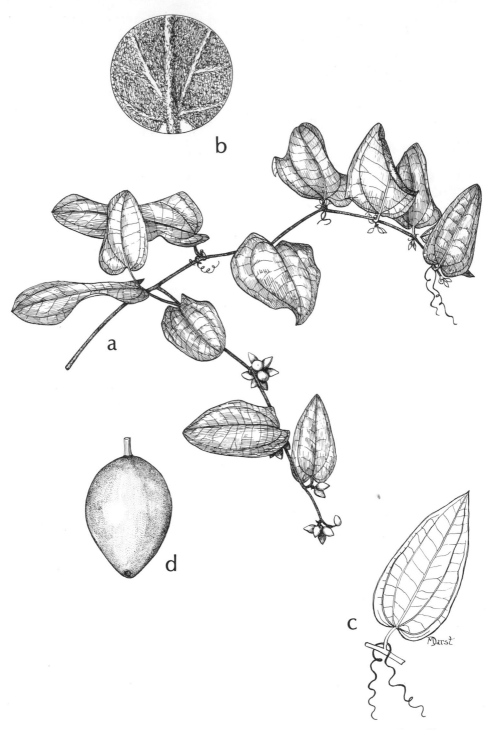

Fig. 24. **Smilax pumila:** a, habit, one branch fruiting; b, lower surface of base of leaf showing pubescence; c, leaf, with tendrils; d, fruit, much enlarged.

Fig. 25. **Smilax rotundifolia.** (From Coker in *Jour. Elisha Mitch. Sci. Soc.* 60, 1944.)

93

from nearly as broad as long to about half as broad as long, for the most part not exceeding 8 cm long; bases usually truncate or rounded, sometimes some shortly and broadly tapered, or some subcordate, apices acute, acuminate, or very abruptly short-pointed at their extremities; edges thin, entire or sometimes with minute, short, flattish, soft, irregular enations.

Stalks of the umbels as long as or to 1.5 times as long as the subtending petioles. Berries bluish black, glaucous, rarely dull reddish brown.

Moist to dryish woodlands and thickets, river banks, natural levees, low woodlands around ponds and lakes; throughout our area, rather infrequent. (N.S., s. Maine, N.H., s. Ont., to s.e. Mo. and Okla., generally southward to n. Fla. and e. Tex.)

7. **Smilax smallii** Morong. JACKSON-VINE, JACKSON-BRIER.   Fig. 26

Glabrous evergreen, with thick, hard, very irregularly branched, tuberous rhizomes to about 6 dm long. Leader stems from old, large rhizomes very vigorous,

Fig. 26. **Smilax smallii:** a, habit, a lateral fruiting branch from high on the stem; b, a small portion of leader-shoot from near its base; c, a node with inflorescence.

94

stout, to 2 cm thick, glaucous at first, bearing scattered, hard, sharp prickles both nodally and internodally, their bases relatively broad and somewhat flattened, gradually tapered distally, their distal halves usually dark brown, nearly black (prickles absent from branches). Stems commonly climbing high into trees, there with many spraylike, festooning branches.

Leaf blades relatively thin and flexible, prevailingly lanceolate, some sometimes lance-ovate (mostly ovate, sometimes hastate, on seedlings), 5–10 cm long and 1–3 cm broad; upper surfaces dark green and lustrous, paler green beneath, tapering basally into the short petioles, apices acute, sometimes tipped by a short mucro.

Stalks of the umbels little, if any, longer than the subtending petioles. Berries glaucous, then becoming dark brownish red to blackish red, usually black when fully mature, ripening about a year after flowering. [*S. lanceolata* of authors not L.]

Rich woodlands, mixed pine-hardwood stands, thickets, fence and hedge rows, most commonly in well-drained places, less frequently where temporary flooding occurs; common in much of our area. (Chiefly coastal plain, s.e. Va. to s. cen. pen. Fla., westward to e. Tex. and s. Ark.)

8. Smilax tamnoides L. BRISTLY GREENBRIER. Fig. 27

Semievergreen, with short, knotty, subterranean, nontuberous rhizomes free of prickles. Lower part of leader-shoots with needlelike, dark brown or nearly black, lustrous, slender, very sharp-pointed prickles about 1 cm long on the internodes, sometimes on the nodes, sometimes such needlelike prickles intermixed with numerous, similar, similarly colored, finer and shorter prickles. Branchlets terete, finely ribbed.

Leaf blades green on both surfaces, not glaucous (often drying ashy gray), those on leader-shoots sometimes much larger than those on the ultimate branches, broadly ovate to subrotund, or hastate to fiddle-shaped (the latter usually low on the stem), to about 12 cm long and some nearly as broad, bases truncate to cordate, apices varyingly with abrupt, short points, or short- to long-acuminate. Blades of branches sometimes nearly or quite as large as those of leader-shoots, more commonly much smaller, mostly ovate; margins flat, sometimes revolute, not banded, entire or with small, soft, flattish enations.

Stalks of the umbels 1.5 or more times as long as the subtending petioles. Berries black at maturity. [*S. hispida* Muhl. in Torr.]

Upland woodlands of various mixtures, old fields, thickets, fence and hedge rows, swales, moist to wet clearings, stream banks, often intermixed with *S. bona-nox* and/or *S. glauca*; throughout our area. (Conn., N.Y., to Minn. to Neb., generally southward to s. cen. pen. Fla. and e. and n. cen. Tex.)

Identifiable in the field by the needlelike, dark brown or nearly black, lustrous prickles on the lower stems.

9. Smilax walteri Pursh. CORAL GREENBRIER. Fig. 28

Deciduous, with slender, nontuberous, nonprickly rhizomes. Stems of leader-shoots slender, climbing over low bushes or to 5–6 m into trees, often forming rather extensive, dense, impenetrable tangles; sparingly prickly, prickles broadest at their somewhat flattened bases, gradually tapered to sharp points (absent from nodes); branchlets usually more or less angled.

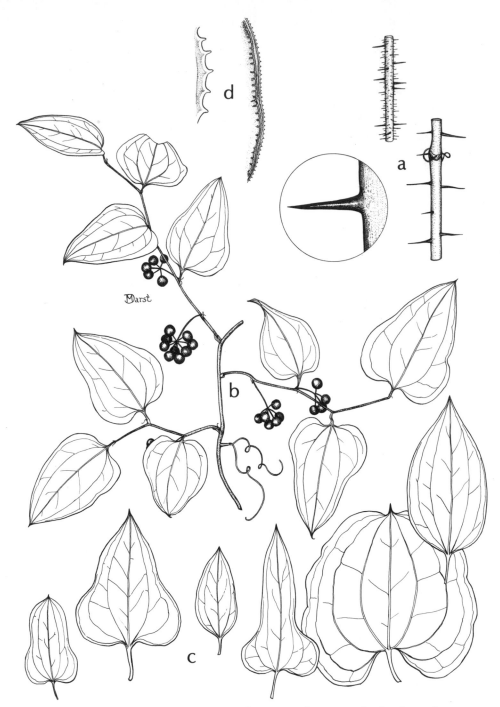

Fig. 27. **Smilax tamnoides:** a, small portions of two main leader-shoots from
near their bases to show variation in prickles and one of the larger prickles
enlarged; b, a fruiting branch from relatively high on the stem; c, variable
leaves; d, enlargement of edges of two leaves to show variation.

Fig. 28. **Smilax walteri. (From Coker in** *Jour. Elisha Mitch. Sci. Soc.* 60, 1944.)

Leaves short-petiolate, petioles usually reddish. Blades ovate to lance-ovate, or ovate-oblong (those of leader shoots sometimes much broader than those of branches), 4–10 cm long and 2–6 cm broad basally; paler green beneath than above, truncate to rounded basally, apices acute to obtuse, often with a cusp at their tips; margins thin, sometimes narrowly and tightly revolute thus simulating a banded edge, usually entire.

Stalks of the umbel not exceeding the subtending petioles. Berries bright red at maturity, usually reaching full maturity after the leaves have fallen in autumn.

Usually in water, at least in places with long periods of inundation, swamps, wet thickets, depressions in pinelands, marshy places; throughout our area. (Chiefly but not exclusively coastal plain, N.J. to cen. pen. Fla., westward to La., cen. Ark., Tenn.)

Plants of *S. walteri* are difficult to distinguish from those of *S. rotundifolia* except by the color of the ripe fruits. The latter species does not to my knowledge inhabit such wet places as does *S. walteri*.

DICOTYLEDONS

Aceraceae (MAPLE FAMILY)

Acer

(Ours) deciduous trees, some kinds flowering when of shrub stature. Leaf scars with 3 vascular bundle scars. Pith homogeneous and continuous. Leaves opposite, petiolate, usually without stipules. Blades simple and palmately veined, or (in one of ours) ternately or pinnately compound and the blades of the leaflets pinnately veined. Inflorescences racemes, panicles, or umbellike fascicles on short, few-leaved, or leafless branchlets developing from apical and/or lateral buds before, with, or after new leafy shoots of the season. Flowers small, radially symmetrical, unisexual by abortion, some sometimes functionally bisexual, plants dioecious or polygamo-dioecious. Calyx of 5 sepals, or sepals united below and 5-lobed, sometimes 4- or up to 12-parted. Petals 5 or of the same number as the calyx parts, inserted on the margin of a nectariferous disc, absent in some species. Stamens 4–12, usually 7 or 8, sometimes fewer than 4, inserted on or within the disc (sterile in the functionally pistillate flowers). Pistil 1 (rudimentary or absent in staminate flowers), ovary superior, 2–locular, each locule 2–ovulate, flattened perpendicular to the partition; style usually very short, stigmas 2, relatively long, stigmatic along their inner surfaces. Fruit a "double samara" (a pair of winged, 1–seeded samaras united at the base).

1. Leaves ternately to pinnately compound, blades of the leaflets pinnately veined.
   1. *A. negundo*
1. Leaves simple, blades palmately veined and for the most part palmately lobed.
   2. Lobes or margins of leaf blades evidently toothed.
      3. The lobes or margins bluntly toothed or bluntly toothed-lobed; central, terminal lobe generally less than half the length of the entire leaf blade. 2. *A. rubrum*
      3. The lobes or margins of leaf blades sharply toothed or toothed-lobed; central, terminal lobe generally more than half as long as the entire blade. 3. *A. saccharinum*

2. Lobes or margins of leaf blades not toothed.
  4. Leaf blades, when fresh, glaucous and grayish beneath; terminal lobes of many or most blades on a given plant (and usually the major lateral lobes) with their sides somewhat tapered downwardly (rather like a keystone).
                   4a. *A. saccharum* subsp. *floridanum*
  4. Leaf blades, when fresh, bright green beneath, not at all glaucous; terminal lobes broadest basally or with parallel sides, the major lateral lobes usually more or less triangular.   4b. *A. saccharum* subsp. *leucoderme*

1. Acer negundo L. BOX-ELDER, ASH-LEAVED MAPLE.   Fig. 29

Small to medium-sized tree, commonly with a short trunk and widely spreading branches. Twigs green, glabrous, sometimes glaucous, older ones with scattered, brown or buff-colored, circular lenticels. Leaf scars narrow, broadly V-shaped, the extremities of the pairs meeting.

Leaves compound, petioles long, slender, often as long as or longer than the blades, their bases dilated and those of pairs connected and thus clasping the stem, axillary buds hidden beneath their bases, each base usually with an adaxial, thin rim bearing a fringe of bristlelike hairs, this usually persistent on the upper side of the leaf scar after leaf-fall. Leaf blades ternate or with 5–7 (9) leaflets pinnately arranged, leaflets pinnately veined; leaflets shortly stalked, stalk of the terminal one usually longest. Blades of leaflets very variable in size and extent of marginal toothing or lobing, mostly 5–10 cm long and 5–7.5 cm broad, although some smaller, ovate, or elliptic, often inequilateral; bases broadly tapered to rounded, apices acuminate; margins with few, coarse, sometimes irregular, serrations mostly from the middle or slightly below upwardly, sometimes more or less lobed-serrate, sometimes toothed or toothed-lobed only on one margin; lower surfaces copiously and densely hairy as the leaves unfurl, the upper surfaces only moderately hairy and soon becoming glabrous, the lower eventually glabrous, pubescent mainly along the veins, or pubescent throughout.

Flowers greenish yellow, usually unisexual (plants dioecious), bisexual flowers rare (see Wagner, 1975); flowers appearing in early spring before or as new shoot growth commences. Staminate flowers fascicled, 10–15 per fascicle, their hairy stalks very lax, filiform, and pendulous, sometimes red. Sepals 4 or 5, unequal, oblong-oblanceolate to obovate, varying to linear. Petals absent. Stamens 4–6, usually 5, their filaments filiform, slightly exserted from the calyx, anthers much longer and broader than the filaments, bananalike in form. Pistillate flowers in lax, stalked, pendulous racemes often having 1–2 unfurling leaves basally, their stalks glabrous, very slender, about 8–12 mm long at anthesis (greatly elongating after anthesis). Fruiting racemes pendent, samaras reaching full size about midsummer, each winged half 2.5–3.5 cm long, falling tardily, many of them remaining on the tree over winter. [*Negundo negundo* (L.) Karst.]

Floodplain woodlands and slopes above them, low, wet woodlands and banks of small streams; spottily distributed throughout our area. (Vt. to Man., generally southward to cen. pen. Fla. and e. half of Tex.; also southward from n. Calif., Nev., Utah, and Colo. to s. Mex. and n. Cen.Am.)

Notable distinguishing characteristics of the box-elder are its green, often glaucous, twigs and compound leaves having dilated petiole bases covering the axillary buds.

National champion box-elder (1976), 16'11" circumf., 110' in height, 120' spread, Lenawee Co., Mich.

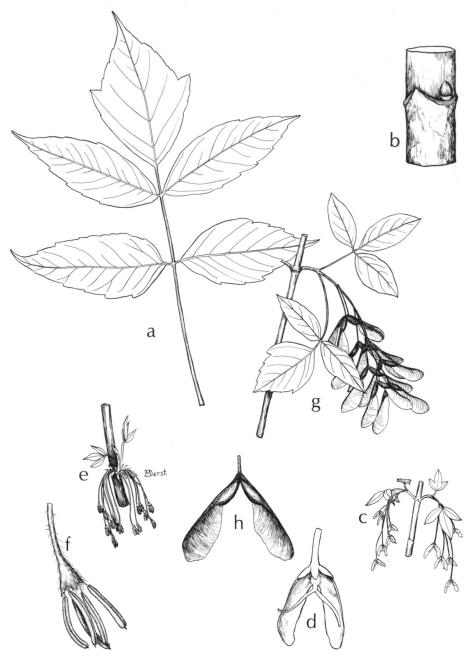

Fig. 29. **Acer negundo:** a, leaf; b, node; c, node with pistillate inflorescences; d, pistillate flower; e, node with staminate inflorescences; f, staminate flower; g, node with infructescence; h, fruit.

100

2. Acer rubrum L. RED MAPLE, SCARLET MAPLE. Fig. 30

Small to large tree, somewhat conspicuous in winter because of its clusters of small red to yellowish flowers which appear long before new shoot growth occurs, following that and still before new shoot growth commences the pistillate or largely pistillate trees much more conspicuous because of the profusion of deep red, scarlet, pink, or yellowish samaras they bear.

Bark of trunks furrowed, ultimately divided into long, narrow ridges covered with scaly plates free at their extremites; bark of branches smooth and gray, that of small twigs usually purplish red. Leaf scars mostly narrow and V- or crescent-shaped.

Leaves simple, slenderly petiolate, petioles to as long as the blades. Blades variable (see illustration), commonly ovate to orbicular in overall outline, sometimes obovate, usually palmately 3-lobed, sometimes with a pair of small, near-basal lobes in addition, unlobed on a rare individual and then ovate-lanceolate to lanceolate; leaf size, principal lobing, and marginal toothing variable; in our area, trees having 3 more or less triangular lobes, the terminal lobe usually shorter than to equaling the body of the leaf below it is perhaps the more common; often, however, the terminal lobe sublobed below its triangular tip and with parallel sides below the sublobing; bases truncate, rounded, subcordate, or U-shaped; margins variably bluntly serrate (in ours) upwardly from somewhat above the base or from above the middle; upper surfaces dark green, glabrous or sparsely short-pubescent along the principal veins, lower surfaces grayish to silvery, on some plants glabrous, and on others pubescent along the midrib, along the midrib and principal lateral veins, or densely felty-pubescent throughout.

Flowers glabrous, functionally unisexual, in fascicles from clustered buds on branches of the previous year, appearing long before new shoot growth commences, the staminate and pistillate commonly on different trees, apparently less frequently in separate clusters on the same tree in which case there are usually many more staminate than pistillate clusters. Bud scales red, flowers, from tree to tree, varying (as do the samaras later) from deep red to pinkish, brownish, or straw-colored. Flower stalks slender, glabrous, 1 cm long or a little more, those of the pistillate flowers quickly elongating to 3–4 (6) cm after anthesis. Sepals and petals 5 each, short-oblong to linear, 1.5–2 mm long, much alike, the sepals often slightly shorter and broader than the petals. Staminate flowers with 5–8 stamens, filaments well exserted, longer than the anthers which are very dark red and held more or less horizontally. Stamens of fertile pistillate flowers short, not exserted, sterile. Samaras ripening and falling by the time new shoot growth occurs, their halves spreading at various angles, each winged half 2 cm long or a little more. [*Rufacer rubrum* (L.) Small; incl *R. carolinianum* sensu Small]

In our area a common and abundant inhabitant of swamps and wet woodlands, less frequent on wooded slopes and in upland woodlands. (E. Can. to Man., generally southward to s. pen. Fla. and e. Tex.)

Red maple exhibits notable variation in size, form, lobing, marginal toothing, and pubescence of leaves. As described above, our plants may be designated as var. *trilobum* K. Koch, or those individuals with leaves densely pubescent (tomentose) beneath as var. *trilobum* forma *tomentosum* (Tausch.) Siebert & Voss.

CARL A. RUDISILL LIBRARY
LENOIR-RHYNE COLLEGE

Fig. 30. **Acer rubrum:** a, tip of branchlet and, below, variable leaves; b, twig with functionally pistillate flowers, at right, and functionally pistillate flower, at left; c, twig with staminate inflorescences, at left, and staminate flower, at right; d, node with infructescences.

3. Acer saccharinum L. SILVER OR SOFT MAPLE. Fig. 31

Medium-sized, fast-growing tree, generally with a relatively short trunk from which several nearly equal, erect branches grow, the crown relatively open. Bark of trunk breaking into long, loose, scaly plates more or less free at their extremities. Young stems glabrous, green at first, later becoming brown or reddish brown and with scattered, buff-colored, vertically lenticular lenticels; leaf scars crescent-shaped.

Leaves simple, slenderly long-petiolate, petioles commonly red. Blades palmately veined and palmately 5–lobed, the terminal lobe and often the 2 laterals below sublobed, ovate to orbicular in overall outline, mostly 6–12 cm long and about as broad across the major lateral lobes; the terminal lobe half or more than half as long as the entire blade, sinuses between the terminal lobe and the major laterals varying from narrowly acute to broadly V-shaped; bases truncate to broadly inverted V-shaped, apices of major lobes long-tapered, margins of lobes generally irregularly, prominently, and sharply toothed; upper surfaces bright green and glabrous, lower surfaces of unfurling leaves densely shaggy-pubescent, pubescence quickly sloughing, at maturity silvery, appearing glabrous but with suitable magnification seen to have very short, appressed pubescence throughout.

Flowers in compacted fascicles seated within the scales of opened buds clustered on wood of the previous season well before new shoot growth commences. Scales of the buds usually red, thickly fringed marginally with pale to reddish shaggy hairs. Flowers unisexual, the staminate and functionally pistillate in separate clusters on the same tree or one or the other on separate trees (staminate trees said to switch to pistillate as they age). Flower stalks so short that, at anthesis, in the staminate flowers only the stamens are exserted from within the bud scales, in the pistillate clusters only the stigmas (and sometimes the tips of the abortive anthers) are evident. In the case of the pistillate clusters the flower stalks quickly elongate, extending the entire flower from the bud, the ovary by this time already expanding, the stalks on ripening fruits eventually 2–4 cm long. Calyx essentially funnellike, barely if at all lobed, glabrous or nearly so, longer in the staminate than in the pistillate flower. Petals none. Stamens 3–7 per flower. Ovary very densely clothed with longish, matted, straw-colored hairs, most of the hairs sloughing as the fruit matures. Samaras mature and falling about the time or shortly after shoot growth commences, each winged half 3–6 cm long. [*Argentacer saccharinum* (L.) Small]

River banks and natural levees, colonizing clearings of bottomlands and adjacent slopes; s.w. Ga., Apalachicola River, and Choctawhatchee River, Panhandle of Fla., s.e. and s.w. Ala. (N.B. and Que. to Minn. and S.D., generally southward to cen. Panhandle of Fla. and La.)

National champion silver maple (1972), 22'7" circumf., 125' in height, 111' spread, Oakland Co., Rochester, Mich.

4a. Acer saccharum Marsh. subsp. floridanum (Chapm.) Desmarais. SOUTHERN SUGAR MAPLE, FLORIDA MAPLE. Fig. 32

Medium-sized tree. Bark gray and smoothish on relatively small trunks and on limbs, grayish, shallowly furrowed and scaly-ridged on large ones; twigs glabrous, green at first, becoming reddish brown and with scattered, paler, vertically lenticular lenticels; leaf scars mostly narrowly V-shaped or crescent-shaped.

Leaves simple, slenderly petiolate, petioles variable in length, glabrous (in

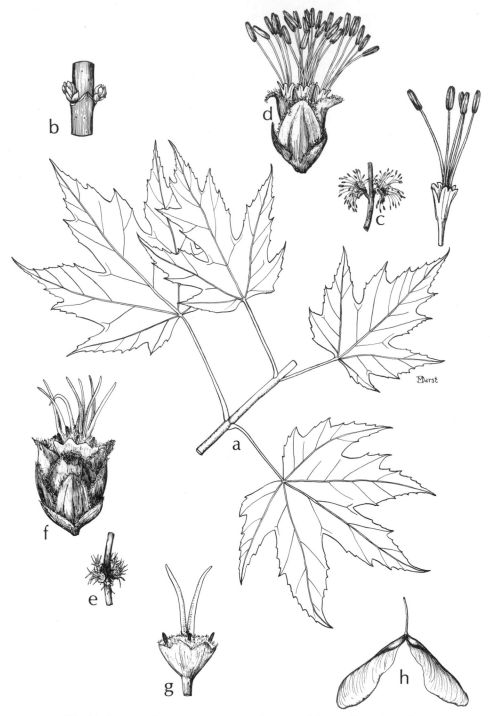

Fig. 31. **Acer saccharinum:** a, portion of stem with leaves; b, node with winter
buds; c, node with a pair of staminate inflorescences and d, one of same much
enlarged; e, node with clusters of functionally pistillate inflorescences and f,
one of same much enlarged; g, functionally pistillate flower; h, fruit.

Fig. 32. **Acer saccharum** subsp. **floridanum:** a, fruiting branch; b, cluster of fruits, enlarged somewhat; c, staminate inflorescence at left, staminate flower at right; d, pistillate inflorescence at left, pistillate flower at right.

ours). Blades generally as broad as or broader than long (breadth measured from tip to tip of the lateral lobes), those on crowns of mature trees often considerably smaller than on saplings, varying from 3 to 9 cm long; palmately veined and palmately (3) 5–lobed, the three major lobes commonly with 3 small, blunt-tipped sublobes distally, the principal lobes of many or most leaves on a given tree evidently to subtly narrowed downwardly below the lateral sublobes, somewhat like a keystone, the major lateral lobes of some leaves broadest basally and more or less triangular; basal pair of lobes, if present, small and bluntly triangular; margins or margins of lobes not toothed; upper surfaces of unfurling leaves essentially glabrous, the lower surfaces copiously pubescent; upper surfaces of *fresh*, *mature* blades bright green and glabrous, the lower glaucous and pale grayish, sparsely to moderately short-pubescent throughout, the hairs along principal veins longer than elsewhere, occasionally pubescent only along the major veins. (Rarely, in our limited range, all the leaves of a given tree may have leaves with 3 lobes, the lobing more or less triangular, bases U-shaped, yet the lower surfaces as described above, blades apparently closely similar to *A. saccharum* subsp. *saccharum* forma *rugelii* (Pax) Palmer and Styerm.) Autumnal coloration bright yellow to scarlet; dead, brown leaves tend to remain on the trees through much of the winter.

Flowers unisexual and staminate or bisexual and functionally pistillate, the anthers sterile, borne in fascicles terminating short-shoots of the season before those shoots have elongated from the bud, or after they have elongated somewhat and have barely evident unfurling leaves basally; staminate and functionally pistillate flowers in the same or separate fascicles on a given tree, or only one or the other on different trees. Flower stalks elongate, filiform, lax and pendent, sparsely pubescent with longish hairs throughout or only distally. Calyx cuplike or saucerlike, the lobes very short, broadly triangular. Petals none. Staminate flowers with 7 or 8 stamens, filaments exserted, interior of the calyx and sometimes the margins of the lobes pubescent, the base of the calyx interiorly densely bristly-hairy, many of the hairs protruding from the calyx giving a ciliate-fringed effect. Functionally pistillate flowers having the filaments of the stamens very short, only the tips of the fully formed but sterile anthers protruding from the calyx; calyx hairy within, ovary copiously bristly-hairy and some of its pubescence extruding from the calyx and contributing to the ciliate-fringed effect. Samaras maturing in midsummer, shriveled calyx often persistent at the base of the "twinned" seed-bearing portion, surfaces glabrous or with a few long hairs persisting, each winged half 2.5–3 cm long. [*A. barbatum* of authors; *Saccharodendron barbatum* (Michx.) Nieuwl.; *S. floridanum* (Chapm.) Nieuwl.]

Mesic wooded bluffs and ravines, sandy or calcareous upland woodlands, wooded natural levees; throughout much of our area, perhaps absent from southeasternmost Ga. and northeasternmost Fla. (Coastal plain and piedmont, s. Va. to cen. pen. Fla., westward and northwestward through much of Ala. and Miss. thence to s.e. Mo., Ark., e. Okla., e. Tex.)

Co-national champion Florida maples (1981), 8'10" circumf., 110' in height, 60' spread, Sumter Co., Ala.; (1981), 7'5" circumf., 126' in height, 57' spread, Stevens Creek Natural Area, S.C.

4b. Acer saccharum Marsh. subsp. **leucoderme** (Small) Desmarais. CHALK MAPLE. Fig. 33

Respecting the maples of our area, this one is most nearly like subsp. *floridanum*. Trees generally of smaller potential stature, the bark of their trunks said to have

Fig. 33. **Acer saccharum** subsp. **leucoderme**.

a chalky surface (perhaps owing to the profusion of chalky lichens sometimes growing on it). Leaves 3–5–lobed, 3–lobed ones more common than in subsp. *floridanum*, the three major lobes commonly tapered upwardly from their bases, especially the laterals, and more or less triangular, but sometimes, especially the terminal one, with parallel sides; tips of the lobes on the average more acuminately pointed than in subsp. *floridanum*. Lower surfaces of *fresh* leaves green, sublustrous, not at all glaucous (with just a hint of yellowish green when pressed and dried), distribution of pubescence as in subsp. *floridanum*. [*A. leucoderme* Small (*Saccharodendron leucoderme* (Small) Nieuwl.)]

Inhabiting wooded river bluffs and ravines, upland calcareous woodlands; in our area, cen. Panhandle of Fla., chiefly Apalachicola and Chipola River drainages of Liberty, Gadsden, and Jackson Cos., s.w. Ga., s.e. Ala. where it often grows intermixed with subsp. *floridanum*. (Locally distributed, e. Tenn., N.C. to cen. Panhandle of Fla., Ala., e. Miss., La., e. Tex., Ark., e. Okla.)

Co-national champion chalk maples (1981), 2'7" circumf., 30' in height, 40' spread, Clemson University Forest, S.C.; (1981), 2'3" circumf., 35' in height, 34' spread, Pickens Co., S.C.

Anacardiaceae (SUMAC FAMILY)

1. Inflorescences paniculate, each panicle borne above the uppermost leaf of a branchlet of the season (or sometimes in *Rhus copallina* a series of panicles terminating a branchlet of the season); fruit copiously red-pubescent. 1. *Rhus*
1. Inflorescence spreading or drooping panicles borne in axils of leaves on shoots of the season; fruit glabrous, straw-colored or ivory-colored, often streaked with purple.
2. *Toxicodendron*

1. Rhus

Shrubs or small trees. Twigs without terminal buds. Leaves alternate, deciduous, petiolate, trifoliolate or odd-pinnately compound, without stipules. Flowers small, radially symmetrical, functionally unisexual, the plants dioecious (said to be polygamous). Calyx united basally, 4–6-lobed, usually 5. Petals 4–6, usually 5. Stamens of the same number as the sepals or petals, inserted on the rim of a disc at the base of the flower. Pistillate flowers with 1 pistil seated on a disc, ovary superior, pubescent, 1–locular, styles 3. Fruit a 1–stoned dry drupe copiously clothed with red pubescence.

1. Leaves trifoliolate; flowers in short spikes or short panicled spikes near or at the ends of branchlets, appearing before or as shoot growth occurs. 1. *R. aromatica*
1. Leaves odd-pinnately compound, leaflets usually 9 or more; flowers in panicles on twigs of the season, wholly above the uppermost leaf or in panicles both in leaf axils and above the uppermost leaf or bracteal leaf.
  2. Twigs copiously shaggy-pubescent; axis of leaf blade, between the leaflets, winged; margins of leaflets usually entire; lateral buds in the leaf axils. 2. *R. copallina*
  2. Twigs glabrous; axis of the leaf blades not winged; margins of leaflets serrate; lateral buds enclosed by the bases of the petioles. 3. *R. glabra*

1. Rhus aromatica Ait. FRAGRANT SUMAC. Fig. 34

Straggly to upright shrub, to about 2.5 m tall, sometimes thicket-forming. Twigs slender, brown and sparsely pubescent at first (ours), becoming glabrous; leaf

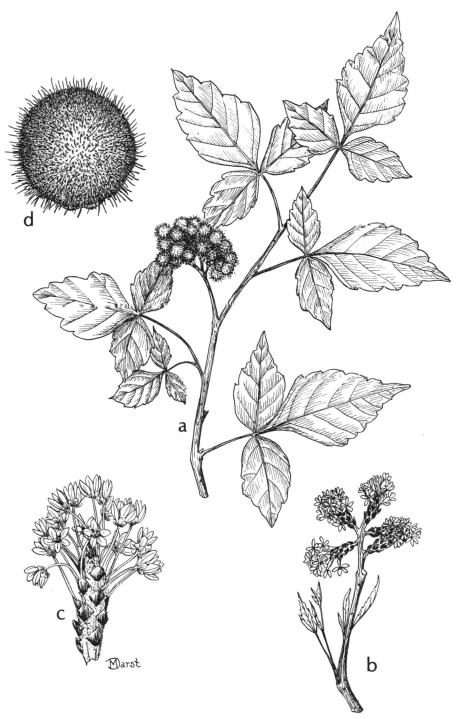

Fig. 34. **Rhus aromatica:** a, fruiting branch; b, inflorescence; c, enlargement
of one branch of inflorescence; d, fruit (most of trichomes of which slough
before fruits are shed).

scars angularly raised, the long-hairy buds partially obscured within the angles formed by them.

Surfaces of developing leaflets on emerging shoots copiously soft-pubescent, especially beneath. Petioles of mature leaves variable in size, mostly 1–3 cm long, usually shorter than the blades, glabrous or sparsely pubescent. Leaflets 3, the lateral pair sessile, equilateral or oblique basally, ovate to oblong-elliptic, or obovate, usually blunt apically, terminal leaflet equilateral, larger than the laterals, 2.5–8 cm long, mostly ovate, acuminate basally to a subpetiole, sometimes blunt apically, more frequently acute or short-acuminate; margins of leaflets coarsely crenate-undulate upwardly from a little below their middles; surfaces variably sparsely short-pubescent to glabrous. Crushed herbage sweet-scented.

Flowers in short, dense, bracteate spikes or racemes, or racemes panicled, appearing near or at the tips of twigs of the previous season before or as new shoot growth commences; axes of the spikes, racemes, or panicles copiously short-pubescent; bracts short, subrhombic, brown pubescent about their bases and along their margins. Flowers sessile or with sparsely pubescent stalks 2–3 mm long. Sepals ovate or oblong-ovate, somewhat cupped, blunt apically, glabrous, little, if any, more than 1 mm long. Petals pale yellow, ovate-oblong or oblong, obtuse apically, about 3 mm long.

Drupe subglobose, 3–5 mm in diameter, bright red, surface with longish, whitish hairs and very short, red, glandular ones intermixed. [*Schmaltzia crenata* (Mill.) Greene]

Dry or rocky woodlands and glades; rare in our range, s. Ala. and w. Panhandle of Fla. (Species exhibiting much variation.) N.E., Que. to Mich. and Iowa, southward to w. Panhandle of Fla., and Tex.)

2. **Rhus copallina** L. WINGED SUMAC OR SHINING SUMAC.   Fig. 35

Deciduous, bushy-branched shrub, commonly colonial and thicket-forming, or a small tree, probably little, if any, exceeding 8 m tall. Twigs relatively stout, copiously shaggy-pubescent with tawny hairs, eventually with numerous roundish or transversely elliptic, somewhat raised lenticels, pubescence gradually sloughing, older stems glabrous, brown, and vaguely fluted; vascular bundle scars obcordate, the copiously hairy axillary buds seated between the cordations; vascular bundle scars usually 9, in 3 groups of 3 each; pith light brown, relatively thick, continuous; sap clear and watery.

Leaves alternate, petiolate, petioles 2–6 cm long, shaggy-pubescent, their bases nearly encircling the axillary buds; blades odd-pinnately compound, 1–3 dm long, leaflets 9–23, sessile or very shortly subpetiolate, lanceolate to oblong-elliptic, some of them, or all of them, on a given leaf inequilateral and more or less falcate, 3–8 cm long and 1–3 cm broad, rarely broader, bases mostly shortly tapered, sometimes rounded, apices acute to rounded; upper surfaces glabrous or minutely pubescent on the major veins, sublustrous, dark green, lower surfaces pubescent, much paler and dull green, margins usually entire, occasional ones few-toothed on one or both sides; axis of the blade, between the leaflets, winged, the wings varying considerably as to their prominence, more conspicuous seen from above than from beneath.

Flowers borne in panicles on twigs of the season, the number of panicles per twig and their distribution relative to fully developed or considerably reduced leaves very variable (even on individual plants). Sometimes there is a single

Fig. 35. **Rhus copallina:** a, fruiting branch; b, piece of stem from current year's growth; c, woody stem, cut below to show relative extent of pith, leaf scar, enlarged, with axillary bud to right; d, small portion of staminate panicle; e, flowers, staminate below, pistillate above; f, fruit.

111

panicle subtended by the uppermost leaf of the branch (there being no terminal bud) and appearing terminal, or there may be few to numerous panicles on a given branch having, often, some fully developed and some reduced leaves not subtending panicles interspersed, such a floriferous branch system approximately conical, sometimes as much as 9 dm long and approximately as broad across its base; inflorescence axes and the short flower stalks shaggy-pubescent with translucent hairs.

Calyx lobes deltoid, erect, about 1 mm long, sparsely pubescent basally; petals about 3 mm long, oblongish, blunt to rounded apically, greenish white, spreading at early anthesis then strongly reflexed and hiding the calyx; anthers exserted. Functionally pistillate flowers with rudimentary stamens; calyx lobes deltoid, pubescent, margins of their distal halves white and fringed with short, translucent cilia; petals oblong, erect, apically obtuse or rounded; styles 3, short, stout, reddish above the middle, each stigmatic at the dark red tip. Drupe dullish red, semiglobular but somewhat asymmetric, 4–5 mm in diameter, its surface pubescent with straight, short, translucent hairs and copious, red, short-stipitate or sessile glands, shriveled calyx persistent at its base, shriveled styles persistent at its summit.

At or just within the borders of upland mixed woodlands, uplands, open pinelands, old fields, fence and hedge rows, highway, railway, and power line rights-of-way; throughout our area. (N.H. to s. Mich. and Mo., generally southward to s. Fla. and e. half of Tex.)

The staminate plants, especially, are attractive during the relatively short period of their flowering, the pistillate rather attractive as their red fruiting panicles mature, both handsome in autumn when the leaves turn red.

The description above is drawn from local plants and may not be appropriate in respect to character of leaves and inflorescences in various other parts of the overall range.

National champion southern winged sumac (1972), 1'1" circumf., 22' in height, 15' spread, Camp Chenyata, Fla.

3. Rhus glabra L. SMOOTH SUMAC.   Fig. 36

Similar to *R. copallina* in general features; usually sparingly branched, rarely with the stature of a small tree, vegetative parts all glabrous, young twigs glaucous, soon having numerous, scattered, raised, corky lenticels, sap milky; leaf scars having the copiously hairy axillary buds occupying most of their surface area, the narrow space around the buds also hairy and obscuring the several vascular bundle scars.

Leaves with larger leaflets than in *R. copallina*, petioles 6–15 cm long, their bases enclosing the axillary buds; blades odd-pinnately compound, leaflets 11–31, lanceolate to oblong-lanceolate, 5–14 cm long, 1–4 cm broad near their usually rounded and slightly oblique bases, often slightly falcate, upper surfaces bright green but not lustrous, lower grayish owing to a waxy bloom, margins serrate; petiole and nonwinged leaf axis glaucous, commonly purplish red on the side toward the light source, becoming fully purplish red toward autumn and before the leaflets turn red.

Panicle conical, wholly above the uppermost leaf of a branch and appearing terminal, not at all leafy, its axes shaggy-pubescent; panicles 1–1.5 dm long, the staminate sometimes longer. Flowers essentially like those of *R. copallina*. Drupes 3–4 mm across, their smooth, brown inner surfaces completely covered

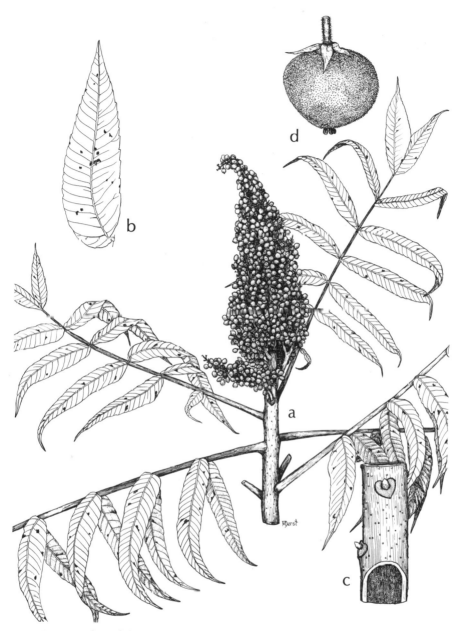

Fig. 36. **Rhus glabra:** a, fruiting branch; b, leaflet; c, small piece of woody stem cut away below to show extent of pith (and note that leaf scar surrounds axillary bud indicating that petiole base encloses axillary bud); d, fruit.

with a profusion of bright red, fleshy, glistening, clavate trichomes, the fruiting panicles very showy and handsome over much of the summer.

The leaflets by maturity often bear irregularly scattered, dark, purplish, "disease blotches" visible on both surfaces. Leaves brilliantly colored in autumn.

Woodland borders, old fields, thickets, fence and hedge rows, rights-of-way; s. cen. and s.w. Ga., local in Fla. Panhandle from Jefferson to Jackson Co., s. Ala. (Widely ranging, across s. Can., generally southward to cen. Fla. Panhandle, n. Mex., Ariz.)

National champion smooth sumac (1976), 1'11" circumf., 42' in height, 23' spread, Akron, Ala.

2. Toxicodendron

Contact with plants of the following three species may cause susceptible persons to become afflicted with a dermatitis. Degree of sensitiveness varies from person to person; some persons are insensitive. Fumes or smoke from burning of such plants is also harmful to those having the sensitivity.

Twigs without terminal buds. Leaves alternate, odd-pinnately compound, without stipules. Inflorescences axillary panicles on developing twigs of the season. Flowers very small, functionally unisexual (plants functionally dioecious). Calyx inserted beneath a central disc, united basally, 5-lobed. Petals 5. Staminate flowers with 5 stamens inserted on the rim of the disc and with a rudimentary pistil. Pistillate flowers with 5 rudimentary stamens, 1 pistil seated upon the disc or its base slightly embedded in it, ovary superior, 1-locular, base of the style short and stout, stoutly 3-forked above, stigmas tipping each fork. Fruit a glabrous, dry drupe, ovoid or somewhat depressed-subglobose, usually somewhat asymmetric, straw-colored, yellowish gray, or ivory-colored, commonly longitudinally streaked with purple, shriveled calyx persistent at the base and shriveled style at the summit, outer wall thin and brittle.

Reference: Gillis, W. T. 1971. "The Systematics and Ecology of Poison-ivy and the Poison-oaks." *Rhodora* 73:72–159, 161–237, 370–443, 465–540.

1. Leaves trifoliolate.
  2. Leaflets acute or acuminate apically.   1. *T. radicans*
  2. Leaflets blunt to rounded apically.   2. *T. toxicarium*
1. Leaves odd-pinnately compound, leaflets 7–15.   3. *T. vernix*

1. **Toxicodendron radicans** (L.) Kuntze. POISON-IVY.   Fig. 37

A trailing, low, weak shrub, or if support available, more commonly climbing by means of aerial roots from the stem; old main stems often attaining a diameter of at least 6–8 cm. Young twigs sparsely to copiously ashy- or rusty-pubescent; twigs of the previous season glabrous or gradually becoming glabrous; leaf scars crescent-shaped to obcordate, each with several vascular bundle scars; buds densely pubescent.

Leaves petiolate, petioles slender, somewhat shorter than the blades, glabrous or variably pubescent; leaflets 3, thinnish, the lateral pair sessile or very shortly stalked, the terminal one with a stalk 1–4 (6) cm long; lateral pair of leaflets generally inequilateral, infrequently equilateral, the terminal one equilateral, all ovate, rounded or shortly tapered basally, acute to acuminate apically, untoothed or unlobed and entire, or teeth or lobes, if present, bluntly triangular, 1–few per side, not infrequently restricted to one side of the leaflet, terminal

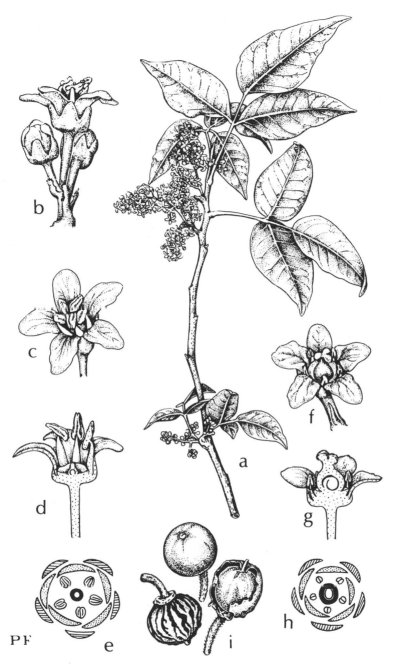

Fig. 37. **Toxicodendron radicans:** a, flowering branch; b, cluster of staminate flowers; c, staminate flower, from above; d, staminate flower longitudinally dissected; e, floral diagram of staminate flower; f, pistillate flower, front-side view; g, pistillate flower longitudinally dissected; h, floral diagram of pistillate flower; i, fruits, from complete to removal of seed coat to show resin canals. (Courtesy of Fairchild Tropical Garden.)

leaflet less frequently toothed or lobed than the lateral ones; size of leaflets varying greatly, largest ones on vigorous specimens or vigorous branches up to about
20 cm long and 12 cm broad; upper surfaces dark green, often with very short
pubescence, especially along the midrib, lower surfaces much paler, sparsely to
moderately pubescent with shaggy hairs, mostly on the veins and veinlets,
sometimes more densely pubescent along the midribs, or tufted in the axils of
the lateral veins, infrequently glabrous (ours).

Panicles sessile or very short-stalked, relatively little to profusely branched,
erect-ascending to spreading, not arching or drooping, axes pubescent. Calyx
lobes erect, short deltoid to oblong and blunt-tipped; petals spreading at first
then reflexed (in the staminate flowers), curvate-spreading in the pistillate,
creamy white with purplish venation, ovate-oblongish with blunt tips, about 2
mm long. Drupes short-pubescent or glabrous, 4–7 mm in diameter, surfaces at
maturity tending to be shallowly grooved longitudinally. [*Rhus radicans* L.]

Essentially ubiquitous in terrestrial, moist to wet, open and wooded habitats
throughout our range. (N.S. to B.C., generally southward to s. Fla. and Ariz.;
Mex.; Berm. and Bah.Is.; Asia.)

The description of *T. radicans* above is drawn from plants of our local area;
they belong presumably to the subsp. *radicans*. In other parts of the total range,
this and other variants occur that are variously treated as varieties, subspecies,
and forms.

2. Toxicodendron toxicarium (Salisb.) Gillis. EASTERN POISON-OAK.   Fig. 38

Generally similar in many characteristics to *T. radicans*. A shrub forming clones
by subterranean runners, aerial stems rather stiffly erect, commonly 5–6 dm
high, less frequently to 1 m or a little more, not climbing. Young twigs brown,
sparsely to copiously rusty-pubescent, older portions of stem gradually becoming glabrous; buds copiously rusty-pubescent; leaf scars crescent-, broadly V-,
or shield-shaped, each usually with 5 vascular bundle scars, these often indistinct.

Leaves with slender petioles, nearly as long or as long as the blades, pubescent
like the young twigs; leaflets firm, 3 per leaf, the lateral pair sessile or short-
stalked, the terminal one with a stalk 1–3 cm long; lateral leaflets usually inequilateral, terminal one equilateral; generally all three leaflets of a given leaf
with a few undulations on a side, with 1–3 undulate or angulate lobes on a side,
or incised-lobed, the lobes oblongish and rounded apically, lateral leaflets frequently with undulations or lobing only on one side, or more on one side than on
the other, their terminal leaflets generally more symmetrically undulate or
lobed; in any case, the toothing or lobing notably more "oaklike" than in *T.
radicans*; upper surfaces relatively sparsely short-pubescent, the lower usually
more densely and uniformly longer-pubescent, sometimes mainly so only along
the major veins, apices of the leaflets blunt to rounded.

Inflorescences like those of *T. radicans*, perhaps never so large and abundantly
floriferous as is often the case for that species; flowers and fruits essentially as in
T. radicans. [*Rhus toxicodendron* L.]

In relatively xeric places, open woodlands of various mixtures, longleaf pine–
scrub oak ridges and hills, open pine-hardwood and hardwood second-growth
woodlands, open banks of highway and railway rights-of-way; throughout our
area of coverage. (N.S. to B.C., generally southward to s. Fla. and Ariz.; Mex.;
Berm. and Bah.Is.; Asia.)

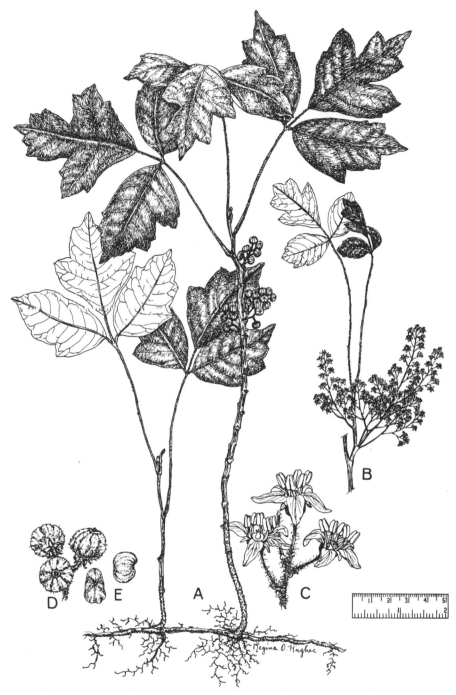

Fig. 38. **Toxicodendron toxicarium:** a, habit; b, inflorescence; c, flowers; d, drupes; e, stones. (From Reed, 1970. *Selected Weeds of the United States.* [As *Rhus toxicodendron*].)

3. Toxicodendron vernix (L.) Kuntze. POISON SUMAC.

Sparingly branched, deciduous shrub or small tree. Young twigs glabrous, usually reddish, later tan, very soon showing numerous lenticels, these often very numerous and many of them contiguous; leaf scars mostly obcordate, with relatively numerous, often irregularly spaced, vascular bundle scars. If a leaf has no inflorescence in its axil, an evident axillary bud develops there; if the leaf has an inflorescence in its axil, the axillary bud develops within the base of the inflorescence stalk, is not evident until that is shed, and then has around it a more or less definite, somewhat raised, ring of residual tissue of the inflorescence stalk; buds pubescent.

Leaves petiolate, petioles 2–10 cm long, leaflets 7–15, very shortly stalked, oblong, elliptic, or ovate, some sometimes inequilateral, 5–10 cm long and 2–5 cm broad, bases mostly shortly tapered, apices short-acuminate, margins entire, upper surfaces dark green and glabrous, lower much paler, usually pubescent on the veins and veinlets. Stalks of the leaflets usually reddish or maroon, petioles and leaf axes often so.

Inflorescences arching or drooping, long-stalked, open, pubescent, axillary panicles, usually developing as the leaves are maturing. Calyx lobes erect, short-deltoid to oblongish-triangular; petals erect, narrowed at the base then oblong, blunt at their tips, cream-colored with purplish venation, about 2 mm long. Drupe 5–6 mm in diameter. [*Rhus vernix* L.]

Nonalluvial swamps, pineland depressions, shrub-tree bogs or bays, wet thickets; nearly throughout our area of coverage. (N.E. to s.w. Que., N.Y., s. Ont., s. Mich., s. Wis., and s.e. Minn., and generally southward to n. Fla. and e. Tex.)

National champion poison sumac (1973), 3'1" circumf., 20' in height, 25' spread, Robins Island, N.Y.

Annonaceae (CUSTARD-APPLE FAMILY)

Asimina (PAWPAWS)

Trees or shrubs. New shoots variously pubescent in most, woody twigs with prominent lenticels; terminal bud none. Pith with transverse diaphragms, eventually becoming chambered between the diaphragms. Leaves short-petiolate, without stipules, blades membranous or leathery, pinnately veined, deciduous, alternate, 2-ranked, margins entire. Flowers bisexual, radially symmetrical, stalked, borne 1–4 in leaf axils or axils of leaf scars, enlarging during anthesis, nodding or erect. Calyx united only at the base, with 3 or 4 prominent, triangular to ovate-deltoid, equal lobes. Petals 3 or 4 in each of 2 unequal cycles, veiny, spreading to erect or recurved, those of the inner cycle saccate-based and interiorly with or without a corrugated nectary zone. Stamens numerous, at first compacted into a tight ball on the elevated receptacle. Pistils 3-several, variously fusiform, variously appressed-pubescent, sessile at the center of the summit of the receptacle. Fruit more or less oblong-cylindric, pulpy, a few- to many-seeded berry. Seeds brown, bean-shaped, laterally compressed, disposed in 2 often irregular rows in the fruit.

In places where habitat disturbance is prevalent and where representatives of two species grow intermixed or in close proximity, hybrids may occur but with one exception they are not accounted for in the descriptions below.

Treatment herein adapted from Kral, Robert. 1960. "A Revision of *Asimina* and *Deeringothamnus* (Annonaceae)." *Brittonia* 12:233–278.

1. Leaf blades usually membranous, long-obovate to long-oblanceolate, at least some of them on a given plant acuminate.
  2. The leaf blades 15–30 cm long (many of them on a given plant over 15 cm); flowers 2–3 cm broad, their stalks 1 cm long or more at anthesis.   1. *A. triloba*
  2. The leaf blades 6–15 cm long (occasional plants with a few to 20 cm); flowers 1–1.7 cm broad, their stalks at anthesis less than 1 cm long, usually so short that flowers appear subsessile.   2. *A. parviflora*
1. Leaf blades leathery, linear, oval, oblong, or oblong-obovate, their apices never acuminate.
  3. Flowers from buds on the previous year's wood prior to or as new shoot growth takes place.
    4. Upper and lower surfaces of newly emergent leaf blades densely tomentose with pale blonde or tan pubescence; mature blades elliptic, ovate, oblong, or obovate, flat to wavy-margined; outer petals white or yellowish white, the inner yellowish white with deep yellow corrugated zone.   3. *A. incarna*
    4. Upper surfaces of newly emergent leaf blades sparsely pubescent, the lower densely so; mature blades evidently revolute, cuneate to oblong; outer petals white, flat to wavy-margined; inner petals white, yellowish white, or pink, with a deep maroon to purple corrugated zone.
      5. Leaf blades pale green to glaucous, cuneate to ovate oblong; bark of upper portion of old wood gray or gray-brown.   4. *A. reticulata*
      5. Leaf blades deep green, oblong to oblong-lanceolate (rarely spatulate); bark of upper portion of older wood reddish brown with raised pale lenticels.   8. *A. X nashii*
  3. Flowers arising after emergence of leaves of the season, either axillary to new leaves or terminating new shoot growth.
    6. New shoots, petioles, veins of lower leaf surfaces, and flower stalks brightly red-pubescent; flowers borne axillary to uppermost leaf on new shoot growth.   5. *A. obovata*
    6. New shoots, etc., glabrous to sparsely hairy; flowers borne axillary to leaves on new shoots.
      7. Stems seldom over 0.5 m tall or long, decumbent to arching, sparsely branched; outer petals pink with maroon streaks or wholly maroon, their tips reflexed, the inner petals one-third to two-thirds the length of the outer, deep maroon, their tips recurved.   6. *A. pygmaea*
      7. Stems 1 m tall or more, erect or suberect, variously branched; outer petals yellowish white or pale pink, inner petals white to pink-streaked or maroon.
        8. Leaf blades oblong to oblong-lanceolate, rarely broadly spatulate.   8. *A. X nashii*
        8. Leaf blades linear, linear-elliptic, oblanceolate, or narrowly long-spatulate.   7. *A. longifolia*

1. Asimina triloba (L.) Dunal.   Fig. 39

Shrub or small tree, 1.5–11 (14) m tall, in our range rarely reaching tree stature. New shoots moderately to copiously dark brown–hairy, aging smooth and gray-brown; winter buds dark brown–hairy, 2.5–5 mm broad.

Leaf blades membranous, mostly longish oblong-obovate, elliptic, or oblanceolate, 15–30 cm long, apices acute to acuminate, gradually attenuate from a little above their middles or from their middles to the base; surfaces of young leaf blades sparsely reddish appressed-pubescent above, densely so beneath, becoming glabrous above and sparsely hairy on the veins beneath.

Flowers maroon, 2–3 (4) cm broad, with a faintly fetid aroma, borne from the axils of leaf scars, their nodding stalks 1.5–2 (2.5) cm long, densely dark brown–

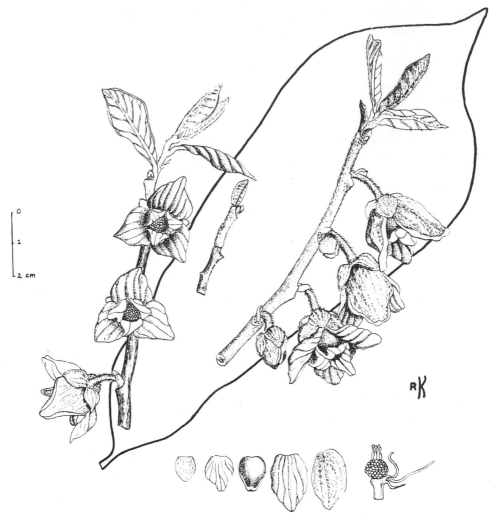

Fig. 39. **Asimina triloba:** [note outline of leaf in background] at left and right, flowering twigs; at center, a winter twig with terminal bud; at bottom, dissected flower. (From Kral in *Brittonia* 12:243. 1960.)

hairy. Calyx 8–12 mm high, with 3 deltoid lobes striate with brown hairs on the outside, glabrous within. Outer petals 1.5–2.5 cm long, oblong-elliptic, their bases ascending, tips recurved, copiously appressed-hairy along the veins exteriorly, glabrous and impressed-veiny interiorly; inner petals one-third to two-thirds the length of the outer, elliptic, saccate-based, recurved tipped, becoming glabrous exteriorly, glabrous and impressed-veiny interiorly. Pistils 3–7 (10), appressed-red-hairy. Berries 5–15 cm long, yellow-green to brownish at ripening.

Mesic woodlands and bottomlands where rarely flooded; cen. Panhandle of Fla., s.e. Ala. (N.Y. and s. Ont. to s. Mich. and e. Neb., generally southward to Fla. Panhandle and n.e. Tex.)

2. Asimina parviflora (Michx.) Dunal. Fig. 40

In general appearance very much like *A. triloba* but generally more slender and potentially not so tall, probably not exceeding about 6 m tall, commonly not over 2–3 m. New shoots reddish brown and reddish-tomentose, becoming gray-brown. Winter buds about 2.5 mm broad, reddish-hairy.

Leaf blades with the texture and form as in *A. triloba*, 6–15 cm long, rarely to 20 cm, densely reddish-tomentose beneath when young, becoming sparsely red-hairy on the veins beneath.

Flowers maroon, 7–15 mm broad, with a fairly fetid aroma, subsessile or slightly nodding on reddish hairy stalks 0.3–0.8 cm long, borne in the axils of leaf scars. Calyx 4–7 mm high, with 3 deltoid sepals striate with brown hairs on the exterior, glabrous within. Outer petals 1–1.3 cm long, fleshy, oblong to ovate, acute apically, tips recurved, moderately reddish-hairy exteriorly, glabrous and impressed-veiny interiorly; inner petals about one-half the length of the outer, fleshy, ovate with strongly recurved tips, saccate bases incised-veiny within. Pistils 5–7, erect, more or less fusiform, appressed-red-hairy. Berries ripening to a greenish yellow, becoming smooth, 3–6 (7) cm long.

Mesic and submesic woodlands, borders of sinkholes, coastal hammocks; throughout our area. (Coastal plain, s.e. Va. to s. cen. Fla., westward to s.e. Tex.)

A. parviflora and *A. triloba* are very much alike in general appearance, differ in overall potential size, quantitative differences in size of leaf blades and flowers, and in length of flower stalks, subtle differences in pubescence. *A. parviflora* is restricted to the coastal plain whereas *A. triloba* is infrequent there.

3. Asimina incarna (Bartr.) Exell. Fig. 41

Shrub to 1.5 m tall, with 1–numerous, stiff, primary shoots. Bark of woody stems brown to grayish brown or gray, relatively thinly pubescent. Young shoots and leaves densely and softly short-pubescent with blond or grayish hairs, the upper blade surfaces less densely pubescent than the lower.

Mature leaf blades narrowly obovate, oblong-obovate, elliptic, some sometimes broadly elliptic-oblong, bases strongly tapered to rounded, apices bluntly obtuse to rounded, the latter often emarginate, 4–10 cm long, 2–6 cm broad, both surfaces eventually sparsely pubescent, reticulate-veiny beneath.

Flowers fragrant, nodding on densely and softly blond-hairy stalks 2–3.5 cm long, 1–4 per node from buds on the previous season's growth before or as new shoot growth takes place. Calyx 8–12 mm long, the prominent lobes 3 (4), deltoid, striate with orange hairs exteriorly, glabrous within. Outer petals 3–7 cm long, oblong to obovate, wavy-margined, white or yellowish white, the outer vein surfaces tan-hairy, interiorly impressed-veiny and glabrous; inner petals one-third to one-half the length of the outer, more or less triangular and revolute above a saccate shortly narrowed base with the inner face yellow-corrugated. Pistils 3–5 (11), narrowly fusiform, with appressed pale pubescence. Berries short-oblong, to 8 cm long, terete or irregularly bulging, ripening yellow-green. [*Pityothamnus incanus* (Bartr.) Small; *Asimina speciosa* Nash]

Inhabiting well-drained sandy soils, longleaf pine–scrub oak ridges, old fields, and pine flatwoods (where little or no saw palmetto occurs); s.e. Ga., cen. n. and n.e. Fla. to s. cen. pen. Fla.

4. Asimina reticulata Shuttlw. ex Chapm. Figs. 42, 43

Shrub to 1.5 m tall, with 1–numerous, stiff, primary shoots. Bark of woody stems gray, that of younger ones brown to grayish brown or gray, thinly and

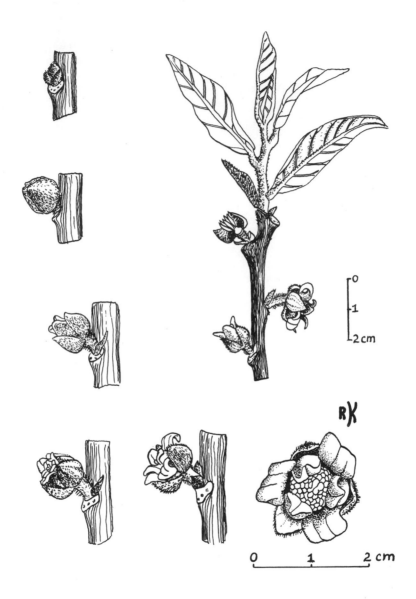

Fig. 40. **Asimina parviflora:** at left, bud and flower development; at upper right, a flowering twig; at lower right, an enlarged flower. (From Kral in *Brittonia* 12:245. 1960.)

Fig. 41. **Asimina incarna:** at left, from top to bottom, inner petal as it infolds the androecium and gynoecium, outer view of inner petal, inner view of inner petal, two longisections of a fruit with extracted seeds; at center, a fruiting twig; at right, a flowering twig. (From Kral in *Brittonia* 12:251. 1960 [as *A. speciosa*].)

Fig. 42. **Asimina reticulata:** at left, and right, mature flowers; at center, a flowering twig. (From Kral in *Brittonia* 12:257. 1960.)

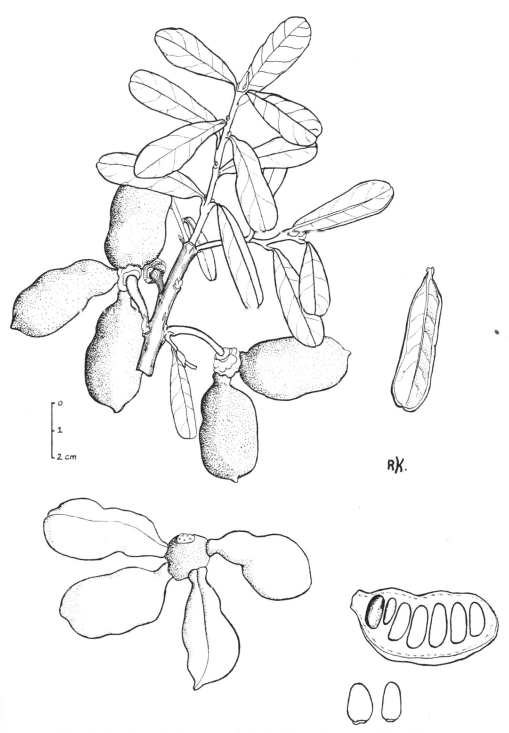

Fig. 43. **Asimina reticulata:** upper left, a fruiting twig; upper right, a lower leaf surface; lower left, top view of torus with four attached fruits; lower right, a longisection of fruit plus two extracted seeds. (From Kral in *Brittonia* 12:258. 1960.)

tightly pubescent distally. Young shoots with a reddish brown to tan, sparse to dense coat of pubescence, upper surfaces of newly emerging leaf blades green, the green somewhat subdued by relatively thin, gray pubescence, lower surfaces densely blondish-pubescent throughout or, almost always, on some blades the midribs densely reddish brown–pubescent.

Mature leaf blades leathery, oblong, narrowly oblong-obovate, oblong-elliptic, or cuneate, 5–8 cm long and 1–3 cm broad, apices rounded, rounded-emarginate, or bluntly obtuse; upper surfaces sparsely short-appressed-pubescent, some eventually glabrous, the lower pale grayish green with the principal veins reddish brown, sparsely pubescent, especially on the veins, veinlets notably reticulate; leaf margins revolute.

Flowers fragrant, nodding on moderately orangish-hairy stalks 2–3.5 cm long, 1–3 flowers per node from buds on wood of the previous season's growth prior to or as new shoot emergence occurs. Calyx 8–10 mm high, the prominent lobes deltoid and with reddish hairy striations exteriorly, glabrous within. Outer petals 3–7 cm long, oblong, oval, or obovate, wavy-margined, white and white-, tan-, or orange-hairy on the exterior veins, glabrous and impressed-veiny within; inner petals one-third to one-half the length of the outer, white, yellowish white, or pink, more or less triangular-hastate and revolute above a narrower saccate base deep purple–corrugated within. Pistils 3–8, fusiform, surfaces appressed-orange-hairy. Berries 4–7 cm long, short-oblong to almost globular, terete or irregularly bulging, ripening to yellowish green. [*Pityothamnus reticulatus* (Shuttlw. ex Chapm.) Small]

Inhabiting poorly drained sandy soils of slash or longleaf pine–saw palmetto flatwoods, coastal dune scrub; chiefly in pen. Fla., barely reaching our range in n.e. Fla.

5. Asimina obovata (Willd.) Nash. Figs. 44, 45

Arborescent shrub to 4.5 m tall, primary shoots 1–few, suberect to arching. Bark of older woody stems gray, glabrous, that of year-old wood reddish to light brown with raised pale lenticels. Young shoots very densely red-pubescent.

Early emerging leaf blades sparsely reddish brown–pubescent above, very densely so beneath, much of the pubescence rather quickly sloughing beneath leaving them, when partially grown, with a rather copious pubescence on the principal veins and on the reticulate veinlets. Mature leaf blades leathery, 4–10 (12) cm long and 2–4 cm broad, oblong-obovate, oblong, oblong-oval, mostly rounded basally and apically, less frequently rounded-emarginate apically, upper surfaces becoming essentially glabrous and lustrous dark green, the lower sparsely reddish brown–pubescent.

Flowers fragrant, subsessile, borne singly (usually) axillary to the uppermost leaves of new shoots of the season. Calyx 0.5–1.5 cm high, the lobes elliptic to ovate, rounded apically, red-hairy on the veins exteriorly, glabrous within; inner petals one-fifth to one-half the length of the outer, oblong-oval, oblong, or occasionally lanceolate, white or occasionally rose-pink and slightly revolute distally, saccate-based and the inner surface of the base purple-corrugated, glabrous. Pistils 3–8 (11), narrowly fusiform, sparsely appressed-red-pubescent.

Berries 5–9 cm long, short-oblong, terete or irregularly bulging, ripening yellow-green. [*Pityothamnus obovatus* (Willd.) Small]

Inhabiting well-drained sandy soils, coastal dunes and hammocks, sand pine– or longleaf pine–scrub oak ridges and slopes; s.e. to n.e. pen. Fla., barely if at all reaching our area.

Fig. 44. **Asimina obovata:** at left, a flowering twig; upper right, an outer petal; right center, a flowering shoot with outer petals of flower removed; lower right, a shoot with young flower. (From Kral in *Brittonia* 12:255. 1960.)

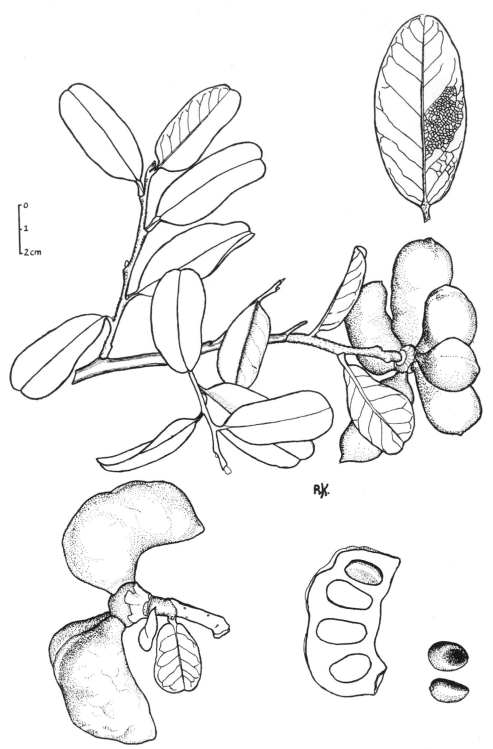

Fig. 45. **Asimina obovata:** at left and center, a fruiting twig; upper right, a summer leaf; lower left, torus with two attached fruits; lower center, a long-isection of a fruit; lower right, two seeds. (From Kral in *Brittonia* 12:256. 1960.)

6. Asimina pygmaea (Bartr.) Dunal.   Fig. 46

Low shrub, to about 3 dm tall, primary shoots 1–several, unbranched or branched. Very young shoots sparsely to densely pubescent with bright reddish brown hairs, the pubescence soon sloughed leaving the stems dark reddish brown. Newly emerging leaf blades usually densely blonde to reddish brown–pubescent beneath, sometimes sparsely so, sparsely pubescent above, both surfaces becoming glabrous rather quickly.

By twisting of the petioles the leaf blades usually held more or less erectly ascending-secund from the twigs. Mature leaf blades leathery, oblanceolate, spatulate, narrowly obovate, or oblongish, mostly somewhat cuneate from above their middles, apices obtuse, rounded or rounded-emarginate, infrequently acute, 4–11 cm long and 1–4 cm broad, dark green and glabrous above, pale green beneath, the veinlets reticulate.

Flowers with a fetid aroma, solitary, nodding-secund from the leaf axils on shoots of the season. Calyx 5 6 mm high, its 3 prominent lobes mostly broadly ovate, blunt-tipped. Outer petals 1.5–3 cm long, oblong to lance-ovate, pink with longitudinal maroon streaks or maroon, tips reflexed; inner petals one-third to two-thirds as long as the outer, ovate-acute to lance-ovate, deep maroon, revolute marginally, the tips abruptly reflexed, the saccate base deeply corrugated within. Pistils 2–5, narrowly fusiform, pubescent. Berries oblong-cylindric, often curved and often bulging, 3–5 cm long, ripening yellowish green. [*Pithyothamnus pygmaeus* (Bartr.) Small]

Low slash or longleaf pine–saw palmetto flatwoods and savannas, adjacent roadsides, intermittently boggy slopes, old fields; s.e. Ga. (Charlton Co.), n.e. Fla. to a little w. of the Suwannee River, pen. Fla. southward to Polk Co.

7. Asimina longifolia Kral.   Figs. 47, 48

Shrub to 1.75 m tall, usually with several rigid primary shoots. Bark of woody stems gray-brown to tan, eventually gray. Very young shoots and lower surfaces of very young emerging leaves densely reddish brown–pubescent, most of the pubescence quickly sloughing, all of it eventually, the young stem then dark reddish or purplish.

Mature leaf blades leathery, elongate linear-elliptic, linear-oblong, or linear-lanceolate, varying to linear-oblanceolate, linear-spatulate, or spatulate, to 15–20 cm long and 0.5 to 3 cm broad, varying from broadest at about their middles in the narrower leaves to broadest well above their middles in the broader leaves; averaging broader above their middles in the western portion of the range than in the eastern portion; surfaces darker green above than beneath, the lower with veinlets reticulate.

Flowers fragrant, solitary, and nodding from leaf axils on shoot growth of the season. Calyx 1–1.5 cm high, lobes elliptic to lance-ovate, apices usually blunt. Outer petals white, rarely pale pink, 3–8 cm long, oblong or obovate, sparsely pubescent along the veins exteriorly, glabrous within; inner petals one-third to one-half the length of the outer, white, infrequently pink, pink-streaked, or maroon, oblong-lanceolate distally and marginally revolute, base saccate, one-third to one-half of the inner face with purple corrugations. Pistils 2–7 (12), fusiform. Berries oblong-cylindric, terete or irregularly bulging, 4–10 cm long, ripening yellowish green. [*Pithyothamnus angustifolius* (A. Gray) Small]

Slash or longleaf pine flatwoods, well-drained sandy soils of longleaf pine–scrub oak ridges and slopes, old fields, pastures, roadsides; s. Ga., s.e. Ala., n. Fla., and pen. Fla. southward to about Lake Co.

Fig. 46. **Asimina pygmaea:** upper left, longisection of fruit and extracted seeds; above, center, a flowering twig; below, center, a fruiting twig; upper right, flowers; lower right, an upper leaf surface. (From Kral in *Brittonia* 12:263. 1960.)

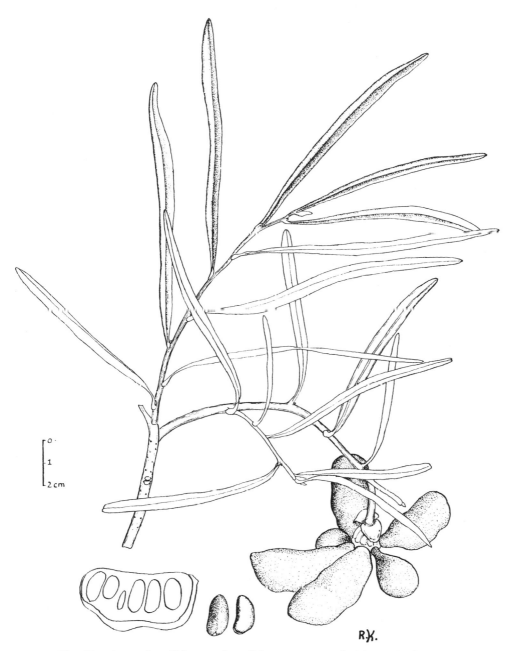

Fig. 47. **Asimina longifolia** var. **longifolia:** at center, a fruiting twig; lower left, a longisection of fruit with two extracted seeds. (From Kral in *Brittonia* 12:267. 1960.)

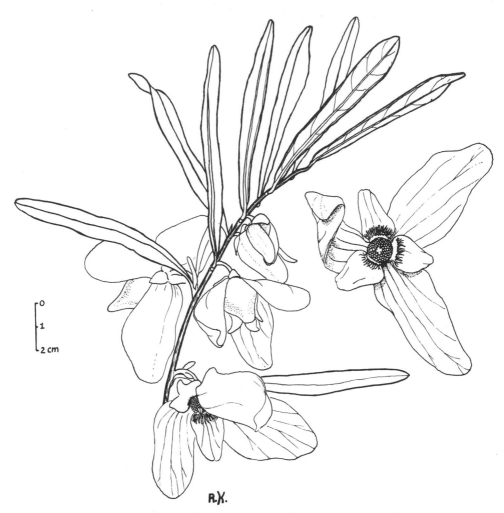

Fig. 48. **Asimina longifolia** var. **spatulata:** a flowering twig; at right, top view
of a flower. (From Kral in *Brittonia* 12:268. 1960.)

Plants representing this species occurring in the more westerly portion of the
range, having the broader leaves, and sometimes pink or maroon petals, have
been designated *A. longifolia* var. *spatulata* Kral; those occurring in the more
easterly portion of the range with narrower leaves and always with white petals,
A. longifolia var. *longifolia*.

8. Asimina × nashii Kral.

Shrubs of frequent occurrence where plants of *A. longifolia* and *A. incanna* grow
intermixed, having intermediate characteristics and presumed to be F_1 hybrids.

Bark of older wood reddish brown. Mature leaf blades dark green above, paler
green beneath, and with scattered hairs along the veins and veinlets, oblong,
oblanceolate, or rarely spatulate, mostly broadest above their middles, apices
rounded or obtuse, infrequently acute, mostly 4–15 cm long and 1.5–3 cm
broad. Flowers fragrant, erect or slightly nodding on red or orange-red stalks,

solitary from buds on wood of the previous season (as in *A. incarna*) or, if cut back or burned back, flowering from axils of leaves on shoots of the season (as in *A. longifolia*). Outer petals white or pale pink, 3–7 cm long (to 10 cm on leafy sprouts), oblong to obovate; inner petals one-half to one-fourth as long as the outer, white to maroon, oblong to hastate-lanceolate, saccate base with deep purple corrugated zone on the inner face.

Well-drained sandy soils, longleaf pine–scrub oak ridges, adjacent old fields and pastures; n. Fla., e. of the Suwannee River.

Apocynaceae (DOGBANE FAMILY)

Trachelospermum difforme (Walt.) A. Gray. CLIMBING DOGBANE.   Fig. 49

Slender, deciduous, twining vine, shoots of the season herbaceous at first, glabrous or pubescent, becoming woody and glabrous. Woody stems reddish purple, papillose at first, some, at least, of the papillae eventually becoming slightly raised lenticels, a line-scar connecting the opposite more or less U-shaped leaf scars with 1 vascular bundle scar, the line scar left after the sloughing of several basally connecting minute stipules.

Leaves opposite, short-petioled or subsessile; blades, even on a single plant, varying considerably in shape and size, lanceolate, elliptic, oval, obovate, or suborbicular, to 14 cm long and 8 cm broad (mostly little more than half that size or less), bases obtuse to rounded, apices abruptly short-acuminate, upper surfaces dark green, glabrous or very sparsely short-pubescent, lower pale green, usually moderately short-pubescent, sometimes glabrous.

Flowers small, in panicled, axillary cymes (only in the axil of 1 of a pair of leaves), cyme axes glabrous, flower stalks slender, glabrous, 4–10 mm long. Calyx glabrous, erect, lobes 5, about one-third as long as the corolla tube, lance-ovate basally, abruptly narrowed to long-acuminate or subulate tips. Corolla funnelform, pale yellow or greenish yellow, lined within with brown streaks, tube 5–7 mm long, lobes 5, obliquely obovate, spreading, 3–4 mm long. Stamens 5, filaments adnate to the corolla tube for most of their length, anthers fused to the stigma. Ovaries 2, their styles united, stigma and anthers not exserted, each ovary ripening into a very slenderly cylindric, glabrous follicle 1–2 dm long. Seeds 10–12 mm long, angled, very narrow, base narrowest, slightly broadening upwardly to a truncate summit bearing a tuft of conspicuous, long-silky hairs.

River banks, floodplain woodlands, clearings, thickets, marshes, sloughs; more or less throughout our area. (Coastal plain and piedmont, Del. to n. Fla., westward to e. Tex., northward in the interior to Mo. and s. Ind.)

Aquifoliaceae (HOLLY FAMILY)

Ilex (HOLLIES)

Deciduous or evergreen shrubs or small trees. Leaves simple, alternate, pinnately veined, commonly varying considerably in size on a given plant, stipules minute, quickly deciduous in most; vascular bundle scar 1 in each leaf scar. Inflorescences axillary to leaves or leaf scars, fascicled, cymose, umbellike, or

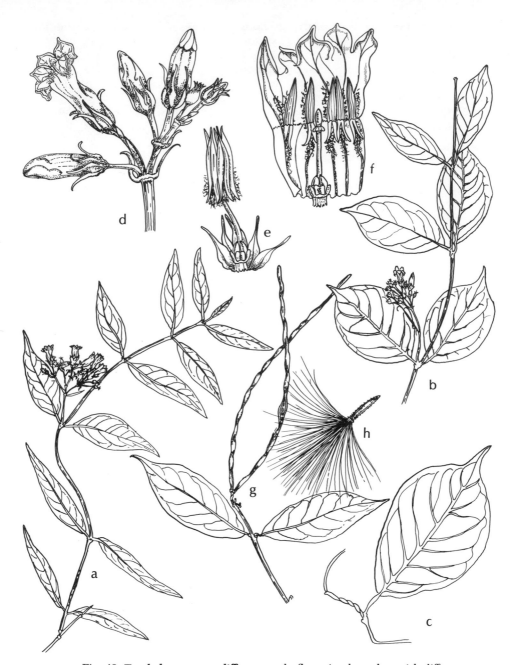

Fig. 49. **Trachelospermum difforme:** a, b, flowering branches with different leaf shapes; c, leaf; d, inflorescence; e, flower with corolla removed and calyx somewhat spread; f, corolla, opened out, and pistil; g, follicles; h, seed. (From Correll and Correll, 1972. *Aquatic and Wetland Plants of Southwestern United States.*)

panicled, sometimes flowers solitary (pistillate more often solitary than the staminate). Flowers mostly unisexual, the staminate and pistillate usually on different individuals, staminate sometimes with an abortive pistil, pistillate with abortive stamens; sometimes a few functionally bisexual flowers intermingled. All flowers small, radially symmetrical. Calyx united below, 4–9-lobed or -toothed. Corolla slightly united basally, 4–9-lobed, lobes spreading radially, usually white, greenish white, or yellowish. Stamens as many as the corolla lobes and alternate with them, the filaments usually adnate basally to the short corolla tube. Pistil 1, ovary superior, 4–9-locular, ovules 1 in each locule, style very short in most, stigma capitate or discoid, usually more or less lobed. Fruit a globular, ellipsoidal, or oblate, berrylike drupe, calyx persistent at its base, 1 nutlet in each locule; nutlets usually with 2 flattish sides and 1 rounded, the latter smooth, longitudinally striate, and/or ridged and grooved, sometimes irregularly reticulate, stones of the nutlet hard in most kinds, soft in a few.

1. Leaves leathery, evergreen.
  2. Leaf blades with marginal, distinctly spine-tipped dentations, or if margins entire, then always with a distinctly spine-tipped dentation apically. (Calyx and corolla 4-lobed)   1. *I. opaca*
  2. Leaf blades entire or toothed, if toothed, then the teeth not spine-tipped, the blades never distinctly spine-tipped apically.
    3. Margins of leaf blades appressed-crenate from base to apex. (Calyx and corolla 4-lobed.)   2. *I. vomitoria*
    3. Margins of leaf blades entire, or if toothed, then the teeth absent from the basal margins.
      4. Lower surfaces of leaf blades punctate-dotted, drupe black.
        5. Margins of some, at least, of the leaf blades with 1–3 appressed-crenate teeth on a side near their apices. (Calyx and corolla 5–8-lobed)   3. *I. glabra*
        5. Margins of leaf blades entire or with a few, spreading, short, bristlelike teeth. (Calyx and corolla 5–9-lobed.)   4. *I. coriacea*
      4. Lower surfaces of leaf blades not punctate-dotted; drupe red, orange-red, or rarely yellow.
        6. Larger leaf blades at least 15 mm broad, commonly much broader; branchlets mostly ascending at angles less than 45 degrees to the branch from which they arise. (Calyx and corolla 4-lobed.)   5. *I. cassine*
        6. Larger leaf blades not exceeding 8 mm broad; branchlets mostly borne at angles of more than 45 degrees to the branches from which they arise, commonly at 90-degree angles. (Calyx and corolla 4-lobed.)   6. *I. myrtifolia*
1. Leaves membranous, deciduous.
  7. Leaf blades predominantly oblanceolate, spatulate, or obovate, most of them broadest above their middles, cuneate-attenuate basally, margins obscurely crenate distally. (Calyx and corolla 5-lobed.)   7. *I. decidua*
  7. Leaf blades predominantly oblongish, elliptic, oval, or ovate, most of them broadest at or below their middles, mostly rounded to broadly tapered basally, or if cuneate then not cuneate-attenuate, margins entire, serrate, or obscurely serrate.
    8. Plants occurring in wet habitats.
      9. Leaf blades predominantly oblongish, few if any acuminate apically, their margins entire or with minute teeth. (Calyx and corolla lobes 4.)   8. *I. amelanchier*
      9. Leaf blades predominantly elliptic or oval, most of them acuminate apically, their margins with small but evident and sharp teeth. (Calyx and corolla lobes 5–7 [8].)
           9. *I. verticillata*
    8. Plants occurring in well-drained places. (Calyx and corolla lobes [3] 4 [5].)
           10. *I. ambigua*

1. Ilex opaca Ait. AMERICAN HOLLY.

Leaf blades bright dark green above, paler dull green beneath, flat, their edges not revolute, larger ones 3.5–5.5 cm broad; inhabiting mesic or submesic woodlands of various mixtures, river banks, and floodplain woodlands subject to brief periods of flooding. 1a. var. *opaca*

Leaf blades dull, somewhat yellowish green above, olive-green beneath, their edges revolute and their sides tending to be rolled downward, larger ones 1–2.5 cm broad; inhabiting sand pine–oak scrub. 1b. var. *arenicola*

1a. Ilex opaca var. opaca. AMERICAN HOLLY. Fig. 50

Evergreen understory tree to about 15 m tall. Young twigs sparsely very short-pubescent, becoming glabrous, older twigs brown, roughish, with circular raised lenticels. Bark of trunks light gray, somewhat roughened by wartlike processes. When growing where not crowded by other plants, the crown symmetrically conical to cylindric, its branches rather rigid and spreading more or less at right angles from the central trunk.

Leaves with sparsely short-pubescent petioles 5–12 (18) cm long. Blades dark green above, pale and dull green beneath, oblong, elliptic, oval, slightly ovate or obovate, flat, 3–10 (12) cm long, 2–5.5 cm broad, stiff-leathery at maturity, usually with spine-tipped marginal dentations, the dentations varying from a single one on a side distally, to all along each side to the base, usually not more than 7 or 8 on a side at most, *the tip of the blade always spine-tipped*; occasional trees with margins wholly entire; midrib on the upper surface of young leaves short-pubescent, usually glabrous in age, lower surface wholly glabrous or with a few short hairs along the midrib. Staminate and pistillate flowers borne similarly in stalked, simple or compound cymes on leafless parts of shoots, in leaf axils on developing shoots, or in leaf axils on wood of the previous season; inflorescence axes and flower stalks short-pubescent. Staminate and pistillate flowers on separate trees. Calyx 4-lobed, lobes triangular, acute or acuminate, their margins usually somewhat erose. Corolla white or cream-colored, 4-lobed. Drupe red, without luster, yellow on a rare individual, globose to oval, 7–10 (12) mm in diameter. Nutlets 4, 6–8 mm long, irregularly ribbed and grooved on the rounded side.

Inhabiting mesic woodlands of various mixtures, floodplain woodlands subject to brief periods of flooding, river banks; throughout our area. (S. Mass. to Md., s.e. N.Y., e. Pa., W.Va., s. Ohio to s.e. Mo., generally southward to cen. pen. Fla. and e. and s. cen. Tex.)

The national champion American holly (1969) 12′1″ circumf., 51′ in height, 45′ spread, St. Mary's College, Md.

1b. Ilex opaca var. arenicola (Ashe) Ashe. SCRUB OR HUMMOCK HOLLY.

Compact shrub or small tree to about 4–5 m tall, in general appearance resembling *I. opaca* var. *opaca*. Branches tending to be rigidly ascending, twigs stiff and crooked, the bark of trunks permanently smooth (West & Arnold, 1956). Leaf blades rigid, dull and somewhat yellowish green above, olive-green beneath, their edges revolute and their sides tending to be rolled downward (inverted boatlike), larger ones 1.5–5 cm long and 1–2.5 cm broad. Drupes orangish red. [Incl. *I. cumulicola* Small]

Inhabiting deep sands of sand pine–oak scrub of central Fla., barely reaching our area, or perhaps only closely peripheral to it.

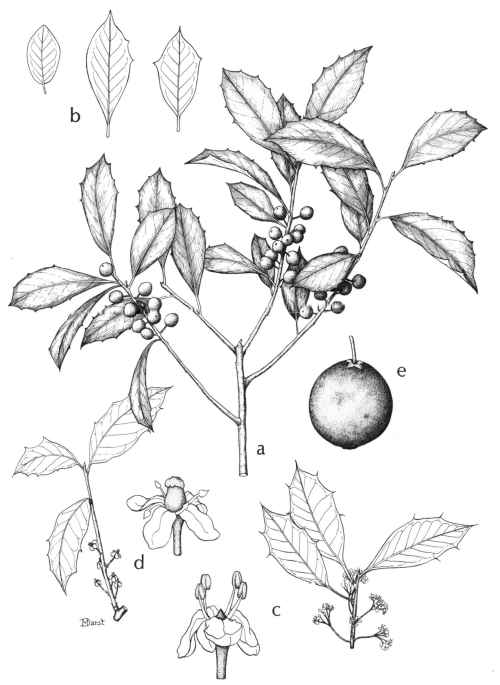

Fig. 50. **Ilex opaca** var. **opaca:** a, fruiting branch; b, variable leaves; c, functionally staminate flowering branchlet to right, flower to left; d, functionally pistillate flowering branchlet to left, flower to right; e, fruit.

Fig. 51. **Ilex vomitoria:** a, piece of a fruiting branch; b, leaf enlarged to accentuate marginal toothing; c, staminate flowering branchlet at left, staminate flower at right; d, functionally pistillate flowering branchlet at right and flower at left; e, fruit, two views.

2. Ilex vomitoria Ait. YAUPON.

Fig. 51

Evergreen shrub or small tree to about 8 m tall, commonly sprouting from the roots and forming dense thickets, especially near the coasts. Young twigs sparsely to densely short-pubescent, the pubescence usually evident throughout the first year, older twigs gradually becoming glabrous and with a waxy gray coating that breaks up into a thin, more or less interlacing surface pattern; woody twigs notably rigid. Bark of trunks thin, gray, smooth.

Leaves with short-pubescent petioles 2–3 mm long, rarely longer. Blades stiff-leathery, oval, elliptic, or oblong-elliptic, infrequently lanceolate, 0.5–3 cm long and 0.5–2.5 cm broad, upper surfaces dark green and lustrous, short-pubescent only along the midrib when young, lower surfaces paler and dull green, glabrous; margins appressed-crenate from base to apex, small glandular mucros at the sinuses and in a shallow notch at the blunt to rounded tip, bases rounded. Functionally staminate and pistillate flowers for the most part on separate plants, infrequently a few functionally bisexual flowers intermixed with the staminate, the staminate flowers much more abundant than the pistillate and much more noticeable from a little distance. Flowers borne in short-stalked or sessile fascicles in the axils of leaves or leaf scars of leafless nodes, the fascicles solitary or 2-several at a given node; flower stalks mostly 1.5–3 mm long, more or less short-pubescent, those of the pistillate usually more evidently so than those of the staminate, calyces similarly pubescent. Calyx lobes 4, broadly short-triangular, obtuse to rounded apically. Corolla lobes 4, white or yellowish, oblongish, commonly with their broad tips curled upwardly. Drupe globose, 4–8 mm in diameter, clear bright red and lustrous, yellow on a rare individual plant. Nutlets 4, 3–4 mm long, irregularly shallowly ribbed on the rounded side.

Coastal dunes and interdune depressions, maritime forests, upland woodlands of various mixtures, pine flatwoods; for the most part inhabiting well-drained places but occurs as well on the edges of stream banks, in wet woodlands and floodplains; throughout our area. (Coastal plain, s.e. Va. to n. cen. pen. Fla., westward to s.e. and s. cen. Tex., s.e. Okla., Ark.)

Yaupon commonly forms shrub thickets on coastal dunes where it is a component of the slanting, salt-spray-pruned dense masses of shrubs so characteristic of seaside communities. The evergreen, dark green foliage is attractive and in the pistillate plants combined with the bright red, shiny fruits is an attractive decorative subject. Moreover, it responds favorably to pruning and is widely used for hedges.

National champion yaupon (1972), 4'1" circumf., 45' in height, 40' spread, near Devers, Tex.

3. Ilex glabra (L.) A. Gray. GALLBERRY, INKBERRY.

Fig. 52

Evergreen shrub to 2–3 m tall, commonly sprouting from subterranean runners and clonal. Twigs of the season green, very finely powdery-pubescent, older twigs glabrous, gray or grayish brown, the lenticels roundish and with vertical slits.

Leaves with powdery-pubescent petioles 3–8 mm long. Blades glabrous, somewhat leathery, elliptic to oblanceolate to slightly obovate, 2–5 cm long, upper surfaces dark green and sublustrous, lower paler and dull green, with scattered punctate glands, these usually reddish in color; margins somewhat thickish-banded, mostly with 1 or 2 (3) small, appressed teeth on a side distally, sometimes entire, bases acute to obtuse, apices blunt, usually with a tiny mucro.

Fig. 52. **Ilex glabra:** a, fruiting branch; b, enlargement of lower leaf surface to accentuate punctations; c, enlargements to accentuate nature of marginal teeth of a leaf; d, flowering branchlet (functionally staminate) to right and flower to left; e, flowering branchlet (functionally pistillate) to left and flower to right; f, fruit.

140

Functionally staminate and pistillate flowers for the most part on separate plants. Staminate flowers in axillary, stalked cymes, stalks mostly 6–10 mm long, individual flower stalks of a given cyme variable in length to about 10 mm, minutely bracted near the base. Pistillate flowers mostly solitary in the leaf axils, sometimes 2 or 3, their stalks 4–10 mm long, with a pair of minute bracts varying from about midway their length to near their bases. Calyx glabrous, lobes 5–8, very short and broadly obtuse. Corolla 5–8-lobed, lobes oblongish, white. Drupe black (very rarely white on an individual plant), dull or only sublustrous, globose, 5–7 mm in diameter, *pulp dryish when fully mature, bitter; drupes persistent throughout the winter.* Nutlets 5–8, 3–4 mm long, smooth on the rounded side.

Ilex glabra and the somewhat similar *I. coriacea* commonly grow intermixed or in close proximity where their ranges overlap. The latter blooms several weeks earlier in the season than the former.

Pine savannas and flatwoods, bogs, seepage areas in woodlands, lower slopes and bottoms of wooded ravines, acid prairies; throughout our area. (N.S. to s. pen. Fla., westward to n.e. Tex., chiefly on the coastal plain.)

4. Ilex coriacea (Pursh) Chapm. LARGE OR SWEET GALLBERRY. Fig. 53

Evergreen shrub, rarely with the stature and dimensions of a small tree, to 5 m tall. Twigs of the season dark brown, finely short-pubescent, older twigs grayish to tan, lenticels nearly circular, with longitudinal slits.

Leaves with short-pubescent petioles 5–10 mm long. Blades leathery, elliptic or oval, occasional ones lanceolate or subobovate, mostly 3.5–9 cm long and 1.5–4 cm broad, usually a few on a given branch smaller; upper surfaces dark green and sublustrous, short-pubescent along the midrib, paler beneath and with scattered punctate glands, glabrous or pubescent, the pubescence often stellate; margins entire or with a few, almost bristlelike, usually spreading teeth; bases acute, apices acute or short-acuminate, rounded or rounded and notched on occasional leaves.

Functionally unisexual flowers on separate plants. Staminate flowers in clusters axillary to leaf scars or leaves on year-old wood, also singly proximally on leafless bases of shoots of the season and upwardly axillary to developing leaves. Pistillate flowers borne axillary to proximal leaves on shoots of the season. Calyx and corolla 5–9-lobed. Calyx more or less tuberculate exteriorly, lobes triangular, acute to acuminate, their margins varyingly entire, irregularly shallowly erose, or finely toothed. Corolla white, lobes more or less oblong. Drupes globose to oblate, 6–8 mm in diameter, at maturity black and lustrous, pulp juicy and sweet, falling shortly after maturing in summer. Nutlets 5–9, smooth on the rounded side.

Commonly growing intermixed with or in close proximity to *I. glabra* where their ranges overlap, *I. coriacea* blooming several weeks earlier than *I. glabra.*

Pine savannas and flatwoods, shrub-tree bogs and bays, open bogs, seepage areas in woodlands, on lower slopes and in bottoms of wooded ravines; throughout our area. (Coastal plain, s.e. Va. to cen. pen. Fla., westward to s.e. Tex.)

National champion large or sweet gallberry (1973), 5″ circumf., 17′ in height, 12′ spread, Hardin Co., Tex.

5. Ilex cassine L. DAHOON, CASSENA. Fig. 54

Evergreen shrub or tree generally to about 10 m tall, branches and branchlets mostly strongly angling-ascending. Young twigs, petioles, and leaf blades varying not a little in vestiture, young twigs and petioles sometimes glabrous, more

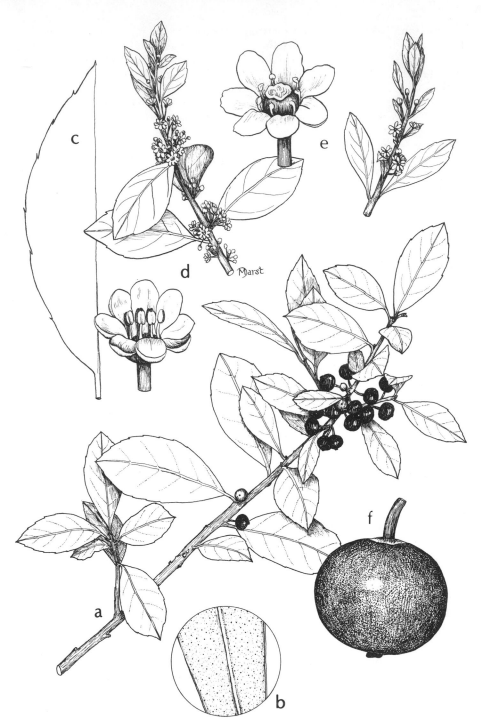

Fig. 53. **Ilex coriacea:** a, fruiting branch; b, portion of lower surface of leaf, enlarged to accentuate punctations; c, enlargement of half-leaf outline to accentuate marginal toothing; d, staminate flowering branchlet above, flower below; e, functionally pistillate flowering branchlet to right, flower to left; f, fruit.

Fig. 54. **Ilex cassine:** a, fruiting branch; b, flowering branchlet (functionally staminate) above and flower below; c, flowering branchlet (functionally pistillate), to left and flower to right; d, fruit.

commonly shaggy-pubescent, less frequently with short, curly pubescence. Bark of trunks gray, very thin, permanently smooth.

Petioles 5–15 mm long. Blades variable in shape and size from plant to plant or even on a single one, oblanceolate, nearly spatulate, oblong, oval, elliptic, or subobovate 2–8 (14) cm long and 0.8–4.5 cm broad; bases narrowly to broadly tapered, apices acute or obtuse, less·frequently rounded, usually tipped with a small mucro; margins mostly entire, some of them or most of them on some plants with a few, short, sharp dentations; upper surfaces glabrous save along the midrib, the lower surfaces varyingly wholly glabrous, copiously shaggy-pubescent only along the midrib and with longish, straight hairs relatively sparsely distributed over the remainder of the surface.

Functionally unisexual flowers on separate plants. Staminate flowers in axillary, usually diffusely panicled cymes to about 5 cm long. Pistillate flowers solitary or in stalked 2–4-flowered axillary cymes, stalks of the cymes about 5 mm long, sometimes with a pair of opposite or subopposite bractlets somewhat above the base; stalks of the flowers 2–3 mm long; occasionally pistillate flowers borne in diffusely panicled cymes similar to the staminate. Inflorescence axes, flower stalks, and calyces generally sparsely to copiously pubescent. Calyx and corolla 4-lobed. Calyx lobes broadly triangular, margins minutely toothed or erose. Corolla lobes white, oblongish, their broadly rounded tips somewhat cupped. Drupes globose, bright red, orangish red, or yellowish red, rarely yellow on an individual plant, 5–8 mm in diameter. Nutlets 4, about 4 mm long, irregularly ribbed-grooved on the rounded side.

Edges of spring-fed rivers and streams and on their floodplains where surface water stands much of the time, cypress-gum ponds or depressions, depressions in flatwoods, wet hammocks with shallow soil over limestone, shrub-tree bogs and bays, banks of fresh or brackish marshes; throughout our area. (Coastal plain, s.e. N.C. to s. Fla., westward to s.e. Tex.)

In the cen. and w. Panhandle of Florida, the dahoon is seldom associated with pond cypress and gum in cypress-gum ponds or depressions in flatwoods, although there *Ilex myrtifolia* occurs in abundance in many of them; eastwardly and especially southeastwardly, the dahoon does occur commonly in such places in pen. Fla. to the exclusion of *I. myrtifolia*.

National champion dahoon (1975), 2'10" circumf., 72' in height, 22' spread, Osceola National Forest, Fla.

6. Ilex myrtifolia Walt. MYRTLE-LEAVED HOLLY OR DAHOON.　　Fig. 55

Evergreen shrub or small, scrubby tree, usually not exceeding 5–6 m tall, often with several to many stems from the base, the branches and branchlets stiff, many of them borne perpendicularly or nearly so to the branches from which they arise, almost all of them at an angle greater than 45 degrees. Young twigs and petioles ashy-pubescent, older twigs becoming glabrous and with a gray, waxy coating which breaks up into a thin, more or less interlacing, surface pattern. Bark of trunks gray, thickish (relative to that of *I. cassine*), and very much roughened by corky excrescences.

Petioles 1–3 mm long. Blades 5–30 mm long, 3–8 mm broad, narrowly lanceolate, oblong, oblanceolate, or elliptic, stiff-leathery, dark green and lustrous above, sometimes with sparse, minute pubescence on or near the midrib, paler beneath, usually wholly glabrous, occasionally with sparse, short pubescence on·the midrib and over the remainder of the surface; margins entire, tips mucronate.

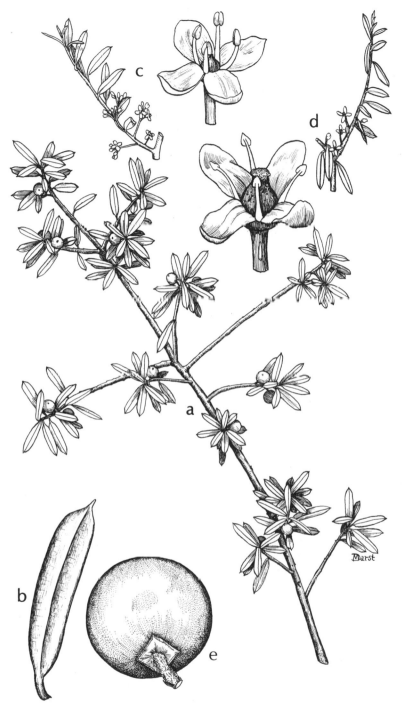

Fig. 55. **Ilex myrtifolia:** a, fruiting branch; b, leaf, much enlarged; c, staminate flowering branchlet to left, flower to right; d, functionally pistillate flowering branchlet, above, flower, to left; e, fruit.

Functionally unisexual flowers on separate plants. Staminate flowers rarely solitary, usually in stalked cymes, not over 8–10 per cyme, borne axillary to leaves on branchlets of the season, stalks pubescent, 2–8 mm long, without bractlets above their bases. Pistillate flowers axillary and solitary, their stalks pubescent, with 2 opposite or subopposite bractlets usually about halfway between their bases and midpoints. Calyx 4-lobed and squarish, sparsely short-pubescent. Corolla lobes 4, white, oblongish, their rounded tips slightly cupped. Drupes globose, 5–8 mm in diameter, red (rarely orange or yellow), without luster. Nutlets 4–4.5 mm long, ribbed-grooved on the rounded side. [*I. cassine* var. *myrtifolia* (Walt.) Sarg.]

Cypress-gum ponds and depressions, outer, sandy rims of ponds, depressions in pine flatwoods; throughout much of our area. (Coastal plain, s.e. N.C. to n. Fla., westward to La. and near the mouth of the Brazos River, Tex.)

Hybridization between *I. myrtifolia* and *I. cassine* is alleged to occur in places where the ranges of the two overlap.

National champion myrtle-leaved dahoon (1972), 5'7" circumf., 46' in height, 35' spread, Bradford Co., Fla.

7. Ilex decidua Walt. POSSUM-HAW.   Fig. 56

Deciduous shrub or small understory tree, often attaining a height of about 10 m. Twigs of the season greenish brown, older ones gray and with raised, nearly circular lenticels. Bark of trunks thin, grayish, smooth or slightly roughened.

Leaves with glabrous to sparsely (rarely copiously) short-pubescent petioles 2–15 mm long. Blades mostly, but not always, broadest somewhat above their middles, oblanceolate, spatulate, obovate, or elliptic, cuneate basally, apices rounded, obtuse or obtusely subacuminate, variable in size from tree to tree, even on a single branch, 1–6 (8) cm long, 0.8–3 (4.5) cm broad, margins appressed-crenate, the teeth gland-tipped, upper surfaces sometimes sparsely pubescent on or near the midrib, at least when young, lower surfaces sparsely to copiously shaggy-pubescent on the elevated midrib, usually on the principal lateral veins as well, sometimes sparsely pubescent between the veins.

Functionally unisexual flowers borne on separate plants. Staminate flowers mostly in fascicles at the junction of spur-shoots of the previous season and the twigs of the current season, also solitary in axils of developing leaves; flower stalks slender, 5–12 (20) mm long. Pistillate flowers solitary or 2 or 3 in the leaf axils, or at nodes on twigs of the previous season; flower stalks mostly 3–5 (8) mm long, on relatively few individuals in our range to about 2 cm long. Calyx glabrous, lobes 5, deltoid. Corolla lobes 5, white to yellowish green, oblong to oblanceolate. Drupes globose or subglobose, 6–8 (10) mm in diameter, red or orangish red, yellow on a rare individual plant, persisting for a considerable period after leaf fall. Nutlets 4.5–5 mm long, irregularly ridged on the rounded side. [Incl. *I. cuthbertii* Small]

Alluvial floodplains, low woodlands astride creeks, wet thickets, infrequently on well-drained wooded slopes or sandy pineland ridges; throughout much of our area, probably not in s.e. Ga. and n.e. Fla. (Coastal plain, Md. to w. cen. pen. Fla., westward to e. and cen. Tex., e. Okla., northward in the interior to s. Ind., s. Ill., s. Mo.)

Plants with the stalks of pistillate flowers (and fruits) 1–2 cm long or a little more, rare in our range, more common and abundant in some parts of the over-all range, may be referred to *I. decidua* var. *longipes* (Chapm. ex Trel.) Ahles. (*I. longipes* Chapm. ex Trel.)

Fig. 56. **Ilex decidua** var. **decidua:** a, fruiting branch of the more common "form" having flowering/fruiting stalks relatively short, to left, and at the right a fruit with its stalk; b, fruiting branch of the less common "form" (in our area) with relatively long flowering/fruiting stalks and which may be referred to as *I. decidua* var. *longipes.*

147

Fig. 57. **Ilex decidua** (the "form" which may be distinguished as *I. decidua* var. *curtissii*): a, staminate flowering branch at the left, fruiting branch to the right; b, leaf, enlarged to accentuate the marginal toothing; c, flower just prior to full anthesis; d, staminate flower to the left, functionally pistillate flower to the right; e, fruit.

Plants generally inhabiting nonalluvial soils, rather thin soils with much fine organic material overlying limestone, somewhat to the east and west of the Suwannee River, have narrower leaves than those of alluvial soils. These have been segregated as *I. decidua* var. *curtissii* Fern. (*I. curtissii* (Fern.) Small). Wunderlin (1977) does not consider this as an entity worthy of recognition as a taxon; he places the names in the synonymy of *I. decidua*. See Fig. 57.

National champion possum-haw (1981), 3' circumf., 42' in height, 52' spread, Congaree Swamp National Monument, S.C.

8. Ilex amelanchier M. A. Curtis in Chapm. SARVIS HOLLY.   Fig. 58

Deciduous shrub, sometimes with several to numerous main stems from near the base, or arborescent and to about 5 m tall. Twigs of the season grayish brown, powdery-pubescent or sparsely short-pubescent and with dotlike blackish lenticels; older twigs smooth, grayish brown, the lenticels horizontally elliptical and pale buff-colored.

Leaves with powdery-pubescent petioles 3–15 mm long. Blades oblong, oblong-obovate, or elliptic, the several shapes often on a single branch, larger ones 5–9 cm long and 3–4 cm broad, bases rounded or shortly tapered, apices rounded or obtuse, rarely abruptly very short-acuminate; upper surfaces glabrous at maturity, dull green, lower surfaces and petioles persistently and rather uniformly clothed with soft, shaggy pubescence, margins entire or with a few small teeth.

Functionally unisexual flowers on separate plants, the staminate numerous in axillary fascicles, the pistillate solitary or few in leaf axils or on leafless portions of twigs of the previous season. Calyx lobes 4, narrowly triangular, calyx apparently deciduous well before maturation of the drupes. Corolla lobes 4, oblongish to oblong-elliptic, white or yellowish. Drupes cherry-red, without luster, subglobose or slightly oblate, 5–10 mm in diameter. Nutlets 4, their rounded sides with 2 broad, longitudinal furrows, sometimes with a rib connecting one or both of the furrows.

Woodlands along creeks, river floodplain forests, cypress-gum swamps. Local, e. N.C. to cen. and w. Panhandle of Fla., s. Ala., s. Miss., s.e. La.

9. Ilex verticillata (L.) A. Gray. BLACK-ALDER, WINTERBERRY.   Fig. 59

Deciduous shrub or small tree, to about 8 m tall, trunk usually branching close to the ground. Twigs of the season sparsely pubescent at first, sometimes glabrous later, greenish; older twigs grayish to brown, with nearly circular, tan lenticels. Bark of trunks smooth, thin, brown to dark gray.

Leaves with pubescent petioles 5–20 mm long. Blades variable in size from plant to plant, often even on a single branch, elliptic, oval, rarely obovate, 2–10 cm long and 1–5.5 cm broad, bases tapered, apices abruptly short-acuminate to prominently acuminate, margins appressed-serrate; upper surfaces sparsely short-pubescent, chiefly along the midrib, often becoming glabrous, lower surfaces varyingly glabrous, pubescent on the veins, or pubescent throughout.

Functionally unisexual flowers borne on separate plants. Staminate flowers usually in shortly stalked, rarely sessile, verticels in the leaf axils, stalks of the verticels, if present, 2–6 mm long, stalks of the flowers 2–5 mm long, both somewhat pubescent. Pistillate flowers solitary, or 2–4 in short-stalked or sessile verticels in the leaf axils. Calyx tube and lobes usually pubescent exteriorly, lobes 5–7 (8), triangular, apically obtuse, ciliate marginally. Corolla lobes 5–7 (8), oblong, white. Drupes 5–7 mm in diameter, globose, bright red, yellow on a rare

Fig. 58. **Ilex amelanchier:** a, fruiting branch; b, leaf and enlargement of small portion of lower surface to right; c, flowering branchlets, functionally pistillate to left, staminate to right; d, flowers, functionally staminate below, pistillate above; e, fruit.

Fig. 59. **Ilex verticillata.**

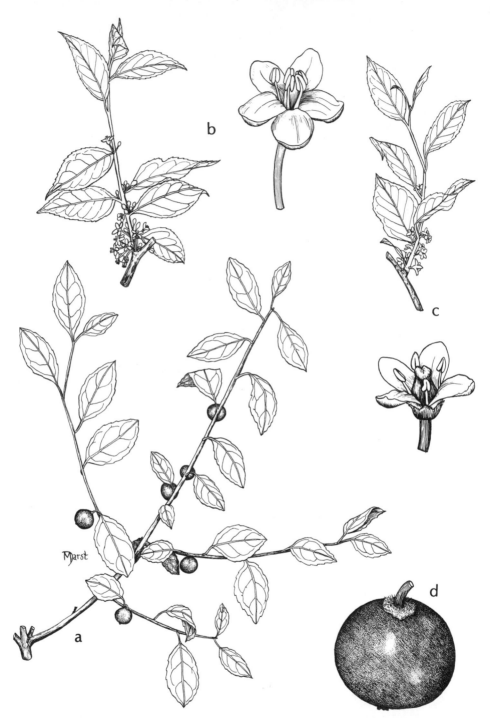

Fig. 60. **Ilex ambigua** var. **ambigua:** a, fruiting branch; b, staminate flowering branchlet to left, flower to right; c, functionally pistillate flowering branch above, flower below; d, fruit.

individual plant, usually persistent until after leaf-fall. Nutlets 5 or 10, 3–4 mm long, smooth on the rounded side.

Swamps, wet woodlands, bogs, seepage areas in woodlands; local, s.w. Ga., cen. and w. Panhandle of Fla., s. Ala. (N.S. to Minn., generally southward to Fla. Panhandle, s.e. Miss. and s.e. La.; cen. w. La. and e. Tex.)

10. Ilex ambigua (Michx.) Torr. var. **ambigua.** CAROLINA OR SAND HOLLY.

Fig. 60

Deciduous, often a low shrub, varying to arborescent and to about 6 m tall. Twigs of the season green or brownish, varying from glabrous to copiously short-pubescent, woody twigs gray or brownish gray.

Leaves variable in size and amount of pubescence from plant to plant, perhaps owing to relative fertility and/or moisture content of the substrate and whether in sun or shade. Petioles slender, sparsely to densely short-pubescent, 2.5–10 mm long. Blades elliptic, oval, ovate, or infrequently obovate, 1.5–8 cm long, and 1–4 (5) cm broad; bases rounded, obtuse, or acute; apices acute, obtuse, or short-acuminate and with acuminations bluntly to sharply tipped but not attenuate; surfaces glabrous or sparsely to moderately short-pubescent, margins entire or appressed-crenate-serrate from about their middles upwardly, the teeth rather obscure to the unaided eye, with short taillike tips, however, as viewed with magnification.

Functionally unisexual flowers on separate plants. Staminate flowers few to numerous in fascicles about the base of petiole or bases of petioles of leaf or leaves at the ends of spur-shoots. Pistillate flowers similarly disposed, varyingly solitary, in pairs, or up to 4 in a fascicle; flower stalks 1–4 mm long (some pistillate flowers sometimes sessile). Calyx and corolla each 4–5 lobed, commonly 4-lobed, rarely 3-lobed, lobes of calyx short-triangular, their margins short-ciliate, lobes of the corolla oblongish to obovate, white. Drupes subglobose to ellipsoid, red, 5–7 (10) mm in diameter. Nutlets 4 or 5, about 5 mm long, furrowed on their rounded sides. [Incl. *I. beadlei* Ashe, *I. buswellii* Small]

Inhabiting a wide variety of well-drained sites, mesic or submesic mixed woodlands, pine–scrub oak sand ridges and flats, sandy hardwood hammocks, sandy old fields; throughout our area. (Coastal plain and piedmont, N.C. to s. cen. pen. Fla., westward to e. Tex., Ark., s.e. Okla.)

Ilex ambigua var. *monticola* (A. Gray) Wunderlin & Poppleton, chiefly occurring in mesic woodlands of the mountains and piedmont northward of our range, may reach our area along the Chattahoochee River. Its leaves are consistently large (especially longer) than those of var. *ambigua*, blades more thinly membranous, the slender petioles to 2 cm long, blades more consistently elliptic in outline, 8–12 (18) cm long and 3–4 (6) cm broad, their apices usually acuminate-attenuate, their margins more conspicuously serrate, the teeth more evident to the unaided eye.

Araliaceae (GINSENG FAMILY)

Aralia spinosa L. DEVIL'S WALKING-STICK, HERCULES'-CLUB.

Fig. 61

A very distinctive shrub, occasionally having the stature of a small tree, perhaps not exceeding 8 m tall, perennating by elongate subterranean runners. Stem unbranched until the season after the first inflorescence is produced, eventually with relatively few ascending branches; abundantly armed with stiff, sharp,

Fig. 61. **Aralia spinosa:** a, tip of twig (in winter); b, portion of a single leaf; c, leaflet (upper) and enlargement of its lower surface (lower); d, small portion of a petiole having prickles, at right, and small portion of leaf rachis, at left; e, very small portion of inflorescence; f, flower; g, very small portion of infructescence; h, fruit.

straight or curved prickles, these sometimes sloughing on older stems, leafless nodes with raised, narrow leaf scars obliquely nearly encircling the stem, vascular bundle scars many in a line the full extent of the leaf scar.

Main stem or branches bearing very large, twice or thrice odd-pinnately compound, closely set leaves disposed so as to give an umbrellalike appearance. Leaves 6–15 dm long including the petioles, nearly triangular in overall outline, alternate, petiolate, petioles stout, to 3 dm long or a little more, their bases dilated, partially fused with stipules and clasping the stem obliquely; on some individual plants none of the parts of the leaves armed with prickles, on other plants the leaf axes, sometimes the petioles, sometimes the midribs on the lower leaflet surfaces, armed with irregularly disposed, slender, usually hooked prickles; basal pair of leaf segments to about 7 dm long, upwardly the pairs of leaf segments gradually shorter, the most distal few leaflets undivided; lateral leaflets of a given leaf segment short-stalked, the terminal one much longer-stalked; main axis of the leaf with a pair of deflexed, "accessory" leaflets subtending each of the pairs of secondary axes, often such an "accessory" pair of leaflets, or a single one, at the juncture of a tertiary axis or subtending a leaflet; blades of ultimate leaflets ovate, variable in size, 3–10 cm long and 1.5–8 cm broad, pinnately veined, bases subcordate, rounded, truncate, or shortly tapered, sometimes inequilateral, apices acuminate, less commonly acute, margins sharply to obscurely serrate, upper surfaces dark green, the lower pale green, sometimes with a few short stiff hairs on the midribs above, beneath, or both.

Inflorescences large, terminal, compound panicles up to 12 dm long and often as broad (main axis of the panicle sometimes short relative to the overall size of the inflorescence), the ultimate flower-bearing divisions slender-stalked, ball-like umbels; secondary and tertiary branches of the panicle and flower stalks sparsely to copiously pubescent with short, stiff hairs. Flowers small, ovary inferior, floral tube surmounted by 5 very small calyx segments, 5 small, white, spreading or reflexed petals, and 5 stamens; styles 4–6, usually 5, fused basally.

Fruit a subglobose, 5-stoned drupe 5–8 mm in diameter, crowned by the persistent styles, purplish black. All axes of fruiting panicles reddish purple and showy although developing fruits remaining green until ripening.

In upland and lowland woodlands, thickets, shrub bogs; throughout our area. (N.J. to s. Ind., Ill., and Iowa, generally southward to n. cen. pen. Fla. and e. Tex.)

The characteristic prickly, canelike stems marked by ring-scars and crowned by an umbrellalike canopy of very large 2–3-pinnately compound leaves makes this plant easily identifiable. Its features make it a very interesting subject for horticultural use although its habit of spreading afar by underground runners is perhaps a disadvantage.

National champion devil's-walking-stick (1976), 1'11" circumf., 51' in height, 16' spread, San Felasco Hammock State Preserve, Fla.

Note: *Tetrapanax papyrifera* (Hook.) K. Koch, rice-paper plant, native to southern China, also a member of Araliaceae, is often planted in at least some parts of our area for its bold effect. It does not become naturalized from seed, insofar as I know, but it may spread a great deal by means of potentially elongate subterranean runners; and then appears to have "gone wild."

It is a shrub with wandlike stems, the pith white and very large relative to the diameter of the stem. Young vegetative growth bears dense but loose buff-colored to grayish stellate tomentum. Leaves with long petioles bearing prominent subulate stipules, the blades variable in size but the large fully grown ones up to 50 cm wide, palmately veined and lobed, lobes 7–12, each usually sublobed; upper surfaces eventually glabrous, the lower retaining a dense grayish tomentum. Inflorescence a very large and striking panicle whose branches bear numerous ball-like stalked umbels 1.5–2 cm across, the stalks subtended by subulate bracts as long as or a little longer than the stalks; all parts of the inflorescence, including the floral tubes, densely tawny-tomentose.

The leaves generally get frost-killed in winter; if the cold is severe, the stems die back to ground level.

Aristolochiaceae (BIRTHWORT FAMILY)

Aristolochia tomentosa Sims. WOOLLY DUTCHMAN'S PIPE, WOOLLY PIPE-VINE.

Fig. 62

A twining vine, sometimes extending to 25 m into the crowns of trees, the main stems softly woody, those of branchlets herbaceous throughout much of the season, copiously tawny-pubescent at first.

Leaves alternate, simple, with densely soft-pubescent petioles usually somewhat shorter than the blades; blades broadly cordate-ovate to orbicular-reniform, palmately veined, 5–15 cm long or a little more, mostly about as broad as long, margins entire, apices broadly obtuse to rounded; both surfaces of unfurling blades densely clothed with silky, tawny pubescence, eventually the upper surfaces becoming green and with relatively sparse pubescence, somewhat scabrid, lower surfaces becoming grayish, relatively thinly hairy but soft to the touch.

Flowers solitary or paired on bractless stalks from leaf axils, the stalks and exterior of floral tube densely soft-pubescent; flower strongly bent a little below the middle, with an oblique, nearly closed, purple orifice and a spreading to reflexed, greenish yellow to purple, 3-lobed limb; flower (straightened out) about 5 cm long when fully developed. Stamens 6, anthers sessile and adnate to the 3–6-lobed style. Pistil 1, ovary partially or wholly inferior, ovary 6-locular, style short and stout.

Fruit a septicidal, longitudinally ribbed capsule, broadly oblong, truncate to rounded at both extremities (rarely beaked apically owing to failure to produce seeds in the distal portion), 4–8 cm long. Seeds numerous, flat, triangular, about 1 cm long, in longitudinal rows.

Alluvial woodlands, river and creek banks, often on natural levees; in our range, s.w. Ga., w. Fla. Panhandle, s. Ala. (S. Ind. to s.e. Kan., generally southward to Fla. Panhandle and e. and n. cen. Tex.)

Distinctive features of this woody, twining vine are its broadly cordate-ovate or orbicular-reniform leaf blades, velvety to the touch beneath, especially when young, dutchman's pipelike flowers singly or paired in leaf axils, cylindrical capsule, broadly oblong in outline, seeds numerous, flat, in longitudinal rows.

Avicenniaceae (BLACK MANGROVE FAMILY)

Avicennia germinans (L.) L. BLACK MANGROVE.

Fig. 63

Evergreen shrub or small tree, sometimes becoming 20–25 m tall (in the tropics). Twigs at first densely very short-gray-pubescent, appearing very finely granular with suitable magnification, becoming grayish brown, then gray; bark

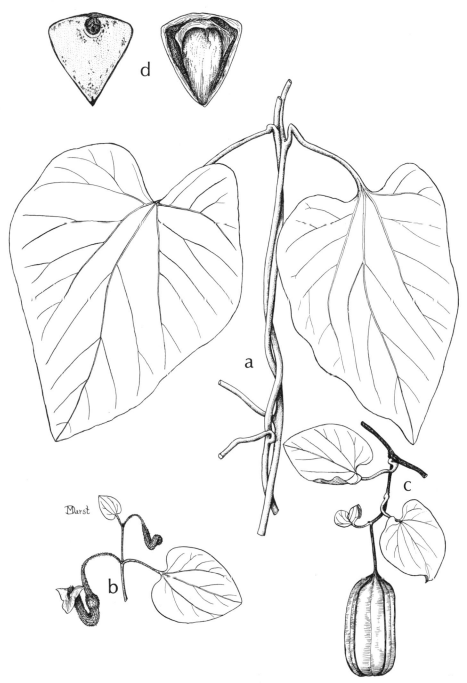

Fig. 62. **Aristolochia tomentosa:** a, small portion of two twining stems each with one leaf; b, branchlet with flower; c, branchlet with fruit; d, seed, viewed from each side.

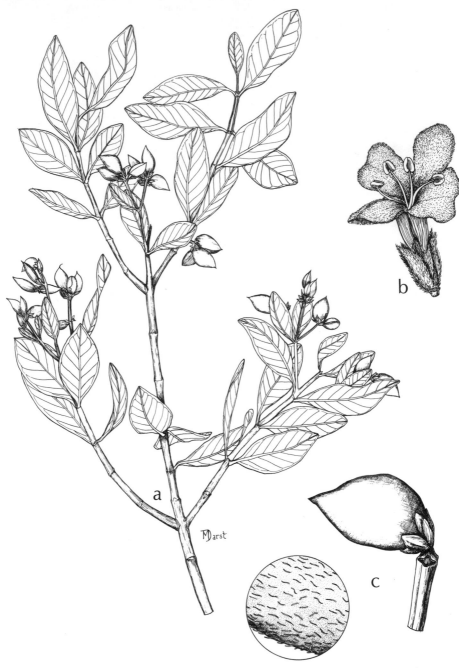

Fig. 63. **Avicennia germinans:** a, fruiting branch; b, flower; c, fruit, at right, enlarged portion of surface of fruit at left.

of large stems becoming irregularly and roughly fissured and broken; leaf scars more or less crescent-shaped, with 3, often indistinct, vascular bundle scars; pith pale tan, continuous.

Leaves simple, opposite, pinnately veined, shortly subpetiolate; blades elliptic, oblong, lanceolate or somewhat oblanceolate, their apices obtuse to rounded, bases tapering, upper surfaces green, glabrous and finely punctate, lower surfaces with very tight, felty, gray pubescence completely obscuring the surface proper; stipules none.

Inflorescences stalked, compact, bracteate cymes borne on twigs of the season from the leaf axils or terminally on the twigs, copiously short-pubescent throughout. Flowers sessile on the cyme branches, bisexual, radially symmetrical. Sepals 5, pubescent, nearly separate, 3–4 mm long, persistent on the fruit. Corolla white, tubular below then flaring and with one lobe above and a 3-lobed lower lip. Stamens 4, the filaments attached to the lower part of the corolla tube, anthers barely exserted. Pistil 1, pubescent ovary superior, style slender, stigma bifid. Fruit a flat, asymmetric, almost velvety, 1-seeded, 2-valved pod much resembling a very short lima bean pod. [*Avicennia nitida* Jacq.]

Essentially maritime, swamps and shores; coasts of Fla. southward from St. Johns Co. on the Atlantic Coast and southward from Levy Co. on the Gulf Coast; sporadic along the Gulf Coast westward to s.e. and s. Tex. where probably not attaining stature of more than 1 m tall and flowering and fruiting during some years, at least; Berm., trop. Am.; W.Afr.

National champion black mangrove (1975), 7'2" circumf., 61' in height, 42' spread, Everglades National Park, Fla.

Bataceae (SALTWORT FAMILY)

Batis maritima L. SALTWORT. Fig. 64

Plant with thickish, cordlike roots, young stems succulent-herbaceous, old stems woody and with a pale buff, very soft, corky bark readily shredding into irregular flaky pieces. Herbage lustrous light green to yellow-green, strongly scented, stems elongate-spreading, prostrate, arching, or creeping, the tips of arching branches commonly rooting, then forming a cluster of new branches, eventually forming large dense clones.

Leaves 1–3 cm long, opposite, sessile, glabrous and succulent, commonly curved, half- to semiterete, margins entire, without stipules.

Flowers unisexual (plants dioecious), small, crowded in fleshy, conelike, axillary, sessile or stalked spikes solitary in the leaf axils, with imbricate, fleshy scales subtending the flowers. Staminate spikes to 1 cm long, oblong- or ovate-cylindric, sessile, with persistent, reniform to suborbicular bracts each subtending a flower; calyx cuplike, 2–lobed, corolla none; stamens 4 or 5, inserted at the base of the calyx, with alternating staminodia. Pistillate spikes on short stalks, 4–12-flowered, bracts deciduous, perianth none, the flower thus of a single pistil, the ovary 4-locular, 1 ovule in each locule, no evident style, a flattened stigma sitting directly on the summit of the ovary. As the 4–12 pistils of each spike ripen (the bracts having sloughed) they fuse to each other into a fleshy, knobby whole 5–15 mm long.

Salt marshes and salt flats, mangrove swamps, muddy tidal shores and flats,

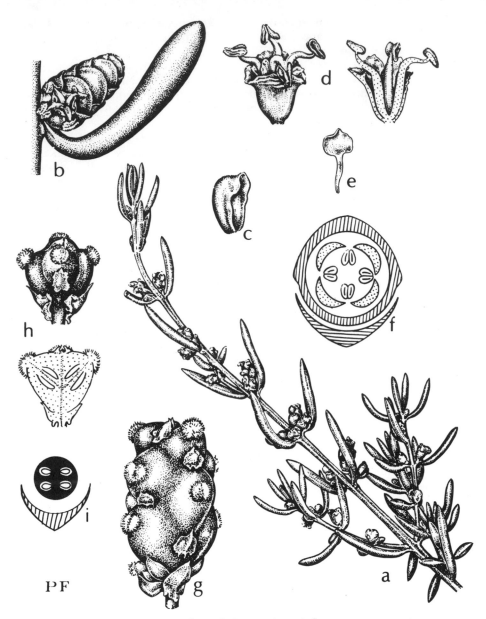

Fig. 64. **Batis maritima:** a, branch; b, staminate inflorescence; c, staminate flower bud; d, on left, staminate flower, on right, staminate flower longitudinally dissected; e, petal; f, floral diagram of staminate flower; g, pistillate inflorescence; h, above, complete pistillate flower, below, pistillate flower longitudinally dissected; i, floral diagram of pistillate flower. (Courtesy of Fairchild Tropical Garden.)

both Atlantic and Gulf coasts of our area. (S.C. southward through the Fla. Keys, Gulf Coast to Tex.; trop. Am.; Haw. Is.)

Berberidaceae (BARBERRY FAMILY)

Nandina domestica Thunb. NANDINA. Fig. 65

Evergreen, glabrous, unarmed shrub to about 2 m tall, with relatively slender, unbranched or few-branched stems, the leaves large, compound, and somewhat lacy; plant often extensively perennating by subterranean runners.

Leaves alternate, closely arranged near the ends of stems or branches; petioles short and relatively rigid, flared basally and clasping the stem. Blades large, the larger ones to 5 dm long and to at least 8 dm across the lateral divisions, 3–4-odd-pinnately compound, with 3 primary divisions arising together from a joint at the end of the petiole, each primary division with opposite secondary divisions, each of the ultimate divisions with 5, 3, or 2 often unequal leaflets, occasionally a leaflet solitary. Lateral leaflets subsessile, terminal ones usually narrowed basally to short stalks. Leaflets narrowly ovate to lanceolate, 2–7 cm long and 1–2 (2.5) cm broad, bases obtusely to acutely tapered, apices with a longer, acute taper; upper surfaces dark green, sublustrous, lower surfaces paler dull green, margins entire. Axes of the leaf often partially or wholly purplish red. Leaves when they fall generally break off at the juncture of petiole and blade, the petiole remaining intact for an indefinite period; the petiole eventually breaks off, not leaving a clear-cut leaf scar but a short, irregular stubble of fibers.

Inflorescence a relatively large, erect, stalked panicle, its axes at least partially purplish red; terminal bud none, panicle axillary to the uppermost leaf of the season, usually a small compound leaf (bracteal leaf) subtending the lowermost branch of the panicle. Flowers small, each axillary to a subulate bract, the flower stalks 7–15 mm long, usually bearing 3–5 alternate bractlets, bracts and bractlets persistent. Perianth parts in numerous series of 3 each, outermost scarious, short-ovate, inwardly in the series becoming gradually longer, thinner, more oblong, and more whitish and petallike. Stamens 6. Pistil 1, ovary superior, 1-locular, 2-ovuled, style very short, stigma more or less 3-lobed. Fruit a bright red, 2-seeded, globose berry 6–12 mm in diameter, the stigma persistent at its apex.

Native to Asia, widely cultivated as an ornamental; in our area sporadically naturalized in woodlands, usually in the vicinity of human habitations.

Betulaceae (BIRCH FAMILY)

1. Leaves in 3 ranks on the stem; buds stalked; fruiting cones becoming woody, persistent overwinter and many of them through at least the next season. 1. *Alnus*
1. Leaves in 2 ranks on the stem; buds not stalked; fruiting catkins not persistent overwinter.
  2. Outer bark of trunks exfoliating in large, thin patches; leaf blades more or less triangular-ovate or subrhombic; fruiting catkins falling soon after maturation in spring.
  2. *Betula*

Fig. 65. **Nandina domestica:** a, leaf with infructescence from its axil; b, flower.

2. Outer bark of trunks smooth, granular, or finely shredding; leaf blades ovate-oblong, oblong, or oblong-elliptic, neither triangular-ovate nor subrhombic; fruiting catkins either leafy-bracted, or its parts inflated-saclike, in either case persisting well into summer.

    3. The outer bark of trunks smooth or granular, grayish, not shredding, the trunks usually somewhat fluted; leaf blades glabrous above, with tufts of hair in the angles formed by the major veins beneath; fruiting catkins leafy-bracted.   *3. Carpinus*

    3. The outer bark of trunks brown, finely shredding, the trunks not fluted; mature leaf blades sparsely pubescent above, downy-pubescent beneath; fruiting catkins with individual fruits enclosed in thin-papery, inflated sacs.   *4. Ostrya*

1. Alnus

Alnus serrulata (Ait.) Willd. HAZEL ALDER.   Figs. 66, 67

Deciduous shrub, occasionally with the stature of a small tree. Young twigs copiously brown-shaggy-pubescent, the pubescence gradually sloughing as the twigs become woody; bark of woody twigs with a gray waxy bloom that eventually shreds irregularly exposing reddish brown bark, lenticels small, scattered, slightly raised; leaf scars triangular to half-round, a stipule scar at either side of their summits, vascular bundle scars 3, often indistinct. Terminal buds none, axillary buds stoutly stalked. Pith pale green, continuous, 3-lobulate in cross-section.

Leaves simple, alternate, 3-ranked on the stems, petioles mostly 1–1.5 cm long, stipules membranous, oblong, truncate basally, blunt apically, 6–10 mm long, soon deciduous. Blades mostly 5–10 cm long and 2.5–5 cm broad, obovate, elliptic, or oblongish, bases rounded or broadly and shortly tapered, apices usually broadly obtuse; venation pinnate, impressed on the upper surfaces, raised on the lower, the lateral veins equal, angling-ascending toward the margins parallel to each other; margins commonly irregularly slightly wavy and irregularly finely toothed; upper surfaces dark green, glabrous or sparsely pubescent, especially along the principal veins, lower surfaces with a brownish cast because the major veins, often the veinlets, are brown, usually sparsely pubescent at least on the veins, tufts of hairs in the axils formed by the midvein and the major laterals.

Flowers unisexual, borne in staminate and pistillate catkins on the same branches, the catkins partially developed and evident well before leaf-fall, reaching anthesis in spring well in advance of new shoot emergence; fully developed staminate catkins, at anthesis, dangling, long-cylindrical, 4–8 cm long and about 5–6 mm in diameter; each staminate catkin bearing many compactly disposed clusters, usually 3 flowers per cluster, each cluster subtended by a fleshy peltate bract to which 5 bractlets are almost imperceptibly fused and really appearing as 1, each flower with a 3–5-parted calyx and 3–5 stamens; pistillate catkins very much smaller than the staminate, 5–10 mm long and about 2 mm in diameter, sticky, held stiffly erect; bracts compactly disposed, fleshy at anthesis (bract with tiny scalelike bractlets adnate to it but the bractlets imperceptible for all practical purposes at anthesis), each bract subtending 2 minute pistillate flowers each composed of the pistil only; styles 2, exserted from between the bracts of the catkin.

Bracts and associated fused bractlets eventually becoming woody, flattish, obovate in outline, undulate across the truncated apex, together forming a short-oblong to ovate-oblong burlike or conelike "fruit" 1–2 cm long and 0.5–1 cm broad, the actual fruits, subtended by the bracts, being laterally winged, chestnut-brown, lustrous nutlets.

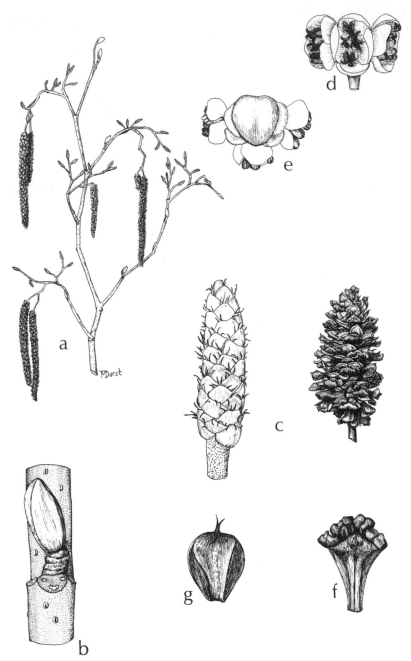

Fig. 66. **Alnus serrulata:** a, branchlet with pendent staminate catkins and erect pistillate catkins; b, node with leaf scar and axillary bud; c, pistillate catkin at anthesis to the left, woody mature fruiting catkin to the right; d, bract complex and staminate flowers from staminate catkin, abaxial view; e, bract complex and pistillate flowers, adaxial view; f, scale and bract complex from mature fruiting catkin; g, fruit.

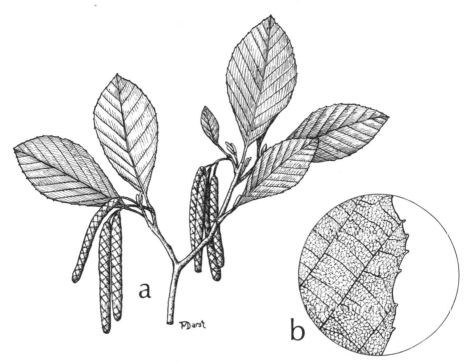

Fig. 67. **Alnus serrulata:** a, branchlet (of late summer) with staminate catkins already formed and which remain dormant until spring; b, small portion of leaf blade enlarged to show margin.

The conelike "fruits" mostly persist over winter, many of them through at least the next season, sometimes for a second season. Their presence together with stalked buds represent field-identifying features for the plant.

In alluvial soils, open stream banks or shores, swales, borders of swamps and wet woodlands, sloughs, and ditches; spottily distributed throughout our area of coverage. (Maine to N.Y., Ohio, Ind., Mo., generally southward to n. Fla. and e. Tex.)

National champion hazel alder (1965), 1'5" in circumf., 40' in height, 22' spread, near Shreve, Wayne Co., Ohio.

2. Betula

Betula nigra L. RIVER BIRCH. Fig. 68

Tree, reaching 30–35 m tall, often flowering and fruiting when of shrub stature. Bark of trunk freely sloughing in curly thin patches of dark coppery brown to buff, exposing yellowish to salmon-colored inner bark. Young branchlets densely pubescent; woody twigs slightly zigzagging, slender, dark reddish brown and with small, tan, horizontally elliptical lenticels; leaf scars vaguely 3-angular to nearly oval, each with 3 vascular bundle scars, stipule scars shortly linelike, often indistinct; terminal buds none, lateral buds sessile; pith tan, continuous.

Leaves alternate, deciduous, 2-ranked on the stem, with narrow, very quickly

165

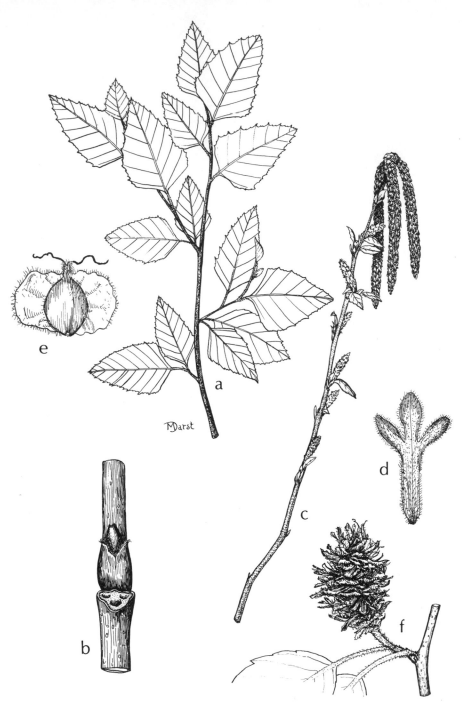

Fig. 68. **Betula nigra:** a, leafy branch; b, piece of woody twig with leaf scar and axillary bud; c, branch with pendulous staminate catkins at tip and erect pistillate catkins below; d, bract of fruiting catkin that subtends the fruit; e, fruit; f, mature woody fruiting catkin after fruits have been shed.

deciduous, hairy stipules 4–5 mm long, petioles short, densely pubescent. Blades triangular-ovate to subrhombic, sometimes a few of them nearly elliptic, 3–10 cm long and 1.5–3 cm broad, bases broadly and shortly tapered to truncate, apices obtuse to acute, margins doubly serrate; emerging blades very densely woolly beneath, moderately woolly above, usually becoming nearly glabrous beneath in age save for pubescence on the principal veins and in the principal vein axils, upper surfaces becoming glabrous save for sparse pubescence along the principal veins; venation pinnate, the lateral veins angularly ascending approximately parallel with each other to the leaf margins.

Flowers unisexual, in catkins, both staminate and pistillate catkins on the same plant. Staminate catkins partially developed in summer, singly or in clusters at or near tips of twigs of the season, remaining dormant over winter, developing fully, elongate and pendent, as the leaves are emerging or just before; staminate flowers 3 above each peltate bract, each of the 2 lateral ones subtended by a bractlet, each with a 2–4-lobed calyx and 4 stamens. Pistillate catkins stalked, developing in spring, usually singly at the ends of short new lateral branchlets, the catkins stalked, thick, short-cylindric; pistillate flowers 2 or 3 subtended by a bract with 3 longish lobes, perianth none.

Fruit a flat, laterally winged nutlet. Scales of the catkin shedding as the nutlets fall early in the season.

Alluvial wooded banks of rivers and streams, floodplain woodlands; throughout our area. (N.H. to N.J., Pa., westward to Ohio, Ind., Ill., Iowa, s.w. Wis., s.e. Minn., generally southward except in the s. Appalachians to n. Fla. and e. Tex.)

National champion river birch, 14′9″ in circumf., 92′ in height, 84′ spread, Milton, N.Y.

3. Carpinus

Carpinus caroliniana Walt. AMERICAN HORNBEAM, IRONWOOD, BLUE-BEECH.

Fig. 69

An understory tree attaining heights of 10 m or a little more, trunk diameters up to 3 dm, sometimes more. Bark of the trunk thin, smoothish or granular, gray, not shredding, the trunks often somewhat fluted, appearing in places somewhat like flexed muscles. Very young branchlets sparsely soft-hairy. Woody twigs slender, smooth, reddish brown and with small, vertically elliptical lenticels; leaf scars horizontally oval to crescent-shaped, each with 3 vascular bundle scars, a small lenticular stipule scar at either side of the summit of the leaf scar. Pith narrow, pale buff-colored, continuous.

Leaves simple, alternate, deciduous, 2–ranked on the stem, petioles short, slender, pubescent. The very small unfolding leaves on emerging branchlets copiously silky-hairy beneath, most of that pubescence sloughed before the blades reach much size; at this stage the stipules evident and fairly conspicuous, narrow, membranous, yellowish or brownish yellow, 6–10 mm long, very quickly deciduous. Mature blades oblong-ovate to oblong, 2–6 cm long and 1–3 (4) cm broad, rounded basally and often slightly asymmetrical, apices acute to acuminate, margins doubly serrate; venation pinnate, the lateral veins angling-ascending parallel to each other from the midrib to the margins, not conspicuous seen from above, much more so seen from beneath; upper surfaces glabrous, the lower with tufts of hairs in the angles formed by the midrib and lateral veins, sometimes pubescent along the principal veins.

Flowers unisexual, borne in catkins, staminate and pistillate in separate catkins on the same plant, the staminate much more conspicuous than the

167

Fig. 69. **Carpinus caroliniana:** top center, flowering branch, staminate catkins on wood of the previous season, pistillate catkins terminating branchlets of the season; center left, staminate flower with its subtending bract; center right, pistillate catkin; center, pair of pistillate flowers with subtending bracts; lower center, fruiting branch; lower right, small medial portion of lower surface of leaf much enlarged; lower left, bract with nutlet basally.

pistillate. Staminate catkins develop from buds at various places on wood of the previous season, each solitary and sessile, pendent, cylindrical, 2–4 cm long, the bracts broadly short-ovate, ciliate marginally, acute or obtuse apically, each subtending a staminate flower composed of 3–20 stamens, the filaments of each divided, each branch bearing an apically pilose half-anther. Pistillate catkins inconspicuous, solitary and softly erect terminating developing branchlets of the season, at anthesis when the leaves of the branchlets are in early stages of unfolding and while the stipules are still present; pistillate flowers in pairs above the scales, each flower subtended by a minute bract joined basally to 2 minute bractlets, the bract and fused bractlets eventually becoming very much enlarged and foliar, the fruits maturing seated on their bases.

Fruit a broadly ovate, longitudinally-ribbed, brown nutlet about 4 mm long, several short, sharp-pointed teeth at its summit. Foliaceous fruiting catkins present on the trees through much of the summer.

Floodplain woodlands, low woodlands, lowermost wooded slopes by streams; throughout our area. (N.S., N.B., Maine to Minn., Iowa, e. Neb., generally southward to cen. pen. Fla. and e. Tex.)

National champion American hornbeam (1975), 7'2" circumf., 65' in height, 66' spread, Milton, N.Y.

4. Ostrya

Ostrya virginiana (Mill.) K. Koch. EASTERN HOPHORNBEAM, ROUGH-BARKED IRON-WOOD. Fig. 70

A small understory tree attaining heights of 7–12 m, occasionally more, trunk diameters up to 25 cm, perhaps more. Bark of trunk thin, the outer bark brown, eventually peeling off or shredding in thin strips. Stems of emerging branchlets silky-hairy, later the woody twigs slender, slightly zigzagging, dull brown, scattered, persistent pubescence on year-old twigs, and with irregularly scattered, tan lenticels; leaf scars half-round to horizontally elliptical, each with 3 vascular bundle scars, linelike stipule scars to each side of the summits of the leaf scars. Terminal bud none, lateral buds sessile. Pith very thin, pale buff-colored, continuous.

Leaves alternate, simple, deciduous, 2-ranked on the stem, very short petiolate, petioles pubescent. Blades of small, developing leaves on emerging branchlets copiously silky-pubescent beneath, sparsely so above but densely short-pubescent along the midrib, much of the pubescence quickly sloughing as the leaves grow, at this stage, membranous, hairy, linear or linear-oblong, brownish stipules 7–8 mm long evident, these very quickly deciduous. Mature leaf blades oblong, ovate-oblong, or elliptic, 5–10 cm long and 2–5 cm broad, bases rounded to subcordate, sometimes shortly and broadly tapered, apices acute to acuminate; margins sharply serrate or doubly serrate; upper surfaces usually short-pubescent along the midribs or often with sparse, short pubescence more or less over the entire surface, lower surfaces generally sparsely short-pubescent throughout.

Flowers unisexual, staminate and pistillate in separate catkins on the same plant. Staminate catkins relatively conspicuous, pendent, sessile, produced singly or 2–3 from buds near the tips of the previous year's branchlets, 2–4 cm long, bracts very broadly ovate, firm-textured, reddish brown, their bodies broader than long, tips abruptly attenuated, each bract subtending a staminate flower composed of several stamens, their filaments usually forked, the forks

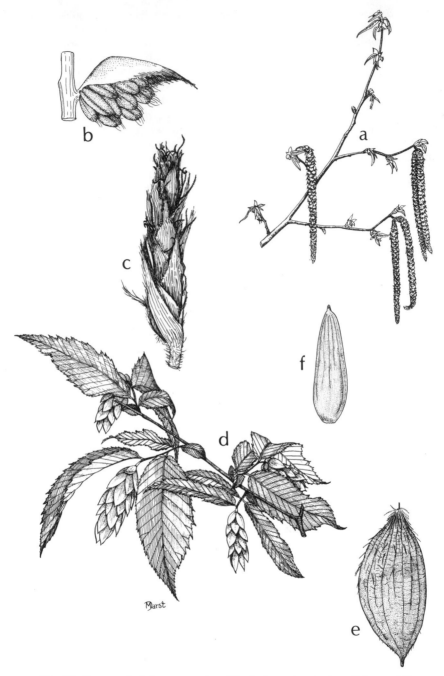

Fig. 70. **Ostrya virginiana:** a, twig with obvious staminate catkins and tiny pistillate catkins; b, bract of staminate catkins with staminate flower; c, pistillate catkin very much enlarged; d, branch with mature fruiting catkins; e, saclike bract of fruiting catkin (nutlet enclosed); f, nutlet.

bearing half-anthers with a tuft of hairs apically. Pistillate catkins inconspicuous, softly silky-pubescent, borne singly and erect at the tips of emerging branchlets, 5–8 mm long, bracts very closely overlapping, subulate, quickly deciduous, each subtending 2 flowers, each composed of a single pistil, its ovary enclosed in a tubular bract; ovary inferior, floral tube bearing at its summit a short-bearded "perianth," styles 2, long-linear, well exserted. Bract enclosing the ovary eventually becoming a thin-papery, inflated, ovate sac, at maturity bearing within its base an oblong-ellipsoid, smoothish, dull light brown nutlet about 5 mm long, the densely overlapping sacs sometimes bristly-hairy at the base, together forming a pale fruiting catkin 4–5 cm long and about 2.5 cm broad. Fruiting catkins persisting through much of the summer.

Well drained mixed mesic woodlands, throughout our area; on lower slopes this and *Carpinus caroliniana* often grow in close proximity. (N.S. to Man., generally southward to n. pen. Fla. and e. Tex.)

National champion eastern hophornbeam (1976), 9'4" circumf., 73' in height, 88' spread, Monroe Center, Mich.

Bignoniaceae (BIGNONIA FAMILY)

1. Plant a tree; leaves simple, opposite or whorled.   1. *Catalpa*
1. Plant a woody vine; leaves compound, opposite.
  2. Leaflets 2 per leaf, their blades with entire margins, rachis of the leaf terminated by a tendril by which climbing is achieved.
    3. Tendrils distally with 3 primary branches, these subbranched, the tendrils twining and also forming tiny adhesive discs at their tips; corolla orange or reddish orange exteriorly, tube yellow to red within.   2. *Bignonia*
    3. Tendrils 3-forked, tips of the forks stiffly hooked and clawlike (older woody stems also producing much branched, aerial adventitious roots which tenaciously grasp the substrate); corolla bright yellow with orange lines in the throat.   4. *Macfadyena*
  2. Leaflets of most leaves 7 or more, margins of the blades serrate; climbing achieved exclusively by aerial, adventitious roots on the stem.   3. *Campsis*

1. Catalpa

Catalpa bignonioides Walt. SOUTHERN CATALPA, INDIAN-BEAN, CIGAR-TREE, CATAWBA-TREE, CATERPILLAR TREE.   Fig. 71

A small to medium-sized tree, often shrubby about rural homesites where fishermen not infrequently plant it and sometimes keep it pruned. Caterpillars, used as fish-bait, commonly feed on the foliage.

Twigs stout, brittle, at first purplish green, becoming grayish brown with relatively large, raised, pale corky lenticels. Terminal buds none. Leaf scars relatively large, elevated, nearly round, vascular bundle scars in a circle but often indistinct. Pith white, continuous. Bark of older trunks grayish brown tinged with red, separating into large, irregular scales.

Leaves opposite or whorled, often some of both on the same branches, simple, deciduous, long-petiolate, blades subpalmately-veined (the midrib more pronounced than the basal lateral veins), broadly ovate, up to about 18 cm broad basally, sometimes as broad as long, sometimes about twice as long as broad, and the distal half long-attenuate, truncate to cordate basally, acuminate apically, sometimes abruptly constricted to a narrowly attenuated tip; margins entire or obscurely wavy, occasionally with one or a pair of acuminate lateral

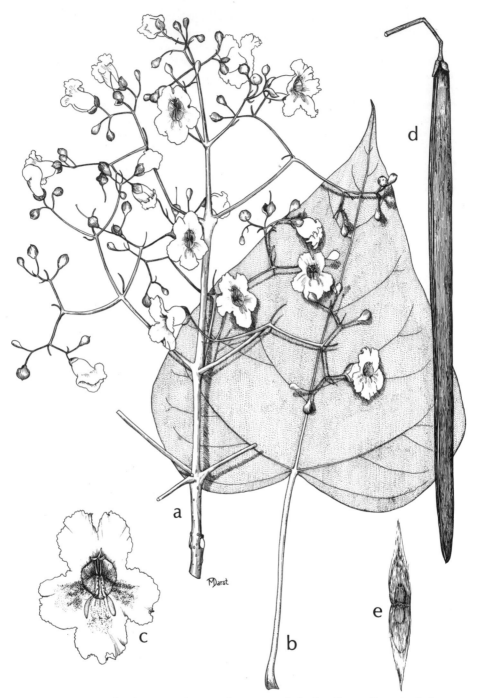

Fig. 71. **Catalpa bignonioides:** a, inflorescence; b, leaf; c, flower (face view); d, fruit; e, seed.

lobes on the upper side(s); upper surface glabrous or sparsely pubescent on the veins, the lower at first copiously soft-pubescent, much of the pubescence sloughing as the leaves age; usually a patch of purplish tissue (more pronounced on the lower surface) in the angles formed by the midvein and major lateral veins.

Inflorescence an open, broadly pyramidal panicle 1–3 dm long above the uppermost leaf of a branchlet of the season. Flowers large, about 4–5 cm long, bilaterally symmetrical, bisexual. Calyx irregularly 2-lobed, 6–10 mm long, glabrous, purplish. Corolla tube broadly and obliquely subcampanulate, expanded at the throat into 4 or 5 spreading, broad, short lobes crisped or erose marginally; corolla white exteriorly, marked within by yellow lines and con-spicuous purple spots. Fertile stamens 2, they and accompanying, usually 3, staminodia inserted near the base of the corolla tube. Pistil 1, ovary superior, 2-locular, ovules many in each locule, style slender, stigmas 2.

Fruit a long, narrow, cylindrical pod 1–3 dm long or longer, somewhat taper-ing from its midportion to either end, loculicidally 2-valved. Seeds numerous, flat, the embryo evident centrally within a narrowly elliptic or oblong wing, its two flat cotyledons "spread-eagled" parallel to the long axis of the wing; body of the wing about 3 cm long and nearly 1 cm broad, the opposite extremities fringed into longish hairlike tails; in each locule of the pod, the seeds longitudi-nally disposed and overlapping each other in 2 ranks.

Inhabiting wooded natural levees, banks, and floodplains of the larger rivers in the Fla. Panhandle and adjacent s.w. Ga., s. Ala., and s.e. Miss.; planted both as an ornamental and as a source of caterpillars, often naturalized in the vicinity of such plantings; often planted n.e. of the range stated above and there sometimes naturalized.

This is our only tree having opposite or whorled leaves with large ovate blades and bearing long, slender pods.

National champion southern catalpa (1981), 22'2" circumf., 80'in height, 60' spread, Henderson Co., Ill.

2. Bignonia

Bignonia capreolata L. CROSS-VINE, TRUMPET-FLOWER.   Fig. 72

A high-climbing, glabrous, woody, evergreen vine.

Leaves opposite, compound, petiolate, the petioles short, about 1 cm long, without stipules; leaflets 2, opposite, their stalks 1–2 times the length of the petioles, a distally branched tendril arising between them, the branches of the tendril twining and also forming minute, adhesive discs at their tips, climbing being achieved by a combination of twining and clinging; blades of leaflets pin-nately-veined, 5–15 cm long and 2–8 cm broad, the larger ones on vigorous shoots varyingly oblong-ovate, oblong-elliptic, lanceolate, rarely obovate, their bases rounded to subcordate or auriculate, apices acute to acuminate, rarely obtuse, margins entire.

Flowers large and showy, bisexual, borne in sessile (infrequently short-stalked) axillary clusters of 2–5, the flower stalks 2–3 cm long. Calyx small rela-tive to the corolla, cuplike, truncate at its rim or with 5 very short blunt lobes. Corolla showy, orange or reddish orange, about 5 cm long, relatively narrowly tubular basally then abruptly flaring to a tube about 2 cm across, limb slightly bilabiate, with 5 broad, short, spreading lobes, the tube yellow to red within. Stamens 4 in 2 pairs, inserted within the narrow base of the corolla tube, the

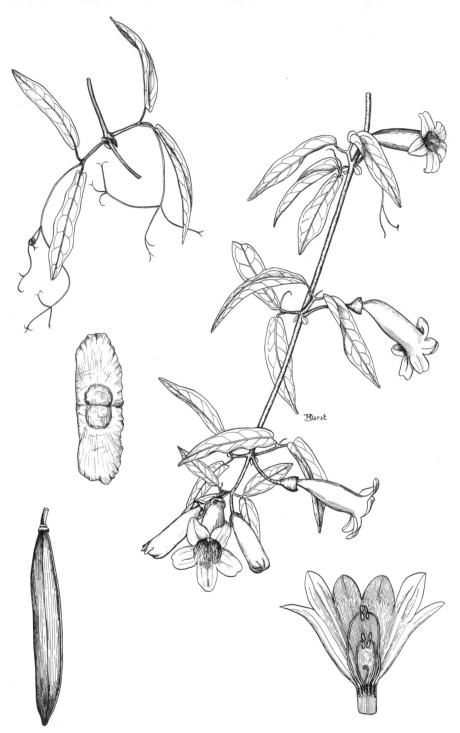

Fig. 72. **Bignonia capreolata:** piece of flowering stem, at right; at lower right, corolla opened out; node from young growing branch with opposite compound leaves, each having a terminal tendril, at upper left; fruit, lower left; and seed, center left.

filaments arching-ascending within the broad portion of the tube and not exserted from it, the filament of one of each pair a little shorter than the others. Pistil 1, ovary superior, 2-locular, ovules numerous in each locule, style slender, reaching nearly to the throat of the corolla, stigma shortly bifid.

Fruit a capsule flattened parallel to the partition, tapered at each extremity, linear between, 1–2 dm long, 2–2.5 cm broad, dehiscing septicidally. Seeds in 4 unequal rows, flat, conspicuously transversely 2-winged, about 3 cm long. [*Anisostichus capreolata* (L.) Bureau]

In a wide range of habitats relative to moisture conditions of the soil, upland woodlands of various mixtures, floodplain and lowland woodlands, edges of evergreen shrub-tree bogs and bays, thickets, fence rows; throughout our area. (E. Md. to s. Ohio and s. Mo., generally southward to s. Fla. and e. Tex.)

3. Campsis

Campsis radicans (L.) Seem. ex Bureau. COW-ITCH-VINE, TRUMPET-CREEPER.

Fig. 73

Woody vine, often trailing in open places where nothing upon which to climb, usually climbing by means of aerial stem-roots, in forests reaching high into the crowns of trees, lower parts of old stems to at least 10 cm in diameter, the bark shreddy, buff-colored.

Leaves deciduous, opposite, petiolate, odd-pinnately compound, leaflets mostly 7–11 (5–15), blades mostly ovate but varying to lanceolate, the terminal one sometimes nearly orbicular, bright green above, paler green beneath, variable in size to about 8 cm long and 4 cm broad, bases rounded or shortly tapered, abruptly narrowed to short stalks, apices acuminate or with an attenuated cusp above the uppermost teeth, margins coarsely serrate; leaf axis glabrous or sparsely pubescent between the pairs of leaflets, upper surfaces glabrous to very sparsely short-pubescent, the lower varying from glabrous to short-shaggy-pubescent, especially along the veins.

Inflorescence a short few-flowered to crowded compound cyme, the flowers of a given cyme up to at least 20, not all at anthesis simultaneously. Flowers bisexual. Calyx leathery, cylindric-campanulate, tube about 1.5 cm long, with 5 short-deltoid, erect lobes at the summit. Corolla thick-tissued, cylindric-tubular and yellow for about 2 cm from the base then flaring and trumpetlike, the dull orangish red funnellike tube about 4 cm long, the similarly colored limb slightly asymmetric, with 5 short, roundish, arching-spreading lobes; interiorly the funnellike portion of the tube striped with numerous dark red lines, lighter yellow-red between them. Stamens 4 in two pairs of somewhat unequal length inserted proximally within the corolla tube, not exserted. Pistil 1, ovary superior, slender, seated on a thickish disc broader than its narrowed base, style slender with 2 slightly unequal dilated stigmas positioned at the level of the uppermost pair of anthers.

Fruit a 2-locular, loculicidally dehiscent, 2-valved, fusiform, often slightly falcate capsule keeled along the sutures, convex to either side of the sutures, 10–20 cm long. Seeds many, flat, the body of each dark brown, short-obovate, a conspicuous pale obliquely transverse wing to either side, about 15–18 mm long overall. [*Bignonia radicans* L.]

In diverse habitats, woodlands of various mixtures, both upland and lowland,

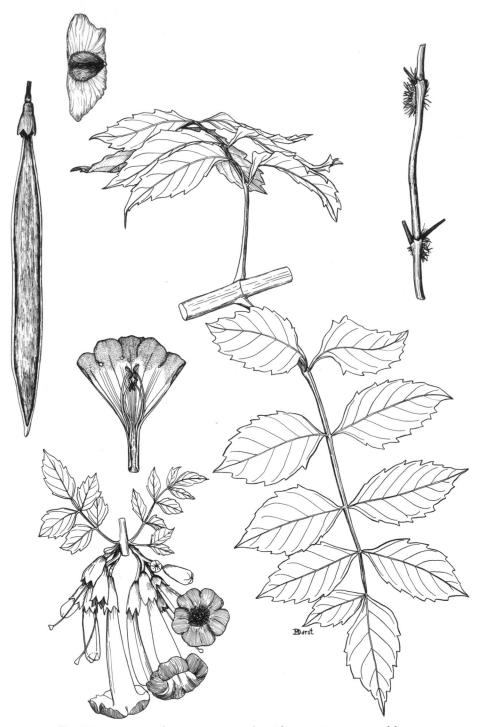

Fig. 73. **Campsis radicans:** center, node with opposite compound leaves; at upper right, piece of stem (leaves cut off) to show position of adventitious roots; lower left, node with inflorescences; center left, corolla opened out, and to its left, fruit; upper left, seed.

thickets, fence and hedge rows, old fields, on utility poles and walls; throughout our area. (N.J. to Iowa, generally southward to Fla. and e. half of Tex.; naturalized northeastward from N.J. to N.E.)

4. Macfadyena

Macfadyena unguis-cati (L.) A. Gentry. CAT'S CLAW VINE.   Fig. 74

High-climbing woody vine, woody stems of old plants attaining a diameter of at least 5–6 cm. Seedling first forms a short stem with few leaves, its further growth then being arrested for a considerable period of time during which the root becomes tuberoid, the tuber more or less fusiform or nearly orbicular; following this stage, elongation of the stem is rapid and very gradually over the years the root becomes elongate-tuberous. If, as the stem elongates, no erect substrate is available, it and its branches will run along the surface of the ground for not inconsiderable distances, its tendrils clinging tenaciously to whatever is available until something is reached upon which to climb; climbing is first achieved by the tendrils hooking very tightly to a substrate, wood, masonry, or stems or trunks of other woody plants; later the woody stems produce much branched but not very long, aerial, adventitious roots which very tightly fasten to the substrate.

Leaves opposite, deciduous or tardily deciduous in our climate, their petioles slender, about 2 cm long, each leaf with a pair of opposite, nearly sessile blades, the rachis extended beyond the blades as a 3-forked tendril, the tip of each fork stiffly hooked and sharply clawlike; blades of leaflets ovate to lanceolate, mostly 3–7 cm long, rounded or tapered basally, apices obtuse to acuminate, venation pinnate, margins entire.

Flowers showy, solitary or in few-flowered, stalked clusters axillary to leaves, bisexual. Calyx campanulate, 1–1.5 cm long, its rim truncate and minutely undulate. Corolla bright yellow with orange lines in the throat, bilaterally symmetrical, tube funnelform, 5–8 cm long, throat oblique, limb 1.5–2 cm across, with 5 somewhat unequal, broad, spreading lobes. Fertile stamens 4, 2 pairs of unequal length, about half the length of the corolla tube, and 1 short, linear staminodium. Pistil 1, ovary superior, style slender, reaching the throat of the corolla. Ovary maturing into a linear, flat capsule to 50 cm long, the calyx persisting at its base. Seeds linear-oblong, membranous-winged at the extremities, 2–3 cm long. [*Doxantha unguis-cati* (L.) Rehd.]

Native to Trop. Amer., cultivated as an ornamental, in our area occasionally naturalizing, chiefly near human habitations.

Calycanthaceae (STRAWBERRY-SHRUB FAMILY)

Calycanthus floridus L. STRAWBERRY-SHRUB, SWEET-SHRUB, CAROLINA ALLSPICE.
Fig. 75

Deciduous shrub, stems to about 3 m tall, extending by root sprouts and somewhat colonial, loosely few-branched, spicy-aromatic when crushed or bruised. Twigs dark reddish brown, at first quadrangular in cross-section, soon becoming terete, with scattered paler lenticels; leaf scars somewhat raised, more or less horseshoe-shaped, each with 3–5 vascular bundle scars.

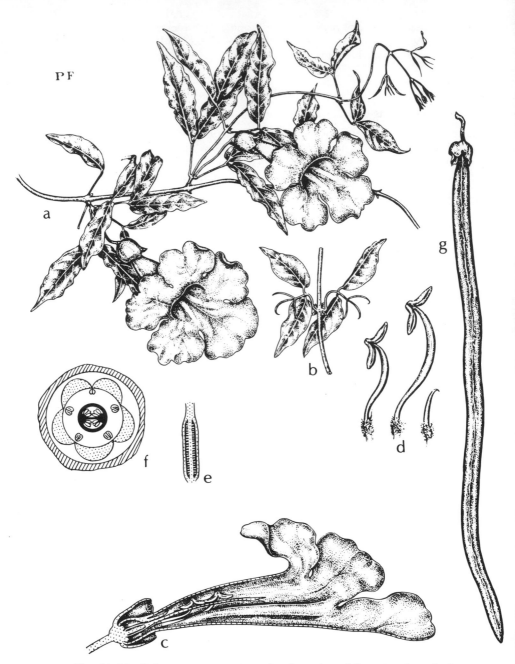

P F

Fig. 74. **Macfadyena unguis-cati:** a, distal portion of flowering branch; b, node to show leaflets and tendrils; c, flower longitudinally dissected; d, two stamens and staminode; e, ovary longitudinally dissected; f, floral diagram; g, fruit. (Courtesy of Fairchild Tropical Garden.)

Fig. 75. **Calycanthus floridus:** flowering/fruiting branch, center; flower, lower
left; longitudinal section of flower, lower right; "fruit" at upper right.

Leaves opposite, short petiolate, the blades elliptic, oval, or ovate, rounded to
broadly cuneate basally, acute to acuminate apically, margins entire, dark green
and lustrous above, somewhat rugose and commonly glaucous or short-pubescent beneath, 5–15 cm long.

Flowers short-stalked, borne singly and terminally on short branches of new
growth, bisexual, radially symmetrical, about 3 cm long, fragrant. Floral tube
cup-shaped, bearing bracts on its outer surface toward the apex, numerous
strap-shaped, somewhat fleshy, greenish brown or brownish purple perianth
parts (tepals) on its much thickened rim. Stamens numerous on the inner rim of
the floral tube, with stout filaments; inner stamens reduced to staminodia.
Pistils distinct, numerous, inserted on the base or sides of the interior of the cup;
styles filiform, protruding. Fruits 1-seeded nutlets enclosed by the fleshy, indehiscent floral tube ("fruit").

Rich woodlands of ravine slopes and bottoms, banks of rivers and small woodland streams; s.w. Ga., Fla. Panhandle from about the Ochlockonee River westward, s. Ala. (S.e. Pa., w. Va., s. W.Va., s.e. Ky., generally southward to Fla. Panhandle and s.e. Miss.)

Two varieties of *C. floridus* are sometimes distinguished: var. *floridus* having
the lower surfaces, twigs, and petioles pubescent; var. *oblongifolius* (Nutt.)
Boufford & Spongberg having lower leaf surfaces glabrous or with few scattered
hairs, twigs and petioles glabrous to slightly pubescent.

179

Caprifoliaceae (HONEYSUCKLE FAMILY)

1. Plant a twining or trailing vine.   1. *Lonicera*
1. Plant an erect shrub or small tree.
  2. Leaves pinnately compound.   2. *Sambucus*
  2. Leaves simple.
    3. Inflorescences short axillary spikes; corolla tubular-campanulate.
         3. *Symphoricarpos*
    3. Inflorescences flat or round-topped, compound cymes terminating branches; corolla barely united basally, lobes rotate.   4. *Viburnum*

1. Lonicera (HONEYSUCKLES, WOODBINES)

Twining or trailing woody vines (ours). Leaves simple, opposite, short-petiolate, without stipules, blades pinnately veined. Inflorescence of axillary or terminal 2–several-flowered, bracteate clusters, the terminal clusters sometimes aggregated in spikes. Flowers bisexual, ovary inferior. Floral tube surmounted by 5 small calyx segments surrounding only the very base of the corolla tube. Corolla longish-tubular below, 5-lobed above, the limb radially symmetrical or strongly bilateral and 2-lipped with the upper lip 4-lobed, lower 1-lobed, a nectariferous area of sessile glands at the base within. Stamens 5, inserted near the summit within the corolla tube, at least a little exserted. Pistil 1, inferior ovary 2–3-locular, each locule with 1–3 ovules, style slender, about equaling the stamens, stigma capitate. Fruit a few-seeded berry.

1. Young stems pubescent; petioles pubescent, leaf blades usually pubescent at least beneath; flowers solitary or more commonly in pairs terminating short axillary shoots; corolla strongly bilaterally symmetrical, generally whitish or cream-colored, becoming yellow with age, or on some plants pinkish or purplish.   1. *L. japonica*
1. Young stems and leaves glabrous, leaf blades glaucous beneath; flowers in terminal clusters or interrupted spikes; corolla long-tubular below and with very short, approximately equal lobes, tube red without, yellow within.   2. *L. sempervirens*

1. Lonicera japonica Thunb. JAPANESE HONEYSUCKLE.   Fig. 76

Twining high-climbing or trailing woody vine. Young stems pubescent. Leaves evergreen, blades ovate, elliptic or oblong, bases rounded, apices obtuse or rounded and mucronate, mostly 4–8 cm long, petioles pubescent, blades usually at least sparsely pubescent along the midrib on both surfaces, the margins usually pubescent, entire save on vigorous new spring shoots, blades of which are often pinnately lobed. Stalks bearing a single flower, more often a pair of flowers, 2–10 mm long, subtended by a pair of bracteal leaves with a pair of minute bractlets in their axils. Flowers very fragrant. Calyx segments minute, narrowly triangular-acute. Corolla 3–5 cm long, whitish or cream-colored, becoming yellow with age (or on occasional individual plants pinkish or purple-tinged), tube narrowly funnelform, abruptly expanded into a 2-lipped limb, the upper lip with 4 lobes, lower with a single lobe. Stamens and style much exserted from the throat of the corolla. Berry black, globose or nearly so, 5–6 mm long. Seeds lustrous black, finely reticulate, 3 mm long or a little more. [*Nintooa japonica* (Thunb.) Sweet]

    Native of eastern Asia, widely naturalized, often perniciously weedy in woodlands, fields, thickets, roadsides, commonly overwhelming and eradicating native flora and difficult to control; throughout our area. (S. N.Eng. to Mo. and Kan., generally southward to Fla. and Tex.)

Fig. 76. **Lonicera japonica:** a, flowering branch; b, flower; c, fruiting branch; d, fruit; e, branchlet, spring growth, with some lobed leaves.

The Japanese honeysuckle is reported to be excellent deer-browse and is sometimes planted for such in our area.

2. Lonicera sempervirens L. CORAL HONEYSUCKLE, TRUMPET HONEYSUCKLE.

Fig. 77

Twining or trailing, glabrous, woody vine, usually not diffusely branched and usually not exceeding about 5 m long. Leaves tardily deciduous, mostly short-petiolate, blades 3–7 cm long, variable in shape, oblong to elliptic, obovate, or suborbicular; uppermost 1 or 2 pairs usually connate-perfoliate, rarely not so; apices obtuse to rounded, bases acute to rounded, margins entire, glaucous beneath.

Flowers terminal on new growth, in interrupted spikes, 2–4 at a node, each cluster subtended by a pair of small bracts and each flower by a pair of minute bractlets. Calyx segments minute. Corolla long-tubular, tube pubescent within, 4–5.5 cm long, with 5 short, blunt-tipped, approximately equal, erect or spreading lobes, red without and often yellow within. Stamens and style barely exserted from the throat of the corolla. Berries red. Seeds lustrous golden-brown, alveolate-reticulate, curvate-oblong, 5 mm long and 3 mm broad. [*Phenianthus sempervirens* (L.) Raf.]

For the most part in upland, well-drained places, open woodlands and their borders, thickets, fence rows, sometimes in moist to wet woodlands but not where water stands for more than short intervals; throughout our area, relatively local. (N.Eng. to Iowa, Neb., southward to Fla. and Tex.)

2. Sambucus (ELDERS, ELDER-BERRIES)

Sambucus canadensis L. ELDER-BERRY, COMMON ELDER.

Fig. 78

Soft-stemmed shrub, to about 4 m tall, with a large, white pith, the herbage with a rather rank odor when crushed or bruised, producing elongate subterranean runners and colonial. Bark grayish brown, with prominent lenticels.

Leaves opposite, 1-odd-pinnately compound *or* (in our range) the lower leaflets, 1 or 2 pairs, irregularly strongly lobed, subdivided, or divided and thus often partially bipinnate; petioles 3–10 cm long, without stipules but sometimes with a line of short bristlelike hairs between the opposite, usually pubescent petioles; leaflets mostly 5–11, lanceolate, elliptic, or ovate, variable in size (those on vigorous shoots much larger than those on branchlets) 5–15 (18) cm long, 2–6 (8) cm broad, serrate; bases rounded to acute, apices acuminate to caudate-acuminate; upper surfaces dark green, short-pubescent on the midrib and the principal veins at least proximally, sometimes sparsely short-pubescent between the veins, lower surfaces paler green, glabrous, pubescent along the midrib and major veins, sometimes downy-pubescent throughout; leaflets sessile or with stalks to about 1 cm long, each leaflet subtended by a subulate stipel.

Inflorescences commonly relatively large (although varying a great deal in size up to 4 dm across) flat-topped to broadly rounded, decompound cymes terminal on branches; since growth may occur over much of the season (here even sometimes in winter) flowering may extend across much of the season although there is a peak of bloom in mid-summer; axes of the cyme, especially distal divisions, mealy and with some hairs. Flowers small, white, somewhat fragrant, 3–5 mm across the corollas. Floral tube surmounted by 5 short-triangular, erect

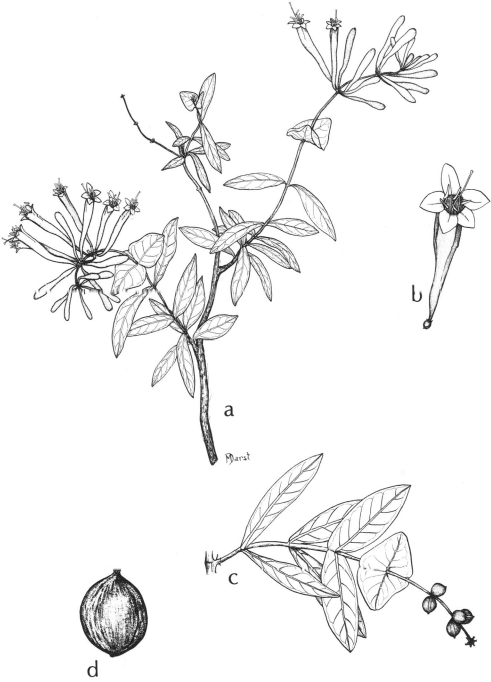

Fig. 77. **Lonicera sempervirens:** a, flowering branch; b, flower; c, fruiting branch; d, fruit.

Fig. 78. **Sambucus canadensis:** a, flowering branch and, in the background, a
bipinnate leaf such as occurs on some individual plants; b, flower; c, fruit.

calyx segments. Corolla with 5 spreading, rounded-ovate, short white petals. Stamens 5. Pistil 1, ovary inferior, 3–5-locular, stigma sessile, 3–5-lobed. Fruit 4–6 mm long, a juicy, purplish black, berrylike drupe containing 3–5 stone-covered seeds.

Commonly in moist to wet open places, borders of swamps and wet wood-lands, ditches, banks of canals and bayous, abundantly colonizing wet clearings and disturbed sites; throughout our area. (N.S. and Que. to Man. and S.D., gen-erally southward to s. Fla. and Tex.; W.I., Mex.)

3. Symphoricarpos

Symphoricarpos orbiculatus Moench. CORALBERRY, INDIAN-CURRANT.   Fig. 79

Deciduous, slenderly branched shrub 5–10 (20) dm tall; young twigs brown to purplish and copiously clothed with short, grayish pubescence; older woody stems with light brown exfoliating bark.

Leaves opposite, simple, very short-petiolate, without stipules; blades pin-nately veined, ovate, oval, broadly elliptic, infrequently suborbicular, 1.5–5 cm long and 1–3 cm broad, bases rounded to broadly short-tapered, apices rounded or obtuse, sometimes acute on smaller ones; upper surfaces dull green and glabrous or very nearly so, the lower paler, somewhat glaucous, short-shaggy-pubescent, margins entire (blades said to be lobed on vigorous shoots). Lateral branchlets generally have the smallest blades at their bases, the size gradually

Fig. 79. **Symphoricarpos orbiculatus:** a, flowering branch; b, flower; c, co-rolla, spread out; d, fruit.

increasing to maximal more or less midway of the branchlet, about equal in size distally from there.

Inflorescences inconspicuous, compacted, short, bracteate, axillary spikes on branchlets of the season, the bracts short-ovate, exteriorly copiously short-pubescent, their margins short-ciliate. Flowers small, bisexual, radially symmetrical, ovary inferior. Floral tube urceolate, the body ellipsoid, 5 deltoid sepals crowning the neck, both the neck and sepals pubescent. Corolla pinkish, 2–3 mm long, campanulate-tubular below, with 5 short ovate-oblongish, erect lobes with rounded tips, villous within the tube. Stamens 5, inserted on the upper margin of the corolla tube, anthers barely exserted. Pistil 1, ovary adnate to the floral tube, 4-locular, an ovule in each locule, but those in two of the locules small and abortive; style short, hairy, included. Fruit an ellipsoid or subglobose, often oblique, 2-stoned drupe with the neck of the floral tube and calyx persisting at its summit, the fruit body 4–5 mm long and about as broad, coral-red or pink, sometimes purple-tinged. [*S. symphoricarpos* (L.) MacM.]

In our area restricted to wooded limestone bluffs of the Marianna Lowlands, Florida Panhandle. (N.Y. to n. Ga. and n. Ala., disjunct in Fla., Jackson and Levy Cos., westward to Tex., Colo.; n. Mex.)

4. Viburnum (viburnums, arrow-woods, haws)

Deciduous (ours) shrubs or small trees. Winter buds with 1 or 2 pairs of outer scales. Leaf scars more or less V-shaped or 3-lobed, narrowly so in some, each with 3 vascular bundle scars. Pith continuous.

Leaves opposite, simple, petiolate, with or without stipules; blades pinnately or palmately veined. Inflorescences stalked or sessile, round-topped cymes terminal on branchlets of the season. Flowers small, radially symmetrical, bisexual (in ours), ovary inferior. Floral tube surmounted by 5 small, persistent, sometimes rudimentary, calyx segments. Corolla white or off-white (often pink-tinged in the swollen bud stage), inserted on the rim of the floral tube, united only basally, with 5 spreading lobes. Stamens 5, inserted at the base of the corolla. Pistil 1, inferior ovary 3-locular, 2 locules abortive, the third 1-ovulate; style short, conical, stigma 3-lobed. Fruit a 1-seeded drupe, the shriveled calyx and the style persistent at its summit.

1. Leaf blades (sometimes all of them on a given plant, usually most of them) palmately 3-veined and lobed.   1. *V. acerifolium*
1. Leaf blades pinnately veined.
   2. The leaf blades with relatively prominent lateral veins angling-ascending parallel to each other and they or their ultimate branches ending in marginal teeth. 2. *V. dentatum*
   2. The leaf blades with relatively inconspicuous lateral veins arching-ascending, branching into veinlets before reaching the margins, veinlets anastomosing.
      3. Upper surfaces of leaf blades glistening-green, the lower with irregular patches of loose, kinky, rusty pubescence mainly along the midrib or both along it and the major veins, not punctate-dotted.   5. *V. rufidulum*
      3. Upper surfaces of leaf blades dull green, lower surfaces, if pubescent, pubescence unlike that described above, punctate-dotted.
         4. Leaf blades mostly oblanceolate or spatulate, sometimes short- or nearly round-obovate.   4. *V. obovatum*
         4. Leaf blades elliptic, lance-elliptic, oval or a few broadly obovate.
                                                                                        3. *V. nudum*

1. Viburnum acerifolium L. ARROW-WOOD, MAPLE-LEAVED VIBURNUM.   Fig. 80

Slender shrub to about 2 m tall, often perennating by subterranean runners and clonal. Young stems copiously pubescent with long, simple hairs and short-stellate ones, the latter much the more abundant, both to a degree sloughing as the stem ages; older woody stems brown, smooth.

Leaves short-petiolate, often (apparently not always) stipulate, stipules soft, acicular, 2–3 mm long, inserted on the petiole base and usually partially adnate to it. Blades mostly palmately 3-veined and -lobed, some sometimes, rarely all, essentially pinnately veined and unlobed on a given plant, membranous, ovate, commonly as broad as long, varying in size, in our range relatively small, larger ones to about 8 cm long and broad, marginal teeth irregular and blunt or margins undulate, lobes blunt to short-acuminate, bases subcordate to truncate, upper surfaces bright green and with few, usually straight hairs, lower pale green, pubescence usually mixed long, simple hairs and short, stellate ones, the latter generally the more abundant, sometimes hairs notably tufted in many of the vein axils, sometimes the long hairs dense along veins and upwardly appressed.

Cymes 2–6 cm across, slenderly stalked, stalks 2–4 cm long. Drupe purplish black, ellipsoid, about 8 mm long. [Incl. *V. densiflorum* Chapm.]

Mesic, mixed woodlands of bluff and ravine slopes, coastal sand pine–sand live oak woodlands; in our area, s.w. Ga., cen. and w. Panhandle of Fla., s.e. Ala. (Que., N.B., to Minn., generally southwesterly in the east to cen. Panhandle of Fla. and in the w. to e. Tex.)

2. Viburnum dentatum L. ARROW-WOOD.   Fig. 81

A polymorphic species complex variously interpreted by authors as composed of several varieties or several species, some of them with varieties. Plants occurring in our range belong to var. *scabrellum* T. & G. (*V. scabrellum* (T. & G.) Chapm.; incl. *V. semitomentosum* (Michx.) Rehd.)

Shrub to about 3 m tall, or infrequently with a single main stem and of low tree stature. Young stems, petioles and cyme axes copiously shaggy-pubescent with stellate hairs, the hairs gradually sloughing leaving their wartlike bases intact, this feature remaining on year-old stems.

Leaves with petioles 1–3 cm long. Blades mostly ovate, some ovate-elliptic or lance-ovate or obovate, variable in size, 3–12 cm long and 2.5–8 cm broad, bases subcordate, rounded, or truncate, apices mostly obtuse; margins dentate-serrate, teeth sometimes sharp, mostly blunt, lateral veins angling-ascending approximately parallel with each other and they or their ultimate branches ending in the marginal teeth

Cymes stalked or sessile, mostly with 6–8 principal branches, 4–10 cm across. Drupes ellipsoid to globose, sometimes obovate, 5–8 mm long, blue-black.

In both well-drained and poorly drained places, river and stream banks and associated woodlands, shrub-tree bogs and bays, pine flatwoods, upland woodlands of various mixtures; throughout our area. (S. cen. and s.w. Ga. to w. cen. pen. Fla., westward to e. Tex.)

3. Viburnum nudum L. POSSUM-HAW.   Fig. 82

Shrub or small tree to about 5 m tall. Young stems, cyme axes, petioles, and buds sparsely to densely clothed with small, rusty-scurfy scales, the buds notably so, the pubescence gradually sloughing leaving the surfaces finely warty.

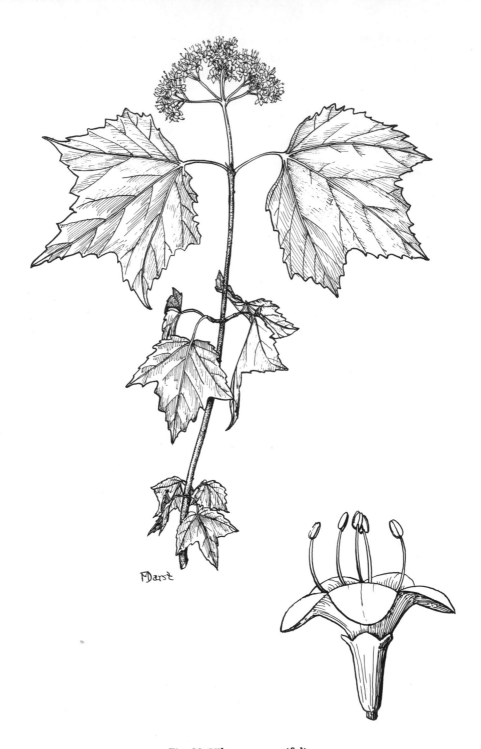

Fig. 80. **Viburnum acerifolium.**

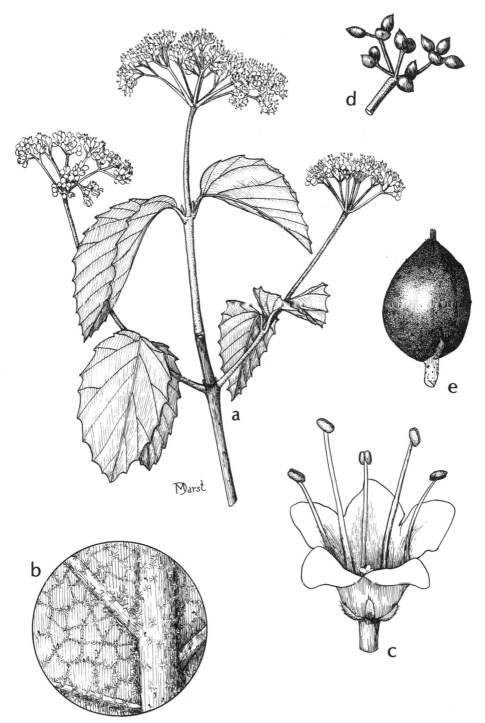

Fig. 81. **Viburnum dentatum:** a, flowering branch; b, enlargement of small portion of lower surface of leaf; c, flower; d, infructescence; e, fruit.

Fig. 82. **Viburnum nudum:** a, flowering branch; b, flowers; c, fruiting branchlet (showing enlargement of leaves subsequent to flowering). (From Correll
and Correll, 1972. *Aquatic and Wetland Plants of Southwestern United States*.)

Leaves with narrowly membranous-winged petioles 5–20 mm long. Blades varying in size and shape, to 15 cm long, narrowly elliptic, elliptic-oblong, lanceolate, lance-ovate, infrequently obovate, bases acute to rounded, apices abruptly short-acuminate, acute, obtuse, or infrequently rounded; margins mostly entire and somewhat revolute, varying to finely and irregularly crenate-undulate, or finely serrate; upper surfaces sometimes sparsely and finely glandular-dotted, lower surfaces relatively copiously so, sometimes, especially when young, scurfy-scaly, mainly along the midribs.

Cymes stalked, stalks 5–25 mm long at anthesis (to 40–50 mm long on the fruiting cymes), mostly with 4–5 primary branches, to about 15 cm across, flowering after shoots of the season are fully or nearly fully expanded. Drupes mostly ellipsoid or ovoid, 6–10 mm long, during maturation their surfaces varying from yellowish, whitish, or pink, then deep blue, glaucous.

Swamps, shrub-tree bogs and bays, wet woodlands and thickets; throughout our area. (Nfld. to Man., generally southward to cen. pen. Fla. and e. Tex.)

4. Viburnum obovatum Walt. SMALL VIBURNUM. Fig. 83

Shrub or small tree, commonly the latter, the branches stiff and often with short, stubby, angling-ascending branchlets. Young stems, petioles, and cyme axes more or less, sometimes abundantly, clothed with rusty-scurfy scales; stems in the first year, usually through the second, narrowly 2-winged.

Leaves sessile or with narrowly winged petioles to about 6 mm long. Blades chiefly oblanceolate to spatulate, occasional ones short- or nearly round-obovate, 2–5 cm long and 1–3 cm broad, apically obtuse or rounded; margins somewhat revolute, entire or finely and irregularly serrate or crenate-dentate upwardly from about their middles, upper surfaces dull green and glabrous, the lower with many small brown glandular dots.

Cymes sessile, with 2–5 primary branches, 4–6 cm across, flowering during emergence of new shoots in early spring. Drupes ellipsoid to spherical, sometimes ovate, 6–10 mm long, passing from red to black during maturation, not glaucous.

Stream banks, wet hammocks, floodplain woodlands, wet pine flatwoods; throughout our area. (Coastal plain, S.C. and Ga. to s. pen. Fla., Panhandle of Fla. westward to about Washington Co., s.e. Ala.)

5. Viburnum rufidulum Raf. SOUTHERN OR RUSTY BLACK-HAW. Fig. 84

Shrub or small understory tree, bark of trunk checkered-blocky. Pubescence of parts stellate, the branches kinky. Buds very densely and tightly clothed with rusty-brown pubescence. Young stems and petioles, especially on the short-shoots having cymes, often densely rusty-pubescent, varying to irregularly patchy-pubescent. Year-old twigs grayish, with few, scattered, raised, corky, brown lenticels.

Petioles of leaves below the cymes about 8 mm long, relatively broadly winged, densely rusty-pubescent abaxially, purple and essentially glabrous adaxially, their blades usually nearly as broad as long, short-oval, short and broadly elliptic, obovate, or nearly rotund, rounded at their extremities, sometimes notched apically; other leaves with more narrowly winged petioles, their blades oval, elliptic, to elliptic obovate, some oblanceolate, their extremities generally shortly tapered but varying to obtuse; all blades having upper surfaces glistening dark green, lower paler and dull, not brown-dotted, with varying amounts of patchy, rusty pubescence, especially along the major veins,

Fig. 83. **Viburnum obovatum:** a, flowering branch, at right, fruiting branch, to the left; b, leaf, enlarged; c, flower; d, fruit.

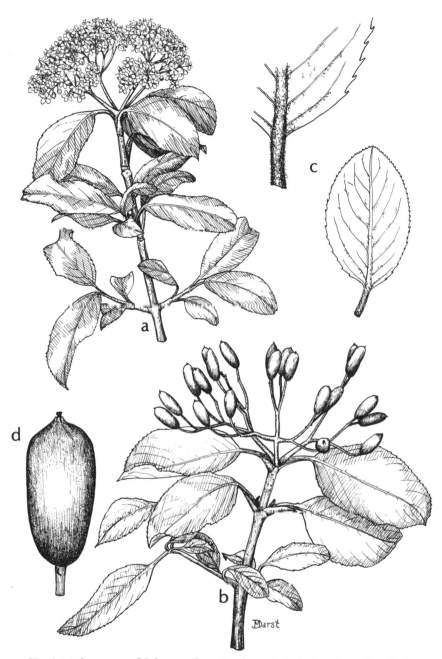

Fig. 84. **Viburnum rufidulum:** a, flowering branch; b, fruiting branch; c, leaf, at lower right and enlargement of abaxial base of leaf showing pubescence; d, fruit.

much of it often eventually sloughing; margins very finely but sharply serrate.

Cymes very attractive, with 2–5, commonly 4, primary branches, 5–10 cm across, at anthesis when leaves (of flowering short-shoots) are about fully developed. Drupes ellipsoid to subglobose, 1–1.5 cm long, dark blue to purple, glaucous.

Upland, well-drained woodlands of various mixtures, occasionally in fence and hedge rows; throughout our area. (Va. to s. Ohio, s. Ind., s. Ill., Mo., generally southward (except in mts.) to cen. Fla., westward to cen. Tex.)

The lustrous foliage and handsome flowering cymes would seem to make it a suitable subject for use as an ornamental.

National champion southern or rusty black-haw (1961), 3'11" circumf., 25' in height, 30' spread, Hempstead Co., Ark.

Celastraceae (STAFF-TREE FAMILY)

Euonymus

Openly branched, (ours) deciduous or tardily deciduous shrubs, rarely with the stature and form of small trees. Twigs green at least the first year; pith white, continuous.

Leaves simple, opposite (some sometimes subopposite), petiolate, margins of blades with minute gland-tipped teeth; stipules minute, quickly deciduous.

Flowers bisexual, borne singly on slender stalks from leaf axils or internodes, or in 2-several-flowered, simple or compound, slenderly stalked, open cymes from leaf axils or internodes on branchlets of the season after the leaves are fully grown. Ovary partly inferior, being surrounded by a broad, flat, nectariferous disc, the tip of the ovary protruding above it, a flat floral tube subtending the disc, petals 4 in one of ours, 5 in the other, attached at the rim of the floral tube. Stamens of the same number as the petals; sepals 4 or 5, clearly or poorly differentiated from the floral tube. Ovary 3–5-locular, ovules usually 2–6 per locule. Fruit a somewhat leathery, purplish or red capsule, its surface smooth or muricate-warty.

1. Stems green throughout; leaves subsessile, petioles no longer than 2–3 mm, mostly shorter; petals 5, yellowish green; surface of fruit muricate-warty, red at maturity.
1. E. americanus

1. Stems green through the first year, brown thereafter; leaves distinctly petiolate, petioles 10–15 (20) mm long; petals 4, maroon, or greenish and partially suffused with maroon; surface of fruit smooth, pinkish purple or purple at maturity. 2. *E. atropurpureus*

1. Euonymus americanus L. STRAWBERRY-BUSH, HEARTS-BURSTING-OPEN-WITH-LOVE. Fig. 85

Openly branched, tardily deciduous shrub to about 2 m tall, branches sometimes divaricate, varying to strongly ascending, sometimes with a notably straggly habit. Stems green throughout, minutely pustulose.

Leaves subsessile or very short-petiolate, petioles 2–3 mm long or less. Blades varying a great deal in shape and size, narrowly lanceolate, lance-ovate, elliptic, broadly ovate, subrotund, or obovate, bases tapered to rounded, apices acute or acuminate, or merely short-cuspidate on the roundish blades; usually sparsely short-pubescent on the midribs of the upper surfaces, sometimes on the lower, margins minutely crenate-serrate, small glands tipping the teeth.

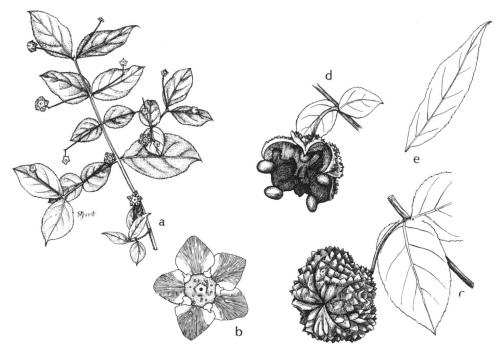

Fig. 85. **Euonymus americanus:** a, flowering branch; b, flower, face view; c, node with fruit; d, node with dehisced fruit; e, narrow leaf characteristic of those of many individual plants.

Flowers yellowish green. Rim of the floral tube slightly undulating, without differentiated sepals, or sometimes the undulations more pronounced and with hyaline edges more suggestive of sepals, the number being 5. Petals 5, borne on the rim of the floral tube, shortly clawed, their blades ovate or suborbicular, corolla 10–12 mm across the spreading petals. Stamens 5, inserted on the rim of the disc, the filaments short, often curved inward so that the anthers lie on the upper edge of the disc. Summit of the ovary (above the disc) flattish, with 5 short, radiating lobes, the stigma appearing flat and nearly, if not wholly, sessile centrally within the lobes. Fruit red at maturity, somewhat oblate, 1–2 cm in diameter, with 3 5 vertical rounded lobes, its surface muricate-warty, splitting when ripe, its segments reflexing, the scarlet seeds then exposed at which time the whole is very ornamental. The first of the common names given above assumably derives from the unopened mature capsule, the second from the opened one.

Chiefly in mesic woodlands, also in floodplains of small woodland streams and in parts of river floodplains subject to temporary flooding; throughout our area. (S. N.Y. to s. Ill. and s.e. Mo., generally southward to pen. Fla. and e. Tex.)

2. **Euonymus atropurpureus** Jacq. WAHOO, BURNING-BUSH.   Fig. 86

Erect, deciduous shrub or small tree, to 6 or 8 m tall. Stems green the first year, grayish brown to brown thereafter, and with vertically elongated, buff-colored lenticels.

195

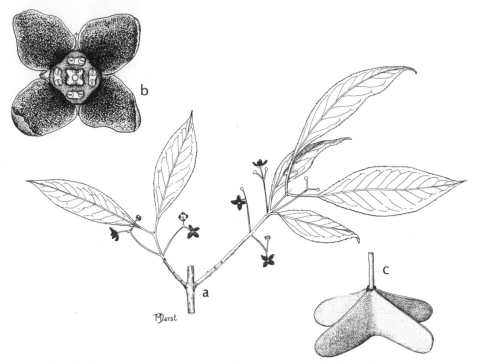

Fig. 86. **Euonymus atropurpureus:** a, flowering branch; b, flower, face view; c, fruit.

Leaves distinctly petiolate, petioles 10–15 (20) cm long. Blades long-elliptic, broadly elliptic, or oblong-elliptic, mostly 6–12 cm long and 2–5 cm broad, bases subrounded to cuneate, apices notably cuspidate, acuminate, less frequently acute, lower leaf surfaces usually sparsely short-pubescent.

Flowers mostly in very slenderly stalked, simple or compound cymes, rarely solitary. Floral tube, sepals, petals, and disc varying from wholly maroon to green or yellowish green and more or less suffused with maroon. Floral tube very shortly funnellike, sepals 4, reniform, spreading beneath and hidden by the corolla, cupped, much shorter than the petals. Petals 4, ovate, about 4 mm long, usually radially spreading from the rim of the floral tube beneath the edge of the disc. Corolla 10 mm across the spreading petals. Disc squarish in outline, with 4 rounded lobes, a very short stamen perched centrally atop each lobe, the stamen very short, fleshy, colorless and somewhat knoblike, the anther sacs within its summit and opening terminally; stamens detaching extremely readily leaving a nipplelike protuberance atop each lobe of the disc. Stigma rather indistinct at the moundlike summit of the ovary (the portion above the disc). Capsule (unless some locules abort) deeply and equally 4-lobed, its surface smooth, pinkish purple or purple, splitting when ripe but not exposing the seeds.

Stream banks, wooded bluffs and ravine slopes, thickets; s.w. Ga., adjacent Panhandle of Fla., s.e. Ala. (Ont. to Mont., generally southward to Fla. Panhandle and cen. Tex.)

196

Chrysobalanaceae (COCO-PLUM FAMILY)

Licania michauxii Prance. GOPHER-APPLE, GROUND-OAK.   Fig. 87

Low shrub, extensively colonial by subterranean, stoutish stems from which arise slender, glabrous, aerial shoots 1–4 dm tall.

Leaves simple, alternate, essentially sessile, stipulate, stipules short-subulate, blades pinnately and notably reticulately veined, 2–10 cm long and 1–3.5 cm broad, lower and upper leaves often much smaller than the median ones, firm and stiffish in texture, mostly oblong-oblanceolate or oblong-spatulate, bases cuneate, apices mostly blunt, rounded or obtuse, rounded ones occasionally emarginate, less frequently some of them acute, upper surfaces glabrous and lustrous, lower glabrous or felty-pubescent, margins irregularly, shallowly undulate to entire, cartilaginous.

Inflorescence a terminal panicle of cymes, often more or less triangular in overall outline, panicles with relatively few to numerous flowers; larger panicles to about 10 cm long and 5–7 cm broad at the base. Flowers small, bisexual, radially symmetrical or nearly so, ovary free within a floral tube. Inflorescence axes sparsely short-pubescent proximally, gradually becoming more densely so and more grayish distally, the short bracts similarly pubescent, flower stalks, floral tube, and exterior of calyx segments densely grayish pubescent, interior of calyx segments pubescent distally. Calyx segments 5, triangular-acute, erect. Petals 5, white, ovate to oblong, spreading radially between and beyond the sepals, pubescent, sometimes unequal. Stamens numerous, the filaments united basally forming a wide tube within the floral tube, free above. Pistil 1, ovary superior, glabrous or pubescent, 1-locular, style slender, arising from the base of the ovary, stigmatic at the tip. Fruit an ellipsoid drupe 2–3 cm long. [*Chrysobalanus oblongifolius* Michx.]

Well-drained sands of pineland ridges and hills, sandy roadside banks bordering pinelands, stabilized coastal dunes; throughout our area. (Coastal plain, s.e. S.C. to s. Fla., westward to La.)

Clethraceae (WHITE-ALDER FAMILY)

Clethra alnifolia L. SWEET PEPPERBUSH.   Fig. 88

Shrub to about 3 m tall, commonly 1–1.5 m. Young branchlets tomentose, woody twigs glabrous, purplish brown; leaf scars obcordate, each with 1 vascular bundle scar.

Leaves deciduous, simple, alternate, short-petiolate, stipules none. Blades pinnately veined, the lateral veins moderately prominent at least beneath, curvate-ascending nearly parallel to each other; variable in size to about 8 cm long and 4 cm broad, averaging about one-half that, mostly narrowly obovate to oblanceolate, sometimes elliptic-lanceolate, cuneate basally, apically acute, acuminate, obtuse, or less frequently rounded, margins usually serrate on the distal halves, occasionally entire, glabrous above, lower surfaces varying from glabrous at full maturity to loosely tomentose or densely tomentose. (Pubescence of various parts, some of it at least, stellate.)

Flowers in terminal or terminal and axillary dense racemes on branchlets of

Fig. 87. **Licania michauxii**: a, habit; b, flower; c, infructescence; d, stone of the fruit.

Fig. 88. **Clethra alnifolia:** a, branch with inflorescence; b, flower; c, fruit.

the season; raceme axes, bracts, flower stalks, calyces, and ovaries densely pubescent. Bracts subtending the flowers 2–10 mm long, linear-oblong to linear, acute apically, shorter than to about as long as the flowers, usually deciduous by the time the flowers open or shortly thereafter; flower stalks 2–10 mm long; calyx 3–4 mm high, oblong or campanulate, united below, the 5 lobes triangular-ovate to oblong, apically rounded, obtuse, or acute; petals 5, separate, white, oblong-obovate, glabrous, about 8 mm long; stamens 10, 5 in each of two whorls, filaments glabrous, anthers sagittate, opening by terminal pores; ovary superior, 3-locular, style slender and elongate, stigma capitate or slightly 3-lobed. Fruit a 3-valved, pubescent, subglobose capsule about 3 mm in diameter, calyx persistent and loosely surrounding it. Flowers emitting a pungent fragrance. [Incl. *C. tomentosa* Lam.]

Wet thickets, wet pine savannas and flatwoods, swamps, wet woodlands, bogs; throughout our area. (S. Maine, s. N.H., chiefly coastal plain southward to n. Fla., westward to s.e. Tex.).

Compositae (Asteraceae) (SUNFLOWER FAMILY)

A highly diversified and large family composed of annual, biennial, and perennial herbs, shrubs, and trees. The few woody kinds in our flora are shrubs. Flowers sessile, in heads, borne on a common receptacle, the heads commonly referred to as flowers by laymen. Heads bearing few to very numerous individual flowers surrounded by an involucre of bracts, these few to numerous and varyingly disposed in a single series to much imbricated. Each flower of the head may be subtended by a bract (chaff or pale), the receptacle then said to be chaffy, or flowers without bracts, the receptacle said to be naked, or the receptacle may be bristly or hairy. Flowers epigynous, the corolla surmounting the inferior ovary, calyx none but exterior to the corolla a pappus may or may not be present; if present, commonly (not always) persistent on the fruit. Pappus may be composed of few to numerous, smooth, barbellate, or plumose capillary bristles in one or more series, or of scales, or a combination of the two, or of awns, teeth, or a ring or crown. Corollas of three general types: (1) tubular, 4–5-lobed distally, or truncate at the summit, radially symmetrical; (2) tubular at base above which flat (ligulate or raylike), commonly bent to one side, sometimes toothed apically; and (3) more or less intermediate between the first and second types, bilabiate, the outer side usually larger. Heads may be composed of all tubular, radially symmetrical flowers, usually called disc flowers (heads discoid); or heads with all ligulate flowers (heads ligulate); or heads may have disc flowers centrally, ray flowers peripherally (heads radiate). Stamens, if present, (4) 5, filaments attached well down the corolla tube, free or united above, anthers elongate, united into a tube in most, merely connivent in a few. Pistil solitary, composed of 2 carpels, ovary 1-locular and with 1 ovule, style columnar, stigma bilobed. Fruit an achene.

1. Heads with both ray and disc flowers.
   2. Rays lavender-pink or pink.   1. *Aster*
   2. Rays yellow.
      3. Leaves alternate; receptacle of the head naked.   4. *Chrysoma*
      3. Leaves opposite; receptacle of the head chaffy, the bracts hard and rigid, with spinose tips.   3. *Borrichia*

1. Heads with disc flowers only.
    4. Flowers bisexual; corollas lavender-pink.   5. *Garberia*
    4. Flowers unisexual; corollas white, greenish, or cream-colored.
        5. Staminate and pistillate flowers in separate heads and only one or the other on an individual plant (plants dioecious); flowers not subtended by chaff.   2. *Baccharis*
        5. Staminate and pistillate flowers both in the same head (plants monoecious); flowers, some of them at least, subtended by chaff.   6. *Iva*

1. Aster

Aster carolinianus Walt. CLIMBING ASTER.   Fig. 89

Principal stem and its branches woody, branchlets herbaceous at least distally; stem usually widely and diffusely branched, arching and scrambling over other vegetation to a height of 4 m or more, or forming a tangle of its own; "climbing" considerably effectuated by strongly divaricate lateral branches which serve in anchoring them as growth through other vegetation takes place; older portions of woody stems to at least 1 cm in diameter, buff or tan, unevenly striate-ribbed.

Leaves simple, alternate, pinnately-veined, the larger ones abruptly constricted proximally to a suprabasal, shortly subpetiolar portion, the latter then flared on either side to a small but distinctly clasping auricle; blades elliptic to lanceolate, acute apically, 2–6 cm long and to 1.5, rarely 2, cm broad; margins entire, surfaces sparsely to copiously short-pubescent.

Heads for the most part disposed on bracteate ultimate branchlets from leaf axils, the flowering branchlets numerous and the plant showy when in bloom. Rays of the head pistillate and fertile, 1–1.5 cm long, lavender-pink or lavender; disc flowers yellow, narrowly tubular below, the tube expanding distally to a funnelform or campanulate throat, 6–7 mm long overall, 5 short lobes at the summit. Involucre hemispheric to turbinate-campanulate, 7–12 mm high, bracts pubescent, strongly imbricated, the outer ones pale and linear-oblong below and with elliptic to subrhombic, dilated, often recurved, green tips, inwardly bracts becoming less dilated at their tips, the innermost linear and attentuated at their less conspicuously green tips, often lavender-pink along their scarious margins a little back of the tips.

Achene narrowly columnar, brown, glabrous, 2.5–3 mm long, surmounted by a ring of numerous capillar bristles 6–7 mm long.

Marshy shores, stream banks, marshes, edges of swamps and wet woodlands, often in water; for our area, near-coastal portions of s.e. Ga., n.e. Fla. to e. part of Fla. Panhandle. (Coastal plain, s.e. N.C. to s. pen. Fla., e. Fla. Panhandle)

2. Baccharis

Tardily deciduous shrubs, stems often resinous. Leaves alternate, simple, 1–3-nerved. Heads discoid, relatively small, flowers relatively numerous per head, solitary or more frequently 2-several in clusters borne from axils of leaves or reduced bracteal leaves of branchlets. Involucre of imbricated, pale, papery bracts. Receptacle of the head pitted, minutely fringed, or smooth, flowers without subtending chaff. Flowers unisexual (pistillate) and functionally unisexual (staminate), plants dioecious. Flowers of staminate heads with abortive ovary; corolla filiform below and dilated a little below the 5-lobed summit; pappus of one series of crisped capillary bristles surmounting the abortive ovary and not evident beyond the involucre. Flowers of pistillate heads with corollas only slightly dilated distally from about the middle, truncate at the summit or with

Fig. 89. **Aster carolinianus:** a, flowering branchlet; b, flower-head; c, achene.

minute teeth, the straight, white, numerous pappus bristles conspicuously extending beyond the involucre and the corollas, the collective pappuses of the numerous pistillate heads making the pistillate plants very much more conspicuous at flowering time than are the staminate plants. Achene cylindric or nearly so, 5–10-ribbed.

1. Leaves nearly linear, almost needlelike, 1–3 mm broad.   1. *B. angustifolia*
1. Leaves much broader, the larger ones on a given plant rarely less than 10 mm broad.
    2. Heads mostly in sessile clusters scattered along stiffish branchlets of a branch and not giving the effect of a paniculiform inflorescence; leaf blades without punctate glands or these inconspicuous.   2. *B. glomeruliflora*
    2. Heads mostly distinctly stalked or in stalked clusters so that the overall flowering branch has a distinctly paniculiform appearance; leaf blades with conspicuous, pale amber punctations.   3. *B. halimifolia*

1. Baccharis angustifolia Michx. FALSE-WILLOW.   Fig. 90

Much branched shrub to about 4 m tall. Older twigs tan, splitting and forming rough, corky, irregular fissures. Leafy twigs slender, striate-angled and channeled, green and with copious, minute, pale surface scales.

Leaves sessile, 1-nerved, bright lustrous green, nearly linear and almost needlelike, tapering only slightly to both extremities, 2–4 (6) cm long, mostly with entire margins, but occasional leaves with 1–3 serrations on a side; surfaces not or inconspicuously punctate, eventually with minute scales, drying wrinkled.

Heads usually borne in abundance, sessile and stalked, stalks to about 15 mm long, distributed on the numerous leafy-bracted branchlets of a branch so that the overall branch has a distinctly paniculiform appearance. Involucre campanulate, about 4–5 mm high, bracts closely imbricated, the outer ones very short-ovate, gradually becoming longer inwardly, the innermost linear-oblanceolate, outer surfaces stramineous or with purple tips, the latter more characteristic of pistillate than of staminate heads. Achenes slightly over 1 mm long, with 10 pale ribs, the intervals between ribs brown, pappus 10 mm long.

Coastal marshes, estuarine shores, banks of ditches and sloughs, shrub-tree islands in marshes, coastal hammocks, interdune swales. Near-coastal, N.C. to s. pen. Fla. and around the Gulf Coast of Fla.

2. Baccharis glomeruliflora Pers.   Fig. 91

In general vegetative appearance closely similar to *B. halimifolia* and often mistakenly identified as such. Leaves tending to be brighter green and to remain so during much or all of the season; blades without punctate glands or these inconspicuous. Heads mostly in sessile clusters scattered on stiffish branchlets of a branch and not giving the effect of a paniculiform inflorescence. Involucres 8–10 mm high, the bracts not resinous, the tips of all bracts rounded or blunt. Achene 1.2–1.5 mm long, with 10 pale ridges, the intervals between the ribs light brown; pappus about 8 mm long.

Commonly occurring in partial to full shade of moist to wet hammocks or on their edges, not aggressively weedy, also in dune swales, on banks of marshes and in hammocks surrounded by marshes; more or less throughout n. Fla., rare in s.w. Ga. (Coastal plain, s.e. N.C. to s. pen. Fla., Fla. Panhandle, s.w. Ga.)

3. Baccharis halimifolia L. GROUNDSEL-TREE, SILVERLING, SEA-MYRTLE, CONSUMPTION-WEED.   Fig. 91

Bushy-branched shrub to about 4 m tall. Young stems with a sticky exudate, later becoming minutely pale-scaly.

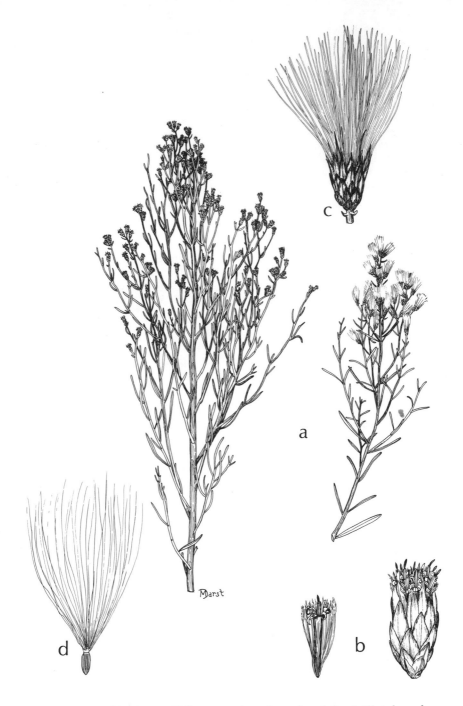

Fig. 90. **Baccharis angustifolia:** a, staminate branch, at left, pistillate branch, at right; b, staminate head, at right, staminate flower, at left; c, fruiting head; d, fruit.

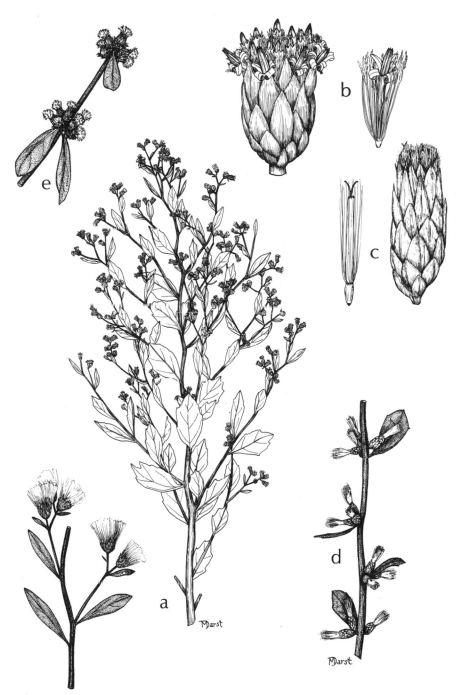

Fig. 91. a–c, **Baccharis halimifolia:** a, flowering branch, staminate plant at right, fruiting branchlet at left; b, staminate head at left, staminate flower at right; c, pistillate head at right, pistillate flower at left; d–e, **Baccharis glomeruliflora:** d, piece of staminate branch; e, piece of pistillate branch, these to show sessile clusters of heads.

Leaves cuneately narrowed basally to short petioles or subpetioles; bladed portions of many larger ones obovate to oblanceolate, sometimes 3-nerved, with 1-several low to salient blunt-tipped teeth upwardly from about their middles, tipped by a similar tooth or tip obtuse to rounded; smaller leaves oblong-oblanceolate to oblanceolate, many with 1–2 teeth on a side distally; both leaf surfaces with relatively large, glandular punctae; in the latter part of the season and in winter the leaves and twigs tend to become dull grayish green giving the plant a decidedly grayish hue as seen from a little distance. (Plants in what appear to be randomly scattered localities of the range may have all leaves narrowly oblanceolate or spatulate with entire or barely toothed margins; these probably represent only sporadic forms.)

Heads mostly stalked or in distinctly stalked clusters on the numerous branchlets of a branch giving the overall branch a decidedly paniculiform appearance. Involucres usually somewhat resinous, 5–6 mm high, campanulate, their tips blunt or commonly the innermost, sometimes the medial, acute. Achene about 1.2 mm long, with 10 pale ribs, the intervals honey-colored, pappus about 10 mm long.

Perhaps formerly restricted to near-coastal areas, marshes, shores, swales, and the like. Now aggressively and noxiously weedy in a wide variety of kinds of disturbed places, even far inland; throughout our area. (Coastal plain and piedmont, Mass. to s. pen. Fla., westward to Tex., Ark.; W.I.)

3. Borrichia

Borrichia frutescens (L.) DC. SEA-OXEYE.   Fig. 92

Subshrubby maritime plant, rhizomatous and forming extensive clones. Stem 1.5–12 dm tall, usually with relatively few strongly ascending branches, these at first clothed with a dense, compact, gray pubescence, eventually becoming glabrous and light brown, finally gray; older stems very irregularly ridged and grooved.

Leaves opposite, simple, subsucculent, gradually tapering basally to petioles or subpetioles whose bases are somewhat dilated and clasping the stem, an acutely triangular projection of stem tissue on either side of the node forming a wedge between the nearly-meeting petiole bases; the wedges commonly becoming free as sharp points after leaf-fall. Leaf blades narrowly to broadly oblanceolate, sometimes some of them elliptic, 2–6 cm long, margins usually entire, some sometimes dentate or spinose-dentate, apices blunt or rounded and mucronate, or shortly spine-tipped, pubescence dense, compact, and gray, obscuring the venation apart from the midrib. Occasional plants (or clones) have glabrous, glistening green leaves.

Heads borne terminally on the branchlets, each flower subtended by a bract (pale), bracts oblong, hard and rigid, with hard, sharp, erect, spinose tips 1–3 mm long. Heads radiate, rays yellow, bisexual and fertile, short tubular at base then abruptly forming a short, oblongish strap; disc flowers bisexual and fertile, brownish yellow, the corolla tube expanding little upwardly, 5-lobed at the summit, lobes erect but slightly curved inwardly. Involucre hemispheric, bracts closely imbricated, the outer ones of different form and color than the inner; outer ones ovate, with texture and pubescence like the leaves; inwardly bracts firmer and harder, brown, much more sparsely pubescent, their tips often spinose.

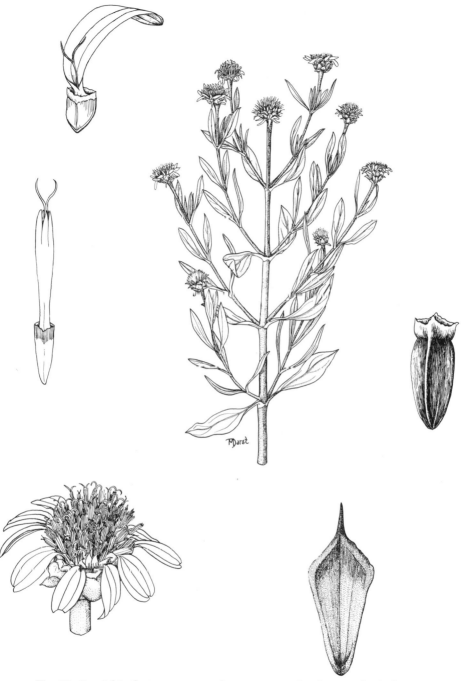

Fig. 92. **Borrichia frutescens:** top of stem, center; head, at anthesis, lower left; ray flower, upper left, disc flower, center left; bract that subtends flower, lower right; achene, center right.

Achene obconic, regularly 3–4-angled, metallic gray, about 3.5 mm long, the pappus a short or low crown having a short tooth above each angle of the achene. In fruit, the heads very "stickery" owing to the hardness, stiffness, and sharpness of the exserted, erect pales, the entire head hard and difficult to disassemble.

Tidal marshes and mud flats, mangrove thickets, coastal and estuarine shores, often weedy in vacant lots or on roadsides near the sea; maritime areas throughout our range. (Va. to Fla. Keys, westward to Tex.; e. Mex., Berm.)

4. Chrysoma

Chrysoma pauciflosculosa (Michx.) Greene. WOODY GOLDENROD.   Fig. 93

Taprooted, glabrous, evergreen, low shrub, eventually with several to numerous, strongly ascending to erect, leafy branches forming a more or less flat-topped crown; younger portions of stems, and to some extent the leaves, clothed with a sticky exudation, older stems grayish brown, not sticky; leafy portion of stems, before flowering season, with numerous, closely approximate leaves, each stem or branch usually producing in autumn a relatively fast-growing shoot with leaves gradually reduced and much less closely disposed, this eventually terminated by a conical or pyramidal inflorescence whose lower, longer branches are leafy proximally, floriferous distally, branches gradually shorter upwardly, the uppermost branches bracteate.

Leaves grayish green, alternate, sessile or their attentuated bases barely subpetiolate; blades firm, narrowly oblong-elliptic, elliptic, or oblanceolate, mostly 2–6 cm long and 0.4–1.2 cm broad, apices blunt, margins entire, midrib evident beneath, scarcely or not at all so above, scarcely a suggestion, if that, of a pair of lateral, strongly ascending nerves from the midrib somewhat above its base beneath; to either side of the midrib beneath and over the entire surface above, surfaces strongly reticular, spaces within the reticulae small, irregularly blocky (the reticulum more evident on dried leaves).

Flowers in numerous involucrate heads more or less paniculately disposed, overall inflorescence size variable, largest one about 2 dm broad basally and to about 3 dm from the origin of lowest branches to the tip. Involucres about 6 mm high, bracts relatively few, at full anthesis of the head bright yellow or the smaller, lower ones with tinges of green, sticky and appearing as though glued together, later becoming stramineous, separating, and more or less disposed in vertical rows, outermost 1–2 mm long, increasing in length inwardly, innermost 5–6 mm long; a few (1–3) irregularly disposed, peripheral flowers of most heads pistillate and fertile, their straplike, yellow rays 4–6 mm long; other flowers 2–5 per head, bisexual and fertile, their corollas radially symmetrical, narrowly funnelform-tubular for about 3 mm from the base, 5 subulate, erect to spreading, yellow lobes 1 mm long at the summit.

Achene pale, narrowly turbinate, pubescent, about 3 mm long, a ring of numerous capillary bristles 4–5 mm long at the summit.

Coastal dunes, sand pine–oak scrub, longleaf pine–scrub oak ridges, Panhandle Fla., Franklin Co., westward, s.w. Ala. (Inner coastal plain, local, s.e. N.C., cen. S.C.; Fla. Panhandle, westward to s. Miss.)

This low shrub of coastal dunes, sand ridges, and hills is distinguished by the grayish green hue of its foliage, its closely approximate, oblong-elliptic, elliptic, or oblanceolate leaves with the only evident venation the midrib beneath, the

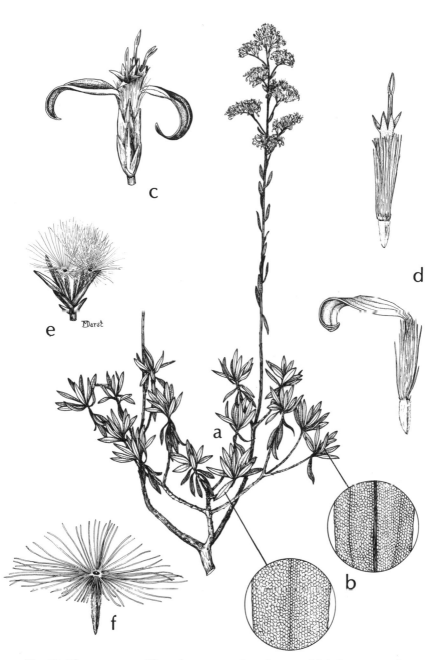

Fig. 93. **Chrysoma pauciflosculosa:** a, portion of stem with inflorescence; b, enlargement of portions of leaf surfaces, upper at left, lower at right; c, flowering head; d, flowers, ray flower below, disc flower above; e, fruiting head; f, fruit.

crown stems producing in autumn relatively fast-growing panicles of numerous, bright yellow heads. On coastal dunes and in thinly wooded pinelands, it sometimes occurs very abundantly so that when flowering, the autumnal near-ground-level aspect is rather like a "sea of gold."

5. Garberia

Garberia heterophylla (Bartr.) Merrill & F. Harper.   Fig. 94

Semievergreen, bushy-branched shrub to 2–2.5 m tall, the main branches with clustered, suberect branchlets, each in autumn terminated by a corymbose cluster of flowering/fruiting heads. Young stems copiously clothed with grayish, finely mealy indument intermixed with amber atomiferous glands and viscid; pubescence of older stems becoming brown and without stickiness, eventually the brown bark finely roughened by low interlacing ridges and grooves; leaf scars pale, obliquely raised, vascular bundle scars 3 in each, indistinct.

Leaves simple, alternate, approximate on the stem, the blades held more or less vertically by a subpetiolar base and overlapping, mostly 1–3 cm long including the subpetiolar base, having a dull grayish green hue; blades mostly obovate, varying to subrotund, pinnate venation indistinct, apices rounded, often slightly retuse, margins entire, surfaces with a mealy indument more scanty than that of the stems and somewhat viscid with scattered to copious, mostly sessile, minute glands, the latter varyingly white, amber, or almost black.

Corymbs with erect branches, lowest longest, progressively shorter upwardly, thus the summit convex to nearly flat. Heads discoid, about 1 cm high, usually 5 flowers in each, receptacle naked. Involucres subcylindric in outline, angular in cross-section because the mealy and glandular, somewhat fleshy bracts tend to be appressed in vertical ranks, commonly 3 in each rank. Corollas lavender-pink, 8–10 mm long, tube narrowly cylindric below, abruptly expanded to a campanulate throat, and with 5 narrow, spreading lobes somewhat longer than the throat. Achene 7–8 mm long, narrowly obconic, 10-ribbed, dark grayish and purplish with pale pubescence; pappus of numerous, barbellate, whitish to purplish, capillary bristles (about equaling the corollas in length) that eventually become tawny, both the pappus and corollas well exserted from the involucre. [*Garberia fruticosa* (Nutt.) A. Gray]

Sand ridges and hills, mostly associated with sand pine–oak scrub, cen. and n.e. pen. Fla.

Distinctive features for this relatively low shrub of the sand pine–scrub oak are its dull grayish, viscid foliage, the alternate, short, mostly erect-overlapping, obovate leaves with indistinct venation and, when in bloom late in the season, its terminal corymbs of lavender-pink heads. Since the flowers of all the heads are at anthesis approximately simultaneously, the plants are then notably attractive suggesting the possibility of use an an ornamental.

6. Iva

Annual or perennial herbs or shrubs. Leaves opposite or opposite below, alternate above. Heads in spicate, spicate-racemose, or paniculate arrangement, each bearing both staminate and pistillate flowers, the former 3–20 and central

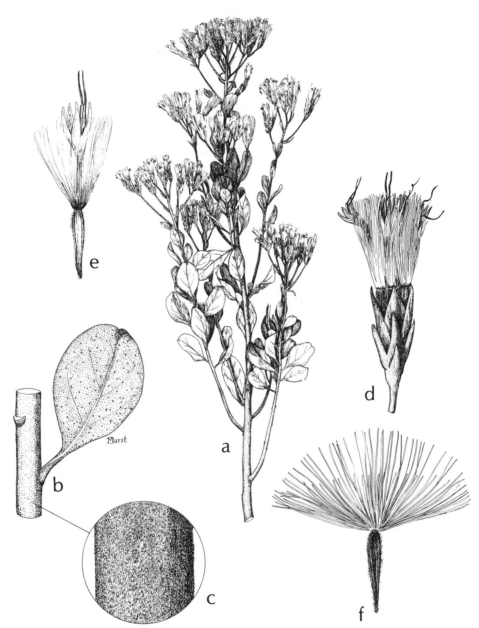

Fig. 94. **Garberia heterophylla:** a, flowering branch; b, enlargement of a small bit of stem and leaf; c, enlargement of a small area of surface of stem; d, head; e, flower; f, fruit.

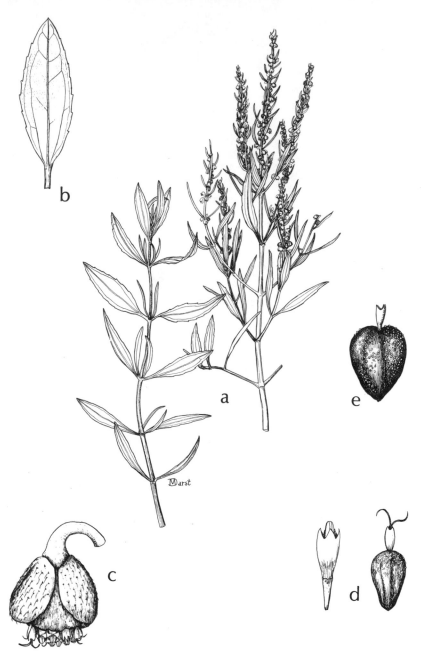

Fig. 95. **Iva frutescens:** a, tips of shoots; b, leaf, enlarged; c, head; d, flowers, staminate to left, pistillate to right; e, fruit.

in the head, the latter 1–9 and peripheral; chaff subtending both staminate and pistillate flowers or not one or the other. Involucres turbinate to hemispheric, bracts 3–9, free or united. Corolla of staminate flowers funnelform, 5-lobed at the summit, that of pistillate flowers tubular and truncate or 2-notched at the summit, absent in some; corolla sometimes persisting at the summit of the achene until it is ripe. Achene cuneate to obovate, somewhat compressed parallel to the involucral bracts, pappus none.

1. Leaves all opposite except the uppermost bracteal ones, not succulent. 1. *I. frutescens*
1. Leaves all alternate except the lowermost, succulent. 2. *I. imbricata*

1. Iva frutescens L. SUMP-WEED, HIGHWATER-SHRUB, MARSH-ELDER. Fig. 95

Maritime plant, usually with numerous branches from near the base, woody below and with more or less herbaceous branchlets, to about 3.5 m tall. Young stems short-pubescent; woody stems tan, shallowly ridged and furrowed; nodes with a line between the opposite leaf scars.

Leaves tardily deciduous, simple, opposite except the upper bracteal ones, with a pair of weak lateral nerves, narrowed proximally to subpetioles somewhat dilated basally and connecting around the stem; blades 4–7 cm long, dull green, lanceolate (smaller ones narrowly so), margins of larger ones serrate from somewhat above the base or often entire distally, bracteal leaves commonly entire, surfaces at first short-pubescent with unbranched hairs and/or hairs irregularly branched from their bases.

Heads short, recurved-stalked, roundish, borne terminally, paniculately and abundantly on leafy, bracteate branchlets. Involucres 2–4 mm high, bracts few, sparsely pubescent, broad and roundish, essentially enclosing the flowers each of which is subtended by a small linear-spatulate bract. Achene about 2 mm long, obovate in outline, somewhat compressed, with 2 angled faces and 1 face broadly rounded, truncate or shallow-notched apically, surfaces dark purplish brown and with pale dots of resin.

Coastal marshes and shores, both Atlantic and Gulf coasts of our area. (N.S. to s. pen. Fla. westward to Tex.)

2. Iva imbricata Walt. DUNE SUMP-WEED. Fig. 96

Glabrous subshrub, commonly bushy-branched from near the base, seldom over 6 dm tall, some branches often decumbent or plants becoming partially buried by shifting sands. In general with features similar to those of *I. frutescens*.

Leaves sessile, subsucculent, only the lowermost opposite; principal leaves broadly spatulate to oblong-spatulate, 2–4 cm long, broadly tapered basally, rounded or obtuse apically, with a pair of weak lateral nerves, margins entire; many or most of the nonbracteal leaves shed by the time the flowering branches are well developed.

Heads numerous, more or less paniculiformly disposed on distal branches, 4–6 mm high at anthesis, increasing in size during maturation of the fruits, becoming as much as 10 mm high, the longer ones roughly oblong. Involucral bracts remaining loose or sometimes completely fused about the achenes, the fruiting heads then knobby and hard. Achenes obovate in outline, plump, 2–4 mm long, reddish brown, minutely resin-dotted, corollas often persistent at their summits.

Coastal dunes, both Atlantic and Gulf coasts of our area. (Va. to s. pen. Fla., westward to La.; Bah.Is. and Cuba)

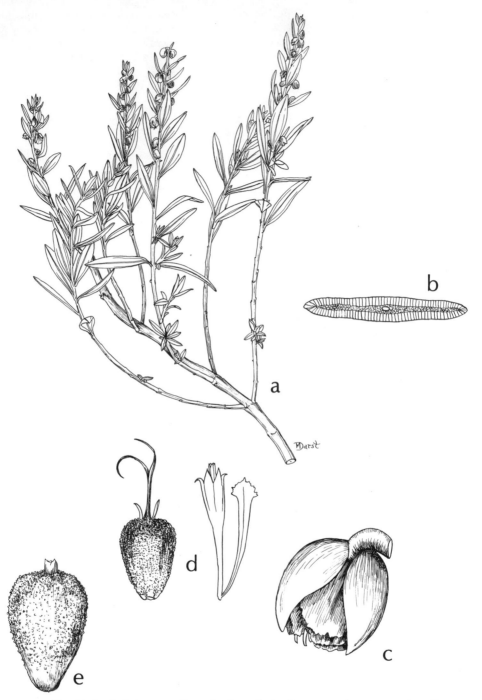

Fig. 96. **Iva imbricata:** a, flowering branch; b, diagrammatic cross-section of leaf; c, head; d, flowers, staminate with subtending pale to right, pistillate to left; e, fruit.

Cornaceae (DOGWOOD FAMILY)

Cornus (DOGWOODS, CORNELS)

Deciduous shrubs or small trees (ours). Leaves opposite (or alternate in one of ours), simple, petiolate, without stipules. Blades pinnately veined, marginally entire (ours). Inflorescence a dichotomously branched, bractless, axillary or terminal cyme or panicle, or (in one of ours) a head surrounded by 4 large, petal-like bracts and an inner ring of small, obtuse, membranous bractlets. Flowers small, bisexual (ours), radially symmetrical, ovary inferior. Floral tube bearing on its rim 4 minute calyx segments. Petals 4, oblong, spreading, inserted marginally on a nectariferous disc. Stamens 4, exserted. Pistil 1, ovary 2-locular, 1 ovule in each locule, 1 ovule usually aborting, style arising from the center of the disc, swollen just beneath the stigma in some species, stigma terminal. Fruit a drupe with a 2- or 1-seeded stone.

Pubescence of parts, when present, may consist of unbranched hairs, hairs 2–forked from the base and with a V-form, 2-forked from the base and with a V-form, a U-form, a Y-form, or attached at the middle and with 2 divaricate, usually appressed, branches. Hairs on a given surface may be mostly one or the other of the kinds but on some parts are commonly intermixed in various proportions. Pubescence, if characteristically present on young stems and on one or the other or both leaf surfaces when young, may eventually be mostly or wholly sloughed.

Plants of the taxa which characteristically become small trees flower and fruit when of small stature.

1. Leaves alternate.   1. *C. alternifolia*
1. Leaves opposite.
  2. Flowers and fruits in open cymes.
    3. Upper surfaces of leaf blades definitely rough to the touch.   2. *C. asperifolia*
    3. Upper surfaces of leaf blades smooth to the touch.
      4. Pith of year-old or older twigs white.   3. *C. foemina*
      4. Pith of year-old or older twigs brown or tawny.   4. *C. amomum*
  2. Flowers and fruits sessile in compact heads, the flowers surrounded by 4 conspicuous, white or cream-white, obovate, petallike bracts.   5. *C. florida*

1. Cornus alternifolia L. f. ALTERNATE-LEAVED OR PAGODA DOGWOOD.   Fig. 97

A small understory tree, the crown somewhat flat-topped, the crown branches more or less spraylike and layered. Bark of trunks relatively smooth, greenish or olive-greenish streaked with brown or gray. Twigs green at first, sparsely if at all pubescent, becoming brown to maroon, axillary buds not evident. Pith white.

Leaves alternate, tending to be closely set near the tips of lateral branchlets, so closely set sometimes that the alternate arrangement is more or less obscured. Petioles slender, 1–6 cm long, glabrous or sparsely pubescent. Blades elliptic, oval, ovate, or infrequently obovate, 5–12 cm long and 2.5–5 cm broad, bases shortly and broadly tapered to rounded, apices short-acuminate, upper surfaces dark green and glabrous, lower grayish green, sparsely appressed-pubescent, sometimes shaggy-pubescent along the major veins as well.

Cymes showy, flat-topped or convex, 3–6 cm across; cyme axes usually sparsely shaggy-pubescent, floral tube densely appressed-pubescent; calyx segments minute; styles not swollen beneath the stigma; corollas cream-colored.

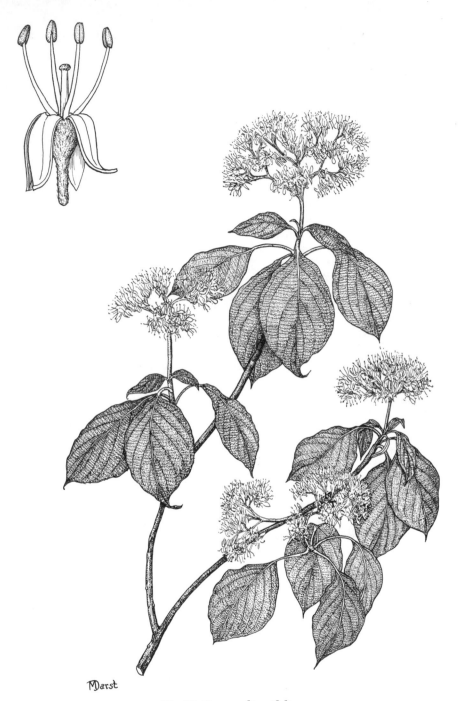

Fig. 97. **Cornus alternifolia.**

Drupes blue, subglobose, 4–7 mm in diameter, surfaces sparsely appressed-pubescent. [*Svida alternifolia* (L. f.) Small]

Mesic woodlands, chiefly in ravines; infrequent, s.w. Ga., s.e. Ala., cen. Panhandle of Fla. (Apalachicola and Chipola River drainages) and near Knox Hill, Walton Co. (Nfld. and N.S. to Minn., southward in the east to Fla. Panhandle and from Minn. to n. Ark., the distribution e., w., and s. of the s. Appalachians very spotty.)

National champion alternate-leaved dogwood (1972), 5'8" circumf., 30' in height, 50' spread, Old Westbury, N.Y.

2. Cornus asperifolia Michx. Fig. 98

Shrub 1–5 m tall. Young twigs short-shaggy-pubescent to appressed-pubescent, brown beneath the hairs, becoming gray by the end of the first year or during the second. Axillary buds evident, stiffly short-pubescent with rust-brown hairs. Pith white.

Leaves short-petiolate, petioles 2–5 mm long, shaggy- or appressed-pubescent, pubescence often gray and rusty brown intermixed. Blades elliptic, ovate, or smaller ones often lanceolate, 2–8 cm long and 1–4 cm broad, rarely larger, bases rounded to cuneate, apices abruptly short-acuminate, the tips of the acuminations blunt; upper surfaces dull green, moderately pubescent with stiffish, spreading to appressed hairs and notably rough to the touch, lower surfaces similarly pubescent, often more densely so than on the upper, hairs or their branches tending to be longer, gray or rusty brown or both.

Cymes flat-topped or convex, half as long to as long as subtending leaves, mostly 3–4 cm across, axes and short flower stalks relatively sparsely pubescent, floral tubes densely appressed-pubescent, small deltoid calyx segments sparsely pubescent; corolla cream-colored, petals appressed pubescent exteriorly; styles a little swollen just beneath the stigmas. Drupes light blue, sparsely appressed-pubescent, subglobose, about 8 mm in diameter. [Not *C. asperifolia* sensu Small, 1933; incl. *C. microcarpa* Nash (*Svida microcarpa* (Nash) Small); *C. foemina* subsp. *microcarpa* (Nash) J. S. Wilson.]

Chiefly inhabiting well-drained, calcareous or circumneutral mixed woodlands, less frequently places in bottomlands that flood infrequently and briefly; throughout our area. (Coastal plain, s.e. N.C. to cen. pen. Fla., westward to Ala., perhaps elsewhere.)

3. Cornus foemina Mill. SWAMP DOGWOOD, STIFF DOGWOOD OR CORNEL.

Fig. 98

Shrub or infrequently of small tree stature (to 8 m tall). Bark of larger stems gray to brownish gray, somewhat plated. Young twigs glabrous, green at first, eventually reddish brown to maroon, sometimes tan. Axillary buds evident, clothed with short, stiff, ascending, rusty brown hairs. Pith white.

Leaves short-petiolate, petioles to about 1 cm long, pubescent with short, grayish, appressed hairs. Blades lanceolate, elliptic, or ovate, generally 2–10 cm long and 1–4 cm broad, somewhat larger on vigorous shoots; bases shortly and broadly tapered, occasionally strongly tapered, sometimes rounded, apices mostly acuminate (acuminations pointed or blunt), less frequently acute; upper surfaces dark green, yellowish green, or grayish green, glabrous or with sparse, very short, appressed hairs *and smooth to the touch*, lower surfaces paler and duller grayish green, glabrous or with sparse, short, closely appressed hairs and smooth to the touch.

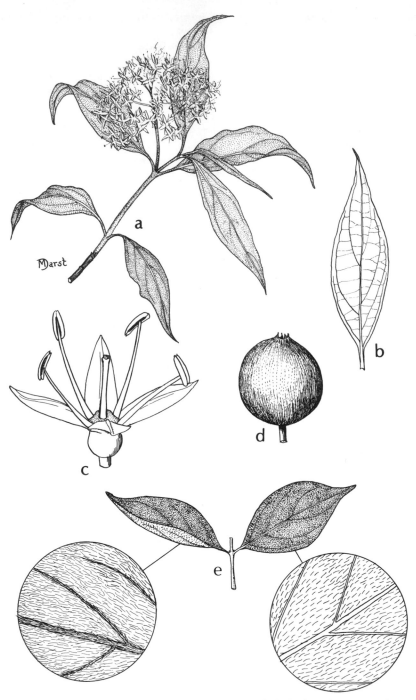

Fig. 98. a–d, **Cornus foemina:** a, flowering branch; b, leaf (surfaces smooth to the touch); c, flower; d, fruit. e, **Cornus asperifolia,** pair of leaves and enlargements of portions of surfaces (rough to the touch), upper surface to right, lower at left.

Cymes having stalks of varying lengths, 2–4 (7) cm, convex, mostly 3–7 cm across, axes and flower stalks glabrous or sparsely pubescent with short, rusty brown hairs, floral tubes densely pubescent with grayish, appressed hairs, small triangular calyx segments sparsely pubescent; corollas cream-colored, petals with few, minute, appressed hairs exteriorly; styles somewhat swollen just beneath the stigmas. Drupes blue, subglobose to ellipsoid, 4–6 mm in diameter. [*C. stricta* Lam. (*Svida stricta* (Lam.) Small)]

Stream banks, banks of marshes, low wet woodlands, marshy shores, in or on borders of cypress-gum ponds or depressions, floodplains, wet thickets and clearings; throughout our area. (Coastal plain, s.e. Va. to s. pen. Fla., westward to e. Tex., in the interior, S.C., Ga., Ala., Tenn., s.w. Ky., s. Ind. and Ill., s.e. Mo., Ark.)

4. Cornus amomum Mill. subsp. amomum. SWAMP DOGWOOD. Fig. 99

Shrub, older plants usually with several erect, arching, or leaning stems from the base, to 5 m tall. Main stem relatively smooth, green, olive-green, reddish brown, or nearly red, with irregular, horizontal flecks of low corky outgrowth, some surrounding lenticels. Young twigs copiously more or less shaggy-pubescent with silvery or rusty brown hairs, much or all of the pubescence sloughing by the end of the growing season and the twigs maroon or green and maroon. Axillary buds evident, copiously pubescent. Pith of year-old and older twigs tawny to brown.

Leaves opposite, slenderly petiolate, petioles 0.5–2 cm long, usually very shaggy-pubescent. Blades mostly ovate, varying to oval or oblong, varying considerably in size, even on a single branch, 3–10 cm long and 2–8 cm broad, bases truncate, rounded, or shortly and broadly tapered, apices abruptly short-acuminate; upper surfaces dark green, glabrous or sparsely pubescent, lower paler and duller green, pubescence appressed or shaggy or both, gray or rusty brown or both, often the principal veins shaggy-pubescent with gray hairs, smaller ones appressed-pubescent with rusty brown hairs.

Cymes flat-topped to hemispherical, slenderly stalked, extending to about half the length of blades of subtending leaves, their axes, flower stalks, floral tubes, and calyx segments shaggy-pubescent, calyx segments subulate, 1 mm long or a little more; styles swollen just below the stigmas. Drupes subglobose, blue with areas of cream-color, 5–9.5 mm in diameter. [*Svida amomum* (Mill.) Small]

River and stream banks, marshy or swampy shores, borders of swamps, wet thickets and clearings; s.w. Ga., s. Ala., cen. and w. Panhandle of Fla. (S. Maine to s. Ohio, s. Ind., Ky., southward to s.w. Ga., Fla. Panhandle, s. Ala., n.e. Miss.)

5. Cornus florida L. FLOWERING DOGWOOD. Fig. 100

A small tree, to about 12 m tall, usually with a short trunk and rounded crown, conspicuous in spring as flower heads reach full development before and as new shoot growth commences, contributing greatly to autumnal coloration, the leaves then variously turning purplish pink to purplish red, popularly used as an ornamental. Bark of trunks gray to nearly black, roughly broken up into scaly blocks. Young twigs very minutely and sparsely appressed-pubescent, eventually becoming glabrous, usually green or red and green through the first year, eventually brown. Axillary buds not or scarcely evident. Pith pale green in 1–2-year-old twigs, eventually white.

Leaves short-petiolate, petioles 0.5–1.5 cm long, sparsely short-pubescent.

Fig. 99. **Cornus amomum** subsp. **amomum:** a, flowering branch; b, node with infructescence; c, flower; d, fruit.

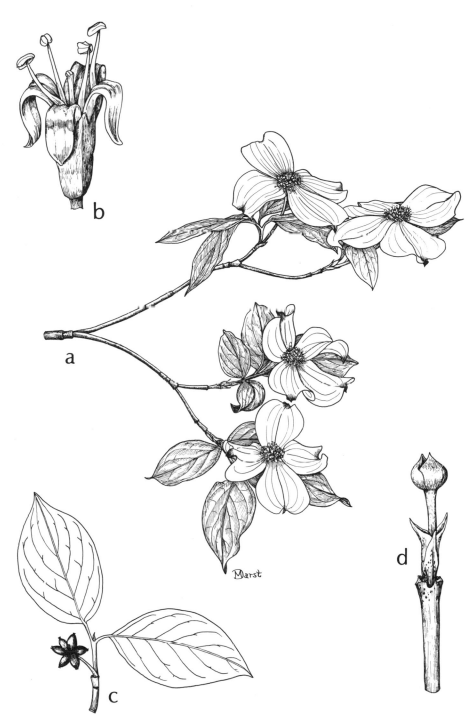

Fig. 100. **Cornus florida:** a, branch with inflorescences; b, flower; c, branchlet with cluster of fruits; d, tip of branchlet (winter condition) with overwintering inflorescence bud.

Blades somewhat decurrent on the petioles, ovate, narrowly to broadly elliptic, even subrotund, or oblong-elliptic, infrequently lanceolate, varying a great deal in size, 3–10 cm long and 2–7 cm broad, bases often somewhat inequilateral and oblique, perhaps more commonly equilateral, acute to rounded, apices abruptly to gradually acuminate, sometimes acute; upper surfaces dark green, with relatively sparse, minute, appressed pubescence, lower similarly appressed-pubescent, often with shaggy pubescence along the major veins. All pubescence grayish.

Small flowers compacted in heads surrounded by 4 conspicuous, petallike, white or pale cream-colored, spreading bracts, the bracts gradually developing before any new shoot growth commences and reaching full size and becoming fully white or cream-white by the time the flowers reach anthesis and by which time new shoot growth may begin; bracts usually obovate and notched apically. Flowers sessile, small, floral tubes and exterior of calyx segments densely clothed with short, appressed hairs. Corolla greenish yellow to yellow, appressed-pubescent exteriorly at least on their upper halves. Styles little if any enlarged below the stigmas. Drupes glabrous, glossy bright red, oblong-ellipsoid, 8–14 mm long, persistent calyces at their apices. [*Cynoxylon floridum* (L.) Raf.]

In our area, a form with pink bracts rather rarely cultivated, presumably not well adapted, much more commonly in cultivation to the north and northeast of our range.

Well-drained woodlands of various mixtures; throughout our area. (Southeasternmost Maine to s. Mich., generally southward to cen. pen. Fla., the w. boundary of the range from Mich. to s. Ill., s. Mo., e. Okla., e. Tex.)

Co-national champion flowering dogwoods (1976), 4'7" circumf., 55' in height, 56' spread, St. Joseph Co., Mich.; (1971), 5'1" circumf., 51' in height, 43' spread, Maclay Gardens State Park, Fla.

Cyrillaceae (CYRILLA FAMILY)

1. Leaf blades with lateral veins scarcely if at all evident on either surface; flowers borne in racemes at or near the tip of branchlets of previous season prior to emergence of new shoot growth in early spring; calyx segments deltoid, short-oblong, or crescent-shaped, obtuse or rounded at the summit, usually pinkish white; stamens 10; fruit markedly 2–5-winged.   1. Cliftonia

1. Leaf blades with lateral veins (and usually veinlets) evident at either side of the midrib on both surfaces; flowers in radiating clusters of racemes from the summit of twigs of the previous season and below leafy shoots of the season; calyx segments lance-ovate, sharply acute to acuminate apically; stamens 5; fruit not winged.   2. Cyrilla

1. Cliftonia

Cliftonia monophylla (Lam.) Britt. ex Sarg. BUCKWHEAT-TREE, BLACK TITI.
Fig. 101

Glabrous, evergreen shrub or small slender tree, very commonly forming dense stands. Leaf scars shield shaped, each with a single, transverse vascular bundle scar which often appears perforated. Leaves simple, alternate, without stipules, sessile or shortly subpetiolate; blades somewhat leathery, mostly elliptic, vary-

Fig. 101. **Cliftonia monophylla:** a, fruiting branch; b, flowering branchlet; c, flower; d, fruit.

ing to elliptic-oblanceolate or oblanceolate, 2.5–10 cm long and 1.2–1.8 cm broad, bases cuneate, apices acute to obtuse, sometimes shallowly emarginate, surfaces glabrous, the upper dark green, the lower paler and with a bluish-whitish bloom; lateral veins of the blades scarcely if at all evident on either surface; margins entire.

Flowers bisexual, radially symmetrical, borne in 1–several racemes at or at and near the tips of twigs of the previous season, each rarely exceeding 5–7 cm long, at full anthesis in early spring before new shoot growth commences; flower stalks mostly about 5 mm long, each bearing 2 small, quickly deciduous bracts at or somewhat below their middles; axis of raceme ridged, each ridge terminating in a small mound of tissue just below a flower stalk and subtending a bract which is deciduous before full anthesis; calyx very much shorter than the corolla and surrounding only the bases of the latter, united below and with 5 broadly deltoid or short-oblong lobes obtuse to rounded at the summit; petals 5, white or less frequently pinkish, 5–8 mm long, clawed at the base, blades spatulate to obovate, rounded apically; stamens 10, the filaments dilated below; pistil 1, ovary superior, 3–5-locular, one ovule in each locule, style short, stigma 2–5-lobed or entire. Fruit a small, indehiscent, dry, 2–5-winged drupe 5–7 mm long, in outline a little longer than broad, calyx persistent at its base.

Acid shrub-tree bogs or wet woodlands along stream courses and in flatwoods depressions; s. cen. and s.e. Ga., Fla. Panhandle from Jefferson Co. westward (disjunct in Clay Co., n.e. Fla.), s. Ala., s.e. Miss., s.e. La.

Massed stands of this plant when in bloom and viewed from a distance appear almost as though a slight mist were hanging in them; at close range, the white racemes are very handsome. To my knowledge, the plant has had no appreciable, if any, use horticulturally; its evergreen habit and floral characteristics would seem to favor its use as an ornamental unless there are problems in its propagation or culture.

Co-national champion buckwheat-trees (1967), 5′6″ circumf., 30′ in height, 21′ spread, Crooked Creek, Fla.; (1981), 49″ circumf., 44′ in height, 31′ spread, Washington Co., Fla.

2. Cyrilla

Cyrilla racemiflora L. TITI, CYRILLA, HE-HUCKLEBERRY.   Figs. 102, 103

Glabrous shrub or low tree, commonly reproducing vegetatively by sprouts from shallow horizontal roots and forming thickets. Young stems angled-ridged, the leaf scars somewhat obliquely raised, each with a single, usually roundish vascular bundle scar.

Leaves simple, alternate, without stipules, tardily deciduous or semipersistent, shortly subpetiolate or sessile; blades stiffish-leathery at full maturity; blades variable in size from about 1 to 10 cm long and 0.5–2.5 cm broad, oblanceolate, narrowly obovate, elliptic-oblanceolate, or elliptic, cuneate basally, apices acute, obtuse, rounded, or emarginate, both surfaces evidently reticulate-veined lateral to their midribs, the upper more pronouncedly so than the lower, margins entire.

Flowers bisexual, radially symmetrical, borne in narrowly cylindric racemes 2–15 cm long, these clustered at the tips of twigs of the previous season and below shoots of the season, raceme axes angled-ridged; flower stalks 1–2.5 mm long, each subtended by a subulate bract usually persisting below the stalk of the fruit, two similar subulate bracts on the flower stalks distally, these not persistent; calyx surrounding only the base of the corolla, its 5 persistent seg-

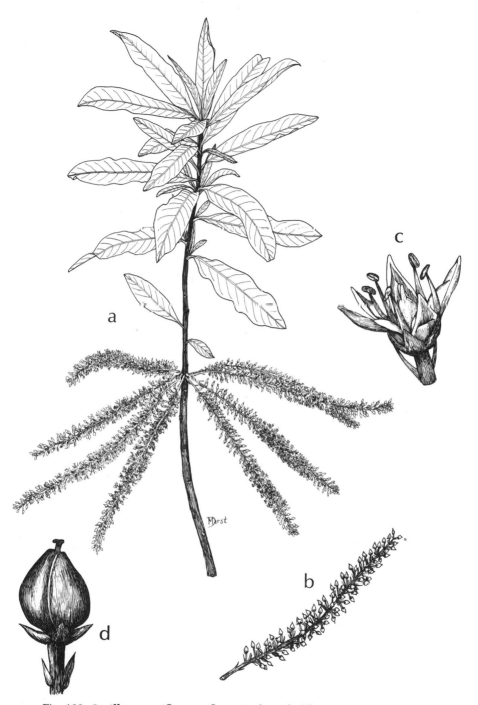

Fig. 102. **Cyrilla racemiflora:** a, flowering branch (illustration representing large-leaved extreme form of general distribution in the coastal plain of s.e. U.S.); b, fruiting raceme; c, flower; d, fruit.

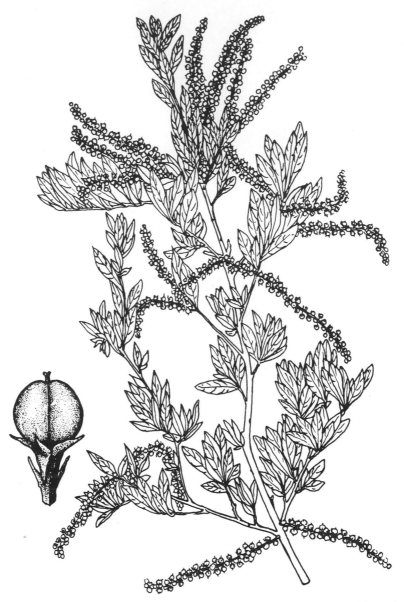

Fig. 103. **Cyrilla racemiflora:** illustration representing the small-leaved extreme often common and abundant in northern Florida, chiefly from Taylor Co. westward. (From Kurz and Godfrey [as C. parvifolia], 1962. *Trees of Northern Florida.*)

ments lanceolate to deltoid, their apices acute to acuminate; petals 5, white or creamy white, about 2.5–3.5 mm long, sometimes longer, broadest basally and narrowed to acute apices; stamens 5, opposite the sepals, filaments broadest basally and gradually narrowed distally. Pistil 1, ovary superior, 2–3 (4) -locular, ovules 1–3 in each locule, style short, sometimes persistent on the fruit, stigma 2-lobed. Fruit 2–2.5 mm long, a small, indehiscent, dry ovoid to subglobose drupe, the calyx persistent at its base.

Acid shrub-bogs, along alluvial and nonalluvial stream courses, flatwoods, and flatwoods depressions, seepage areas in woodlands; throughout our area. (Coastal plain, s.e. Va. to cen. pen. Fla., westward to s.e. Tex.; s. Mex., Belize, W.I., n. S.Am.)

Thomas (1960), having studied *Cyrilla* from its extensive geographic range, concluded that the numerous described taxa, showing local patterns of variation, intergrade to such an extent that but a single species can be recognized. In our area, occurring in the outer or lower coastal areas of the Fla. Panhandle, the extreme small-leaved form has been known as *C. parvifolia* Raf. See illustration.

National champion swamp cyrilla or titi (1981), 3'10" circumf., 52' in height, 28' spread, Washington Co., Fla.

Ebenaceae (EBONY FAMILY)

Diospyros virginiana L. PERSIMMON.   Fig. 104

Medium-sized tree, in open areas sometimes spreading by root-sprouts and thicket-forming. Bark of older trunks brown to blackish, roughly broken into small blocks and cross-fissures; wood hard. Terminal buds absent, uppermost buds frequently abort following which shoots develop from several axillary buds in such a way as to form a bushy or spraylike branch pattern. Pith of the twigs homogeneous through most, if not all, of the first year, in year-old twigs becoming transversely diaphragmed, the diaphragms very thin, chambered between them. Leaf scars half-moon- to approximately shield-shaped, with 1 narrowly crescent-shaped vascular bundle scar.

Leaves deciduous, simple, alternate, without stipules, short-petiolate, adaxial surface of the petiole often (not always) having irregularly distributed, few to numerous minute sessile glands, these also extending proximally along the midrib of the upper surface of the blade. Blades broadly lanceolate, ovate-oblong, or widely elliptic, mostly 7–15 cm long and 3–7 cm broad; bases mostly rounded, sometimes broadly and shortly tapered, apices usually short-acuminate, surfaces usually glabrous but sometimes pubescent beneath, upper surface dark green, often with blackish blemishes, lower pale, somewhat grayish green, margins entire.

Flowers greenish and inconspicuous, functionally unisexual and the plants functionally dioecious. Staminate flowers smaller than the pistillate, in few-flowered axillary clusters; pistillate solitary in leaf axils. Calyx united below, 4 (5) -lobed above, lobes of the staminate flower much smaller than those of the pistillate. Corolla 1–1.5 cm long, very much exceeding the calyx, greenish yellow, campanulate-urceolate, lobes 4 (5), recurved. Staminate flowers with 16 stamens in 8 pairs, 1 stamen above the other in a pair, pubescent proximally, collectively the hairs forming a roof over a cavity containing a pistillode. Pistillate flower with 8 hairy staminodes in 2 whorls, 1 pistil, with 4–8 locules in the superior ovary, the styles variable according to the number of locules.

Fig. 104. **Diospyros virginiana:** a, staminate flowering branchlet; b, pistillate flowering branchlet; c, longitudinal section of year-old woody stem (at which time pith is diaphragmed and chambered between the diaphragms); d, longitudinal section of functionally staminate flower; e, longitudinal section of functionally pistillate flower; f, fruit.

Fruit a several-seeded, globose or depressed-globose berry to 4–5 cm in diameter, an enlarged persistent calyx at its base; in ripening its color turning from green to yellow, to orange, surface glaucous. Pulp of the fruit very astringent during maturation, becoming sweet and with a distinctively delicious flavor when fully ripe, the palatability varying, however, from tree to tree.

Inhabiting a wide range of sites of varying moisture and fertility conditions, fields, upland woodlands of various mixtures, river and stream bottomlands, pine flatwoods; throughout our area. (S. N.Eng., e. N.Y., Pa., W.Va., Ohio, Ind., Ill., s.e. Iowa, e. Kan., generally southward to s. pen. Fla. and e. Tex.)

Sterile specimens of persimmon and those of black gum (*Nyssa sylvatica*) are sometimes difficult to distinguish, one from the other. Pith of year-old twigs of persimmon are diaphragmed, chambered between the diaphragms; petiole of the leaf having minute sessile glands on the adaxial surface and usually extending proximally along the adaxial midrib. Pith of the year-old twigs of the black gum diaphragmed and homogeneous and solid between the diaphragms; adaxial petiole surface and midrib without glands.

Co-national champion persimmons (1972), 12′2″ circumf., 60′ in height, 58′ spread, Leon Co., Tex.; (1977), 6′9″ circumf., 131′ in height, 40′ spread, Big Oak Tree State Park, Mo.

Elaeagnaceae (OLEASTER FAMILY)

Elaeagnus

Shrubs (ours), stems, petioles, one or both surfaces of leaf blades, flower stalks, exterior of floral tube and calyx segments, and fruits clothed with silvery or brown, peltate or stellate-peltate scales. Leaves alternate, simple, without stipules, short-petiolate. Flowers fragrant, radially symmetrical, bisexual or unisexual (plants polygamodioecious), mostly borne in short, congested racemes, infrequently solitary, axillary to leaves on shoots of the season, some sometimes axillary to leaf scars on wood of the previous season. Floral tube having its base closely surrounding the ovary then flaring-obconical upwardly, 4 small, deltoid calyx segments at its summit. Corolla none. Stamens 4, anthers short, sessile, inserted at the throat of the floral tube and below the sinuses between the calyx segments. Pistil 1, ovary 1-locular, 1-ovululate, style 1, slender, stigma capitate. Fruit drupelike, the outer layers of the part of the floral tube surrounding the ovary becoming fleshy during maturation, the inner layer hard and wholly enclosing the achene.

1. Leaves evergreen; leader-shoots of the season forming axillary, divaricate, very sharp, rigid thorns 4–5 cm long; upper surfaces of mature leaf blades dark green, punctate, without scales, lower surfaces densely scaly, scales mostly silvery, a few russet ones interspersed thus appearing silvery and only slightly speckled with russet; flowers produced in autumn on mature, lateral branchlets of the season, fruits maturing the following spring.

1. E. pungens

1. Leaves deciduous; leader shoots of the season usually not forming axillary thorns; upper surfaces of mature leaf blades moderately green, bearing tiny, silvery scales separated from each other, lower surfaces densely and compactly scaly, scales metallic-silvery; flowers produced in spring on short, lateral, developing branchlets of the season, sometimes a few from axils of leaf scars on wood of the previous season, fruits maturing in late summer or autumn.

2. E. umbellata

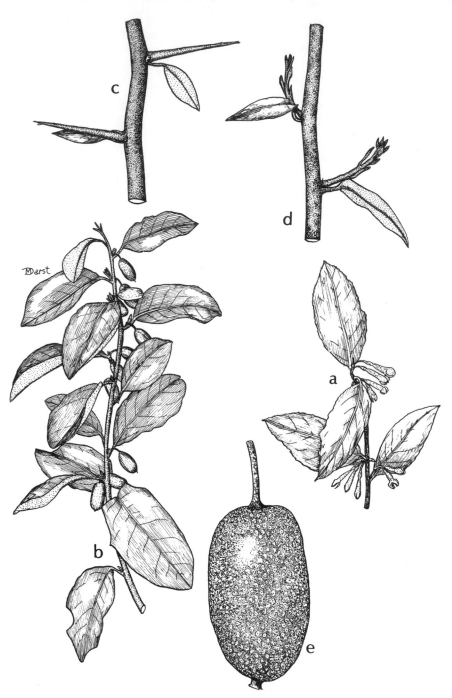

Fig. 105. **Elaeagnus pungens:** a, flowering branchlet; b, fruiting branchlet; c, piece of elongation shoot, first year with thorns; d, piece of elongation shoot, spring of second year at which time thorns soften and become a leafy short-shoot; e, fruit.

1. Elaeagnus pungens Thunb. Fig. 105

Evergreen shrub well-established specimens of which produce long, pliant, brown, fast-growing shoots bearing small leaves most of which have in their axils rigid, very sharp, divaricate thorns. In spring of the following season, the thorns become soft, produce buds which by further development produce leafy branches and these in autumn bear floral racemes in the leaf axils. Successive seasons of such growth may yield a tangle of stems which may scramble through other vegetation in a somewhat vinelike manner, the scrambling being effected by the divaricate branches holding or anchoring, as it were, as they grow into and through branches of other plants.

Blades of unfurling leaves copiously scaly on both surfaces, at this stage numerous russet scales overlie silvery ones so that the surfaces are distinctly russet-speckled; as the leaves mature, the scales all slough from the upper surfaces which are then dark green, glabrous, and punctate, almost all of the russet scales slough from the lower surfaces leaving them copiously and compactly metallic-silvery and only very slightly speckled with russet; mature blades variable in size, 3–10 cm long and 2–5 cm broad, in outline mostly oval, elliptic, or slightly ovate, sometimes a few of them suborbicular; margins variably entire, somewhat crisped, or irregularly very shallowly undulate.

Flowering in autumn on branchlets of the season; flower stalks 5–8 mm long, flowers about 10 mm long. Fruits maturing in early spring, ellipsoid or oblong-ellipsoid, about 15 mm long, the flared portion of the shriveled perianth usually persisting at their summits until just before full maturity.

Native to China and Japan, here cultivated as an ornamental, infrequently naturalized.

2. Elaeagnus umbellata Thunb. Fig. 106

Bushy-branched, deciduous shrub to 3 m tall or a little more. In spring, lateral buds on wood of elongation shoots of the previous season give rise to relatively short shoots producing flowers as they elongate and as the leaves are unfurling.

Upper surfaces of mature leaf blades grayish green, the color yielded by a combination of many tiny silvery scales separated from each other and the light green color of the surface between them; lower surfaces densely and compactly covered with metallic-silvery scales; blades variable in size, 1–8 cm long and 1–3 cm broad, largest leaves on elongation shoots of the season; in outline, oval, elliptic, or oblanceolate, some of the smaller ones subrotund.

Flower stalks 3–6 mm long, flowers about 10 mm long. Fruits subglobose to ellipsoid, 6–8 mm long, the flared portions of the floral tubes dehiscing soon after anthesis.

Native to China, Korea, and Japan, cultivated as an ornamental, infrequently naturalized in our area.

Empetraceae (CROWBERRY FAMILY)

Ceratiola ericoides Michx. ROSEMARY. Fig. 107

Evergreen, bushy-branched shrub with a distinctive rounded form. Young twigs densely and compactly gray-pubescent; leafless twigs roughened by the numerous, small, raised leaf scars, gray bark sloughing in small, thin plates exposing brown inner bark that itself eventually becomes gray.

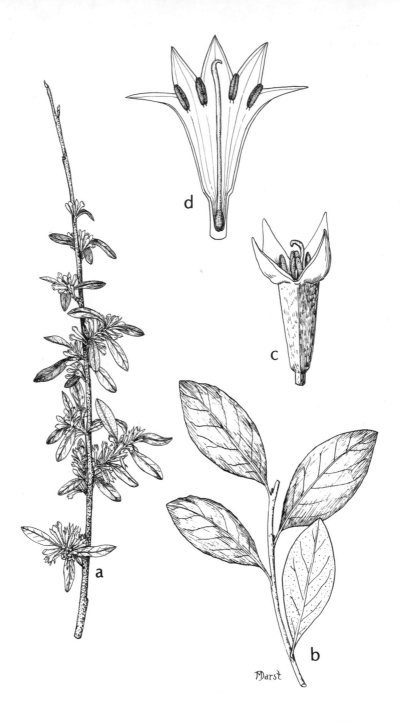

Fig. 106. **Elaeagnus umbellatus:** a, flowering branch; b, piece of elongation
shoot; c, flower; d, flower, opened out.

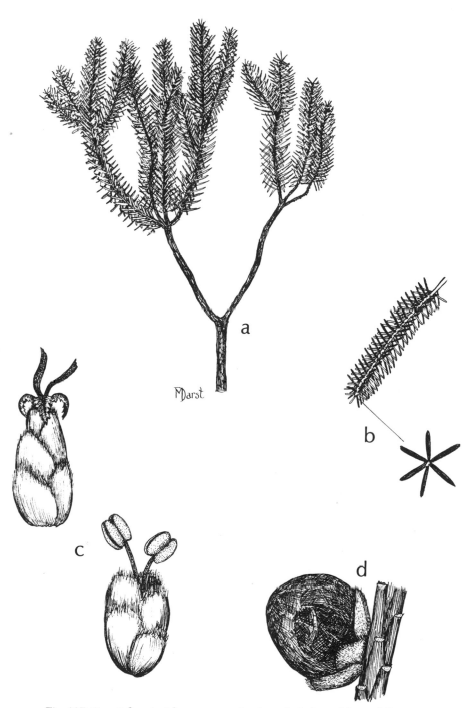

Fig. 107. **Ceratiola ericoides:** a, vegetative branch; b, branchlet and diagrammatic sketch to show ranking of leaves; c, flowers, pistillate above, staminate below; d, fruit.

Leaves simple, very closely set in opposite pairs at right angles to each other and appearing essentially whorled; blades needlelike, dark green, thickish, more or less square in cross-section, the adaxial sides of the blade very strongly revolute, turning under and meeting beneath thus completely obscuring the abaxial surface; blades 8–12 mm long and 1 mm broad or slightly more, surfaces not punctate.

Flowers very small, sessile, unisexual (plants dioecious), borne singly in leaf axils in autumn, fruits maturing overwinter, "perianth" of 5–6 bracts and sepals; staminate flower with 2 exserted stamens, pistillate with 1 pistil, ovary superior, 2-locular, style 1, splitting distally into 4 spreading, incised-pinnatifid lobes from which 2 stigmas protrude. Fruit a greenish yellow, subglobose, 2-stoned drupe 2–3 mm in diameter.

Usually locally abundant, coastal stabilized dunes and hammocks, deep sands of pineclad ridges and hills; n. Fla., s.w. Ala. (S.e. S.C. to s. pen. Fla., westward to s. Miss.)

Ericaceae (HEATH FAMILY)

1. Plant with slender, rooting stems creeping shallowly within the substrate, producing short, leafy branches elevated just above or partially within and partially above the substrate.   3. *Epigaea*
1. Plant with erect stems (or in the case of *Pieris phillyreifolia* if climbing then with scale-leaved stems within the bark of trunks of trees and these producing leafy branches emanating from the bark of the "host").
  2. Larger leaf blades on a given plant 15–18 cm long and 2.5–7.5 cm broad.
                                                   8. *Oxydendrum*
  2. Larger leaf blades on a given plant much smaller.
    3. Ovary superior, fruit capsular.
      4. Calyx and corolla 7-merous; petals separate.   2. *Befaria*
      4. Calyx and corolla 5-merous; petals partially united.
        5. Corolla tube campanulate or cylindric-tubular below and flaring above, in either case the aperture relatively broad.
          6. Corolla tube campanulate, anthers in the bud or at early anthesis seated in pouches in the corolla tube; capsule little, if any, longer than broad.   5. *Kalmia*
          6. Corolla tube cylindric-tubular below, notably flaring above, anthers not seated in pouches of the corolla tube; capsule much longer than broad. 10. *Rhododendron*
        5. Corolla tube essentially urceolate, its aperture narrow.
          7. Flower stalks with 2 opposite or subopposite, subulate bractlets a little below the calyx.   9. *Pieris*
          7. Flower stalks with a pair of ovate bractlets either basally or just below the calyx, or if bractlets minute or narrow then borne approximately basally on the flower stalks.
            8. Flowers or fruits in definite racemes; capsule without thickened ridges over the sutures.
              9. Flower stalks stoutish, 1–4 mm long; filaments straight; pith homogeneous.
                                    6. *Leucothoe*
              9. Flower stalks slender, 7–10 mm long; filaments strongly curvate (nearly looped) just below the anthers; pith transversely diaphragmed, chambered between the diaphragms.   1. *Agarista*
            8. Flowers or fruits in umbelliform clusters or short fasciclelike racemes; capsule with prominently thickened ridges over the sutures, these usually separating and falling at dehiscence.   9. *Lyonia*
    3. Ovary inferior; fruit a berrylike drupe with 10 hard nutlets or a many-seeded berry.

10. Leaf blades with sessile, amber, glandular dots beneath; fruit a berrylike drupe with 10 hard nutlets.							4. *Gaylussacia*
10. Leaf blades without sessile, amber glandular dots beneath, if glandular-pubescent then the hairs stipitate-glandular or clavate; fruit a many-seeded berry.

11. *Vaccinium*

1. Agarista

[All members of Ericaceae lack stipules, although this is not noted in the descriptions below.]

### Agarista populifolia (Lam.) Judd.							Fig. 108

Evergreen shrub to about 4 m tall, the branches usually weakly arched-ascending or leaning. Young twigs light brown, sometimes sparsely short-pubescent and/or with few irregularly scattered stipitate glands; pith transversely diaphragmed, chambered between the diaphragms.

Leaves short-petioled; blades ovate, lance-ovate, or lanceolate, 4–10 cm long and 1–4 cm broad, bases mostly rounded, some broadly short-tapered, apices acute to acuminate, reticulately veined to either side of the midrib, glabrous or sparsely short-pubescent on the midrib beneath, margins narrowly cartilaginous-banded, entire or obscurely wavy, sometimes obscurely to sharply serrate, serrations tipped by gland-headed hairs.

Flowers in short, axillary, shortly stalked racemes. Flower stalks slender, 7–10 mm long, they and raceme axis sparsely to moderately pubescent with short, curvate hairs, sometimes having irregularly scattered stipitate glands intermixed; several alternately disposed, small, narrowly triangular bracts on the stalk of the raceme and a similar, slightly smaller bract subtending each flower, usually a pair of similar but still slightly smaller bractlets basally on the flower stalk. Calyx united for about half its length, 5-lobed, lobes broadly ovate-triangular, about 1 mm long, sometimes stipitate-glandular and margins finely short-ciliate. Corolla white, tube more or less urceolate, about 8 mm long, somewhat constricted and necklike at the summit, with 5 short, broad lobes whose edges recurve. Stamens 10, not exserted, filaments strongly curvate (nearly looped) just below the anthers, relatively long-pubescent below the curvature, anthers awnless. Pistil 1, ovary superior, style slender, stigma capitate, barely extending beyond the aperture of the corolla. Capsule depressed-ovoid or depressed-globose, 5–6 mm broad, 5-valved, dehiscing loculicidally. [*Leucothoe populifolia* (Lam.) Dippel; *L. acuminata* (Ait.) D. Don]

Local, moist to wet hammocks and wet woodlands, often about large springs and along spring-runs; n. cen. pen. Fla. to s.e. S.C.

2. Befaria

### Befaria racemosa Vent. TAR-FLOWER, FLY-CATCHER.							Fig. 109

Evergreen, slender shrub with relatively few stiffly erect branches, 1–2.5 m tall. Twigs powdery-pubescent and with few to numerous long, spreading hairs.

Leaves sessile or subsessile, narrowly to broadly elliptic or ovate, occasional ones oblanceolate or obovate, mostly 2–4 cm long, 0.6–2 cm broad, cuneate basally, acute to rounded apically, margins entire, lateral veins usually faint, upper surfaces powdery-pubescent, lower pale, glabrous or with few to numerous longish hairs along the midrib.

Flowers showy, fragrant, in racemes or panicles terminal on shoots of the sea-

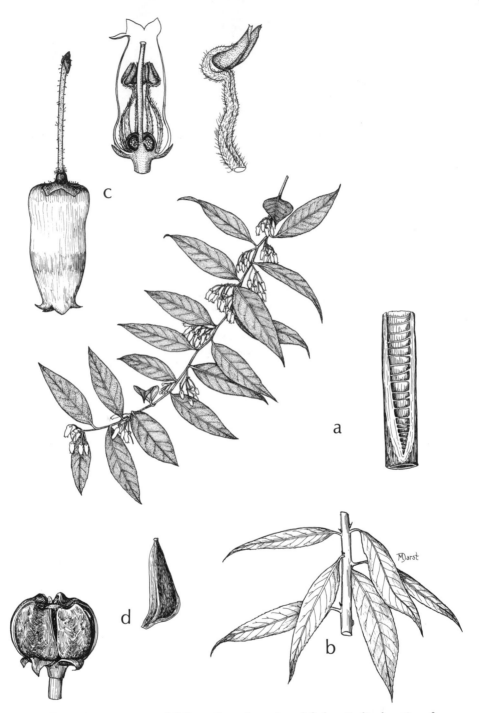

Fig. 108. **Agarista populifolia:** a, flowering twig at left, longitudinal section of twig at right; b, piece of elongation growth; c, flower at left, longitudinal section of flower at center and stamen at right; d, capsule at left, seed at right.

Fig. 109. **Befaria racemosa:** flowering branch at right; fruiting branch, center; flower, upper left; capsule, lower left; seed, center left.

son. Flower stalks 1–3 cm long, subtended by reduced or bracteal leaves, or the uppermost by subulate bracts, each stalk bearing 1, or usually 2, subulate bractlets closely below or at a little distance below the calyx. Calyx persistent, about 5 mm high, tube short-campanulate, surmounted by 7 erect, ovate-triangular lobes, their inner surfaces nectariferous. Petals 7, separate, white and pink-tinged, sometimes wholly pink proximally, spreading, elongate, 2–3 cm long, tapering from a narrowly clawed base to a narrowly spatulate or oblanceolate blade, glutinous exteriorly. Stamens 12–14, filaments long, one-half to two-thirds as long as the petals, dilated below and cottony-pubescent, anthers short, dehiscing by 2 oblique terminal pores. Ovary superior, 7-locular, style slender, 1.5–2 cm long, stigma capitate.

Fruit a subglobose, 7-valved, septicidal, glutinous capsule 6–8 mm broad. Seeds narrowly elongate, somewhat falcate, 1–1.5 mm long, amber-colored.

In both poorly drained pine flatwoods and in well-drained sand-scrub. (Coastal plain, s.e. Ga. to s. pen. Fla.)

Rather a distinctive evergreen shrub of flatwoods or sand-scrub, irregularly branched, the branches stiffly erect, the alternate leaves relatively small, white to pink, fragrant flowers borne in racemes or panicles terminating branches of the season, calyx nectariferous within, 7 narrowly oblanceolate or spatulate petals greatly exceeding the calyx, capsules glutinous.

3. Epigaea

Epigaea repens L. TRAILING-ARBUTUS. Fig. 110

Evergreen subshrub with elongate, branching, slender stems shallowly running in the soil, the leafy branch tips protruding through leaf litter. Leafy stems and petioles abundantly clothed with longish, spreading, rust-colored hairs; old leafless stems generally bearing petiole bases having small, soft, whitish buds in their axils.

Leaves simple, alternate, petiolate, petioles 2–5 cm long. Blades stiffish, oblong, elliptic-oblong, infrequently ovate, smaller ones sometimes subrotund, bases usually rounded or cordate, seldom shortly and broadly tapered, apices mostly rounded, sometimes obtuse, their tips often minutely mucronate, 2–9 cm long and 1.5–6 cm broad, pinnately-veined, veins all impressed on the upper surfaces, beneath the midvein strongly elevated, the major laterals only slightly so at most, the veinlets impressed, veinlets appearing notably reticulated on both surfaces; surfaces and edges bearing numerous, stiffish and longish, rust-colored hairs, these sometimes long-persistent, sometimes most of them breaking off near their bases leaving the surfaces rough to the touch.

Inflorescences short, few-flowered spikes in the axil(s) of the uppermost leaf or leaves of a branchlet before new growth commences in spring, each flower subtended by an ovate basal bract and with 2 similar but somewhat larger bracts immediately above, the bracts obscuring much of the calyx. Plants functionally dioecious; flowers on some plants structurally bisexual, having functional stamens and a pistil, the stigma of the latter, however, not receptive to pollen, therefore the pistil sterile; flowers on other plants with a functional pistil but the stamens rudimentary, usually represented only by filaments, these sometimes pink-petaloid. Calyx united only basally, with 5 ovate-acuminate lobes 5–6 mm long. Corolla pink to white, its tube cylindric, the limb with 5 ovate to short-oblong, spreading, sometimes slightly recurved lobes; corollas about twice as large in functionally staminate as in functionally pistillate flowers; co-

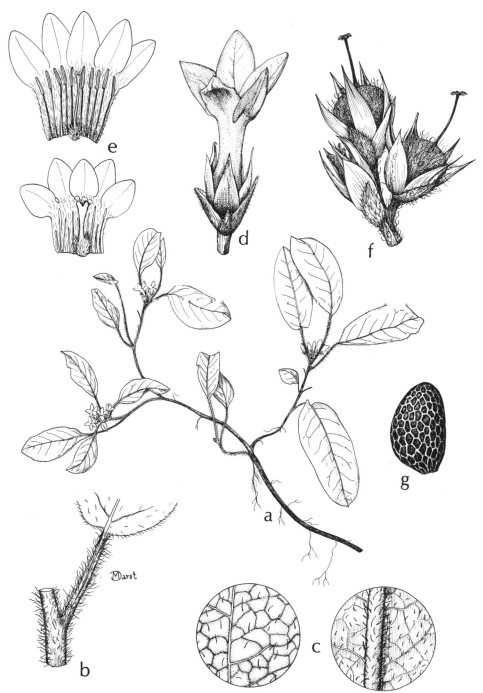

Fig. 110. **Epigaea repens:** a, portion of stem with flowers; b, piece of stem with petiole and leaf blade enlarged to accentuate pubescence; c, small portion of leaf surfaces, upper surface on left, lower on right; d, flower; e, flowers opened out (calyces removed), functionally staminate flower above, functionally pistillate flower below; f, cluster of fruits; g, seed.

rolla tube pubescent interiorly. Stamens 10 in the functionally staminate flowers, inserted on the base of the corolla tube, half as long to as long as the tube. Pistil 1, ovary superior, outer surface glandular-pubescent, 5-locular, style slender, half as long to as long as the corolla tube, elongating somewhat after anthesis, stigma glandular, 5-lobed, lobes radially spreading in the functionally pistillate flowers, lobes erect and dry in the functionally staminate flowers.

Fruit a small, depressed-globose, septicidal capsule, its surface shaggy-pubescent with rust-colored hairs; bracts and calyces persistent on the infructescence, the style and stigma persistent at the summit of the capsule until it ripens. Seeds numerous, minute, hard, lustrous-brown.

Wooded slopes of ravines and bluffs, local from Liberty Co. westward in the Fla. Panhandle, s.w. Ala. (Widespread northward of our range.)

4. Gaylussacia (HUCKLEBERRIES)

Deciduous shrubs (ours), often with short to elongate subterranean runners and colonial. Leaves alternate, simple, short-petioled, blades pinnately veined. Flowers in axillary bracteate racemes on wood of the preceding season, flower stalks with 1 or usually 2 bractlets. Floral tube adnate to the ovary. Sepals 5, persistent. Corolla white, greenish, greenish red, or pink, tubular below, tube campanulate to urceolate. Stamens 10, inserted on the base of the corolla, filaments short, anthers without appendages on the back, each half narrowed upwardly into a tube, opening by a terminal pore or slit, awnless. Ovary 10-locular, each locule with 1 ovule. Fruit a berrylike drupe with 10 smooth, 1-seeded nutlets.

1. Bracts of racemes not foliaceous, deciduous soon after flowering; floral tube and fruit glaucous, glands, if any, sessile. *2. G. frondosa*
1. Bracts of racemes subfoliaceous, persistent until the fruits mature; floral tube and young fruit not glaucous, glands, if any, stipitate.
  2. Young twigs, raceme axes, flower stalks, and floral tubes with short curly hairs, with or without short stipitate-glandular ones intermixed. *1. G. dumosa*
  2. Young twigs, raceme axes, flower stalks, and floral tubes bearing long, spreading, silvery-silky, longish hairs, minutely glandular at their tips. *3. G. mosieri*

1. Gaylussacia dumosa (Andr.) A. Gray. DWARF HUCKLEBERRY.   Fig. 112

Shrub 1–5 dm tall, forming clones by subterranean runners. Young twigs usually copiously pubescent with short curly hairs, with or without short stipitate-glandular hairs intermixed.

Leaf blades oblanceolate or short-obovate, some sometimes elliptic, bases mostly strongly cuneate, apices rounded or infrequently acute, mucronate, variable in size on a given stem, the larger ones 2–3 cm long and 5–20 mm broad, margins sometimes with barely a suggestion of glandular crenulations; upper surfaces of young leaves usually short-pubescent at least on the veins and sparsely glandular-punctate, usually becoming glabrous and lustrous, edges often remaining pubescent, lower surfaces pubescent at least on the veins and gland-dotted. (Rarely, in our range, leaf surfaces and margins more or less stipitate-glandular.)

Racemes 5–40 cm long, mostly 4–10-flowered, bracts subfoliaceous and persistent, axes and flower stalks with somewhat matted curly hairs, short-stipitate glands intermixed. Floral tube usually abundantly clothed with both sessile and short-stipitate glands, the sepals hairy on their margins. Corolla white or pinkish, campanulate when fully expanded, 6–9 mm long, constricted at the summit, lobes short-ovate.

Fruit black, not glandular by full maturity, about 8 mm in diameter. [*Lasiococcus dumosus* (Andr.) Small]

Commonly in well-drained pinelands or pine-oak woodlands, also in areas transitional from high pinelands to shrub-tree bogs and in sphagnous bogs; throughout our area. (Nfld. to cen. pen. Fla., westward to Miss.; e. cen. Tenn.; local in mts. of W.Va., s.w. N.C., n.w. S.C.)

2. Gaylussacia frondosa (L.) T. & G. DANGLEBERRY.

Shrub to 2 m tall, with sessile, amber, glandular dots on various of its parts. Young twigs often glaucous, glabrous or sparsely short-pubescent to densely short-pubescent, glandular-dotted.

Leaf blades oval, elliptic, oblanceolate, or obovate, cuneate basally, rounded to obtuse apically, sometimes retuse; margins entire, upper surfaces glabrous or very sparsely short-pubescent, lower moderately to strongly glaucous, less frequently without glaucescence, sparsely to densely short-pubescent.

Racemes loosely 5–12-flowered, bracts not foliaceous, axes and flower stalks glabrous varying to copiously short-pubescent, sparsely gland-dotted. Floral tube glaucous, sometimes gland-dotted, calyx lobes triangular, about 1 mm long, obtuse to acute, gland-dotted. Corolla greenish white, usually suffused with purplish pink, dull and inconspicuous, tube cuplike, 2–4 mm long, a little longer than broad to broader than long, lobes very small, triangular, obtuse to acute, recurved.

Fruit 5–8 mm in diameter, glaucous-blue. [*Decachaena frondosa* (L.) T. & G.]

Populations northeastward of our range (N.H., Mass., s.e. N.Y., southward chiefly on the coastal plain to S.C.) attaining a height of 2 m, having widely spreading branches, glabrous twigs, leaves 3–6 cm long and 2–3.5 cm broad, corolla 3–4 mm long, usually a little longer than broad, constitute *G. frondosa* var. *frondosa*.

Two varieties occur in our area of coverage:

a. G. frondosa var. tomentosa A. Gray.

Relatively loosely branched, to about 1 m tall. Young twigs copiously pubescent with short, curly hairs and glandular-dotted. Larger leaves from about 3 to 6 cm long, sparsely to densely pubescent beneath and not glaucous. Floral tube and calyx glabrous, not glaucous, glandular-dotted. [*G. tomentosa* (A. Gray) Small; *Decachaena tomentosa* (A. Gray) Small]

Borders of shrub-tree bogs or bays, wet pinelands, well-drained pinelands; local in our area. (S.e. S.C. to cen. pen. Fla., Fla. Panhandle, s. Ga. and s. Ala.)

b. G. frondosa var. nana A. Gray.   Fig. 111

Stems with relatively short branches, with a columnar aspect, 2–6 (–10) dm tall. Young twigs copiously pubescent with short, curly hairs, glandular-dotted, and glaucous. Larger leaves 2–4 cm long, sparsely short-pubescent and strongly glaucous beneath. Floral tube and calyx glabrous, glaucous, with or without glandular dots. [*Decachaena nana* (A. Gray) Small]

Both wet and well-drained pinelands, borders of shrub-bogs or bays and cypress depressions; relatively frequent in our area. (S. Ga., s. Ala., across n. Fla. and to cen. pen. Fla.)

Perhaps the two taxa of *Gaylussacia* occurring in our area of coverage and treated just above as varieties of *G. frondosa* are best considered to be specifically distinct from each other and from the tautonomous variety. Duncan and Brittain (1966) came to this conclusion; Walter Judd (personal communication)

Fig. 111. **Gaylussacia frondosa** var. **nana:** a, flowering branch to left, fruiting branch to right; b, portion of lower leaf surface enlarged; c, flower; d, fruit.

agrees. If so, then the names are *G. tomentosa* (A. Gray) Small and *G. nana* (A. Gray) Small.

3. Gaylussacia mosieri (Small) Small.   Fig. 112

In general aspect similar to *G. dumosa* but its branches more spreading, its potential height about 1.5 m. Young twigs, raceme axes, flower stalks, and floral tubes bearing longish spreading, silvery-silky hairs minutely glandular at their tips, the glandular tips, at least some of them, red.

Leaf surfaces usually somewhat pubescent when young but the pubescence variable as to density and the relative number of stipitate-glandular and non-glandular hairs.

Mature fruit black, not glaucous, bearing gland-tipped hairs, 8–10 mm in diameter. [*Lasiococcus mosieri* Small]

Pitcher plant bogs, shrub tree bogs and bays (mainly on their borders), seasonally wet pine savannas and flatwoods; in our area, s. cen. and s.w. Ga., Fla. Panhandle, s. Ala. (Also Miss., La.)

5. Kalmia

Evergreen (ours) shrubs or small trees. Leaf scars shield-shaped, each with 1 vascular bundle scar; winter buds very small, usually with 2 outer scales; stipules none. Leaves simple, alternate (in ours), pinnately veined, petioled or sub-

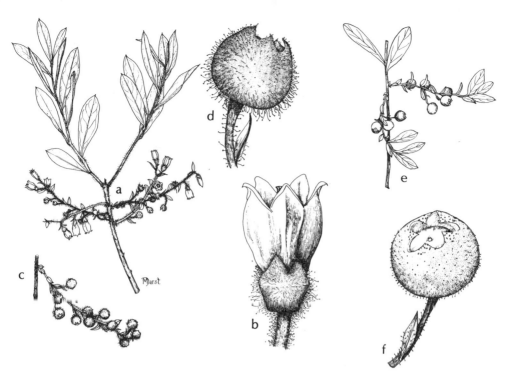

Fig. 112. a–d, **Gaylussacia mosieri:** a, flowering branch; b, flower; c, fruiting raceme; d, fruit. e–f, **Gaylussacia dumosa:** e, fruiting raceme; f, fruit.

sessile. Flowers borne in terminal or axillary panicles, lateral racemes or fascicles, or solitary in leaf axils; radially symmetrical, bisexual, each subtended by a bract and the flower stalks with a pair of bractlets at its base. Calyx united below, deeply 5-lobed above, persistent on the fruit or tardily deciduous. Corolla saucer- or bowl-shaped, shallowly 5-lobed above, the bowl with a red ring or a ring of red dots within, at, or near its base and having 10 pockets more or less medially interiorly in which the anthers are held in the bud. Stamens 10, filaments slender, anthers in the bud lodged in the pockets of the corolla, their filaments tripped (presumably by an insect visitor), the pollen being catapulted out in the process, anthers unappendaged, opening by terminal slits Pistil 1, seated on a 10-lobed disc, ovary superior, 5-locular, ovules several to numerous in each locule; style slender, straight or bent, stigma capitate. Fruit a septicidally 5-valved capsule, usually with a distinct depression apically.

1. Leaves well distributed along very slender stems, relatively small, 5–15 mm long and 2–8 mm broad, sessile or subsessile, edges revolute; stems mostly 1–6, rarely to 10, dm tall; flowers or fruits borne singly, less frequently 2–3, in leaf axils all along segments of the present season's growth.   1. *K. hirsuta*
1. Leaves relatively large, in compact clusters near the tips of branchlets, blades 2–10 cm long and 3–5 cm broad, not revolute, petioles 1–4.5 cm long; stems coarse, 1–6 m tall or more; flowers or fruits in panicles terminally at the tips of branchlets of the previous season.   2. *K. latifolia*

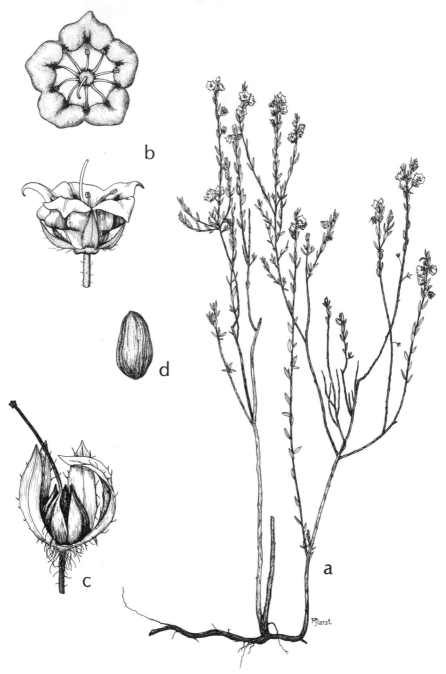

Fig. 113. **Kalmia hirsuta:** a, habit; b, flower, side view below, face view above; c, capsule, with subtending persistent sepals (one removed); d, seed.

1. Kalmia hirsuta Walt. HAIRY WICKY.

Fig. 113

Straggly, evergreen shrub 1–6, rarely to 10, dm tall, commonly with several to numerous slender stems from a hard, stout, subterranean base. Twigs light brown, bearing numerous long, spreading hairs.

Leaves small, light green, sessile or subsessile, variable in shape, even on a single stem, ovate, elliptic, oblong, oblong-elliptic, oval, oblanceolate, or subrotund, 5–15 mm long, 2–8 mm broad, bases rounded to truncate, apices acute, obtuse, infrequently rounded, surfaces glabrous to moderately pubescent with scattered long hairs, almost always with scattered long hairs on or near the revolute margins.

Flowers solitary, less frequently 2–3, in leaf axils along most of the present year's growth, their stalks 3–10 (–20) mm long, hirsute, each with a pair of elliptic-oblong, foliaceous bractlets just above the subtending leaf. Calyx lobes lanceolate, acute, 5–8 mm long, margins markedly ciliate, often a few long hairs on the midrib above. Corolla campanulate, pale to deep pink or rose, less frequently white, marked interiorly with red or purple around the anther-pockets and with a ring of red spots lower down, limb 5–6 mm high.

Capsule ovoid, 2–3 mm across or a little more, pointed apically, tan, the persistent sepals exceeding it; style persistent, after dehiscence of the capsule usually remaining attached to the tip of one of the valves. Seeds minute, mostly somewhat asymmetrical, a little longer than broad, to about 0.5 mm long, amber, surfaces finely more or less striate-lined and reticulate (as seen with suitable magnification). [*Kalmiella hirsuta* (Walt.) Small]

Pine savannas and flatwoods, borders of shrub-tree bogs or bays, hillside bogs in pinelands; throughout our area. (Coastal plain, s.e. S.C. to n. cen. pen. Fla., westward to s.e. La.)

Although the hairy wicky commonly has a disheveled appearance owing to its low, straggly habit, and lacks showiness from afar because its flowers are so small, it is nonetheless attractive close-up. The flowers, having the distinctively attractive corollas of the genus, the more attractive since it has a long blooming period, are both pretty and unusual. Isolated individual specimens are sometimes especially handsome suggesting its being given more attention than it gets as an ornamental subject, perhaps for moist, acidic rock gardens.

2. Kalmia latifolia L. MOUNTAIN-LAUREL, CALICO-BUSH, MOUNTAIN IVY.

Fig. 114

Commonly a shrubby plant with multiple stems from the base or near the base, occasionally single-trunked specimens attain small tree stature. In many places, plants grow so thickly as to be almost impenetrably coppicelike. Trunks and branches characteristically crooked and contorted, the ultimate branchlets many and with closely set foliage effecting a leafy crown. Young twigs reddish green, somewhat viscid and short-pubescent; older bark reddish brown and with long narrow furrows and ridges, sloughing in narrow strips or flakes. Leaf scars impressed, having rims around them, the buds above them with a visorlike protuberance over them.

Leaves sometimes so closely approximate that they appear opposite, petioles 1–4.5 cm long. Blades leathery, lateral veins not evident, 2–10 cm long and 3–5 cm broad, lanceolate, elliptic, elliptic-lanceolate, or oblanceolate, cuneate basally, apically obtuse, acute, or short-acuminate, the extreme tips bluntly callused; upper surfaces dark green and glabrous, lower paler green, sometimes

245

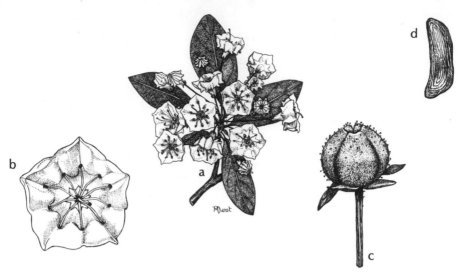

Fig. 114. **Kalmia latifolia:** a, flowering branchlet; b, flower, face view; c, capsule; d, seed.

with small, reddish, clavate glands or shortly stipitate glands, sometimes finely and granularly punctate, margins entire.

Flowers borne in showy panicled corymbs at the ends of leafy branchlets of the previous season. The individual flowers are exquisite both in the bud stage and at full anthesis, the buds notable for their usually deep coloration and their ridged and pebbled surface texture. Inflorescence axes, flower stalks, and often the calyces viscid, stipitate-glandular. Flower stalks slender, each subtended by a short-ovate bract, and each with a pair of shorter bractlets at the base of the stalks, these partially obscured by the primary bract. Calyx lobes oblong, blunt apically, about 3 mm long. Corollas angled saucer- or bowl-shaped, resembling, old-timers might say, inverted, starched, calico skirts, their color from rose-pink to pale, delicate pink, or almost white, often varyingly thus as they age. Interiorly, the corolla having a pink or reddish band or ring of dots at its near-basal throat and pink, reddish, or purplish markings at the anther pouches.

Capsule depressed-globose or oblate in outline, usually longitudinally with 5 low rounded lobes, surface stipitate-glandular, apex with a small depression, the persistent sepals spreading across its base, not exceeding the periphery of the base, the long slender style often persistent until dehiscence, sometimes attached to the tip of one of the valves after dehiscence. Seeds small, mostly about 1 mm long, about one-fourth as broad, tending to be falcate, opaque, pale-amberish, surface finely striate-lined and reticulate (as seen with suitable magnification).

In our area, in woodlands on ravine slopes, along small woodland streams, wooded bluffs along the larger rivers; s.w. Ga., s. Ga., s. Ala., Fla. Panhandle from Leon Co. westward. (N.B. to s. Ont. and s. Ind., generally southward to w. Fla. Panhandle and La.)

The mountain laurel has long been prized as a garden ornamental; however, although it occurs as a native shrub in parts of our area of coverage, its use as an ornamental here appears to be very much restricted. I am aware of but a very

246

few specimens in my home city, Tallahassee, Fla., but these are very vigorous and handsome ones.

6. Leucothoe

Evergreen or deciduous shrubs, some colonial. Leaves simple, alternate, short-petiolate, blades pinnately veined. Flowers in axillary racemes from buds on wood of the previous season. Flower stalks each subtended by a bract and each having a pair of bractlets either basally or beneath the calyx. Calyx united basally, 5-lobed, persistent. Corolla white or white tinged with pink, tube urceolate, with 5 small lobes at the summit. Stamens 10, not exserted, filaments straight, anthers opening by terminal pores, awnless or 2–4-awned apically. Pistil 1, ovary superior, 5-locular, style straight, included or exserted. Fruit an oblate 5-valved, loculicidal capsule. Seeds numerous.

1. Leaves evergreen; bract subtending the flowers broadly short-ovate, usually broader than long, firm-textured; bractlets borne basally on the flower stalks, ovate, tips blunt, completely hidden by the bract.   1. *L. axillaris*
1. Leaves deciduous; bract subtending the flower deciduous long before anthesis, bractlets borne just below the calyx, ovate-alternate.   2. *L. racemosa*

1. Leucothoe axillaris (Lam.) D. Don. DOG-HOBBLE.   Fig. 115

Evergreen, loosely branched shrub, leafy branches commonly more or less arching, to 1.5 m tall. Young branchlets green, usually very short-pubescent, twigs eventually becoming brown.

Leaves with short-pubescent petioles 2–10 mm long, blades leathery, lustrous dark green above, duller and paler green beneath, ovate, elliptic, elliptic-oblong, or oval, 5–14 cm long, 1.5–5 cm broad, obtusely short-tapered basally, apices short-acute, short-acuminate, or abruptly mucronate, upper surfaces glabrous, lower minutely pubescent when young, often glabrous in age, margins cartilaginous (more obvious after drying), sometimes variously serrate, the teeth few and mostly on the distal margins, or more or less throughout the margins, or margins sometimes entire.

Flowers in sessile racemes 2–7 cm long, borne in axils of leaves of the previous season; axes of racemes pubescent, flowers not secund; flower stalks stoutish, 1–4.5 mm long, pubescent, each subtended by a broadly short-ovate or suborbicular persistent bract, 2 smaller bractlets closely set above the bract, and completely hidden by it. Calyx united basally, lobes 5, persistent, ovate or lance-ovate, bluntish at their tips. Corolla white or white tinged with pink, 6–8 mm long, tube oblong-urceolate, the 5 lobes short-deltoid, recurved. Stamens not exserted from the corolla, anther halves awnless at the apex.

Capsule oblate, slightly if at all lobed, about 5 mm across, dark brown with buff sutures. Seeds 1.5–2 mm long, irregularly compressed-angular, lustrous amber, surfaces reticulate (as observed with suitable magnification).

Wet woodlands astride small streams, seasonally wet depressions in woodlands, floodplain forests, wooded stream banks, swamps, shrub-tree bogs; throughout our area. (Coastal plain, s.e. Va. to Fla. Panhandle, westward to La.)

2. Leucothoe racemosa (L.) Gray. FETTER-BUSH.   Fig. 115

Deciduous or tardily deciduous shrub to about 4 m tall. Twigs of the season copiously short-pubescent.

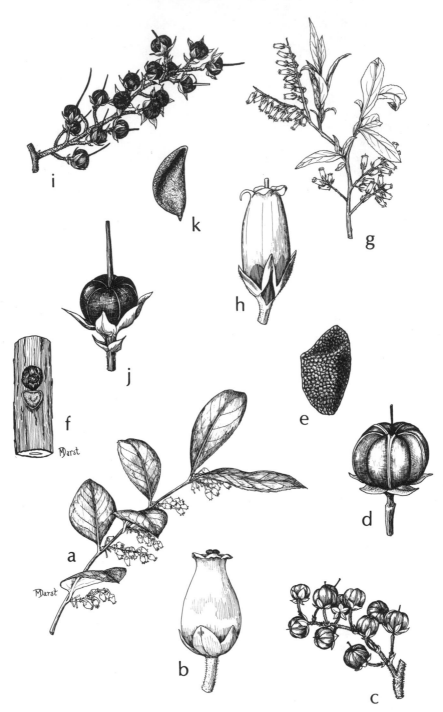

Fig. 115. a–e, **Leucothoe axillaris:** a, flowering branch; b, flower; c, fruiting raceme; d, fruit; e, seed. f-k, **Leucothoe racemosa:** f, small portion of twig with leaf scar and bud; g, flowering branchlet; h, flower; i, fruiting raceme; j, fruit; k, seed.

Leaves with very short, pubescent petioles; blades membranous, mostly elliptic, varying to lance-elliptic or oval, variable in size on a given branch, 1–5 cm long, 6–30 mm broad, cuneate to rounded basally, acute or short-acuminate apically, surfaces short-pubescent at least when young, usually remaining so, margins shallowly and obscurely crenate-serrate.

Racemes sessile, partially developing from leaf axils during the summer before flowering, at anthesis 2–10 cm long, axes densely short-pubescent, flowers secund; flower stalks about 2 mm long, subtending bracts linear, 8–10 mm long, deciduous long before anthesis of the flower, bractlets short, ovate-attenuate, borne just below the calyx, persistent. Calyx lobes lanceolate to lance-ovate, acuminate apically, a little less than half as long as the corolla. Corolla white, sometimes tinged with pink, tube oblong-cylindric, about 8 mm long; lobes 5, very short, broadly rounded, recurved; anther halves 2-awned at the summit.

Capsule oblate, not lobed, 4–5 mm broad, dark brown with buff sutures. Seeds angled, with 2 flat faces, the back rounded, about 0.3–1.0 mm long, light brown. [*Eubotrys racemosa* (L.) Nutt.]

Swamps, shrub-tree bogs, open bogs, cypress-gum depressions, shoreline thickets; throughout our area. (Chiefly but not exclusively coastal plain, e. Mass. to n. cen. Fla., westward to s.e. La.; cen. Tenn.)

7. Lyonia

Evergreen or deciduous shrubs, rarely of small-tree stature. Bark thin, grayish or reddish brown, thinly longitudinally ridged and furrowed. Terminal buds none, axillary buds ovoid, conical, or flattened and triangular in outline, usually with 2 outer scales; leaf scars half-round, each with a single, central vascular bundle scar. Leaves simple, alternate, short-petiolate, blades pinnately-veined; stipules none. Flowers in umbelliform clusters borne on leafy or leafless branches, forming racemes or panicles; each flower stalk subtended by a bract and each with bractlets near its base. Perianth 5–merous (in ours). Calyx united basally, 5–lobed, persistent in most, deciduous in fruit in some. Corolla tube cylindric-campanulate, urceolate, or globose-urceolate, the 5 lobes short. Stamens 10; filaments flattened, S-shaped, with or without a pair of spurlike appendages below anther-filament junction; anthers awnless, opening by terminal pores. Pistil 1, ovary superior, 5-locular (in ours), style straight, not exserted, stigma truncate or capitate. Fruit a 5-valved, loculicidal capsule thickened along the paler sutures which appear as ridges before dehiscence. Seeds numerous, appearing like fine sawdust (scobiform).

1. Plant evergreen, some leaves present on wood of a previous season.
  2. Lower leaf surfaces (especially when leaves young) bearing stipitate-peltate or shieldlike scales.
    3. Flowers (and subsequent fruits) borne on twigs of the previous season essentially prior to production of any new shoot growth.　　　　　　　1. *L. ferruginea*
    3. Flowers (and subsequent fruits), with rare exception, borne on twigs of the season after the twigs are nearly fully developed.　　　　　　　2. *L. fruticosa*
  2. Lower leaf surfaces glabrous.　　　　　　　4. *L. lucida*
1. Plant deciduous, no leaves present on wood of the previous season.
  4. Young twigs terete; leaf margins minutely serrate; corolla ovoid-globose, 3–4 mm long.　　　　　　　3. *L. ligustrina*
  4. Young twigs angled; leaf margins entire; corolla oblong-cylindric, 7–14 mm long.　　　　　　　5. *L. mariana*

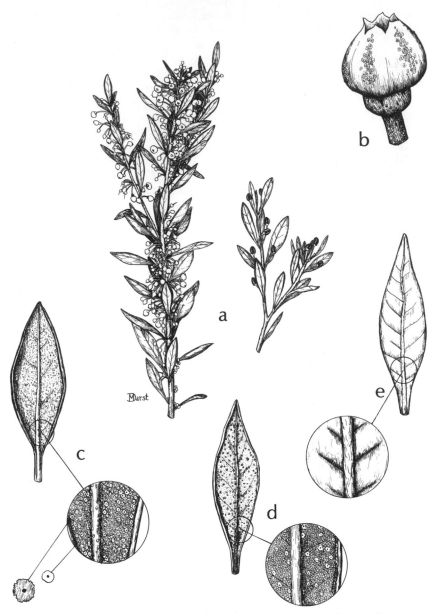

Fig. 116. **Lyonia ferruginea:** a, flowering branchlet to left and fruiting branchlet at right; b, flower; c, lower surface of fresh leaf, an enlargement of a small portion of it, and further enlargement of scales representing two sizes present; d, lower surface of dried leaf with enlargement of portion of same; e, upper surface of dried leaf with similar enlargement. (Flowering takes place before new growth of the season commences.)

1. Lyonia ferruginea (Walt.) Nutt. STAGGER-BUSH, POOR-GRUB.   Fig. 116

Evergreen shrub, often clonal, to 6 m tall or a little more, laxly branched, larger specimens, especially, with crooked trunks, spreading crowns, irregularly spreading branchlets. Young twigs with moderate to dense, short, gray pubescence and stipitate-peltate or shieldlike scales, becoming much less scaly in age, somewhat less pubescent, and dirty-grayish.

Leaf blades when only half or less than half developed wholly rusty-colored with stipitate-peltate scales on both surfaces, the scales quickly deciduous from the upper surfaces leaving them glossy green and usually with pale, short pubescence along the midribs; lower surfaces usually with persistent, short, and inconspicuous pubescence and with at least some rust-colored and some gray scales, the scales usually of two size classes, the larger rust-colored and irregularly margined, the smaller entire-margined and gray, both more or less persisting or only the smaller one persisting; leaves short-petioled; blades with entire margins, elliptic or oblanceolate to narrowly obovate, 1–7.5 cm long, 0.5–3 cm broad, at full maturity commonly with a downward curvature and revolute, principal lateral veins, at least parts of them, impressed on the upper surfaces and sometimes in troughlike depressions.

Flowers in axillary fascicles on twigs of the previous season, their stalks loosely scaly and sometimes with nonglandular pubescence, each with a pair of bractlets at the base. Calyx persistent, lobes triangular, 1–2 mm long, outer surfaces scaly, inner glabrous or pubescent. Corolla subglobose-urceolate, 2–4 mm long, usually broader at the base than long, lobes very short, white. Capsule ovoid, scaly and pubescent, 5-angled, 3–6 mm long, with five thickened sutures which separate as a unit from the valves in dehiscence. [*Xolisma ferruginea* (Walt.) Heller]

In both poorly drained pine flatwoods (where commonly intermixed with *L. fruticosa*) and in well-drained sand pine–oak scrub (where *L. fruticosa* is present less frequently); s.e. and s. cen. Ga., n. Fla., westward to about Bay Co. (S.e. S.C., southward to about the latitude of n. Lake Okeechobee in pen. Fla., and Fla. Panhandle.)

National champion stagger-bush (1971), 2′5″ circumf., 40′ in height, 21′ spread, Orange Home, Fla.

2. Lyonia fruticosa (Michx.) G. S. Torr. in Robins. STAGGER-BUSH, POOR-GRUB.

Fig. 117

Evergreen shrub, usually clonal, to 1.5, rarely to 3, m tall, branches usually rigidly erect; young twigs often densely pubescent and abundantly clothed with rust-colored scales, becoming glabrous or sparsely pubescent in age.

Leaf blades when only half or less than half developed wholly rust-colored with stipitate-peltate scales of one size class; scales soon deciduous on upper blade surfaces leaving them dull green and with short, grayish pubescence along the midrib; scales on lower leaf surfaces mostly eventually sloughing or turning gray leaving the surfaces dull grayish; mature leaf blades flat or their edges sometimes slightly curved upwardly, margins rarely revolute, obovate to oblanceolate, oval, or elliptic, 0.5–5 cm long and 0.3–2.8 cm broad, veins not impressed on the upper surfaces, usually raised and often notably reticulate.

Flowers in axillary fascicles, borne on newly developed twigs of the season (rarely a few flowers on an occasional branch of an occasional plant on twigs of a previous season). Calyx persistent, lobes triangular, 1–1.5 mm long, the outer surfaces scaly-pubescent, the inner sparsely to moderately nonglandular pubes-

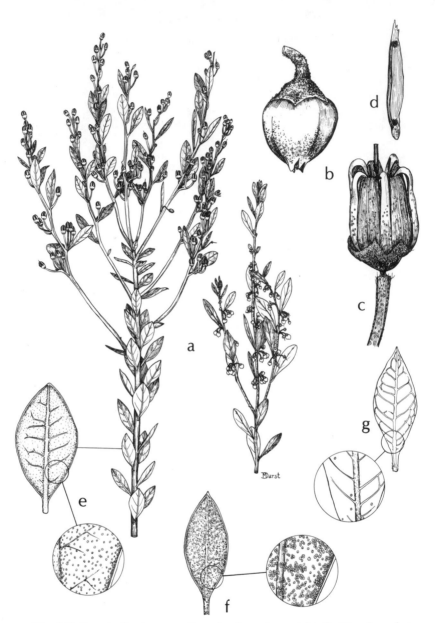

Fig. 117. **Lyonia fruticosa:** a, flowering branch to right, fruiting branch to left; b, flower; c, fruit; d, seed; e, lower surface of fresh leaf (from previous season's growth) with enlargement of small portion of same; f, lower surface of dried leaf (from current season's growth) with enlargement of small portion of same; g, upper surface of dried leaf from current season's growth with similar enlargement. (Flowering is almost exclusively on the current season's growth.)

cent. Corolla ovoid-urceolate, 2.5–5 mm long, usually about as broad at the base as long, scaly without, lobes very short, white.

Capsule ovoid, 5-angled, 3–5 mm long, with thickened sutures which separate as a unit from the valves in dehiscence. [*Xolisma fruticosa* (Michx.) Nash]

Poorly drained pine flatwoods, shrub-tree bogs and bays, less frequently in sand pine–oak scrub; s.e. and s. cen. Ga., n. Fla. westward to Liberty and Franklin Cos. (S.e. S.C. to s. pen. Fla. and e. half of Panhandle of Fla.)

3. Lyonia ligustrina (L.) DC. var. foliosiflora (Michx.) Fern. MALEBERRY, HE-HUCKLEBERRY. Fig. 118

Deciduous shrub 0.4–4 m tall. Young twigs terete, sparsely short-pubescent at first, eventually becoming glabrous; older twigs purplish brown, leaf scars raised; fully developed buds conical.

Leaves short-petiolate; blades variable in size and shape, lanceolate, elliptic, oval, oblanceolate, or obovate, to 8 cm long or a little more and 1–5 cm broad, cuneate basally, mucronate at the rounded, obtuse, or acute apices, margins minutely serrate, upper surfaces pubescent chiefly along the midrib, lower more conspicuously pubescent along the principal veins, petioles pubescent.

Flowers in fascicles proximally, singly distally, borne on irregularly leafy-bracted, racemolike, angling ascending shoots arising from axils of leaf scars on distal portions of wood of the previous season; vigorous new shoots of the season sometimes well developed from below the inflorescence and ascending so as partially to obscure it; flowering branches often numerous from the old wood giving the effect of a panicle. Calyx united about half its length, the 5 lobes deltoid, 0.5–1 mm long, pubescent. Corolla white or with a faintly pinkish tinge, tube globose-ovoid, 3–4 mm long and as broad, the 5 very small lobes curving outwardly. Stamens not at all exserted from the corolla, filaments flat, S-curved, without appendages, anthers curved, terminal pores oblique.

Capsule oblate to globose, sparsely short-pubescent, the sutures prominently pale-cartilaginous, the leathery calyx persistent about its base. Seeds numerous, variable in shape and size, narrow, oblongish and irregularly angled longitudinally, often falcate, amber-colored, 1–2 mm long. [*Arsenococcus frondosus* (Pursh) Small]

Thickets and woodlands along streams, shrub thickets in pineland depressions, shrub-tree bogs and bays, swamps, cypress-gum depressions; throughout our range. (Chiefly coastal plain, s.e. Va. to cen. pen. Fla., westward to e. Tex., Okla., Ark.)

4. Lyonia lucida (Lam.) K. Koch. FETTERBUSH, STAGGER-BUSH, HURRAH-BUSH.
Fig. 119

Evergreen shrub, commonly very showy and handsome when in flower; flowering specimens varying very greatly in stature from a few dm tall to at least 4 m; the large ones robustly branched from the base, their crowns about as broad as the height of the plant, usually occurring where surface water is present much or nearly all of the time. Young twigs strongly angled, green flecked with dark, loose, soon deciduous, narrow scales.

Very young, partially grown leaves minutely scaly on both surfaces (similar scales on flower stalks and calyces); mature leaf blades leathery, glabrous, dark, glossy green above and with irregularly scattered, very minute punctate dots and somewhat paler beneath, there the dots much more abundant and more evenly distributed (dots purple on fresh leaves, usually brownish on dried ones);

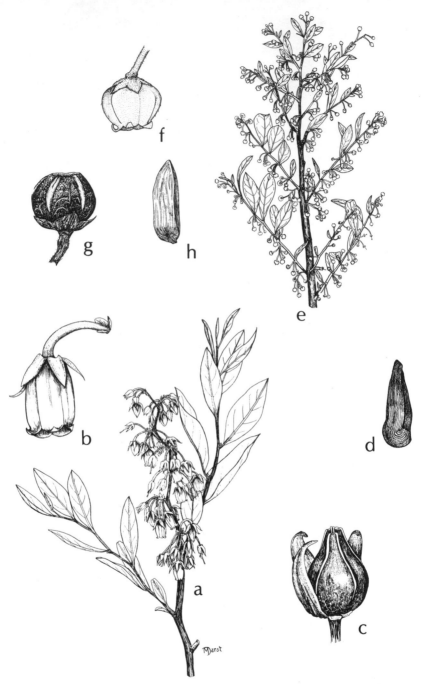

Fig. 118. a–d, **Lyonia mariana:** a, flowering branch; b, flower; c, capsule, two sepals removed; d, seed. e–h, **Lyonia ligustrina:** e, flowering branch; f, flower; g, capsule; h, seed.

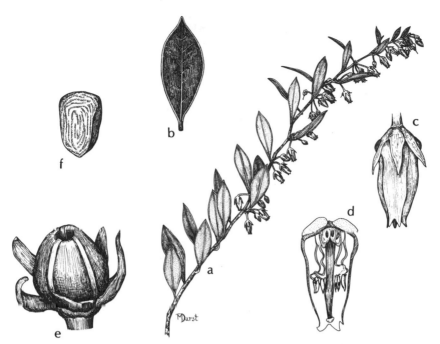

Fig. 119. **Lyonia lucida:** a, flowering branch; b, leaf, lower surface; c, flower; d, flower (corolla) in longitudinal section; e, capsule; f, seed.

blades narrowly to broadly elliptic, varying to oval or obovate, 2–8 cm long, 1–4 cm broad, bases cuneate, apices short-acuminate or acute, margins entire, a vein running parallel to and near each margin, not revolute when fresh, revolute when dry.

Flowers in fascicles, mostly to 10 (15) per fascicle, in leaf axils on wood of the previous season before new growth commences in spring; flower stalks 2.5–12 mm long, usually bearing few, scattered, inconspicuous, short-knobby trichomes. Calyx united only basally, 5–lobed, the lobes elongate-triangular, mostly acute apically, 2–5 (7.5) mm long, very much shorter than the corolla, persistent on the fruit, sometimes a little shorter than the fruit, sometimes a little exceeding it. Corolla pink, varying to nearly white or red, swollen basally when fresh, more or less cylindrical distally from the swollen base, 5–9 mm long, with 5 minute, rounded lobes apically. Filaments flattened, 3–5 mm long, usually having 2 spurs below anther-filament junction.

Capsule ovoid to globose-ovoid, sometimes urnlike, 3–5 mm long and as broad, with pale, slightly to strongly thickened sutures which remain attached to an adjacent valve at dehiscence. Seeds many, minute, mostly thinly wedge-shaped, amber-brown. [*Desmothamnus lucidus* (Lam.) Small]

Shrub-tree bogs and bays, seasonally wet pine savannas and flatwoods, cypress-gum ponds, wet woodlands, sometimes in Florida scrub; throughout our range. (Coastal plain, s.e. Va. to s. cen. pen. Fla., westward to La.; Cuba.)

5. Lyonia mariana (L.) D. Don. STAGGER-BUSH.   Fig. 118

Shrub with subterranean runners and to some extent colonial, 2–10, sometimes to 20, dm tall, usually not branched below, branches above strongly ascending. Young twigs angled, glabrous (in ours).

Leaves deciduous (some sometimes overwintering, deciduous in spring), very short-petioled, blades elliptic, oblong-elliptic, or narrowly obovate, bases cuneate, apices obtuse to rounded, rarely acute, sometimes mucronate, the larger ones 4–10 cm long, 1.5–4.5 cm broad, margins entire; upper surfaces glabrous, lower glabrous or short-pubescent along the principal veins (in ours).

Flowers nodding, borne in fascicles in axils of leaf scars on wood of the preceding season. Calyx purplish red, united only basally, the lobes 5, lance-subulate or lance-oblong, 3-nerved, 6–8 mm long, deciduous by full maturity of the capsule. Corolla white or pinkish, tube oblong-cylindric, 7–14 mm long, lobes very short and broad, outwardly recurved. Stamens a little more than half as long as the corolla tube, filaments flattish, broadest basally, arching, white, pubescent.

Capsule 5–7 mm long, ovoid-obpyramidal, truncate apically, sutures prominently pale-cartilaginous. Seeds many, mostly irregularly angled and obconic in outline, summits truncate or obliquely truncate, amber-colored, 1–1.5 mm long. [*Neopieris mariana* (L.) Britt.]

Pine savannas and flatwoods, hillside bogs in pinelands, edges of cypress-gum ponds or depressions, borders of shrub-tree bogs or bays, open, mixed, well-drained woodlands; in our range, s. cen. Ga., n. Fla., westward to cen. Fla. Panhandle. (Coastal plain and adjacent piedmont, s. R.I., s. Conn., N.J. and e. Pa., thence to cen. pen. Fla. and cen. Fla. Panhandle; w. La., e. Tex., s.e. Okla., Ark., s.e. Mo.)

8. Oxydendrum

Oxydendrum arboreum (L.) DC. sourwood.   Fig. 120

A medium tall, slender, deciduous tree with sour- or acid-tasting sap, commonly flowering/fruiting when of low shrublike stature. Young twigs green or reddish green, eventually with relatively conspicuous, vertically lenticular lenticels. Leaf scars hemispheric in outline, each with 1 vascular bundle scar. Bark of trunk gray tinged with red, longitudinally furrowed between scaly ridges.

Leaves alternate, simple, slenderly petioled, petioles usually reddish green, mostly 1.5–2 cm long. Blades pinnately veined, lateral veins arching-ascending and anastomosing well inside the margins; variable in length to about 15–18 cm long and 2.5–7.5 cm broad, long elliptic-oblong to elliptic, bases rounded, apices acute or acuminate; margins often reddish, entire or finely and irregularly doubly serrate; upper surfaces glabrous, brightly medium green, lower dull green, varyingly glabrous, or mainly along the midribs ashy-pubescent or with sparse relatively coarse straight hairs, or both. Leaf blades in autumn turning orange-yellow, then scarlet or crimson, the trees then notably contributing to autumn coloration.

Inflorescence composed of longish racemes emanating from a central axis, diverging bowlike from the axis more or less in one plane, spraylike overall and the sprays arching-declining from the tips of branchlets of the season in a very attractive manner. Flowers individually small, shortly stalked, vertically declining from the lower side of the raceme axes; after flowering the stalks of the developing fruits turn upward, the fruits eventually held erect above the axes; raceme axes, flower stalks, and flowers white; axes, bracts, bractlets, flower stalks, calyces, and corollas minutely but copiously pubescent with white, curly hairs. Flowers bisexual, radially symmetrical, each flower stalk subtended by a minute, white bract deciduous well before the flower is fully grown, in addition

256

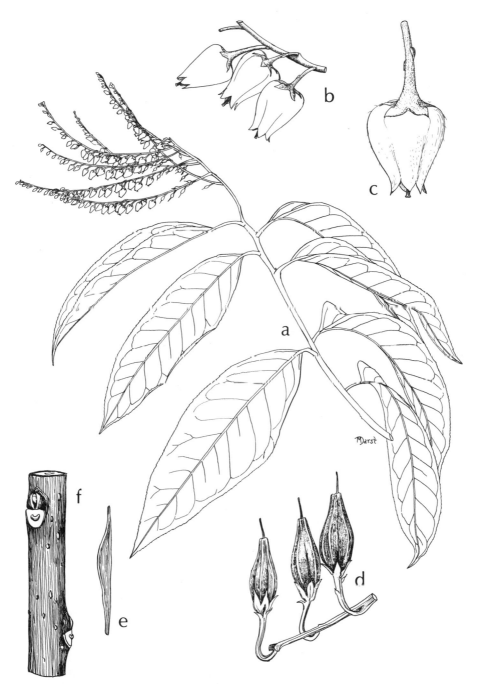

Fig. 120. **Oxydendrum arboreum:** a, branchlet with inflorescence; b, portion of raceme (flowers descending); c, flower; d, portion of fruiting raceme (fruits erect); e, seed; f, piece of woody twig.

with 1 or 2 minute, persistent bractlets, usually alternately disposed if 2, more or less midway of the length of the stalk. Calyx composed of 5 small, deltoid sepals, united only basally, embracing the base of the corolla. Corolla 6 mm long or a little more, tube ovoid-urceolate, not constricted at the throat, distally with 5 obtusely triangular, short, erect lobes. Stamens 10, about two-thirds as long as the corolla tube, filaments white, somewhat fleshy, broadest basally, pubescent distally, appressed against the ovary, anthers awnless, orange, as long as the filaments, opening by terminal slits. Pistil 1, ovary superior, ovoid, rounded at the summit, pubescent, 5-locular, style thickish, stigma capitate.

Fruit a 5-valved, loculicidal, oblongish capsule 5–8 mm long, the persistent calyx subtending it, the style persistent until dehiscence. Seeds numerous, with a more or less ellipsoid, amberish-opaque body and taillike extremities, about 3 mm long overall, surface of the body finely striate-reticulate (observed with suitable magnification).

Sparsely scattered in well-drained woodlands of bluffs, ravines, and hills; in our area, s. Ga., Fla. Panhandle, s. Ala. (s. N.J. and Pa., s. Ohio, s. Ind., s.e. Ill., to n. Fla. on the eastern boundary, on the western boundary from s.e. Ill. to n.e. La.; e. Tex.)

National champion sourwood (1972), 6'5" circumf., 118' in height, 25' spread, Nantahala National Forest, N.C.

9. Pieris

Pieris phillyreifolius (Hook.) Small. VINE-WICKY. Fig. 121

Evergreen shrub; sometimes with weakly erect stems from subterranean runners; perhaps more commonly ascending the trunks of pond-cypress (*Taxodium ascendens*) and white cedar (*Chamaecyparis thyoides*), its main, somewhat branched, stem growing beneath the outer bark, the stem strongly flattened, bearing minute, brown scale-leaves, the growing tips calloused, and having lateral "normal" leafy branches exserted at intervals through the cypress bark; also sometimes growing amongst persisting, decaying petiole bases of cabbage palm (*Sabal palmetto*), or if old petiole bases have fallen off and the trunks of the palms well covered with moss, ascending the trunks amongst or beneath the moss mats.

Leaves alternate, very short-petiolate, blades glabrous, pinnately veined, elliptic, oblong-elliptic, or lanceolate, 1.5–7 cm long, 5–20 mm broad, margins revolute, entire or with a few, small, very blunt teeth distally, less frequently toothed from somewhat below their middles, surfaces punctate, the punctae with minutely stalked red glands, lower surfaces more conspicuously so than the upper.

Flowers in 3–9-flowered racemes borne in the axils of leaves near branch tips well before new growth of the season commences; raceme axis and flower stalks powdery-pubescent, stalks 3–5 mm long, thickish, each subtended by a subulate bract about as long as the stalk and bearing a pair of subulate bractlets at a little distance below the calyx, both bracts and bractlets glandular-pubescent. Calyx persistent, united only basally, lobes 5, lance-triangular, about half as long as the corolla tube, margins stipitate-glandular above the middle. Corolla white, somewhat translucent, 7–9 mm long, tube ovoid-urceolate, narrowed at the throat and then with 5 very short, somewhat recurved lobes. Stamens 10, filaments somewhat S-curved, anthers with a pair of deflexed spurs on the back above the junction with the filament, each anther-half opening by an oval or

Fig. 121. **Pieris phillyreifolia:** a, habit of nonvining plant; b, portion of vining plant in bark of tree, bark partially pulled away; c, d, stems from beneath bark of tree showing scale leaves; e, enlargement of small section of d; f, flower; g, corolla opened out with stamens and pistil; h, stamen.

V-shaped pore. Ovary superior, 5-locular, fruit a loculicidally 5-valved, subspheroidal capsule. Seeds brown, irregularly angled longitudinally, more or less obpyramidal in outline, about 1 mm long.

The vine-form of this plant most commonly on pond-cypress, but occasionally on associated shrubs or trees. [*Ampelothamnus phillyreifolius* (Hook.) Small]

Cypress-gum ponds and depressions and their pineland borders; throughout our range. (Coastal plain, S.C., Ga., to cen. pen. Fla., Fla. Panhandle, s. Ala.)

10. Rhododendron (RHODODENDRONS, AZALEAS)

Evergreen or deciduous shrubs or small trees, often perennating by subterranean runners. Leaves simple, alternate, short-petiolate, without stipules, blades pinnately veined. Buds with numerous, imbricated scales, the inflorescence buds larger than the shoot buds. Leaf scars shield-shaped to more or less obcordate, each with 1 vascular bundle scar. Flowers showy, in corymbose clusters, rarely solitary, from buds terminally or subterminally on shoots of the previous season, produced before or as shoots of the season develop or after shoots of the season are essentially fully formed. Each flower subtended by a bract, a pair of elongate bractlets at the base of the flower stalk. Calyx small, united below, 5-lobed or -toothed above, persistent. Corolla cylindric-tubular below or cylindric-tubular proximally and flaring distally, the 5-lobed limb spreading, weakly bilaterally symmetrical, the largest lobe often spotted or differently colored than the others. Stamens 5 or 10, occasionally of an intermediate number, filaments longish, usually unequal in length, well exserted, anthers unappendaged, opening by terminal pores. Pistil 1, ovary superior, 5-locular, ovules numerous in each locule; style slender, curved, stigma capitate. Fruit a septicidal capsule. Seeds numerous, minute, flat and more or less winged to fusiform.

Walter S. Judd kindly permitted me to see his manuscript on this genus prepared for the forthcoming flora of the southeastern United States and to some extent I have adapted from it. For that I am greatly in his debt.

1. Leaves evergreen, their lower surfaces with rust-colored, peltate scales.   4. *R. minus*
1. Leaves deciduous, their lower surfaces without scales.
  2. Corollas red, orange, orange-red, or yellow.
    3. Scales of winter buds glabrous exteriorly, their margins minutely ciliate; flowers appearing after the shoots of the season are mature; corolla tube glabrous or with few stipitate glands.   5. *R. prunifolium*
    3. Scales of winter buds densely to moderately pubescent exteriorly, margins minutely ciliate; flowers appearing before or as new shoot growth occurs; corolla tube usually moderately to densely stipitate-glandular or hirsute.   2. *R. austrinum*
  2. Corollas white to pink.
    4. Flowers appearing in summer after shoots of the season are mature. 6. *R. viscosum*
    4. Flowers appearing before or as the shoots of the season are developing.
      5. Corolla white and with a blotch of yellow on the upper lobe; scales of winter buds glabrous exteriorly (apart from ciliate margins).   1. *R. alabamense*
      5. Corollas deep pink to pale pink or white with a tinge of pink, without a blotch of yellow on the upper lobe; scales of winter buds pubescent exteriorly. 3. *R. canescens*

1. Rhododendron alabamense Rehd. ALABAMA AZALEA.   Fig. 122

Deciduous shrub to about 4 m tall. Twigs of the season with scattered, enlarged-based, longish hairs and otherwise uniformly covered with numerous, short, curvate hairs, both gradually sloughing. Exposed faces of scales of winter buds essentially glabrous, their margins minutely ciliate.

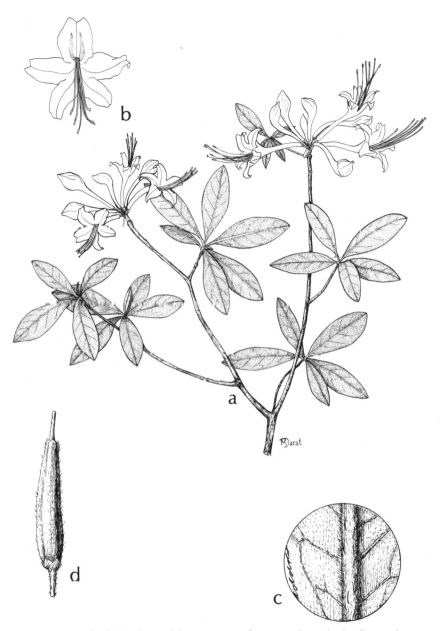

Fig. 122. **Rhododendron alabamense:** a, flowering branch; b, flower, face view; c, enlargement of portion of lower surface of leaf; d, capsule.

Petioles 3–6 cm long; blades mostly elliptic, oblanceolate, or narrowly obovate, 2.5–7.5 cm long and 1–3.3 cm broad, bases cuneate, apices obtuse to rounded, less frequently acute, sometimes shortly mucronate; margins entire but short-ciliate, both surfaces relatively sparsely pubescent, hairs much like those on the young twigs, the midveins more copiously pubescent, usually only with short hairs.

Flowers appearing before or as new shoot growth occurs in spring, their stalks 5–10 mm long, with longish hairs usually mixed with stipitate-glandular ones; bracts usually longer than the flower stalks, markedly dilated distally, bractlets shorter and little dilated distally. Calyx lobes with few hairs on their faces, margins fringed with longish hairs, usually with a few stipitate-glandular ones as well. Corolla 2.5–4 cm long, tube narrowly funnelform, sparsely stipitate-glandular, gradually expanded into the limb, limb white with a blotch of yellow on the upper lobe, lobes about 1–2 cm long. Stamens 5, much exserted, about twice as long as the corolla tube. Capsules 1.5–2 cm long, subcylindric, straight or falcate, longitudinally 5-lobed, surfaces with few longish hairs, often some stipitate-glandular ones, and many short-curvate ones, stub of the style protruding from the persistent spreading calyces. [*Azalea alabamensis* (Rehd.) Small]

Usually in mixed, well-drained woodlands, occasionally where poorly drained; apparently rare in our area of coverage: Leon Co., Fla., Decatur Co., Ga., Monroe and Clarke Cos., Ala. (Otherwise s. to n. in w. Ga., throughout much of Ala. and cen. Tenn.)

Where *R. alabamense* and *R. canescens* grow intermixed or in close proximity, hybridization between them apparently occurs; the flowers of the putative hybrids are generally pale pink and have 1, 2, or 3 of the lobes blotched with yellow.

2. Rhododendron austrinum (Small) Rehd. YELLOW AZALEA, FLORIDA AZALEA.

Fig. 123

Deciduous shrub to 3–6 m tall. Young twigs and petioles stipitate-glandular and with nonglandular pubescence as well, most of the pubescence sloughed by the end of the first year and twigs brown. Exposed faces of scales of winter buds copiously short-pubescent over much of the surface, copiously short-ciliate marginally and margins of some scales with small reddish glands as well.

Petioles 2.5–14 mm long; leaf blades elliptic to obovate, mostly 3–11 cm long and 1.5–4.5 cm broad, bases cuneate, apices acute to rounded with a very small mucro. Mature blades pubescent on both surfaces, the lower more abundantly so than the upper, midveins usually with longer hairs than the surfaces otherwise; margins minutely dentate, the dentations each bearing an antrorsely inclined, transparent, spiculelike trichome.

Flowers appearing before or as shoots of the season develop, sometimes fragrant, their stalks 5–15 mm long, usually notably stipitate-glandular, glands pale to reddish, sometimes the hairs nonglandular, both conditions sometimes in the same inflorescence. Calyx similarly pubescent, margins of the lobes longish-ciliate, the hairs gland-tipped or not. Corolla 2.5–4.5 cm long, the tube 1.5–2.5 cm long, yellow to orange, red, or orange-red, exteriorly with scattered stipitate glands, these usually diminishing in number distally, sparsely to moderately nonglandular-pubescent as well, interiorly nonglandular-pubescent; limb more or less abruptly expanded from the tube, lobes about 1–2 cm long, usually pale to golden yellow, sometimes orange or orange-red. Stamens 5, much exserted, yellow to reddish, 2–3 times as long as the corolla tube, filaments non-

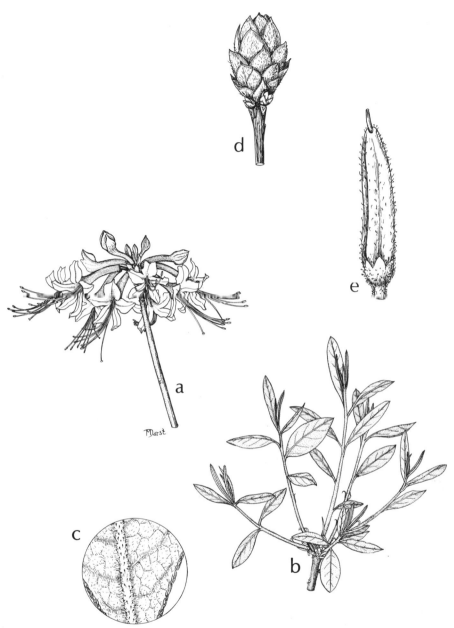

Fig. 123. **Rhododendron austrinum:** a, flowering branch; b, leafy branch; c, enlargement of portion of lower surface of leaf; d, winter (inflorescence) bud; e, capsule.

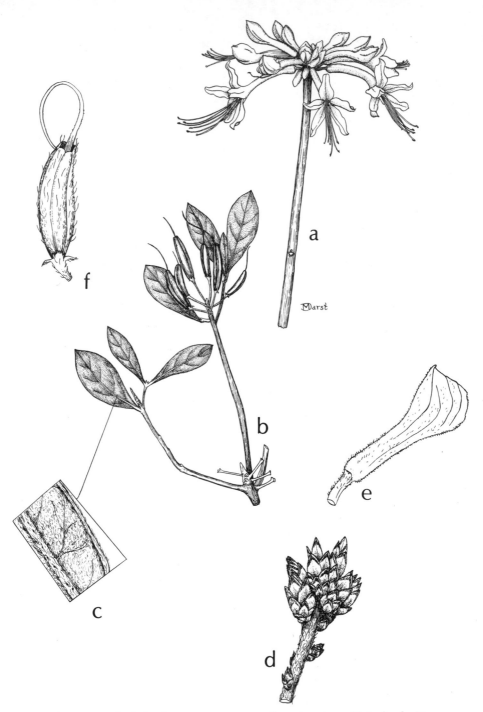

Fig. 124. **Rhododendron canescens:** a, flowering branchlet; b, fruiting branch; c, enlargement of a portion of lower surface of leaf; d, end of twig with winter floral buds; e, flower just prior to anthesis; f, capsule (fully grown but not yet mature enough for persisting style to have dehisced).

glandular pubescent proximally. Capsules 1.5–2.5 cm long, straight or falcate, subcylindric, longitudinally 5-lobed, stipitate glandular pubescence often intermixed with both longish and short nonglandular hairs, persistent stubs of styles often extruding from the persistent, erect calyces. [*Azalea austrina* Small]

Wooded bluffs and slopes, banks of small woodland streams; Panhandle of Fla., Leon and Liberty Cos. westward, s.w. Ga., s. and cen. Ala.

3. Rhododendron canescens (Michx). Sweet. SOUTHERN PINXTERBLOOM, PIEDMONT AZALEA, HOARY AZALEA. Fig. 124

Deciduous shrub to 5 m tall. Young twigs with moderate to copious, shaggy, long hairs or more frequently a mixture of them and short, tangled, cottony hairs, sometimes with some stipitate-glandular ones as well. Exposed faces of scales of winter buds densely short-pubescent medially and on their apical mucros, their margins usually copiously very short-ciliate.

Petioles 2–10 mm long, pubescent like the stems; mature blades mostly oblanceolate or narrowly obovate, varying to elliptic, 2–10 cm long, bases cuneate, apices obtuse to acute, minutely mucronate, sparsely to densely pubescent beneath, often with stouter, stiffer hairs along the midrib, margins bristly-ciliate with short, curved ascending hairs.

Flowers appearing before or with new shoots of the season, fragrant, their stalks 4–12 (17) mm long, densely soft-pubescent, rarely stipitate-glandular. Calyx densely soft-pubescent, the lobes pectinate-ciliate and/or stipitate-glandular. Corollas 2.5–4.5 cm long, tube funnelform, 1.3–1.7 cm long, pink to nearly white, exteriorly densely pubescent with soft hairs and with sparse to copious stipitate-glandular ones intermixed, short-pubescent within; tube more or less abruptly expanding into the deep to pale pink or nearly white limb, its lobes 10–18 mm long. Stamens 5, much exserted, about 3 times as long as the corolla tube. Capsules 1.2–3.2 cm long, lance-falcate in outline, pubescent with scattered longish hairs, sometimes stipitate-glandular ones, and short, dense hairs. [*Azalea canescens* Michx.; incl *A. candida* Small]

Wet woodlands, springy places in woodlands, borders of shrub-tree bogs, and bays, pine flatwoods, well-drained woodland slopes, banks of woodland streams; throughout our area. (Coastal plain and piedmont, Del., Md., to n. cen. pen. Fla., westward to e. Tex., s.e. Okla., Ark., w. Tenn.)

Where *R. canescens* grows intermixed with or in close proximity to *R. alabamense*, hybridization between them apparently occurs; the flowers of the putative hybrids are generally pale pink and have 1, 2, or 3 of the lobes blotched with yellow.

4. Rhododendron minus Michx. var. chapmanii (A. Gray) Duncan & Pullen. CHAPMAN'S RHODODENDRON. Fig. 125

Evergreen shrub to about 3 m tall, branching relatively open and generally stiffly ascending. Young twigs, petioles, and lower surfaces of leaf blades markedly dotted with rust-colored, peltate scales, the upper blade surfaces much less conspicuously dotted with much smaller dark dots. Exposed surfaces of scales of winter buds also with rust-colored scales, sometimes short-pubescent as well, their margins short-ciliate.

Flowers appear in early spring, usually, if not always, before new shoot growth commences, their stalks 5–15 mm long, copiously scurfy-dotted as are the calyces, the minute lobes of the latter ciliate. Corollas 1.5–3.5 cm long, pink to rose-pink, tube funnelform, 1–2 cm long, rather gradually flaring into the limb, exteriorly both the tube and the limb irregularly scaly, limb 3–4 cm across.

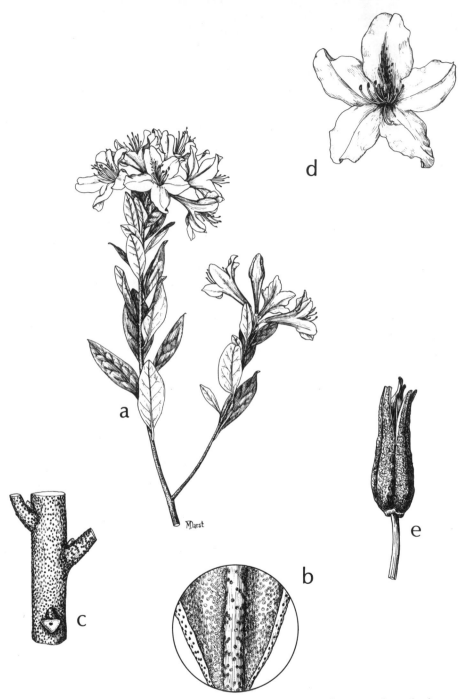

Fig. 125. **Rhododendron minus** var. **chapmanii:** a, flowering branch; b, enlargement of a portion of lower surface of leaf; c, piece of woody stem; d, flower, face view; e, capsule, dehisced.

Stamens 10, relatively little exserted from the throat of the corolla, somewhat longer than the style. Capsule 6–14 mm long, longitudinally 5-lobed, dark rusty-scaly, slightly urceolate in outline. [*R. chapmanii* A. Gray]

Seasonally wet pine flatwoods, borders of shrub tree bogs and bays, transition zone between the latter and sand ridges; n.e. Fla., cen. Panhandle of Fla.

R. minus var. *minus* (incl. *R. carolinianum* Rehd.), of general distribution in the s. Appalachians and extending southward to s.w. Ga. and s.e. Ala., just to or closely peripheral to our area, has a branching habit more compact or more straggling, branches not so rigidly erect-ascending, as in var. *chapmanii*; its petioles longer, 2–7 cm, and its leaf blades for the most part averaging longer, 2.5–13 cm, their apices more generally acute to short-acuminate; its flowering may occur before new shoot development but, apparently, frequently as shoots of the season are about two-thirds mature.

5. Rhododendron prunifolium (Small) Millais. PLUMLEAF AZALEA.

Deciduous shrub to 4–6 m tall. Young twigs glabrous or with a few longish hairs. Exposed faces of scales of winter buds glabrous, their margins short-ciliate.

Petioles 3–11 mm long, blades elliptic to obovate, 3–13.5 cm long and 1–5 cm broad, bases cuneate, apices rounded, obtuse, less frequently acuminate; upper surfaces glabrous or rarely sparsely pubescent with longish hairs, the midveins short-pubescent, lower surfaces glabrous or with a few longish hairs on the mid-vein and sometimes on the lateral veins, margins ciliate.

Flowers appearing in summer after shoots of the season are mature, not fragrant, mostly 4–5 cm long or a little more, their stalks 5–10 mm long, sparsely shaggy-pubescent. Calyx lobes fringed with longish, nonglandular or long-stipitate glandular hairs. Corolla tube funnelform, gradually dilated into the limb, 2.3–3.5 mm long, tube and limb pink to crimson, glabrous to sparsely stipitate-glandular, sometimes having short, nonglandular pubescence as well, short-pubescent within, lobes of the limb 1.3–2.2 cm long. Stamens 5, much exserted, about 3 times as long as the corolla tube. Capsules 1–2 cm long, densely pubescent with longish hairs and sparsely with short ones. [*Azalea prunifolia* Small]

Chiefly in forested ravines along streams; s.e. Ala. and s.w. Ga., perhaps only peripheral to our area of coverage.

6. Rhododendron viscosum (L.) Torr. SWAMP OR CLAMMY AZALEA.   Fig. 126

Deciduous shrub to 3–5 m tall, the branching relatively open and stiffly ascending-erect. Young twigs sparsely to densely shaggy-pubescent with long hairs, sometimes with short ones as well. Exposed faces of scales of winter buds glabrous to partially or wholly short-pubescent, often with dark brown bands within the distal margins, the latter short-ciliate, less frequently glandular; scale apices rounded or with mucros arising somewhat back of the rounded or obtuse tips and extending a little beyond the tips.

Leaves subsessile or with petioles to 6 mm long; blades oblanceolate to narrowly obovate, varying to elliptic, 1–7 (9) cm long and 0.7–3 cm broad or a little more, bases cuneate, apices rounded, obtuse, or acute, usually minutely mucronate; both surfaces green, or green above, glaucous beneath, upper glabrous or with sparse longish hairs, sometimes also short ones, stout appressed-ascending hairs along the midribs beneath and otherwise more or less short-pubescent throughout; margins bristly-ciliate, the trichomes usually curved beneath the leaf edge and often scarcely visible from above.

Fig. 126. **Rhododendron viscosum:** a, flowering branch; b, enlargement of portions of leaf surfaces, lower surface to right, upper to left; c, overwintering inflorescence buds (with leafy branches removed—flowering occurs after new growth of the season is well developed); d, flower just prior to anthesis; e, fruiting branch; f, capsule.

Flowers appearing in summer after shoots of the season are mature, very fragrant, their stalks 5–20 mm long, usually with some sparse, cottony pubescence intermixed or overlaid with long-spreading, nonglandular and stipitate-glandular hairs. Calyx with some thin-cottony pubescence and sparsely stipitate-glandular, margins of the lobes usually copiously clothed with elongate, stipitate-glandular hairs. Corolla white, white with pink stripes, or pink-tinted; tube cylindric below, abruptly flared distally into the limb, 1.5–2.5 cm long, lobes about 1 cm long, spreading, shorter than the tube, exteriorly with some soft, short, pubescence, the tube with scattered stipitate-glandular hairs, the latter also in a band along each side of the middle of each lobe, tube short-pubescent interiorly, stamens 5, exserted, 1–2 times as long as the corolla tube, filaments pubescent from their bases to about two-thirds their length. Capsules mostly 1–2 cm long, usually lanceolate in outline, shallowly longitudinally 5-lobed, abundantly clothed with elongate stipitate-glandular hairs (the glands tiny and sometimes sloughed) and very fine, inconspicuous, short hairs as well. [*Azalea viscosa* L.; incl. *A. serrulata* Small (*R. serrulatum* (Small) Millais, *R. viscosum* var. *serrulatum* (Small) Ahles)]

In our area, mostly in swamps, wet woodlands, bogs, shrub-tree bogs and bays, seasonally wet pine flatwoods; nearly throughout. (Maine to Ohio, generally southward to s. cen. pen. Fla. westward to s. Miss.)

11. Vaccinium

Low shrubs to small trees; some, beyond our range, trailing shrubs or epiphytic. Leaves simple, alternate, short-petiolate, blades pinnately veined. Flowers bisexual, ovary partly to completely inferior, borne singly or in few-flowered clusters from buds on wood of the previous season, solitary in the axils of leaves on branchlets of the season, or in racemes or sprays on branchlets of the season. Flower stalks usually with 2 bractlets, these commonly quickly deciduous. Floral tube surmounted by 4 or 5 persistent calyx segments, 5 in ours. Corolla white, greenish, varying to red, cylindric, urceolate, or campanulate, shallowly to deeply 4–5-lobed, 5-lobed in ours. Stamens 8 or 10, anthers with or without spurs on the back, the anther-halves narrowed distally into tubes opening by terminal pores. Pistil 1, ovary 4–5 locular, or falsely 10-locular above, 5-locular below; style slender, usually exceeding the stamens, stigma small, sometimes minutely capitate. Fruit a 5-many-seeded berry crowned by the persistent calyx.

A large genus, mostly of the northern hemisphere, composed of 8 subgenera which are by some authors segregated as genera. Representatives of 3 subgenera occur in our area.

Taxonomically, the genus is, in general, difficult, not a little of the difficulty being attributed to hybridization. For eastern North America, a plethora of names exists for variant taxa at one or another level and there is little consensus on the part of authors in their respective treatments.

1. Corolla open as flowers develop from the early bud stage to full anthesis; anthers well exserted from the corolla at full anthesis; flower stalks not jointed just below the floral tube. 1. *V. stamineum*
1. Corolla closed in the bud; anthers not exserted at full anthesis; flower stalks jointed just below the floral tube.
  2. Small tree, to 8–10 (14) m tall (although flowering and fruiting when of low stature); many, even most, leaf blades on a given plant rounded apically apart from a minute mucro at their tips, blade margins essentially entire and bearing minute sessile glands,

at least when young; corolla open-campanulate at anthesis; anthers with awns on the back, these about half the length of the tubular extensions of the anther-halves.

                                             2. *V. arboreum*

2. Shrubs to about 4 m tall; most leaf blades on a given plant pointed, blade margins entire or toothed, without glands, or if with them then the glands stipitate, tipping teeth, or sessile in the sinuses of low, rounded teeth; corolla urceolate, usually somewhat constricted at the throat; anthers without awns on the back.

    3. Evergreen shrub (leaves present on wood of the previous year).

        4. Lower surfaces of leaf blades with stalked or clavate glandular hairs; floral tube and fruit not or only slightly glaucous.   3. *V. myrsinites*

        4. Lower surfaces of leaf blades without pubescence; floral tube and fruit conspicuously glaucous.   4. *V. darrowi*

    3. Deciduous shrub (leaves shed in autumn or if some persist overwinter, these shed in spring).

        5. Leaf blades (the larger ones on a given plant) exceeding 3.5 cm long, marginally entire or, if toothed, the teeth not gland-tipped.   5. *V. corymbosum*

        5. Leaf blades (the larger ones on a given plant) not exceeding 3 (3.5) cm long, minutely toothed marginally, the teeth tipped by gland-tipped hairs.

            6. Lower surfaces of leaf blades glabrous or pubescent, if the latter then the hairs nonglandular; stems to 40 dm tall.   6. *V. elliottii*

            6. Lower surfaces of leaf blades with at least some of the hairs gland-tipped or clavate-glandular; stems to 3.5–4 dm tall.   7. *V. tenellum*

1. Vaccinium stamineum L. DEERBERRY, SOUTHERN-GOOSEBERRY, SQUAW-HUCK-LEBERRY, BUCKBERRY.   Fig. 127

A complex exhibiting numerous enigmatic variations. The "elements" have been treated, historically, as parts of the genus *Vaccinium* or as composing the genus *Polycodium* (contemporaneously considered as a subgenus of *Vaccinium*). Baker (doctoral dissertation, 1970) presented the results of a careful and exhaustive analysis of the variation and concluded that the subgenus is composed of a single species. Hitherto, some authors had considered rather poorly defined variants at specific and intraspecific levels and there is a plethora of epithets (too numerous to be cited here as synonyms) in combination with *Vaccinium* or *Polycodium* or both.

Shrub to about 5 m tall, largest stem diameter about 16 cm, generally with several to numerous stems from the base, branching diffuse, often, however, where subject to frequent burning or severe habitat disturbance, having subterranean runners and forming small to large clones, the aerial stems of such clones 1–6 dm high; branches terete, glabrous to moderately pubescent, sometimes stipitate-glandular; older woody stems brownish gray, scaly. Hairs in general, of two basic types, uniseriate and either straight or recurved, or biseriate and stipitate-glandular.

Leaves deciduous or tardily deciduous, slenderly short-petiolate. Blades membranous to somewhat leathery, variable, on sterile leader shoots 2–8 cm long and from about 1 to 3 cm broad, elliptic, ovate, oblong, obovate, or suborbicular; bases acute to rounded or subcordate, apices acute to short-acuminate with small mucros or apicules at their tips; margins mostly entire, sometimes appearing ciliate from above because of marginally spreading pubescence on their lower surfaces, rarely irregularly finely serrate, often revolute if somewhat leathery, often with stipitate glands on basal margins; surfaces with or without varying degrees of glaucescence on both surfaces or only on the lower surfaces with varying densities of pubescence, the glaucescence and pubescence, if present, generally more pronounced or denser, respectively, on lower surfaces than

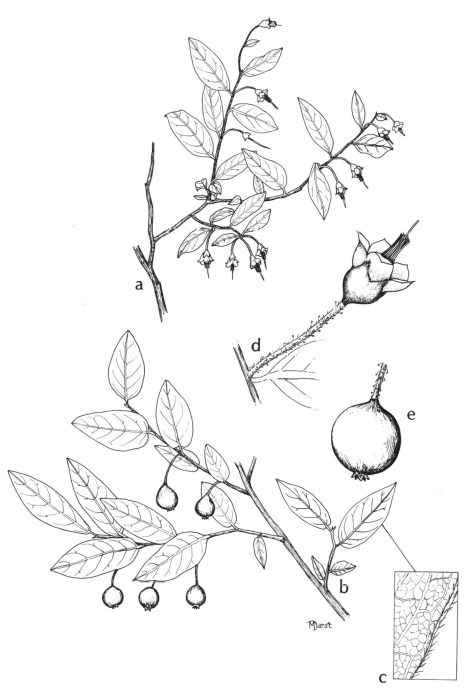

Fig. 127. **Vaccinium stamineum:** a, flowering branch; b, fruiting branch; c, enlargement of small portion of leaf; d, flower; e, fruit. (Illustration is representative of one of the several variants.)

upper; stipitate-glandular hairs may be present on stem, flower stalks, floral tubes, or calyces, variably present or disposed even on different branches of the same plant.

Buds, in spring, are of two kinds, turgid, ovoid ones which give rise to the flowering branches, and narrowly ovoid ones which give rise, usually later, to the season's leader branches. Since the flowering branches arise from buds distinctive from those for the vegetative leader shoots, the flowering ones varying from having flowers solitary in the axils of only moderately reduced leaves (bracts), relative to leaves of vegetative branches, to relatively proliferously flowered, small-bracteate, spraylike branches; such branches may be termed inflorescences with a racemose to paniculately racemose form.

Flowers open from the early bud stage until reaching full anthesis, then 7–13 mm high, stalks more or less pendulous, often stipitate-glandular; floral tube surmounted by a greenish calyx united below and with 5 semicircular, obtuse, acute, or deltoid lobes 1–2.5 mm long, glabrous except for ciliate margins. Corolla, at full anthesis, whitish to greenish, sometimes with dark veins, occasionally tinged with pink, united below, 5-lobed (rarely 4-lobed), lobes 1–3 mm long, one-half to one-third the length of the corolla, their apices rounded to broadly acuminate. Stamens 10, distinct, 5 in each of 2 whorls, those of outer and inner whorls respectively opposite the calyx lobes and corollas lobes; filaments short, arising from the floral disc, each anther sac with a long tubular projection extending beyond the corolla. Style filiform with a capitate stigma, exserted beyond the stamens.

Fruits mostly a little broader than long, varying from 5–16 mm in diameter, oblate to globose, ovoid, or slightly pear-shaped; at maturity (if they persist until maturity) variously whitish, yellowish, light to dark reddish, blue, purple-blue, dark blackish purple, reddish black to dark purple, glaucous or nonglaucous, glabrous or sparsely pubescent, juicy, somewhat bitter and sourish. [*Polycodium stamineum* (L.) Greene]

Inhabiting well-drained places, upland mixed woodlands, pine woodlands, open longleaf pine–scrub oak hills and ridges; throughout our area. (W. Mass. to s. Ont., generally southward to s. cen. Fla. easterly, the western boundary roughly from e. Ohio southwesterly to s. Ind., s. Mo., e. Okla., e. Tex.)

2. Vaccinium arboreum Marsh. SPARKLEBERRY, FARKLEBERRY, TREE-HUCKLE-BERRY, WINTER-HUCKLEBERRY. Fig. 128

Small tree (flowering and fruiting when of small shrublike stature), to 8–10 (14) m tall, lower trunks to about 3 dm in diameter. Bark of larger stems relatively thin, grayish brown, sometimes with a purplish tinge, sloughing in irregular thin plates or flakes exposing smooth, reddish brown inner bark, usually with a pleasingly mottled effect. Crown branching tending to be crooked or somewhat contorted and yielding a rounded crown. Twigs of the season shaggy-pubescent, year-old twigs glabrous, with a thin gray coating which breaks up revealing spots of reddish brown bark.

Leaves commonly tardily deciduous, often pinkish to reddish in autumn, short-petiolate or blades tapering below to subsessile bases. Mature blades stiffish and firm, obovate to short-elliptic, oval, or less frequently suborbicular, varying very much in size on a given plant, the larger ones 2–4 (7) cm long and 1–2.5 (4) cm broad, bases mostly cuneate (rounded on suborbicular

Fig. 128. **Vaccinium arboreum:** a, flowering branch; b, flowers; c, stamen; d, fruiting branch; e, fruit.

ones), apices rounded to obtuse, tipped by minute apiculations; upper surfaces glossy green, usually pubescent along the midrib, lower paler and dull green, generally pubescent along the midrib, along it and the major lateral veins, sometimes pubescent throughout; margins bearing small glands, these sometimes sloughed or dried-up in age.

Flowers borne in profusion in bracted or leafy-bracted racemes or sprays 2.5–7.5 cm long emanating from buds on old wood at the same time that nonflowering shoots of the season are developing, or even have about fully developed. Flower stalks slender, more or less pendulous, with 2 quickly deciduous bractlets a little below their middles, jointed below the flower. Floral tube with its calyx segments campanulate, tube glabrous, about 1.5 mm high, segments deltoid, pubescent near their tips. Corolla closed in the bud, at anthesis white, tube campanulate, 3–4 mm long, lobes about as long, oblongish, their narrowed tips usually recurving. Stamens reaching the throat of the corolla, filaments pubescent, anthers with awns a little more than half the length of the tubular extensions of the anther-halves; style extending beyond the anthers. Berries black, dryish, 5–8 mm in diameter, often some of them persisting into the winter. [*Batodendron arboreum* (Marsh.) Nutt.]

Acid to calcareous xeric woodlands, thickets, clearings; throughout our area. (Coastal plain and piedmont, Va. to n. Fla., westward to e. and s. cen. Tex., northward to s.e. Kan., Mo., s. Ill., s. Ind., Ky.)

National champion sparkleberry (1977), 3'7" circumf., 30' in height, 30' spread, Pensacola, Fla.

273

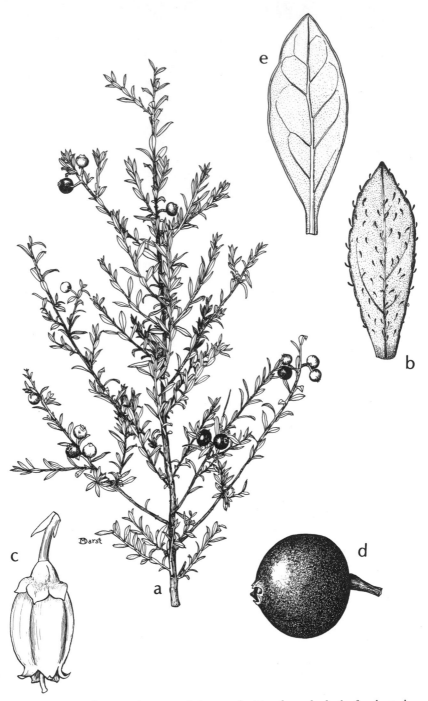

Fig. 129. a–d, **Vaccinium myrsinites:** a, fruiting branch; b, leaf enlarged, lower surface; c, flower; d, fruit. e, **Vaccinium darrowi:** leaf enlarged, lower surface.

3. Vaccinium myrsinites Lam. SHINY BLUEBERRY.   Fig. 129

Low, evergreen shrub, to about 6 dm tall, with stout and elongate subterranean runners and colonial. Young twigs glabrous or sparsely to densely pubescent with short, curvate hairs.

Leaves sessile or subsessile, commonly glossy green, varying to glaucous and grayish green, blades oblanceolate, short-obovate, or elliptic, mostly 5–15 mm long, 2–10 mm broad, glabrous above, sparsely beset with stalked or clavate glandular hairs beneath, these usually sloughed on mature leaves, occasionally sparsely pubescent beneath with nonglandular hairs as well, margins not revolute, with minute, appressed-ascending gland-tipped teeth, the glands eventually deciduous.

Flowers 2–8 in axillary fascicles or short racemes from buds on wood of the previous season, usually appearing before new shoot growth of the season commences. Floral tube glabrous, sometimes glaucous, calyx segments broadly triangular-obtuse, deltoid, or very broadly rounded, about 1 mm long, usually glabrous, sometimes short-ciliate marginally. Corolla closed in the bud, white to deep pink, urceolate, 6–8 mm long, lobes short-triangular or oblongish-triangular, apices mostly obtuse to rounded. Anthers not awned on the back. Berries black, less frequently blue-black, 6–8 mm in diameter. [*Cyanococcus myrsinites* (Lam.) Small]

In seasonally wet to well-drained pinelands, scrub, prairies, borders of shrub-tree bogs or bays; throughout n. Fla. and s. Ga., s.w. Ala. (Coastal plain, s.e. S.C. to s. pen. Fla., and s.w. Ala.)

4. Vaccinium darrowi Camp. GLAUCOUS BLUEBERRY.   Fig. 129

Evergreen shrub with habit, leaf size and shape, and general appearance of *V. myrsinites* save that it is more markedly glaucous and bluish green, the two often sharing the same habitat where their ranges overlap. Young twigs moderately to densely pubescent with short, curvate, gray hairs.

Leaves commonly glaucous on both surfaces at first, usually remaining glaucous beneath, upper surfaces glabrous, short-pubescent along the midrib, or not infrequently the entire surface clothed with very fine, short, appressed gray hairs, lower surfaces glaucous and glabrous, sparsely pubescent proximally along the midrib or along all the midrib, without glandular hairs, thickened margins usually somewhat revolute, entire or with low, appressed-ascending, gland-tipped teeth. Fruit glaucous-blue, 4–6 mm in diameter.

Habitats essentially as for *V. myrsinites*; throughout our area. (Coastal plain, s. Ga. to s. cen. pen. Fla., westward to s.e. Tex.)

5. Vaccinium corymbosum L. HIGHBUSH BLUEBERRY.   Fig. 130

The highbush blueberries, an important small-fruit crop in eastern and midwestern North America, have long been subject to various taxonomic interpretations. "Elements" include diploids, tetraploids, hexaploids, and various hybrid combinations. A high degree of hybridization and consequent morphological diversity amongst what have been considered species is at a maximum from Charleston, S.C., to the south of Jacksonville, Fla., and westward to n.e. Tex. (Vander Kloet, 1980). Vander Kloet, in his investigation, culminating in the 1980 paper, sought "to assess morphological variation, habitat diversity, and niche specilization in 56 highbush blueberry sites located in 48 regions scattered throughout eastern North America." From his data, he concluded that the complex should be considered as one species, namely *V. corymbosum*, including (*for our area of coverage*) the following named taxa at the species level: *V. arkansanum* Ashe; *V. ashei* Reade; *V. atrococcum* (A. Gray) Heller (*Cyanococcus atrococcus* (A. Gray) Small); *V. australe* Small; *V. elliottii* Chapm.; *V. fuscatum* Ait. (*C. fuscatus* (Ait.) Small); *V. virgatum* Ait. (*C. virgatus* (Ait.) Small).

It may be noted in addition that "elements" of this complex are also alleged to produce natural hybrids with the lowbush blueberry *V. darrowi*.

Inasmuch as I myself have not managed (with one exception) to conceptualize entities in the highbush blueberries occurring in the local area, I choose here (with not a little reluctance) to follow Vander Kloet in large part, the exception being *V. elliottii*, which seems to me to have such distinctiveness as to be recognizable in the field at a glance.

The variation in characteristics attributable to *V. corymbosum* sens. lat. is so great, the varying way characters combine from plant to plant is so diverse, that it is virtually impossible to make more than a very generalized description (such as that which follows).

Deciduous or tardily deciduous shrub to 3 (5) m tall, stems solitary or up to at least 12 from a single bole, often suckering or root-sprouting where disturbed or where having been burned. Twigs green, yellow, or reddish, glabrous and sometimes glaucous to densely pubescent, rarely glandular.

Mature leaf blades ovate, oblongish, elliptic, lanceolate, less frequently spatulate to obovate, varying considerably in size on most individuals, up to 8 cm long and to 4.5 cm broad (largest ones on any given plant generally considerably larger than the largest ones on specimens of *V. elliottii*); surfaces glabrous, lower sometimes thinly to moderately glaucous, or variously pubescent, sometimes glandular beneath, margins usually entire.

Flowers borne in short fascicles or racemes from buds on the previous year's wood, appearing before or as shoot growth of the season commences; corolla closed in the bud, white, white tinged with pink, or pink, subcylindric to urceolate, varying from 5 to 10 mm long. Anthers without awns on the back. Berries dull black, shining black, or blue and glaucous, 4–12 mm in diameter. [*Cyanococcus corymbosus* (L.) Rybd.]

Bogs, swamps, shrub-tree bogs or bays, wet to well-drained pinelands, upland mixed woodlands; throughout our area.

6. Vaccinium elliottii Chapm. MAYBERRY. Fig. 130

Deciduous shrub to 4 m tall, often with a broad, bushy-branched crown. Young twigs varyingly glabrous, powdery-pubescent, or short shaggy-pubescent, the latter sometimes with a few stipitate-glandular hairs intermixed; woody stems green.

Petioles short, pubescent, blades usually thin, glossy green above, narrowly to broadly elliptic, oblanceolate, or oval, 1–2 (3) cm long, 5–10 (15) mm broad, usually more or less pubescent along the midrib both above and beneath, sometimes sparsely pubescent over the entire surface beneath, margins minutely serrate, each serration of a developing leaf usually tipped by a quickly deciduous, stipitate gland.

Flowers in fascicles of 2–6 from buds on the previous season's wood, appearing before or as the shoots of the season develop. Flower stalks 2–5 mm long, glabrous or pubescent, floral tube and calyx segments glabrous, rarely glaucous, the segments broadly triangular, tips obtuse to rounded. Corolla closed in the bud, white or pinkish, narrowly urceolate, 5–7 mm long. Berries black, rarely glaucous, 5–10 mm in diameter.

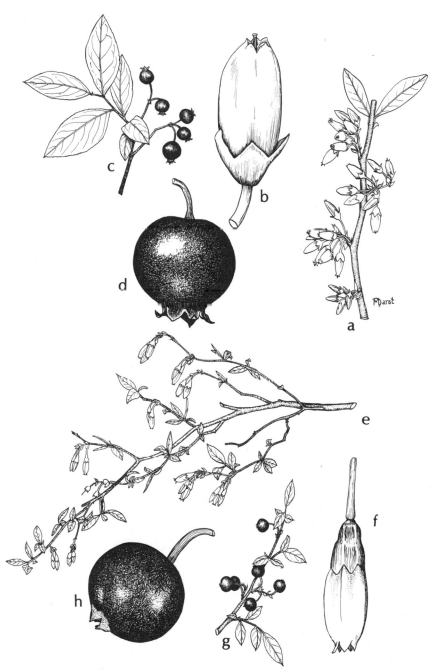

Fig. 130. a–d, **Vaccinium corymbosum:** a, piece of flowering branch; b, flower; c, fruiting branchlet; d, fruit. e–h, **Vaccinium elliottii:** e, flowering branch; f, flower; g, fruiting branchlet; h, fruit.

Flowering and fruit ripening occurring earlier than for *V. corymbosum*. [*Cyanococcus elliottii* (Chapm.) Small]

In a wide variety of habitats, from wet to very well drained, various upland woodland mixtures, stream banks, wet thickets and clearings, bottomland woodlands; common throughout our area and often abundant locally. (Chiefly coastal plain, s.e. Va. to n. Fla., westward to e. Tex. and Ark.)

7. Vaccinium tenellum Ait.

Shrub, perennating by subterranean runners and forming clones, erect stems to about 3.5–4 dm tall. Young twigs densely pubescent, much of the pubescence persisting on year-old woody twigs.

Leaves deciduous, oblanceolate to obovate, infrequently elliptic, 1.5–3.5 cm long, tips minutely apiculate, margins minutely toothed, teeth usually tipped by stipitate glands, upper surfaces pubescent with nonglandular hairs, chiefly along the midribs, lower surfaces usually bearing both nonglandular and stipitate-glandular or clavate-glandular hairs.

Flowers in fascicles or short racemes from buds on year-old wood, appearing before or as new shoot growth commences in spring. Floral tube glabrous or pubescent, calyx segments ovate to triangular. Corolla pink to whitish, tube urceolate, 5–8 mm long, lobes minute, oblongish or triangular. Berries black, 5–8 mm in diameter. [*Cyanococcus tenellus* (Ait.) Small]

Xeric, sandy, thin woodlands and their borders, clearings, well-drained pinelands; rare in our area of coverage or perhaps overlooked. (Chiefly coastal plain, s.e. Va. to s. cen. and s.w. Ga., n.e. Fla. (Clay Co.), not reported for Ala. by Clark (1971), Miss., Tenn.)

Euphorbiaceae (SPURGE FAMILY)

1. Leaf blades mostly as broad as long or broader than long.
    2. The leaf blades palmately veined or lobed.
        3. Blades of leaves cordate-ovate, unlobed, or if lobed, the lobes not at all leafletlike; inflorescences in early spring borne terminally on short shoots of the season as leaves are unfolding.   1. *Aleurites*
        3. Blades of the leaves deeply divided into 7–12 leafletlike lobes; inflorescences borne above the uppermost leaves of fully developed branches of the season.   2. *Manihot*
    2. The leaf blades pinnately veined. (BEWARE: The milky sap of this plant has poisonous properties.)   3. *Sapium*
1. Leaf blades much longer than broad.
    4. Wood of the stem hard, not extremely light in weight; margins of leaf blades entire; plants occurring in mesic or moist woodlands, if on floodplains then where flooded only temporarily.   4. *Sebastiania*
    4. Wood of the stem very light in weight, approximately as light as cork; margins of leaf blades minutely appressed-crenate, the teeth tipped by tiny callosities; plants growing where surface water stands much of the time.   5. *Stillingia*

1. Aleurites

Aleurites fordii Hemsl. TUNG-TREE.   Fig. 131

The tung-tree, native to China, a small tree, was until relatively recently extensively cultivated, principally in northern Florida and westward perhaps to east-

Fig. 131. **Aleurites fordii:** a, tip of twig with emerging leaves and inflorescence, a leaf from a seedling/sapling in background; b, leaf from mature tree and to the right enlargement to show glands at junction of petiole and blade; c, node with leaf scar; d, pistillate flower; e, staminate flower.

279

ern Louisiana, for its seeds which yield an oil used in commerce. It became naturalized in the vicinity of plantings and has spread from there.

Twigs glabrous, relatively stout and stiff, at first with a waxy bloom which as it sloughs leaves them brown, eventually having scattered, elliptic to oblong lenticels. Leaf scars subrotund, with several vascular bundle scars disposed more or less in a ring. Sap milky.

Leaves simple, alternate, deciduous, long-petiolate, stipulate, the stipules linear-sublate, about 1 cm long, quickly deciduous. Petioles bearing a pair of conspicuous red glands on their upper sides immediately below the bases of the palmately-veined blades. Blades of seedlings, sprouts, and young saplings relatively very large, to 30 cm long and broad, cordate basally, distally with a prominent terminal, acuminate lobe and one or a pair of smaller acuminate lobes to either side of it; at the base of the sinus of the lobe or lobes (at the end of a veinlet) there is a small rounded red gland. Saplings gradually, over several years, produce fewer and fewer lobed leaves, eventually all are unlobed and of smaller size, their blades ovate, about as broad as long, cordate to truncate basally, short-acuminate apically, margins not toothed. Petioles of young leaves sparsely pubescent, the blade surfaces with more or less matted, brown pubescence, denser beneath than above; most or all pubescence eventually sloughing from petioles and upper surfaces and becoming relatively sparse, not at all matted, on the lower.

In spring short shoots from buds terminally on the twigs bear developing leaves and relatively large, open, compound, paniclelike cymes of showy flowers, development of the cymes occurring more rapidly than that of the leaves. Flowers unisexual, usually the central flower of a given cyme pistillate, the others staminate; occasionally a cyme may bear more than one pistillate flower or a given cyme may have all staminate flowers. Receptacle with a 5-segmented disc. Calyx of 2 oblong-ovate sepals fused only basally. Corolla commonly having 5 oblong-obovate, curvate, spreading, separate petals, most corollas about 4 cm across; petal number may, however, vary to 6, 7, or 8, the variable number more characteristic of pistillate than of staminate flowers; petals nearly white exteriorly, interiorly red-veined near the base, the red color becoming fainter upwardly on the veins. Staminate flower with 8–16 stamens in 2 series, the outer shorter than the inner, filaments fused below, free above. Pistillate flowers with 1 pistil, ovary superior, 3–5 locular, one ovule in each locule, style short, stigmas 3–5. Fruit green and red, ovoid to globose, depressed globose, or obovoid, thick fibrous, 4–8 cm in diameter, tardily dehiscing by 3–5 valves, containing 3–5 large seeds each with 2 flat faces, 1 rounded.

Borders of woodlands, fence and hedge rows, sporadic throughout our range. (Ga., n. Fla., westward to s. Miss. and s. La.)

The tung tree is the only tree in our area having long-petioled leaves with broadly ovate, palmately-veined blades, a pair of glands on the upper side of the petiole immediately below the base of the blade. The Chinese tallow-tree has petioles similarly glandular, but its leaf blades are pinnately-veined.

2. Manihot

Manihot grahamii Hook.   Fig. 132

Arborescent, to at least 7 m tall, basal woody stems of larger specimens to at least 2 dm in diameter. Growth of the season vigorous, herbaceous until relatively late in the season, stem glabrous, green, slightly glaucous; second-year

Fig. 132. **Manihot grahamii:** a, habit (with some parts removed); b, flowers, functionally staminate above, pistillate below; c, infructescence.

woody stems stout, often 1.5–2 cm in diameter, light brown, with scattered, more or less circular, buff-colored lenticels; leaf scars subhemispheric, vascular bundle scars indistinct; sap milky, pith reticulately diaphragmed peripherally and pale greenish, solid centrally and white.

Leaves alternate, deciduous, with filiform, toothed stipules 8–30 mm long, usually quickly deciduous leaving minute stipular scars; petioles to about 2 dm long; blades palmately deeply divided into 5–11 leafletlike segments (lobes), the basal portion of the leaf undivided for about 1 cm and its basalmost portion (from which the veins radiate) low moundlike and somewhat pinkish pubescent; leafletlike segments pinnately veined, veins much more conspicuous on the upper, dark green surfaces, lower surfaces slightly glaucous and with a bluish-green cast, in shape oblanceolate, some or most of the medial ones usually fiddle-shaped above their middles, apices acuminate, lateral, basal segments smallest, about 8 cm long and often declined, gradually larger medially and to 15 cm long; overall the leafletlike segments of larger leaves radiating to about a full circle, in outline broader than long.

Inflorescence above 3 (4) very closely approximate leaves at the terminus of a shoot of the season (lower leaves of which are well separated); lowermost branches of inflorescence usually 3–5 in number, these spreading, stalked racemes, occasional ones subpaniculate, 7–10 cm long including stalks, lowermost 2–3 flowers of the racemes pistillate, those above staminate; central axis of the inflorescence more or less erect, loosely paniculate, 7–10 cm long or more, its flowers often all staminate, sometimes with some pistillate flowers. During the time that such an inflorescence is itself developing and during the period of its anthesis, another fast-growing branch arises from the axil of each of the 3 or 4 very closely approximate leaves subtending the inflorescence just described and terminally on each of these branches the process is repeated.

Perianth consisting of a calyx only, about 1.5 cm high, united only basally in the pistillate or to about half its length in the staminate, with 5 oblongish, apically blunt lobes, tube and lobes yellowish green without, interiorly with 2–4 conspicuous purplish red stripes extending from just below the throat to about two-thirds the length of the lobes; seated within the perianth is an orange-yellow disc. In the staminate flowers the disc having 10 rounded lobes marginally, a filament of each of the 10 stamens arising beneath the disc and each one curving around and then upward between 2 lobes of the disc, eventually the filaments and anthers more or less interlacing within the perianth. In the pistillate flowers the disc is somewhat thinner, is not lobed marginally, is often pink tinted; pistil 1, ovary seated on the disc, 2-locular, 1 ovule in each locule, style short, stigma with 3 radially spreading, fleshy, crinkled sublobes. Calyx open, lobes radially spreading only in the morning when flower at full anthesis. Fruit a round capsule, both extremities truncate, about 1.5 cm long and as broad, septicidally dehiscent.

Native to s.e. Brazil, Uruguay, Paraguay, n. Argentina, sparingly cultivated as an ornamental in our range and naturalized mostly in vacant lots, waste places, thickets, sometimes in hedges and amongst shrubbery in gardens.

The palmately divided, rather large and numerous leaves give this plant a distinctive and handsome appearance. It is one of many species of the genus native to the American tropics the most notable of which, *M. esculenta* Crantz and its cultivar varieties, is widely grown in most of the world's lowland tropical areas, the root being a carbohydrate source; it is variously known as tapioca plant, cassava, yuca, manioc.

3. Sapium

1. Sapium sebiferum (L.) Roxb. CHINESE TALLOW-TREE, POPCORN-TREE.

Fig. 133

The Chinese tallow-tree, a small to medium-sized tree, is native in eastern Asia, is frequently planted in our area as an ornamental, and in recent years has become naturalized in numerous places. In autumn its leaves are colorful and attractive, varying from orange-yellow to orange-yellow suffused with varying amounts of pinkish or purplish red to wholly dark red. Also in autumn the capsular fruits dehisce, the capsular valves shedding and leaving 3 conspicuous, dull white seeds attached for a while, these giving somewhat the effect of popped corn, hence the vernacular name, popcorn-tree. The sap is milky.

Usually fast-growing, glabrous throughout, the young branches slender, commonly arching or somewhat drooping. Potentially a tree, but flowering and fruiting from the time it is about a meter high. Leafy twigs more or less herbaceous early in the season, glaucous, the waxy surficial layer flaking as they become woody, bark tan; twigs bearing scattered, nearly circular, rough, slightly raised, orangish brown lenticels; leaf scars shield-shaped, with 3 vascular bundle scars, a leathery, persistent stipule at either side of the base of a naked, leathery, short triangular bud.

Leaves simple, alternate, deciduous, stipules short-ovate; petioles slender, mostly 2–5 cm long, blades pinnately-veined, broadly subrhombic-ovate or ovate, as broad as or broader than long (excluding their acuminate tips), mostly 3–6 cm broad, bases broadly rounded, nearly truncate, or with but a short, broad taper, apices abruptly acuminate, margins entire, a pair of glands at the summit of the petiole on the upper side and immediately below the base of the blade.

Inflorescences on some trees solitary, terminating branchlets of the season, each narrowly cylindrical and spikelike or tassellike, to 20 cm long, bearing pistillate flowers proximally, each in the axil of a bract, and staminate flowers distally in many short fascicles of up to 15 flowers per fascicle, their stalks unequal, each fascicle in the axil of a bract having a pair of glands basally, each flower with a small 2–3-lobed calyx, 2–3 stamens; on other trees, the inflorescence is branched, the central terminal branch elongate and usually wholly staminate, 2–3 much shorter, lateral branches below, each with a few pistillate flowers proximally and staminate fascicles distally. Pistillate flower with a small calyx united only basally and with 3 triangular lobes, pistil 1, ovary superior, 3-locular, 1 ovule in each locule, styles 3. Staminate flower with a small cuplike calyx about 1 mm across, stamens 2. Fruit a 3-lobed capsule about 1 cm long and as broad, upon dehiscing the capsular walls fall leaving 3 dull white seeds intact for a time. [*Triadica sebifera* (L.) Small]

Often in low, swampy or submarshy places, shores of streams, ponds, lakes and impoundments, sometimes on floating islands; also in upland, well-drained places, especially near human habitations, throughout our range. (S.e. S.C. to e. Tex.)

A combination of easily observed characteristics which serve to give distinctiveness to this plant: milky sap, alternate leaves whose blades are broadly subrhombic-ovate or ovate and with acuminate tips, 3–6 cm broad, as broad as or broader than long, a pair of glands at the summit of the petiole on the upper side and immediately below the base of the blade. (The milky sap of *Sapium* has poisonous properties!)

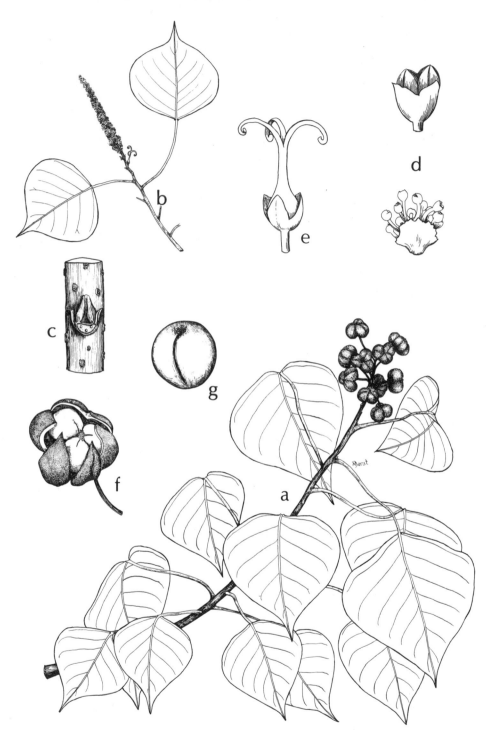

Fig. 133. **Sapium sebiferum:** a, fruiting branch; b, flowering branch; c, node with leaf scar, stipular remnants to either side, and axillary bud; d, fascicle of staminate flowers with subtending bract, below, and staminate flower above; e, pistillate flower; f, dehisced capsule, seeds within; g, seed.

National champion tallow-tree (1978), 10' circumf., 52' in height, 86' spread, Travis Co., Tex.

4. Sebastiania

Sebastiania fruticosa (Bartr.) Fern. SEBASTIAN-BUSH.   Fig. 134

Loosely branched, essentially glabrous subevergreen shrub 1–3 m tall. Leafy twigs herbaceous until about midseason, stipules 1.5–2 mm long, subulate-triangular, less frequently somewhat spatulate, apparently with abscission zones somewhat above their bases, the distal portions detaching fairly soon leaving a somewhat leathery stub at either side of the petiole; woody twigs grayish brown, with raised leaf scars, the stubs of the stipules persisting for a relatively extended period, vascular bundle scars scarcely if at all evident.

Leaves simple, alternate, petioles 2–10 mm long, blades pinnately-veined, elliptic, oval, or lanceolate, cuneate basally, acute to acuminate apically, variable in size on most branches, 2–7 cm long, 1–2 (–3) cm broad, margins entire, pubescence, if any, minute on the petioles.

Inflorescences very narrow spikelike racemes 1–4 cm long, borne in the axils of uppermost leaves of branchlets, green or greenish yellow. Flowers small, unisexual, without petals, each flower subtended by a broad, short bract with a gland on either side. Pistillate flowers on the lower portion of the raceme, staminate above. Staminate flower with 3 small sepals united basally, 3 exserted stamens. Pistillate flower with 3 nearly separate sepals, sometimes 2 tiny additional ones in 2 of the sinuses between; ovary 3-locular, each locule with 1 ovule, stigmas 3, recurved.

Capsule ovoid, about 8 mm long, dehiscing septicidally and loculicidally, the septae remaining attached to one side of each of the dehisced valves. Seed terete, ovoid-oblong, brown with a somewhat silvery cast, about 5 mm long. [*Sebastiania ligustrina* (Michx.) Muell.-Arg.]

River and stream banks, floodplain woodlands, richly wooded adjacent slopes, throughout our area. (Coastal plain, s.e. N.C. to cen. pen. Fla., westward to e. and s.e. Tex.)

A relatively inconspicuous, loosely branched essentially glabrous shrub, chiefly of streambanks, alluvial woodlands and adjacent wooded slopes, the alternate leaves relatively small, with entire margins, yellowish green, narrowly spikelike inflorescences solitary in the axils of uppermost leaves of branchlets.

5. Stillingia

Stillingia aquatica Chapm. CORKWOOD.   Fig. 135

Glabrous shrub to 12 or 15 dm tall, with a short taproot bearing numerous small rootlets. Stem single, terete, tapering gradually from the base upwardly, eventually leafy only on the branches. Wood of stem relatively soft, not dense, very light in weight. Leafy twigs herbaceous at least during the early part of the season, glaucous, reddish or purplish, the waxy surficial layer soon sloughing, becoming brown as they age, eventually often grayish brown; lenticels widely scattered, inconspicuous, nearly circular; leaf scars crescent-shaped, often becoming raised, with 3 vascular bundle scars, these often obscure; axillary buds naked; stipules variable, sometimes minutely subulate, sometimes divided from very near the base into 2–3 filiform divisions 2–3 mm long.

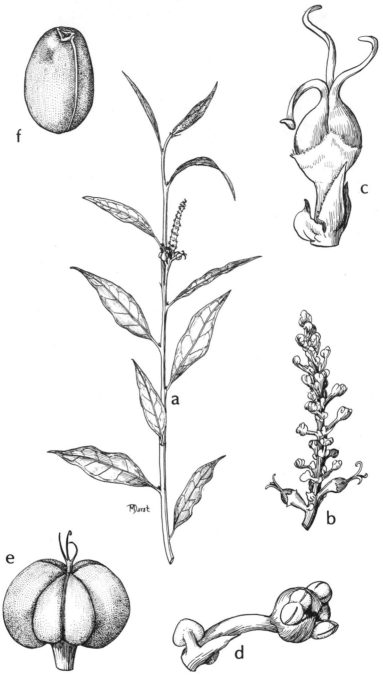

Fig. 134. **Sebastiania fruticosa:** a, branchlet with an inflorescence; b, inflorescence, enlarged; c, pistillate flower; d, staminate flower; e, capsule; f, seed.

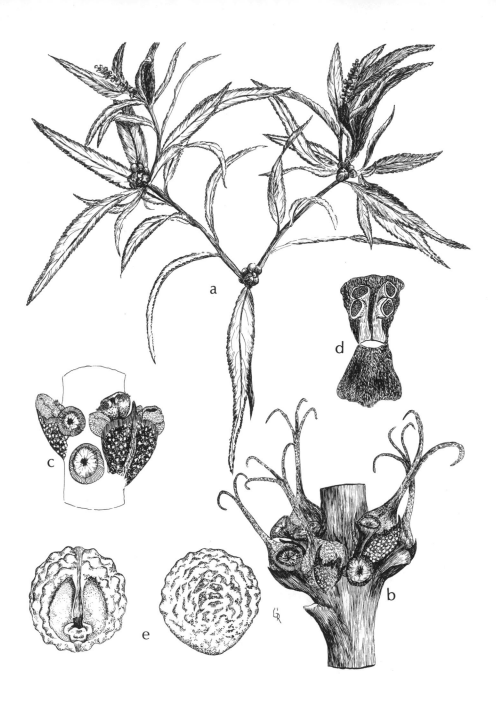

Fig. 135. **Stillingia aquatica:** a, top of plant, flowering and fruiting; b, portion of inflorescence with pistillate flowers; c, portion of inflorescence with staminate flowers; d, staminate flower (sepal spread back); e, seed, two views.

Leaves simple, alternate, closely set on the twigs, some of them usually over-wintering; petioles 1–4 mm long, blades pinnately-veined, lanceolate to linear-lanceolate, sometimes narrowly elliptic or rhombic, 3–8 cm long, cuneate at both extremities, margins minutely appressed-crenate, the teeth tipped by tiny callosities; other than the midvein, venation (on lower surfaces) not or only faintly evident.

Narrowly cylindrical, spikelike, inflorescences terminal on the branchlets, green, yellow, or red. Flowers unisexual, 1 per bract (pistillate) or 1–several in a fascicle subtended by a bract (staminate), bracts with 2 marginal glands. Pistillate flowers borne on the lower portion of the inflorescence, sepals 3 or 2, usually separate, sometimes united, sometimes absent; petals none; ovary (2) 3-locular, each locule with a single ovule, stigmas (2) 3. Staminate flowers above the pistillate on the axis, calyx 2-lobed, always present; petals none, stamens 2, exserted, filaments joined at the base.

Fruit a short-ovate capsule, about 1 cm broad, septicidally dehiscent. Seed about isodiametric, 4–4.5 mm across, silvery gray, with low, pinnaclelike tubercles or wrinkled-tuberculate.

Flatwoods ponds and depressions, drainage ditches and canals, old borrow-pits, in general in shallow water or where surface water stands much of the time, throughout our range. (Coastal plain, s.e. S.C. to s. pen. Fla., westward to s.w. Ala.)

Distinctive features of this aquatic or semiaquatic shrub are its wandlike stem with leafy branches only at the summit, the wood of the stem very light in weight, leaves usually lanceolate or linear-lanceolate, their margins with appressed crenations tipped by tiny callosities.

Fagaceae (BEECH AND OAK FAMILY)

1. Winter buds ovoid, 2–4 mm long; inflorescences narrowly spikelike, axillary to leaves on shoots of the season when all but the distal leaves on a given shoot are fully grown; fruit a conical, chestnut-brown nut enclosed in a 2-valved spiny bur.   1. *Castanea*
1. Winter buds narrowly lanceolate or awl-shaped, 10–20 mm long; staminate inflorescences in ball-like heads borne on pendent stalks both from proximal, bracteate, leafless portions of young developing branchlets and singly from axils of some developing leaves; pistillate flowers, usually singly or paired, borne in axils of uppermost unfurling leaves of developing shoots; fruit a 3-winged or 3-edged nut, usually 1–3 of them enclosed in a 4-valved husk covered with rusty brown pubescence and numerous spinelike enations.
    2. *Fagus*
1. Winter buds ovoid or oblongish, seldom exceeding 10 mm long, usually less; staminate inflorescences dangling, cylindric catkins borne singly or more commonly in clusters at the base of emerging shoots of the season; pistillate flowers borne singly or 2–several on short spikes in the axils of some of the unfurling leaves on developing shoots of the season; fruit a terete nut, seated in a many-bracteate, saucer- or bowllike involucre (in only one of ours, the involucre embracing more than one-half of the nut).   3. *Quercus*

1. Castanea

Castanea pumila (L.) Mill. var. pumila. CHINQUAPIN.   Fig. 136

Deciduous shrub, or tree to about 20 m tall (the shrubby form commonly with subterranean runners and forming extensive clones where the site is burned-over annually or at frequent, short intervals). Stems of young, developing shoots

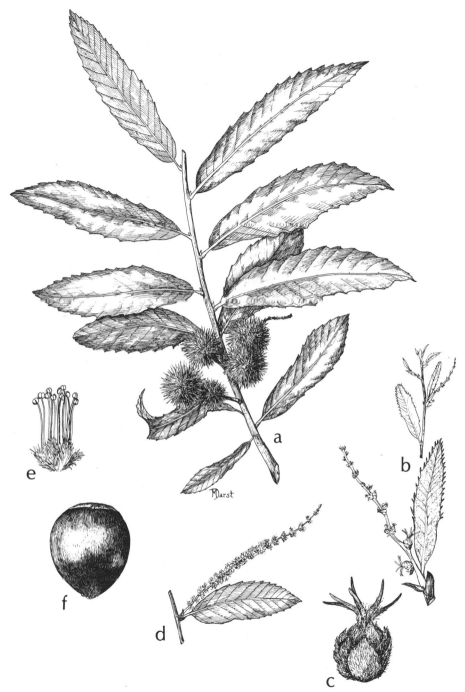

Fig. 136. **Castanea pumila:** a, fruiting branch; b, pistillate flowering spikes, the lower one enlarged; c, pistillate flower; d, staminate flowering spike; e, staminate flower; f, nut (drawn inverted!).

varying from densely gray-tomentose to thinly pubescent with longish hairs, to ashy-pubescent, to glabrous. Woody twigs round or slightly fluted, brown and dotted with small, paler lenticels; axillary buds 2–4 mm long, ovoid, dark brown, few-scaled; terminal bud none. Trunks of small trees gray and smooth, becoming slightly to moderately dark and shallowly furrowed as they enlarge.

Leaves alternate, usually 2-ranked, stipulate, stipules membranous, ovate, linear, lanceolate, or oblong (often all of these shapes on a single branch), quickly deciduous. Petioles 0.3–1.2 cm long. Blades pinnately-veined, the major lateral veins angling-ascending parallel to each other and ending in the marginal teeth; variable in shape, elliptic, oblong-elliptic, lance-elliptic, oblong, oblanceolate or obovate, rarely ovate, predominantly, however, broadest above their middles, proximally mostly cuneate to a usually rounded base, apically rounded, rarely emarginate, to acute or short-acuminate, from about 4.5–15 (18) cm long and 1.5–8 cm broad; margins serrate or dentate-serrate, teeth often abruptly narrowed to bristlelike tips; upper surfaces usually glabrous, infrequently sparsely pubescent along the midrib, lower surfaces densely and compactly tomentose, densely but more loosely tomentose and velvety to the touch, varying to glabrous.

Flowers fragrant, unisexual, plants monoecious. Inflorescences axillary to leaves on shoots of the season, produced in late spring or early summer when all but the distal leaves on a given shoot are fully grown; most inflorescences bearing many staminate flowers, these in axils of leaves proximally and medially on the shoots, relatively elongate and narrow, erect, ascending, or spreading; other inflorescences essentially similar but usually shorter and more distal on the branch and bearing a few pistillate flowers at the base; other inflorescences less conspicuous, usually in axils of less fully developed leaves distally on a branch, the flowers pistillate. Corolla none. Staminate flowers mostly in clusters of 3–7 (some solitary on the axis), the cluster or flower subtended by a minute, ovate bract; calyx of (5) 6 (8) ovate sepals in 2 series, stamens usually 12 in one cycle. Pistillate flowers 1–3 at a given position on the axis, surrounded by bracts, each flower with an urnlike calyx with 6 short lobes, the pistil (4) 6 (9) carpellate, with as many spreading styles, each hairy below and tipped by a minute, punctiform stigma. Fruit a conical, chestnut-brown nut about 7–20 mm long and about as broad basally, enclosed in a 2-valved globose or ellipsoid bur densely to openly covered with branched, short-pubescent, sharp spines 4–17 mm long. [Incl. *Castanea ashei* Sudw. in Ashe (*C. pumila* var. *ashei* Sudw.); *C. floridana* (Sarg.) Ashe (*C. alnifolia* var. *floridana* Sarg.); *C. alnifolia* Nutt.]

Locally often very abundant as a low, clonal shrub on longleaf pine–scrub oak sand ridges and hills that are burned frequently; similarly in open stands of planted pine on ridges and hills; less frequent in sand pine–oak scrub; in railroad rights-of-way, fence and hedge rows, old fields; local and scattered in xeric to mesic mixed woodlands; throughout our area. (N.J., s. and e. Pa., W.Va., e. Ky., Tenn., Ark., s.e. Okla., generally southward to n. cen. Fla. and e. Tex.)

2. Fagus

Fagus grandifolia Ehrh. AMERICAN BEECH.   Fig. 137

A medium to large deciduous tree, sometimes to about 35 m tall and to 1 m d.b.h. or a little more. Winter buds chestnut-brown, narrow, lance- or awl-shaped, 1–2 cm long; terminal bud none; leaf scars half-round, vascular bundle scars usually 3; pith continuous. Young emerging branchlets and petioles co-

Fig. 137. **Fagus grandifolia:** a, tip of winter twig (on increment of previous season's growth); b, node with leaf scar and bud on lateral short twiglet from increment of growth 2 years old; c, twig with emerging flowering branchlets; d, ball-like staminate catkin; e, pair of pistillate flowers nearly surrounded by bracts and involucre; f, branchlet with fruits; g, seed.

291

piously pubescent with blond, silky, antrorsely disposed hairs; stipules conspicuous but quickly deciduous, linear-subulate to linear-oblong, some slightly broadest at their middles, mostly 2–4.5 cm long, chestnut-brown to the unaided eye, with magnification seen to be conspicuously parallel-veined, the veins darker than the intervening tissue; upper surfaces of unfurling leaves with sparse silky pubescence, the lower surfaces with copious silvery-silky pubescence antrorsely disposed along the major veins and shorter and more nearly dull greenish, soft pubescence between the veins; emerging twigs, many or most of them bear pendent staminate catkins from their proximal, bracteate but leafless portions and usually from the axils of the lowermost of the emerging leaves as well, some of the uppermost leaves subtend single or paired pistillate flowers. Twigs of the season becoming dark reddish brown as they mature, are dotted with buff-colored lenticels, retaining some of the pubescence through the season; older twigs and branches gray. Bark of trunks ashy-gray, smooth, not sloughing, often mottled and a little roughened by lichens or moss growing upon it.

Leaves simple, alternate, pinnately veined, essentially 2-ranked, petioles 2–10 mm long. Mature blades stiffish, oval, elliptic, ovate, or slightly obovate, 4–12 (15) cm long and 2–7 cm broad; lateral veins prominent and straight, anglingascending parallel to each other and ending in the marginal teeth or in veinlets at the margin; bases mostly rounded, some very slightly cordate, sometimes oblique and the blades inequilateral, apices mostly abruptly short-acuminate; margins shallowly dentate or dentate-serrate; upper surfaces dark glossy green, usually pubescent along the midrib, lower surfaces paler and duller green, pubescent with long hairs along the major veins, usually with short, soft pubescence between the veins, glabrous between the veins on an occasional tree.

Flowers unisexual, plants monoecious. Staminate flowers individually small, numerous, and borne in globular balls dangling on slenderly flexuous, silkyhairy stalks. Flower stalks short, silky-pubescent; calyx 5–8-cleft, copiously clothed with long, silky, white hairs; corolla none; stamens 8–16, anthers exserted. Pistillate flowers inconspicuously borne singly or in pairs on short, stoutish, silky-hairy stalks, invested by numerous subulate bracts, the innermost bracts coherent at base and eventually forming a 4-lobed involucre, its surface with dense, rusty-brown pubescence and numerous, relatively soft, hairy, subulate, usually curvate, spinelike enations; involucre enclosing 1–3, 3-winged or 3-edged nuts, splitting by 4 valves at maturity.

In autumn the leaves of the American beech turn yellow, then russet-brown, finally brown, and after dying tend to remain on the trees into the winter season.

Inhabiting mesic woodlands; in our area mainly in s.w. Ga., Fla. Panhandle, s. Ala.; also in Alachua and Columbia Cos., n.e. Fla. (N.S. and N.B. to Minn., generally southward to n. Fla. (apparently absent in s.e. Ga.), and e. Tex.)

National champion American beech (1976), 13'11" circumf., 161' in height, 105' spread, Berrien Co., Mich.

3. Quercus (OAKS)

Shrubs or trees, trees of some species potentially of large size. Winter buds tend to be clustered at or near the ends of twigs. Pith continuous and homogeneous. Leaf scars half-round for the most part, vascular bundle scars scattered. Leaves alternate, simple, at least short-petioled, 5-ranked, deciduous in autumn or in some species for the most part over-wintering and falling just before or as new growth of the season commences; blades pinnately veined, margins entire, variously toothed, or pinnately lobed; stipules present as new shoots emerge but quickly deciduous as the leaves develop. Flowers unisexual, plants monoecious. Staminate flowers on pendent catkins borne singly or more commonly in clusters at the base of emerging shoots of the season, usually from the axils of the inner scales of the opening bud, sometimes before any of the shoot is evident, sometimes perhaps from buds from which no shoots are produced; each flower solitary in the axil of an inconspicuous bract, calyx bowllike below, (4) 5-lobed above, 3–12 stamens seated within. Pistillate flowers inconspicuous, borne singly or 2-several in short spikes in the axils of developing leaves; each flower consisting of a single pistil with 3 styles, the ovary seated within and adherent to a 6-lobed calyx which in turn is enclosed in an involucre of overlapping scales, these by subsequent development eventually forming an involucral cup around the base of or in some almost wholly around the matured ovary, the nut or acorn which is generally apiculate apically. (In the descriptions of the species following, the involucral cup together with the acorns may be, for the sake of brevity, referred to as the fruit.) Ovary 3-locular, 2 ovules in each locule only one of which develops into a seed. Inner surface of the acorn glabrous in some kinds of oaks, pubescent in others, if pubescent, the pubescence usually copiously woolly.

Plants of some species of oaks exhibit a considerable degree of variability of leaf blades. In some kinds, at least, the leaves of seedlings, sprouts, root-sprouts, or sucker shoots may differ notably from those of crown branches of a mature tree. In some, leader shoots produced in summer have leaves differing considerably from the leaves on branches below that were produced in spring. In some kinds of oaks, lower branches of the crown may have leaves not a little different from those on upper parts of the crown. Given these kinds of variability, the task of preparation keys and descriptions is unusually difficult. In order not to muddle the language of keys and descriptions inordinately, one tends to use characterisitcs of leaves of fertile, crown branches and to encourage the novice to restrict his observations, *at first*, to crown branches and to learn eventually to make the necessary correlations of the variables.

Identification of oaks is to some extent complicated because hybridization occurs between plants of various species growing in close proximity, the hybrid progeny having characteristics intermediate between those of the parents. Hybrids are perhaps not so frequently produced as is sometimes alleged, the apparent frequency of their occurrence attributable to the relatively long life of individual plants. In general, hybrid individuals tend to be more prevalent on sites where mechanical disturbance of the soil has occurred and where, therefore, "normal" community structure no longer obtains. The following treatment does not take hybrids into account.

1. Lower surfaces of mature leaf blades wholly glabrous.
  2. Leaf blades of crown branches lobed, the sinuses between the major lobes predominantly extending halfway or more than halfway to the midrib.   1. *Q. alba*
  2. Leaf blades of crown branches, if lobed, lobes not reaching halfway to the midrib.
    3. Lobes short and rounded, not bristle-tipped.
      4. The leaf blades relatively uniform in size and the lobing on each side regular (immature specimens, perhaps to 6–8 m tall).   1. *Q. alba*
      4. The leaf blades on a given tree tending to be relatively variable in size and their lobing irregular.   2. *Q. austrina*
    3. Lobing none on the blades, or if leaves on increments of summer growth lobed, then the lobes bristle-tipped.

5. The leaf blades predominantly oblanceolate but some narrowly elliptic or narrowly lanceolate, margins entire (or lobed only on increments of summer growth); plants potentially medium-sized to large trees.   13. *Q. hemisphaerica*

5. The leaf blades oval, short-elliptic, short-oblong, or obovate, margins entire; plants thicket-forming shrubs or small scrubby trees.   16. *Q. myrtifolia*

1. Lower surfaces of mature leaf blades pubescent throughout, only along the major veins, or only in the axils formed by the midrib and major lateral veins.

  6. Lateral veins of leaf blades conspicuously angling-ascending approximately parallel to each other and extending into marginal teeth or undulations.

    7. Leaf blades lanceolate to elliptic, sometimes some obovate; marginal teeth generally sharply pointed, their tips callused.   3. *Q. muehlenbergii*

    7. Leaf blades predominantly obovate, often broadly so; margins varying from undulate to definitely toothed, the teeth rounded to obtuse, if callused, calluses minute.
                    4. *Q. michauxii*

  6. Lateral veins of leaf blades not conspicuous angling-ascending parallel to each other.

    8. Leaf blades of crown branches lobed, the sinuses between the major lobes predominantly extending halfway or more to the midrib.

      9. Lobes of the leaf blades not bristle-tipped.

        10. Woody twigs of the season persistently very densely stellate-pubescent.
                    5. *Q. stellata*

        10. Woody twigs of the season glabrous or with few, scattered hairs 2-forked basally most of which are eventually sloughed.

          11. Woody twigs of the season glabrous; involucre bowllike, embracing about one-third of the acorn; shrubs or small trees of well-drained pine—scrub oak sand ridges and hills.   6. *Q. margaretta*

          11. Woody twigs of the season with few, scattered hairs mostly 2-forked from the base, these usually gradually sloughed as the season advances; involucre wholly or nearly wholly enclosing the acorn; potentially large trees of river and stream bottomlands.   8. *Q. lyrata*

      9. Lobes of leaf blades (or teeth on the lobes, if any) attenuated to bristle-tips.

        12. Lower surfaces of mature leaf blades with a dense, tawny or dull gray mat of short, stellate pubescence, *or* a thinnish mat of orangish yellow pubescence.

          13. Pubescence of lower blade surfaces tawny or dull gray; petioles not twisted and plane surfaces of leaf blades not oriented vertically to the ground.

            14. Base of blades of crown leaves predominantly U-shaped; pubescence of lower blade surfaces usually tawny (when fresh).   22. *Q. falcata*

            14. Bases of leaf blades not U-shaped, usually obliquely truncate, truncate, or shortly tapered; pubescence of lower surfaces of blades dull gray. 23. *Q. pagoda*

          13. Pubescence of lower blade surfaces orangish yellow; petioles, for the most part, twisted so that the plane surfaces of the blades are oriented vertically to the ground.   21. *Q. laevis*

        12. Lower surfaces of mature leaf blades not having densely matted pubescence, the pubescence, if any, in axils formed by the midrib and major lateral veins, along the veins, or thinly distributed so that surface tissue is evident.

          15. Petioles short, 0.5–1 (1.5) cm long, predominantly twisted so that plane surfaces of blades are oriented vertically to the ground.   21. *Q. laevis*

          15. Petioles 2–5 (6) cm long, not twisted.

            16. Involucre enbracing one-third to one-half of the acorn.

              17. Winter buds densely pubescent, upper scales of the involucre of the fruit loose, forming a fringe at the rim.   24. *Q. velutina*

              17. Winter buds glabrous or thinly pubescent; upper scales of the involucre not loose and not forming a fringe.

                18. Woody twigs of the season dull gray or brown; involucre abruptly narrowed basally to a stalklike portion; lowest sinuses of leaf blades relatively broad, on some leaves, at least, truncate basally and paralleling the midrib.
                    27. *Q. nuttallii*

18. Woody twigs of the season bright reddish brown; involucre rounded basally; lowermost sinuses of the leaf blades relatively narrow and rounded. 28. *Q. coccinea*

16. Involucre saucerlike, embracing the base of the acorn.

19. Leaf blades prevailingly with 3 lateral lobes per side, major lobes divaricate or nearly so. 26. *Q. shumardii*

19. Leaf blades prevailingly with 4 or more lobes per side, major lobes ascending. 25. *Q. rubra*

8. Leaf blades of crown branches unlobed, or if lobed then the sinuses between the lobes shallow.

20. The leaf blades variable as to marginal features, some entire, some with 1–few sinuate teeth or short, rounded lobes, in size, (2) 4–8 (10) cm long and (1) 2–6 (8) cm broad; in outline, oblong, elliptic, obovate, or spatulate; leaves mostly persisting overwinter; (shrubs or small, scrubby trees of sand pine–oak scrub). 7. *Q. chapmanii*

20. The leaf blades not having the above combination of characteristics.

21. Blades of mature crown leaves broadened toward or at their summits, often slightly or shallowly 3–5-lobed near the tip, or with 3 broad lobes or shoulders distally.

22. Leaf blades mostly 1.5–3 times as long as broad at their broadest places, generally longish-cuneate below the broadest places. 18. *Q. nigra*

22. Leaf blades mostly as broad as long at their broadest places.

23. Blades predominantly 2 5 cm long and 1–3 cm broad; leaves overwintering. 16. *Q. myrtifolia*

23. Blades of most leaves on a given plant very much larger than in no. 23 above; leaves deciduous in autumn.

24. Twigs relatively slender; leaf blades thinnish and somewhat pliable, green beneath and any pubescence tawny and sparse, sometimes only in the major vein axils. 19. *Q. arkansana*

24. Twigs relatively stout; leaf blades stiff or rigid, clothed beneath with thin to copious, glandlike pubescence, this yellow to olive-green (and usually desiccated after leaves are fully mature). 20. *Q. marilandica*

21. Blades of mature crown leaves not predominantly and significantly broadened toward their tips.

25. Margins of leaf blades entire or irregularly only slightly wavy (on increments of spring growth but sometimes shallowly toothed or toothed-lobed if on increments of summer growth).

26. Lower surfaces of blades, for the most part, clothed with densely matted pubescence throughout (the pubescence in some so tightly matted as to appear glabrous to the unaided eye).

27. Mature leaf blades with a bluish green color, thus the crowns with a bluish green hue as seen from a little distance. 17. *Q. incana*

27. Mature leaf blades (upper surfaces) dark green, or in *Q. pumila* sometimes graylsh green but without a bluish hue.

28. Plants low shrubs which usually spread by subterranean runners and form extensive low clones, stems of which rarely exceed 1 m high (to 1.5–2 m if the site not recently burned).

29. Leaves all deciduous in autumn or only a few overwintering; margins of blades entire, or sometimes crisped or with a few rounded undulations; fruits sessile or very shortly stalked; acorns little longer than broad, mostly 0.8–1.2 cm long, their exposed outer surfaces light brown. 12. *Q pumila*

29. Leaves wintergreen (usually with a more leathery texture than in the preceding and their upper surfaces darker green), commonly some leaves having the margins with spinose-tipped teeth and prickly; fruits sessile but spicate on longish stalks (peduncles), only 1–2 near the tips of the stalks or up to 5 or 6 along the stalks; acorns twice as long as broad, mostly 2–2.5 cm long, their exposed outer surfaces dark brown to black. 11. *Q. minima*

(Note: *Q pumila* and *Q. minima*, in much of our range, frequently grow intermixed.)

28. Plants potentially trees; however, *Q. geminata*, particularly when growing in sand pine—oak scrub, or in coastal shoreline scrub, may form thickets of low stature but many of the stems considerably exceeding 1 m in height.

30. Leaf blades flat, their margins slightly thickened but scarcely revolute; midvein and major lateral veins not, or very little, impressed on upper surfaces, on the lower surfaces the major lateral veins very slightly raised, if at all, and not very observable to the unaided eye; pubescence on lower surfaces so compactly matted as to appear glabrous to the unaided eye, the hairiness only just perceptible with 10× magnification; "fruits" usually only 1 or 2 per stalk (peduncle).   9. *Q virginiana*

30. Leaf blades revolute and the sides of the blades commonly slightly to conspicuously rolled downward so that the blades have the appearance of shallow, inverted boats; midvein and major lateral veins of the upper surfaces impressed, on the lower surfaces the major lateral veins and, to a lesser extent, the anastomosing veinlets between them, raised so that to the unaided eye the surfaces appear to some extent rugose-veiny; hairiness of the lower surfaces more perceptible than in the preceding; "fruits" may be 1 or 2 per stalk but frequently 3 or more.   10. *Q. geminata*

26. Lower surfaces of blades not having densely matted pubescence throughout.

31. Mature leaf blades on a given tree prevailingly lanceolate but some varying to elliptic or narrowly oblong, only an occasional one oblanceolate; lower surfaces with tufts of hairs in some or most of the axils formed by the midvein and major lateral veins, sometimes with patches of pubescence scattered beside the midrib, less frequently with patches along the lateral veins; leaves deciduous in autumn.   15. *Q. phellos*

31. Mature leaf blades on a given tree prevailingly spatulate, oblanceolate, subrhombic, or obovate; lower surfaces with tufts of hairs in some or most of the axils formed by the midrib and major lateral veins (late in the season, most of this pubescence may be sloughed, but there is usually some evidence of it persisting); leaves, for the most part overwintering and dropping just before or as new shoot growth commences in spring.   14. *Q. laurifolia*

1. Quercus alba L. WHITE OAK.   Fig. 138

Medium to large, deciduous tree. Woody twigs purplish gray to greenish red. Winter bud or buds at tip of twig reddish brown, ovoid-conic to globose, usually vaguely, obtusely, and longitudinally angular, margins of scales with a fringe of hairs, sometimes also sparsely and patchily short-pubescent distally on their backs. Bark of trunks light ashy gray, that of medium-sized trees broken into small blocks vertically aligned, that of larger trunks irregularly plated or divided into broad, flat, flaky or scaly ridges.

Young developing stems heavily clothed with pinkish, shaggy, stellate pubescence, this completely sloughing rather quickly. The short petioles and surfaces of unfurling leaf blades densely pubescent with stellate pubescence, pink at first, soon becoming grayish, the pubescence all sloughed by the time the leaves reach full size, during the sloughing the pubescence becoming loosely cottony and patchy.

Blades of mature leaves exhibiting a considerable variability in size and degree of lobing, 5–15 (20) cm long and 3–10 cm broad, usually broadest above their middles, sometimes at their middles; with respect to lobing, those of saplings, even trees of moderate size or of branches low on relatively large trees, generally having relatively sinuous, short lobes, the sinuses between them shallow, while those of middle and upper crown branches of trees of flower-

Fig. 138. **Quercus alba.**

ing/fruiting size have much longer, oblongish or sometimes unevenly forking lobes with V- to U-shaped sinuses between them reaching halfway to the midribs or more; lobes mostly 7–10 in all, rounded apically, not bristle-tipped; bases cuneate to U-shaped; surfaces glabrous, the upper bright green, sometimes shiny, the lower dull green to gray or whitish.

Fruits maturing in one season, sessile or stalked, their involucres bowllike, 10–12 mm deep, mostly 15–20 (22) mm across from rim to rim, embracing about one-fourth to one-third of the acorns, scales dark brown, pubescent, thickflattened, outwardly disposed, more or less fused basally, their rounded apices not appressed, thus the overall surface very roughly warty or knobby; acorns glabrous, shiny tan to brown, oblong to narrowly ovoid or obovoid, 1.5–3 cm long, flat basally, rounded apically, inner surfaces glabrous.

Inhabiting mesic or submesic hardwood forests; s. cen. and s.w. Ga., s. Ala., n. Fla. from about the Suwannee River westward. (Maine to Mich. and Minn., generally southward to the Panhandle of Fla. and e. Tex.)

National champion white oak (1972), 29'8" circumf., 102' in height, 158' spread, State Park, Wye Mills, Md.

2. Quercus austrina Small. BLUFF OAK.   Fig. 139

A medium-sized to large deciduous tree, often with a narrow crown. Winter bud or buds at tips of twigs reddish brown or brown, ovoid-conic, vaguely, obtusely, and longitudinally angled, median and upper scales mostly uniformly and moderately pubescent with longish hairs. Bark of trunks light gray and scaly, similar to that of the white oak.

Young developing stems usually green around one half, reddish around the other half, clothed with pale pubescence; petioles reddish on their edges, pubescent with pale hairs. Unfurling leaf blades pinkish to the unaided eye; with 10X magnification, upper surfaces are seen to be finely dotted with an abundance of minute, dark red, short-stipitate glands and sparsely clothed with pale pubescence, lower surfaces nonglandular, major veins green, pink between the veins, pale-pubescent throughout, the pubescence more abundant than on the upper surfaces. (All of the pubescence softly stellate.)

Mature leaves with short, glabrous petioles, their blades not so dissimilar on saplings, young trees, and crowns of large older trees as in white oak, yet variable in size and degree of lobing on individual plants, the lobing tending to be very irregular and variable throughout; commonly on a given fairly large branch, blades varying from about 3 cm long and 2 cm broad, unlobed or with few, irregular, short lobes, to about 10 (15) cm long and 4–8 cm broad across the widest parts, the lateral lobes even on the largest leaves oftentimes only 2–4 (in truth the lobing so variable as to be difficult to describe in concise terms, see illustration); in general the bases cuneate, or if rounded, then narrowly so, the larger blades mostly broadest above, less frequently at, their middles; both surfaces glabrous, the upper dark green and moderately shiny, the lower dull, more or less olive-green.

Fruits maturing in one season, their involucres (very much unlike those of white oak) saucerlike or shallowly bowllike, 6–8 mm deep, embracing about one-fourth of the acorn, scales brown, densely short-pubescent, lanceolate or more or less triangular, apices pointed or blunt, closely appressed throughout; acorns glabrous, shiny tan to brown, oblong or narrowly ovoid-oblong, 1.2–1.5 cm long, bases somewhat rounded, apices rounded and apiculate, inner surfaces glabrous.

Fig. 139. **Quercus austrina.**

Inhabiting submesic, mixed deciduous forests of river bluffs, ravine slopes, high areas above or near rivers and streams, usually where calcareous, or perhaps at least where the soil is circumneutral; local, coastal plain, s.e. N.C. to n. cen. pen. Fla., s.w. Ga., Panhandle of Fla., s. Ala., s.e. Miss.

Elias (1980), via a distributional map, attributes the related Durand oak, *Q. durandii*, to our area of coverage; however, he used the distributional map for *Q. durandii* from Little (1976) who does not recognize *Q. austrina* as distinct from *Q. durandii* as does Elias, thus that map shows the combined ranges of the two. Superimpose Elias' distributional map for *Q. austrina* over that of Little for *Q. durandii* (including *Q. austrina*) and our area is blocked out, as it were. Whether that would reflect the real distributions of the two, I am not, of course, sure. In any event, I have not, as yet, encountered in our area plants of the alliance with leaf blades copiously stellate-pubescent beneath, a characteristic of *Q. durandii*.

Florida champion bluff oak (1983), 8'5" circumf., 105' in height, 78' spread, Chiefland, Fla.

3. Quercus muehlenbergii Engelm. CHINQUAPIN OAK, YELLOW CHESTNUT OAK.
Fig. 140

A medium-sized deciduous tree. Woody twigs of the season slender, brown, usually sparsely pubescent. Winter bud or buds at tips of twigs 4–5 mm long, ovoid, bluntly pointed apically, their distal scales with patches of short pubescence on their backs. Bark of trunks relatively thin, light gray, breaking into narrow scales.

Young developing stems and petioles sparsely pubescent, hairs pale, unbranched, 2–3-branched from their bases, or minutely starlike, commonly a mixture of these; upper surfaces of unfurling blades green, the pubescence much as on the stems and petioles, lower surfaces on some individuals green and sparsely pubescent, more frequently gray, the major lateral veins and the surfaces between them clothed with minute, starlike, gray hairs.

Fully mature leaves with slender, usually glabrous, petioles 1–3 cm long. Blades (5) 8–15 cm long, (2) 4–10 cm broad at their broadest places, obovate, elliptic, or broadly lanceolate, tapered to wedgelike or rounded bases, apices generally short-acuminate; lateral veins angling-ascending approximately parallel to each other and extending into the marginal teeth; toothing variable from tree to tree, rarely little more than undulations, generally definite and somewhat curving upwardly with their tips somewhat incurved, tips salient and acute, each tooth tipped by a glandular-callused mucro; upper surfaces dark green, glabrous, and at least sublustrous, lower surfaces generally grayish-glaucous, between the lateral veins with sparse to moderately dense, minute, stellate pubescence.

Fruits maturing in one season, their involucres thin, narrowly bowllike, 7–8 mm deep, 10–15 mm across from rim to rim, embracing one-third to one-half of the acorns, scales more or less obovate, longitudinally moderately ridged or humped on their backs, tightly gray-pubescent, the cup surface finely knobby; acorns ellipsoid to slightly ovoid, 1.5 cm long or a little more, 1–1.4 cm in diameter, outer surfaces brown to almost black, densely ashy-pubescent distally and on the apicules, inner surfaces glabrous. [*Q. prinoides* Willd. var. *acuminata* (Michx.) Gl.]

Locally inhabiting submesic woodlands, principally calcareous places, bluffs and slopes, glades; s.w. Ga., Panhandle of Fla. from n. Leon Co. to Jackson Co., perhaps in s.e. Ala. along the Chattahoochee River (although local in much of

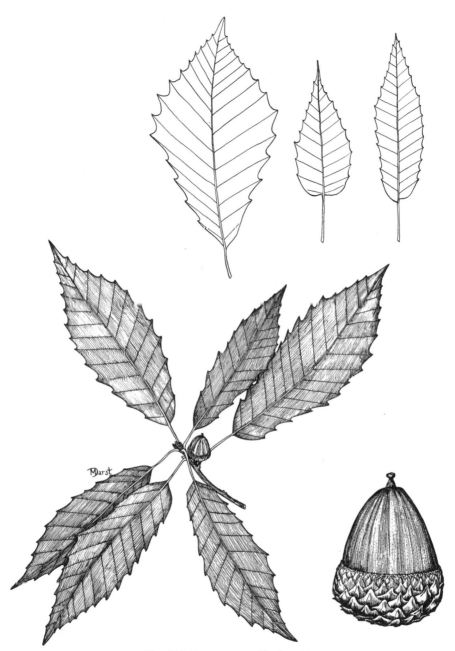

Fig. 140. **Quercus muehlenbergii.**

Ala. except the southernmost portion). (Vt., n. N.Y., s. Ont., s. Mich., Wis., and Minn., s.e. Neb., generally southward to n.w. Fla. and cen. Tex.)

4. Quercus michauxii Nutt. SWAMP-CHESTNUT, COW, OR BASKET OAK. Fig. 141

A medium-sized to large deciduous tree. Woody twigs of the season relatively stout, brown, glabrous, older twigs gray. Winter bud or buds at tips of twigs conic-ovoid, often vaguely, obtusely, and longitudinally angled, brown, scales variously glabrous to copiously grayish-pubescent on their backs, sometimes with a fringe of hairs around their apical margins. Bark of larger trunks thick, light gray, shaggy, irregularly furrowed and broken into long, narrow plates that flake off exposing brownish gray inner bark.

Young, developing stems bearing scattered, shaggy, pale hairs, some of them unbranched, others 2–3-branched from the base, these hairs quickly sloughed; hairs of unfurling leaves similar, shaggy, much more abundant on the petioles than on the stems and generally persistent there; pubescence on upper surfaces of unfurling blades thin throughout, that on lower surfaces of blades of some trees very dense and silvery, velvety to the touch, that on lower surfaces of blades of other trees very much less dense throughout, making the surfaces neither velvety to the touch nor silvery.

Mature leaves with petioles 0.5–2.5 cm long, occasional ones sessile. Blades relatively large, larger ones 10–22 cm long and 7–10 (15) cm broad (vigorous sprouts or suckers may have much larger blades); most blades broadest above their middles and obovate, tapering gradually from their broadest places to obtuse or rounded bases, apices with small acuminations; lateral veins angling-ascending approximately parallel to each other, each extending into a marginal tooth or undulation; marginal teeth or undulations, a few of them, from a little above the base of the blades small and inconspicuous, others then becoming more uniformly sized and spaced distally, slightly angling-ascending, on some leaves short and merely rounded (undulate) varying to more or less shortly triangular and with rounded to obtuse tips, the latter, if callused, calluses minute; upper surfaces darkish green, usually glabrous by full maturity, sometimes pubescent along the midribs, lower surfaces usually dull green, sometimes gray-glaucous, varyingly sparsely pubescent mainly along the veins, moderately pubescent throughout, less frequently densely pubescent throughout.

Fruits maturing in one season, varying in average size from tree to tree, involucres bowllike, relatively thin to very thick, 1.5–2 cm deep, 2.5–3.5 cm across the cup from rim to rim, embracing one-third to one-half of the acorns, scales more or less triangular-acute, very irregularly but notably humped on their backs, thus the surfaces of the cups rough-knobby; acorns ovoid or short-ellipsoid, infrequently subglobose, 2.5–4 cm long and 2–3 cm broad, bases flat, apices rounded, outer surfaces dull brown and with buffish powdery pubescence throughout, or glabrous and sublustrous proximally and pubescent distally and on the apicules, inner surfaces glabrous, but innermost tissue very thin, brittle, and easily broken revealing pubescence beneath. [*Q. prinus* L.—*nomen ambiguum*]

Occurring in river bottom forests where inundation is of short duration, in mesic, mixed hardwood or hardwood-pine stands of bluffs, ravine slopes, and flatwoods where there is subsurface limestone; throughout our area. (Coastal plain and piedmont, s. N.J. and s.e. Pa., southward to n. cen. pen. Fla., westward to e. Tex., northward in the interior, s. and e. Ark., w. Tenn., s.e. Mo., s. Ill. and Ind., w. Ky.)

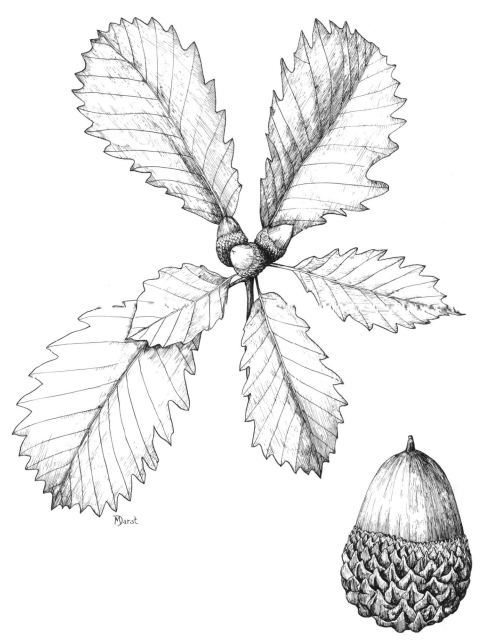

Fig. 141. **Quercus michauxii.**

The leaves of *Q. michauxii* and *Q. muehlenbergii* are more closely similar than those of either are to those of other oaks of our area, especially with respect to venation pattern and marginal toothing, sometimes to shape as well. In general, the leaf blades of *Q. michauxii* average larger in size, more of them obovate, the marginal teeth less sharply pointed than in *Q. muehlenbergii*; trees of the former perhaps much less frequently glaucous and grayish beneath than those of the latter which are almost always thus; hairs of lower leaf surfaces between the lateral veins, if any, in the former, generally longish, unbranched or 2–3-branched from the base, easily visible with 10× magnification, usually hairs of the latter in the same places are minute, starlike, and appressed, visible only with higher magnification. Cups and acorns in *Q. michauxii* larger than in *Q. muehlenbergii*. Trees of the latter generally grow in less mesic and more calcareous places.

National champion swamp chestnut oak (1974), 22′7″ circumf., 122′ in height, 123′ spread, Talbot Co., Md.

5. Quercus stellata Wang. POST OAK.   Fig. 142

Small to medium-sized deciduous tree. Twigs of the season densely (rarely sparsely) clothed with tawny, short-stellate pubescence, older twigs dark gray to blackish owing to shriveled, persisting, discolored pubescence. Winter bud or buds at the tips of twigs brown, globose to broadly and shortly ovoid, vaguely, obtusely, and longitudinally angled, margins of scales with a fringe of buff-colored hairs, usually some of them with patchy, appressed pubescence on their backs. Bark of trunks moderately thick, reddish brown to dark gray, forming relatively deep longitudinal furrows separating rounded to broad, scaly ridges.

Young developing stems and petioles clothed with dense tawny to olivaceous pubescence; upper surfaces of unfurling leaf blades pinkish, sparsely clothed with a mixture of pale, short-shaggy, stellate hairs and few to many very minute reddish glandlike ones, lower surfaces with dense, shaggy, yellowish to orangish stellate hairs.

Mature leaves with petioles 1.5–2 cm long, some, at least, of the pubescence persisting; blades 10–15 (18) cm long, often very nearly as broad, or as broad, measured across the major lateral lobes; pattern of lobing varying, but that of many or most blades on any given tree as follows: 3 prominent distal lobes, a terminal one and 2 diverging laterals (sometimes all three distinctly to vaguely sublobed), the 3 giving the effect of a cross; below the cross the blade narrowed to an isthmus below which usually 2 much shorter, less conspicuous lobes rounded at their tips occur (occasional blades with 1 or 2 still smaller lobes below them); bases cuneate to bell-shaped, sometimes oblique at the extreme base; uppermost 3 lobes variable in outline, often oblongish or squarish, often irregularly so, usually more or less rounded apically; lobes not bristle-tipped; upper surfaces dark green, at least sublustrous, glabrous and smooth or essentially glabrous but very slightly roughened by persisting bases of hairs, lower surfaces dull green, usually most of the pubescence sloughed save along the veins.

Fruits maturing in one season, sessile or shortly stalked, their involucres bowllike, 6–8 mm deep, embracing nearly one-third of the acorns, scales very tightly appressed, more or less elongate-triangular, pubescent on their backs; acorns 1.5–2 (2.5) cm long, oblongish in outline but slightly broadest distally, bases rounded, apices rounded and with densely short-hairy apicules, lower two-thirds of outer surface tan to dark brown, smooth and somewhat shiny,

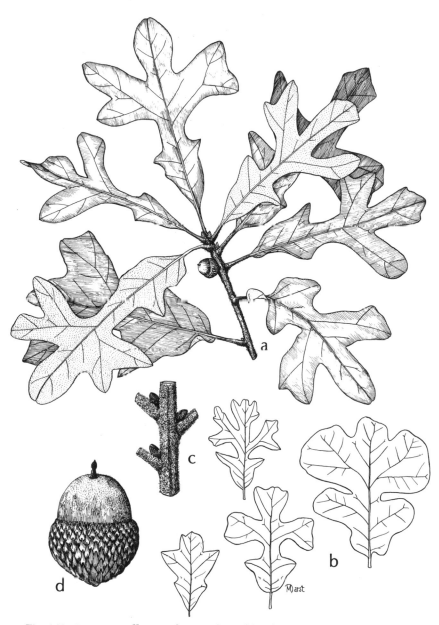

Fig. 142. **Quercus stellata:** a, fruiting branchlet; b, variable leaves, drawn ×
⅝ actual size; c, portion of twig of the season (leaves cut off) to show pubescence; d, fruit.

upper one-third dull-tawny pubescent, *or* entire surface dull-tawny and with sparse powdery pubescence, inner surfaces glabrous, but pubescent beneath a very thin, innermost, parchmentlike covering which easily peels away.

Inhabiting upland mixed hardwood or hardwood-pine forests where (in our area) the soils are sandy-clayey or gravely, often where there is subsurface limestone, frequently associated with *Q. velutina* and *Q. falcata*; local, s. cen. and s.w. Ga., n. Fla. to the w. of the Santa Fe River, s. Ala. (S.e. Mass. to s.e. Pa., W.Va., s. Ohio, Ind., Ill., and Mo., generally southward to n. cen. Fla. and e. and cen. Tex.)

Co-national champion post oaks (1980), 12'10" circumf., 94' in height, 92' spread, campus of Univ. of N.C. at Chapel Hill; (1981), 14'8" circumf., 88' in height, 113' spread, Hampton Co., S.C.

6. Quercus margaretta Ashe. SAND POST OAK.   Fig. 143

Generally a small, scrubby, deciduous tree, often forming small thickets or groves by subterranean runners. Woody twigs brown, *glabrous*, relatively slender. Winter bud or buds at tips of the twig reddish brown, very slightly angled (as in *Q. stellata*), scales with a very minute fringe of hairs marginally and some of them with a little pale pubescence distally on their backs. Bark of trunks rough, brown or reddish brown.

Young developing stem and petioles with sparse, pale, stellate pubescence; upper surfaces of unfurling leaf blades yellowish green, with sparse, pale, stellate pubescence, lower surfaces gray with dense, soft pubescence.

Mature leaves with sparsely pubescent petioles 0.3–1.0 cm long; blades variable in size and form, 2–12 (15) cm long and up to 12 cm broad across the largest lateral lobes; sometimes a few of the smallest leaves on a given branch unlobed, oblong or elliptic, in general, however, 1–few sinuate-margined to shallowly to deeply 3–5 (7) -lobed, the pattern of the lobing of some blades somewhat as in *Q. stellata* although few of them with the distal 3 lobes suggesting a cross, the major lateral lobes more angling-ascending, shorter and narrower than in *Q. stellata* (see illustration for some, at least, of the variability); upper surfaces dull medium-green, slightly roughened by persisting bases of sloughed hairs, lower surfaces much paler and duller green, sparsely shaggy-pubescent throughout or sometimes most of the hairs sloughed save along the veins.

Fruits maturing in one season, their involucres bowllike, about 8 mm deep, embracing about one-third of the acorn, scales appressed, more or less triangular, pubescent on their backs; acorns oblong-ellipsoid, sometimes slightly broadest basally or apically, 1.5–2 cm long, rounded basally and apically, outer surfaces brown, with tawny, powdery pubescence about their summits and on their apicules, inner surfaces glabrous, but pubescent beneath a thin parchmentlike covering which peels easily. [*Q. stellata* Wang. var. *margaretta* (Ashe) Sarg.]

Chiefly with longleaf pine–scrub oak or sand pine–scrub oak on deep, well-drained sands (or where the pines have all or nearly all been harvested, or in pine plantations on such sites); much less frequently and abundantly in mixed pine-hardwood second-growth woodlands where there is subsurface limestone, some of the tree associates there: *Q. falcata, Q. stellata, Q. tomentosa, Q. shumardii, Q. hemisphaerica, Q. nigra, Carya tomentosa, Liquidambar styraciflua, Cornus florida, C. asperifolia, Pinus palustris* or *P. taeda*; throughout our area. (Chiefly coastal plain, s.e. Va. to cen. pen. Fla., westward to e. and cen. Tex., northward to Okla., Ark., southeasternmost Mo.)

Fig. 143. **Quercus margaretta:** drawing at center right is tip of twig of the season (leaves cut off) to contrast with "c" in Fig. 142.

Fig. 144. **Quercus chapmanii:** drawing at lower left: enlargement of small, medial portion of lower surface of leaf blade.

National champion sand post oak (1979), 11'1" circumf., 70' in height, 79' spread, Chowan Co., N.C.

7. Quercus chapmanii Sarg. CHAPMAN OAK. Fig. 144

A shrub or a small, scrubby tree. Woody twigs of the season tan, usually densely stellate-pubescent, older twigs gray. Winter bud or buds at the tips of twigs short-oblong to obovoid, rounded to obtuse apically, chestnut-brown, scales as broad as or broader than long, very broadly rounded apically, usually minutely lacerate on the distal margins, with or without some loose pubescence on their backs. Bark of trunks of larger specimens thickish, grayish brown, its surface irregularly breaking into flat strips or plates.

Young developing stems and petioles densely clothed with blond, stellate pubescence. Upper surfaces of unfurling leaf blades with a pinkish hue, relatively sparsely clothed with pale, stellate hairs, lower surfaces with very dense yellowish pubescence, much of the pubescence of upper surfaces sloughing as the leaves mature to full size, that of lower surfaces becoming relatively sparse and shriveled as the blades expand to full size; pubescence of the young developing stems and leaves generally having so much pollen adhering to them as to obscure its nature.

Leaves in most years more or less persistent over the winter, sometimes one by one turning yellow or reddish and falling gradually during the winter, sometimes remaining green until late winter then turning yellow to brown and falling just before new growth commences.

Mature leaves short-petiolate, petioles usually with persistent, shriveled pubescence. Blades varying in size, shape, and marginal features; margins unlobed, to irregularly wavy-margined, to irregularly few-lobed, lobes short and rounded or just shoulderlike; in size (2) 4–8 (10) cm long and (1) 2–6 (8) cm broad; in outline, oblong, elliptic-oblong, elliptic, obovate, or spatulate, bases broadly and shortly tapered to rounded, some of the obovate ones cuneate from above their middles; upper surfaces dark green, lustrous or sublustrous, with little or no persisting pubescence, lower surfaces dull green, sparsely pubescent throughout or only along the veins.

Fruits maturing in one season, their involucres bowllike, 6–8 mm deep, embracing nearly one-half of the acorns, scales more or less triangular, their apices truncate to rounded, pubescent on their backs; acorns oblong to slightly ovoid, about 1.5 cm long, bases rounded, apices rounded or narrowly crateriform, with brown outer surfaces having sparse, short pubescence below and usually more densely pubescent about their apices, hairs buff-colored; inner surfaces glabrous but the innermost, glabrous layer very thin and brittle, breaking up easily, pubescent within that.

Inhabiting sand pine–oak scrub where associated oaks are, principally, *Q. myrtifolia* and *Q. geminata*, or a mixture of these and *Q. laevis*, *Q. margaretta*, *Q. incana*, sometimes also *Q. hemisphaerica*; in our area, s.e. Ga., n. Fla., s.w. Ala. (Outer coastal plain, S.C. to n.e. Fla., in the scrub of coastal and interior of pen. Fla., chiefly near-coastal in the Panhandle of Fla. and s.w. Ala.)

8. Quercus lyrata Walt. OVERCUP OAK. Fig. 145

Large deciduous tree. Woody twigs of the season pubescent at first, becoming glabrous as the season advances. Winter bud or buds at tips of twigs short and round, appearing almost knobby, usually some persistent stipules below them,

Fig. 145. **Quercus lyrata.**

scales thickly clothed with tawny hairs. Bark of trunks brownish gray, forming thick, irregular plates covered with thinner scales.

Young developing stems and petioles sparsely pubescent, the hairs forked basally, the branches longish; unfurling leaves more or less olive-green on their surfaces and with sparse pubescence, the hairs 2-forked basally or stellate with few branches, lower surfaces copiously pubescent throughout.

Mature crown leaves with petioles 1–1.5 cm long. Blades obovate or oblong in overall outline, (7) 12–20 (25) cm long and (3) 5–10 (12) cm broad (measuring across the largest and opposite or subopposite lobes), generally broadest above their middles, mostly irregularly 5–9-lobed, the lateral lobes below the terminal one usually largest and spreading nearly laterally, downwardly the lobes smaller, the lowest usually short-triangular; lobes usually with rounded, less frequently with acute apices, these not bristle-tipped, sinuses between the lobes rounded basally; bases cuneate; upper surfaces dark green and glabrous, lower surfaces varyingly copiously and compactly grayish-pubescent to green and sparsely pubescent (or lower surfaces of leaves of trunk sprouts or seedlings completely glabrous and lustrous); margins or margins of lobes entire.

Fruits maturing in one season, their involucres stalked, oblate to subglobose, enclosing two-thirds to almost all of the acorn, scales broadest basally, medially with thick, hard knoblike, lateral protuberances, apices flat, pointed, appressed between protuberances of other scales; acorns slightly ovoid to oblongish, bases flat, apices rounded to flattish, about 2.5 cm long and broad, outer surfaces dark brown, short-pubescent, inner surfaces glabrous.

River banks and floodplain forest for the most part, sometimes on adjacent lower slopes, commonly associated with *Q. laurifolia* and *Q. nigra*; n. Fla. from about the Suwannee River westward, s. Ga. save for a small area in the southeasternmost part, s. Ala. (Chiefly coastal plain and piedmont, s. N.J. to Suwannee River, Fla., westward to e. Tex., northward in the interior, s.e. Okla., s. and e. Ark., s.e. Mo., w. Tenn. and Ky., s.e. Ind., s. Ill.; and locally in n.w. Ga., cen. Tenn., w. cen. Ill., s. cen. Mo., e. Iowa.)

Co-national champion overcup oaks (1976) 22′ circumf., 123′ in height, 48′ spread, Cogaree Swamp, S.C.; (1977), 21′5″ circumf., 116′ in height, 118′ spread, Tuckahoe State Park, Md.; (1973), 21′5″ circum., 116′ in height, 118′ spread, Tuckahoe State Park, Md.

9. Quercus virginiana Mill. LIVE OAK. Fig. 146

A tree that, when growing relatively isolated, does not attain great height but nonetheless may attain picturesque proportions, the trunks short and bulky, the crowns broadly rounded and with very large principal limbs that spread horizontally, the lower ones often descending and their extremities touching the ground; when growing in forests, the trunks may be relatively tall, the crowns high and irregular. Woody twigs of the season tan to gray, varyingly essentially glabrous or with some, even most, of the desiccated pubescence persisting even through the first winter. Bark of older trunks dark, thick, roughly ridged and furrowed, eventually becoming blocky. Winter bud or buds at tips of twigs minute, 1–2 mm long, globose or short-oblong or -ovoid, chestnut-brown, tips rounded, lowermost scales pubescent on their backs, others with a minute fringe of hairs marginally.

Leaves, for the most part, overwintering, falling just before or as new growth of the season commences. (Leaf fall, flowering, and new shoot growth appears to occur considerably prior to that of *Q. geminata*.)

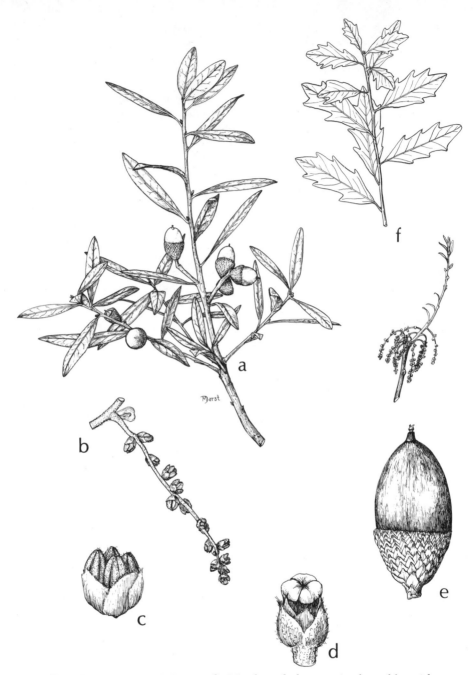

Fig. 146. **Quercus virginiana:** a, fruiting branch; b, emerging branchlet with staminate catkins; c, staminate catkin; d, pistillate flower with surrounding involucre; e, fruit; f, elongation shoot of summer, leaves of which are often toothed-lobed.

Young developing stems and petioles sparsely to densely patchy-pubescent with gray stellate hairs. Unfurling leaf blades flat, their margins not revolute, upper surfaces green but moderately pubescent with gray, stellate hairs, lower surfaces gray, the stellate pubescence dense and, at this stage, relatively loose.

Mature leaves with short, usually pubescent, petioles 2–6 mm long. Blades stiff-leathery varying in size and shape, even on the branchlets of a single branch, or in average size and shape on different trees in a local population, or from one local population to another; margins of blades usually not revolute, on increments of spring growth usually entire, infrequently slightly wavy or with 1–few undulations on a side; on increments of summer growth, however, some, even most of the blades with 1–few blunt teeth on a side, thus such shoots notably contrasting with the spring shoots; in size, in general, 2–12 (15) cm long and 1–3 (5) cm broad; in shape, narrowly to broadly oblong, elliptic, oblong-elliptic, cuneate-spatulate, rarely cuneate-obovate; upper surfaces glabrous, dark green and lustrous, principal veins not impressed on upper surfaces, lower surfaces dull, grayish green, principal lateral veins very little raised, surfaces to either side of the midrib with stellate pubescence so dense and so compacted as to appear glabrous to the unaided eye, usually only just perceptibly pubescent with 10× magnification.

Pistillate flowers mostly 1 or 2, sessile near the tips of stalks (peduncles), the latter usually elongating somewhat after anthesis, reaching 0.5–2.5 cm long. Fruits maturing in one season; involucres turbinate or like bowls of goblets, about 8 mm deep, embracing one-third of the acorns or a little more; scales broadest and thickest near their bases, distally thinner and tapered to more or less triangular, obtuse to nearly truncate tips, their backs densely covered with grayish brown pubescence throughout, or their tips essentially glabrous and medium brown to dark brown; acorns variable in size, oblong-ellipsoid, to slightly ovate or obovate, 1–2.5 cm long, outer surfaces glabrous, usually light brown proximally (within the cups), at full maturity (before they drop), usually dark brown or black distally, sometimes with a grayish-glaucous bloom, inner surfaces glabrous.

Inhabiting a wide variety of sites having relatively heavy and fertile, well-drained to seasonally wet soils, sometimes in almost pure stands, or scattered in mixed woodlands, hammocks, flatwoods, borders of salt marshes, roadsides, city lots, commonly scattered (residual) in pastures; throughout our area. (Outer coastal plain, s.e. Va. to s. pen. Fla., westward to e. and s. Tex.)

National champion live oak (1976), 36'7" circumf., 55' in height, 132' spread, near Louisburg, La.

Florida champion live oak (as given in list for Florida, revised in 1984), 340" circumf., 83' in height, 150.5' spread, Alachua, Fla.

The live oak is the State Tree of Georgia.

10. Quercus geminata Small. SAND LIVE OAK. Fig. 147

Plants generally growing on deep, relatively infertile sands, their form varying a great deal in differing environments. In coastal scrub where subject to salt spray from the seas, commonly forming impenetrable thickets of varying heights, the "crowns" of the thickets appearing as though clipped or sheared; in sand pine–oak scrub, occurring as individual, small, scrubby trees, or together with other kinds of scrubby oaks forming thickets; in longleaf pine forests or where longleaf pine occurred formerly (residual in pastures, for example), com-

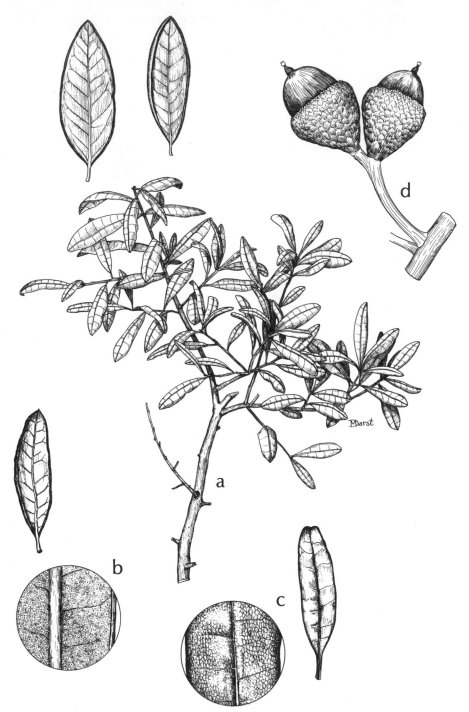

Fig. 147. **Quercus geminata:** a, leafy branch (two separate leaves above from two other trees, drawn to same scale as those in "a"; b, leaf, abaxial view, at left above and small portion of abaxial surface enlarged; c, leaf, adaxial view, at right, and small portion of abaxial surface enlarged; d, fruiting spike.

314

monly reaching moderate to fairly large trees with a form not unlike that of *Q. virginiana*. Twigs, buds, and bark of trunks as in *Q. virginiana*.

Leaves, for the most part, overwintering and falling just before or as new growth of the season commences. (Leaf fall, flowering, and new shoot growth appears to occur considerably later than in *Q. virginiana*.)

Young developing stems and petioles uniformly densely pubescent with gray to somewhat tawny stellate hairs. Unfurling leaf blades having their sides conspicuously rolled downward and more or less folded toward the lower midribs, thus obscuring some or nearly all of the lower surfaces at this stage; upper surfaces green but moderately pubescent with gray, stellate hairs, lower surfaces gray, the pubescence dense, some of it, at least, relatively loose.

Mature leaves with densely pubescent petioles 2–10 mm long. Blades stiffleathery, varying in size and shape, even on branchlets of a single branch, or in average size and shape on different trees in a local population, or from one local population to another; margins of blades commonly revolute and often the sides rolled downward so that the blades have an overall appearance of an inverted, shallow boat; in size 2–12 cm long, 0.5–4 cm broad; in shape, narrowly to broadly oblong, oblong-elliptic, oblanceolate, or cuneate-spatulate; upper surfaces glabrous, dark green and lustrous, generally the midribs and major lateral veins somewhat impressed; lower surfaces dull grayish or somewhat tawny, midveins notably raised, major lateral veins moderately raised and evident, surfaces to either side of the midrib with dense, stellate pubescence which, in general, is less compacted than in *Q. virginiana*.

Pistillate flowers generally 2–several-spicate distally on stalks (peduncles), the latter usually elongating after anthesis, eventually reaching 1.5–5 cm long; fruiting stalks sometimes maturing only one fruit, more commonly 2, often up to 5 or 6. Fruits maturing in one season. Involucres usually turbinate, sometimes bowllike, 7–10 mm deep, 10–15 mm across from rim to rim, embracing about one-third of the acorn; scales as in *Q. virginiana*; acorns variable in size, ellipsoid, oblong-ellipsoid, or slightly broadest distally, 1–2 cm long, outer surfaces glabrous, light brown proximally (within the cups), at full maturity (just before they drop) usually black distally, inner surfaces glabrous. [*Q. virginiana* Mill., var. *maritima* (Michx.) Sarg., misapplied.]

Inhabiting sites having relatively deep, infertile sands, coastal dunes, sand pine–oak scrub, beneath and/or about slash pine in near-coastal areas, longleaf pine–scrub oak ridges and hills, commonly scattered (residual) in pastures; throughout much of our area. (Outer coastal plain, s.e. Va. to s. cen. Fla., westward to s. Miss.)

Florida champion sand live oak (as in Florida list, revised in 1984), 107" circumf., 58' in height, 72.5' spread, High Springs, Fla.

11. Quercus minima (Sarg.) Small. DWARF LIVE OAK. Fig. 148

Shrub, forming extensive clones by subterranean runners; stems to 1 m tall, infrequently to about 2 m, profusely sprouting from their bases after burning of pinelands. Woody twigs of the season dark brown, glabrous. Winter bud or buds at tips of twigs minute, 1–1.5 mm long, globose or subglobose, scales powdery-pubescent.

Leaves overwintering, falling just before, as, or just after new growth commences in spring. Stems of young developing shoots sparsely or more commonly densely stellate-pubescent, the hairs pale gray, sloughing during the first season; unfurling leaves usually flat (their edges sometimes revolute but not

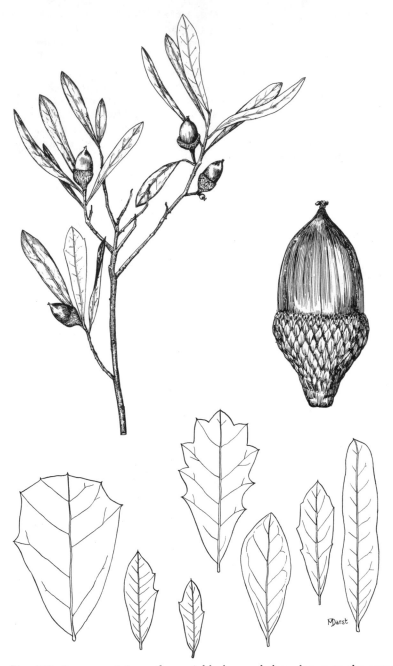

Fig. 148. **Quercus minima:** the variable leaves, below, drawn to the same scale as those leaves on the fruiting twig, above (from sprouts after the site had been burned).

recurving so as to cover edges of lower surfaces); upper surfaces green, very sparsely and finely stellate-pubescent, pubescence rather quickly sloughing as leaves mature, the lower surfaces usually gray owing to being copiously stellate-pubescent, pubescence rather quickly sloughing as the leaves mature, the lower surfaces usually tan owing to being copiously stellate-pubescent although the midribs and some lateral veins not obscured by the pubescence.

Mature leaves very short-petiolate. Blades stiff and leathery, variable in size, shape, and marginal features; sometimes most or all of them entire-margined, mostly 3–6 cm long and 1–2.5 cm broad, oblong-oblanceolate, narrowly oblong, or narrowly cuneate-obovate; sometimes blades similarly sized and shaped but regularly or irregularly toothed-lobed marginally; sometimes (presumably on sprouts after fire) most or all leaves larger, 4–8 (10) cm long and 2–4 (6) cm broad, most of them regularly or irregularly toothed-lobed marginally, narrowly to rather broadly cuneate-obovate in outline; teeth or lobes, when present, often spinose-tipped and prickly; upper surfaces of mature blades glabrous, dark green and lustrous or sublustrous; lower surfaces generally densely and compactly gray-pubescent, sometimes the pubescence so compacted that the surfaces appear glabrous, or sometimes some leaves very nearly or wholly glabrous beneath.

Fruits maturing in one season, rarely sessile or paired and sessile in a leaf axil, usually (owing to the pistillate flowers being borne on a short- to long-stalked spike which elongates after anthesis) the fruits are sessile but borne singly, in pairs, or up to several distally on a fruiting stalk (peduncle), the stalks of various lengths. Involucres broadly turbinate, less frequently cuplike or bowllike above but abruptly tapered proximally, scales closely appressed, broadest and thickest at the basal one-third to one-half then tapered to rounded or truncate tips, their lower portions clothed with buff-colored pubescence, sometimes the base of the distal tapers as well, all of the taper on only its tip with little pubescence and dark brown or brownish red in color, thus the involucre, to the unaided eye, appearing dark-spotted; acorns about twice as long as broad, mostly 2–2.5 cm long, bases more or less rounded, only slightly broader near the base than above and oblongish ellipsoid, outer surfaces glabrous, brown to dark brown or blackish, inner surfaces dark brown and glabrous.

Inhabiting open, sandy, usually flat pinelands; s. Ga., n. and cen. pen. Fla., s. Ala.? (included for s. Ala. by Small, 1933).

Q. minima and *Q. pumila* commonly grow in close proximity or intermixed where their ranges overlap. Their growth habit is similar and it is sometimes difficult to distinguish one from the other, the more so if, in the case of *Q. minima*, none of the leaf blades are toothed-lobed. The pubescence of the lower leaf surfaces of *Q. minima* is much more compact than that of *Q. pumila*. Only rarely are 1–few leaf blades of *Q. pumila* toothed-lobed as they commonly are in *Q. minima*. Winter buds of *Q. minima* are essentially globose, those of *Q. pumila* ovoid-conic and pointed apically.

12. Quercus pumila Walt. RUNNING OAK. Fig. 149

Shrub, commonly forming extensive clones by subterranean runners; stems to 1 m tall, infrequently to 1.5–2 m if the site not recently burned, profusely sprouting from their bases after burning of pinelands, the leaves of sprouts often larger than those of stems not having been burned for several years. Woody twigs of the season grayish brown, usually much of their pubescence persisting through the first year. Winter bud or buds at the tips of twigs ovoid-conic, 3–5 mm long, brown, scales mostly with a fringe of minute hairs around their apical margins.

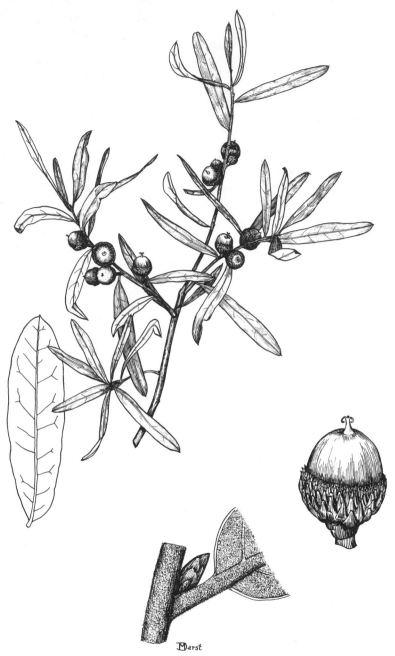

Fig. 149. **Quercus pumila:** leaf at left, drawn to same scale as leaves on fruiting branch, characteristic of those on vigorous sprouts.

Leaves all deciduous in autumn or a few of them overwintering and falling just before or as new growth commences in spring. Stems of young developing shoots moderately to densely stellate-pubescent; edges of unfurling leaf blades folded downward and recurved so as to cover about half of the lower surfaces, their upper surfaces with sparse, pale, stellate pubescence, the lower surfaces densely and compactly pale-gray-pubescent; as the blades enlarge most of the pubescence quickly sloughs from the upper surfaces.

Mature leaves very short-petiolate, petioles stellate-pubescent. Blades mostly 3–10 cm long and 0.7–2 (3) cm broad (to 15 cm long and to 5 cm broad on some sprouts), oblanceolate or spatulate, narrowly elliptic, elliptic-oblong, or lanceolate, often with short bristle tips; bases cuneate to narrowly rounded, apices rounded to acute; upper surfaces glabrous, dark green and lustrous or sublustrous, sometimes dull green, lower surfaces densely and compactly grayish pubescent (or sparsely pubescent to essentially glabrous and bright green on large leaves of sprouts); flat and with entire margins, sometimes their edges somewhat crisped, only rarely with a few, low, rounded undulations.

Fruits maturing in one season, sessile or shortly stalked, their involucres bowllike, 4–5 mm deep, embracing about one-third of the acorns, scales tightly appressed, grayish brown, broadest basally where many or most of them are humped or bulged, gradually narrowed distally to truncated, flat tips; acorns ovoid, subglobose, or somewhat oblate, 8–12 mm long and broad, flat basally, apically rounded to nearly truncate, outer surfaces light brown, glabrous or faintly and sparsely very short-pubescent near their summits, inner surfaces loosely pale-pubescent, the hairs blond to tawny.

Inhabiting open, sandy, usually well-drained pinelands, commonly intermixed with or in close proximity to the dwarf live oak, *Q. minima*; see *Q. minima* for comparative features; throughout our area. (Coastal plain, s.e. N.C. to cen. pen. Fla., westward to s.e. Miss.)

13. Quercus hemisphaerica Bartr. ex Willd. DARLINGTON OAK, LAUREL OAK.
Fig. 150

A relatively large tree. Woody twigs of the season brown to gray, usually slender. Winter bud or buds at the tips of twigs dark brown, short-ovoid, sometimes vaguely, obtusely, and longitudinally angled, scales glabrous on their backs, sometimes the lower and medial ones with very short hairs fringing their margins, the distal ones usually with much longer hairs around the apical margins, less frequently they appear not to be fringed at all. Bark of smaller trunks gray and smoothish, that of larger ones blackish or gray, irregularly and shallowly ridged and furrowed, not blocky, or appearing somewhat crustose.

Unfurling leaves more or less olive-green, very sparsely short-pubescent on and near the midribs of the upper surfaces, very short-pubescent throughout on the lower surfaces. Leaves tardily deciduous, some falling over winter, most falling in spring just before or as flowering and new growth commence (the latter usually occurring somewhat later than in *Q. laurifolia*).

Mature leaf blades oblanceolate, varying to narrowly elliptic or lanceolate, mostly 2.5–8 cm long and 1.5–3 cm broad; cuneate proximally to a subsessile base or only slightly tapered to a narrowly rounded base, a short petiole evident, apices narrowly blunt to sharply acute, if the latter, then commonly bristle-tipped; margins entire; upper surfaces dark green and sublustrous, lower surfaces somewhat paler but bright green, both surfaces wholly glabrous. Leaves on increments of summer growth often having several irregularly dis-

319

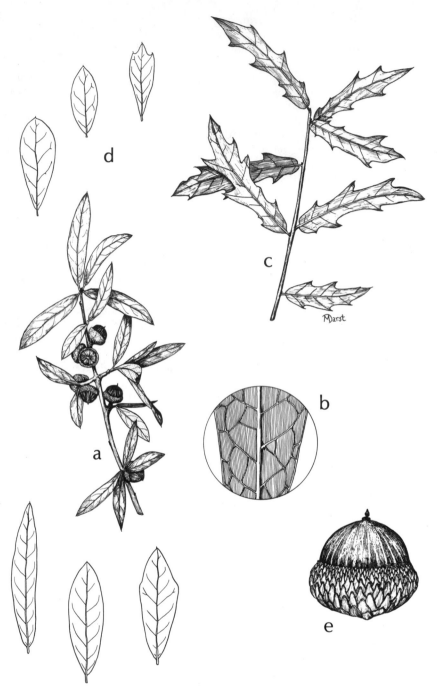

Fig. 150. **Quercus hemisphaerica:** a, fruiting branch; b, enlargement of small portion of lower surface of leaf blade; c, elongation shoot of summer (leaf blades of such shoots or of saplings or sprouts often are toothed-lobed); d, variable leaves, top and bottom, left; e, fruit.

320

posed, triangular-ascending, bristle-tipped lobes on each side. Low root-sprouts often abundant beneath and around mature trees in some mixed woodlands, their leaves so different from those of crowns of the "parent" tree as to be difficult to associate with them; they are commonly oblong or oblong-obovate, to 15 cm long and 6 cm broad at their broadest places, varyingly with entire margins or with several more or less dentate, short, bristle-tipped lobes on each side, overall very much resembling leaves of root-sprouts of *Q. virginiana* save that their surfaces are bright green and wholly glabrous whereas those of *Q. virginiana* are grayish-green beneath and are very compactly and densely pubescent (as viewed with suitable magnification).

Fruits maturing the second season, sessile or short-stalked, their involucres shallowly saucerlike, 2–2.5 mm deep, embracing little more than the bases of the acorns, scales sparsely pubescent, closely appressed, triangular, apices blunt; acorns ovoid, nearly flat basally, apically rounded, mostly about 10 mm long, outer surfaces tan, very short pubescent throughout to only sparsely pubescent distally, inner surfaces copiously pubescent with blond hairs. [*Q. laurifolia* sensu Small (1933); in part *sensu* West and Arnold (1956), Radford *et al.* (1964), Long and Lakela (1971), Clark (1972), Elias (1980), Lakela and Wunderlin (1980), Wunderlin (1982)]

Common in well-drained woodlands of various mixtures, longleaf pine–scrub oak sand ridges, slash pine plantations where well-drained, natural levees and terraces along rivers and streams, maritime woodlands, often abundantly colonizing and eventually forming groves in old fields, in fence and hedge rows, commonly planted as a shade tree; throughout our area. (Coastal plain, s.e. Va. to cen. pen. Fla., westward to s.e. Tex.)

See discussion at end of account for *Q. phellos* for some features distinguishing *Q. hemisphaerica* from it.

14. **Quercus laurifolia** Michx. DIAMOND-LEAF OAK, SWAMP LAUREL OAK.

Fig. 151

Large tree (in floodplain forests, largest ones often with more or less planklike, flying buttresses to 2 m high). Woody twigs of the season dark brown, glabrous, slender or stoutish. Winter bud or buds at the tips of the twigs ovoid-conic, obscurely, obtusely, and longitudinally angled, dark reddish brown, scales glabrous on their backs, the distal ones, especially, with a fringe of hairs on their apical margins.

Unfurling leaves bronze-colored to purplish red, both surfaces finely and sparsely short-pubescent and with dense tufts of hairs in at least some of the axils formed by the midvein and major lateral veins.

Leaves tardily deciduous, falling gradually over winter or, more commonly, most of them retained until just before new growth commences in spring. (In most years leaf fall and flowering with concomitant initiation of new growth in spring occurring prior to that in *Q. hemisphaerica*.)

Mature leaf blades exhibiting considerable variation on a given tree or from tree to tree (3) 5–10 (14) cm long and (1.5) 2–4 (5) cm broad; prevailingly spatulate, oblanceolate, subrhombic, or obovate, these generally with cuneate bases, sometimes tapering indistinctly into the short petioles, sometimes the short petioles distinct, apices rounded, bluntly tapered, rarely acute and only the latter occasionally bristle-tipped; occasional blades with 1–2 lateral lobes distally, lobes usually rounded, infrequently some blades lanceolate or narrowly ovate; *at least a few leaves on nearly all trees subrhombic; occasionally blades on incre-*

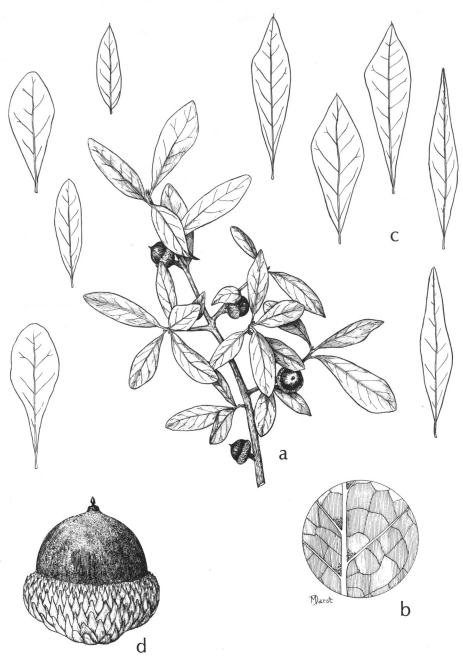

Fig. 151. **Quercus laurifolia:** a, fruiting branch; b, enlargement of small portion of lower surface of leaf showing tufts of hairs in vein axils; c, variable leaves; d, fruit.

ments of summer growth irregularly and shortly 1-, 2-, or 3-lobed on each side; upper surfaces glabrous, dark green and slightly lustrous, lower surfaces duller but not significantly paler green, with tufts of hairs in some or most of the axils formed by the midrib and major lateral veins; margins or margins of lobes entire. ·

Fruits maturing the second season, usually shortly stalked, their involucres and acorns varying a great deal in size from tree to tree, shallowly to deeply bowllike, 5–10 mm deep, their bases abruptly short-tapered, embracing slightly less than one-third to nearly one-half of the acorns, scales pubescent, closely appressed, ovate-triangular, their apices blunt; acorns slightly ovoid to oblong, nearly flat basally, rounded apically, 10–20 mm long, outer surfaces tan to dark brown, densely to sparsely pubescent with fawn-colored hairs, inner surfaces densely pubescent with blond hairs. [Not *Q. laurifolia* sensu Small (1933); *Q. obtusa* (Willd.) Ashe]

Floodplain forests and bottomland woodlands generally (where often associated with *Q. lyrata*), river and stream banks, shores of lakes and impoundments, bayheads; throughout much of our area. (Coastal plain and piedmont, s. N.J. to s. pen. Fla., westward to e. Tex. and s.e. Ark.)

15. Quercus phellos L. WILLOW OAK. Fig. 152

Medium-sized to large deciduous tree. Woody twigs of the season brown, glabrous. Winter bud or buds at the tips of twigs lance-ovoid to ovoid, usually obscurely, obtusely, and longitudinally angled, brown, medial and upper scales with longish, antrorsely disposed hairs on their backs. Bark of smaller trunks grayish and smoothish, that of larger ones dark gray, shallowly fissured and splitting into irregular small plates or scales.

Unfurling leaf blades more or less olive-green, upper surfaces very sparsely short-pubescent, the sides strongly rolled downwardly or even folded over the lower surfaces, obscuring much or even most of the lower surfaces, the latter sparsely and minutely stellate-pubescent along the midribs to copiously white-stellate-pubescent throughout.

Mature blades predominantly lanceolate, some narrowly elliptic-lanceolate, elliptic, or narrowly lance-oblong, an occasional one oblanceolate, mostly 6–12 cm long and 0.7–2 cm broad; usually short-tapered basally, acute apically and bristle-tipped; upper surfaces dark green and glabrous, lower surfaces somewhat paler green, with tufts of hairs in some or most of the axils formed by the midrib and major lateral veins; margins entire; petioles very short.

Fruits maturing the second season, sessile or very shortly stalked, their involucres shallowly saucerlike, about 3 mm deep, embracing about one-fourth of the acorns, scales grayish tan, closely appressed, ovate-triangular to subrhombic, appressed-pubescent, apices blunt; acorns slightly ovoid or subglobose, slightly convex basally, apices rounded, outer surfaces grayish brown, clothed with short, grayish pubescence, inner surfaces densely pubescent with blond or pale buff-colored hairs.

Chiefly in bottomland or lowland woodlands, sometimes spreading to fence or hedge rows and into upland old fields; in our area, disjunct in n.e. Fla., local in the Panhandle of Fla. from Leon Co. westward, s. cen. and s.w. Ga., s. Ala. (S. N.J. and s.e. Pa., generally southward, mainly on the coastal plain and piedmont, to the Fla. Panhandle (apparently not in s.e. Ga.), westward to e. Tex., northward in the interior, s.e. Okla., s. and e. Ark., s.e. Mo., s. Ill., Ky., and Tenn.)

Q. phellos and *Q. hemisphaerica* have leaves, blades of which, at a glance,

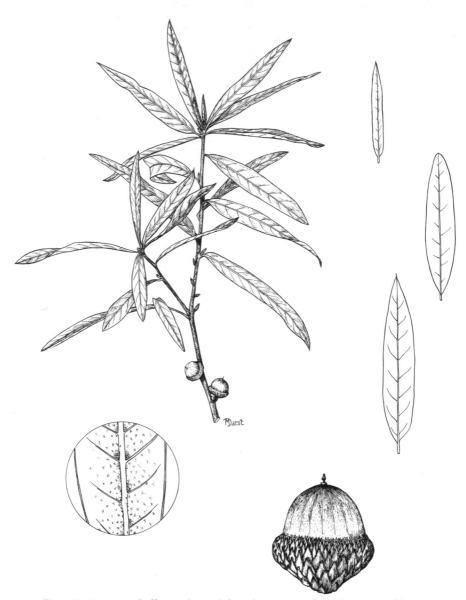

Fig. 152. **Quercus phellos:** at lower left, enlargement of small portion of lower
surface of leaf to show distribution of pubescence.

appear much alike; on a given tree, those of the former are predominantly lanceolate and with pubescence beneath mainly in major vein axils and sometimes in patches along the midrib and major lateral veins, those of the latter are predominantly oblanceolate and wholly glabrous.

Co-national champion willow oaks (1972), 24' circumf., 112' in height, 125' spread, Taliaferro Co., Ga.; (1972), 23'6" circumf., 125' in height, 106' spread, Queenstown, Md.

16. Quercus myrtifolia Willd. MYRTLE OAK. Fig. 153

Commonly a thicket-forming shrub, less frequently a small, scrubby tree. Woody twigs of the season brown, usually some of the dessicated pubescence persisting. Winter bud or buds at tips of twigs ovoid, brown, uppermost scales pubescent distally, or at least with a fringe of hairs around their distal margins. Bark of branches and small trunks light gray, on older trunks dark gray and roughish, commonly, however, more or less coated with lichens.

Leaves persisting over winter, most or all of them dropping immediately before or as flowering and growth of the season commences. Stems of young, developing shoots very densely stellate-pubescent; surfaces of unfurling leaves clothed with a mixture of pale stellate hairs and orange, glandlike hairs, the density of the mixture varying (depending in part upon the extent of expansion of the blades), tending to be very dense at first, particularly on the lower surfaces; in addition lower surfaces may have puffs of stellate pubescence elevated above the other pubescence at the position of some of the axils formed by the midrib and lateral veins (the axils themselves obscured by pubescence); much of the pubescence of the leaf surfaces sloughs during maturation of the leaves, usually all of it from the upper surfaces, the lower surfaces may retain some of the glandlike hairs (now desiccated and not evident to the naked eye), may retain some of the tufts of hairs in the vein axils, may be very finely stellate-pubescent throughout, or may be wholly glabrous.

Mature leaves short-petiolate, blade tissue slightly winging the petioles. Blades stiff-leathery, flat or their sides often cupped downwardly, variable in size and shape, as broad as long to about twice as long as broad, 2–5 (8) cm long and 1–3 (5) cm broad, oval, elliptic, oblong, or obovate, bases broadly short-tapered to rounded, sometimes cordate or subcordate (probably thus only on sprouts), apices mostly rounded, infrequently bristle-tipped; upper surfaces dark green and lustrous, lower surfaces paler and duller green, sometimes yellowish green or somewhat orangish brown, margins entire, rarely sinuate (perhaps only on sprouts).

Fruits maturing the second season, short-stalked, involucres shallowly saucerlike to somewhat more bowllike, 3–4 mm deep, embracing about one-fourth of the acorns or slightly more, scales grayish brown, broadest basally, tapering from about their middles to rounded tips, pubescent; acorns ovoid to subglobose, flat basally, apically rounded, outer surfaces tan, sparsely pubescent, inner surfaces densely and compactly pubescent with blond hairs.

A major component of sand pine–oak scrub, less frequently present in longleaf or slash pine–scrub oak communities on well-drained sandy soils; s.e. Ga., n.e. Fla., Fla. Panhandle, s.w. Ala. (Coastal plain, S.C. to cen. pen. Fla., westward to s.e. Miss.)

National champion myrtle oak (1972), 4'3" circumf., 26' in height, 33' spread, Fort Clinch State Park, Fla.

Fig. 153. **Quercus myrtifolia:** lower center, enlargement of small portion of
lower surface of leaf; upper left, variable leaves.

17. Quercus incana Bartr. BLUEJACK OAK.

Fig. 154

A small, often shrubby tree, sometimes forming thickets by subterranean runners, crowns open and irregular, trunk usually short. Twigs grayish. Winter bud or buds at tips of twigs conic-ovoid, dark reddish brown, a sparse mixture of longish and short hairs on their backs.

Young developing stems densely stellate-pubescent, hairs gray, much of the pubescence retained through at least the first year; unfurling leaves densely short-stellate pubescent on both surfaces, the venation obscured by the pubescence, upper surfaces pinkish so that the crowns have a definite pinkish cast during leaf emergence.

Mature leaves deciduous in autumn, tardily so if autumn and winter temperatures are mild. Petioles short, pubescent. Blades, on increments of spring growth, generally unlobed, margins entire, mostly 3–10 (12) cm long and 1–3 cm broad, lanceolate, oblanceolate, oblong, elliptic-oblong, or slightly obovate, bases cuneate to rounded, apices rounded, obtuse, or acute, sometimes with minute bristle-tips; blades on increments of summer growth, if any, and of sprouts, averaging somewhat larger, often some of them irregularly 1–few-lobed on each side, lobes short and rounded to more or less short-triangular and their tips often bristle-tipped; upper surfaces wholly glabrous or with only a few short hairs basally, lower surfaces with dense and compact gray pubescence. Crown, overall, having an ashy green or slightly bluish green hue as seen from a little distance.

Fruits maturing the second season, sessile, their involucres shallowly bowllike, embracing about one-fourth of the acorn, scales grayish brown, tightly appressed, broadest basally and bases somewhat bulged or humped, tapered to blunt or rounded tips, appressed-pale-pubescent; acorns ovoid to subglobose, mostly 10–14 mm long, some longer than broad, some broader than long, flat or convex basally, outer surfaces brown, unevenly and finely short-pubescent, inner surfaces tawny-pubescent. [Q. cinerea Michx.]

Occurring on well-drained sandy ridges and flats of pineland; throughout our area. (Coastal plain, s.e. Va. to n. cen. pen. Fla., westward to e. and cen. Tex., s.w. Ark., s.e. Okla.)

National champion bluejack oak (1972), 7' circumf., 51' in height, 56' spread, Freestone Co., Tex.

18. Quercus nigra L. WATER OAK.

Fig. 155

Medium-sized tree. Woody twigs of the season dark brown, glabrous. Winter bud or buds at the tips of twigs lance-ovoid to ovoid, obscurely, obtusely, and longitudinally angled, brown, at least the medial and upper scales thickly clothed on their backs with long, antrorsely disposed hairs. Bark of trunks at first smoothish, grayish brown, that of large older ones blackish, with close, irregular patching and eventually shallowly and irregularly scaly-ridged and furrowed.

Young, developing stems and petioles clothed with sparse, pale, stellate pubescence. Both surfaces of unfurling leaf blades bearing sparse, short, pale, appressed pubescence along the major veins, lower surfaces with tufts of white hairs in the axils formed by the major lateral veins and the midrib.

Leaves tardily deciduous, usually falling gradually over winter. Mature blades exhibiting notable variation in size and shape from tree to tree; in outline broadest distally, spatulate to cuneate-obovate, the broadened apices unlobed varying to 3-lobed, long-cuneate basally and narrowed gradually into a short

Fig. 154. **Quercus incana.**

Fig. 155. **Quercus nigra:** upper left, small portion of lower surface of leaf showing tufts of hairs in vein axils; two leaves at lower left from seedling or small sapling, other separate leaves to show some of the variation.

petiole, the latter scarcely distinguishable as such, or gradually narrowed to a narrowly rounded base and the short petiole evident; blades 5–10 (15) cm long and 2–5 (10) cm broad distally, bristle tips sometimes evident on the lobes; upper surfaces dullish dark green and glabrous, lower surfaces paler green, with tufts of hairs in some or all of the axils formed by the midrib and major lateral veins, hairs of the tufts more or less sloughing as the leaves age but usually to some degree evident. Leaves of seedlings, saplings, and vigorous sprouts usually differing markedly from those of crowns of well-developed trees.

Fruits maturing the second season, sessile or nearly so, their involucres shallowly saucerlike, thin, 1.5–2 mm deep and with rounded bases, enclosing little more than the bases of the acorns, scales pubescent, ovate-triangular, tightly appressed, blunt apically. Acorns ovoid to subglobose, nearly flat basally, rounded apically, mostly 8–10 mm long, outer surfaces tan, finely short-pubescent, inner surfaces densely pubescent with fawn-colored hairs.

In a wide variety of habitats, mixed upland woodlands, locally in pine flatwoods, river and stream banks, floodplain woodlands, wet hammocks, old fields, fence and hedge rows; throughout our area. (Coastal plain and piedmont, rarely in the foothills of the Blue Ridge, Del. to cen. pen. Fla., westward to n.e. Tex., northward in the interior to s.e. Okla., s.e. Mo., and Tenn.)

National champion water oak (1980), 21'7.5" circumf., 105' in height, 117' spread, Itawamba Co., Miss.

19. Quercus arkansana Sarg. ARKANSAS OAK.   Fig. 156

Generally a rather small deciduous tree. Woody twigs of the season dark brown, glabrous or much of the pubescence persisting. Winter bud or buds at the tips of twigs ovoid, most of them obscurely, obtusely, and longitudinally angled, brown, medial and upper scales usually pubescent on their backs and the distal margins with a fringe of minute hairs. Bark of trunks thickish, nearly black, with long, narrow ridges and furrows.

Young developing stems, petioles, and both surfaces of unfurling leaf blades moderately densely to relatively sparsely pubescent with minute, pale, stellate hairs, the lower surfaces also having tufts of hairs in the axils formed by the major lateral veins and the midrib, sometimes especially hairy along the midrib between the lateral veins and along the major lateral veins; upper surfaces dark green, the lower more or less olive-green.

Mature leaves with petioles 8–20 mm long, desiccated pubescence persisting on them. Mature blades on increments of growth produced in spring generally obovate, varyingly unlobed or with a pair of rounded distal lobes, irregularly with a few short bristles marginally extending beyond vein endings, or on tips of lobes if lobes present; size variable, width at widest part commonly approximately equal to the length although some leaves longer than broad, 6–15 cm long. Mature blades on increments of growth produced in summer, if any, in outline obovate to broadest at about the middle, with 3–5 more definite lobes, the lateral lobes more or less triangular, the terminal one broader and more pronounced. Upper surfaces dark green, somewhat shiny, varying from glabrous to sparsely pubescent throughout, the hairs stellate and tiny, lower surfaces paler, less shiny, with shaggy, tawny stellate pubescence in the major vein axils, sometimes along the lateral veins as well, between the veins glabrous or with sparse pubescence persisting.

The outline shape of blades of leaves of *Q. arkansana* (on increments of spring growth, especially) is not unlike that of many crown leaves of *Q. marilandica*.

Fig. 156. **Quercus arkansana:** leaves of fruiting twig drawn life-size. Variable leaves drawn × ⅝; two leaves at top center from increment of summer growth.

The blades of the former, however, are more pliable, not so stiff-leathery as in *Q. marilandica*, and have none of the orangish, glandlike pubescence on their surfaces as in *Q. marilandica*.

Fruits maturing the second season, their involucres shallowly saucerlike, embracing only one-fourth of the acorn, scales brown, triangular, tightly compressed, finely pubescent on their backs but their thin margins glabrous; acorns subglobose to oblate, 6–10 mm long, outer surfaces pubescent with longish-forked, gray pubescence, inner surfaces blond-hairy. [Incl. *Q. caput-rivuli* Ashe]

Inhabiting well-drained, sandy or rocky, mixed woodlands or ecotones between them and branch bays; in our area, local, perhaps only in the w. half of the Panhandle of Fla. and s.w. Ga. (Local, s. cen. Ga., Fla. Panhandle, cen. Ala., s.e. La., s. Ark.)

National champion Arkansas oak (1972), 11'6" circumf., 62' in height, 66' spread, Howard Co., Ark.)

20. Quercus marilandica Muenchh. BLACKJACK OAK. Fig. 157

Small to medium-sized tree, often scrubby and with an irregular crown. Woody twigs of the season stout, dark brown or grayish brown. Winter bud or buds at tips of twigs narrowly conical, scales densely covered with tawny or rusty brown hairs. Bark of larger trunks thick, blackish, deeply fissured and broken into irregular blocks.

Stems of young developing shoots and petioles copiously stellate-pubescent, hairs tawny or yellowish brown; upper surfaces of unfurling leaf blades moderately pubescent with a mixture of pale stellate hairs and minute, orangish, glandular or glandlike hairs, the lower surfaces densely pubescent with similar hairs, the orangish ones very much more numerous than the stellate ones and rendering the lower surfaces orangish throughout; in addition lower surfaces sometimes having tufts or puffs of pale stellate hairs in axils formed by the midrib and major lateral veins.

Blades of mature leaves stiff, broadly obovate to triangular-obovate, 7–20 (25) cm long and 5–20 cm broad, unlobed or more commonly with 3–5 broadly rounded lobes distally, lobes usually shortly bristle-tipped; from the broad summits abruptly or gradually narrowed to narrowly rounded or subcordate bases; upper surfaces green, glabrous, usually shiny, the lower surfaces shaggy stellate-hairy in the vein axils, surfaces otherwise thinly to copiously clothed with the now desiccated glandlike pubescence, this yellow to olive-green if the pubescence abundant. Mature leaves of seedlings and saplings commonly more distinctly and prominently 3-lobed distally.

Fruits maturing the second season, their involucres, *on some trees*, narrow basally and expanding upwardly to distinctly rounded rims, about 10 mm deep and 20 mm broad, embracing about one-third of the slightly obovate to short-oblong acorns, these 15–20 mm long; apparently less frequently, *on other trees*, the cups rounded basally, more saucerlike, without a rounded rim, about 5 mm deep and 15 mm broad, embracing about one-fourth of the more nearly oblate acorns, these about 10 mm long; scales of the cups brown, relatively coarse on the larger cups, relatively fine on the smaller cups, in either case pubescent on the back; acorns tawny-pubescent within.

Generally in open woodlands, often locally abundant, usually in clayey soil or where top-soil is sandy and with clayey soil or limestone beneath; in our area, s.e. Ga. and w. of the Okefenokee Swamp and the Suwannee River. (S.e. N.Y., N.J., s.e. Pa., n.e. W.Va., s. Ohio, Ind., and Ill., s.e. Iowa, generally southward to n. Fla. and e. and w. cen. Tex., also s.e. Mich.)

Fig. 157. **Quercus marilandica.**

National champion blackjack oak (1981), 13′9″ circumf., 48′ in height, 78′ spread, Grant Co., Okla.

21. Quercus laevis Walt. TURKEY OAK.

Fig. 158

Usually a small, commonly scrubby, deciduous tree. Twigs relatively stout, brown the first year, gray when older. Winter bud or buds at tips of twigs long-conical, brown, scales pubescent on their backs. Bark of trunks blackish, deeply corrugated and blocky.

Young developing stems sparsely pubescent with pale or yellowish stellate hairs and a few yellow glandular or glandlike hairs; midrib and proximal portions of major lateral veins of upper blade surfaces with mostly grayish stellate hairs, the surfaces otherwise densely clothed with greenish yellow (when fresh) chiefly glandlike hairs; midribs and major lateral veins of lower surfaces with a mixture of longish pale grayish hairs and greenish yellow glandlike ones, the surfaces otherwise densely clothed chiefly with glandlike ones. From a little distance the unfurling leaves have a distinctive chartreuse hue.

Mature leaves with petioles predominantly 1 cm long or less, these, for the most part, twisted so that the plain surfaces of the blades are oriented more or less vertically, a characteristic not shared by other kinds of our oaks having a similar pattern of lobing of the leaves. Blades stiffish, 7–20, rarely to 30, cm long, cuneate basally, irregularly lobed, on a given tree with the size varying and the primary lobes varying from 3 to 7; primary lobes relatively prominent, few-sublobed or toothed, mostly spreading laterally or slightly ascending, tips of lobes or teeth pointed and bristle-tipped, sinuses between the lobes usually open, rounded basally and reaching more than halfway to the midrib; upper surfaces dark green, glabrous, and lustrous, lower surfaces lighter green, or often yellowish owing to the persistence of desiccated pubescence which is sometimes patchy and scalelike, or late in the season, if all or most of the pubescence has sloughed, glabrous save for tufts of pubescence in the major vein axils.

Fruits maturing the second season, their involucres narrow basally, flaring above the base and bowllike, about 10 mm deep and 20 mm wide distally, embracing one-third to one-half of the acorn, uppermost several series of scales rolled inward, then downward and appressed against the upper inner face of the cup, thus the upper edge of the cup appearing like a terrace around the acorn; scales of the cup relatively coarse, brown, tightly appressed, oblongish-triangular, short-pubescent on their backs; acorns ovoid-oblong, mostly about 2 cm long, copiously blond-pubescent within.

In the autumn of most years, leaves of the turkey oak turn red and contribute significantly to autumnal coloration, later turn brown and gradually drop. [Q. catesbaei Michx.]

Common on well-drained sandy ridges with longleaf pine (or where longleaf pine has all or nearly all been harvested), or with planted slash pine, and associated with Q. incana, Q. margaretta, Q. geminata, with Q. marilandica in some places; with sand pine and associated Q. geminata, Q. myrtifolia, Q. chapmanii; less frequently in well-drained mixed pine-hardwood stands, including Q. falcata, Q. stellata, Q. margaretta, Q. hemisphaerica, Q. nigra, Carya glabra or C. tomentosa, Liquidambar styraciflua, and others; throughout our area. (Coastal plain, s.e. Va. to s. cen. Fla., westward to La.)

Co-national champion turkey oaks (1972), 6′9″ circumf., 83′ in height, 67′ spread, near Branford, Fla.; (1979), 8′8″ circumf., 67′ in height, 70′ spread, Pierce Co., Ga.

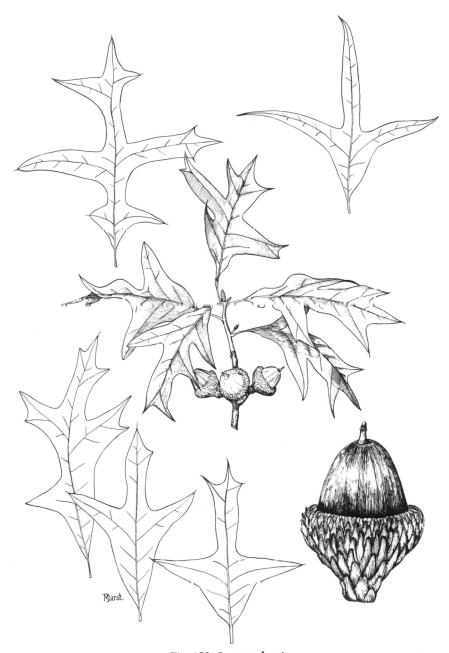

Fig. 158. **Quercus laevis.**

Fig. 159. **Quercus falcata:** variable leaves at top drawn × ⅝ relative to leaves of fruiting branch at × 1.

22. Quercus falcata Michx. SOUTHERN RED OAK, SPANISH OAK.   Fig. 159

Medium-sized to large deciduous tree. Twigs brown by the end of the first year and then varyingly pubescent to essentially glabrous. Winter bud or buds at tips of twigs lance-ovoid to ovoid, brown, scales with longish, antrorsely disposed hairs on their backs. Bark of trunks nearly black, deeply fissured between flat ridges.

Young, developing stems clothed with abundant, tawny, stellate hairs; petioles of unfurling leaves densely shaggy, stellate-pubescent with tawny hairs; upper surfaces of blades with obscure venation, surfaces moderately densely clothed with a mixture of small, pale stellate hairs or minute, amberish glandlike ones, the surfaces appearing brown to the unaided eye, soft to the touch; lower surfaces having prominent, somewhat raised, major veins, these covered mainly with rust-colored, stellate pubescence, surfaces between the veins densely clothed with (at this stage) grayish stellate pubescence, thus the rust-colored veins and grayish areas between clearly evident to the unaided eye.

Mature leaves with petioles 1–4 (5) cm long, relatively densely pubescent. Blades of leaves of saplings, sprouts, and relatively young, small trees generally with 3 primary lobes distally, narrowing below the lobes to rounded or, less frequently, cuneate bases, the bulk of the blades below the lobes U- or V-shaped. Blades of most leaves on mature crowns 12–20 cm long or a little more, prominently lobed, bases rounded or broadly tapered, most of them U- or V-shaped; primary lobes 3–7, triangular to triangular-attenuate, lowermost usually triangular, angling-ascending, the terminal one commonly oblongish and sublobed or toothed, sinuses between the lobes open, mostly broadly rounded, and those of the major lobes reaching more than halfway to the midrib, tips of lobes pointed, sometimes attenuate, and they and the sublobes or teeth bristly-tipped; upper surfaces dark green, glabrous, and sublustrous, lower surfaces densely short-pubescent and more or less tawny (when fresh), soft to the touch.

Fruits maturing the second season, their involucres shallowly bowllike, 4–5 mm deep, 12–14 mm broad, embracing about one-third to one-half of the acorn, scales grayish brown, oblongish, short-pubescent on the back, tightly appressed; acorns subglobose or slightly oblate, blond- or tawny-pubescent within. [*Quercus rubra* sensu Small, 1933]

In our range, in mixed, well-drained, relatively fertile woodlands, often associated with *Q. stellata, Q. velutina, Q. virginiana, Carya tomentosa, Pinus taeda,* and/or *P. echinata;* throughout much of our area. (S.e. N.Y., s. N.J. and s.e. Pa., W.Va., s. Ohio, Ind. and Ill., s.e. Mo., generally southward to cen. pen. Fla. and e. Okla. and Tex.)

National champion southern red oak (1971), 27'3" circumf., 128' in height, 149' spread, Harwood, Md.

23. Quercus pagoda Raf. CHERRYBARK OAK, SWAMP RED OAK.   Fig. 160

Relatively large deciduous tree. Woody twigs of the season reddish purple and grayish-pubescent, often glabrous by autumn. Winter bud or buds at tips of twigs lance-ovoid to ovoid, often vaguely, obtusely, and longitudinally angled, brown, scales, at least the upper ones, with longish, antrorsely disposed hairs on their backs, usually with a fringe of short hairs marginally. Bark of trunks with narrow, flaky or scaly ridges, reddish-tinged.

Young developing stems relatively densely pubescent with gray stellate hairs (hairs later becoming tawny, then dirty-gray by the end of the season); upper surfaces of blades of unfurling leaves with evident venation clothed with long-

337

Fig. 160. **Quercus pagoda:** variable leaves at upper right drawn × ⅝ actual size.

ish, purple-tinged, stellate hairs, areas between the veins relatively thinly clothed with tiny, purple-tinged, stellate hairs, the whole surface with a purplish hue (to the unaided eye), but the hue of the veins more pronounced; lower surfaces with evident venation, but the veins not sharply defined relative to the remainder of the surface, the entire surface being uniformly and densely gray-pubescent and soft to the touch, axils formed by the midrib and major lateral veins with small, short tufts of hairs. (The purplish hue of the upper surfaces of unfurling leaves is subtly evident even on the crown of the tree as viewed from a little distance.)

Mature leaves with slender petioles 3–6 cm long, most of the pubescence sloughing by the end of the season. Blades 8–15 (20) cm long, prominently 5–7-lobed (more regularly lobed than in *Q. falcata*), lobes oblongish to triangular, on the average tending to diverge more nearly at right angles to the midrib than in *Q. falcata*, sinuses between lobes open and broadly rounded at their bases, the larger lobes sometimes sublobed or toothed, lobes or teeth mostly bristle-tipped; bases of larger blades varyingly truncate or obliquely truncate to broadly and shortly tapered; upper surfaces dark green, glabrous, and lustrous, lower surfaces usually permanently densely gray-pubescent, occasionally only sparsely pubescent and more green by midseason or later. Crown leaf blades of the cherry bark oak do not have U-shaped bases such as are characteristic of many crown leaves of any given tree of *Q. falcata*.

Fruits as in *Q. falcata*. [*Q. falcata* Michx. var. *pagodaefolia* Ell.]

In our area, inhabiting floodplain forests where only periodically and briefly flooded, river banks, and in woodlands well elevated above major rivers but apparently not at great distances from the rivers, probably where the soils are circumneutral; s.w. Ga., along or near the Chattahoochee River, perhaps along the same river in s.w Ala., in Fla. along or near at least the upper Apalachicola River, possibly elsewhere westward of it. (Chiefly coastal plain, s.e. Va. to cen. Panhandle of Fla., westward to e. Tex., northward in the interior, e. Okla., s.e. Mo., s.e. Ill., Ala. to w. Tenn. and Ky.)

National champion cherrybark oak (1975), 29′ circumf., 120′ in height, 126′ spread, Perquimons Co., N.C.

24. Quercus velutina Lam. BLACK OAK, YELLOW-BARKED OAK.   Fig. 161

Medium-sized to large deciduous tree. Woody twigs of the season reddish brown, glabrous. Winter bud or buds at tips of twigs ovoid or lance-ovoid, longitudinally obtusely angled, scales densely pubescent. Bark of trunks thick, dark brown to almost black, with broad scaly ridges and deep vertical furrows, inner bark yellowish orange.

Young developing stems and petioles densely pubescent with tawny stellate hairs; upper surfaces of unfurling blades densely pubescent throughout with short, grayish stellate hairs, the surface tissue beneath the hairs pinkish, lower surfaces densely pubescent with a mixture of tawny, short-stellate hairs and longer straight ones, the latter more abundant along the major veins than between them.

Mature leaves with petioles (2) 4–7 cm long, sometimes longer, glabrous or sparse pubescence persisting. Blades of crown branches of mature trees variable, larger ones 10–20 cm long and 8–15 cm broad, with 5–7 primary lobes, lowermost triangular, others more or less oblongish, these few-sublobed and/or toothed, lobes or teeth bristle-tipped, sinuses between lobes varyingly less than

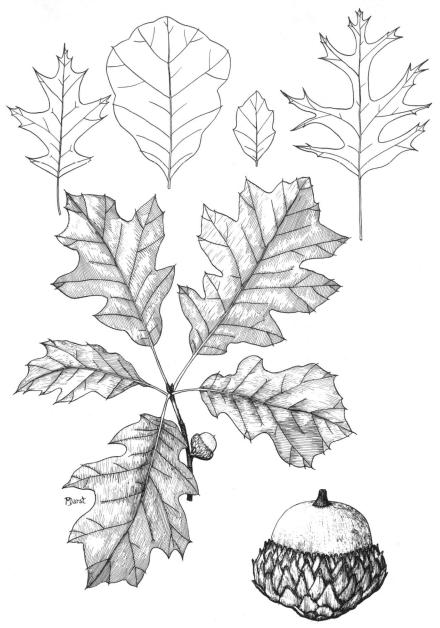

Fig. 161. **Quercus velutina.**

half to more than half the distance to the midrib, open and more or less U-shaped; bases truncate, subtruncate, or broadly and shortly tapered, often inequilateral; upper surfaces dark green and glabrous or sometimes with a few hairs along the midribs, lower surfaces paler and duller green or brownish green, usually thinly and finely short-pubescent throughout when young, sometimes even in age, in age, however, more frequently hairy only along the major veins or only with tufts of hairs in the major vein axils. Blades of saplings, and not infrequently those of lower crown branches, with much shorter and less conspicuous lobes.

Fruits maturing the second season, their involucres bowllike, 10–12 mm deep and 15–20 mm broad, gradually tapered to abruptly constricted basally, embracing about one-half of the acorn, scales more or less triangular, the uppermost erect and forming a fringe around the cup margin or sometimes folded inward and downward, pubescent on their backs, loosely to tightly appressed; acorns oblongish in outline, sometimes slightly broadest distally, 12–15 mm long and 10 mm broad or a little more, outer surfaces finely grayish-pubescent, inner surfaces tawny-pubescent.

Inhabiting mixed, upland woodlands, usually with clay subsoils, frequently associated with *Q. falcata* and *Q. stellata*; n. cen. Panhandle of Fla., adjacent s.w. Ga. and s.e. Ala. (S.e. Maine to s. Mich. and Wis., s.e. Minn., much of Iowa, generally southeastward to cen. Fla. Panhandle and e. Tex.)

Co-national champion black oaks (1981), 19'9" circumf., 117' in height, 129' spread, St. Clair Co., Mich.; (1981), 21'2" circumf., 107' in height, 122' spread, Monroe Co., Mich.

25. Quercus rubra L. RED OAK.

Medium-sized to large deciduous tree. Woody twigs of the season moderately stout, glabrous, dark brown. Winter bud or buds at tips of twigs ovoid, pointed or blunt apically, scales dark reddish brown, somewhat shiny, glabrous on their backs, the distal ones sometimes with ciliate apical margins. Bark of smaller trunks smoothish, gray to grayish brown, that of larger older ones becoming dark grayish brown to black, with shallow vertical furrows and low flat to rounded ridges that become checkered with age.

Young developing stems, petioles, and both surfaces of unfurling leaf blades copiously shaggy-pubescent, the blades with a pink hue.

Mature leaves of crown branches with slender, glabrous petioles 2–5 cm long. Blades mostly 10–22 cm long and 6–15 cm broad (across the largest lobes), usually broadest just above their middles, larger ones usually with 8–9 ascending lobes, sinuses between the major lobes reaching about halfway to the midribs (not so deep on blades of saplings, sprouts, or lower crown branches), major lobes usually toothed-lobed distally, the teeth attenuated into bristle-tips; bases truncate, oblique-truncate, or with a short, broad taper; upper surfaces dull dark green and glabrous, the lower surfaces a little paler green, glabrous save for tufts of hairs in the angles formed by the midrib and major lateral veins. Commonly leaf blades turn orangish red in autumn.

Fruits maturing the second season, their involucres saucerlike or shallowly bowllike, embracing only the base of the acorn or to about one-fourth of it, scales closely appressed, more or less triangular with truncate to rounded tips, brown but with appressed gray pubescence covering most of their backs; acorns slightly ovoid, mostly 1.5–2 cm long, slightly longer than broad, outer surfaces brown, with patches of sparse, fine, gray pubescence, inner surfaces copiously shaggy-pubescent with tawny hairs. [*Quercus maxima* (Marsh.) Ashe; *Q. borealis* Michx. f.; *Q. rubra* var. *borealis* (Michx. f.) Farw.]

Inhabiting mesic forests on fertile, loamy soils; in our area restricted to and infrequent in s. Ala. (P.E.I., N.B., s. Que., s. Ont., Minn., generally southwestward in the East to middle s.w. Ga. and s. Ala., westward of that southward to n.e. Miss., n. half of Ark., e. Okla., and with a few outlying stations in w. Miss. and adjacent e. La.)

26. Quercus shumardii Buckl. SHUMARD OAK. Fig. 162

Large deciduous tree, to 40 m tall. Twigs of the season dark reddish brown, glabrous. Winter bud or buds at tips of twigs dull tan or grayish tan, lance-ovoid to ovoid, tips acute, scales glabrous. Bark of trunks thick, with more or less interlacing, grayish scaly ridges and dark furrows, inner bark tan or buff, sometimes pinkish.

Young developing stems, petioles, and blade surfaces loosely pubescent with a mixture of short-stellate and much longer forked hairs, those of the blades usually purplish red at first, especially the short-stellate and much longer forked hairs; in addition on the lower surfaces of the blades in the axils formed by the midrib and the principal lateral veins conspicuous tufts of hairs appear after the blades have expanded considerably.

Mature leaves with slender, glabrous petioles 2–6 cm long. Blades of most or many of the crown leaves of mature trees 7–15 cm long and 6–12 cm broad, prevailingly, besides the terminal lobe, with 3 lateral, divaricate or nearly divaricate lobes on a side, each usually with several acute sublobes or teeth with conspicuous bristle-tips, sinuses between the major lobes commonly open and more or less U-shaped, reaching more than halfway to the midrib, sometimes adjacent lobes touching or nearly touching distally and the sinuses teardrop-shaped, bases mostly truncate or nearly truncate, truncation often slightly oblique; upper surfaces dark green, glabrous and lustrous, lower surfaces paler but bright green, glabrous, save for tufts of pubescence in main vein axils; lobes such as to give the leaves, collectively, a lacy appearance. Blades of saplings and always a few of those on crown branches smaller, irregularly fewer- and shorter-lobed, their bases usually much narrower and more oblique; shaded leaves of lower crown branches often with similar form, but not lacy, and relatively large; moreover, if large, their form much like that of many leaves of *Q. velutina*.

Fruits maturing the second season, their involucres bowllike, about 8 mm deep and 16 mm broad, embracing about one-third of the acorn, scales brown, more or less triangular, slightly humped medially below their middles, closely appressed, irregularly pubescent on their backs, innermost series of scales behind and obscured by the next below; acorns ovoid-oblong (only slightly broadest basally), about 2 cm long and 1.5 cm broad, outer surfaces with grayish pubescence which appears, to the unaided eye, to be in longitudinal bands or lines, with buff-colored pubescence within.

Occurring in mixed, mesic woodlands of hammocks, bluffs, ravines, and river banks, often where limestone underlies the surface soil; local in much of our area. (Local, s. Pa. to Va., s. Ohio, e. Tenn., more general, coastal plain and piedmont, N.C. to n. cen. pen. Fla., Panhandle of Fla., westward to e. and cen. Tex., northward in the interior to Ind., s. Ill. and Mo., s.e. Kan., e. Okla.)

National champion Shumard oak (1975), 21'9" circumf., 97' in height, 105' spread, Lake Providence, La.

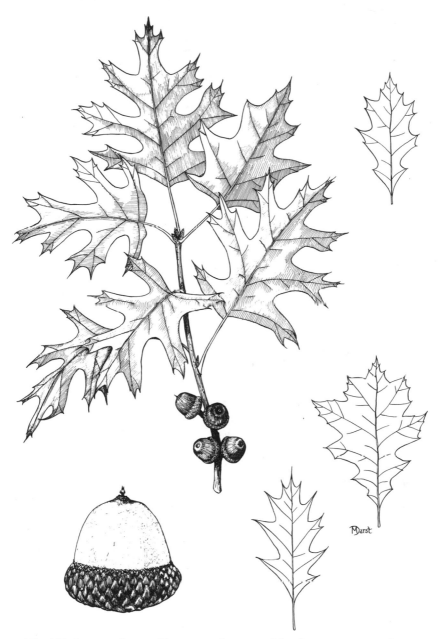

Fig. 162. **Quercus shumardii:** separate leaves at right, drawn to same scale as those on fruiting branch, characteristic for saplings.

27. Quercus nuttallii Palmer. NUTTALL OAK.

Medium-sized to large deciduous tree. Woody twigs of the season grayish brown to dark reddish brown, glabrous. Winter bud or buds at tips of twigs lance-ovoid to ovoid, usually vaguely, obtusely, and longitudinally angled, dull brown, scales with antrorsely appressed hairs on the back and with a marginal fringe of hairs, sometimes glabrous. Bark of larger trunks shallowly furrowed between thin scaly ridges.

Mature leaves with slender, glabrous petioles 2–5 cm long. Blades 5–14 cm long and 4–8 cm broad, with 5–7 prominent primary lobes, these sometimes sublobed or toothed, lobes or sublobes, for the most part, having attenuated apices and bristle tips, sinuses between the lobes broadly U-shaped, lowermost sinuses of some blades, at least, truncate, their bases paralleling the midrib, some sinuses reaching more than halfway to the midrib; bases equilateral or inequilateral and oblique, shortly and broadly tapered to obliquely truncate; upper surfaces bright green, shiny, lower surfaces more brownish green, and glabrous save for tufts of pubescence in the axils formed by the major lateral veins and the midrib, early in the season, at least, the tufts sometimes more or less covering and obscuring portions of the veins, late in the season much of the tufted pubescence sloughed.

Fruits maturing the second season, their involucres bowllike, about 1–1.5 cm deep and 1.4–2 cm broad distally, embracing one-third to one-half of the acorn; the cup abruptly narrowed at the base of the bowl to a thickish stalklike portion about 7–8 mm long; acorn oblong-ellipsoid, 2.5–3 cm long and 1.3–2 cm broad medially, outer surface tawny-pubescent, faintly striated, inner surface with buff-colored pubescence. [*Q. palustris* Muenchh. forma *nuttallii* C. H. Mueller]

Inhabiting alluvial woodlands of river and stream bottoms; in our area, known to me only from along the Tombigbee River, s.w. Ala. (Cen., w., and s.w. Ala. to La. (e. Tex. according to range map of Little, 1971, but not treated by Correll and Johnson, 1970, as occurring in Tex.), s. and n.e. Ark., s.e. Mo., westernmost Tenn., and Ill.)

Elias (1980) does not treat *Q. nuttallii*. One assumes he does not distinguish it from *Q. palustris* although he does not give the name, *Q. nuttallii*, in synonymy. If, however, he does consider the two indistinguishable, then the distributional map he uses for *Q. palustris* (that of Little, 1971, map 177E) does not reflect the distribution of the combined pair (see Little's maps 177E and 177E on facing pages for the two distributions separately).

Co-national champion Nuttall oaks (1972), 16'11" circumf., 130' in height, 80' spread, Delta National Forest, Miss.; (1978), 20'7" circumf., 85' in height, 76' spread, Horseshoe Lake Island, Ill.

28. Quercus coccinea Muenchh. SCARLET OAK.   Fig. 163

Generally a medium-sized, deciduous tree. Woody twigs of the season relatively slender, reddish brown or purplish, glabrous. Winter bud or buds at tips of twigs narrowly ovoid to ellipsoid, blunt apically, lowermost scales brown or dark reddish brown, glabrous or slightly pubescent distally, other scales densely silky-pubescent with tawny hairs. Bark of trunks brown to nearly black, thickish, with shallow fissures and irregular ridges that sometimes become scaly.

Young developing stems and petioles densely stellate-pubescent, hairs tawny; both surfaces of unfurling blades densely tawny-pubescent, mostly longish hairs along the major veins, shorter, tangled, stellate hairs between them.

Mature leaves of crown branches with slender, glabrous petioles 3–6 cm long.

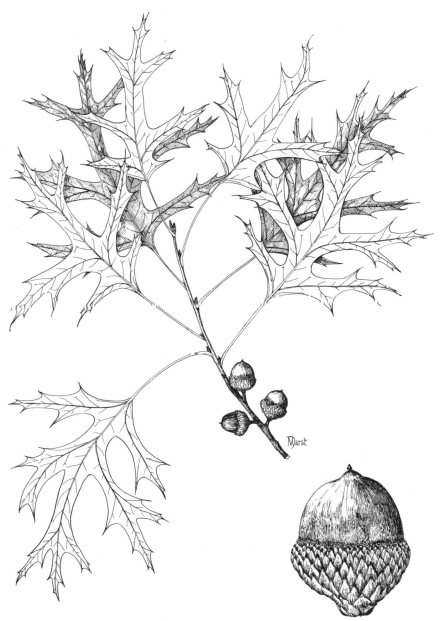

Fig. 163. **Quercus coccinea.**

Blades 7–15 cm long and 6–15 cm broad, prominently 5–7 (9) -lobed, lobes mostly spreading laterally or slightly ascending, lowermost ones short-triangular, others squarish basally and few-sublobed or toothed distally, lobes or teeth prominently bristle-tipped; sinuses between the lobes mostly rounded basally; bases often inequilateral and obliquely truncate, sometimes truncate and not oblique; upper surfaces bright green, glabrous, and shiny, lower surfaces lighter green but somewhat shiny and sometimes with small tufts of hairs in at least the proximal axils formed by the major lateral veins and midrib, often wholly glabrous. Leaf blades turning bright red or scarlet in autumn.

Fruits maturing the second season, their involucres bowllike, rounded basally, 10–15 mm deep and 20 mm broad, embracing one-half to one-third of the acorn, scales brown, triangular, somewhat humped basally, tightly appressed, finely short-pubescent distally on their backs; acorns ovoid, slightly broadest basally, 1.5–2 cm long or a little more and about 1.8–2 cm broad basally, outer surfaces shortly woolly-pubescent, inner surfaces pale tawny-pubescent.

Inhabiting well-drained, often infertile, upland, mixed woodlands; in our area local in s.w. Ga. and s. Ala. (Maine and s. Ont. to s. N.Eng., N.Y. to s. Mich., Ind., s. Ill., s.e. Mo., generally southward to s.w. Ga., Ala., n. Miss.)

National champion scarlet oak (1978), 17'10" circumf., 150' in height, 128' spread, Maud, Ala.

Guttiferae (Hypericaceae)
(ST. JOHN'S-WORT FAMILY)

Hypericum (ST. JOHN'S-WORTS, ST. PETER'S-WORTS)

Annual or perennial herbs or shrubs. Shrubs (of our area) evergreen, the stems, leaves, sepals, and petals having translucent or transparent punctate secretory glands (showing as dots or short lines). Leaves opposite, sessile, without stipules, with or without an articulation at the base, one or both surfaces punctate-dotted. Sepals 4 or 5 (rarely 3 or 6), persistent below the fruit, in some deciduous at time of capsule dehiscence. Petals 4 or 5 (rarely 3 or 6), inequilateral, yellow or orange-yellow, deciduous soon after anthesis or long persistent as withered remains. Stamens 5–numerous, separate or filaments united below and in 3–5 clusters, withered filaments commonly persistent around the fruits. Pistil 1, 2–5-carpellate, the ovary 1–locular with 2–5 parietal placentae, or 3–7-locular by intrusion of the placentae. Styles 2–5, separate or united for part of their length, stigmas minute. Fruit capsular, mostly septicidally dehiscent. Seeds numerous.

To learn to recognize woody representatives of *Hypericum* species, nos. 7–14 in the following treatment, requires patient and diligent study, especially since representatives of two or more of them not infrequently grow intermixed or in close proximity, particularly where habitat disturbance or modification has taken place. In attempting to achieve recognition of the kinds, it is necessary to familiarize oneself with general habit, characteristics of the bark on larger portions of stems, relative leaf size, disposition of flowers, capsule size and form, characteristics of seeds (with magnification). Identification using only small pieces of branches is virtually impossible.

1. Sepals and petals 4 each (in species nos. 4 and 5 the two inner sepals so reduced as to be scarcely evident).
   2. Leaves subcordate to cordate basally and somewhat clasping.   1. *H. tetrapetalum*

2. Leaves not subcordate or cordate basally, not clasping.

  3. Styles 3 or 4.

    4. Outer sepals much larger and more conspicuous than the inner. 2. *H. crux-andreae*

    4. Outer sepals somewhat larger than the inner but not markedly so.

          3. *H. microsepalum*

  3. Styles 2.

    5. Stalks of the flowers 6–12 mm long, becoming reflexed after anthesis, bearing a pair of bractlets near their bases.   4. *H. suffruticosum*

    5. Stalks of the flowers 2–3 mm long, remaining erect after anthesis, bearing a pair of bractlets just below the calyx.   5. *H. hypericoides*

1. Sepals and petals 5 each.

  6. Midstem leaves ovate-triangular, subcordate to cordate basally and somewhat clasping.   6. *H. myrtifolium*

  6. Midstem leaves not ovate-triangular.

    7. Leaves all linear-subulate and needlelike, their margins essentially parallel.

      8. Main stem leaves (not those of short axillary branches) not exceeding 11 mm long.

        9. Plants usually decumbent (main stems rooting) and becoming matted, erect portions of stems usually not over 5 dm tall; internodes 6-angled in cross-section, especially just below the nodes; capsules mostly 6–7 mm long; seed usually black at full maturity, its outer surface coarsely alveolate, the alveolae squarish and in longitudinal rows.   8. *H. reductum*

        9. Plant with erect stem attaining as much as 1.5 m tall; internodes 2-winged in cross-section and flattened (best observed on young stems); capsules mostly 4–5 mm long; seed very dark brown, surface very finely alveolate, the alveolae nearly round and not obviously in longitudinal rows.   7. *H. brachyphyllum*

      8. Main stem leaves (not those of axillary branches) usually 13 mm long or more.

        10. Young stems, leaves, and sepals strongly glaucous; bark of upper larger stems smooth, metallic-silvery, that of older lower portions sloughing in large, thin, curled plates; seeds 1.5 mm long or a little more.   9. *H. lissophloeus*

        10. Young stems, leaves, and sepals not at all or scarcely glaucous; bark variously roughened, sloughing in irregular strips or flakes; seeds less than 1 mm long.

          11. Stem slender and wandlike, limber, unbranched or with few long-ascending branches from about the middle or above, these paniclelike in flower; stem seldom exceeding 8 mm in diameter near the base.   10. *H. exile*

          11. Stem not limberly wandlike, bushy-branched above, eventually 1–several cm thick near the base.

            12. Bark of older stems thin and relatively tight, reddish brown, brown, or grayish, sloughing in narrow, thin flakes or strips.   11. *H. nitidum*

            12. Bark of older stems thin-corky to thick-corky and spongy, sloughing in sheets, the portions exposed buffish or cinnamon-colored.

              13. Flowers or fruits commonly solitary in leaf axils or in 3–flowered cymes, usually relatively few flowers at anthesis simultaneously on a given plant so it does not appear abundantly floriferous; bark of lower stems soft and spongy, eventually becoming as much as 3–4 cm thick, with conspicuous vertically aligned cordlike lactifers.   13. *H. chapmanii*

              13. Flowers or fruits solitary in leaf axils, more commonly in 3–flowered or compound cymes, generally many flowers at anthesis simultaneously and the plant appearing abundantly floriferous; bark of lower stems thin-spongy if at all so, not eventually attaining notable thickness owing to sloughing, the lactifers threadlike and inconspicuous.   12. *H. fasciculatum*

    7. Leaves not with parallel sides although if very narrow then at least somewhat dilated distally.

      14. The leaves with an articulation at the extreme base, this showing as a narrow horizontal line or groove.

        15. Bark of larger stems smooth, eventually sloughing in large, thin, somewhat curled plates; leaves oblong, elliptic, or lance-ovate, 3–6 cm long and 1–2 cm broad;

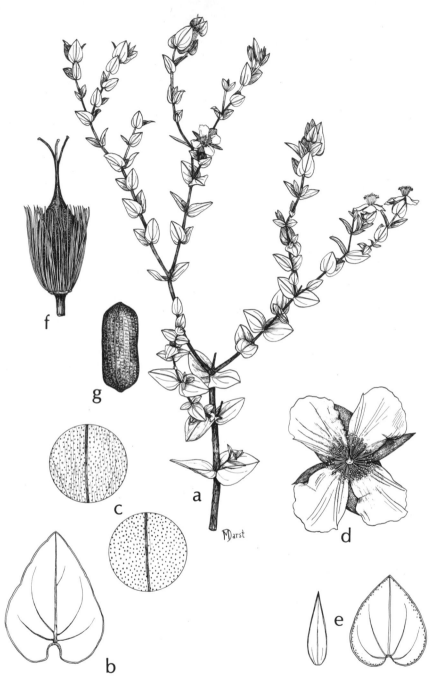

Fig. 164. **Hypericum tetrapetalum:** a, habit; b, leaf; c, enlargement of leaf surfaces, upper to left, lower to right; d, flower; e, two sepals, one of outer pair to right, one of inner pair to left; f, capsule (sepals removed); g, seed.

flowers solitary or in 3–flowered cymes terminating branchlets, individually showy, 2.5–4 cm across the corollas when the petals are spreading (they usually become reflexed).   18. *H. frondosum*

15. Bark of larger stems roughish, eventually sloughing in small, thin, roughish flakes; leaves narrowly oblanceolate, 2–3 cm long and about 7 mm broad distally; flowers in axillary, relatively abundantly floriferous, often paniclelike cymes, each flower about 1.5 cm across the corolla.   14. *H. galioides*

14. The leaves without an articulation at the base.

16. Main stem leaves usually with short branches in their axils, successive pairs disposed conspicuously at right angles to each other, linear-lanceolate, or linear-oblong, mostly broadest near the base and tapering to an acute or blunt tip, thickish and stiffish, less than 10 mm broad; seeds pale brown, dull, faintly alveolate-reticulate, 0.4–0.5 mm long.   15. *H. cistifolium*

16. Main stem leaves without short branches in their axils, elliptic, lance-elliptic, or lanceolate, 1–3 cm broad, thinnish and flexuous, rounded to broadly obtuse apically; seeds dark brown, conspicuously alveolate-reticulate, 1.5–2.0 mm long.

17. Flowers usually in many-flowered compound cymes terminating branches; sepals 1.5–2 mm long, usually triangular-acute; capsules (excluding styles) ovoid to subglobose, 4–5 mm long and about as broad; seeds 1.5–1.8 mm long, usually falcate-cylindric, dark purplish brown and lustrous at full maturity.

16. *H. nudiflorum*

17. Flowers in simple 3-flowered cymes or in compound cymes with up to about 8 flowers; sepals 3 mm long, oblong, obtuse apically; capsules ovoid, 8–10 mm long (excluding styles) and mostly 5–7 mm broad; seeds 1.8–2.0 mm long, cylindric, sometimes slightly falcate, dull brown.   17. *H. apocynifolium*

1. **Hypericum tetrapetalum** Lam.   Fig. 164

Shrub 1–10 dm tall. Stem slender, older portions with thin, reddish brown bark exfoliating in thin, grayish, irregular strips or flakes.

Leaves ovate to oblong-ovate, the larger ones 1–2 cm long, subcordate or cordate basally, obtuse apically. Openly dichotomously branched above, solitary flowers terminating the branchlets.

Flowers 2.5–3 cm across or a little more. Sepals 4, punctate-dotted on both surfaces, the outer 2 broader than the inner, varying considerably in size, the larger ones 15–20 mm long and about 12 mm broad, sometimes as broad as long, ovate, cordate basally and broadly obtuse to rounded apically; inner sepals elliptic, acute apically. Petals 4, bright yellow, obovate, much exceeding the sepals. Stamens numerous.

Capsules about 6 mm long, oval-elliptic in outline, styles 3, diverging. Seeds oblong, 0.8–1.0 mm long, dark blackish brown, surfaces reticulate, sublustrous. [*Ascyrum tetrapetalum* (Lam.) Vail in Small]

Pine flatwoods and pond margins; in our area s.e. Ga., n. Fla. westward to Okaloosa Co. (Also pen. Fla.; Cuba.)

2. **Hypericum crux-andreae** (L.) Crantz. ST. PETER'S-WORT.   Fig. 165

Shrub 3–10 dm tall, not colonial. Stem slender, older portions with reddish brown bark exfoliating in thin strips or flakes. Leaves lanceolate, oblong, oval, or oblanceolate, the larger 2–3 cm long, bases mostly rounded or subtruncate, sometimes very slightly clasping, rarely cuneate, apices obtuse or rounded.

Flowers 2–3 cm across, solitary and terminal or axillary, or in small cymes on the branchlets. Sepals 4, punctate-dotted on both surfaces, the outer 2 much broader than the inner, ovate, 1–1.5 cm long and commonly as broad basally, bases cordate, apices broadly obtuse to rounded; inner sepals lanceolate or nar-

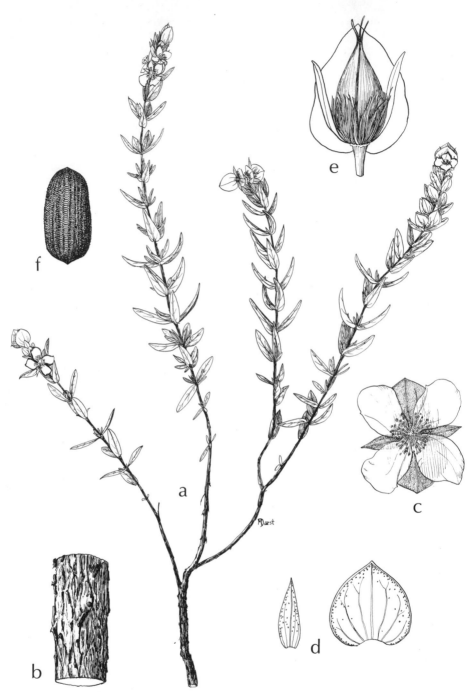

Fig. 165. **Hypericum crux-andreae:** a, flowering branch; b, section of lower stem; c, flower, face view; d, at right, outer sepal, at left, inner sepal; e, capsule, one outer sepal removed; f, seed.

rowly elliptic, acute apically. Petals 4, bright yellow, 10–18 mm long, obovate. Stamens numerous.

Capsules narrowly ovate or oval, varying to subglobose, 7–10 mm long, styles 3, rarely 4, diverging. Seeds oblong, about 0.8 mm long, brown, reticulate, sublustrous. [*Hypericum stans* (Michx.) Adams & Robson; *Ascyrum stans* Michx.]

Pine savannas and flatwoods, adjacent ditches, cypress depressions, swales, marshy or boggy shores, meadows, sometimes in well-drained upland mixed woodlands and on deep sands of longleaf pine–scrub oak hills and ridges; throughout our area. (L.I., N.J., e. Pa., southward to cen. pen. Fla., westward to e. Tex., s. and cen. Ark., e. Okla.; Cumberland Plateau, n. Ala. to s.e. Ky.)

3. Hypericum microsepalum (T. & G.) Gray ex S. Wats.   Fig. 166

A low, straggly to bushy-branched shrub, not exceeding 10 dm tall, commonly much lower, flowering in very early spring. Branches very leafy, flowers solitary at tips of branchlets or in leaf axils near the branch tips.

Principal leaves mostly 5–10 (–15) mm long, 1.5–2.0 mm wide (commonly with very short branchlets bearing much smaller leaves in their axils), linear-oblong, oblanceolate, elliptic, or obovate, bright green above, pale green below, margins revolute, basally scarcely narrowed to cuneate, apices rounded to obtuse.

Flowers showy, 1.5–2.0 cm across. Sepals usually 4, subequal, linear-elliptic to oblong-elliptic, obtuse to acute apically. Petals usually 4, yellow, subequal, spreading, usually asymmetric, obovate, broadly rounded apically (occasional flowers on some plants may have 5-merous calyces and corollas, or 3-merous calyces and 5-merous corollas, or 5-merous calyces and 3-merous corollas). Stamens numerous.

Capsules conic, narrowly ovate, or elliptic-oblong, 3–4 mm long, styles 3. Seeds oblong, about 1 mm long, dark brown and lustrous, finely pitted-reticulate. [*Crookea microsepala* (T. & G.) Small]

Moist to wet pine flatwoods; common and abundant in the Fla. Panhandle from Madison and Taylor Cos. westward to Walton Co. and in s.w. Ga.

4. Hypericum suffruticosum Adams & Robson.   Fig. 167

Low shrub, 5–15 cm tall, stem slender, variably few-branched and erect, multibranched, erect, and often cushionlike, or decumbent and more or less matted. Bark reddish brown, sloughing in thin strips or flakes.

Leaves of the main axes mostly as long as or a little longer than the internodes (often subtending very short, axillary branches with smaller leaves), 4–8 mm long and 2–3 mm broad, variably elliptic, oblong, narrowly obovate, or nearly spatulate, sometimes thus variable on a single specimen; lowermost leaves on some stems linear, but those commonly shed early; margins entire, surfaces distinctly punctate-dotted beneath, sometimes also above.

Flowers solitary at nodes ending ultimate branchlets, their stalks 6–12 mm long, bearing a pair of narrowly subulate bractlets near the base, mostly eventually reflexed. Outer sepals conspicuous, ovate, broadly elliptic, or nearly orbicular, 5–8 mm long, broadly obtuse to rounded apically, broadly rounded to subcordate basally, punctate-dotted on both surfaces; inner sepals minute, scarcely if at all evident. Petals 4, pale yellow, narrowly obovate, 5–7 mm long. Stamens numerous.

Capsule ovate or elliptic, about 4 mm long, styles 2. Seeds oblong, terete, about 1 mm long, very finely cancellate-reticulate. [*Ascyrum pumilum* Michx.]

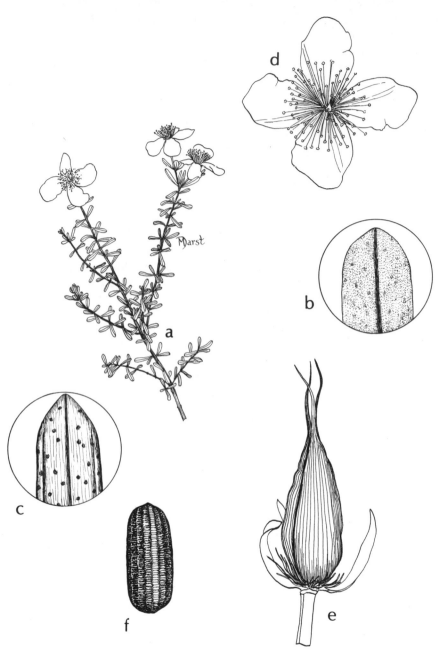

Fig. 166. **Hypericum microsepalum:** a, flowering branchlet; b, portion of upper surface of leaf; c, portion of lower surface of leaf; d, flower, face view; e, capsule (one sepal removed); f, seed.

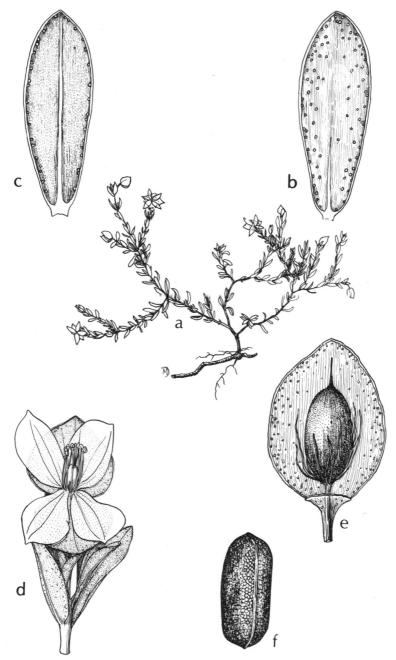

Fig. 167. **Hypericum suffruticosum:** a, habit (plant diminutive); b, upper surface of leaf; c, lower surface of leaf; d, flower; e, capsule (one sepal cut off); f, seed.

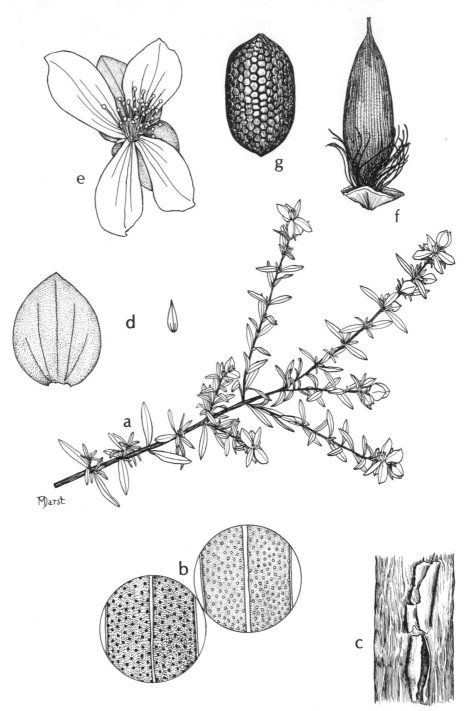

Fig. 168. **Hypericum hypericoides:** a, flowering branch; b, portion of leaf surfaces enlarged, lower surface to left, upper surface to right; c, piece of lower stem; d, sepals, an outer large one to left, an inner tiny one to right; e, flower; f, capsule (outer large sepals cut off just above the base); g, seed.

354

Well-drained sands of pinelands; throughout our area. (Coastal plain, s.e. N.C. to n. Fla., westward to s.e. La.)

5. Hypericum hypericoides (L.) Crantz. Fig. 168

Plant erect with a single main stem freely branched above, sometimes with several principal, erect stems from the base, each freely branched above, commonly with short, leafy axillary branches, 3–15 dm tall; bark of older portions of stems reddish brown, sloughing or shredding in thin plates or strips. Leaves variable, narrowly oblanceolate, linear-oblong, or linear, the larger mostly 8–25 mm long, 1.5–6 mm wide (considerably larger if plants in deep shade), narrowed basally, obtuse to rounded apically, punctate-dotted on both surfaces.

Flowers usually solitary above the pair of leaves terminating an ultimate branchlet, their stalks erect, 2–3 mm long, bearing a pair of narrowly subulate bractlets just below or very little below the calyx. Sepals 4, outer herbaceous, conspicuous, ovate to broadly elliptic, rounded or subcordate basally, obtuse apically, variable in size, 5–12 mm long, 4–10 mm wide, punctate-dotted on both surfaces; inner sepals minute. Petals 4, pale yellow, narrowly oblong-elliptic, mostly 8–10 mm long. Stamens numerous.

Capsules ovate to elliptic in outline, 4–9 mm long, styles 2. Seeds oblong, 1 mm long, black, reticulate. [*Ascyrum hypericoides* L.]

In a wide variety of habitats, open, well-drained upland woodlands, pine barrens on sand hills and ridges, floodplain and wet woodlands, hammocks, moist to wet thickets, pine flatwoods, cypress-gum depressions, bogs; throughout our area. (Va. to s.e. Mo. and e. Okla., generally southward to s. Fla. and e. third of Tex.; Berm., Bah.Is., Cuba, P.R., Haiti, Dom.Rep., Jam., e. Mex., Guatem., Hond.)

6. Hypericum myrtifolium Lam. Fig. 169

Shrub to about 1 m tall. Stems slender, glaucous, mostly solitary, erect, loosely branched above, bark on older portions glaucous, grayish, sloughing in thin plates, becoming slightly corky if submersed.

Leaves numerous, mostly ovate-triangular, 1–3 cm long, basally subcordate to cordate and somewhat clasping, obtuse apically, light green above, paler and slightly glaucous below, margins usually becoming narrowly revolute in drying, lower surface finely punctate-dotted.

Flowers in cymes terminating the branchlets, 2–2.5 cm across. Sepals 5, glaucous below, subequal, ovate to elliptic, 8–10 mm long, acute apically, punctate-dotted on both surfaces. Petals 5, yellow, obovate, 10–12 mm long. Stamens numerous.

Capsules ovate or ovate-conic, very dark brown to almost black, about 8 mm long, broadly depressed along the 3 sutures and convex between the depressions, 3 persistent styles erect. Seeds narrowly oblong, about 1 mm long, apiculate at both extremities, surfaces very dark brown, reticulate.

Pine flatwoods, cypress-gum depressions and ponds, borders of ponds (on exposed shores to high-water mark or in shallow water depending upon water level), bogs; throughout our area. (Coastal plain, Ga. to s. cen. pen. Fla., westward to s.e. Miss.)

7. Hypericum brachyphyllum (Spach.) Steud. Fig. 170

Shrub, stem usually solitary, erect, bushy-branched above, usually 5–10, sometimes to 15 dm tall. Stem diameter near the base to about 1.5 cm, bark relatively thin and tight, exfoliating in small strips or plates. Principal leaves with

Fig. 169. **Hypericum myrtifolium:** a, base of plant; b, top of fruiting stem; c, flower, face view; d, capsule, dehisced.

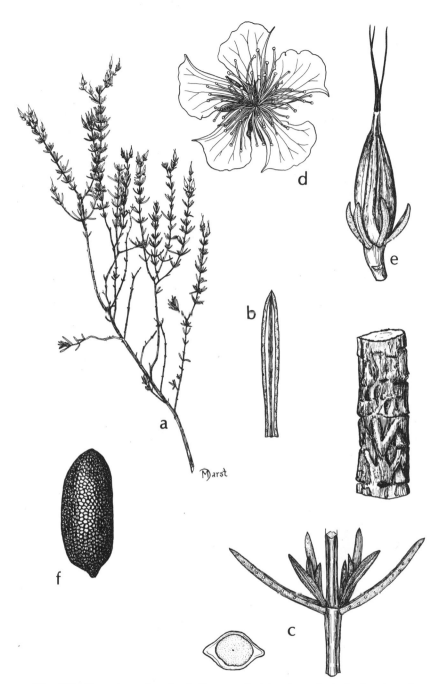

Fig. 170. **Hypericum brachyphyllum:** a, fruiting branch; b, leaf, enlarged, from below; c, node with pair of leaves and short branches in their axils, a diagrammatic cross-section of twig to left; d, flower, face view; e, capsule; f, seed.

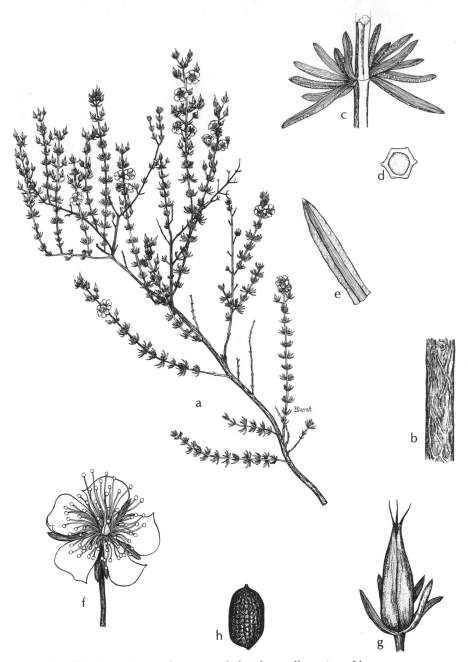

Fig. 171. **Hypericum reductum:** a, habit; b, small portion of lower stem; c, node with pair of leaves and short branch axillary to each one; d, diagrammatic cross-section of stem; e, leaf, abaxial view; f, flower; g, capsule; h, seed.

very short axillary branchlets having smaller leaves giving a fasciculate appearance to the nodes.

Leaves needlelike, very strongly revolute, the margins usually folded tightly around and against the lower surface and together covering about two-thirds of the lower surface, the revolute margins (from beneath) with relatively large punctate dots, these often appearing there in a longitudinal line, some punctate dots scattered on the rest of the upper surface; larger leaves 6–10 mm long.

Flowers sometimes solitary in leaf axils, mostly in short, leafy cymes on branchlets, overall abundantly floriferous. Sepals 5, essentially like the leaves in form, 2–3.5 mm long. Petals 5, bright yellow, spatulate to obovate, mostly about 8 mm long. Stamens numerous.

Capsules narrowly conical, narrowly depressed along the sutures, 4–5 mm long excluding the styles; styles 3 (rarely 4), usually erect, 3 mm long. Seeds short-oblong, about 0.4 mm long, very dark brown, surface very finely alveolate, the alveolae round or nearly so and not obviously in longitudinal rows.

Wet pine flatwoods and savannas, pond margins, cypress-gum ponds and depressions, borrow pits, wet ditches; throughout our area. (Coastal plain, s. Ga. to s. Fla., westward to s. Miss.)

8. Hypericum reductum P. Adams. Fig. 171

Shrub, usually bushy-branched from the base, often with numerous decumbent stems rooting and more or less mat-forming, a single plant as much as 1 m across; erect stems usually not exceeding 5 dm tall, often 1–2 dm. Stem diameter seldom reaching 1 cm, bark relatively thin and tight, exfoliating in thin plates and strips.

Leaves as in the preceding, the larger ones usually not exceeding 5 mm long, the fasciclelike appearance of the nodes generally more pronounced. Branches abundantly floriferous, more flowers solitary in the leaf axils than in the preceding, many of the branchlets thus appearing racemose, some flowers in short cymes as well. Sepals 5, 2–3 mm long. Petals 5, bright yellow, obovate, 7–8 mm long. Stamens numerous.

Capsules narrowly oblong but shortly tapered at the tip, 4–8 mm long excluding the styles; styles 3, straight, 1–1.5 mm long. Seeds short-oblong, about 0.4 mm long, usually black when fully mature, alveolate, the alveolae relatively fewer and larger than in the preceding, squarish, in longitudinal rows.

For the most part inhabiting well-drained sandy soils, longleaf pine–scrub oak hills and ridges, sand pine–scrub oak hills and ridges, old stabilized coastal dunes, moist to wet interdune hollows or swales, outer sandy shores of ponds where sometimes in water at least for short periods. Local, coastal plain, s.e. N.C. to cen. pen. Fla., Fla. Panhandle, s.e. Ala.

9. Hypericum lissophloeus P. Adams. Fig. 172

Shrub to 4 m tall, the stems limber, to 4.5 cm thick near the base. Bark smooth, chestnut-brown, shiny at first, metallic silvery later, exfoliating in large, thin, curled plates exposing shiny, chestnut-brown inner bark. Seedlings occur on exposed shores; older plants tolerating water to a depth of about 1.5 m during which time few to numerous, unbranched or branched, descending and arching, woody prop-roots may be produced, these arising from the stem from all along the extent of their submersion, probably from the base upwardly as water levels gradually rise. If numerous individual plants grow thickly, their respective prop-roots interlace in the same manner as do those of the red mangrove.

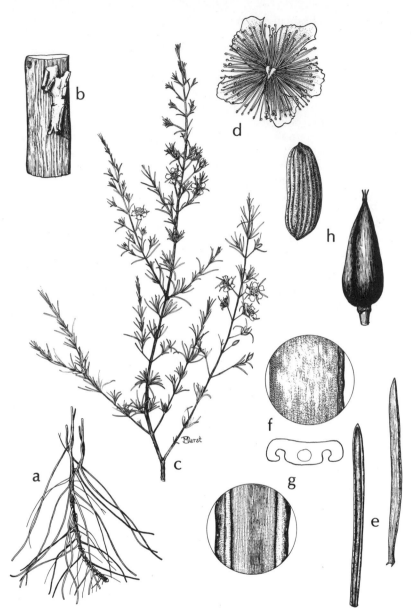

Fig. 172. **Hypericum lissophloeus:** a, lower submersed stem with prop-roots (drawn very much reduced from life); b, small portion of medial part of stem showing how outer bark sloughs in papery curls; c, flowering branch; d, flower, face view; e, upper side of leaf, to right, lower side to left; f, enlargement of portion of upper surface of leaf; g, enlargement of portion of lower leaf surface, below, and diagrammatic cross-section of leaf above; h, capsule to right, seed to left.

Leaves articulate at the base, linear-subulate and needlelike, the larger ones mostly 12–17 mm long, occasionally about 10 mm, strongly glaucous, the edges strongly curved downward, in drying flattened against the lower surface, punctate-dotted mostly along the revolute edges.

Floriferous branchlets thin-flexuous, ascending, the flowers mostly solitary in leaf axils, occasionally in 3-flowered cymes. Sepals 5, articulate at the base, subulate, glaucous, 7–8 mm long, revolute like the leaves, deciduous by the time the fruits are fully ripe. Petals 5, bright yellow, asymmetrical, obovate, with a tooth obliquely placed to one side of the summit, 10–12 mm long. Stamens many. Styles 3, rarely 4, connivent at anthesis.

Capsules ovate-conic, narrowly depressed along the sutures, lustrous reddish brown, 6–8 mm long, styles slender, erect, about 4 mm long, often breaking off near the base by maturity of the capsules. Seeds oblong, tan to brown, 1.5 mm long or a little more, surfaces longitudinally markedly striate, with faint cross-striae between.

Locally abundant, restricted to shores of sinkhole ponds and lakes, on exposed shores to the high-water mark and at high-water levels in water to 1.5 m deep. Endemic to Bay and Washington Cos., Fla. Panhandle.

10. Hypericum exile P. Adams. Fig. 173

Slender shrub, stems wandlike, limber, usually less than 10 dm tall, with 2–few ascending branches above. Bark thin, reddish brown, exfoliating in very thin, small, irregular flakes or strips. Leaves articulate at the base, spreading, linear-subulate and needlelike, (10–) 15–25 mm long, the edges strongly rolled downward, on drying tightly folded against the lower surface and punctate-dotted.

Distally, the stems or main long branches, with short floriferous branchlets, the flowers in 3–7-flowered cymes, the inflorescence overall having the aspect of a compact narrow panicle. Sepals 5, articulate at the base, subulate, not revolute, 6–7 mm long, usually deciduous at about the time the fruits are fully mature. Petals 5, bright yellow, asymmetrical, nearly elliptic, rounded at the tip, a tooth on one side somewhat below the tip, 6–7 mm long. Stamens many. Styles 3, rarely 2 or 4, connivent at anthesis.

Capsules about 7 mm long excluding the beak, lustrous reddish brown, long-conical in outline, deeply depressed along the sutures, the styles persistent as a beak or spreading, 3 mm long. Seeds reddish brown, lustrous, oblong, 0.4–0.6 mm long, surface very finely reticulate.

Pine savannas and flatwoods, usually where soils remain water-saturated only for short periods, occasionally where wet for extended periods. Endemic to Liberty, Franklin, Gulf, Bay, and Washington Cos., Fla. Panhandle.

11. Hypericum nitidum Lam. Fig. 174

Plants commonly with several to numerous stems from the base, each bushy-branched above, overall relatively broad and dense, the main stems to 3–4 cm thick near the base, potentially 2–3 m tall. Bark thin and tight, brown, reddish, or gray, exfoliating in small thin flakes or strips.

Leaves articulate at the base, linear-subulate, needlelike, spreading, the larger ones 15–25 mm long, the edges strongly curved around and in drying becoming flush against the lower surface, punctate-dotted.

Flowers solitary or in small cymes, plentifully distributed on the profusion of branchlets commonly present, individual plants, or thickets of them, often presenting a billowing mass of yellow when in full bloom. Sepals 5, articulate at

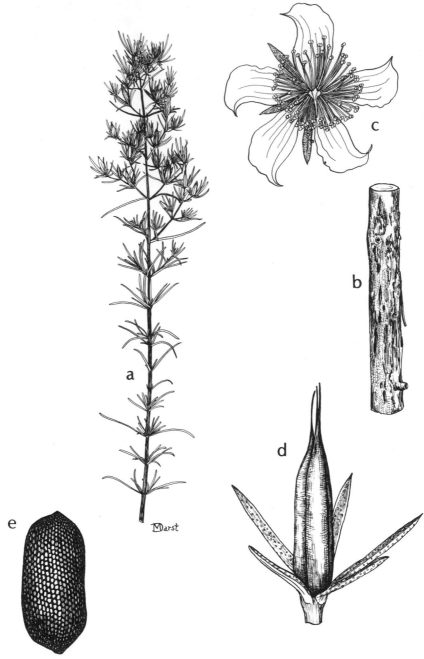

Fig. 173. **Hypericum exile:** a, upper part of wandlike fruiting stem; b, piece of lower stem; c, flower, face view; d, capsule; e, seed.

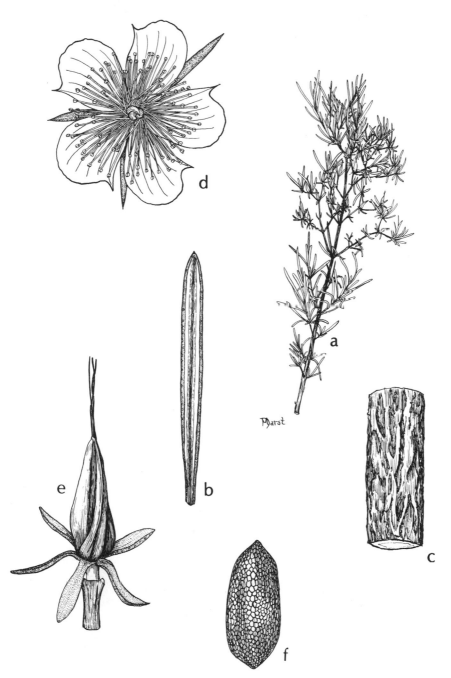

Fig. 174. **Hypericum nitidum:** a, fruiting branch; b, leaf, from below; c, piece of lower stem; d, flower, face view; e, capsule; f, seed.

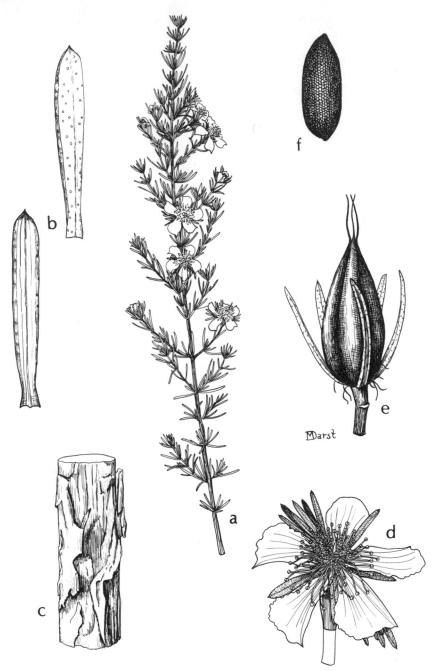

Fig. 175. **Hypericum fasciculatum:** a, flowering branch; b, leaf, two views; c, piece of lower stem; d, flower; e, capsule; f, seed.

the base, linear-subulate and needlelike, strongly revolute like the leaves, 3–4 mm long, usually deciduous by the time the fruits are fully mature. Petals 5, bright yellow, 5–6 mm long, obovate, asymmetrical, a tooth obliquely placed to one side of the rounded summit. Stamens many. Styles 3, sometimes connivent at anthesis.

Capsules oblong-conic, 3–4 mm long, deeply depressed along the sutures, dull reddish brown, the slender styles about 4 mm long, usually breaking off about 1 mm from the base by maturity of the capsules. Seeds dull brown, 0.4–0.5 mm long, surfaces finely alveolate-reticulate, the alveolae roundish, not in obvious rows.

Wet pine flatwoods, particularly open shores of blackwater streams draining flatwoods or shrub-tree bays or bogs, occasionally in masses in freshwater at shores of upper reaches of blackwater estuaries and their backwater basins, ditches, borrow pits, bogs; s. Ga., n. Fla., s. Ala. (Also Lexington Co., S.C., and Brunswick Co., N.C.)

12. Hypericum fasciculatum Lam. Fig. 175

Erect shrub, to 1.5–2 m tall, much branched above, not usually conspicuously dwarf-treelike as is usually the case with older specimens of the following species, but occasionally so. Older plants, if their bases inundated for considerable periods, forming prop-roots. Youngest internodes with a winged ridge on each side, not terete; bark on older stems corky, exfoliating in tissue-thin sheets, exposed portions buff or cinnamon-colored; bark sloughing freely and never attaining great thickness or notable soft-thickness on old basal portions. Around ponds and lakes in which water levels fluctuate considerably, when stems of this plant are submersed significantly, masses of fibrous roots may form; when only the bases of the plants are submersed and remain so for extended intervals, the lower stems may form woody, adventitious prop-roots essentially like those figured in the drawings for *H. lissopheloeus*.

Principal leaves with short branchlets with smaller leaves in their axils giving a fasciculate appearance to the nodes. Leaves spreading to ascending, articulate at the base, linear-subulate and needlelike, the larger ones 13–26 mm long, revolute, punctate-dotted above and on the revolute portions below.

Flowers mostly in 3–26-flowered cymes terminal or terminal and axillary on the branchlets, some sometimes solitary in the leaf axils. Sepals 5, similar to the leaves, (3–) 4.5–7 mm long, usually persistent below the fruits. Petals 5, bright yellow, 6–9 mm long, asymmetrical, obovate, a tooth obliquely to one side of the tip. Stamens many. Styles 3, sometimes connivent at anthesis.

Capsules ovate-conic in outline, deeply depressed along the sutures, dark reddish brown, 3–5 mm long, the slender styles spreading, about 4 mm long, usually breaking off near the base by the time the capsules are fully ripe. Seeds oblong, about 0.4 mm long, finely alveolate-reticulate, alveolae roundish, not in obvious lines.

Shores of ponds and lakes, cypress-gum ponds and depressions, wet flatwoods, ditches, borrow pits, bogs, open banks of streams, commonly in water; throughout our area. (Coastal plain, e. N.C. southward from about the Neuse River, southward to s. Fla., westward to s. Miss.)

13. Hypericum chapmanii P. Adams Fig. 176

Erect shrub, to 2–3 (–4) m tall, usually with a single main stem, bushy branched above, older plants with a markedly dwarf-treelike appearance if growing in

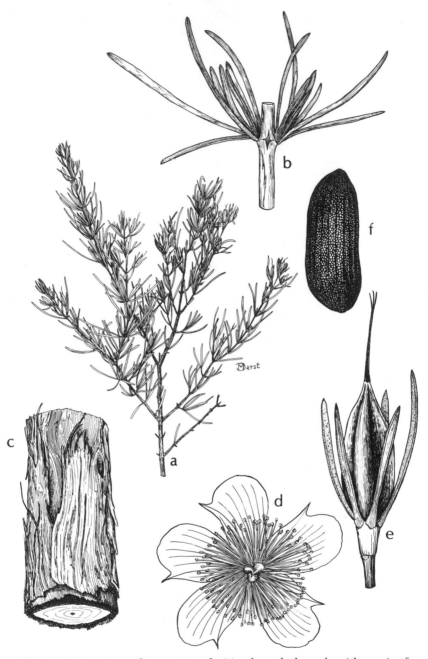

Fig. 176. **Hypericum chapmanii:** a, fruiting branch; b, node with a pair of leaves with short-shoots in their axils; c, piece of lower stem; d, flower, face view; e, capsule; f, seed.

open places, slender, tall, and straggly if surrounded by dense vegetation. Bark soft, spongy, with conspicuous vertical, resin-filled laticifers (resin-filled canals) at discrete levels between cork layers 3–4 mm thick, large size of laticifers giving torn bark a striated appearance; cork buffish, reddish brown, or cinnamon-colored within, weathering grayish, bark often attaining 3–4 cm in thickness on old stems, soft-corky even on small stems. Leaves and sepals much as in *H. fasciculatum.*

Principal leaves linear-subulate and needlelike, angling-ascending, 8–25 mm long, articulate at the base, strongly revolute, the revolute edges in drying appressed against about one-half of each side of the lower surface, the upper surface, including the revolute portions, punctate-dotted; most of the principal leaves having very short, short-leaved branchlets in their axils.

Flowers relatively few, solitary, or in 3-flowered cymes in the axils of leaves near the tips of branchlets.

Capsules ovate, about 6 mm long, depressed along the sutures, styles about 1 mm long, usually breaking off by the time capsules are fully ripe. Seeds oblong, 0.6–0.8 mm long, brown, surface very finely alveolate-reticulate, alveolae roundish, not in evident lines.

Wet pine flatwood depressions, cypress-gum ponds or depressions, borrow pits. Endemic to coastal portion of Fla. Panhandle from a little e of the Ochlockonee River to Santa Rosa Co.

14. **Hypericum galioides** Lam.   Fig. 177

Shrub, mostly 1–1.5 m tall, bushy branched, branchlets slender, flexuous, and usually spreading. Bark of larger stems roughish, eventually sloughing in small, thin, roughish flakes.

Leaves articulated at the base, narrow, and the larger ones mostly 2–3 cm long and 7 mm wide or less, narrowly oblanceolate (sometimes the sides rolling strongly downward in drying and then appearing linear), tapered below to a short-petiolate or subpetiolate base, apically rounded or blunt, upper surfaces punctate-dotted, lower sometimes with evident lateral veins. Main stem leaves with short branchlets in their axils.

Flowers about 1.5 cm across, mostly in short axillary cymes, sometimes the distal portions of branches not very leafy, heavily floriferous, and paniclelike, often some flowers solitary in leaf axils. Sepals and petals 5. Sepals narrowly oblanceolate, 3–4 mm long, articulated at the base. Petals bright yellow, obovate, with a small tooth at one side of the rounded apex. Stamens numerous.

Capsules conic, 4–5 mm long, the 3 persistent styles 2 mm long. Seeds oblong, 0.1 0.8 mm long, dark brown, faintly reticulate.

Cypress-gum swamps and depressions, floodplain forests, river and stream banks, wet woodlands, pine savannas and flatwoods, moist to wet clearings, shores of ponds and lakes, ditches; throughout our area. (Coastal plain, N.C. to n. Fla., westward to s.e. Tex.)

15. **Hypericum cistifolium** Lam.   Fig. 178

Shrub with slender, simple stems, or with relatively few, erect-ascending branches, to about 1 m tall. Successive pairs of main stem leaves conspicuously at right angles to each other, usually strongly ascending, for the most part with short axillary branches in their axils.

Leaves sessile, not articulated at the base, 1-nerved, firm and somewhat leathery and stiffish, lanceolate, linear-lanceolate, lance-oblong, or oblong, the larger

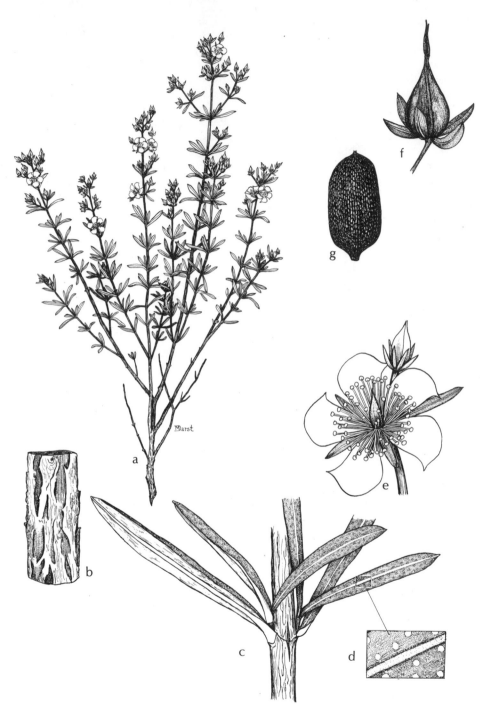

Fig. 177. **Hypericum galioides:** a, flowering branch; b, small portion of lower stem; c, node with opposite leaves and short branch axillary to each; d, enlargement of very small, medial portion of lower leaf surface; e, flower, face view; f, capsule; g, seed.

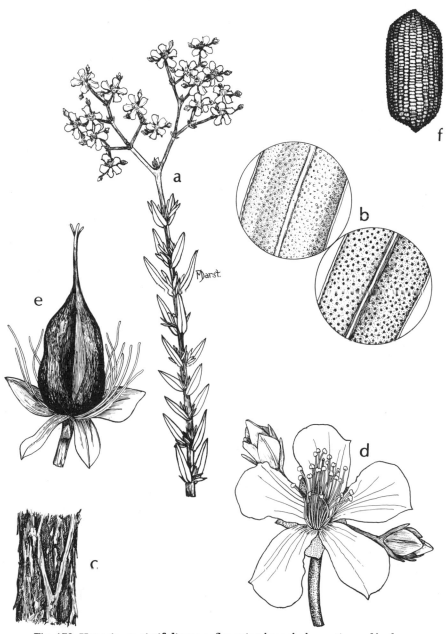

Fig. 178. **Hypericum cistifolium:** a, flowering branch; b, portions of leaf surfaces enlarged, lower surface to right, upper surface to left; c, piece of lower stem; d, flower; e, capsule; f, seed.

ones mostly 1.5–3 cm long and 6–7 mm wide, rarely somewhat wider, broadest near the base, the extreme base somewhat auriculate, the auricles declining, most of them tapering to a point distally, sometimes blunt, punctate-dotted on both surfaces, the midrib below continuous into a wing of the stem.

Flowers 10–12 mm across, in compound terminal cymes. Sepals and petals 5. Sepals 2–3 mm long, oblong, oblong-elliptic, or less frequently ovate or obovate, the shape often differing in a single flower. Petals 5–8 mm long, bright yellow, obliquely cuneate-obovate. Stamens numerous.

Capsules ovate to short-oblong, or shortly elliptic-oblong, 4–5 mm long, persistent styles 3, wholly connivent or separated at their extremities, 2 mm long. Seeds light brown to brown, short-oblong, 0.4–0.5 mm long, shallowly alveolate-reticulate. [Incl. *Hypericum opacum* T. & G.]

Wet pine flatwoods and savannas, adjacent ditches, bogs, boggy seepage slopes of pinelands, cypress-gum depressions, interdune swales, banks of marshes; throughout our area. (Coastal plain, N.C. to s. pen. Fla., westward to e. Tex.)

16. Hypericum nudiflorum Michx. ex Willd. Fig. 179

Slender erect shrub, 5–20 dm tall, usually loosely branched. Bark of older stems more or less cinnamon-brown, sloughing in thin flakes.

Leaves not articulated at the base, membranous, lanceolate, lance-oblong, narrowly elliptic, or oblanceolate, the larger ones 3–6 cm long and 1–2.5 cm broad, without short branches in their axils; abruptly narrowed from the base of the green blade to a short, appressed, winged portion of different texture, apices mostly rounded, some broadly obtuse, both surfaces finely punctate-dotted, midrib beneath continuous into the wing of the stem.

Flowers about 1.5 cm across, in compound cymes terminating branches, cymes usually many-flowered. Sepals and petals 5. Sepals triangular-acute, 1.5–2 mm long. Petals 6–8 mm long, oblong-obovate, copperish yellow. Stamens numerous.

Capsules ovoid to subglobose, 4–5 mm long, persistent styles 3 but connivent and appearing as 1, sometimes tardily separating, 2–3 mm long. Seeds very dark purplish brown, almost black, lustrous, oblong-cylindric, usually falcate, 1.5–1.8 mm long, finely but markedly alveolate-reticulate.

Banks of rivers and streams, bottomland woodlands, wet hammocks, wet thickets; s.w. Ga., Fla. Panhandle, s. Ala. (Coastal plain and piedmont, s.e. Va. to Fla. Panhandle, westward to s. Miss.; Cumberland Plateau, Tenn.)

17. Hypericum apocynifolium Small. Fig. 179

Vegetatively like *H. nudiflorum*. Terminal cymes commonly only 3-flowered, if cymes compound, up to 8 flowers. Sepals oblong, obtuse apically, 3 mm long. Petals 8–10 mm long, obovate, copperish yellow. Capsules ovoid, 8–10 mm long and 5–7 mm broad. Seeds dull brown, not lustrous, 1.8–2.0 mm long, cylindric, sometimes slightly falcate.

Mesic or submesic wooded ravine slopes and bluffs, ravine bottoms, natural levees and other elevated places in floodplains; in our area, s.w. Ga. and s.e. Ala., cen. Fla. Panhandle, perhaps elsewhere. (S.w. Ga., s.e. Ala., cen. Fla. Panhandle; w. La., n.e. Tex., s. Ark.)

18. Hypericum frondosum Michx.

Openly branched shrub 1–3 m tall. Bark of larger stems smooth, eventually sloughing in large, thin, curled plates.

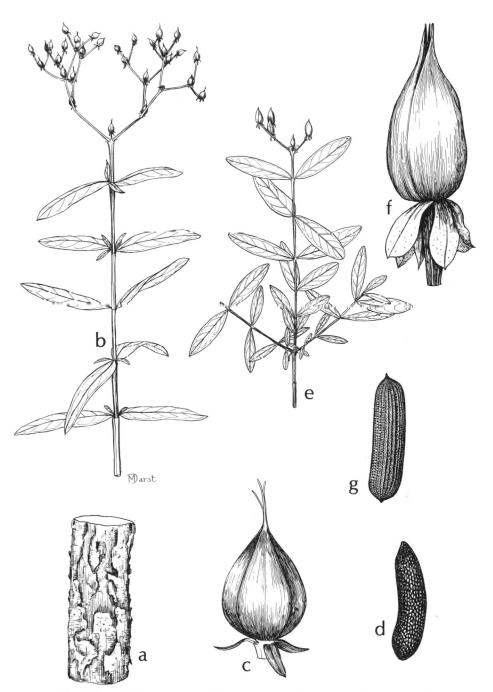

Fig. 179. a–d, **Hypericum nudiflorum:** a, small portion of lower stem; b, upper part of stem with infructescence; c, capsule; d, seed. e–g, **Hypericum apocynifolium:** e, upper part of stem with infructescence; f, capsule; g, seed.

Leaves articulated at the base, oblong, elliptic, or lance-ovate, mostly 3–6 cm long and 1–2 cm broad, shortly tapered basally, usually rounded apically.

Flowers showy, 2.5–4 cm across when the petals are spread, solitary or in 3-flowered cymes terminal on branchlets. Sepals foliaceous, articulated at the base, very unequal, the longer ones to 2 cm long or a little more. Petals bright yellow, obliquely obovate, 15–20 mm long, usually becoming reflexed. Stamens very numerous.

Capsules conic, 9–12 mm high, their tips attenuate into the persistent styles. [*Hypericum aureum* Bartr.]

In our area, I am aware of its occurrence at only one station, that in a woodland close to but well elevated above the Chattahoochee River, Jackson Co., Fla. (Cedar glades, river bluffs, rock outcrops, s. Ind., Ky., Tenn., Ga., Ala., sometimes naturalized from cultivation elsewhere.)

Hamamelidaceae (WITCH-HAZEL FAMILY)

1. Leaf blades pinnately veined, unlobed, stellate-pubescent at least beneath when young, usually sparsely so even in age; flowers in stalked, few-flowered clusters or in dense, globose to oblong spikes.
　　2. The leaf blades crenate usually only distally from somewhat above their middles; flowers numerous in globose to oblong, compact spikes terminating branches or axillary to leaf-scars and appearing in spring before or as new shoot growth commences; petals none; stamens numerous.　　　　　　　　　　　　　　　　　　1. *Fothergilla*
　　2. The leaf blades with undulate margins distally from about their middles or a little below; flowers few in short-stalked, loosely spicate clusters axillary to leaf scars on twigs of a previous season, usually in autumn, sometimes in winter or spring; petals 4, stamens 4.　　　　　　　　　　　　　　　　　　　　　　　　　2. *Hamamelis*
1. Leaf blades palmately veined and lobed, glabrous; flowers and fruits in dense, globose heads.　　　　　　　　　　　　　　　　　　　　　　　　　　3. *Liquidambar*

1. Fothergilla

Fothergilla gardeni Murray. WITCH-ALDER.

Slender shrub to about 1 m tall, sprouting from subterranean runners and colonial. Buds, young twigs, petioles, and both leaf surfaces densely stellate-pubescent; older leaves varying from pubescent to nearly smooth, usually a sparse pubescence on lower leaf surfaces even in age. Buds naked, the terminal one stalked, shaped like a scalpel blade, lateral ones borne singly, short, sessile.

Leaves 2-ranked, short petiolate, blades pinnately veined, principal lateral veins usually 3–5 on a side, these angling-ascending parallel to each other to the leaf margin; blades sometimes about as broad as long, mostly obovate, less frequently broadly elliptic, 2–6 cm long and 1.5–3 cm broad, rounded, truncate, or broadly cuneate basally, apices obtuse, rounded, or nearly truncate, margins usually crenate near their apices, sometimes entire; stipules about 2 mm long, ovate-oblong, blunt apically, densely stellate-pubescent.

Inflorescences dense, oblong spikes terminal on the twigs, overlapping, ovate, copiously hairy bracts relatively conspicuous. Flowers bisexual, or the lower ones in the spikes male-sterile; floral tube cuplike, truncate at the summit or with 5–7 minute, stubby calyx segments. Petals none. Stamens numerous, borne at the rim of the floral tube, filaments relatively long, somewhat dilated distally. Pistil 1, ovary 2-locular slightly sunken in the base of the floral tube; styles 2, slender at anthesis, in fruit becoming thick basally, hornlike, their tips curving outward. Fruit a capsule more or less fused to about the middle with the floral tube, one oblong, lustrous seed in each locule.

Pine savannas, hillside bogs, shrub bogs or bays; in our area known to me only from s.e. Ala. and Fla. Panhandle. (Coastal plain, N.C. to Ala., Fla. Panhandle.)

2. Hamamelis

Hamamelis virginiana L. WITCH-HAZEL.   Fig. 180

Shrub or small tree, the buds, young twigs, and petioles densely stellate-pubescent with buff-colored hairs and appearing somewhat scurfy. Buds naked, the terminal one with a longish stalk and shaped like a scalpel blade, the lateral buds in pairs, one beside the other, one of them stalked and relatively large, shaped much like the terminal bud, the other sessile and relatively small; leaf scars mostly shield-shaped, each with 3 vascular bundle scars, these often indistinct.

Leaves 2-ranked, short-petiolate, blades pinnately veined, principal lateral veins usually not over 6 on a side, these angling ascending parallel to each other to the leaf margin; blades often as broad as long, obovate, oval, or broadly elliptic, variable in size, to 15 cm long and 10 cm broad, bases broadly tapered to rounded, often oblique, apices rounded or obtuse, infrequently acute, margins distally from the middle or a little below with broad undulations; leaf surfaces sparsely to moderately stellate-pubescent, some but not all of the pubescence usually sloughing, stipules lanceolate, 6–8 mm long, densely stellate-pubescent, early deciduous.

Flowers slightly fragrant, bisexual, radially symmetrical, borne from the axils of leaf scars usually in a stalked cluster of 3 subtended by a closely appressed involucre of 3 scales. Floral tube cuplike, with 2 or 3 bractlets at its base, 4 ovate or oblong calyx segments about 2.5 mm long on its rim, their exterior surfaces densely stellate-pubescent, their tips usually reflexed. Petals 4, spreading, yellow, sometimes suffused with red, linear, 1–1.5 cm long. Fertile stamens 4, alternate with the petals, a scalelike staminodium between each of the fertile stamens and opposite the petals. Pistil 1, ovary densely pubescent, partially inferior, 2-locular, styles 2, anthers and styles at about the same level, not exserted. Fruit a hard, ovoid or thickly ellipsoid capsule, its base fused with the floral tube, the persistent stylar beaks recurved, a single, hard, brown seed in each locule. Flowering occurs mostly in late autumn, sometimes in winter or spring southward in the range, fruit maturation during the following summer.

In various upland woodland mixtures, wooded slopes and ravines, in the southeastern part of the range, at least, in floodplain forests, evergreen shrub bogs or bays, stream banks; throughout our area. (Que. and N.S. to n. Mich. and s.e. Minn., generally southward to s. cen. pen. Fla. and Tex.)

Distinctive features of the witch-hazel are the alternately arranged leaves with undulate-margined blades, the stalked, asymmetrically curved (like a scalpel blade), tawny-hairy terminal buds. At time of flowering in autumn, the open clusters of flowers with narrowly linear yellow petals are helpful.

National champion witch-hazel (1976), 1′5″ circumf., 43′ in height, 41′ spread, Wooded Dune State Park, Mich.

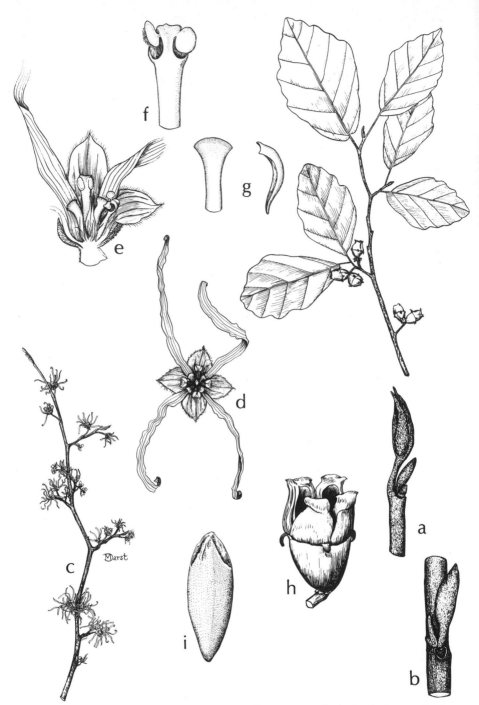

Fig. 180. **Hamamelis virginiana:** a, tip of winter twig; b, node of winter twig; c, twig with buds and flowers; d, flower, face view; e, flower, side view (some parts removed); f, stamen; g, staminodium, two views; h, capsule, dehisced; i, seed.

3. Liquidambar

Liquidambar styraciflua L. SWEETGUM.   Fig. 181

A handsome tree 40 m tall or more, with balsamic juices, its foliage somewhat aromatic when crushed. Twigs stoutish and stiff, glabrous, often (not always) beset with corky outgrowths taking the form of wings, ridges, or warts. Terminal buds ovoid, pointed apically, having a shiny, resinous coating. Leaf scars half-round to elliptical, with 3 vascular bundle scars. Pith continuous, streaked white and tan, angularly lobed in cross-section. Bark of trunk dark gray, with many vertically interlacing ridges and furrows.

Leaves simple, alternate, deciduous, petioles slender and pliable, commonly as long as the blades, even longer, slightly channelled on the upper side, more or less dilated at the base; petioles of developing leaves bear a pair of linear-oblanceolate stipules 8–10 mm long on the upper side somewhat above the base, these soon deciduous leaving a pair of roundish, glandlike scars evident, usually, through the remainder of the season. Blades palmately-veined, ovate in overall outline, nearly as broad as long, with 5–7 prominent, marginally serrate, acute or acuminate lobes giving a starlike effect; bases of blades commonly truncate or the lowermost lobes more or less curvate-declined from either side of the summit of the petiole; upper surfaces lustrous-green and glabrous, the lower paler and duller, pubescent at least in the basal vein-axils, often in most of the major vein-axils, sometimes with sparse pubescence along the major veins. As the new leaves unfurl, their surfaces, especially the lower, are densely pubescent, most of the pubescence quickly sloughing.

Flowers very small, unisexual (the plant monoecious), perianth none. Inflorescence borne on the same shoot axis as the leaves, bursting from the bud prior to the unfolding leaves or with them, several leaves closely set at the base of the shoot, the staminate inflorescence comprising the terminus of the shoot; staminate flowers greenish yellow, in sessile or short-stalked, compact, ball-like or oblongish masses for 3–6 cm along the terminal portion of the axis; pistillate flowers very numerous, in 1-several dense and compact ball-like heads on dangling, slender stalks from the axils of unfolding leaves. Staminate flower with a minute floral tube with, usually, 4 stamens on its rim. Entire staminate portion of the inflorescence falling after the pollen is shed. Pistillate flowers pale green, with a minute floral tube adnate to the lower part of the 2-locular ovary; styles 2, stigmatic on the inner faces of their incurving tips; ovules many, more or less in 4 longitudinal rows in each locule, mostly aborting, only the basal 1–2 ovules functional.

Capsular fruits consolidated by their bases into a hard ball with "spiny" periphery, the "spininess" deriving from the hardness, stiffness, and sharpness of the many persistent styles whose stigmatic tips have coalesced during maturation of the fruits. Capsules dehisce septicidally below the style base. Viable, fully developed seeds usually 1–2 per locule, winged, the embryo-bearing portion dark brown, attached medially on one side, the other side convex, the tan, thin-papery wing terminal, the whole about 10 mm long and about 3 mm broad medially; the very numerous aborted ovules eventually become hard, very irregularly angled, often slightly winged, reddish brown, and reminiscent of tiny grapenuts. Both the seeds and the grapenutlike "abortions" are eaten by mourning doves, goldfinches, white-throated and chipping sparrows, etc.

The sweetgum inhabits a variety of kinds of wooded mesic sites, old fields, fence and hedge rows, in our area in wet or swampy woodlands as well, even on

Fig. 181. **Liquidambar styraciflua:** a, tip of twig with infructescence; b, emerging branchlet with pendent pistillate inflorescence and erect staminate inflorescence; c, enlargement of one unit of staminate inflorescence; e, to right, small portion of twig with leaf scar and bud, and to left a small portion of twig having corky wings; f, individual fruits, before and after dehiscence; g, winged seed to left, and aborted ovules to right.

sites where water stands much of the time. (S. Conn. to s. Ill., generally southward to s. Fla., e. and s.e. Tex., s.e. Okla.; Mex. and C.Am.)

The sharply serrated, regularly 5 (7) lobed, semi-starlike leaf blades, leaves alternately arranged, suffice to identify this tree. Should there be a tendency to mistake it for a maple, bear in mind that the leaves of the maple are oppositely arranged.

National champion sweetgum (1970), 19'8" circumf., 125' in height, 100' spread, Richland Co., S.C.

Hippocastanaceae (BUCKEYE FAMILY)

Aesculus

Deciduous small shrubs to large trees, ours shrubs or small trees. Twigs tan, brown, or reddish brown and with pale lenticels; leaf scars shield-shaped or obcordate, each with 3 or more vascular bundle scars disposed in a V- or U-shaped line. Leaves opposite, long-petioled, palmately compound, each with 5–11, usually 5–7 leaflets; lateral veins of leaflets relatively pronounced, angling-ascending approximately parallel with each other from the midrib toward the blade margins. Inflorescence a more or less cylindrical panicle terminating a shoot of the season, the panicle with a single central axis bearing lateral, much shorter panicles or clusters. Flowers unisexual and staminate, or bisexual, usually both in the same inflorescence. Calyx united for much of its length, distally with 4–5 usually somewhat unequal lobes. Petals 4 or 5, subequal or unequal, clawed basally. Stamens 6–8, rarely 5, an extrastaminal, usually 1-sided disc present. Pistil 1 (rudimentary in the staminate flowers), ovary superior, 3-locular, rarely 2- or 4-locular, ovules 2 in each locule; style slender and elongate, often arcuate, stigma terminal.

Fruit a loculicidal, subglobose or obovoid, often asymmetric, brown capsule with large seeds, each chestnutlike, usually 1–3, rarely 4–6, per capsule.

1. Young twigs green; leaflets caudate-acuminate apically; calyx with white ground color and with some fine flecks of brown; corolla and filaments white (drying brownish).
    1. *A. parviflora*
1. Young twigs dark reddish brown; leaflets, most of them, acute apically; calyx, corolla, and filaments red or yellowish red (drying purplish red).   2. *A. pavia*

1. Aesculus parviflora Walt. BOTTLEBRUSH BUCKEYE.   Fig. 182

An understory shrub or small tree, forming extensive clones by subterranean runners. Twigs green at first, glabrous, eventually becoming gray and smooth; pith of twigs large, brown, continuous.

Petioles 5–20 cm long, stalks of the leaflets 0.5–3 cm long. Leaflets usually 5, sometimes 7, infrequently 3 or 4, elliptic, oblong-elliptic, or obovate, overall size of leaves and their respective leaflets variable; central leaflet largest, to 20 cm long or a little more, usually obovate, outer laterals smallest and usually inequilateral, mostly elliptic or oblong-elliptic; bases of the 3 central leaflets usually longish-tapered, those of the outer laterals mostly obliquely obtuse, apices caudate-acuminate; upper surfaces dark green and glabrous, the lower paler and duller, with pale, soft, short-shaggy pubescence, this sometimes sloughed in age; margins irregularly doubly serrate.

Panicle long-stalked, stalk 8–15 cm long, panicle itself 15–40 cm long and 4–5

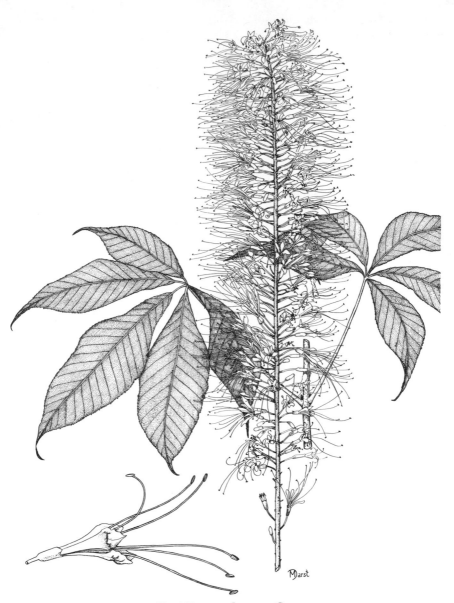

Fig. 182. **Aesculus parviflora.**

cm broad, elongate cylindric, the lateral branches very close together, short, few-flowered, thus the panicle relatively dense, its axis sparsely to copiously short-pubescent as are the flower stalks. Calyx tube cylindric, 4–5 mm long, exterior ground color white with short white pubescence some of which becomes brown giving the surface a brownish-flecked appearance; calyx lobes usually 4, pubescent, short-oblong, rounded to truncate apically. Petals usually 4, essentially equal, white, extending about 10 mm beyond the calyx, each composed of more or less folded and channellike claws, bladed only at the tip, the blades oblanceolate, quickly becoming much crinkled. Stamens usually 6, filaments slender, white, often of unequal length, the longest about 40 mm, some straight although flexuous, some arching, often forming a complete or nearly complete circle; anthers small, pinkish orange in color, glabrous.

Capsules 2.4–5 cm in diameter, surface smooth, pale brown, containing 1 or 2 chestnut-brown seeds, the hilum tan.

Woodlands of river bluffs, ravines, hillsides, sandy or rocky banks; in our area of coverage, s.w. Ga., s. Ala. (Otherwise locally in Ala. save perhaps for the northernmost and cen. s. parts.)

2. Aesculus pavia L. RED BUCKEYE. Fig. 183

Shrub or small understory tree, infrequently to about 12 m tall. Twigs dark reddish brown at first and with sparse, irregularly scattered, very short, grayish pubescence, later becoming gray and glabrous; pith relatively small, white, continuous.

Petioles mostly 8–15 cm long, usually dark reddish brown and with pubescence as on the twigs, leaflets sessile or subpetiolate for the most part. Leaflets usually 5, sometimes 7, infrequently 3 or 4, long-elliptic to long-oblanceolate, infrequently obovate, overall size of leaves and their respective leaflets variable; central leaflet usually largest but frequently not much exceeding the size of the pair to each side of it, the outer laterals smallest, varyingly slightly inequilateral to equilateral, bases of all leaflets usually tapered, apices mostly acute, sometimes slightly acuminate; upper surfaces dark green and glabrous or sparsely short-pubescent throughout or only along the main veins at least when young, the lower surfaces only little paler green than the upper if pubescence is lacking or sparse, much paler and grayish or brownish if pubescence dense; amount of pubescence, if any, varying greatly; margins irregularly serrate or doubly serrate, sometimes teeth on a given margin notably irregular.

Panicle sometimes sessile, usually stalked, the stalk to about 5 cm long, axes and flower stalks dark reddish brown, varying from essentially glabrous to densely short-pubescent; panicle itself 10–25 cm long and to about 12 cm broad, usually loosely flowered, flowers few to relatively numerous. Calyx usually red (drying purplish red), sometimes yellowish red, the tube cylindric or long-campanulate, often gibbous at the base, 10–18 mm long, varying from sparsely to densely short-pubescent, sometimes with sessile and/or stipitate glands; lobes 5, short-oblong, obtuse, rounded, or truncate apically, pubescent and usually also stipitate-glandular at least on their apical margins. Petals usually 4, the yellowish-pubescent claws of the 2 upper ones well exserted at an angle from the calyx, their spatulate blades relatively small; other 2 petals (laterals) with claws wholly within the calyx, their blades oblong-obovate, usually not wholly exserted, erect; blades of the petals red, dotted with darker glands. Stamens usually 6–8, filaments red, unequal in length, the longest exserted only a little beyond the upper petals.

Fig. 183. **Aesculus pavia:** a, flowering branch; b, node with leaf scar and axillary bud; c, flower; d, fruit; e, seed.

Capsules 3.5–6 cm in diameter, light brown, surface finely pitted. Seeds 1–3, rarely 4–6, per capsule, dark chestnut brown, the hilum pale.

Mesic woodlands of ravine slopes and bottoms, bluffs, natural levees, parts of floodplains rarely and briefly flooded, stream banks, hammocks; throughout our area. (S.e. N.C. to cen. pen. Fla., westward to s. cen. Tex., northward in the interior to s.e. Mo. and s. Ill.; n.w. Ga.)

Both of the buckeyes treated here would seem to be excellent candidates for horticultural use (save for their being deciduous), the foliage and inflorescences of each being especially attractive, the bottlebrush being the more so, in my opinion.

National champion red buckeye (1981), 3'9" circumf., 35' in height, 33' spread, Georgetown Co., S.C.

Illiciaceae (STAR-ANISE FAMILY)

Illicium floridanum Ellis. FLORIDA ANISE, STINK-BUSH, STINKING LAUREL, POLE-CAT-TREE.   Fig. 184

An irregularly branched, glabrous, evergreen shrub or small understory tree. Tips of leafy twigs red (early in the season), brown below, older twigs gray, without lenticels; outline of leaf scars roughly U-shaped, each with a single vascular bundle scar.

Leaves simple, alternate, but those distally on the twigs so closely set as to appear whorled; petioles 1–2 cm long, suffused with red pigment, early in the season and in winter, at least; blades slightly leathery, long-elliptic, infrequently oblong-elliptic, cuneate basally, acute to acuminate apically, 6–15 cm long and 2–6 cm broad, usually broadest very slightly above the middle, margins entire, minutely but copiously gland-dotted beneath, emitting a curious odor, obnoxious to some persons, perhaps pleasant to others, to me the odor somewhat reminiscent of that of fresh fish.

Flowers bisexual, solitary from leaf axils, their stalks 1–3 (4) cm long (usually a little longer on the fruits than on the flowers), individually rather showy but tending to be held in such a way as to be to some extent hidden amongst the foliage, having a marked fragrance like that of the leaves. Sepals 3–6, gland-dotted exteriorly and partially suffused with red, broadly ovate with obtuse tips, a fringe of minute pubescence on their margins, falling as the buds open or shortly thereafter. Petals numerous, 21–33, narrowly oblong and straplike above their somewhat dilated bases, deep red or purplish red, mostly 1.5–2 (2.5) cm long, laxly spreading radially. Stamens numerous, (30–40), somewhat imbricated and spreading in a circle within the corolla, filaments slightly unequal, only a little longer than the anthers and about as broad, colored approximately like the petals, the anthers a little darker than the filaments, pollen white. Pistils superior, separate and numerous, usually about 11–15 in a circle, commonly 13, infrequently 17, the ovary of each maturing into a chartaceous, oblique, flattened, 1-seeded follicle beaked by the persistent style, explosively dehiscing along the upper side, body of the follicle about 1 cm long. Seed obliquely ellipsoid, about 8 mm long, surface light brown, smooth and shiny.

Moist wooded ravines and steep-heads, along or even in the small streams or seepage areas of ravines. S.w. Ga. (rare), inner Fla. Panhandle from the Ochlockonee River westward, throughout the coastal plain of Ala., westward to s.e. La.

Fig. 184. **Illicium floridanum.**

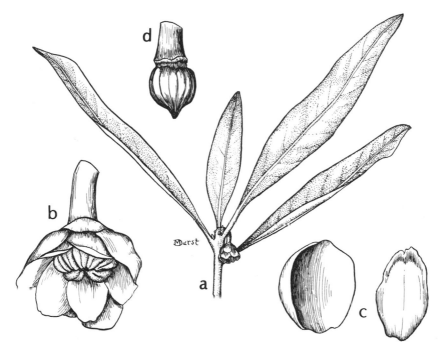

Fig. 185. **Illicium parviflorum:** a, branchlet with flower; b, flower, enlarged; c, sepal on left and petal on right; d, flower with perianth and stamens removed.

This evergreen shrub is unique among shrubs of our flora in having flowers with numerous, narrow, dark red, radially disposed petals, fruits arranged in a whorl, and all parts emitting an odd, somewhat aniselike odor. It has more potential for use as an ornamental than has hitherto been exploited but it requires a moist, shaded place in which to grow.

In 1982, Steven W. Leonard and Wilson Baker, while very carefully exploring the numerous ravines east of the Apalachicola River in Liberty and Gadsden Cos., Fla., discovered a few plants of this species with pink flowers and a few with white flowers growing intermixed with the usual red-flowered ones.

Illicium parviflorum Michx. ex Vent., star anise (see Fig. 185), often cultivated as an ornamental in our area of coverage, occurs naturally peripheral to our area, to the south in Florida (Marion, Volusia, and Lake Cos.), chiefly in hammocks along relatively large spring-fed streams and in bayheads where it is associated with *Chamaecyparis thyoides, Agarista populifolia, Sabal palmetto, Sabal minor, Rhapidophyllum hystrix, Salix floridana, Magnolia virginiana, Persea palustris*, etc. Vegetatively it is very much like *I. floridanum* save that the tips of the leaf blades are obtuse and blunt or even rounded. The flowers are very much smaller and very much less conspicuous than in *I. floridanum*, the perianth segments (sepals and petals) not so clearly differentiated. Sepals unequal, ovate, broader than long, the larger ones 3–4 mm broad, apices rounded, margins minutely fringed with pubescence. Petals yellow, broadly rounded-obovate, 4–5 mm long. Stamens 6–7, colored approximately like the petals. Pistils 10–13.

I. parviflorum is much more drought resistant than *I. floridanum*.

Juglandaceae (WALNUT OR HICKORY FAMILY)

1. Pith of twigs transversly partitioned, chambered between the partitions; leaves pinnately compound, the terminal leaflet usually much reduced or completely aborted and leaves appearing evenly pinnate (commonly odd-pinnate on seedlings and young saplings); husk of the fruit indehiscent; shell of the nut rough-corrugated.   1. *Juglans*
1. Pith of twigs continuous; leaves odd-pinnately compound; husk of the fruit splitting by valves; shell of the nut essentially smooth.   2. *Carya*

1. Juglans

Juglans nigra L. BLACK WALNUT.   Fig. 186

Potentially a large tree, but few large ones have survived commercial exploitation in our range. Bark of trunks dark gray to nearly black, roughened by relatively deep furrows interlacing between narrow ridges. Bud scales 4, in 2 pairs, densely clothed with blond pubescence. Twigs dark brown, copiously pale-pubescent (much of the pubescence persistent into the second year) and with dotlike, somewhat raised lenticels, the latter orangish at first, usually becoming buff-colored; leaf scars more or less obcordately 3-lobed with 3 clusters of vascular bundle scars; pith with traverse diaphragms, chambered between the diaphragms.

Leaves alternate, deciduous, blades large, pinnately compound, sometimes the terminal leaflet much reduced, more often completely aborted so that the leaves appear evenly pinnate (usually odd-pinnate, the terminal leaflet not reduced or aborted, on seedlings and young saplings); leaflets 8 (9) to 22 (23) per leaf, sessile or nearly so, pinnately veined, ovate, lance-ovate, or oblong, medial ones usually largest, mostly inequilateral and somewhat falcate, bases obliquely rounded to subcordate, apices longish-acute to acuminate. Young developing stem and leaf axes relatively densely pubescent with fascicled or tufted hairs and an abundance of both sessile and short-stipitate glands; upper surfaces of unfurling leaflets with fascicled hairs along the major veins and for a short while, at least, sessile glands chiefly along both the major veins and the veinlets, lower surfaces similarly pubescent, the pubescence more conspicuous, glandularness similarly distributed but many more of the glands stipitate. At maturity, leaf axes and major veins beneath the leaflets short-shaggy-pubescent, some glands persisting as well, sparse pubescence between the major veins, upper surfaces usually sparsely pubescent only along the midrib, margins finely serrate; overall length of mature leaves, including petioles, 2–6 dm.

Flowers unisexual, plants monoecious. Staminate flowers in sessile, pendent catkins 7–15 cm long (catkins protruding from buds on wood of the season in autumn, elongating and reaching anthesis as new shoots of the season are developing); each flower sessile or shortly stalked and subtended by a short-ovate, wooly bract; calyx cuplike, 3–6-lobed, anthers sessile, 8–40 per flower. Pistillate flowers solitary or in a few-flowered, densely woolly spike terminating short branches of the season when the leaves are only partially developed; involucre urnlike, densely woolly, sepals barely protruding from its orifice, the style protruding slightly beyond the calyx, stigmas 2, spreading.

Fruit globose or nearly so, 5–8 cm in diameter, husk green, thick, indehiscent, turning black as it disintegrates; nutshell rough-corrugated externally, dark brown, 2-valved but indehiscent; kernel oily, sweet. [*Wallia nigra* (L.) Alef.]

Of limited occurrence in our range in woodlands of bluffs, ravines, river

Fig. 186. **Juglans nigra.**

banks, floodplains where flooding is of brief duration, calcareous upland woodlands; s.w. Ga., s.e. Ala., cen. Panhandle of Fla. (W. N.Eng. to s. Mich., s. Wis., s. Minn., s.e. S.D., generally southward to Panhandle of Fla., n.w. La. and e. Tex.)

National champion black walnut (1975), 22' circumf., 132' in height, 133' spread, Humboldt Co., Calif.

2. Carya (HICKORIES)

Trees with hard, heavy wood. Pith continuous in the twigs. Leaf scars variable in shape, vascular bundle scars 7, more or less scattered, often indistinct. Terminal buds markedly larger than the axillary ones. Leaves deciduous, alternate, petiolate, without stipules. Blades odd-pinnately compound, leaflets pinnately veined, often glandular- or scaly-dotted, lateral ones usually sessile, herbage pungently aromatic when crushed or bruised. Flowers unisexual, plants monoecious, the staminate in pendent catkins, usually in a stalked fascicle of 3 borne from the summit of the twig of the previous year and/or from the base of the developing shoot of the season, the pistillate solitary in 2–6-flowered spikes terminal on developing shoots of the season. Staminate flower with a 2- or 3-lobed calyx subtended and usually exceeded by a bract, stamens 3–8 per flower, anthers sessile, pubescent. Pistillate flower with 1 pistil, the ovary surrounded by and adnate to an involucre 4-lobed at the summit; involucre together with the exocarp of the ovary developing into the husk of the fruit which at maturity separates from the enclosed hard shell of a 1-seeded nut, the husk wholly or partially splitting into (3) 4 (5) valves.

1. Bud scales 4–6, disposed margin to margin (valvate); bud scale scars wider medially than at their extremities, staggered and separated from each other and not forming a close band of scars around the twig; husk of the fruit ridged or keeled along the sutures, at least distally.
  2. Leaves with 5–17 leaflets but many leaves on a given tree with 11 or more; husk of the fruit more or less flattened between the sutures.
    3. Tree inhabiting woodlands of natural levees, banks of rivers and streams, floodplains; nut angled in cross-section, its kernel bitter, each half deeply 2-lobed apically.   1. *C. aquatica*
    3. Tree (in our range) sporadically naturalized from cultivars, occurring in fence and hedge rows, borders of upland woodlands, old fields, vacant lots; nut round in cross-section, its kernel sweet, each half barely notched apically.   5. *C. illinoiensis*
  2. Leaves, most of them on a given tree with 7–9 leaflets; husk of the fruit not at all flattened between the sutures.   2. *C. cordiformis*
1. Bud scales more than 6, imbricated; bud scale scars very narrow, collectively forming a close band of scars encircling the twig; husk of the fruit not ridged or keeled along the sutures.
  4. Margins of leaflets with tufts of pubescence on or on and between the teeth.
                                                  6. *C. ovata*
  4. Margins of leaflets without tufts of hairs on or on and between the teeth.
    5. Pubescence of leaf parts, if any, not fascicled or tufted excepting in the vein axils.
      6. Bud scales clothed with rust-colored, resinous granules or by a mixture of granules and hairs.   3. *C. floridana*
      6. Buds clothed with short hairs, without resinous granules.   4. *C. glabra*
    5. Pubescence of leaf axes and lower surfaces of leaflets notably fascicled or tufted.
      7. Buds more or less covered by amberish, resinous, disclike scaliness; twigs slender; lower surfaces of leaflets with thin, disclike scales.   7. *C. pallida*
      7. Buds very densely and finely pubescent; twigs stout; lower surfaces of leaflets with tiny, gritlike or granular beads of resin.   8. *C. tomentosa*

1. Carya aquatica (Michx. f.) Nutt. WATER HICKORY, BITTER PECAN. Fig. 187

Medium to large tree, to 35 m tall and 1 m d.b.h. Bark of trunks grayish or light brown, eventually splitting freely into platelike, shaggy scales. Buds with valvate, flattened scales covered with an abundance of small, amberish glandular scales and some yellowish, short-shaggy pubescence; terminal bud stalked, lance-attenuate. Very young developing shoots, petioles, leaf axes, and unfurling leaflets densely woolly-pubescent, much of the pubescence quickly sloughing exposing tiny, pale surface scales. Woody twigs slender, dark brown, with scattered pale lenticels; leaf scars light brown, obcordate.

Leaflets 7–17, commonly 9 or 11 (occasionally an even number because of developmental abortion of the terminal one), most of them lanceolate, occasionally oblanceolate, the lateral ones usually inequilateral and falcate, their bases obliquely acute to obtuse, apices long-tapered-acute (sometimes rounded if leaflets oblanceolate), terminal leaflet essentially equilateral, lanceolate or oblanceolate, shortly stalked; at maturity the overall leaf length, including petioles, 15–30 cm long; upper surfaces of mature leaflets generally glabrous and dark green, sometimes retaining some of the minute scales, lower surfaces paler and duller green, short-pubescent along the midvein, often on the lateral veins as well, degree of retention of the scales variable, sometimes none retained; margins with few, remote, very small, blunt, scarcely perceptible serrations, varying to evidently serrate and the teeth blunt.

Staminate catkins slender, about 4–5 mm across, bracts linear-oblong, surfaces of bracts and calyces clothed with transparent scales and a few hairs. Fruit with 4 low-ridged sutures, somewhat flattened on the faces of the valves, broadly oval in outline, abruptly narrowed to bluntly pointed tips, 2.5–4 cm long; husk thin, valves splitting to the base; nut angled, surface usually wrinkled, kernel bitter, each half deeply 2-lobed apically. [*Hickoria aquatica* (Michx. f.) Britt.]

Inhabiting natural levees and banks of rivers and streams, floodplain forests where the duration of flooding is relatively brief; throughout our area. (Chiefly, but not exclusively, coastal plain, s.e. Va. to s. cen. Fla., westward to e. Tex., northward in the interior, s. and e. Ark., w. Miss., s.w. Mo., s. Ill., w. Tenn.)

National champion water hickory (1967), 22'2" circumf., 150' in height, 87' spread, Blountstown, Fla.

2. Carya cordiformis (Wang.) K. Koch. BITTERNUT HICKORY. Fig. 188

Medium-sized tree, to about 30 m tall and to 12 dm d.b.h. Bark of trunks light brownish to gray, smooth until trunks are moderately large, eventually with shallow furrows between interlacing ridges, sloughing in small flakes. Buds valvate, slender, covered by yellow or copperish resinous scales with some pubescence intermixed (the color especially noticeable on the larger terminal buds). Very young developing stems and leaf axes relatively sparsely powdery-pubescent, lower surfaces of leaflets short-pubescent along the veins, sparsely to densely clothed with disclike scales between the veins. Woody twigs of the season greenish brown to dark brown and dotted with buff-colored lenticels; vascular bundle scars shield-shaped or obcordate.

Leaflets 7–9 (11), the terminal one often about the same size as the upper lateral ones, sometimes larger, mostly lanceolate, some sometimes oblanceolate or obovate, lowermost smallest and sometimes ovate, the laterals relatively slightly or not at all inequilateral, bases slightly oblique or symmetrical, more

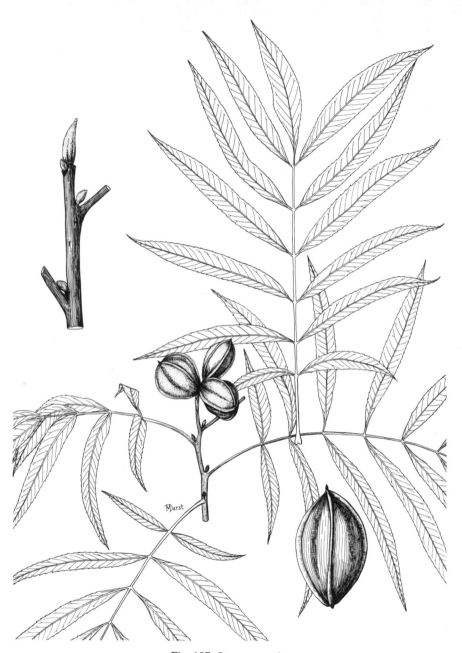

Fig. 187. **Carya aquatica.**

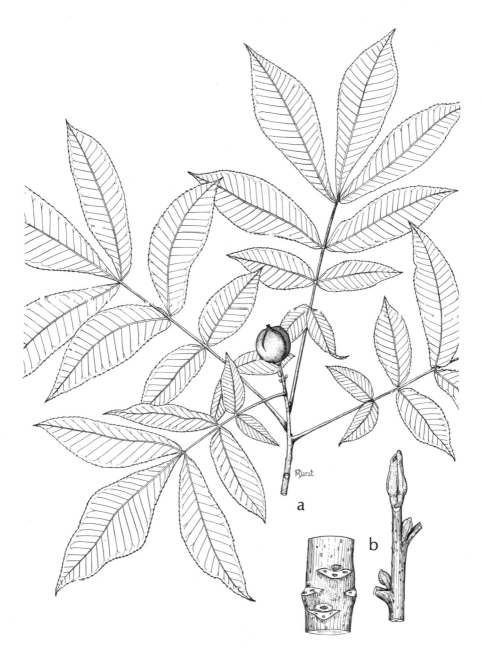

Fig. 188. **Carya cordiformis:** a, fruiting branch; b, at left, annual node showing relatively broad, widely separated bud-scale scars, and at right, tip of branchlet with winter buds.

or less rounded, sometimes cuneate, apices acute, or obtuse to short-acuminate if obovate; terminal leaflet equilateral, cuneately narrowed basally to a sub-petiole; upper surfaces dark green, sometimes sparsely pubescent along the midrib, lower surfaces paler and duller, sparsely to densely clothed with resinous scales, these varying from grayish to rusty brown to purplish, pubescent along the veins, sometimes with tufts of pubescence in the axils formed by the midvein and lateral veins; leaf axes with evenly distributed short-shaggy pubescence; margins of leaflets finely serrate, teeth blunt.

Staminate catkins in slenderly stalked fascicles, stalks about 2 cm long, catkins slender, 6–10 cm long, about 4 mm across, bracts little if any longer than the calyces, bracts and calyces resinous-scaly, anthers pubescent distally. Fruit globular or nearly so, short-beaked, 1.5–2.5 (3) cm in diameter; husk at midseason thickly coated with olivaceous scales, these relatively few and darkened later; husk thin, only slightly ridged along the sutures from about the middle upwardly, splitting tardily to the base; nut usually obscurely 4-angled, shell thin; kernel bitter. [*Hicoria cordiformis* (Wang.) Britt.]

Low woodlands near shores of lakes, floodplain woodlands, natural levees, river banks, upland woodlands; s.w. Ga. and s.e. Ala., Ochlockonee, Apalachicola, and Chipola River drainages in the Panhandle of Fla. (N.H., s.w. Que., s. Ont. to s.e. Minn., Iowa, s.e. Neb., generally southward to cen. Fla. Panhandle and e. Tex., apparently absent from much of s. Ala., s. Miss., s.e. La.)

National champion bitternut hickory (1975), 14′4″ circumf., 120′ in height, 81′ spread, Springfield, Va.

3. Carya floridana Sarg. SCRUB OR FLORIDA HICKORY.   Fig. 189

Small to medium-sized tree, sometimes to about 25 m tall and 8 dm d.b.h. Bark of trunk pale gray to gray-brown, somewhat roughened by flat, interlacing ridges. Buds with imbricated scales, outer ones with their surfaces abundantly clothed by small rusty resinous granules or by a mixture of granules and hairs, surfaces of inner ones densely pubescent. Young developing twigs rusty-scaly, sometimes with a few scattered hairs as well, both rather quickly sloughing. Woody twigs reddish brown with flecks of gray; leaf scars variable in shape, obovate, nearly circular, horizontally elliptic, or somewhat crescent-shaped.

Leaflets 3, 5, or 7, mainly 5 or 7, lanceolate, lance-ovate, or elliptic, some of them usually somewhat inequilateral, the terminal one usually largest and sometimes obovate, sometimes petiolate; at maturity overall leaf length, including petiole, 8–20 cm; bases of laterals obtuse to rounded, apices acute to acuminate, base of terminal one usually cuneately tapered; upper surfaces glabrous and green (rusty-scaly when young), permanently more or less scaly and rust-colored beneath; margins of young, developing blades with bulbous, green "teeth" upwardly from somewhat above their bases, the serrations eventually blunt and somewhat calluslike.

Staminate catkins slender, 2–2.5 mm across, bracts subulate, bracts and calyces rusty-scaly and with a few scattered hairs, anthers pubescent. Fruit usually somewhat obovoid, about 2.5–3 (4) cm long, the husk thin, its surface rusty-scaly at least until near maturity, (3) 4-valvate, valves splitting to the base; shell of the nut hard, kernel sweet. [*Hicoria floridana* (Sarg.) Small]

Restricted, for the most part, to sand pine–oak scrub on stabilized dunes and interior sand ridges, endemic to cen. pen. Fla., barely reaching our range in Marion Co.

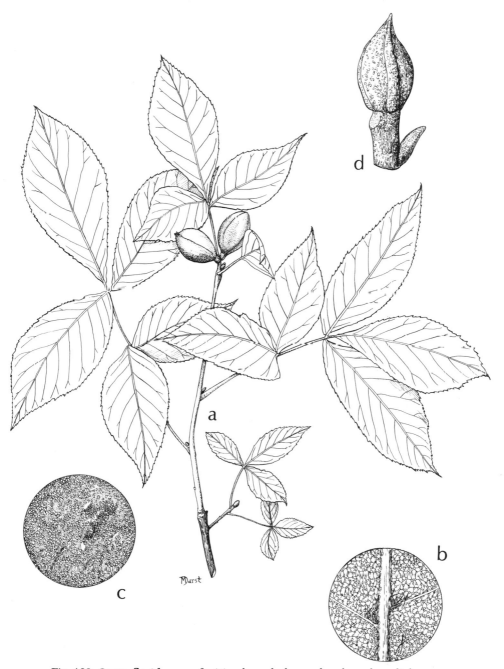

Fig. 189. **Carya floridana:** a, fruiting branch; b, much enlarged medial portion of leaflet showing scaliness and tufts of hairs in vein axils; c, much enlarged small portion of surface of fruit; d, tip of branch with winter bud.

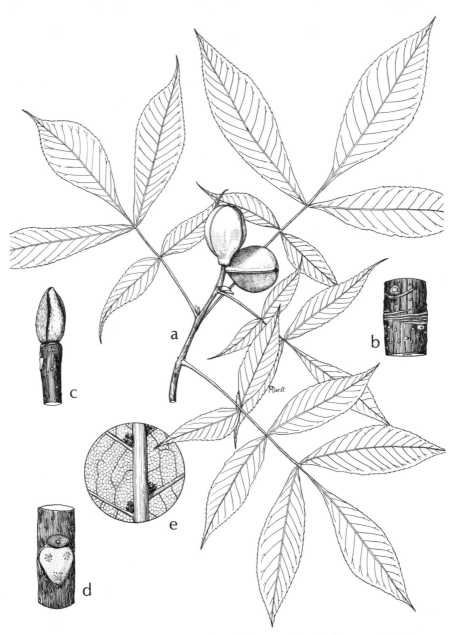

Fig. 190. **Carya glabra:** a, fruiting branch; b, annual node showing relatively narrow bud-scale scars forming a band around the stem; c, tip of stem with winter bud; d, node of twig with leaf scar and axillary bud; e, much enlarged medial portion of lower surface of leaflet.

4. Carya glabra (Mill.) Sweet. PIGNUT HICKORY.   Fig. 190

Variably a small to large tree, larger ones to 40 m tall and to 1.3 m d.b.h. Bark of trunk gray, tight, usually eventually shallowly ridged and furrowed into a diamondlike pattern. Buds brown, with imbricated, very densely and compactly hairy scales. Young, developing stems, petioles, leaf axes, and upper and/or lower surfaces of leaflets often, if not always, with some tiny, sessile or stalked, resinous glands, or scales, these usually sloughed rather quickly but may, rarely, persist on leaf surfaces. Woody twigs usually dark brown; leaf scars lighter brown, shield-shaped to obcordate.

Leaflets (3) 5 or 7, showing much variation in size and shape, lanceolate, elliptic, oblong, lance-oblong, lance-ovate, or ovate, often some of them falcate, upper pair and terminal one (and sometimes the medial pair if leaflet number 7) equal or subequal, bases usually obliquely obtuse, apices usually long-tapered, terminal leaflet usually equilateral and acuminate at both extremities, basally shortly stalked; at maturity overall length, including petioles, 15–40 cm or a little more; upper surface green and glabrous, lower paler green, sometimes with patches of hairs in the axils formed by the midvein and lateral veins, infrequently with tiny resinous glandular dots or scales. Staminate catkins slender, about 5 mm across, variable in length, bracts lance-subulate, glabrous or nearly so *or* shaggy-pubescent with longish hairs, calyces glabrous or nearly so, anthers pubescent. Fruit size and shape variable, often thus from tree to tree where numerous trees grow in close proximity, globose, ellipsoid, obpyriform, or obovoid, 2.5–6 cm long, 2–5-valvate, commonly 3- or 4-, valves splitting tardily and not uniformly, husks relatively thin; shell of the nut thick, kernel sweet to bitter. [*Hicoria glabra* (Mill.) Britt.; incl. *H. austrina* Small]

Inhabiting a wide variety of kinds of sites, well-drained to poorly drained, mesic to xeric mixed woodlands, bottomland woodlands, wet hammocks with shallow soil over limestone; throughout our area. (E. Mass., Vt., to s. Mich., Ill., s.e. Iowa, e. Kan., generally southward to cen. pen. Fla. and e. Tex.)

5. Carya illinoiensis (Wang.) K. Koch. PECAN.   Fig. 191

Potentially a large tree in its native habitats, to 60 m tall and to 2.5 m d.b.h. (trees naturalized from cultivars generally smaller). Bark of trunk grayish or light brown and reddish tinged, eventually dividing irregularly into narrow, forked ridges broken surficially into appressed scales. Buds valvate, at first clothed with small, yellowish or orangish, resinous scales, later thinly to copiously short-pubescent, terminal bud not stalked. Very young shoots and their unfurling leaves with soft, whitish pubescence and copious sessile and short-stipitate, resinous glands. Woody twigs of the season becoming purplish brown and marked by vertically lenticular to narrowly elongate, buff-colored lenticels, usually some scattered, shaggy hairs remaining.

Leaflets 5–17, commonly 9–17, lanceolate to lance-ovate, the lateral ones mostly inequilateral and falcate, sessile or very shortly stalked, bases obliquely rounded to obliquely acute, apices generally long-tapered-acute or -acuminate, usually the median and upper laterals a little larger than the terminal one; at maturity overall length, including petioles, 10–50 cm (largest ones on vigorous saplings or sprouts); upper surfaces dark green, usually glabrous, lower surfaces paler and duller green, glabrous or with scattered, tufted, shaggy hairs mainly on veins, sometimes with minute pale to reddish, resinous surface scales as well;

Fig. 191. **Carya illinoiensis:** a, branch at time of anthesis with staminate catkins from previous season's wood and, at tip of new shoot pistillate flowers; b, mature leaf; c, staminate flower; d, pistillate flower; e, fruits.

leaf axis varyingly glabrous or with scattered, tufted, shaggy pubescence; margins with small, blunt serrations, sometimes serrate only on one margin.

Staminate catkins in sessile fascicles, slender, 12–15 cm long, 5–6 mm across, bracts linear-oblong, bracts and calyces sparsely resinous-scaly, anthers sparsely pubescent. Fruit 3.5–5 cm long, sutures narrowly ridged, ovate, elliptic, or elliptic-oblong in outline, rounded basally, abruptly short-pointed apically; husk relatively thin, covered with small scales, splitting to the base; nut rounded in cross-section, surface smooth, tawny and with mahogany-colored stripes or patches; kernel sweet, each half barely notched apically. [*Hicoria pecan* (Marsh.) Britt.]

The pecan is cultivated for its nuts in our area (native chiefly in the Mississippi River valley) and "wild seedlings" naturalized from cultivars occur sporadically in fence and hedge rows, old fields, vacant lots, and at the borders of upland woodlands.

6. Carya ovata (Mill.) K. Koch. SHAGBARK, SHELLBARK, OR SCALYBARK HICKORY.

Moderately large tree, for the most part to 20–25 m tall and to 1 m d.b.h., sometimes larger. Bark of trunks light to dark gray, eventually broken into loose, thick plates having their extremities curving outward giving a shaggy appearance. Buds with imbricated, appressed- to loose-pubescent scales. In earliest stages of unfurling leaves, petioles, leaf axes, and lower surfaces of leaflets very densely clothed with blond pubescence; on later expanding stages of young stem growth and leaf axes, the pubescence gradually more sparse (as surfaces expand), pubescence of upper leaflet surfaces sparse along the veins, uniformly moderately dense over lower surfaces of leaflets and margins ciliate; in addition, leaf axis and both surfaces of leaflets with sparse to copious, minute, resinous granules, those on the lower surfaces somewhat obscured by the pubescence (pubescence referred to throughout this description tufted). Woody twigs of the season dark reddish brown with raised, buff-colored to orangish lenticels; leaf scars obcordate to 3-lobed.

Leaflets (3) 5 (7), lateral ones essentially sessile, terminal ones with stalks to 2 cm long; at maturity, the uppermost pair often, not always, about equaling the terminal one in length, the latter often, not always, a little broader, lowermost pair considerably smaller than the others, usually ovate, the others broadly elliptic, broadly oblanceolate, or obovate, some usually slightly falcate; bases somewhat inequilateral and obliquely rounded or bluntly tapered, apices obtuse, acute, or acuminate; terminal leaf equilateral, base gradually longish-tapered, apex usually more abruptly narrowed to a point; upper surfaces paler green to olive-green or light brown, sparsely short-pubescent along the major veins, sometimes a few scattered hairs between the veins, surficially with few to numerous, tiny, usually dark red, resinous glandular dots or scales; margins serrate from somewhat above the base, teeth blunt, persistent tufts of hairs on or on and between the teeth.

Staminate catkins 4–10 cm long or a little more, about 5 mm across, bracts subulate, varying from a little longer than to a great deal longer than the calyces, bracts and calyces resinous-granular, anthers pubescent. Fruit subglobose to obovoid, 3–5 cm long, more or less crateriform at the apex, sutures depressed; husk very thick, 10 mm or more, valves splitting to the base; nut 4-ribbed, thin-shelled, kernel sweet. [*Hicoria ovata* (Mill.) Britt.]

Rich, mesic woodlands, stream banks, lower slopes of wooded bluffs or ravines, moist, alluvial, well-drained bottomland woodlands, upland slopes as

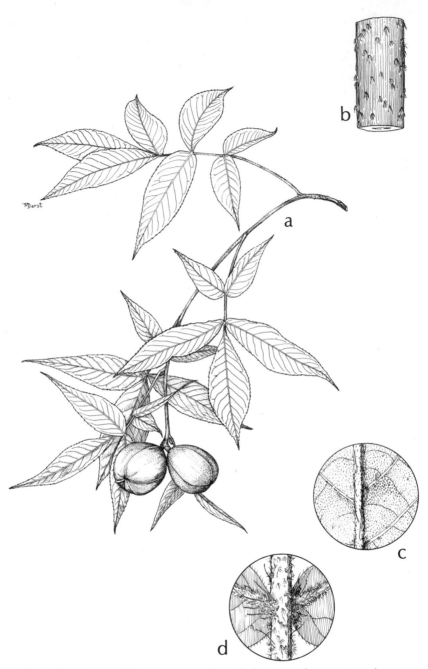

Fig. 192. **Carya pallida:** a, fruiting branch; b, piece of current year's stem showing vestiture; c, much enlarged medial portion of lower surface of leaflet; d, much enlarged bit of leaf rachis and lower surface of bases of leaflets to show vestiture.

well; s.w. Ga., s. Ala. (S. Maine, s.w. Que., s. Ont. to s. Minn.; in the East, generally southward to N.C., s.w. Ga., s. Ala. and s. cen. Miss.; the w. boundary ranging from s. Minn., Iowa, s.e. Neb., to e. Tex. and w. La.)

National champion shagbark hickory, 10'7" circumf., 151' in height, 56' spread, Abbeville, S.C.

7. Carya pallida (Ashe) Engl. & Graebn. SAND OR PALE HICKORY. Fig. 192

Generally a small to medium-sized tree, sometimes (rarely?) to 35 m tall and to 1 m d.b.h. Bark of trunk pale to dark gray, roughish with interlacing ridges and furrows forming more or less a diamondlike pattern. Buds with imbricated scales, their surfaces more or less clothed with amberish, resinous disclike scaliness. Young, developing shoots brown, rusty-scaly and more or less pubescent with some longish, shaggy hairs. Woody twigs reddish or purplish, eventually gray to nearly black; leaf scars shield-shaped, obcordate, or 3-lobed.

Leaflets 3, 5, 7, or 9, commonly 7, lanceolate, elliptic, ovate, oblong-lanceolate, or obovate, some sometimes moderately inequilateral and falcate, the uppermost pair of laterals often as large as or a little larger than the terminal one (the latter often shortly stalked); at maturity overall leaf length, including petioles, 10 25 (30) cm; bases of lateral leaflets mostly rounded, or those of the uppermost pair and the terminal one moderately to strongly cuneate, upper surfaces green and glabrous (brown and with minute, pale scales when very young), lower surfaces more or less olive-green to brownish, permanently with small, disclike, pale, whitish, or reddish scales between the veins, the principal veins (and petioles and leaf axis as well) bearing hairs in tufts; margins finely serrate upwardly from near their bases, teeth blunt-tipped.

Staminate catkins slender, about 4 mm across, bracts subulate and quickly deciduous, calyces and anthers pubescent. Fruit usually somewhat obovoid, 1.5–4 cm long, the husk relatively thin, usually 4-valvate, one valve usually much wider than the other three, valves splitting to the base; shell of the nut thin, kernel small, sweet. [Hicoria pallida Ashe]

In our area generally inhabiting well-drained, sandy, relatively infertile soils of longleaf pine–scrub oak woodlands and mixed woodlands of upper slopes of ravines and outer rims of steepheads; s. cen. Ga., cen. Panhandle of Fla. westward, s. Ala. (S. N.J. to Ky., s.w. Ind. and Ill., generally southward to Fla. Panhandle and e. La.)

National champion sand hickory (1980), 11'1" circumf., 94' in height, 86' spread, Vineland, N.J.

8. Carya tomentosa (Poir. in Lam.) Nutt. MOCKERNUT HICKORY. Fig. 193

Medium-sized to large tree, to 35 m tall and to 1.5 m d.b.h. (perhaps none so large in our area at present). Bark of trunk gray, tight, shallowly and narrowly ridged and furrowed into a diamondlike pattern. Buds with imbricated scales (the terminal bud notably larger than for our other kinds of hickories, about 1.5 cm long and 1 cm across the base); scales light brown, very densely and compactly pubescent. Very young shoots very densely to sparsely pubescent with tufted hairs and with pale amber, sessile glands as well although these obscured if the pubescence is very dense. Woody twigs notably stout, dark gray or dark brown, some pubescence sometimes persistent through the first season although the tuftedness not then apparent; leaf scars prominent, buff-colored, mostly obcordate or 3-lobed with the lower lobe longest.

Leaflets 5, 7, or 9, commonly 7 (the lateral ones longest), lowermost pair usu-

397

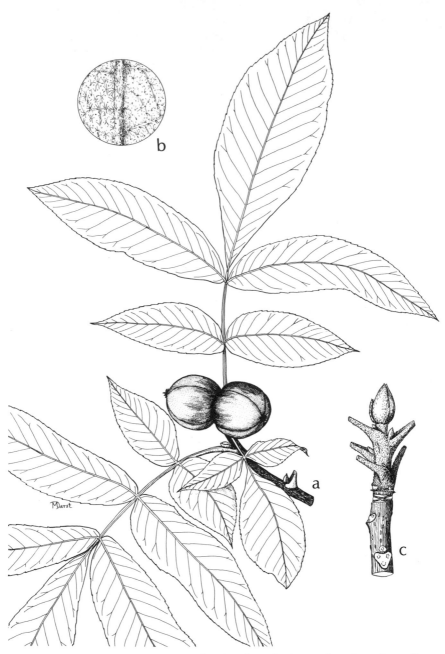

Fig. 193. **Carya tomentosa:** a, fruiting branch; b, much enlarged small medial portion of lower surface of leaflet; c, twig, lower part some of previous year's growth, upper part current year's growth (leaves excised).

ally ovate and much smaller than the others, others lance-oblong, broadly elliptic, or obovate, uppermost pair nearly the same size as the terminal one, mostly
moderately inequilateral, bases of lateral ones obliquely rounded, base of the
terminal one cuneately narrowed to a short stalk, apices long-acute to abruptly
short-acuminate; at maturity overall length, including petioles, 2–4 dm; upper
surfaces of young, unfurling leaflets olive-green, copiously shaggy-pubescent
along the veins, sparsely pubescent and finely gland-dotted between the lateral
veins, hairs tufted, lower surfaces very densely pubescent throughout, hairs
buff-colored; upper surfaces of mature leaflets sparsely pubescent with tufted
short hairs, the lower surfaces with tufted hairs throughout, these much more
numerous than on upper surfaces but not dense, surfaces also dotted with tiny,
granular, amberish or reddish, resinous glands or scales; petiole and leaf axis
permanently tufted hairy; margins generally serrate upwardly from a little
below their middles.

Staminate catkins usually elongate, to 15–20 cm long, 6–8 mm across, bracts
subulate, bracts, calyces, and anthers copiously short-shaggy-pubescent. Fruit
subglobose, oblongish, or slightly obovoid, 4–6 cm long, husk hard and thick,
4-valvate, valves splitting to the base; shell of the nut hard and relatively thick,
kernel sweet.

Inhabiting well-drained, fertile soils of upland woodlands; throughout our
area. (Mass., Vt., N.Y., s. Ont. to s. Mich., Ill., s.e. Iowa, generally southward to n.
cen. pen. Fla. and e. Tex.)

National champion mockernut hickory (1971), 13'5" circumf., 110' in height,
96' spread, Lakeland, Fla. (Species not included in Lakela & Wunderlin, 1980, or
in Wunderlin, 1982.)

Labiatae (Lamiaceae) (MINT FAMILY)

1. Leaves with dense, compact, short pubescence on the lower or on both surfaces; corolla
tube (above the calyx) strongly bent backward then arching forward.   2. *Conradina*
1. Leaves glabrous or with relatively sparse, spreading pubescence; corolla tube not bent
backward.
    2. Leaves and bracts not hiding the flower stalks; fertile stamens 4.   1. *Calamintha*
    2. Leaves and bracts hiding the flower stalks; fertile stamens 2.   3. *Stachydeoma*

1. Calamintha (BASILS)

(Ours) minty-aromatic shrubs. Leaves simple, opposite, sessile or short-petiolate. Flowers solitary in the leaf axils or in short, few-flowered, axillary cymes,
bisexual, bilaterally symmetrical. Calyx tube cylindric or slightly contracted
above the middle then expanded again, limb 2-lipped, aperture of tube with a
ring of inwardly pointing hairs. Corolla tubular below, the limb 2-lipped. Stamens 4, in 2 pairs of somewhat unequal length. Pistil 1, ovary superior, 4-lobed,
style slender, stigma bifid, extending a little beyond the longer stamens. Fruit
splitting into four 1-seeded nutlets.

1. Corolla 3–5 cm long, scarlet (yellow on a rare individual plant).   1. *C. coccinea*
1. Corolla about 1.5 cm long overall, limb lavender-pink.
    2. Leaves sessile.   2. *C. dentata*
    2. Leaves petiolate.   3. *C. georgiana*

Fig. 194. **Calamintha coccinea:** flowering branch, center, and leaf, enlarged, center left; lower right, flower, fruiting calyx above it; nutlets, lower left.

1. Calamintha coccinea (Nutt.) Benth. SCARLET BASIL. Fig. 194

Shrub to about 1 m tall, with relatively few, strongly ascending branches, usually carrying some or all of the leaves on woody portions of the stems through the winter. Youngest stems nearly terete to indefinitely 4-angled, brown and copiously short-pubescent, sometimes atomiferous-glandular; older woody stems with brown, shreddy, pubescent bark; oldest stems grayish brown and with tighter, irregularly fissured bark.

Leaves cuneate below to subpetiolate bases, only the midrib evident on either surface, mostly oblanceolate, spatulate, or obovate, 5–20 mm long and to about 10 mm broad, mostly narrower, glabrous or sometimes very short-pubescent beneath, both surfaces glandular-punctate, the upper more conspicuously so than the lower when fresh, margins entire, edges rolled downward.

Flowers borne singly in the leaf axils distally on the branches, often only a few at anthesis at one time; flower stalks 3–4 mm long, copiously very short-pubescent, with a basal elliptic-oblong bract about half as long as to as long as the stalk. Calyx about 1 cm long, minutely pubescent and sparsely atomiferous-glandular, strongly ribbed, slightly contracted above the middle then expanded again, the limb 2 lipped, upper lip with 3 short-triangular, acuminate lobes, the lower of 2 longer, subulate lobes. Corolla showy, scarlet within and without (yellow on a rare individual plant), its funnelform tube 3–5 cm long, sparsely pubescent exteriorly, lower lip of the limb yellowish basally-centrally and speckled scarlet, with 3 oblongish lobes, upper lip oblongish, shallowly to conspicuously notched apically. Nutlets subglobose, dull brown, about 1 mm in diameter. [*Clinopodium coccineum* (Nutt.) Kuntze.]

Open, well-drained pineland ridges and hills; Fla. Panhandle from Wakulla Co. westward, s.w. Ala. (Cen. s.e. Ga.; pen. Fla., w. Fla. Panhandle to s.e. Miss.)

2. Calamintha dentata Chapm. Fig. 195

Low, suffruticose, strongly aromatic, slenderly branched, at least partially evergreen shrub, the woody lower portions of older plants often gnarled or twisted; herbaceous main branches of the season strongly erect, their secondary branches angled-ascending, generally reaching to about 5 dm tall by the end of the blooming season; branchlets with very finely stipitate-glandular, translucent pubescence; bark of old stems tan, exfoliating in irregular strips.

Leaves sessile, both surfaces conspicuously glandular-punctate and with translucent, very finely stipitate-glandular pubescence (glandular tips of hairs easily detached); lower leaves cuneate-obovate, 8–10 mm long and 7–8 mm broad at their broadest places, irregularly with 1, 2, or 3 dentate, obtuse teeth on their distal margins (teeth rather obscure to the unaided eye); upwardly leaves gradually becoming a little smaller, oblanceolate to spatulate, margins mostly without teeth.

Flowers borne in sessile, leafy-bracted cymes axillary to median and upper leaves, flowers of a given cyme not all developing or at anthesis simultaneously; stalks about 5 mm long fully developed. Calyx tube turbinate-cylindric, 5 mm long, strongly ribbed, atomiferous-glandular between the ribs and with minute glandular pubescence especially on the ribs and lobes at the summit; upper 3 calyx lobes broadly triangular-obtuse, slightly flared, the lower 2 lobes attenuate, incurved, with strongly white-ciliate margins; calyx more or less suffused with purplish red pigment, especially on the lobes interiorly. Corolla tube whitish and with spreading white pubescence on its distal half, at the orifice of

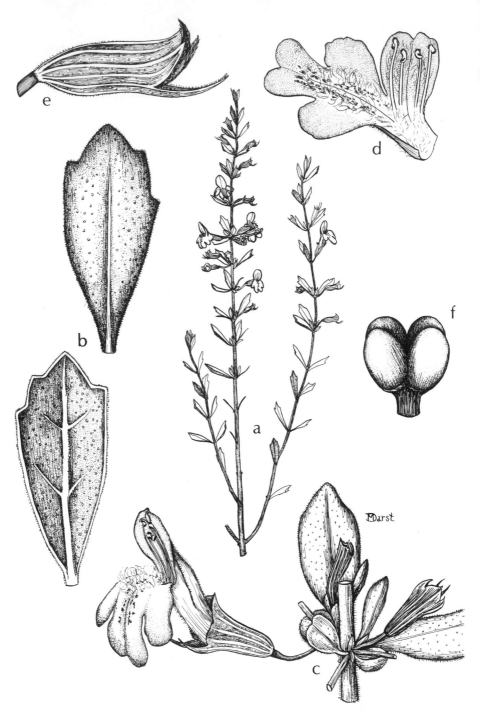

Fig. 195. **Calamintha dentata:** a, flowering branch; b, enlargement of leaf surfaces, upper surface above and lower surface below; c, node with a flower at anthesis; d, corolla, opened out; e, fruiting calyx; f, nutlets.

402

the calyx abruptly expanding and markedly 2-lipped, the upper lip a single short-oblong, angled-ascending lobe, uniformly lavender-pink, pubescent medially, the lower lip arching, having 2 rounded lateral lobes uniformly lavender-pink, a terminal lobe broadly parallel-sided basally, flaring distally, emarginate apically, lavender-pink, centrally the lower lip whitish with purple speckles. Nutlets globular, light brown, about 0.5 mm in diameter. [*Clinopodium dentatum* (Chapm.) Kuntze]

Occurring locally on longleaf pine–scrub oak sand ridges, sometimes in open stands of slash pine or sand pine, also in open sandy woodlands developed on former cultivated fields, restricted to cen. portion of Fla. Panhandle.

3. Calamintha georgiana (Harper) Shinners. Fig. 196

Low, slenderly branched, pleasantly aromatic, deciduous shrub, branches and branchlets commonly spreading laterally. Branchlets at first green, becoming light brown, abundantly clothed with very short, curvate hairs; bark of older stems tan, exfoliating in irregular, thin strips or scales.

Leaves petiolate, blades of principal ones 1–3.5 cm long, ovate, lance-ovate, or oval, sometimes subrotund, shortly tapered basally, blunt apically, margins inconspicuously, often irregularly bluntly toothed, or entire; upper surfaces often, not always, sparsely short-pubescent, lower surfaces pubescent along their midribs; both surfaces relatively indistinctly punctate. Leaves of short axillary branchlets in leaf axils or on short branchlets at leafless nodes very much smaller than the principal stem leaves and appearing somewhat fascicled.

Flowers usually 3 in a cyme in the axil of each of pairs of leaves distally on the branchlets, the flower stalks stoutish, about 2 mm long, the stalk of the central flower without a subtending bract and without a pair of bractlets at the base of the stalk, each of the lateral two flowers subtended by a foliose, narrowly oblanceolate bract about equaling the calyx and with a pair of linear-oblong bractlets at the bases of their stalks, the bractlets as long as the flower stalks or slightly longer; flowers of a given cyme not at anthesis simultaneously. Calyx tube nearly cylindrical, about 4 mm long, strongly ribbed, finely and sparsely atomiferous-glandular; limb of the calyx 2-lipped, the upper lip broadly based and with 2 short, nearly deltoid lobes, the lower lip consisting of 2 longer, setaceous lobes. Corolla 1.5 cm long overall, narrowly cylindric and whitish within the calyx, beyond the latter flaring-funnelform, whitish proximally, pale lavender-pink distally; corolla limb 2-lipped, the upper lip broadly oblong, notched apically, relatively bright lavender-pink, lower lip with 3 subequal, oblongish lobes, the central one notched, the lip below the central lobe brightly purple-speckled; exterior of that portion of the corolla beyond the calyx pubescent Nutlets dull light brown, subrotund, about 1 mm in diameter. [*Clinopodium georgianum* Harper; *Satureja georgiana* (Harper) Ahles]

Xeric, sandy or rocky, semiopen woodlands, wooded, ancient levees along rivers and streams; s.w. Ga., Holmes Co., Fla. Panhandle, s. Ala. (S.e. and s. cen. N.C. to s.w. Ga., cen. Panhandle of Fla., thence westward to La.)

2. Conradina (MINTY ROSEMARYS)

Evergreen, bushy-branched, strongly minty-aromatic shrubs to about 8 dm tall. Twigs 4-sided. Leaves simple, opposite, sessile, linear-oblong to linear-oblanceolate, rounded apically, margins strongly revolute, surfaces glandular-punctate; leaves often with leafy, short branchlets in their axils, leaves of the branchlets

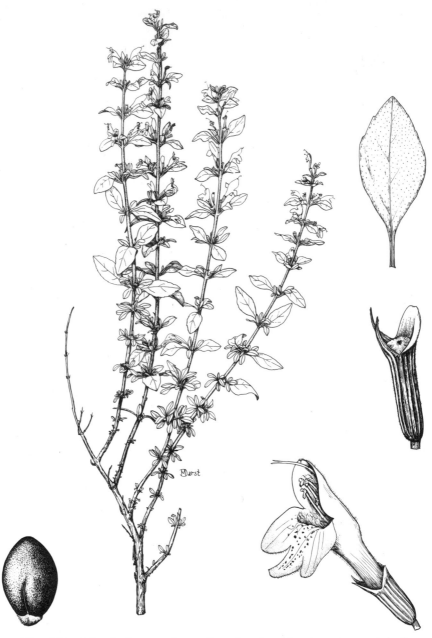

Fig. 196. **Calamintha georgiana:** flowering branch at center, and leaf, enlarged, to its right; flower at lower right; fruiting calyx, right center; a nutlet at lower left.

shorter than those subtending them; leaves of primary branches 10–15 mm long and 1–2 mm broad. Flowers in reduced cymes axillary to leaves on shoots of the season, each stalked, each stalk subtended by a linear bract and with a pair of bractlets basally on the stalk, bisexual, bilaterally symmetrical. Calyx tube cylindric or narrowly funnelform, longitudinally ribbed, glandular-punctate, the lobed summit 2-lipped, lower lip with 2 erect, or curved, subulate lobes, upper lip with 3 spreading, shorter, deltoid lobes, all lobes ciliate marginally, the throat of the calyx with a ring of cilia within that more or less closes the throat. Corolla rather showy, the ground color varying from nearly white to lavender, lavender-pink, or nearly purple; corolla slenderly tubular within the calyx, expanded-tubular above the calyx where bent backward then arching forward, 2-lipped above the throat, the upper lip unlobed and somewhat obovate, the lower lip with a broad, purple-speckled, central portion bearing 2 conspicuous lateral lobes proximally and 1 distally; lowermost portion of the corolla tube pubescent without and within, expanded exterior portion of the tube and the lobes variously glabrous or pubescent. Stamens 4 in 2 somewhat unequal pairs, the anthers of the longer ones reaching the summit of the upper lip of the corolla, each of the pair of filaments arching in a groove of the upper lip. Pistil 1, ovary superior, very short, seated at the very base of the flower, with 4 nipplelike lobes distally; style very slender, stigma bifid, reaching a little beyond the uppermost anthers. Fruit splitting into 4 globular, brown nutlets about 1 mm in diameter.

1. Calyx tube notably pubescent with dense, short hairs and usually numerous, long spreading ones; both leaf surfaces densely short-pubescent and gray.   1. *C. canescens*
1. Calyx tube glabrous or with sparse, short, inconspicuous pubescence; upper leaf surface bright green and sparsely and inconspicuously pubescent or glabrous, the lower gray-pubescent.   2. *C. glabra*

1. **Conradina canescens** (T. & G.) Gray.   Fig. 197

Leaves narrowly linear-oblong, the larger ones mostly not exceeding 10 mm long, conspicuously, densely, short gray-pubescent on both surfaces, the glandular punctae obscured by the pubescence.

Cymes usually 5–1-flowered, some of a given cyme in early bud stage when others are at anthesis or beyond, still others in intermediate stages of development; flower stalks 1 mm long or slightly longer. Calyx about 5 mm long, densely and copiously short-pubescent and usually with numerous, long, spreading hairs as well. Central portion of the lower lip of the corolla usually sparsely purple-speckled. Stamen pairs subequal in length. [Incl. *Conradina puberula* Small]

Sandy soils, stabilized dunes, sand pine–scrub oak ridges, open longleaf (or planted slash) pine ridges. Fla. Panhandle from Liberty and Franklin Cos. westward, common near the coast, less frequent inland, s.e. and s.w. Ala., s.e. Miss.

2. **Conradina glabra** Shinners.   Fig. 197

Leaves narrowly linear-oblong to narrowly oblanceolate, 10–15 mm long, upper surfaces bright green and glabrous, sometimes sparsely and inconspicuously short-pubescent, conspicuously glandular-punctate, the lower usually very compactly gray-pubescent, the punctae evident but much less conspicuous than on the upper surface.

Cymes usually with 3–4 flowers, sometimes only 1 or 2, flower stalks 2–3 mm

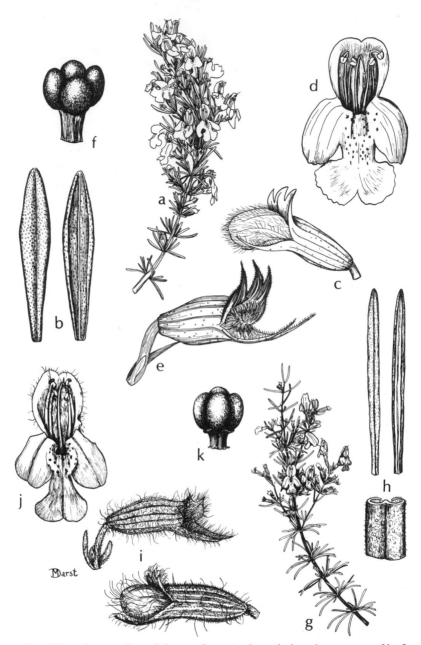

Fig. 197. a–f, **Conradina glabra:** a, flowering branch; b, enlargements of leaf, upper surface on left, lower surface to right; c, flower, just before anthesis; d, flower, face view; e, fruiting calyx; f, nutlets. g–k, **Conradina canescens:** g, flowering branch; h, above, enlargement of leaf, upper surface to left, lower to right, and below, enlargement of medial portion of leaf, adaxial view; i, below, flower, just before anthesis, and above, fruiting calyx; j, flower, face view; k, nutlets.

long. Calyx about 8 mm long, sparsely and inconspicuously short-pubescent, or glabrous save for the ciliate margins of the lobes. Stamen pairs of unequal length.

Longleaf (or planted slash) pine sand ridges, rarely on upper wooded slopes of steepheads, most common and abundant somewhat northward of Bristol, Liberty Co., Fla.; occurring also in ecotones between pinelands and *Chamaecyparis* stands more or less lining the banks of streams in the Blackwater State Forest, Santa Rosa Co., Fla., apparently not in the intervening area.

The two rosemary species treated above are unlike other mint-odored shrubs of our area in having dense, compact, gray pubescence on one or both leaf surfaces; in others the leaves are glabrous or have spreading hairs. The strongly bent-up corollas are peculiar to them as well.

The "true" rosemary of Europe, *Rosmarinus officinalis* L., is the one commercially produced and used as a flavoring herb; ours have very similar properties and perhaps could be a satisfactory substitute.

3. Stachydeoma

Stachydeoma graveolens (Chapm. ex Gray) Small. MOCK-PENNYROYAL.

Fig. 198

An aromatic, suffruticose, low plant, sometimes with few, slender, often straggly, stiffish branches from the base, often with very numerous, slender, stiffly ascending branches forming a compact, more or less rotund crown, the leaves and bracts usually dull purplish-tinted distally on the branches giving an overall pleasing hue. Older woody stem brown, younger herbaceous ones pale or greenish brown, both pubescent with relatively long, shaggy hairs and short, flat, scalelike hairs intermixed with minute gland-tipped hairs.

Leaves numerous, opposite, sessile, the pairs not overlapping on the leafy proximal portions of the stems, but usually overlapping on the floriferous, distal portions, with few, evident pinnate veins, these impressed above, raised beneath; blades ovate, oblong-ovate, or oblong, 1 cm long or less, to 5–6 mm broad, thickish-textured, bases subcordate, rounded, or shortly tapered, largest leaves sometimes obscurely and bluntly few-toothed, others untoothed, sides of the blades rolled downward, margins (of fresh leaves) not revolute; surfaces with amber (becoming dark brown with age) atomiferous glands and minute, translucent, stalked glands upwardly on the stem, the upper surfaces both of leaves and foliose floral bracts with numerous, long translucent hairs as well, particularly near and at the margins.

Flowers usually borne singly from the leaf axils, their stalks stoutish, short, each bearing basally a pair of bracts essentially like, but smaller than, the subtending leaves, the leaves and bracts hiding the flower stalks; flower stalks and calyces pubescent like the leaves and bracts although the atomiferous glands fewer. Calyx tube cylindric-campanulate, 2–2.5 mm long, strongly 10-ribbed, limb 2-lipped, upper lip short-oblong basally and with 3 short-deltoid, erect lobes, the lower lip consisting of 2 slightly arching, subulate lobes, their length about equal to that of the whole upper lip, a ring of inwardly spreading hairs within the orifice. Corolla about 1 cm long, tube cylindrical, nearly white, expanding at about the tips of the calyx lobes to a short throat and a markedly 2-lipped limb, the upper lip roseate, arching forward, oblongish, the sides declined, tip rounded, lower lip spreading outward or bent slightly downward, medially with a yellowish white band mottled with purple, 3-lobed, medial lobe

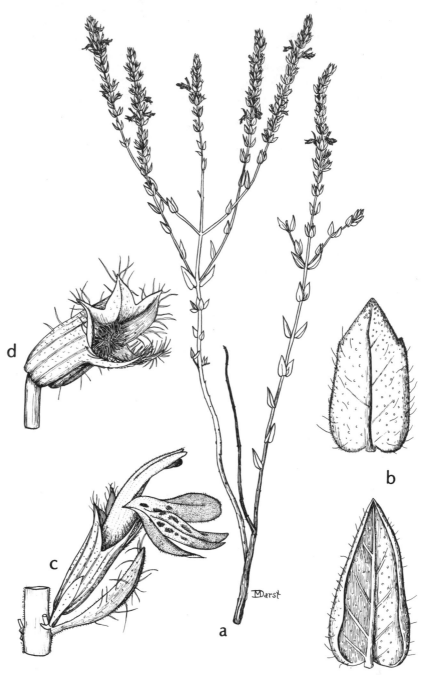

Fig. 198. **Stachydeoma graveolens:** a, flowering branch; b, leaf, much enlarged, upper surface above and lower surface below; c, flower, side view; d, fruiting calyx.

largest, itself 2-lobed, lobes rose-purple. Fertile stamens 2, attached to the corolla tube just below the sinuses of the 2 corolla lips, arching beneath the upper lip of the corolla and extending just beyond it. Pistil 1, deeply 4-lobed above a broad disc, style slender, arching beneath the upper corolla lip, unequally bifid at the tip. Fruit splitting into 4 subrotund, dull brown nutlets about 1 mm in diameter. [*Hedeoma graveolens* Chapm. ex Gray]

Low sand ridges, slightly elevated places in pine flatwoods, slopes from ridges or flatwoods nearly to the edges of shrub-tree bogs or bays; Fla. Panhandle from Leon and Wakulla Cos. westward to Bay Co.

Lauraceae (LAUREL FAMILY)

1. Leaves evergreen; flowers bisexual.
    2. Leaf blades 3-nerved, having 2 major laterals ascending from the midrib somewhat above the base.   1. *Cinnamomum*
    2. Leaf blades not at all 3-nerved.   4. *Persea*
1. Leaves deciduous; flowers unisexual.
    3. Leaf blades narrowly elliptic or oblong, 1–3 cm long and 5–10 mm broad. 3. *Litsea*
    3. Leaf blades variously shaped, most of them very much larger than the above.
        4. The leaf blades, for the most part, with the lowest lateral veins ascending or arched-ascending, relatively more pronounced than the other laterals, thus tending to be 3-nerved; most individual plants with some of the leaves, at least, markedly lobed.
                                                                                    5. *Sassafras*
        4. The leaf blades not at all 3-nerved, never lobed.   2. *Lindera*

1. Cinnamomum

Cinnamomum camphora (L.) Nees & Eberm. CAMPHOR-TREE.   Fig. 199

Native to eastern Asia, the evergreen camphor tree is widely planted in our area as an ornamental and shade tree and is commonly naturalized. Usually its potentially large main trunk is short and gives rise to several ascending secondary trunks, the crown supporting dense handsome foliage. Twigs green or green suffused with red. Cut stems and bruised leaves highly aromatic with the odor of camphor. All vegetative parts glabrous.

Leaves simple, alternate, slenderly petioled, petioles shorter than the blades; stipules none. Blades somewhat leathery at maturity, mostly ovate, 4–10 cm long and 2–5 cm broad, pinnately veined, bases cuneate to rounded, apices acuminate, margins entire and with an opaque-cartilaginous edge, the blade itself somewhat curled toward the margins and appearing wavy; axils formed by the 2 or 3 lowermost lateral veins and the midrib having impressed glands beneath and opaque-cartilaginous callosities above; upper surfaces bright green and lustrous, the lower dull and slightly grayish green, not glaucous.

Inflorescences few-flowered, slender-stalked panicles on branchlets of the season, those panicles proximally on the shoots axillary to scars of early-deciduous bracts, those distally on the shoots axillary to and shorter than the leaves. Panicle branches and the short flower stalks subtended by small linear bracts that are deciduous before full anthesis. Flowers very small, bisexual, radially symmetrical, cream-colored, with a small cuplike floral tube bearing 6 spreading perianth parts, 3 each in 2 series, on its rim, each segment about 1.5 mm long, ovate, pubescent interiorly. Fertile stamens 9, in 3 series, the bases of the fila-

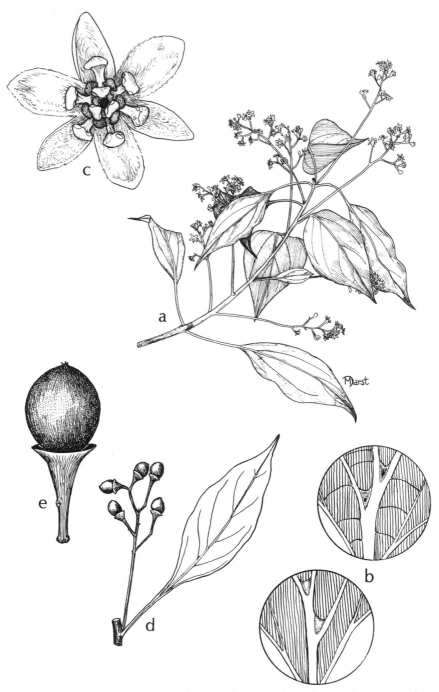

Fig. 199. **Cinnamomum camphora:** a, flowering branch; b, enlargement of proximal medial portion of leaf surfaces to show callosities in the vein axils; c, flower, face view; d, infructescence; e, fruit.

ments of the innermost series flanked by a pair of very shortly stalked, irregularly knoblike, orange-yellow glands; filaments short, anthers 4-locular, the 2 locules of each half-anther dehiscing by a flap-valve. Pistil 1, ovary 1-locular, l-ovuled, style short, stigma capitate. Fruit a black subglobose drupe about 8 or 9 mm in diameter, the persistent leathery floral tube loosely funnellike about its base. [*Camphora camphora* (L.) Karst.]

Throughout our area, naturalized in waste places, vacant lots, various mixtures of upland woodlands, fence and hedge rows.

National champion camphor tree (1977), 30'8" circumf., 72' in height, 102' spread, Hardee Co., Fla.

2. Lindera

Deciduous, aromatic shrubs to 6 m tall. Leaves simple, alternate, short-petiolate, blades pinnately veined, margins entire; stipules none. Flowers unisexual (plants usually dioecious), borne in sessile, roundish clusters of 1–4 per inflorescence, each cluster with 2–6 short-stalked flowers. Perianth segments 6, 3 similar ones in each of 2 series. Staminate flowers "typically" with 9 stamens in 3 series, each of the inner 3 with a pair of conspicuous glands basally (sometimes 10 or 11 stamens, as many as 4 with basal glands), anthers 2-locular, each anther half dehiscing by a terminally hinged flap-valve leaving the whole of each side a pore. Pistillate flowers with variously reduced nonfunctional stamens, pistil 1, ovary superior, 1-locular, style slender, stigma obliquely terminal. Fruit a lustrous, bright red, ellipsoid drupe.

1. Leaf blades, many of them on a given plant at least, obovate, their bases mostly tapered; venation on lower blade surfaces indistinctly reticulate; crushed leaves with a spicy aroma but not that of sassafras.   1. *L. benzoin*
1. Leaf blades narrowly elliptic-oval or oblong, bases rounded; venation on lower blade surfaces distinctly reticulate; crushed leaves with their own peculiar odor.

2. *L. melissaefolia*

1. Lindera benzoin (L.) Blume. SPICEBUSH.   Fig. 200

Shrub or infrequently a small understory tree with spicy aromatic twigs and leaves. Stems of the first year branchlets pubescent, older twigs becoming glabrous, grayish to brown, bearing slightly raised, paler lenticels; leaf scars crescent-shaped to hemispheric or obcordate, each with 3, often indistinct, vascular bundle scars; during the latter part of the season on first-year twigs a very short, essentially sessile, supra-axillary branch develops above some of the axillary buds, this in the winter condition (if perfectly formed) bears a central, smaller, and 1–2 lateral, larger buds; in early spring the laterals developing into sessile inflorescences, the central one into a leafy branchlet either during or after anthesis.

Petioles usually pubescent; blades, many of them on a given branch, obovate, varying to oval or elliptic, those proximally on a branchlet smaller than distal ones; larger blades 6–12 cm long and 3–5 cm broad, bases acuminately, acutely, or obtusely tapered, apices similarly tapered, or less frequently rounded or emarginate; upper surfaces glabrous and dark green, the lower paler, grayish green, pubescent along the midrib varying to uniformly pubescent, lowest 2 pairs of lateral veins diverging from the midrib and ascending approximately parallel to each other, venation on the lower surface indistinctly reticulate.

Flowers yellow, the staminate clusters larger than the pistillate and more

411

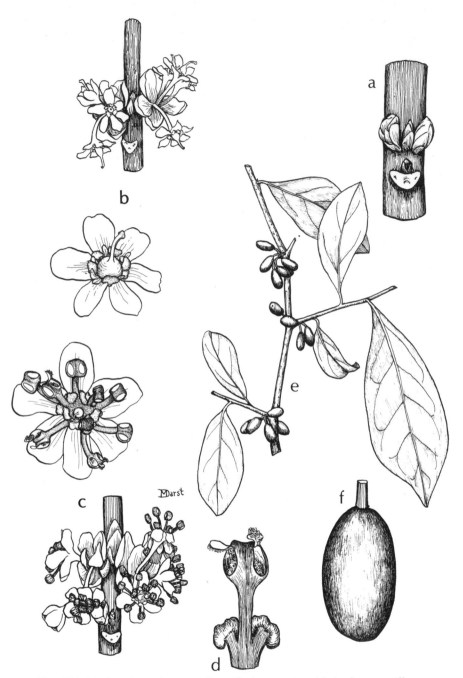

Fig. 200. **Lindera benzoin:** a, a piece of winter twig with leaf scar, axillary bud, and supra-axillary branch buds; b, pistillate flower cluster, above, and pistillate flower, below; c, staminate flower cluster, below, and staminate flower, above; d, stamen; e, portion of fruiting branch; f, fruit.

easily detected at a little distance. Stalks of the fruits slender, 2–4 mm long, little enlarged at the summit. Drupe ellipsoid, 8–10 mm long, its thin flesh spicy-aromatic. [*Benzoin aestivale* (L.) Nees]

Moist wooded banks of streams, alluvial woodlands, mesic wooded slopes, s.w. Ga., adjacent Fla. Panhandle, s.e. and s. cen. Ala., n.e. Fla. (S.w. Maine, s. Ont., s. Mich. to e. Kan., generally southward to e. Fla., Fla. Panhandle, and Tex.)

2. Lindera melissaefolia (Walt.) Blume. JOVE'S-FRUIT, PONDBERRY.

Shrub to 2 m tall, its potential height considerably less than that of *L. benzoin*. Young stems sparsely shaggy-pubescent, soon becoming glabrous. Leaves slightly drooping, crushed ones with their own peculiar odor; petioles short, shaggy-pubescent, blades membranous, elliptic-oblong, oval, or lance-ovate, 5–10 (16) cm long and 1.5–3.5 cm broad, bases mostly rounded to shortly and broadly tapered, apices acute or acuminate, both surfaces pubescent, the lower sometimes densely so; lowest 2 pairs of lateral veins diverging from the midrib and not ascending parallel to each other, venation of lower surfaces distinctly reticulate; both leaf surfaces about the same color. Drupe ellipsoid, 10–11.5 mm long, stalks of the fruit 9–12 mm long, erect, enlarged at the summit. [*Benzoin melissaefolium* (Walt.) Nees]

Sandy sinks, swampy depressions, pond margins, open bogs, rare and local, coastal plain, s.e. N.C., S.C., Ga., Fla. Panhandle, s.w. Ga., s.w. Ala., s.e. Mo. Apparently not collected in the Fla. Panhandle in well over a century.

Reference: Steyermark, Julian A. 1949. "*Lindera melissaefolia*." *Rhodora* 51:153–62.

3. Litsea

Litsea aestivalis (L.) Fern. POND-SPICE.   Fig. 201

Deciduous, much-branched shrub to about 3 m tall, sometimes appearing arborescent, the branchlets, some of them at least, conspicuously zigzagging.

Leaves simple, alternate, with slender, very short petioles; stipules none. Blades pinnately veined, narrowly elliptic, lanceolate, or narrowly oblong, stiff-membranous, 1–3 cm long and 5–10 mm broad, glabrous, mostly obtuse basally and obtuse to acute apically, margins entire.

Flowers unisexual (plants dioecious), borne in few-flowered subumbels axillary to leaf scars prior to new shoot emergence. Perianth yellow, 6–parted, segments nearly free to the base, not persistent. Staminate flowers with 9 or 12 fertile stamens in 3 or 4 series, those of series 3 (and 4, if present) with a pair of stipitate glands basally; each anther-half dehiscing by a terminally hinged flap-valve; pistillate flowers with 9 or 12 staminodia, those of series 3 (and 4, if present) with a pair of glands basally, pistil 1, ovary superior, attenuated into the style, stigma dilated. Fruit a red globose drupe 4–6 mm in diameter. [*Glabraria geniculata* (Walt.) Britt. in Britt. & Brown]

Pond and swamp margins, or in shallow pineland ponds, relatively rare and local. (Coastal plain, s.e. Va. to n. Fla., w. to La.; Tenn.)

4. Persea

Aromatic evergreen shrubs or trees. Leaves simple, alternate, short-petioled, blades leathery, pinnately veined, pubescent at least beneath, margins entire;

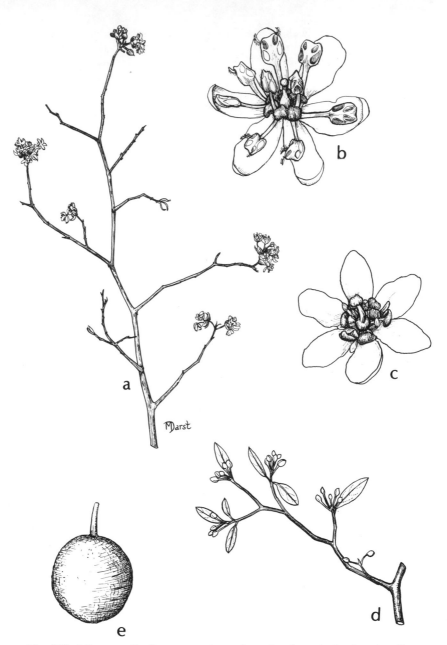

Fig. 201. **Litsea aestivalis:** a, staminate flowering branch; b, functionally staminate flower; c, functionally pistillate flower; d, branch with flower buds (taken from plant in November, buds remain dormant until spring); e, fruit.

stipules none. In general, many leaves on plants of *Persea* have conspicuous insect galls.

Inflorescences stalked, small cymes borne on branchlets of the season, those proximally on the branchlets subtended by small scales, those distally on the branchlets axillary to leaves. Flowers small, bisexual, radially symmetrical, with a short cuplike floral tube bearing 6 perianth segments on its rim, 3 in each of 2 series, all small, the outer shorter than the inner. Fertile stamens 9, three in each of 3 series, filaments pubescent, those of series 3 each flanked by a pair of glands, a fourth series is composed of 3 small staminodia. Pistil 1, ovary 1–locular, 1–ovulate. Fruit a globose to ellipsoid, blue-black to black drupe (ours), the persistent floral tube and persistent perianth segments loosely saucerlike beneath it; perianth segments at this stage essentially woody and usually differently shaped than at anthesis.

1. Pubescence of lower surfaces of mature leaf blades very short and closely appressed, not evident to the unaided eye, with suitable magnification usually appearing as very tiny, glistening-gold flecks.   1. *P. borbonia*
1. Pubescence of lower surfaces of leaf blades uniformly dense over the entire surface, very compact, silky-smooth to the touch, shining chestnut-brown during most of the first season.   2. *P. humilis*
1. Pubescence of lower surfaces of mature leaves relatively long and shaggy, dull brown, moderately abundant on the principal veins, more sparse over the remainder of the surface, much of it evident to the unaided eye.   3. *P. palustris*

1. Persea borbonia (L.) Spreng. RED BAY.   Fig. 202

Usually a medium-sized tree, commonly shrubby, occasional old trees to 1 m d.b.h. and with large crowns. Young twigs sparsely pubescent with short, appressed, glistening gold-colored hairs, appearing glabrous to the unaided eye. Bark of larger trunks brownish gray, roughened by more or less vertical interlacing fissures.

Leaf blades mostly short- to long-elliptic or elliptic-oblong, smaller ones sometimes lanceolate or oblanceolate, varying in size on a given plant, varying not a little respecting length-width ratio from plant to plant; blade lengths varying from 2–15 cm, widths from about 1.5–6 cm; bases and apices vary from rounded to acute or acuminate. Mature leaf blades glabrous above, with very short, appressed glistening gold-colored hairs beneath, these not evident to the unaided eye.

Inflorescence axes, flower stalks, floral tubes, and both exterior and interior of perianth segments with copious, appressed, short, brown pubescence, that of the axes mostly soon sloughing. Outer persisting perianth segments usually shorter than the inner, apices of both very broadly rounded, sometimes nearly truncate. [*Tamala borbonia* (L.) Pax; incl. *T. littoralis* Small]

Mesic woodlands to xeric sandy hammocks, stabilized coastal dune-scrub, rarely in really wet places, throughout our area. (Coastal plain, s. Del. to s. pen. Fla., westward to s.e. Tex.)

National champion red bay (1972), 13′8″ circumf., 58′ in height, 68′ spread, Randolph Co., Ga.

2. Persea humilis Nash. SILK BAY.   Fig. 202

A shrub or small tree. Young twigs with copious, short, somewhat appressed, chestnut-brown pubescence, that remaining on year-old woody twigs grayish, the twigs, however, appearing dark brown or blackish.

Fig. 202. a–c, **Persea humilis:** a, fruiting branch and at right enlargement of small portion of lower leaf surface; b, piece of twig to show pubescence; c, fruit. d–f, **Persea palustris:** d, fruiting branch and at left enlargement of a small portion of lower leaf surface; e, piece of twig to show pubescence; f, fruit. g–h, **Persea borbonia:** g, leaf and at right very much enlarged portion of lower leaf surface; h, piece of twig to show pubescence.

Leaf blades lanceolate to narrowly elliptic, infrequently broadly elliptic, sometimes oblanceolate, 3–8 cm long, 1–3 cm broad, bases acutely tapered, apices tapered but the extreme tips usually blunt; upper surfaces glabrous or with sparse, indistinct pubescence along the midrib, lower surfaces with dense, compact, silky-smooth, chestnut-brown pubescence, this eventually becoming dull and sooty-grayish or -brownish, scarcely evident to the unaided eye, much of it sloughing before the leaves finally fall.

Inflorescence axes, flower stalks, floral tubes, and perianth segments pubescent like the lower leaf surfaces. Outer perianth segments broadly ovate and broadly obtuse to rounded apically, a little more than 1 mm long, the inner twice to thrice as long, ovate to oblong, obtuse apically. Drupes globose, 1–1.5 cm in diameter. [*Tamala humilis* (Nash) Small]

Chiefly inhabiting sand pine–oak scrub, cen. pen. Fla., probably not quite reaching northward into our area, but close to it and included here for that reason.

3. Persea palustris (Raf.) Sarg. SWAMP RED BAY.   Fig. 202

In general habital features closely resembling *P. borbonia*. Young stems densely shaggy-pubescent; year-old woody stems nearly glabrous, dark brown.

Leaf blades frequently somewhat falcate, sometimes notably so, narrowly long-elliptic or long-oblong-elliptic, sometimes moderately broadly elliptic and only about twice as long as broad, bases generally acutely tapered, less frequently broadly short-tapered, apices varyingly acute, acuminate, or obtuse, sometimes the extremities rounded; upper surfaces of mature blades glabrous or shaggy short-pubescent along the midrib, lower surfaces with longish shaggy pubescence, this longer and more abundant along the midrib and major lateral veins, but the entire lower surface at least sparsely shaggy-pubescent; petioles usually densely shaggy-pubescent.

Inflorescence axes, floral tubes, and perianth segments densely pubescent, the pubescence of the inflorescence stalks loose and shaggy, that of the other parts more compact and appressed. Outer perianth segments broadly ovate-obtuse, about 2 mm long, the inner oblong-elliptic, twice as long as the outer. Drupes short-ellipsoid to subglobose, about 1 cm long. At the ripe-fruiting stage, the persisting outer perianth segments are broadly ovate and broadly obtuse apically, somewhat longer than the inner ones which are very broadly rounded. [*Tamala pubescens* (Pursh) Small]

Swamps, wet woodlands, wet pine flatwoods and savannas, branch bays, banks of drainage canals and sloughs, in or at the edges of marshes, throughout our area. (Coastal plain, e. Va. to s. pen. Fla., westward to s.e. Tex.)

5. Sassafras

Sassafras albidum (Nutt.) Nees. SASSAFRAS.   Fig. 203

Potentially a tree to 30 m tall, flowering and fruiting from the time it is of shrub stature, often forming shrubby thickets in open places, all parts aromatic. Twigs at first yellowish green, often copiously short-pubescent, later glabrous and reddish brown and with scattered dark lenticels; leaf scars crescent- to half-moon-shaped, each with a single transverse vascular bundle scar; pith white, continuous; bark of large, older trunks thick, roughly fissured and ridged.

Leaves deciduous, simple, alternate, short-petiolate; stipules none; blades unlobed, lobed right- or left-handed mittenlike, or 3-lobed (rarely 5-lobed), all

Fig. 203. **Sassafras albidum:** a, fruiting branch; b, node with leaf scar and axillary bud; c, variable leaves; d, flowering branch; e, flowers, staminate below, functionally pistillate above; f, fruit.

three types often on a single plant, but plants with all unlobed leaves occasional, edges entire; unlobed blades elliptic, lanceolate, ovate, or obovate, bases mostly broadly short-tapered, apices obtuse to rounded, less frequently acute; lobed leaves tending to an overall obovate outline, bases acutely or acuminately tapered, apices of the lobes mostly blunt but often mucronate; pubescence very variable, emerging branchlets, petioles, and lower blade surfaces commonly densely velvety- or silky-pubescent, upper surfaces moderately pubescent; upper surfaces of mature blades vary from glabrous to moderately pubescent, the lower from glabrous, often glaucous, to permanently velvety-pubescent.

Inflorescences clusters of short racemes at the tips of twigs either before or as new shoots emerge. Flowers unisexual (plants dioecious), radially symmetrical; perianth of 6 greenish yellow, ovate or oblong-ovate segments about 4 mm long; staminate flowers with 9 greenish yellow stamens, 3 each in 3 whorls, the innermost 3 with a pair of dull yellow, short-stalked glands on either side of the filament bases, each anther half opening by an uplifted valve; pistillate flowers with 6 short, yellow staminodia, 1 pistil with superior ovary, slender style and terminal stigma. Fruit a dark blue, ellipsoid drupe about 8–10 mm long, loosely subtended by the enlarged cuplike summit of its stalk surmounted by the blunt basal remains of the perianth segments, tip of the fruit usually with the very short basal remains of the style. [*Sassafras sassafras* (L.) Karst.]

Well-drained sites, commonly in old fields, fence and hedge rows, less commonly in mixed woodlands; throughout our area. (S.w. Maine to Iowa and s.e. Kan., generally southward to cen. pen. Fla. and e. third of Tex.)

National champion sassafras (1972), 17'3" circumf., 100' in height, 68' spread, Owensboro, Ky.

Leguminosae (Fabaceae) LEGUME FAMILY

1. Plants twining vines.
  2. Leaves compound and 3-foliolate.   9. *Pueraria*
  2. Leaves 1-pinnately compound.   12. *Wisteria*
1. Plants not vining.
  3. Leaves simple.
    4. Plant suffrutescent, all parts densely clothed with soft, woolly, grayish pubescence.
            8. *Lupinus*
    4. Plant a small tree, none of the parts woolly-pubescent.   4. *Cercis*
  3. Leaves compound.
    5. The leaves 3-foliolate.   7. *Lespedeza*
    5. The leaves 1- or 2-pinnately compound.
      6. Leaves, all of them, 1-pinnately compound.
        7. The leaves evenly pinnately compound.   11. *Sesbania*
        7. The leaves odd-pinnately compound.
          8. Woody twigs, some of them at least, armed with hard, sharply pointed, stipular spines.   10. *Robinia*
          8. Woody twigs unarmed.
            9. Leaflets not glandular-punctate beneath; pubescence of all parts consisting of hairs attached medially and with 2 strongly divaricate branches.   6. *Indigofera*
            9. Leaflets glandular-punctate beneath; pubescence of unbranched hairs.
                     3. *Amorpha*
      6. Leaves, some or all of them on a given plant, 2-pinnately compound.
        10. The leaves, some of them on an individual plant, 1-pinnately compound, some 2-pinnately compound.   5. *Gleditsia*

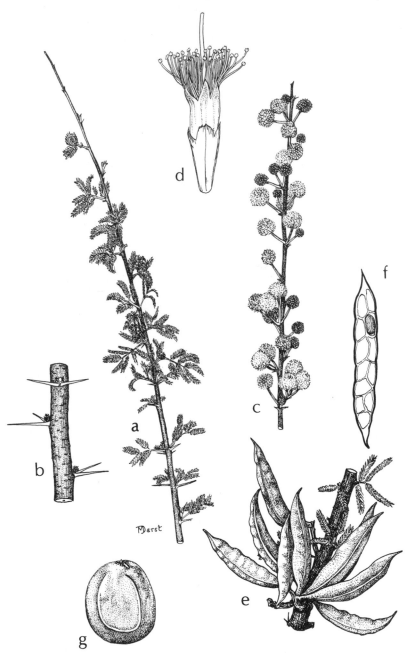

Fig. 204. **Acacia smallii:** a, leafy branch; b, piece of woody stem showing
pairs of infrastipular spines; c, flowering branch; d, flower; e, cluster of fruits;
f, interior of half-pod, 1 seed intact; g, seed.

10. The leaves, all of them, 2-pinnately compound.

    11. Leaflets extremely inequilateral; overall length of leaves 10–30 cm or a little more, few of them as little as 10; twigs unarmed.   *2. Albizia*

    11. Leaflets equilateral or very nearly so; overall length of leaves 7 cm or less; woody twigs armed with sharp stipular spines.   *1. Acacia*

1. Acacia

Acacia smallii Isely. small's acacia. (See Addendum at end of Leguminosae treatment.)   Fig. 204

Deciduous shrub or small tree 2–4 m tall, larger specimens intricately branched from the base. Woody twigs reddish brown and with scattered roundish (at first) or horizontally oblongish lenticels, armed with paired, terete, very sharp, stipular spines 3–10 mm long (Isely, 1969, reports that armed and unarmed plants are sometimes intermixed); axillary buds clustered.

Leaves alternate, petiolate, a gland adaxially on the petiole more or less midway of its length; blades of leaves on leader-shoots evenly bipinnately compound, overall length 3–5 cm and breadth to 3–4 cm, commonly with 8 shortly stalked pinnae 1–3 (3.5) cm long, each having about 30 leaflets; leaflets sessile, linear-oblong, the larger ones 4–5 mm long, very slightly inequilateral, nearly truncate basally, obtuse to rounded apically, sometimes with a small apical cusp; in the axils of leader-shoot leaves and in the axils of leaf scars on year-old twigs occur (usually) 1 or a few short-shoots whose leaves have fewer pinnae and smaller leaflets.

Flowers tiny, borne in stalked, many-flowered, orange-yellow, compactly ball-like heads, these in clusters from axils of leaf scars on wood of the previous season before new shoot growth commences; stalks about 1 cm long, resinous-glandular and minutely pubescent, heads about 1 cm in diameter. Calyx funnelform, with 5 erect, triangular-obtuse lobes, about half as long as the corolla; corolla tube cylindric, with 5 erect, triangular-acute lobes, about 2.5 mm long overall. Stamens many, free, exserted. Pistil 1, short-stipitate, ovary oblongish, style very slender, stigmas minutely capitate, exserted beyond the stamens. Fruit (legume) linear-oblong in outline, 4–6 cm long, straight or falcate, turgid, short stipitate, tapered apically into a short beak. Seeds about 5 mm long. [*Vachellia densiflora* Alexander ex Small]

In scattered bayfront areas, sandy vacant lots, spoil-flats, appearing "weedy," Pensacola area, Fla. (W. Panhandle of Fla. to w. Tex., sporadically to s. Calif.; n.e. Mex.)

2. Albizia

Albizia julibrissin Durazz. silk-tree, mimosa.   Fig. 205

A small deciduous, relatively short-lived tree with handsome feathery or lacy foliage and showy flower clusters. Trunks generally short, the main branches elongate-ascending and forming a somewhat unbrellalike crown.

Leaves alternate, evenly 2–pinnately compound, overall the blades 10–30 cm long or a little more and to 15 cm broad, with 6–25 pinnae which bear, in all, 200–1200 leaflets or thereabouts. Petioles 3–5 cm long, swollen at the base (pulvinate) as are the very short stalks of the pinnae, a conspicuous circular gland adaxially on the petiole somewhat above the swollen base, usually, in addition, smaller glands on the rachis and on the short stalks of the pinnae; leaflets 5–12 mm long and 2–4 mm broad, those near the base and the tip of a pinna some-

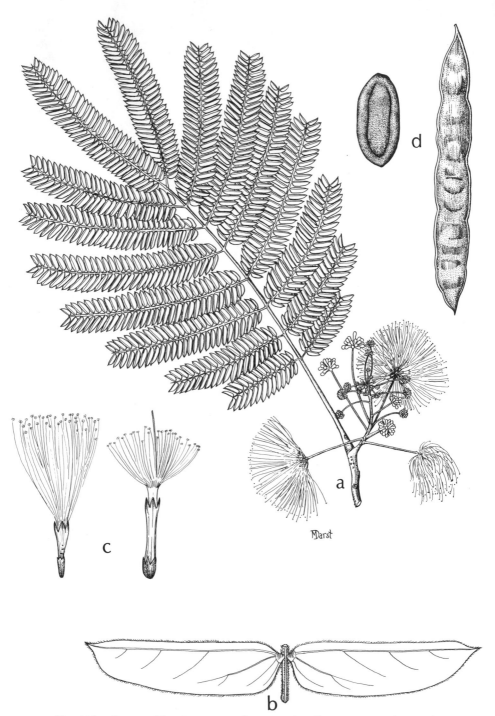

Fig. 205. **Albizia julibrissin:** a, tip of twig with inflorescences and a single leaf; b, a pair of leaflets from a pinna; c, staminate flower to left, bisexual flower to right; d, fruit and seed.

what smaller, notably inequilateral, the midrib close to one margin of a leaflet, each thus like a somewhat falcate half-leaflet, apices pointed, the points at the end of the midrib thus at one side of the tip; margins entire; petioles, leaf axes, and stalks of the leaflets sparsely short-pubescent, lower leaflet surface sometimes sparsely pubescent throughout or only the midrib and edge adjacent to it and/or the opposite margin.

Flowers individually small, radially symmetrical, many in a compacted globose head, the heads relatively long-stalked, solitary (rarely), usually in racemes or irregular panicles axillary to the uppermost leaf on a branchlet of the season. The central flower of the head bisexual, larger than the others which are staminate. Calyx very small and surrounding only the base of the corolla tube, tubular at base, 5-lobed above, lobes triangular, erect, smaller in the staminate flowers than in the bisexual one. Corolla tube funnelform in the staminate flower, more nearly cylindric in the bisexual one, 5 erect, triangular lobes at the summit. Stamens many, filaments united into a tube below, free above, the tube somewhat exserted from the corolla in the bisexual flower, barely if at all so in the staminate flowers, free portions of filaments very long-exserted, with a silky texture, usually pink, giving most of the showy color to the heads. Pistil 1, ovary superior, 1-locular, style slender, elongate, extending a little beyond the anthers. Pod (legume) yellowish, thin, flat, linear-oblong and straplike, to 15 cm long and 2–3 cm broad, the outline of the seeds evident surficially, indehiscent, stipitate at base, persistent base of the style forming an apical beak. Seeds somewhat flattened, oblongish in outline, smooth, brown, mostly about 8–10 mm long and half as broad or nearly so.

Native of Asia, cultivated as an ornamental in our area and elsewhere, commonly naturalized in waste places, roadsides, fence and hedge rows, borders of upland woods, clearings.

The silk-tree is susceptible to a fungal blight which enters the roots and destroys the sap wood. After a given plant is infected it usually dies rather quickly; fortunately, it appears that individuals become infected only after several years of their rapid growth so only the older, larger trees "bite the dust."

National champion silk-tree (1971), 9'8" [at 3'] circumf., 41' in height, 60' spread, Upshur Co., Tex.

3. Amorpha

Amorpha fruticosa L. BASTARD-INDIGO. Fig. 206

Deciduous shrub 1–4 m tall, often bushy-branched, variable with respect to pubescence and size and shape of leaflets. Very young stems, all parts of unfurling leaves, and early-developing racemes copiously pubescent with silky, tawny pubescence, the pubescence on the various parts sloughing to varying degrees as the parts become fully developed. Woody twigs light brown with slightly paler, nearly round lenticels; leaf scars slightly raised, horizontally elliptic to subhemispheric, vascular bundle scar 1, rather indistinct; terminal bud none; lateral (overwintering) buds 3, superposed, uppermost largest, upper 2 always evident, the lowermost formed beneath the base of the petiole and evident if and when it breaks through the leaf scar; pith continuous.

Leaves 1-odd-pinnately compound, 1–3 dm long, stipules setaceous, 4–5 mm long, soon deciduous; leaflets 9–35, very shortly stalked, stipels like the stipules and very quickly deciduous; blades symmetrical or rarely somewhat asymmetrical basally, oblong, oblong-elliptic, ovate-oblong, less frequently elliptic or

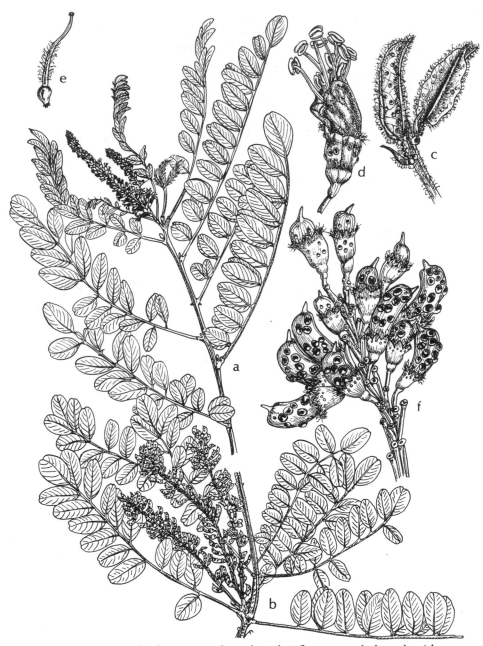

Fig. 206. **Amorpha fruticosa:** a, branch with inflorescence; b, branch with infructescence; c, young leaflets; d, flower; e, pistil; f, infructescence, enlarged. (From Correll and Correll, 1971. *Aquatic and Wetland Plants of Southwestern United States.*)

lanceolate, 1–5 cm long, 0.5–3 cm broad, rounded basally and apically, infrequently acute, apices sometimes emarginate, sometimes apiculate; margins entire; surfaces of mature leaflets varying from glabrous to copiously pubescent and glandular-punctate beneath.

Inflorescence a dense, spikelike raceme or panicle of racemes terminating a branchlet of the season; raceme in the early stages of its development conspicuously bracteate, each bract pubescent, subulate to narrowly lance-attenuate and as long as the bud it subtends, bracts deciduous before flowers are at anthesis. Calyx tube obconic, funnelform or campanulate, 2–3 mm high, glabrous to moderately pubescent, distally punctate-glandular with small and inconspicuous to large amber, resinous glands, occasionally without glands; calyx lobes 5, with short-ciliate margins, all much shorter than the tube, sometimes the upper and lateral ones hardly more than undulations, the lower one usually acute. Corolla consisting of but 1 erect, indistinctly clawed petal (the standard), 5–6 mm long, its blade obovate, dark reddish purple, purple, or blue. Stamens 10, filaments united below, distinct above, extending beyond the calyx, usually beyond the petal. Pistil 1. Fruit an indehiscent, 1-seeded, moderately to strongly curved pod 5–8 mm long, its surface with few to numerous, conspicuous, resinous glands, rarely without them.

Shores and banks of rivers and streams, floodplain woodlands, wet woodlands, marshes and thickets of shorelines, moist to wet clearings, spoil banks and flats; throughout our area. (Maine and s. Que. to N.D., s. Man., S.D., e. Wyo., generally southward to s. pen. Fla., e. Colo., N.Mex., Ariz., s. Calif.; n. Mex.)

4. Cercis

Cercis canadensis L. var. **canadensis**. EASTERN REDBUD, JUDAS-TREE.   Fig. 207

A small tree, often flowering/fruiting when of low, shrublike stature. Smaller twigs usually slightly zigzagging, bark brown, smooth, with scattered small lenticels; terminal buds none; leaf scars more or less obcordate, each with 3 vascular bundle scars; bark of trunk often divided into narrow scaly ridges.

Leaves simple, alternate, deciduous, 2-ranked on the stems, prominently petiolate, petioles dilated (with pulvini) at both extremities. Blades 7.5–12.5 cm long, palmately-veined, broadly ovate to ovate-reniform, about as broad basally as long, bases cordate to truncate, apices short-acuminate to rounded, margins entire; surfaces dark green and glabrous above, pale green beneath and uniformly pubescent, pubescent mainly along the veins, or glabrous.

Flowers appearing in spring before new shoot growth, borne from winter buds in sessile, tight to loose clusters of 4–8, anywhere on the stems except the extremities of the twigs. Flowers bisexual, about 10–12 mm long, their slender stalks as long as or a little longer than the flowers and colored like the calyx. Calyx dark magenta, tube constricted basally then abruptly flaring into a strongly asymmetrical bowllike portion with 5 short, rounded lobes at the summit. Corolla strongly bilateral, composed of 5 light to dark pink or magenta petals, each clawed basally, the claws within the calyx, blades of the 3 upper ones turned (above the calyx) and more or less erect, the lateral pair behind the central one, the 2 lower petals longitudinally arching forward and oriented so as together to form a keel. Stamens 10, free, in 2 series, anthers of one series discharging pollen before those of the other series, filaments arching forward within the keel, their pubescent bases inserted on a nectariferous disc seated beneath and around the base of the ovary. Style slender, the capitate stigma even with the anthers.

425

Fig. 207. **Cercis canadensis.**

Fruits flat, oblong pods (legumes) 6–10 cm long, purplish red at maturity, eventually turning brown, often persisting on the plant into winter; shriveled calyx and sometimes filaments persistent at the base of the pod. Seeds flattened, 6–8 mm long.

Moist woodlands throughout our area. (Conn., Pa. to s. Mich., Iowa, and e. Neb., generally southward to n. Fla. and e. and n. cen. Tex.)

In flower, the redbud is a harbinger of spring, commonly used as an ornamental, the only one of our trees having clusters of pink to magenta flowers (and later fruits) almost anywhere on the stems except tips of twigs. The broadly ovate to ovate-reniform, palmately-veined leaf blades and petioles with pulvini at both their extremities, taken together, are features peculiar to this tree (for our area).

National champion eastern redbud (1976), 8'2" circumf., 47' in height, 36' spread, Springfield, Mo.

5. Gleditsia

Deciduous trees or shrubs, twigs, branches and trunks commonly armed with stout, simple or branched thorns developed from buds above the leaf axils or adventitiously; however, the thorns are by no means uniformly distributed on individual specimens and some may have none at all. Leaves evenly 1- or 2-pinnately compound, or partially 1-compound and partially 2-compound, the 1-compound ones generally solitary or fascicled on essentially sessile short-shoots from axillary buds on branches of previous seasons, the partially or wholly bipinnate ones generally on elongation shoots of the season; leaflets opposite or subopposite, commonly 9–20 pairs on 1-pinnate leaves or on individual pinnae of 2-pinnate ones, lanceolate, lance-oblong, or infrequently ovate-oblong, often slightly inequilateral, 1.5-3 cm long and 0.5–1.2 cm broad, bases rounded or very shortly tapered, apices blunt, infrequently acute, margins very obscurely crenate and with minute glands in the notches of the teeth; stipules minute and deciduous before the unfurling leaves have reached much size, stipels none. Flowers small, radially symmetrical or very nearly so, sessile or shortly stalked, borne on short-stalked, pendulous, narrowly cylindric spikes or racemes 2–8 (10) cm long arising from amongst the 1-pinnate fascicle of leaves on spur shoots well after the leaves are fully developed, occasionally on leafless spur shoots. On some trees, all flowers apparently structurally and functionally staminate; on other trees, most or all flowers structurally bisexual but functionally pistillate, the stamens sterile, an occasional flower may be functionally bisexual, however, although I have not observed the latter. On the staminate spikes, usually the flowers are in sessile clusters of 2-several, few if any of them solitary, spikes appearing densely flowered. On the functionally pistillate spikes, the flowers are generally solitary although some may be closely approximate, the spikes having fewer flowers more loosely disposed. Floral tube narrowly obconic, bearing at its summit perianth parts in 2 series, individual flowers variably with each series having 3, 4, or 5 parts, the outer, the sepals, narrowly awl-shaped, the inner, the petals of equal length or longer, oblongish to elliptic and somewhat cupped, exteriorly usually olivaceous, interiorly yellowish green. In the staminate flowers, the sepals and petals spreading laterally at full anthesis, stamens the same number as the perianth parts, borne on the rim of the floral tube, one opposite each sepal and petal, anthers well exserted; aperature of the floral tube and proximal portions of the filaments pubescent, anthers

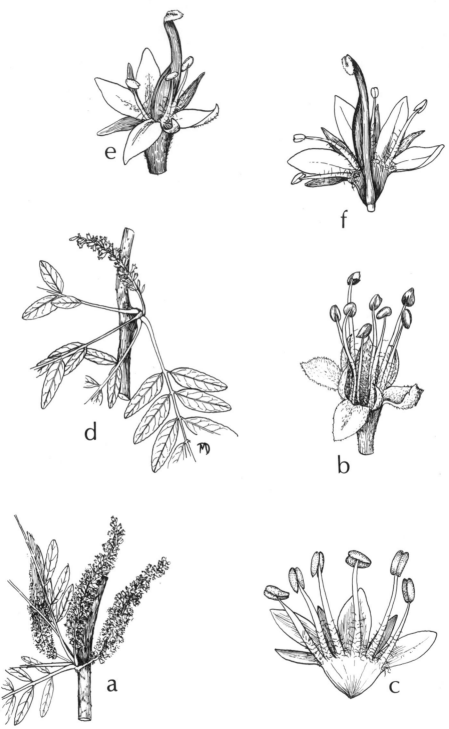

Fig. 208. **Gleditsia aquatica:** a, spur-shoot with spikes of staminate flowers; b, staminate flower; c, staminate flower with perianth opened out; d, spur-shoot with spike of functionally pistillate flowers; e, functionally pistillate flower; f, functionally pistillate flower with perianth opened out.

pinkish. In the functionally pistillate flowers, perianth parts diverging but not spreading laterally, abortive stamens inconspicuous, only as many as the sepals and opposite them (in flowers I have examined). Pistil 1, short-stipitate (the stipe elongating greatly as the fruit develops), ovary 1-locular, style usually bent or curved, stigma curved. Pod (legume) conspicuously stipitate, body flat or nearly so, 1– to many-seeded. Seeds flattened, ovate to suborbicular in outline.

1. Mature leaves with their axes, stalks of the leaflets, and surfaces of the leaflets glabrous or with but a few hairs; pod elliptic-, oval-, or ovate-oblique, 5 cm long or less, 1–3-seeded.
1. G. aquatica

1. Mature leaves with their axes and stalks of the leaflets pubescent, lower surfaces of leaflets pubescent along their midribs at least proximally and margins of leaflets usually short-pubescent; pod long-oblong, commonly falcate, 1–4 dm long and 2–3.5 cm broad.
2. G. triacanthos

1. Gleditsia aquatica Marsh. WATER-LOCUST. Figs. 208, 209

Tree to about 25 m tall, sometimes shrubby, with grayish brown to blackish, smoothish, narrowly furrowed, or warty bark. Thorns simple or few-branched, 7–14 cm long and up to 1.5 cm wide across the basal branches, if any. Axes of mature leaves glabrous or sparsely pubescent adaxially, stalks of the leaflets glabrous or very sparsely short-pubescent, their surfaces glabrous. Ovary stipitate, glabrous or only sparsely ciliate along the sutures. Pod elliptic-, oval-, or ovate-oblique, flat and thin, tardily dehiscent, 2–5 cm long, 2–3.5 cm broad, apiculate apically. Seeds 1–3, not surrounded by pulp, nearly flat, suborbicular, 1–1.5 cm across.

River swamps and floodplains, river banks, wet hammocks astride small streams; generally distributed in n. Fla., less so in s. Ga. and s. Ala. (Coastal plain S.C. to cen. pen. Fla., westward to e. and s.e. Tex., northward in the interior to s.e. Mo., s. Ill., s. Ind.)

National champion water-locust (1980), 7'1" circumf., 91' in height, 40' spread, Pulaski Co., Ill.

2. Gleditsia triacanthos L. HONEY-LOCUST. Fig. 209

Similar to *G. aquatica* in general features, attaining larger stature, perhaps to 45 m tall. Bark with deep fissures, long, narrow scaly ridges between them. Thorns simple or 3- to many-branched, the latter, particularly on older stems or trunks, 6–15 (40) cm long. Mature leaves with their petioles, axes, and stalks of the leaflets pubescent, lower surfaces of the leaflets pubescent along their midribs at least proximally, sometimes other veins as well, margins short-pubescent. Ovary sessile or nearly so, densely pubescent. Pod long-oblong, 1–4 dm long, mostly 2–3.5 cm broad, flat and thickish, indehiscent, the thickened margins usually contracting during maturation causing the pod to curve or coil. Seeds numerous, each surrounded by a sugary pulp, compressed but not flat, irregularly oblong in outline, dark brown, about 0.8 cm long.

In general inhabiting well-drained sites, upland woodlands and their borders, old fields, fence rows, less frequently on sites flooded for periods of short duration, river floodplains and hammocks, sometimes intermixed with *G. aquatica* where their ranges overlap; Fla. Panhandle from Jefferson and Taylor Cos. westward, s. cen. and s.w. Ga., s. Ala. (W. N.Y. and Pa. to e. S.D., generally southwesterly in the east to Fla. Panhandle and southerly in the west to e. half of Tex.)

National champion honey-locust (1972), 17' circumf., 115' in height, 124' spread, Wayne Co., Mich.

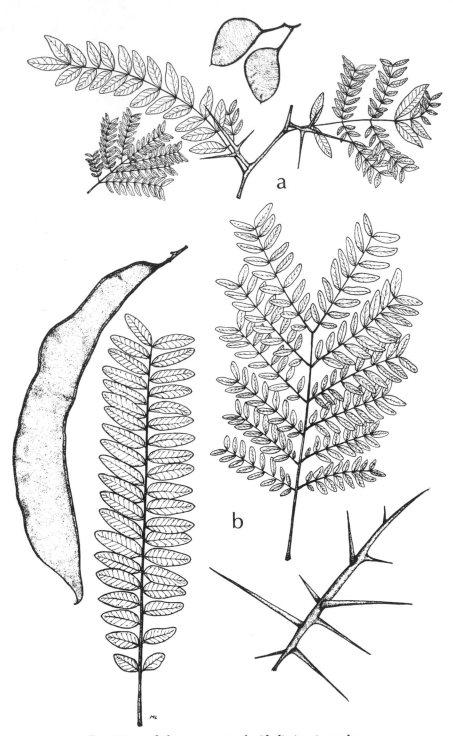

Fig. 209. a, **Gleditsia aquatica**; b, **Gleditsia triacanthos**.

6. Indigofera

Indigofera suffruticosa Mill. INDIGO.

Deciduous shrub to 2 m tall. Pubescence of all parts consisting of hairs attached medially and with 2 strongly divaricate branches. Young stem several-angled, gray-pubescent, the hairs abundant, closely appressed, gradually sloughing as stem ages, eventually with light brown, smooth bark.

Leaves petioled, petioles slender, pubescent, 1–2 cm long, stipules setaceous, pubescent, about 2 mm long, soon deciduous; blades 1–odd-pinnately compound; leaflets 7–15, lateral ones minutely stalked, stalk of terminal one 8–10 mm long, mostly stipellate; as new leaves unfold, the petiole, rachis, and lower leaflet surfaces are densely clothed with blond, appressed, satiny pubescence, that of lower leaflet surfaces becoming more sparse but persistent as the leaflet enlarges, upper surfaces sparsely and persistently pubescent; mature lateral leaflets mostly oblanceolate to elliptic, terminal one often obovate, most of them 2–3 cm long and 1–1.5 cm broad at their broadest places; margins entire, bases cuneate, apices rounded and minutely apiculate.

Flowers small, borne in axillary, sessile, or shortly stalked, spikelike racemes, each flower subtended by a small, subulate, quickly deciduous bract; raceme axis, short flower stalks, and calyces pubescent. Calyx short, 5-toothed at the summit. Corolla papilionaceous, reddish, pinkish, or purplish, quickly shriveling and falling, often leaving the withering stamens exposed. Stamens 10, diadelphous (9 and 1), the filaments of the 9 coalesced into a tube most of the way, but free distally. Pistil 1, ovary linear, pubescent, 1–locular, style relatively short, curvate, stigma capitate. Usually, apparently, fruits are developed only from those flowers on about the basal one-third of the raceme, the pods (legumes), then, appearing in relatively short, bananalike clusters, each strongly cylindric-falcate, 1–1.5 cm long, shortly beaked at the summit, surface dark brown, sparsely pubescent. Seed squarish, dark brown, about 1.5 mm long and broad.

"Weedy" in vacant lots, old fields, by railroads, openly wooded, well-drained pinelands; in our area in n.e. Fla., possibly s.e. Ga. (Nativity questionable, perhaps Asia, perhaps trop. Am., formerly cultivated as a source of chemical precursors of the dye indigo, locally naturalized in the s. U.S.)

7. Lespedeza

The species of *Lespedeza* native to North America are perennial herbs. Two introduced and naturalized species are annual herbs. The two shrubby introduced species treated here are rather widely planted in the southeastern United States for game-food production (chiefly for quail) and for erosion control. In my very limited experience with them, they appear to be variable and some specimens intermediate. In eastern Asia from which these and related shrubby species come, the populations are highly variable and their classification in dispute (Clewell, 1966). It is not unlikely that the original introductions were from diverse sources exhibiting notable variation, the original variation perhaps enhanced by subsequent hybridization. The following generic description is abridged from one that would be more generally applicable. The treatment of the species, it is hoped, will be helpful for purposes of distinguishing the two taxa, so many features of which are similar.

Stems few to numerous from a root-crown, 1–2 (3) m tall, woody at least below, the distal portions often winter-killed. Leaves slenderly petiolate, with persistent or eventually deciduous, small, subulate stipules, pinnately 3-foliolate, leaflets without stipels, pinnately veined, equilateral, margins entire, the terminal leaflet longer-stalked and larger than the laterals. Flowers in stalked racemes from the axils of leaves of upper branches, racemes generally exceeding the subtending leaves and usually numerous giving the floriferous branches a paniclelike appearance; flower stalks subtended by small bracts and with a pair of bractlets immediately beneath the calyx. Calyx tubular below, with 5 narrow lobes, 4 of the lobes nearly equal, the 5th (lower) one longer than the others. Corolla papilionaceous, rose-purple to purple, the standard clawed, its blade oblong-obovate, the wings clawed, curved-oblong, keel petals incurved-obovate. Stamens equal and diadelphous (9 and 1). Ovary slightly stalked, style long and filiform, often persistent on the fruit, stigma minutely capitate. Fruit a 1-seeded, indehiscent, nearly flat pod, its surface reticulately veined and upwardly appressed-pubescent; pods 5–8 mm long, broadly elliptic or less frequently suborbicular.

1. Lobes of the calyx obtuse to acute, the longer (lower) one usually about as long as the calyx tube, the others mostly shorter than the tube; seeds olive green or mottled purple on green.   1. *L. bicolor*
1. Lobes of the calyx acute to acuminate, generally all longer than the calyx tube; seeds dark purple.   2. *L. thunbergii*

1. Lespedeza bicolor Turcz.

Leaflets of lower and medial portions of the stem mostly elliptic or oval, less frequently ovate, variable in size, (1) 2–5 (7) cm long and (0.8) 1–2 (3.5) cm broad, obtuse to rounded basally, mostly rounded apically, sometimes obtuse, usually apiculate, infrequently retuse; mature leaflets with upper surfaces glabrous or with very few, scattered short hairs, lower glabrous, appressed-pubescent on the veins, varying to appressed-pubescent throughout; young stems, petioles, and stalks of the leaflets appressed-pubescent.

Racemes generally longer than the subtending leaves, their stalks, flower stalks, bracts, bractlets, and calyces densely to sparingly, upwardly appressed-pubescent. Flowers 8–12 mm long. Calyx lobes subulate, obtuse to acute apically, the shorter 4 generally shorter than the calyx tube, the longer (lower) one generally equaling to very little longer than the tube and long-acute apically. Seeds olive green or green mottled with purple.

Fields, open woodlands, clearings, fence and hedge rows, roadsides, locally naturalized, probably mostly in the vicinity of plantings.

2. Lespedeza thunbergii (DC.) Nakai.

Leaflets of lower and medial portions of stem averaging larger than in *L. bicolor*, 3–6 cm long and 2–4 cm broad, ovate, lance-ovate, elliptic, or lanceolate, bases rounded or broadly short-tapered, apices obtuse to rounded, infrequently acute, some of them apiculate; surfaces of mature leaflets uniformly short-appressed-pubescent; young stems, petioles, and stalks of leaflets appressed-pubescent.

Racemes generally longer than the subtending leaves, pubescent as in *L. bicolor*. Flowers 12–18 mm long, more showy than those of *L. bicolor*. Calyx lobes linear to narrowly elliptic, all of them longer than the calyx tube. Seeds dark purple.

Apparently less frequently planted than *L. bicolor* and accordingly less frequently naturalized, occurring in the same kinds of places.

8. Lupinus

Lupinus westianus Small. SMALL'S LUPINE.　　　　　　　　Fig. 210

Suffruticose, older plants to about 1 m tall, branches numerous, to some extent decumbent, forming a compact crown to 1 m or more in diameter; lower branches perennial and woody, to at least 2 cm in diameter but most not over 1.5 cm; leafy stems herbaceous. Herbaceous stems (and younger parts of woody stems), petioles, leaf surfaces, inflorescence axes, flower stalks, and calyces very densely clothed with soft, woolly, translucent pubescence rendering them gray-green (often becoming golden brown in drying).

Leaves simple, alternate, petiolate, without stipules, petioles 1–4 cm long. Blades elliptic, the larger ones mostly 7–8 cm long and 3–4 cm broad, very pliable, lateral veins faintly evident, tapered to rounded basally, apices rounded and with a small mucro, margins entire.

Inflorescences erect racemes 1–3 dm long terminating many, or most, of the branches of the season. Flowers each subtended by a subulate bract which is deciduous in the late bud stage, flower stalks very short and thickish, a smaller subulate bract between the base of each lip of the 2-lipped calyx; upper calyx lip 2-forked apically, the lower unforked. Corolla papilionaceous, standard broadly ovate, curved upwardly, its blue sides bent backward, a prominent dark reddish purple spot on its face centrally, wings blue, vertically disposed almost flat one against the other and thrust forward beneath the standard, forming more or less of an oblique envelope enclosing the strongly arched-falcate keel the lower half of which is white, the hornlike distal portion brownish yellow. Stamens 10, monadelphous, wholly within the keel. Pistil 1, the silky-pubescent ovary within the filament tube, style slender and elongate, the capitate stigma reaching the level of the anthers.

Fruit a pod (legume), elliptic-oblong and somewhat oblique, 2–2.5 cm long, beaked by the upturned remains of the style, surface copiously silky-pubescent.

Coastal dunes and dune-blowouts (where any dunes remain), open sands resulting from mechanical disturbance of coastal sand pine–oak scrub, chiefly from Franklin Co. westward to Santa Rosa Co. in the Panhandle of Florida, to some extent inland from there especially on open sands where the longleaf pine–scrub oak sand ridges have been mechanically much disturbed in site preparation for pine plantations.

Another simple-leaved, woolly lupine, closely resembling *L. westianus*, *L. diffusus* Nutt., occurs rather more widely in the w. Panhandle of Florida. It is a biennial herb with decumbent branches. Its corollas are blue, but the central spot on the standard is cream-colored and relatively inconspicuous.

9. Pueraria

Pueraria lobata (Willd.) Ohwi. KUDZU-VINE.　　　　　　Fig. 211

Deciduous, twining vine, annually forming extensive, long, herbaceous, often impenetrable masses high into and completely engulfing woody vegetation or completely covering extensive unwooded areas; medium-sized stems ropelike, woody, brown, tough and leathery, pliable; larger, older stems very much more rigid, somewhat fluted, to at least 5 cm in diameter, the bark roughish, dark brown; herbaceous stems, petioles, stipules, and stipels bearing long hairs and much more numerous short ones, the latter spreading, the former spreading or variably antrorse or retrorse.

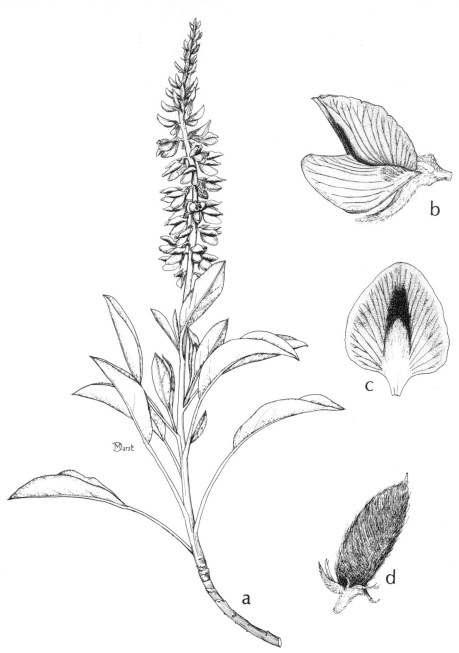

Fig. 210. **Lupinus westianus:** a, flowering branch; b, flower; c, standard or banner (petal), flattened out; d, fruit.

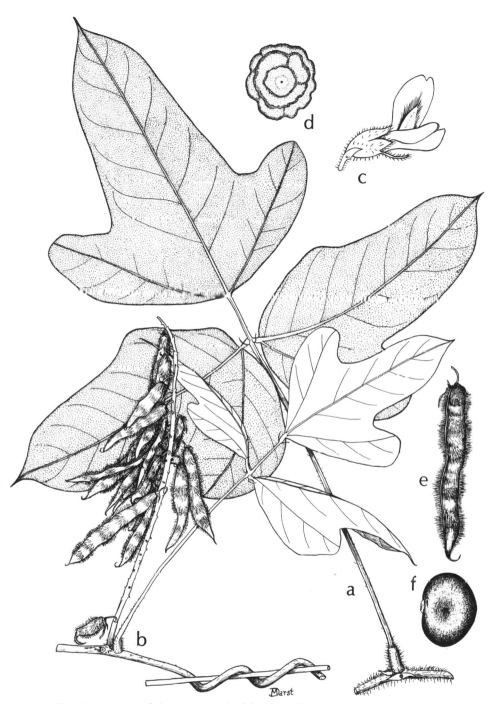

Fig. 211. **Pueraria lobata:** a, large leaf; b, piece of vine with leaf and infructescence; c, flower; d, diagrammatic cross-section of woody stem 5 cm in diameter; e, fruit; f, seed.

Leaves alternate, widely spaced from each other, compound and 3-foliolate, long-petiolate, petioles of mature leaves as long or a little longer than the pinnately veined blades; stipules perfoliate, broadest medially (5 mm) then attenuate to both extremities, about 2 cm long. *Young* developing leaflets copiously pubescent with silky hairs, pubescence much denser beneath than above, the major veins, however, with pubescence as described for the stems, petioles, etc.; some or sometimes much of the pubescence of all parts sloughed as they enlarge and mature; *mature* leaves usually having petioles 25–30 cm long; stalks of lateral leaflets about 8 mm long (consisting of pulvini only), axis of the blade (beyond the lateral leaflets and to the pulvinus of the terminal leaflet) usually 4 cm long; stipels of leaflets thin, narrowly subulate, 10 mm long; blades of lateral leaflets markedly inequilateral, obliquely ovate in overall outline, about 15 cm long and 12 cm broad, unlobed or more frequently the broader side of each 1-lobed, less frequently both sides 1-lobed; terminal leaflet usually equilateral and usually 3-lobed, infrequently unlobed and inequilateral; lateral lobes of all leaflets broadly rounded, apices of all leaflets acuminate. (Measurements for parts of mature leaves of vigorous branches vary somewhat from those given above but not a great deal; "tag-end" branches and flower- or fruit-bearing branchlets have smaller leaves with the same general form.)

Inflorescence a compactly-flowered, axillary, nearly sessile raceme; lowermost flowers at anthesis first, flowering occurring distally as the raceme continues development, maximal length for most racemes being 20 cm or a little more; as the raceme elongates the early flowers, then those blooming sequentially from the base upwardly may be shed until all are shed leaving the knobby axis, or some or all flowers may have produced developing fruits; raceme axis copiously short-pubescent, flowers commonly in pairs (or threes) from a given raised node of the axis; a small, triangular, yellowish green, very quickly deciduous bract subtends each flower and a similar bractlet subtends and is appressed to the calyx. In our range, at least, production of inflorescences appears to be very sporadic from clone to clone and within clones; when flowering does occur, fruit production is very erratic.

Flowers bisexual, bilaterally symmetrical. Calyx with a campanulate, dull purplish, short-pubescent tube, a short, broad, pale, deltoid lobe as an upper lip, and three longer, narrower, triangular-acute, pale lobes, center one longest, as a lower lip, upper lip pressed against the reflexed base of the standard of the corolla, the lower loosely extended beneath the base of the keel of the corolla. Corolla papilionaceous, standard reflexed-erect, broadly oblongish, emarginate or incised apically, wine-colored, paler on the back than on the front, the latter with the bases arched centrally, the medial portion of the arch with a differently colored, conspicuous splotch, its lower half lavender, the upper half bright yellow, its distal edge jagged and purple-margined; base of the keel below the splotch invaginated, a short auricle on either side of the invagination; wing petals clawed basally, the blades elliptic-oblanceolate, bright maroon, flanking the upper portion of the paler, wine-colored keel. Stamens 10, diadelphous, (9 and 1), filament of the 1 free at the base, connivent above with the tube of filaments of the other 9, all free distally, the free portions arching within the arching keel, not exserted. Fruit a generally flattened (bulging above the seeds), falcate, undulate-margined pod 3–8 cm long and about 0.8 cm broad, surface itself dark, dull brown, abundantly clothed with bright, almost golden brown, long-spreading hairs.

Native of e. Asia, introduced for agricultural purposes and erosion control, now out of hand and a significant pest plant throughout much of the s.e. U.S.

10. Robinia

Deciduous shrubs or trees, branchlets often zigzagging. Terminal buds none; lateral (winter) buds (said to be 3 or 4 superposed) naked and formed within or within and beneath the swollen base of the petiole, the cavity lined with copious, bristlelike hairs all pointed inwardly and surrounding the buds; leaf scars raised, 3-lobed (eventually at least), each lobe having an indistinct vascular bundle scar, the tissue between them breaking up and exposing some of the hairs around the buds. Leader shoots of the season, especially those of saplings, sprouts, or suckers, may elongate during most of the growing season by growth from consecutively-produced pseudoterminal buds, branches of these, if any, from axillary, naked buds. Leaves 1-odd-pinnately compound, petiolate, stipules awllike or setaceous at first, becoming hard, sharply pointed, straight or slightly curved spines, persistent unless sloughed or otherwise destroyed during the herbaceous stage; leaflets with very short stalks, stipels like the herbaceous stipules but quickly deciduous; blades pinnately veined, margins entire.

Inflorescences stalked, nodding to pendulant racemes borne axillary on shoots of the season when the leaves are half or a little more than half grown. Flowers showy, slenderly stalked, subtended by small bracts deciduous well before anthesis. Calyx tube campanulate, more or less 2-lipped, 5-lobed, upper 2 lobes shortest and united for part of their length, lower 3 longer and more deeply divided, about equal. Corolla papilionaceous, petals clawed, white, pink, roseate, or purple; standard obcordate to nearly orbicular, more or less reflexed, wings obliquely oblong to obovate, with a basal auricle, keel petals incurved, united above, auriculate basally. Stamens 10, diadelphous (9 and 1), the filaments of the 9 united below, free above, filament of the 10th coherent to the filament tube for about two-thirds its length, free below and above. Pistil 1, ovary short-stipitate, oblong, 1–locular, style slender, incurved, pubescent on the inner side near the apex, stigma minutely capitate. Fruit an oblong compressed, thinly 2–valved pod (legume). Seeds several.

1. Stems of the current season's growth having an abundance of minutely gland-tipped, spreading, stiffish hairs of various lengths to 3 mm long or a little more; corollas purple, reddish purple, or roseate; flowers not aromatic.   1. *R. hispida*
1. Stems of the current season's growth with very short, subappressed or appressed, soft pubescence at first, this all sloughed as the stem becomes woody; corollas white, rarely white tinged with pink; flowers pleasantly aromatic.   2. *R. pseudoacacia*

1. Robinia hispida L. BRISTLY LOCUST.   Fig. 212

Shrub 3–30 dm tall, often much branched, perennating by subterranean runners. Current year's stems having an abundance of minutely gland-tipped, often purple or purplish, spreading, stiffish hairs of various lengths to 3 mm long or a little longer, the hairs on year-old woody stems relatively hard and brown, their glandular tips sloughed, often many of the hairs partially to completely broken off, stipular spines, where present, 3–8 mm long.

Leaves 1–3 dm long, petioles about the length of the lowest leaflets, petiole and at least the proximal portions of the rachis sparsely to densely pubescent with hairs as on the stems; leaflets 7–13, sometimes to 19, per leaf, their blades oblong, ovate, ovate-oblong, or oblong-elliptic, smaller proximal ones sometimes orbicular, 1–5 cm long and 0.8–3.5 cm broad, extremities rounded to obtuse, rarely acute, tips apiculate or minutely mucronate, at maturity sparsely pubescent beneath.

Racemes to 10 cm long or a little more, bearing 3–10 inodorous, purple, red-

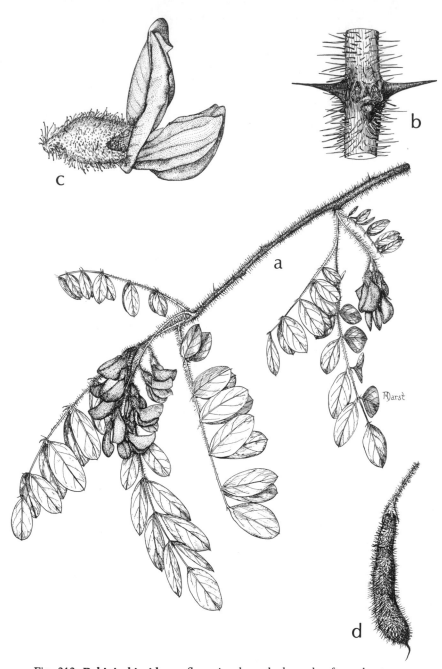

Fig. 212. **Robinia hispida:** a, flowering branch; b, node of woody stem; c, flower; d, fruit.

dish purple, or roseate flowers 2–3 cm long; stalk of raceme, its axis, flower stalks and calyces with pubescence as on young stems. Pods 5–8 cm long, abruptly beaked apically, densely glandular-hispid.

In our range, occasionally cultivated, persisting about abandoned homesites and sparingly naturalized in their vicinity. (Presumably native in or near the mountains, Va. to Ky., southward to Ga. and Ala.)

2. Robinia pseudoacacia L. BLACK LOCUST. Fig. 213

Tree to about 25 m tall, sometimes flowering/fruiting when of shrub stature. Bark of trunk thickish, rough, dark brown or blackish, ridged and furrowed. Very young, developing branchlets terete, sparsely ashy-pubescent, quickly becoming round-ridged and grooved and gradually all pubescence sloughed; stipular spines about 1 cm long, to 1.5 cm on vigorous sprouts or suckers.

Leaves 2–3.6 dm long, leaflets 7–25, longest leaves with the most leaflets on vigorous sprouts or sucker-shoots. Unfurling leaves with petioles and axes copiously short-subappressed-pubescent, the leaflets tightly folded upwardly, their exposed lower surfaces notably silky-appressed-pubescent, much of the pubescence very quickly sloughing. Blades of mature leaflets oblong-elliptic, elliptic, sometimes ovate, mostly 2–6 cm long and 1–2.5 cm broad (larger ones on vigorous shoots), bases rounded to shortly and broadly tapered, apices mostly rounded or obtuse and shortly apiculate, surfaces glabrous, upper dull green, lower grayish green.

Raceme axis, flower stalks, and calyces finely short-pubescent. Flowers delightfully fragrant. Corollas white or creamy-white, the standard with a pale yellow patch, 1.5–2.5 cm long. Pods 5–10 cm long and 1–1.2 cm broad, glabrous, the base of the slender, persistent style usually present.

Cultivated as an ornamental and locally naturalized in our area, in some places abundantly, in fence and hedge rows, in and on the borders of upland woodlands of various mixtures, in urban environments. (Originally native to the s. Appalachian and Ozark Mts., now naturalized throughout much of the eastern U.S., Pacific States, and in Eur.)

National champion black locust (1974), 23'4" circumf., 96' in height, 92' spread, Dansville, N.Y.

11. Sesbania (SESBANS)

Deciduous subshrubs (those treated here), stems and branches with more or less continuous growth over much of the season, the early seasonal growth very slowly becoming woody, in our range some or much of the late seasonal growth becoming winter-killed, lower older stems giving rise to the following season's growth. The more or less continuously growing shoot tips with their unfolding leaves and budding racemes appressed silky-pubescent, much of the pubescence quickly sloughing, all of it on upper leaf surfaces.

Leaves sessile or very short-petiolate, alternate, evenly 1–pinnately compound, length of fully developed ones mostly 2–3 dm long, pairs of leaflets 10 or a few more; stipules subulate, deciduous before the leaves have attained much size; mature leaflets oblong or lance-oblong, 1–3 cm long and 0.5–1 cm broad, medial ones largest, apiculate at the obtuse to rounded apices, margins entire.

Flowers borne in axillary, stalked racemes shorter than the leaves, stalks 2–5 cm long, axis of the flower bearing portions 5–10 cm long; flower stalks slender, at anthesis declining but strongly bent just below the calyx, it being oriented

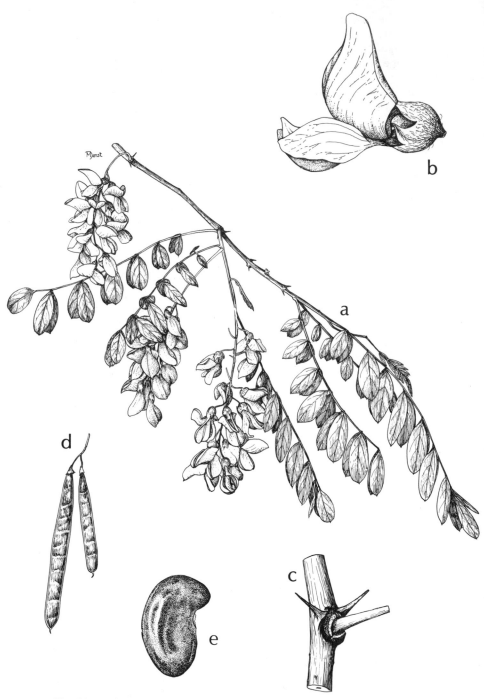

Fig. 213. **Robinia pseudoacacia:** a, flowering branch; b, flower; c, node showing stipular spines; d, fruits; e, seed.

more or less vertically, the corolla curving outward; in the early stage of development of the raceme each developing flower stalk is in the axil of a bract, a pair of lateral bractlets just at the junction of stalk and calyx; as the buds grow and the flower stalks elongate the subtending bracts fall first, the bractlets a bit later, both before full anthesis. Calyx obliquely bowllike and with 5 low lobes broader than high. Corolla papilionaceous, petals clawed, blade of the standard reflexed, much broader than the other petals, suborbicular to subreniform; blades of wing and keel petals oblique. Stamens 10, diadelphous, 9 forming an arched tube below, free distally, the 10th free from the base but arching closely adjacent to the others. Pistil 1, ovary superior, 1-locular. Fruit a legume (pod), stipitate basally, beaked apically, walls thick, hard, and fibrous, interiorly with cross partitions between the seeds.

Prof. Duane Isely, Iowa State University, very kindly permitted me to see his prepared manuscript on *Sesbania* for his forthcoming treatment of the genus for the U.S. This was especially helpful to me in understanding nomenclatural problems, and I am grateful to him.

In general appearances and with respect to vegetative characters, the three species treated below are very much alike.

1. Calyces usually dull reddish; corollas orange-red (usually drying purplish).
1 S punicea

1. Calyces green, corollas yellow.
   2. Legume with 2 thin lateral wings on each edge, a deep, narrow trough between them.
2. S drummondii
   2. Legume with 2 thick, more or less horizontally lined lateral ridges on each edge, a broad shallow trough between them.
3. S. virgata

1. Sesbania punicea (Cav.) Benth. in Mart. RATTLE BOX, BLADDERPOD, PURPLE SESBAN. Fig. 214

Calyces usually with much dull reddish pigment. Blade of the standard bright orange-red, roundish, about 2 cm broad; wing and keel petals a little less brightly colored than the standard.

Legume darkish brown, often with darker speckles, oblong, 5–8 cm long, 1–1.5 cm broad, flattened, not bulged on the surfaces over the seeds, conspicuously 2–winged on each side, wings irregularly slightly wavy or notched, a conspicuous trough between them. Seeds nearly reniform in outline, plump, 6–8 mm long and about 4 mm broad, dull reddish brown, purple at the micropyle. [*Daubentonia punicea* (Cav.) DC.]

Native of S.Am., abundantly naturalized locally throughout our range in sloughs, swales, clearings, vacant lots, along shores. (Coastal plain, N.C. to cen. pen. Fla., westward to e. and s.e. Tex.)

2. Sesbania drummondii (Rydb.) Cory. RATTLEBUSH, RATTLEBOX. Fig. 215

Calyces green. Blade of the standard clear yellow or yellow with reddish lines and fine speckles, subreniform, 1.5–2 cm broad; wing and keel petals only slightly if at all paler yellow than the standard.

Legume tan or light brown, its surfaces bulged over the seeds, usually oblong and with parallel sides, 2–6 cm long and 1–1.5 cm broad or a little more, sometimes abruptly narrowed and isthmuslike between the seeds, conspicuously 2-winged, on each side, wings usually thin, irregularly slightly wavy or notched, a conspicuous trough between them. Seeds reniform to squarish in outline, 5–6 mm long and as broad or slightly broader, dull light brown at preripening, eventually dull reddish purple. [*Daubentonia drummondii* Rydb.]

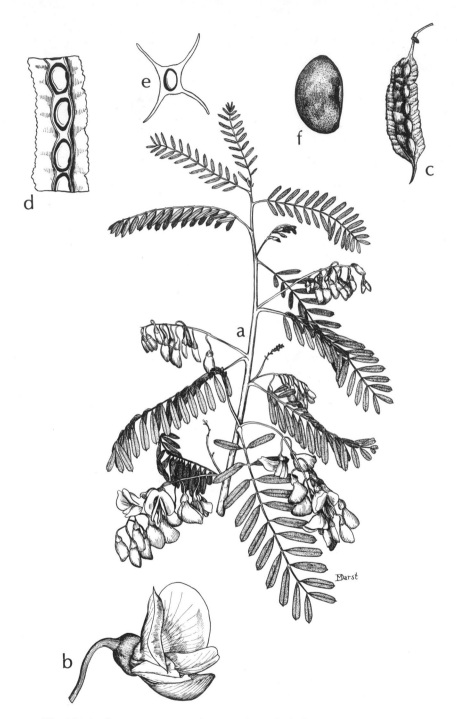

Fig. 214. **Sesbania punicea:** a, flowering branch; b, flower, side view; c, fruit;
d, diagrammatic sketch of portion of half a longitudinal section of fruit; e,
diagrammatic sketch of cross-section of fruit; f, seed.

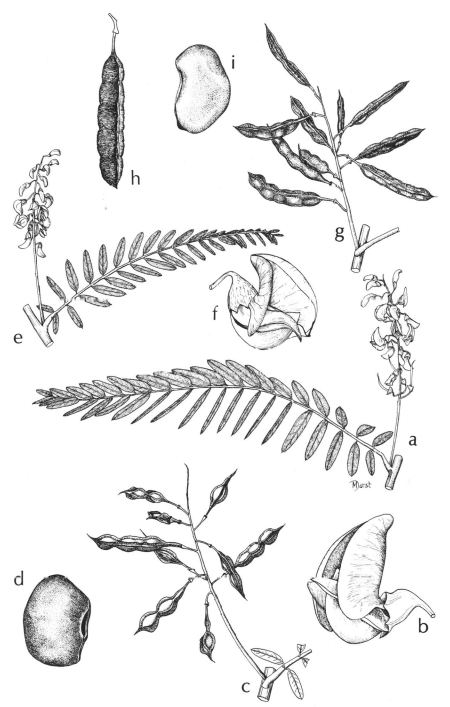

Fig. 215. a–d, **Sesbania drummondii:** a, node with leaf and axillary raceme of flowers; b, flower, side view; c, node with raceme of fruits; d, seed. e–i, **Sesbania virgata:** e, node with leaf and axillary raceme of flowers; f, flower, side view; g, node with raceme of fruits; h, fruit; i, seed.

443

Local, coastal or near coastal, shores, vacant lots, waste places, in or near brackish marshes; w. Panhandle of Fla., s.w. Ala. (thence coastal plain westward to Tex., Mex.)

3. Sesbania virgata (Cav.) Poir. in Lam.   Fig. 215

Flowers essentially like those of *S. drummondii*, averaging somewhat smaller.

Legumes dark reddish brown, for the most part oblong, occasional ones with 1–2 constrictions between seed-bearing portions, 3–6 cm long and mostly about 1 cm broad, surfaces not bulged over the seeds at maturity, laterally bluntly 2-ridged, the ridges more or less horizontally undulate, a broad shallow groove between the ridges. Seeds reniform about 8 mm long and 4 mm broad, plump, purplish brown or reddish brown, tan at the micropyle.

Native to Mex., locally naturalized, coastal or near-coastal, shores, levees, waste places, w. Panhandle of Fla., s.w. Ala. (thence coastal plain to La., Tex.)

12. Wisteria

Deciduous, woody, twining vines. Leaves alternate, 1-odd-pinnately compound, petiolate, stipules and stipels quickly deciduous. Inflorescences bracteate, nodding racemes, bracts falling by or before full anthesis of the flowers. Flowers showy, bisexual, bilaterally symmetrical. Calyx short-campanulate, asymmetrical, unequally 5-lobed. Corolla papilionaceous, blue, lilac, purplish, or white, petals nearly equal in length; standard reflexed, shortly clawed, blade suborbicular and usually with 2 hornlike appendages near its base and at the apex of the claw; wings auriculate basally; keel petals clawed, upwardly curved-falcate, each with an auricle basally, united above. Stamens 10, diadelphous (9 and 1). Pistil 1, ovary stipitate, the stipe surrounded by a collarlike, intrastaminal glandular ring, style inflexed, stigma minutely capitate. Pod (legume) stipitate, flattened, with several seeds, tardily dehiscent.

1. Ovary and pod glabrous; flower stalks 4–6 mm long; flowering in late spring after leaves are nearly fully grown; racemes about 1.2 dm long; old woody stems slender.
                                                                    1. *W. frutescens*
1. Ovary and pod densely velvety-pubescent; flower stalks 10–20 mm long; flowering in early spring before leaves have emerged or when they are just emerging; old woody stems to at least 1.5–2 dm in diameter.   2. *W. sinensis*

1. Wisteria frutescens (L.) Poir. in Lam. AMERICAN WISTERIA.   Fig. 216

Plant climbing into and through shrubs and small trees, stems relatively slender; sparingly perennating by subterranean runners or by shoots running along the ground and adventitiously rooting. Herbaceous branches of the season at first clothed with silky, shaggy hairs, less frequently with short, appressed pubescence, gradually becoming glabrous.

Mature leaves 1–2 dm long, commonly about 1.5 dm; petioles 2–6 cm long, with small swollen bases (pulvini); leaflets 9–15 per leaf, very shortly stalked, their blades lanceolate to lance-ovate, 2–6 cm long and 1–2.5 cm broad, bases broadly rounded or obtuse, apices blunt to acuminate, margins entire; in early stages of leaf expansion, all parts densely clothed with silky pubescence, eventually the upper surfaces sparsely pubescent or glabrous, the lower at least sparsely pubescent; stipules lanceolate, stipels setaceous, both about 2 mm long.

Racemes compactly flowered, to about 15 cm long, mostly pyramidal at mid-

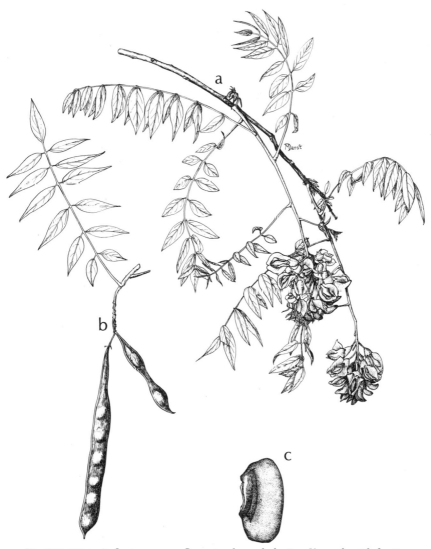

Fig. 216. **Wisteria frutescens:** a, flowering branch; b, tip of branch with fruits; c, seed.

Fig. 217. **Wisteria sinensis.**

development, flowering gradually from base to apex; borne above the upper-most leaf of a branchlet when leaves are nearly fully mature; flower stalks 4–6 mm long, subtended by elliptic to lance-ovate bracts 7–12 mm long; flower stalks, bracts, and calyces densely pubescent, the hairs usually with at least some clavate glands intermixed. Calyx tube 5–6 mm high, not as broad as high. Corolla purplish blue to lilac, 1.5–2 cm long, basal auricle of wing petals linear and 5 mm long or longer. Ovary glabrous. Pod (legume) linear-oblong or often constricted between the seeds, 5–10 cm long, cuneate basally, acuminate apically, glabrous. Seeds few, reniform, 6–8 mm long, brown. [*Kraunia frutescens* (L.) Greene]

Wet shores of streams, ponds, lakes, bayous, in and on borders of wet wood-lands and swamps, moist to wet thickets, often covering fences at such places; throughout our area. (Coastal plain, s.e. Va. to cen. pen. Fla., Ala.)

2. Wisteria sinensis (Sims) Sweet. CHINESE WISTERIA. Fig. 217

Plant climbing into and to the tops of large trees, often from tree to tree, old stems to at least 1.5–2 dm in diameter; perennating extensively by subterra-nean runners and often by very long basal, trailing sprouts rooting adven-titiously. Herbaceous stems of the season's shoots at first densely short-pubes-cent, gradually becoming glabrous, bark of woody stems pale brown or grayish brown.

Mature leaves 1.5–4 dm long; petioles 5–10 cm long, with rather prominently swollen bases (pulvini); leaflets 7–13 per leaf, commonly 9–11, ovate, lance-ovate, or elliptic, 4–10 cm long, mostly 6–8 cm, and 2–6 cm broad, bases ob-tuse, apices acuminate; in early stages of leaf expansion, leaflets densely silky-pubescent, eventually becoming slightly appressed, short-pubescent or nearly glabrous.

Racemes loosely flowered, 1.5–4 dm long, flowers at anthesis approximately simultaneously on a given raceme, racemes borne terminally on short-shoots from old wood, for the most part before leaves have begun to show (save, rarely, for 1–few occasional racemes on leafy branches later in the season); rachis densely short-pubescent, flower stalks 10–20 mm long, flexible, short-pubes-cent, each subtended by an early deciduous arching and hoodlike, tawny bract 1–1.5 cm long. Calyx tube campanulate, 3–5 mm high, densely short-pubescent, upper 2 lobes minutely triangular, others 3–4 mm high, triangular acute or acuminate. Corolla bluish violet (much less frequently white); blade of the stan-dard often with a purple spot adaxially near the base; basal auricle or lobe of wing petal triangular and less than 2 mm long. Ovary densely short-pubescent. Pod (legume) linear-oblong to oblanceolate, 10–15 cm long, velvety-pubescent, often with constrictions between the seeds.

Native of China, widely cultivated; in our area commonly escaped or natu-ralized, usually, but not always, in the vicinity of human habitations or aban-doned homesites, where it may inhabit rather large areas of woodlands.

Addendum

Acacia farnesiana (L.) Willd. SWEET ACACIA.

After this book went to press, my attention was directed to a distributional statement for this species (for s.e. U.S.) by Isely (1969) as follows: "Southern peninsular Florida and Keys, sporadically eastern Florida to southeast Georgia (introduced)." I had not realized the plant occurred in our area of coverage.

In a general way, plants of *A. farnesiana* are similar to those of *A. smallii*. The length of the stipular spines, although variable, is often much longer than in the latter, to about 4 cm rather than up to 1 cm. Petioles, leaf rachi and flower stalks are pubescent, at least at first, glabrous in *A. smallii*. Flowering in *A. farnesiana* occurs in spring (or after drought periods later in the year) from buds on old wood when leaves are present, prior to leaf emergence in spring in *A. smallii*; flower stalks longer and more flexible than in *A. smallii*, 1.5–2 cm as opposed to 1 cm or a little less. [*Vachellia farnesiana* (L.) Wight & Arn.]

In our area, probably naturalized in sandy disturbed places; besides e. Fla., and s. (southernmost?) Ga., a single plant is known to occur on St. Vincent Island, Fla. Panhandle. (Trop. Am.; trop. Asia and Australia where whether it is naturalized from cultivation or is indigenous appears moot.)

Leitneriaceae (CORKWOOD FAMILY)

Leitneria floridana Chapm. CORKWOOD.   Figs. 218, 219

Wandlike or treelike, deciduous shrub to 6 m tall, spreading by shoots from extensive subterranean runners; wood very, very light in weight. Bark reddish brown, smooth, with conspicuous buff-colored lenticels. Twigs at first densely pubescent, becoming smooth with age; leaf scars 3-lobed each with 3 vascular bundle scars; pith white, continuous.

Leaves simple, alternate, petiolate, petioles 2–4 cm long; blades pinnately veined, elliptic or lance-elliptic, 5–17 cm long and 2–5 cm broad, bases and apices acute; emerging leaves sparsely silky-pubescent above between the veins, more densely so on the veins, densely silky-pubescent beneath, at maturity the upper surfaces glabrous, the lower softly pubescent, chiefly along the larger veins, margins entire; larger lateral veins prominent and extending approximately parallel to each other from the midrib, veinlets forming a fine network visible from either surface but more distinct beneath.

Flowers borne in catkins produced from buds on wood of the previous year before the leaves emerge, staminate and pistillate usually on separate plants, plants thus dioecious (rarely the plants polygamodioecious). Staminate catkins erect to spreading, flexuous, often nodding distally, brownish, 2–5 cm long, 1–1.5 cm broad; pistillate catkins stiffly erect, much less conspicuous than the staminate, reddish, mostly 1–2 cm long, about 1 cm broad. Stamens free, mostly in clusters of 10–12, sometimes fewer or more in a cluster, in the axils of spirally arranged, close-set, deltoid-ovate scales, perianth lacking. Pistillate flowers sessile, usually solitary in the axils of spirally arranged primary bracts, each with 2 bractlets at the base and surrounded by a perianthlike structure of (3) 4 (8) segments, two often larger than the others. Fruit an erect, smooth, long-ellipsoid or oblong-ellipsoid drupe, usually brownish at maturity, ripening and falling in early spring.

Of local occurrence. Sawgrass or sawgrass–cabbage palmetto marshes, tidal, estuarine shores, swampy woodlands, swampy prairies. (s. Ga., n. and n.w. pen. Fla., westward to e. Tex., e. Ark., s.e. Mo.)

The only other plant in our flora having wood so light in weight is *Stillingia aquatica*. Legend has it that pieces of stems of both were at one time used as floats for fish nets.

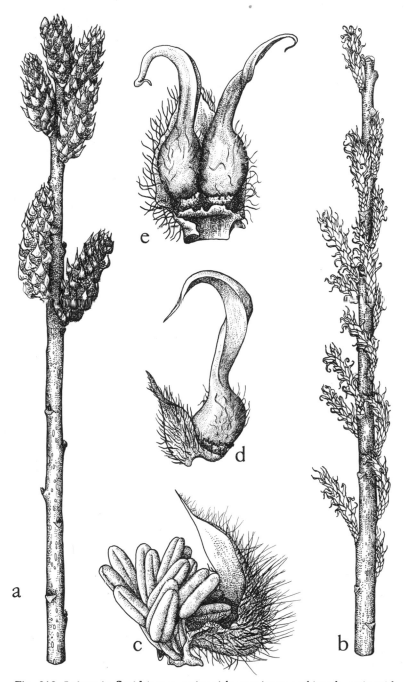

Fig. 218. **Leitneria floridana:** a, twig with staminate catkins; b, twig with pistillate catkins; c, staminate flower with subtending bract; d, usual pistillate flower with subtending bract; e, unusual pistillate flower with subtending bract.

Fig. 219. **Leitneria floridana:** a, top of stem with infructescences; b, piece of twig with leaf scar, axillary bud, lenticels; c, leaf with small section of lower surface enlarged.

Loganiaceae (LOGANIA FAMILY)

Gelsemium

Twining or trailing woody, evergreen vines. Leaves simple, opposite, short-petioled, the petioles usually pubescent on the adaxial side, blades pinnately veined, glabrous, stipules minute, quickly shed as the new shoot develops leaving a stipular line-scar between the leaves. Flowers axillary, solitary or in few-flowered cymes, showy, bisexual, radially symmetrical. Sepals 5, very small relative to the corolla size, appressed to the base of the corolla tube, united only at the base. Corolla funnelform-tubular, the tube about 2 cm long, 5 short and broad spreading lobes at the summit, the tube deeper yellow within the throat and the tube than on the lobes or without. Stamens 5, adnate to the corolla tube below, much exceeding the style and stigma on the flowers of some plants, the stigma being located at about half the length of the tube, the anthers reaching the throat of the corolla; in the flowers of other plants, the style and stigma reach the throat and the stamens reach about halfway the length of the tube. Stigmas 2, each with 2 linear divisions. Fruit a 2-valved capsule, much flattened contrary to the septum. Seeds brown, several to many, flat.

1 Sepals obtuse and blunt to rounded apically, not persistent on the fruit, capsule oblong, very abruptly narrowed to a point 1.5–2 mm long; seed with a prominent membranous wing which is sharply distinguishable from the body of the seed; flowers delicately fragrant.   1. *G. sempervirens*
1. Sepals acuminate apically, persistent on the fruit; capsule elliptical, the tapering tip bearing a definite beak about 3 mm long; seed wingless; flowers without fragrance.
                                                                        2. *G. rankinii*

1. Gelsemium sempervirens (L.) Jaume St. Hil. CAROLINA JESSAMINE, POOR MAN'S ROPE.   Fig. 221

Leaf blades mostly lanceolate, varying to ovate, occasional ones suborbicular, 6–9 cm long, to 1.5 cm broad. Flowers delicately fragrant. Sepals obtuse and blunt to rounded apically, not persistent on the fruit. Capsule broadly oblong, 1.5–2.5 cm long, 8–12 mm broad, abruptly narrowed apically to a point 1.5–2 mm long. Seeds to 1.5 mm long, pale brown, flat, conspicuously and asymmetrically thin-winged.

Commonly in well drained upland woodlands, thickets, fence and hedge rows; also in savannas, pine flatwoods, or lowland woodlands where water stands only for brief periods; throughout our area of coverage. (Coastal plain and piedmont, Va. to s. cen. pen. Fla., westward to e. Tex. and Ark.)

2. Gelsemium rankinii Small. SWAMP JESSAMINE.   Fig. 220

Leaf blades variable as in *G. sempervirens* but on any given plant usually a greater proportion of them ovate or lance-ovate. Flowers not fragrant. Sepals acuminate apically, persistent on the fruit. Capsule elliptical in outline, 1–1.6 cm long, 6–8 cm broad, the tapering tip with a beak about 3 mm long. Seeds full-brown, flat, 3–4 mm long, not winged.

Occurring much more locally than *G. sempervirens*. Swamps, wet woodlands, bogs, in places where the soil may be water-logged for extensive intervals; s. cen. and s.w. Ga., Fla. Panhandle from Leon and Wakulla Cos. westward, s. Ala. (Coastal plain, s.e. N.C. to Fla. Panhandle, westward to La.)

In the spring season of most years, the flowering period of *G. sempervirens*

Fig. 220. **Gelsemium rankinii:** a, flowering branch; b, variable leaves; c, flower; d, e, flowers opened out to show heterostyly; f, flower bud; g, stamen; h, capsule; i, seed.

Fig. 221. **Gelsemium sempervirens:** a, flowering branch; b, variable leaves; c, flower; d, e, flowers opened out to show heterostyly; f, stamen; g, capsule; h, seed.

preceeds that of *G. rankinii*, the former at about the end of flowering when the latter commences.

G. sempervirens has much potential for horticultural use (where it is advantageous to use a vine) since it is a very hardy evergreen bearing beautiful, yellow, delicately fragrant flowers in early spring. *G. rankinii* is equally handsome but may or may not be sufficiently drought resistant for use in well-drained places.

Loranthaceae

Phoradendron serotinum (Raf.) M. C. Johnston. MISTLETOE.

Evergreen shrub parasitic on limbs of a considerable number of kinds of dicotyledonous trees (rarely shrubs) where forming more or less ball-like, bushy-branched growths not infrequently eventually reaching 1 m in diameter. Stems dull green, jointed, brittle at the joints.

Leaves simple, opposite, thickish and leathery, tapered basally to short sub-petioles; blades elliptic, oval, oblanceolate, spatulate, narrowly obovate, or suborbicular, varying in size from 1.5 to 6 cm long, rarely longer, and 1–4 cm broad, apices mostly rounded, margins entire.

Inflorescences axillary, thickish, jointed spikes from a few mm to a few cm long, the very small flowers in clusters depressed into the opposite sides of the joints, opposite pairs alternating at right angles with those below or above. Flowers unisexual (plants dioecious); staminate flowers with a shallow floral tube bearing 3 deltoid sepals, a sessile, orbicular anther seated at the base of each sepal; pistillate flowers with an inferior ovary, the floral tube surmounted by 3 deltoid sepals, the ovary 1-locular, stigma sessile. Fruit a globose, 1-seeded, berrylike, white or slightly yellowish drupe 4–6 mm in diameter. [*Phoradendron flavescens* of authors]

Throughout our area. (N.J. to s.e. Kan., southward to Fla. and e. Tex.).

Note: Krut, Job (1982) treats *Phoradendron* as belonging to the family Viscaceae, formerly considered by some authors to be a subfamily, Viscoideae, of the Loranthaceae. See Krut, "The Viscaceae in the Southeastern United States." *Jour. Arn. Arb.* 63:401–410.

Lythraceae (LOOSESTRIFE FAMILY)

1. Aquatic shrub; leaves opposite or whorled; flowers in axillary cymes.   1. *Decodon*
1. Terrestrial shrub or small tree; leaves alternate; flowers in terminal racemes or panicles.   2. *Lagerstroemia*

1. Decodon

Decodon verticillatus (L.) Ell. SWAMP LOOSESTRIFE, WATER-WILLOW, TIE-DOWN.
Fig. 222

Colonial shrub, older stems woody, the bark if beneath the water surface very soft-corky, above water exfoliating in long, cinnamon-colored, thin strips, young branches herbaceous; main branches strongly arching, some of them, at least,

Fig. 222. **Decodon verticillatus:** a, tip of stem having arched, rooted at the tip, and from which new shoots are being produced; b, flowering branchlet; c, flower; d, seed. (From Correll and Correll, 1972. *Aquatic and Wetland Plants of Southwestern United States.*)

455

rooting at the tips and then forming new stems, this commonly continuing indefinitely; young stems densely short-pubescent, older ones glabrous.

Leaves opposite or whorled, short-petioled, stipules none; blades pinnately-veined, lanceolate or elliptic-lanceolate, the larger ones to about 20 cm long and 5 cm broad, acute at both extremities, sparsely short-pubescent above, more copiously so beneath, margins entire.

Inflorescences short-stalked axillary cymes; some flowers of a given cyme, usually not all, subtended by an oblanceolate, foliose bract and with a pair of smaller, less foliose bractlets at the base of the flower stalk. Flowers bisexual, with a floral tube but the ovary free within it. Floral tube campanulate to hemispheric, inconspicuously striate, pubescent. Sepals 4–7, commonly 5, triangular, slightly spreading, nearly glabrous, alternating with them and arising just below their sinuses are as many terete, incurved-hornlike, somewhat longer, copiously short-pubescent appendages. Petals 4–7, commonly 5, magenta, usually unequal, clawed basally, their blades lanceolate, usually with irregularly crinkled or erose-crinkled edges, 10–14 mm long overall. Stamens usually 10, arising about midway of the floral tube, filaments of 3 possible lengths, in any one flower 5 short ones alternating with the other 5 longer ones. Pistil 1, ovary superior, 3–5-locular, style (of 3 possible lengths) slender, stigma capitate. Fruit a globose, loculicidal capsule closely enveloped by the floral tube, the calyx segments and appendages persisting. Seeds somewhat obpyramidal, slightly rounded across the top and irregularly narrowed and angled to a blunt base, about 2 mm broad across the summit, surfaces very finely reticulate, olive-green or tan, usually with a brown spot on one side.

Swamps, swamp clearings, swampy or marshy shores of streams, pools, ponds and lakes; throughout the area of our coverage. (N.S. to Ont. and Minn., generally southward to cen. pen. Fla. and e. Tex.)

2. Lagerstroemia

Lagerstroemia indica L. CRAPE-MYRTLE.   Fig. 223

Deciduous shrub or small tree, trunks and branches twisted and irregularly fluted, bark buff-colored and sleek, wood very heavy and hard; trunks of larger specimens, when leafless in winter, bring to mind pieces of sculpture; young twigs with a pair of narrow wings extending downwardly from a given node, generally one of the pair extending to the second node below, the other to the third node below; bark of year-old twigs sloughing in narrow, thin strips; leaf scars small, roundish, vascular bundle scars single; terminal bud none.

Leaves alternate, sessile or nearly so; blades pinnately veined, mostly elliptic to slightly obovate, some nearly orbicular, 2–6 cm long and 1–4 cm broad; extremities broadly and shortly tapered to rounded, sometimes emarginate apically; surfaces glabrous or very sparsely and finely short-pubescent along proximal portions of the principal veins; margins entire; stipules minute, conical, dark purplish-tipped at first, becoming brown and eventually drying-up.

Inflorescences foliose-bracted racemes or panicles of reduced cymes terminating branchlets of the season, varying a great deal in amount of branching and in size, to about 2–3 dm long and as broad. Flowers shortly stalked above a small-bracted joint of an ultimate axis. Floral tube thick-walled, campanulate, bearing (usually) 6 triangular, incurving calyx segments 4 mm long on its rim. Petals 4–7, usually 6, each with a very slender claw 1 cm long inserted on the interior rim of the floral cup, extruding from between 2 calyx segments and angling-

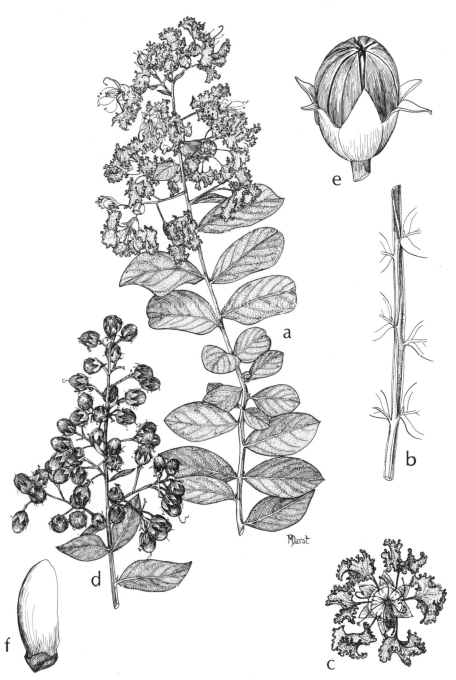

Fig. 223. **Lagerstroemia indica:** a, flowering branch; b, segment of young stem (before it becomes woody) to show wings; c, flower, face view; d, fruiting branch; e, fruit; f, seed.

ascending, abruptly dilated into a broad, very much crinkled and curled, orbicular-cordate blade, the aggregated petal blades of a flower and of all the flowers of an inflorescence very attractive indeed; petal color on different plants white, various shades of pink, watermelon-red, lavender, or purple. Stamens inserted around the base of the free ovary within the floral cup, about 40 in number, consisting of an outer series (same number as the petals) with stouter and longer filaments radiating from the orifice of the floral tube then curling slightly inwardly, *and* a much larger number with thin, relatively short filaments, their anthers more or less filling the center of the flower; anthers of the outer series shedding pollen first. Ovary globose, usually 6-locular, style slender, exserted in a diagonal fashion beyond any of the anthers, stigma capitate. Fruit a hard, globose-ellipsoid capsule 10–12 mm long, about half enclosed by the persistent floral tube and calyx, dehiscing loculicidally. Seed body angular, about 2 mm long, with a terminal, somewhat oblique wing 5–6 mm long.

Native to s.e. Asia and n. Australia, widely cultivated throughout our range, persisting about abandoned homesites and occasionally naturalizing in the relatively near vicinity, in fence and hedge rows, old fields, occasionally in woodlands.

Magnoliaceae (MAGNOLIA FAMILY)

1. Leaf blades as broad as or broader than long, apically broadly V-notched or broadly truncate.   1. *Liriodendron*
1. Leaf blades much longer than broad, unlobed or auriculate-lobed basally, obtuse to acuminate apically.   2. *Magnolia*

1. Liriodendron

Liriodendron tulipifera L. YELLOW-POPLAR, TULIP-TREE.   Fig. 224

A large deciduous tree, to 60 m tall and 3 m in diameter. (In our area relatively few old and relatively large trees extant, many, however, from small saplings to trees about 30 years old.) Bark of trunks with interlacing furrows and ridges. Young stems glabrous, slightly glaucous. Year-old twigs brown, sublustrous, older ones dull grayish, narrow stipular ring-scars encircling the nodes, leaf scars shield-shaped to nearly circular, vascular bundle scars several and scattered, usually obscure. Pith white, diaphragmed, homogeneous between the diaphragms.

Leaves simple, alternate, long-petiolate, stipules foliaceous, conspicuous, oblongish to spatulate, persistent through much of the season on vigorous summer-growing shoots. Blades pinnately veined, variable in size (up to 15 cm broad) and in lobing, as broad as or broader than long, many of them 4-lobed, some sometimes 6-lobed, some 2-lobed, lobes rounded to obtuse, or on some trees abruptly acuminate and tips of the acuminations blunt to sharply pointed, upper pair of lobes, if present, forming a V-shaped notch, truncate if without an upper pair of lobes, bases truncate to broadly rounded to either side of the petioles, margins entire other than the lobing; upper surfaces dark green and glabrous, lower grayish green, glabrous or loosely cobwebby-pubescent along the major veins.

Flowers bisexual, radially symmetrical, conspicuous, borne singly and terminally above the uppermost leaf of branchlets after leaves are fully grown. Sepals

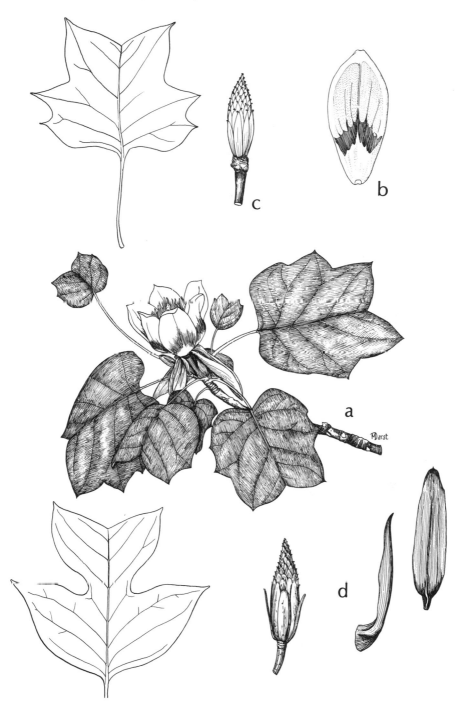

Fig. 224. **Liriodendron tulipifera:** a, flowering branchlet and, to left above and below it, characteristic leaf shapes; b, petal; c, "cone" of pistils of flower after perianth and stamens have fallen; d, "cone" of fruits to left and two views of a fruit at right.

3, green and glaucous, reflexed, oblong to lanceolate, 3–4 cm long, blunt to acute apically, deciduous about the time the flower is fully open. Corolla inverted bell-like or tuliplike, composed of 6 petals in two whorls of 3 each, spatulate, a little shorter than the sepals, green or yellowish green and with a conspicuous blotch of orange basally. Stamens numerous, aggregated below the "cone" of numerous pistils, anthers linear, much longer than the filaments. Pistils numerous, closely spirally imbricated on an elongated conical axis, each ripening into a basally 2-seeded samara, each falling separately at maturity leaving the persistent axis.

In our range inhabiting bottomland woodlands including shrub-tree bogs or bays, and on lower slopes along streams; n.e. Fla., Fla. Panhandle from Jefferson Co. westward, southeasternmost Ga., and from about Lowndes Co. westward, s. Ala. (Vt. to s. Mich., generally southward to n. Fla. and La.)

National champion yellow-poplar (1972), 30′3″ circumf., 124′ in height, 122′ spread, Bedford, Va.

2. Magnolia

Shrubs or trees, some deciduous, some evergreen. Buds covered with spathelike structures each consisting of 2 united stipules. Twigs marked by narrow ring-scars formed by the deciduous bud spathes and by the similar paired stipules shed from developing leaves; stipules large and membranous, free from or adnate to the proximal portions of the petioles, quickly deciduous. Pith indistinctly diaphragmed, homogeneous between the diaphragms. Leaves simple, alternate, petioled, blades pinnately veined, with entire margins. Flowers large and showy in most, appearing terminally on branchlets of the season after the leaves are well developed (in ours). Perianth parts 9–15 in whorls of 3, the outer 3 sometimes differentiated from the others. Stamens and pistils numerous, both in close spirals on a long conical axis, stamens below the pistils. Filaments shorter than the anthers. Pistils collectively mature into a "fruiting cone," each producing a follicle dehiscing abaxially, seeds 2, scarlet to pink, for a while after dehiscence each hanging outside the follicle by a threadlike structure attached inside.

1. Leaf blades (at maturity) conspicuously gray or chalky-gray-glaucous beneath.
  2. Leaves evergreen (wintergreen)   7. *M. virginiana*
  2. Leaves deciduous.
    3. Shrub or small tree, usually with leaning main stems, branches crooked, individual shoots sometimes short-lived and sprouting from behind dying parts; "fruiting cones" subcylindric, mostly about 3 cm in diameter.   1. *M. macrophylla* subsp. *ashei*
    3. Tree to about 18 m tall, with a straight trunk; "fruiting cones" broadly oval to subglobose or globose, 5–6 cm in diameter.   2. *M. macrophylla* subsp. *macrophylla*
1. Leaf blades not conspicuously gray or chalky-gray-glaucous beneath.
  4. Leaves evergreen, blades leathery, lustrous dark green above, lower surfaces densely to thinly rusty-pubescent.   6. *M. grandiflora*
  4. Leaves deciduous, blades not leathery, not lustrous dark green above and not rusty-pubescent beneath.
    5. Leaf blades auriculate- or subauriculate-lobed basally.   3. *M. pyramidata*
    5. Leaf blades without basal lobing.
      6. Leaves closely set terminally on branchlets and collectively umbrellalike, blades 20–40 cm long, their bases long-tapered.   4. *M. tripetala*
      6. Leaves not disposed in an umbrellalike fashion at tips of branchlets, 10–25 cm long, their bases broadly rounded to broadly short-tapered.   5. *M. acuminata*

1. Magnolia macrophylla Michx. subsp. ashei (Weatherby) Spongberg. ASHE MAGNOLIA. Fig. 225

Deciduous, generally a shrub, seldom with the form and stature of a small tree when growing in the forest understory, horticultural specimens may be so if undisturbed and under favorable cultural conditions. In forests, for the most part, behaving as a shrub competing for light by leaning toward openings in the canopy or by having a loose, open, often more or less horizontal growth form. Branches tend to be crooked and with individual shoots short-lived, maintaining its form by sprouting from behind the dying or dead parts. Twigs rather stout, young ones densely pubescent with longish appressed-ascending hairs, eventually becoming green and glabrous during the first year. Older woody twigs brown, terminal winter buds 3–5 cm long, clothed with dense and compact pale gray pubescence, larger leaf scars buff-colored, shield shaped, relatively large, vascular bundle scars relatively numerous and scattered.

Leaves with relatively short and stout petioles, stipules (on emerging leaves) conspicuous, adnate to the petioles proximally, membranous, pale buffish in color, 8–10 cm long, usually longer than the petioles, quickly deciduous; emerging, *partially grown leaf blades* with pubescent petioles, upper surfaces of blades bright green, prominent midrib brown and pubescent, sparse, short pubescence near the margins and more or less ciliate on the margins, lower surfaces green subdued by pale gray, shaggy pubescence throughout; *mature blades*, although varying in size, notably large, 20–60 cm long and 10–30 cm broad at their broadest places, obovate or spatulate, generally narrowed from about their middles toward their bases, bases rounded to auriculate-lobed to either side of the petioles, apices usually broadly obtuse; upper surfaces green, essentially glabrous or pubescent along the major veins, lower surfaces with much of the pubescence sloughed, the surfaces very conspicuously chalky-whitish or -grayish.

Flowers large, fragrant, at anthesis at about the time the leaves are more or less half grown, about 2.5–3 dm across when at anthesis; perianth with a basal urn shape, the tips of the 9 segments arching-recurved prior to full anthesis, all arching-spreading at full anthesis, each segment creamy white and with an adaxial purplish blotch basally. "Fruiting cones" generally subcylindric, 4–5 cm long and mostly 3 cm in diameter, rosy red during early maturation, brown at maturity; seeds red (often drying brown). [*M. ashei* Weatherby]

Inhabiting deciduous or mixed woodlands, ravine slopes, steepheads, bluffs, sometimes on level wooded uplands; endemic to Fla. Panhandle from the Ochlockonee River to Santa Rosa Co.

The Ashe magnolia is very similar in many characteristics to the bigleaf magnolia which has a very much wider, although spotty, range, is much more abundant in many of the places in which it does occur, and has, by and large, a dramatically different growth habit. The more detailed description is given for the subspecific Ashe magnolia simply because in our limited area of geographic coverage it is the one more likely to be encountered, the subspecific bigleaf magnolia reaching our area only in southwesternmost Ala.

National champion Ashe magnolia (1971), 1'6" circumf., 35' in height, 18' spread, Torreya State Park, Fla.

2. Magnolia macrophylla Michx. subsp. macrophylla. BIGLEAF MAGNOLIA.

Tree to about 18 m tall, perhaps taller, commonly with a straight, upright trunk not unlike that of relatively young specimens of tulip-poplar, its crown elevated

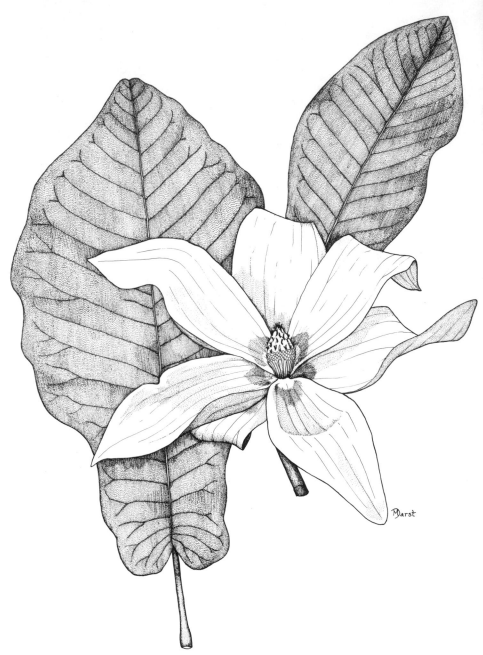

Fig. 225. **Magnolia macrophylla** subsp. **ashei**.

into the canopy until taller trees overtop it. Twigs, stipules, and leaf size and shape as in the subspecies *ashei*, the lower surfaces of mature leaf blades having the same conspicuous, chalky-white or -grayish appearance, the pubescence of lower surfaces generally denser and more of it persistent as the leaves age. Flowers average somewhat larger than in the Ashe magnolia, to about 4 dm across at full anthesis, the perianth parts longer and some of them sometimes much broader, "fruiting cones" broadly oval, subglobose, or globose, 5.5–8 cm long and 5–6 cm broad.

In rich low or upland woodlands of slopes or ravines; in our area, s.w. Ala. (Distribution rather interrupted or spotty, s. Ohio, Ky., Tenn., cen. N.C., s.e. S.C., n. cen. and n.w. Ga., cen. and w. Ala., n. La., n.e. Ark.)

National champion bigleaf magnolia (1972), 9'3" circumf., 59' in height, 62' spread, Baltimore, Md.

3. Magnolia pyramidata Bartr. PYRAMID MAGNOLIA. Fig. 226

A slender, deciduous tree to about 20 m tall and at least 3 dm d.b.h., glabrous throughout. Twigs relatively slender, year-old ones purplish brown, terminal buds slenderly nearly fusiform, leaf scars relatively small, subrotund, vascular bundle scars several and scattered, often obscure or not evident.

Leaves closely set near the tips of branchlets, spreading in a small umbrella-like fashion, slenderly short-petiolate, stipules adnate to the petioles proximally, conspicuous, membranous, buff-colored, more or less oblong, to about 8 cm long and 2 cm broad, quickly deciduous. Blades variable in size, 10–15 (20) cm long and 5–12 cm broad at their broadest places, tapered from above their middles to auriculate or subauriculate bases, the auricles rounded and slightly spreading if fairly pronounced (1–2 cm long), smaller ones not spreading, most blades with abruptly spreading, rounded shoulders or sublobes above the taper then narrowed to blunt tips, the overall shape somewhat kitelike; upper surfaces of mature blades dark green, lower pale green but not at all chalky.

Flowers fragrant, urnlike or cuplike before full anthesis, the 9 creamy white or yellowish, oblanceolate to elliptic perianth parts fully spreading at full anthesis, 12–18 cm across the open flower. "Fruiting cones" oblong to elliptic oblong in outline, mostly 4–9 cm long and about 3 cm broad, more or less rose-colored during preripening, turning purplish to tawny at maturity. Seeds red.

In mesic or submesic woodlands of bluffs, ravine slopes, slopes of steepheads, and level uplands; s.w. Ga., Fla. Panhandle from the Ochlockonee River westward, s. Ala. (With interrupted distribution, cen. S.C., upper s.e. and s.w. Ga., cen. and s. Ala., to e. Tex.)

National champion pyramid magnolia (1972), 6'4" circumf., 57' in height, 37' spread, Newton Co., Tex.

4. Magnolia tripetala L. UMBRELLA MAGNOLIA. Fig. 227

Deciduous small tree, 10–15 m tall and 15–40 cm d.b.h., sometimes with 2–several trunks from the base, trunks straight or often leaning if more than one, branches often somewhat contorted. Young stems stout, green, glabrous or pubescent, eventually becoming reddish brown, brown during the second year, gray during the third; leaf scars somewhat obtriangular in outline, the angles rounded; terminal bud about as thick as the stem below, tapering upwardly from the base, tip usually pointed, glabrous and glaucous, purple, about 3 cm long, sometimes falcate.

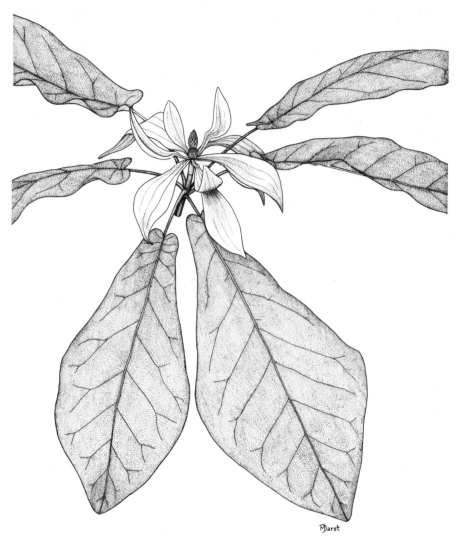

Fig. 226. **Magnolia pyramidata**.

Fig. 227. **Magnolia tripetala**.

Leaves closely clustered on branchlets and disposed in an umbrellalike fashion, with stoutish short petioles (stipules not seen by me). Unfolding blades glabrous above or nearly so, abundantly silky-pubescent beneath, some of the pubescence eventually sloughing; mature blades large, 20–40 cm long and 10–20 cm broad at their broadest places, elliptic to elliptic-obovate, broadest medially or somewhat above their middles and tapering to both extremities, the basal taper narrower than the apical, tips usually cuspidate.

Flowers ill-scented at close range, broadly urnlike before full anthesis, perianth parts 9 or 12, outer 3 greenish white and reflexed at full anthesis, the others white or cream-white, spreading or spreading and ascending. "Fruiting cones" oblong in outline, 8–12 cm long and about 4 cm broad, pink to red at maturity. Seeds dark pink to red.

For our area, known in only one locality, Okaloosa Co., Panhandle of Fla., on a north-facing wooded bluff and in a contiguous ravine system; also in a few localities in s. Ala. but these probably only peripheral to our range. (Irregularly distributed, e. Pa., s.e. Ohio, s. Ind., generally southward from Pa. to s.e. Va., thence to upper s.e. and s.w. Ga., the western boundary roughly cen. Ky., cen. Tenn., s. Ala., and s.e. Miss. with outliers in n.w. Ark., s.e. Okla.)

National champion umbrella magnolia (1969) 9′8″ [at 2′] circumf., 45′ in height, 48′ spread, Lumberville, Bucks Co., Pa.

5. Magnolia acuminata L. CUCUMBER-TREE. Fig. 228

Deciduous, medium-sized tree, to 25 m tall and to 1 m d.b.h. or a little more. Lateral branchlets of the season usually very short, bearing 2–3 leaves, sparsely to densely pubescent as are the petioles and stipules, the latter about 1.5 cm long, slender, adnate to the petioles for three-fourths of their length, quickly deciduous. Year-old woody twigs reddish brown, dotted by numerous, buff-colored, small lenticels, becoming gray after the first year; leaf scars narrow, broadly V-shaped, vascular bundle scars several, disposed in a line; terminal bud broadest basally and tapered upwardly, narrow, 1–2 cm long, densely clothed with long silky-silvery hairs.

Emerging leaf blades sparsely short-pubescent on their upper surfaces, densely silky-silvery-pubescent beneath, much of the pubescence sloughing, sometimes quickly, sometimes tardily. Mature blades broadly oval, elliptic or oblong-elliptic, ovate, or obovate, commonly relatively broad, 10–25 cm long and 8–15 cm broad at their broadest places, broadly rounded to broadly tapered basally, mostly short-acuminate apically, sometimes relatively long-acuminate, lower surfaces moderately short-shaggy-pubescent.

Flowers appearing on short branchlets (as the latter start to develop and when their leaves are small and young), more or less inverted bell–shaped, perianth parts 9, green, greenish yellow, brownish yellow, or orange-yellow, glaucous, outer 3 small, 2–3 cm long and reflexed, others variable in size, 4–8 cm long, concave, oblanceolate to obovate. "Fruiting cones" often with a cucumber shape, often irregularly knobby, sometimes curved, 3–8 cm long, dark red at maturity. Seeds red to reddish orange. [*Tulipastrum acuminatum* (L.) Small; incl. *M. cordata* Michx. (*T. cordatum* (Michx.) Small)]

Rich, mesic woodlands of slopes, ravines, and flats along stream courses; local, cen. panhandle of Fla., s. Ala. (Generally distributed, w. N.Y., w. Pa., e. Ohio, southward to s. S.C. and e. Tenn. and n. and n.e. Ga.; spottily distributed, s. Ont., e. Pa., cen. Va., w. cen. Ga., s. Ind. and s.e. Mo. to Ala., Miss., La., Ark.)

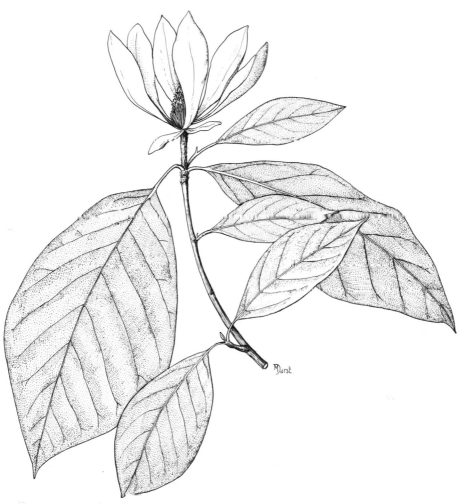

Fig. 228. **Magnolia acuminata**.

Fig. 229. **Magnolia grandiflora.** (Courtesy of the artist, Melanie Darst.)

6. Magnolia grandiflora L. SOUTHERN OR LOBLOLLY MAGNOLIA, BULLBAY.

Fig. 229

A handsome evergreen tree of moderately large proportions, 20–30 (40) m tall and 1 m d.b.h. or a little more. Young twigs clothed with a rusty brown, feltlike pubescence, the petioles and stipules similarly pubescent; year-old twigs dark brown, some of the pubescence (now dirty-brown) persisting, and with scattered, vertically lenticular, small, light brown lenticels; leaf scars hemispherical, relatively large, vascular bundle scars obscure or not evident; terminal buds narrowly conical, clothed with tight, brown, velvety tomentum.

Leaves short-petiolate, stipules free from the petioles, foliaceous, very quickly deciduous as leaves emerge. Emerging blades thin and flaccid, glossy green and glabrous above, usually densely rusty brown, heavily felty pubescent beneath, or on some trees thinly so. Mature blades stiff-leathery, oblong, broadly elliptic, or oval, infrequently narrowly obovate, most of them 10–20 cm long (to 30 cm on vigorous shoots or on sprouts) and 6–10 cm broad, bases acutely to obtusely tapered, apices obtuse, less frequently rounded, frequently abruptly short-acuminate; upper surfaces lustrous dark green, lower persistently and densely rusty-felty pubescent on some trees, on others various degrees of sloughing of the pubescence occurs so that surfaces are varyingly greenish and thinly rusty-brown pubescent, sometimes mottled green and thinly rusty-brown, sometimes dull green.

Flowers large and showy, very fragrant, cream-white, appearing after shoots of the season are well developed, blooming a few at a time over a rather extended period, each with a beautifully urnlike form prior to full anthesis after which the thickish perianth parts irregularly curved-spreading, outer 3 small and quickly deciduous, others, some of them at least, spoon-shaped, 6–15 in number, many of them as broad as long, 15–20 cm across the open flower. "Fruiting cones" oblong, ellipsoid, or subglobose, 5–10 cm long, follicles densely shaggy-pubescent with tawny hairs, with some pinkish or reddish coloration during maturation. Seeds bright, lustrous red.

Upland well-drained woodlands, mesic forests generally on slopes, ravines and steepheads especially, more formerly than now on or surrounded by shifting coastal dunes; in our area abundantly "seeding-in" in secondary mixed upland forests with closed canopies (many or most of them developed on old fields), also becoming prevalent in pine flatwoods where no longer burned, the seed source from shade trees about occupied or abandoned homesites; throughout our area. (Chiefly coastal plain, N.C. to cen. pen. Fla., westward to e. Tex.)

National champion southern magnolia (1978), 20'3" circumf., 86' in height, 96' spread, Bladen Co., N.C.

7. Magnolia virginiana L. SWEETBAY, SWAMPBAY, WHITEBAY.

Fig. 230

Tree, in our range evergreen and reaching about 30 m tall and to about 1 m d.b.h. Young stems and petioles copiously silvery silky-pubescent, pubescence, or much of it, becoming brown, persisting through the first year, sometimes through the second, or most of it or all of it sloughing and stems greenish. Older woody twigs relatively slender, dark brown, with widely scattered, small, buff-colored lenticels, leaf scars relatively small, more or less hemispherical to horizontally oval, vascular bundle scars several, irregularly scattered. Terminal buds finely silky-pubescent and with a sheen, about 2 cm long.

Leaves short-petiolate, stipules loosely silky-pubescent, adnate to the petioles

469

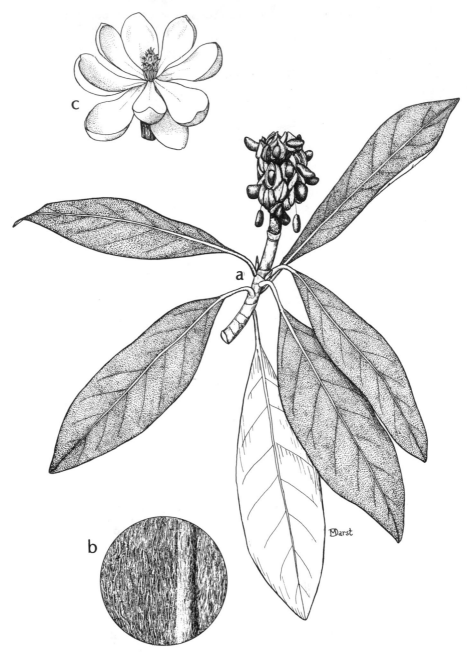

Fig. 230. **Magnolia virginiana:** a, branchlet with fruiting "cone"; b, small portion of lower surface of leaf blade, much enlarged; c, flower.

470

for most of the length of the petioles, the free portions extending well beyond, very quickly deciduous from emerging leaves. Emerging blades silvery silky-pubescent above mainly along the midrib, densely silvery silky-pubescent and with a sheen beneath; mature blades leathery, long-elliptic for the most part, 6–15 cm long and 2–6 cm broad; bases acute to rounded, apices usually blunt; pubescence of lower surfaces becoming dull silvery gray and as though tightly impressed into the heavy gray glaucescence of the surface, upper surfaces glabrous and rather a dull moderate green.

Flowers appearing after the leaves are fully grown, creamy white, fragrant, perianth parts 6–15, the outer 3 reflexed, oblong-spatulate, soon deciduous, others more or less erect, varyingly narrowly spatulate to obovate or suborbicular, 4–7 cm across the open flower. "Fruiting cones" glabrous, often knobby, ellipsoid, ovoid, or subglobose, 2–5 cm long, pinkish red, becoming dark purplish or reddish brown. Seeds red.

Plant occurring in our area of coverage and as described above is referable to var. *australis* Sargent; however, I do not know the precise limits of the respective ranges of var. *virginiana* and var. *australis*.

In swamps, bogs, shrub-tree bogs or bays, savannas, wet ravine and creek bottoms, wet flatwoods; throughout our area. (Chiefly coastal plain and outer piedmont, e. Mass. to s. pen. Fla., westward to e. Tex., s. Ark., w. Tenn.)

National champion sweetbay (1971), 13'1" circumf., 91' in height, 46' spread, Leon Co., Fla.

Meliaceae (MAHOGANY FAMILY)

Melia azedarach L. CHINABERRY, UMBRELLA-TREE. Fig. 231

Potentially a medium-sized, deciduous tree, often flowering and fruiting when of low, shrublike stature. Twigs stout, their bark purplish, smooth and dotted with small, scattered, buff-colored lenticels, terminal buds none; leaf scars prominent, raised, 3-lobed, vascular bundle scars 3, often indistinct, the axillary buds moundlike, densely stellate-pubescent; pith continuous, white.

Leaves long-petiolate; blades twice or thrice odd-pinnately compound, varying considerably in size on a given plant, mostly large, up to 5 dm long, larger ones sometimes with basal leaflets 2-pinnate, or they and those above 1-pinnate, but the distal 3–5 leaflets simple; primary leaflets with stalks 1–2 cm long, the distal simple leaflets sessile or nearly so excepting the terminal one which is acutely tapered to a somewhat longer stalk; some blades usually inequilateral, proximal ones ovate, passing upwardly to elliptic or lance-elliptic, the terminal one narrowly obovate, apices acuminate, margins serrate, irregularly incised-serrate, some sometimes 1–2-incised-lobed basally; upper surfaces dark green, often sparsely short-pubescent along the principal veins, lower a little paler and duller, generally glabrous.

Inflorescences loose, stalked panicles to 20 cm long from axils of lower leaves or leafless nodes on developing shoots of the season. Flowers bisexual, fragrant. Sepals 5 (6), green, pubescent, ovate or oblong, erect or incurved, about 2 mm long. Petals 5 (6), spreading or recurved, attractively pinkish purple or lilac, spatulate, 8–12 mm long, their distal portions somewhat spoonlike. Stamens 10–12, their filaments united into a dark purple cylindrical tube about 8 mm long and crowned by a fringe of twice as many, similarly colored, subulate

Fig. 231. **Melia azederach:** a, flowering branch; b, flower; c, two infructescences; d, fruit; e, node of woody twig with leaf scar and bud.

clefts, the off-white anthers within the fringe of the clefts. Ovary superior, surrounded at the base by a ringlike disc, 5–6-locular, style stout, stigma capitate, obscurely lobed, at the level of the anthers.

Fruit a thinly fleshy, subglobose drupe, yellow or greenish yellow at maturity, 1–1.5 cm in diameter, a single seed within the inner bony walls of each of the locules.

In woodlands and on woodland borders, fence and hedge rows, thickets, disturbed places about human habitations. Native of Asia, formerly planted to give shade, especially in rural areas, now so abundantly naturalized it is probably seldom planted. (E. Va., southward to s. Fla., westward to e. half of Tex., Okla.)

Features distinctive, amongst our woody plants, are the large, 2–3-odd-pinnately compound leaves with leaflets serrate, irregularly incised-serrate, or basally 1–2-incised-lobed, all parts lacking spines, prickles, or thorns. It is to be noted that the fruits, commonly persistent after leaf fall in autumn, have toxic properties if eaten.

National champion chinaberry (1967), 18'6" circumf., 75' in height, 96' spread, Koahe, South Kuona, Hawaii.

Menispermaceae (MOONSEED FAMILY)

1. Leaves not peltate.
   2. Flowers with 6 cream-colored, petaloid sepals, petals none; stamens 12 or more in staminate flowers, staminodes 12 in pistillate flowers; fruit black at maturity, about 20 mm long.   1. *Calycocarpum*
   2. Flowers with 6 sepals and 6 petals; stamens 6 in staminate flowers, staminodia 6 in pistillate flowers; fruit red at maturity, 6–8 mm long.   2. *Cocculus*
1. Leaves peltate, the peltation near the base of the blade.   3. *Menispermum*

1. Calycocarpum

Calycocarpum lyonii (Pursh) A. Gray. CUPSEED.   Fig. 232

A twining, scrambling, deciduous vine, sometimes wholly herbaceous (aboveground parts dying back in winter), sometimes eventually producing long woody stems to at least 1 cm in diameter, climbing into shrubs or trees, the seasonal growth from which is largely herbaceous; herbaceous stems rounded-ridged and -furrowed, glabrous or sparsely shaggy-pubescent (as are the petioles); woody stems brown; irregularly low-warty.

Leaves alternate, simple, without stipules; petioles slender, about equaling to considerably longer than the blades; blades varying greatly in shape and size, in overall outline broadly ovate, cordate basally or less frequently nearly truncate, as broad as long or broader than long, the largest ones to 20–25 cm long, palmately veined, edges entire or with few, remote, small dentations, unlobed or 3–5-lobed, infrequently 7–lobed, the sinuses between the lobes varying from U-shaped to urnlike in outline, apices of the lobes acuminate, surfaces usually bearing irregularly scattered spiculelike hairs.

Inflorescences mostly supra-axillary, minutely bracteate, narrow panicles, the flower stalks short, with 1–3 small bractlets, axes of the panicles and flower stalks pubescent. Flowers small, radially symmetrical, unisexual (plants dioecious), with 6 cream-colored sepals, 3 in each of two series, no petals; stami-

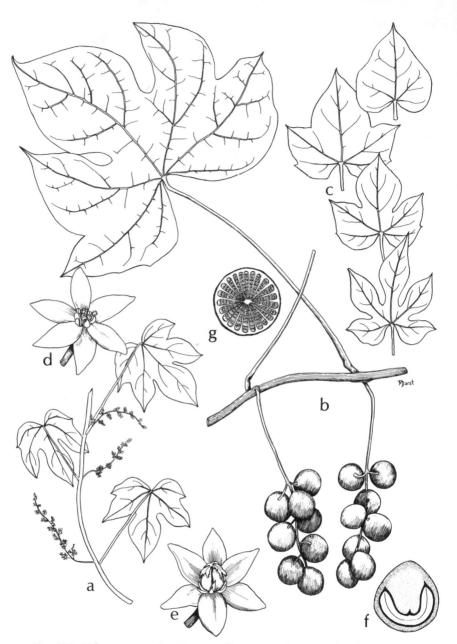

Fig. 232. **Calycocarpum lyonii:** a, small portion of stem with inflorescences; b, small portion of stem with infructescences and characteristic large leaf; c, leaves to show variation in form; d, staminate flower; e, pistillate flower; f, diagrammatic cross-section of fruit; g, diagrammatic cross-section of stem.

nate flowers with 12 stamens or 6 stamens and 6 staminodia; pistillate flowers with 12 very short staminodia, 3 pistils, ovaries superior, style none, stigma irregularly cleft-laciniate.

Usually 1 ovary per flower, sometimes more than 1, ripening into a globose or obovate-globose drupe about 2 cm long, green or greenish yellow then black when fully ripe (drying flattish on one side), surface smooth.

In alluvial woodlands, adjacent wooded slopes and ravines; Fla. Panhandle from about Jefferson Co. westward, adjacent s.w. Ga. and Ala. (S.w. Ga., Fla. Panhandle, westward to e. Tex., northward in the interior to Ky., s. Ill., s. Mo., e. Kan.)

2. Cocculus

Cocculus carolinus (L.) DC. REDBERRY MOONSEED, CAROLINA MOONSEED, SNAIL-SEED, CORALBEADS.   Fig. 233

A twining, scrambling deciduous vine, sometimes only woody basally, or woody stems long, reaching high in trees, the young branches herbaceous; oldest elongate woody stems to at least 2 cm in diameter, brown and conspicuously warty-roughened.

Leaves alternate, simple, without stipules, petioles slender, a little shorter than to about equaling the blades, sparsely to densely shaggy-pubescent; blades varying in shape and size, commonly as broad basally as long, to about 14 cm long but mostly shorter, palmately veined, edges entire, broadly ovate to triangular-ovate, unlobed, regularly or irregularly with 1–2 broadly sinuate lobes on a side, sometimes hastate, bases truncate to subcordate, apices broadly rounded to acute, commonly minutely mucronate, surfaces sparsely to relatively densely pubescent, in the latter case soft to the touch.

Inflorescences mostly supra-axillary, minutely bracteate panicles of variable length, the staminate to 15 cm long, the pistillate mostly shorter, panicles sometimes terminating branches, their axes, flower stalks, and bracts shaggy-pubescent. Flowers small, greenish, unisexual (plants dioecious), radially symmetrical, ovaries superior, flower stalks with 1–3 bractlets. Sepals 6, pubescent exteriorly, in 2 series of 3 each, the outer lance-acute, very much shorter and narrower than the broadly ovate, blunt or rounded inner ones. Petals 6, shorter and narrower than the inner sepals, each inflexed below and loosely clasping the base of a filament (in staminate flower) or a staminode (in pistillate flower). Pistils 6, disposed in a ring upon a short, stout stalk, each pistil shaped somewhat like a tenpin with a strongly bent neck.

One or more ovaries per flower maturing into a bright red drupe 6–8 mm across, round in outline but concave centrally on each face; faces and edges drying roughened ridged-wrinkled save for the central depression, presumably because of the bony endocarp being somewhat sculptured and coiled.

In woodlands of various mixtures, thickets, hedge rows, on fences, throughout our area. (Va. to Ind., Kan., generally southward to cen. pen. Fla. and Tex.; n.e. Mex.)

3. Menispermum

Menispermum canadense L. MOONSEED.   Fig. 234

Twining vine with ribbed stem, the lower stem smooth, slender, brown, woody, upper stem herbaceous.

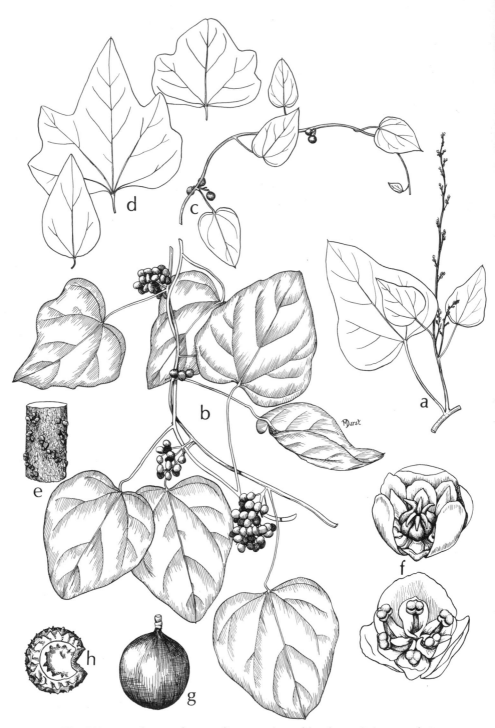

Fig. 233. **Cocculus carolina:** a, flowering branchlet; b, medial piece of vine with infructescences; c, tip of vine with infructescences; d, variable leaf blades; e, small portion of woody stem (piece drawn was 1.5 cm in diameter); f, staminate flower below, pistillate flower above; g, fruit; h, seed.

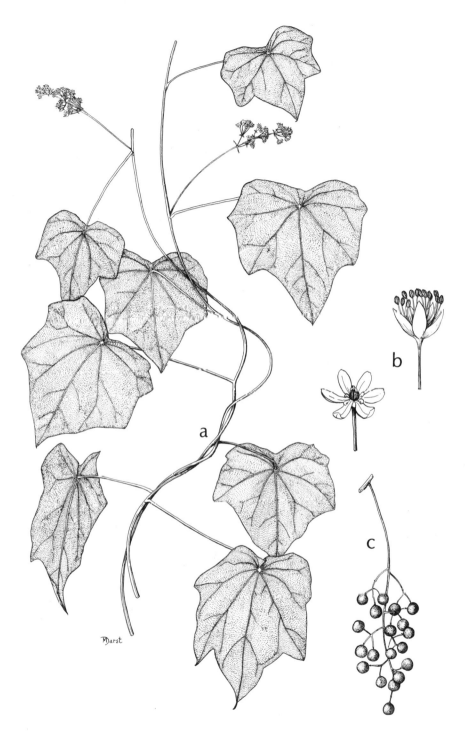

Fig. 234. **Menispermum canadense:** a, pieces of vine with inflorescences; b, flowers, staminate to right, pistillate to left; c, infructescence.

Leaves simple, alternate, slenderly long-petioled; blades thin-membranous, palmately veined, peltate, the peltation near the base of the blade, sometimes very close to it; usually some blades broadly cordate to subreniform and unlobed (or only a suggestion of lobing), their bases subtruncate, apices obtuse but with a small mucro; most leaves cordate basally, broadly ovate to sub-reniform in outline and with 3 broadly triangular, blunt, sometimes minutely mucronate lobes, central lobe largest, sometimes with a suggestion of an additional pair of rounded lobes toward the base; upper surfaces bright green and with a few widely scattered hairs, the lower surfaces grayish green and glaucous, sparsely shaggy-hairy; blades varying considerably in size, 3–15 cm long, often broader than long.

Inflorescences slenderly longish-stalked, supra-axillary, and sometimes terminal panicles 3–15 cm long. Flowers minute, greenish yellow or greenish white, radially symmetrical, unisexual (plants dioecious). Sepals 4–8, often 6, in 2 series, the outer oblongish and a little smaller than the inner elliptic ones. Petals 6–8, usually considerably smaller than the sepals, their sides involuted in the staminate flowers. Staminate flowers with 12–24 exserted stamens. Pistillate flowers with petals somewhat larger than in the staminate, short staminodia 6–12, pistils usually 3, borne on a shortly raised receptacle, styles short, stigmas somewhat dilated, spreading outwardly; ovaries superior. (Flowering infrequently observed in our area.)

Fruit a 1-stoned, shortly stipitate, somewhat obliquely globose, blue-black drupe about 1 cm in diameter.

Rich woodlands, bluff woodlands, streambanks, thickets. Infrequent in n. Fla., s.w. Ga., s.e. Ala. (Que. to Man., generally southward to n. Fla., s.e. Ala., Tenn., Ark., Okla.)

Moraceae (MULBERRY FAMILY)

1. Leaf blades pinnately veined, their margins entire; usually, not always, some twigs armed with axillary spines; "fruit" orangelike.   2. *Maclura*
1. Leaf blades subpalmately veined (having 2 lateral veins and the midrib subequal and arising together from the base of the blade), margins toothed; twigs never armed.
  2. Leaves alternate, opposite, or whorled (often all three conditions on the same plant, especially on saplings or sprouts); twigs of the season and petioles with long, spreading, transparent-glassy hairs.   1. *Broussonetia*
  2. Leaves all alternate; twigs of the season and petioles glabrous, or if pubescent the hairs short and soft.   3. *Morus*

1. Broussonetia

Broussonetia papyrifera (L.) Vent. PAPER MULBERRY.   Fig. 235

Deciduous, fast-growing small tree, commonly producing root-sprouts and thicket-forming, flowering when of shrublike stature. Bark of trunk tan and smooth or only moderately furrowed. Bark of woody twigs brown, with scattered, circular, slightly raised corky lenticels. Terminal buds none, axillary buds dark brown, short-pubescent, scales 2–3. Leaf scars somewhat elevated, nearly circular, vascular bundle scars several, usually indistinct. Sap milky. Twigs of the season and petioles having long, spreading, transparent-glassy pubescence. Pith homogeneous and continuous.

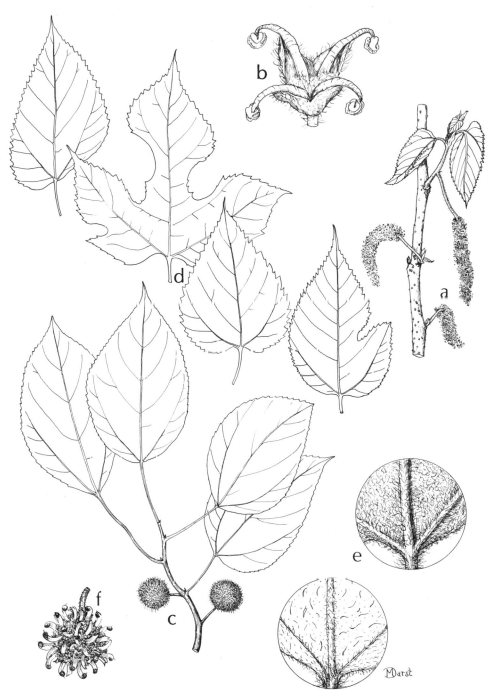

Fig. 235. **Broussonetia papyrifera:** a, portion of branch from staminate plant with catkins; b, staminate flower; c, portion of branch from pistillate plant with ball-like pistillate catkins; d, variable leaves from young plant; e, enlargement of small portion of base of leaf, lower surface above, upper surface below; f, mature aggregation of fruits.

Leaves simple, alternate, opposite, or in whorls of 3 (often all three conditions on the same plant, especially on saplings or sprouts), stipules conspicuous but quickly deciduous, ovate or ovate-oblong, attentuate-tipped, petioles of varying lengths, shorter than to as long as the blades. Mature blades ovate in overall outline, unlobed or variously lobed (see illustration), the lobed leaves more commonly on saplings or sprouts; size of blades variable, 6–20 cm long and 5–15 cm broad or more, often as broad as long, largest leaves with longest petioles usually on saplings and sprouts; bases truncate to broadly rounded, sometimes subcordate, occasional ones oblique, apices or apices of lobes acuminate; margins or margins of lobes, regularly or irregularly serrate above the base; upper surfaces of unfurling blades green and relatively sparsely pubescent, lower surfaces thickly and compactly grayish-pubescent; upper surfaces of mature blades dark green, notably scabrid with persistent bases of hairs, lower surfaces velvety-pubescent.

Flowers small, unisexual, plants dioecious (in much of our area of coverage, all plants staminate). Staminate flowers many, borne in stalked, pendent, flexuous, cylindric catkins, some catkins arising from buds axillary to leaf scars on wood of the previous year, some on developing shoots of the season, either below the leaves or axillary to them, each flower with a deeply 4-lobed calyx and 4 stamens. Pistillate flowers many, coherent on a common ball-like axis together with which forming a dense, globose, stiffly short-stalked, axillary globe, each flower with a 4-lobed calyx closely surrounding the ovary, style elongate-filiform. Collective "fruit" ball-like, 2–3 cm in diameter, the individual parts mainly enlarged calyces enclosing red or orange achenes which protrude at full ripening.

Native of Asia, now widely naturalized in our area and beyond, principally in the vicinity of human habitations.

National champion paper mulberry (1972), 26'7" [at 3.5'] circumf., 27' in height, 32' spread, Wood Co., Tex.

2. Maclura

Maclura pomifera (Raf.) Schneid. OSAGE-ORANGE.   Fig. 236

Deciduous shrub or small tree, often root-sprouting and forming thickets. Trunk often short, malformed, and supporting stout, ascending, often contorted branches collectively forming an open, nearly spherical crown. Bark of trunk orangish brown, thickly and irregularly furrowed between interlacing ridges. Woody twigs green to orangish brown, smooth, some of them having short spur-shoots, some on most trees having axillary sharp spines to 3.5 cm long. Terminal buds none, axially buds inconspicuous, often somewhat sunken. Leaf scars triangular to reniform, vascular bundle scars 3–5. Pith homogeneous and continuous.

Leaves simple, alternate (those on short spur-shoots sometimes terminally fascicled), stipulate, stipules small and very quickly deciduous, petioles slender 3–5 cm long. Blades pinnately veined, mostly ovate, 7–15 cm long and 5–8 cm broad, occasionally lanceolate and 3–4 cm broad, bases truncate, rounded, sometimes tapered, apices strongly acuminate, margins entire, upper surfaces dark green, glabrous, and sublustrous, lower surfaces pale and dull, sparsely pubescent.

Flowers unisexual, plants dioecious, appearing after the leaves are well developed. Staminate flowers numerous in loose, globose to oblong, stalked heads

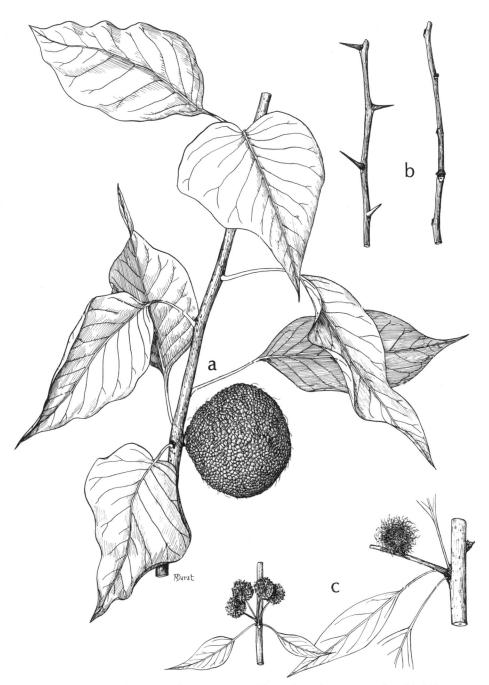

Fig. 236. **Maclura pomifera:** a, piece of branch with a mature "multiple" fruit; b, twigs with and without thorns; c, staminate inflorescences to left and a pistillate one to the right.

axillary to leaves on short spur-shoots, each flower with a 4-lobed calyx and 4 stamens. Pistillate flowers many, in stalked, axillary, spherical heads, each flower with a 4-lobed calyx, a pistil with a nearly spherical, sessile ovary and a greatly elongated filiform style, collectively the styles imparting to the inflorescence a peripherally long-hairy aspect. The collective "fruit" green, orangelike in form, surface wrinkled, composed of the many very much enlarged, fleshy, and compacted calyces, each bearing an achene within it.

Native to s. Ark., e. Okla., and n.e. Tex., widely naturalized eastward, abundantly so in some places; in our area uncommon and sporadic, in hedge rows and woodlands, mostly near human habitations, s.w. Ga., cen. Panhandle of Fla., s. Ala.

National champion osage-orange (1972), 24'6" circumf., 51' in height, 93' spread, Patrick Henry Estate, Brookneal, Va.

3. Morus (MULBERRIES)

Deciduous, unarmed trees, often flowering or fruiting when of shrublike stature. Terminal bud none. Bark of twigs tan, with scattered lenticels, bark of trunks brown or grayish brown, irregularly plated and separating surficially into appressed scales. Leaf scars half-round or oval, several vascular bundle scars in each. Pith homogeneous and continuous. Sap milky, at least early in the season. Leaves alternate, simple, conspicuously long-petiolate, stipulate, stipules quickly deciduous. Blades broadly ovate in overall outline, unlobed or variously lobed, subpalmately veined, margins toothed. Flowers unisexual, borne in stalked staminate or pistillate catkins, one or the other on different branches of the same plant or separately on different plants, mostly the latter; catkins borne on emerging shoots of the season, singly in axils of developing leaves or, usually, 1 or 2 catkins below the lowest leaf, the latter usually emerging from the bud just before any appreciable elongation of the shoot occurs. Staminate catkins elongate-cylindric, lax or pendent, relatively loosely-flowered; pistillate catkins oblongish, not pendent, the flowers compacted. Staminate flowers composed of 4 sepals and 4 stamens; in the bud the sepals short and clasping the stamens filaments of which at that stage are short; at full anthesis the sepals elongated and narrow, enfolding about two-thirds of the filaments, the upper parts of filaments and anthers exserted. Pistillate flowers composed of 4 sepals of 2 sizes, these so closely embracing the ovary as to be discernible only with difficulty. Ovary 2-locular, one locule small and disappearing during further development, stigmas 2, linear. Fruit an achene surrounded by the calyx which has become succulent and juicy during development, the collective fruits forming the mulberry.

Plants of the two species of mulberry occurring here are very similar in general appearance, have leaves similar respecting variation in size and form (see illustration of *M. alba*), but in general, differing in the character and distribution of pubescence. Where they grow intermixed (usually in places where naturalized from plantings), individuals occur which appear to show intermediacy in the nature of pubescence on leaf blades suggesting the possibility of hybridization.

1. Upper surfaces of leaf blades glabrous and smooth to the touch; lower surfaces glabrous, pubescent along the major veins, or the pubescence only in tufts in axils formed by principal lateral veins and the midrib.   1. *M. alba*

1. Upper surfaces of leaf blades usually having short-stiff, antrorsely subappressed hairs and scabrid, or if the hairs eventually sloughed, usually their somewhat enlarged bases persisting, thus the scabrity persisting; lower surfaces of unfurling leaf blades densely soft-pubescent, those of mature blades shaggy-pubescent, soft to the touch, the hairs on both the major veins and the veinlets, usually on the surfaces of islets within anastomosing veinlets as well.   2. *M. rubra*

1. Morus alba L. WHITE MULBERRY.   Fig. 237

Very young twigs and petioles irregularly softly and sparsely pubescent or glabrous, the hairs, if any, sloughing quickly. Leaf blades mostly ovate in overall outline, sometimes broadly oblong, sometimes approximately orbicular, unlobed or variably 2–9-lobed (lobed leaves, in general, much more characteristic on saplings and sprouts than of crown branches although on some individuals some or many of the crown blades lobed); blades often as broad as long, very variable in size, from a few cm long or broad to about 20 cm, bases truncate, rounded, or subcordate, sometimes oblique, apices or apices of lobes abruptly short acuminate or cuspidate, conspicuously acuminate less frequently, sometimes acute or obtuse; margins serrate; upper surfaces dark green and sublustrous, lower slightly paler and duller—see key above for nature and distribution of pubescence. "Fruits" mostly cylindric-oblong, 1–2 cm long or a little more, white, pinkish, pale purple, or nearly black.

Native of e. Asia, formerly cultivated, now naturalized in moist secondgrowth woodlands of stream bottoms, river banks, waste places and vacant lots, fence and hedge rows; throughout our area (and in much of N.Am. and Eur.)

National champion white mulberry (1976), 16'1" circumf., 82' in height, 93' spread, Kalamazoo Co., Mich.

2. Morus rubra L. RED MULBERRY.

Very young twigs and petioles uniformly to irregularly minutely pubescent or glabrous, hairs on petioles, if any, sometimes persisting. Leaf form as in *M. alba*, size perhaps averaging larger, apices or apices of lobes more consistently and prominently acuminate. See key above for nature and distribution of pubescence. "Fruits" 2.5–5 cm long, dark red or dark purplish red.

A native forest tree, usually widely scattered in rich woodlands, rises in bottomland woodlands and floodplain woodlands, much less frequently in waste places and vacant lots than is *M. alba*; throughout our area. (Vt. to S.D., generally southward to s. pen. Fla. and e. and cen. Tex.)

Reports of black mulberry, for example, in Small (1933) and Long & Lakela (1971), probably based on black-"fruited" race of *M. alba*, often distinguished as *M. alba* var. *tartarica* (L.) Ser.

National champion red mulberry (1981), 18'8" circumf., 72' in height, 98' spread, Berrien Co., Mich.

Myricaceae (BAYBERRY FAMILY)

Myrica

Evergreen (ours) shrubs or small trees. Leaf scars small, variable in shape, more or less triangular (angles rounded), half-moon-shaped, obcordate, or crescent-shaped, each with 3 vascular bundle scars. Leaves simple, alternate, pinnately veined, one or both surfaces of blades finely resinous-punctate-dotted; stipules

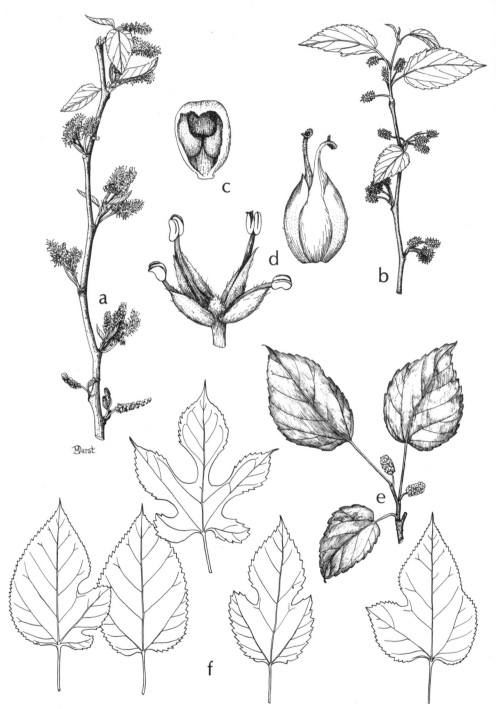

Fig. 237. **Morus alba:** a, twig with staminate catkins; b, twig with pistillate catkins; c, sepal in bud stage with stamen; d, flowers, staminate at left, pistillate at right; e, fruiting branchlet; f, variable leaves of young plants or sprouts.

none (in ours), crushed foliage aromatic save in *M. inodora*. Flowers unisexual (plants dioecious), without perianths, borne in short, scaly-bracted, erect catkins axillary to leaf scars or to leaves on wood of a previous season in early spring prior to development of new shoots of the season. Staminate flowers solitary in the axils of primary, usually ovate bracts, stamens 2–20, usually 4–8, the short filaments free or united basally. Pistillate flowers subtended by an ovate primary bract and usually by 2–4 bractlets in addition, these sometimes absent; pistil 1, ovary superior, 1-locular, 1-ovulate, style short, stigmas 2, linear. Fruit a nutlike drupe, in most covered by small protuberances and heavily coated with gray wax.

1. Leaves not aromatic when crushed, margins of their blades entire; drupes a little longer than broad, short-ellipsoid.   3. *Myrica inodora*
1. Leaves aromatic when crushed, margins of some blades on a given plant, usually on most, with at least one or a few remote teeth on the sides distally; drupes as broad as long, globose or short-ovoid.
  2. Leaf blades finely punctate on both surfaces.   1. *M. cerifera*
  2. Leaf blades finely punctate only on their lower surfaces.   2. *M. heterophylla*

1. Myrica cerifera L. SOUTHERN BAYBERRY, WAX-MYRTLE.   Fig. 238

Aromatic shrub or small tree to about 12 m tall, commonly with several trunks from near the base, on some sites (presumably where burned frequently) a low, slender shrub with subterranean runners and clonal. Leafy twigs brown, glabrous or sparsely to densely shaggy-pubescent, scaly-glandular; older twigs becoming grayish and having nearly circular, corky lenticels.

Leaves thin, oblanceolate (narrowly so in dwarfed specimens), occasional ones narrowly obovate, mostly 3–8 (15) cm long and 1–2 cm broad (averaging smaller on dwarfed specimens), cuneately narrowed from somewhat above their middles to short petioles or subpetioles, apices acute, tapered to blunt tip, or rounded; some leaves, sometimes most of them, on a given plant with few, remote serrations distally, usually some entire; both surfaces punctate, commonly abundantly so, usually with amber to brown dots or scales of glandular exudate, both surfaces glabrous, or the midribs of one or the other or both surfaces sparsely short-pubescent, occasionally sparsely hairy along the margins, occasionally so along the principal lateral veins beneath. Leaves on branchlets of the season tending to be reduced in size on the distal portions of the branchlets. Mature drupes globose or short-ovoid, 2–3 mm across, the surficial tubercles usually compacted, the entire outer surface white- or gray-waxy; wax, or some of it, tending to be very thinly stringy and with magnification giving the effect of pubescence. [*Cerothamnus ceriferus* (L.) Small]

Inhabiting a wide variety of kinds of sites, fresh to slightly brackish banks, shores, flats, and interdune swales, pine savannas and flatwoods, cypress-gum ponds and swamps, wet hammocks, bogs, upland mixed woodlands, old fields, fence and hedge rows; throughout our area. (Coastal plain, s. N.J. to s. pen. Fla. and Fla. Keys, westward to e. Tex., s.e. Okla., Ark., Berm., W.I., Cen.Am.)

Several characteristics combine to render the wax-myrtle a popular ornamental: it is a fast grower, is heavily clothed with attractive evergreen foliage, and responds well to pruning.

The dwarfed, clonal form is recognized as varietally distinct from *M. cerifera* by some authors, as specifically distinct by some: as *Cerothamnus pumilus* (Michx.) Small by Small (1933); as *Myrica pusilla* Raf. by Fernald (1950), by Gleason (1952), and by Long & Lakela (1971); as *M. cerifera* var. *pumila* Michx. by Radford et al (1964).

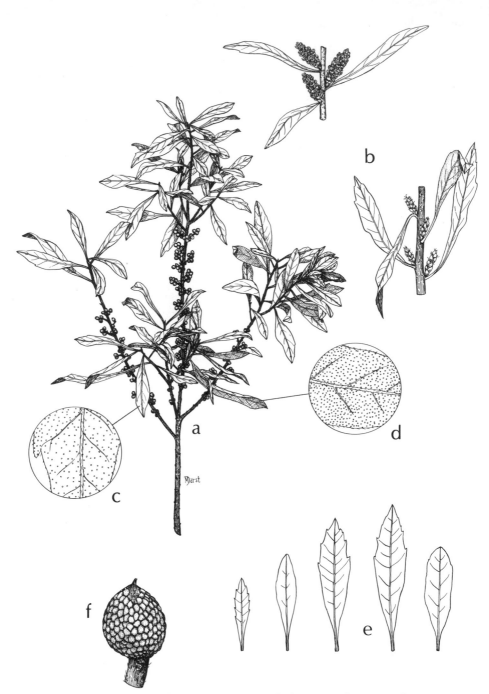

Fig. 238. **Myrica cerifera:** a, fruiting branch; b, pieces of twigs, with staminate catkins above, with pistillate catkins below; c, small portion of upper
leaf surface enlarged; d, small portion of lower leaf surface enlarged; e, variable leaves; f, fruit.

486

National champion southern bayberry (1978), 3′ circumf., 28′ in height, 33′ spread, Bradenton, Fla.

2. Myrica heterophylla Raf. SWAMP CANDLEBERRY. Fig. 239

Aromatic shrub to about 3 m tall. Young stems and petioles glabrous on some plants, generally, however, shaggy-pubescent and much or all of the pubescence persistent during the first year. Stems dark brown to almost black when fully mature, glandularness usually relatively sparse; older portions of twigs gray.

Leaves thickish and subleathery, mostly oblanceolate or narrowly obovate, some sometimes elliptic, variable in size to 12 cm long and 3 cm broad, cuneately narrowed below to short petioles or subpetioles, apices obtuse to rounded, tips often minutely apiculate; margins sometimes entire, for the most part with 1–few remote serrations on each side distally, occasionally only 1 serration on one side; upper surfaces glabrous or sparsely short-pubescent along the proximal portion of or all of the midrib, often on or near the edges as well, not glandular-punctate, lower surfaces punctate, with amber or brownish dots of glandular exudate in the punctae or above them, usually sparsely pubescent along the midrib, sometimes along the major lateral veins as well. Drupes globose or short-ovoid, 2–4.5 mm across, surficial tubercles compacted or not, sometimes thickly coated with white or grayish wax, sometimes with little wax and dark brown, wax tending to be very finely stringy and with magnification giving the effect of pubescence. [*Cerothamnus carolinensis* sensu Small (1933) in part]

Evergreen shrub-tree bogs and bays, depressions in flatwoods, hillside bogs in pinelands, wet pine savannas and flatwoods; throughout our area. (Coastal plain and piedmont, s. N.J. to cen. pen. Fla., westward to e. Tex.)

Leaves of *M. heterophylla*, when fully mature on branchlets of the season, do not tend to be reduced on the distal portions of the branchlets as in *M. cerifera*. Most of the leaves tend to be more conspicuously oblanceolate than those of *M. inodora* which, above the cuneate bases, are elliptic-oblanceolate. They differ from both *M. cerifera* and *M. inodora* in lacking punctations on their upper surfaces. *M. inodora* differs from both in that its crushed leaves are not aromatic.

3. Myrica inodora Bartr. ODORLESS BAYBERRY OR WAX-MYRTLE. Fig. 240

Nonaromatic shrub or small tree to about 7 m tall. Young twigs rust-colored, without hairs, copiously scaly-glandular; older woody twigs grayish and smoothish.

Leaves leathery, narrowed below to short petioles or subpetioles, expanded portion of blades mostly elliptic or elliptic-oblanceolate, seldom narrowly obovate, commonly somewhat cupped, the entire margins somewhat revolute, mostly 4–8 cm long and 2–3 cm broad, apices obtuse or rounded; both surfaces very finely and abundantly glandular-punctate; petioles with inconspicuous short-pubescence on their adaxial sides, this often extending the length of the adaxial midrib (pubescence not at all evident without magnification). Drupes seldom globose, usually oblong-oval in outline, mostly 6–7 mm in diameter, tubercles somewhat pillared, dark brown or nearly black, sometimes with white waxy covering, sometimes wax evident only in the interstices between the tubercles, sometimes not evident at all. [*Cerothamnus inodorus* (Bartr.) Small]

Shrub-tree bogs and bays, nonalluvial acid swamps, pine flatwoods, hillside bogs in pinelands, semiopen boggy places by small streams, cypress ponds; Panhandle of Fla. from Leon and Wakulla Cos. westward, s.w. Ala., s. Miss.

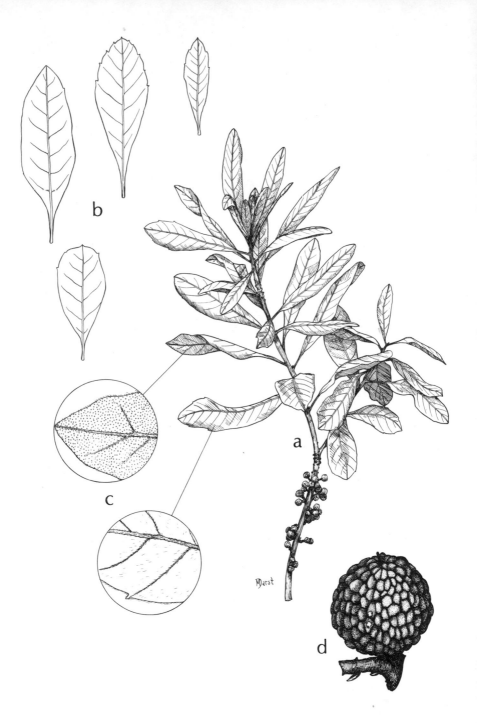

Fig. 239. **Myrica heterophylla:** a, fruiting branch; b, variable leaves; c, small portions of leaf surfaces enlarged, upper surface below, lower surface above; d, fruit.

Fig. 240. **Myrica inodora.**

Nyssaceae (SOUR GUM FAMILY)

Nyssa

Trees, individuals sometimes flowering and fruiting when in a shrublike stage. Leaf scars with 3 vascular bundle scars. Pith with transverse diaphragms, homogeneous between the diaphragms. Leaves simple, alternate, deciduous, petiolate, without stipules; blades pinnately veined, margins entire or with 1–few coarse dentations. Flowers radially symmetrical, unisexual or bisexual, plants androdioecious, that is, some individual plants having only staminate flowers and others having all bisexual ones (at least structurally bisexual although stamens may not always be fertile in a given flower); staminate flowers much smaller than either the pistillate or bisexual ones. Staminate flowers with a very short floral tube bearing perianth segments on its rim, 5 minute sepals or none, 5 or 10 petals or none, the throat of the floral tube closed by a fleshy, nectariferous disc, stamens 5–10, their filaments inserted at the edge of the disc. Pistillate and bisexual flowers with a much more pronounced floral tube wholly adnate to the ovary, a fleshy nectariferous disc capping it, the stamens, if present, borne at the edge of the disc, sepals and petals, if any, on the rim of the floral tube; petals quickly deciduous or very easily detached. Flowers or inflorescences borne proximally on shoots of the season. Ovary 1-locular and 1-ovulate. Fruit a drupe.

1. Petioles of most mature leaves on a given tree 3–6 cm long.   1. *N. aquatica*
1. Petioles of mature leaves on a given tree 0.5–2 (2.5) cm long.
  2. Lower surfaces of leaf blades with a decidedly grayish green hue and a texture soft to the touch; staminate flowers sessile in compact ball-like heads; pistillate flowers solitary; drupes red at maturity, 2–4 cm long.   2. *N. ogeche*
  2. Lower surfaces of leaf blades paler green than those of the upper, but not with a decidedly grayish hue and without a soft texture; staminate flowers mostly stalked and loosely aggregated in short racemes; pistillate flowers usually 2 or more at or near the end of a stalk although some on a given plant may be solitary; drupes blue-black at maturity, not exceeding 1 cm long.
    3. Leaf blades membranous and pliable, mostly obovate to elliptic-oblong and most of them acuminate apically; bark of trunks relatively rough and notably blocky with interlacing ridges and fissures.   3a. *N. sylvatica* var. *sylvatica*
    3. Leaf blades mostly narrowly elliptic to lance-elliptic, lanceolate, or oblanceolate, subleathery and stiffish, most of them acute to rounded apically; bark of trunks usually not notably blocky, but with interlacing fissures and ridges.
        3b. *N. sylvatica* var. *biflora*

1. Nyssa aquatica L. WATER TUPELO.   Fig. 241

A tree attaining about 30 m in height, with swollen buttressed base. Twigs of the season pubescent or glabrous.

Petioles 3–6 cm long, sparsely pubescent; blades variable in size, even on a single branch, ovate or ovate-oblong, varying from about 6 to 30 cm long, bases rounded to subcordate or broadly tapered, apices mostly short-acuminate, occasionally acute, rarely rounded; margins entire or usually some on a given tree with 1–several coarsely dentate teeth or lobes on a side; blades of emerging leaves densely soft-pubescent beneath, becoming sparsely pubescent only along the veins in age, their upper surfaces at first pubescent on the veins, later glabrous, lower surfaces of mature leaves much paler beneath than above.

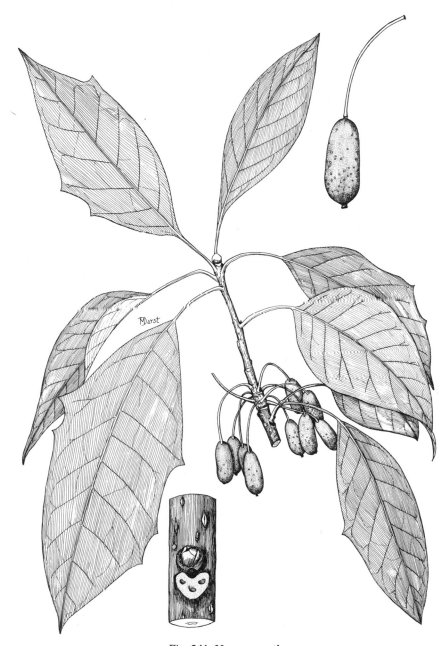

Fig. 241. **Nyssa aquatica.**

Staminate flowers numerous, sessile, in compact, ball-like clusters; pistillate or bisexual solitary, subtended by 2–several unequal bracts at the ends of flexuous stalks 2–4 cm long.

Drupes dark blue or dark purple at maturity, elliptic-oblong to narrowly obovate in outline, 1.5–3 cm long, usually somewhat shorter than their stalks, the latter drooping. Stone of the drupe with 8–10 sharp, longitudinal ridges. [*Nyssa uniflora* Wang.]

Floodplain forests, swamps, pond and lake margins, usually where the bases are inundated during extended intervals and where the soils are clayey or finely silty; in our area, cen. Panhandle of Fla. westward, s.w. Ga., s. Ala.; Duval Co. in n.e. Fla. and lower Suwannee River drainage. (Coastal plain, Va. southwesterly to n. Fla., westward to e. Tex., northward in the interior to s.e. Mo., s. Ill., w. Ky., w. Tenn.)

National champion water tupelo (1972), 27'1" circumf., 105' in height, 58' spread, Kinder, La.

2. Nyssa ogeche Bartr. ex Marsh. OGEECHEE-LIME, OGEECHEE TUPELO.

Fig. 242

Tree to about 20 m tall; frequently with numerous erect branches from a leaning trunk, more commonly with several upright, often crooked trunks from near the base, in some places where original single trunks died or were broken off several to numerous sprouts grew outward from the base, these eventually growing together to form a very large, more or less coalesced or composite base with separate upright trunks from it. Twigs of the season pubescent.

Petioles mostly 1–2 cm long, rarely 3 cm, pubescent; blades varying in shape and size, even on an individual branch, elliptic or subelliptic, narrowly obovate, or oblong-oval, to 1.5 dm long and 8 cm broad; bases mostly cuneate, infrequently rounded or subcordate, apices acute, barely acuminate, obtuse, or rounded; margins usually entire, very rarely with 1–several coarsely dentate teeth on a side; emerging leaf blades sparsely pubescent above, becoming glabrous in age, the lower surfaces very densely soft-pubescent, becoming relatively sparsely soft-pubescent in age, much paler beneath than above, having a decidedly grayish hue and soft to the touch.

Staminate flowers numerous, sessile, in compact ball-like clusters; pistillate or bisexual ones solitary, subtended by 2–several unequal bracts at the ends of stalks 1.5–2 cm long.

Drupes red at maturity, sometimes with only a red blush, oblong to obovate, 2–4 cm long, at full maturity equaling or longer than their stalks, the latter stiffish and not drooping. Stone of the drupe with papery wings extending outward through the flesh to or nearly to the skin.

River swamps, bottomland woodlands along small streams, sloughs, bayous, pond and lake margins. Coastal plain, southeasternmost S.C., s.e. and s. cen. Ga., across n. Fla. from the e. to e. Jackson and Bay Cos., from thence in a wedge to southernmost Okaloosa Co., a tongue from Jackson Co., Fla., into Seminole Co., s.w. Ga.

National champion ogeechee tupelo (1976), 8' circumf., 60' in height, 64' spread, St. Vincent National Wildlife Refuge, Fla.

3. Nyssa sylvatica Marsh. SOURGUM, BLACKGUM, PEPPERIDGE.

The "*Nyssa sylvatica* complex," as it may be called, is relatively wide-ranging, exhibits considerable morphological variation, and inhabits in general two

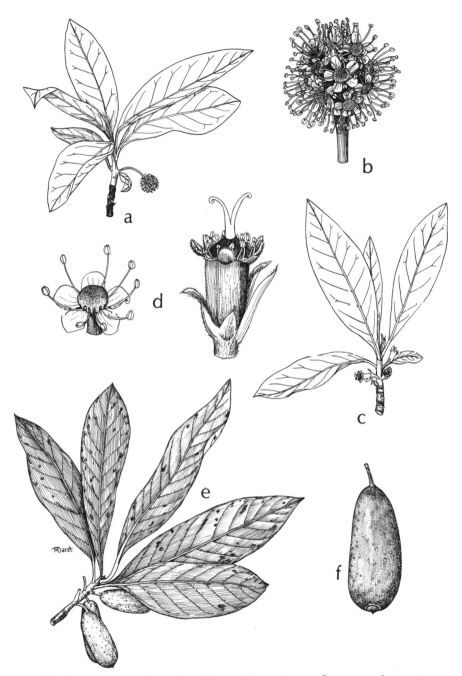

Fig. 242. **Nyssa ogeche:** a, branchlet with staminate inflorescence; b, staminate inflorescence enlarged; c, branchlet with pistillate flowers; d, flowers, staminate to left, pistillate at right; e, fruiting branchlet; f, fruit.

kinds of sites: (1) wetlands, where it is commonly gregarious, and (2) well-drained uplands, where it occurs as scattered individuals in woodlands or forests. Taxonomically it has been treated as comprising three species; one species with four varieties; two species, one of them with three varieties; or one species with two varieties. For our area of coverage, for the most part, I think that two entities can be distinguished, and I choose to treat them as varieties of a single species. I have particular difficulty distinguishing whether sapling individuals belong to one variety or the other in ecotonal situations, places where upland, well-drained sites give way as gentle slopes to bottomlands.

3a. Nyssa sylvatica var. sylvatica. SOUR GUM.

In general a tree of forests on well-drained sites, occurring as single, scattered individuals. The branches tend to be stiffly horizontal, lower ones very often descending. Bark of larger trunks rough and notably blocky with interlacing ridges and fissures.

Leaf blades relatively membranous and pliable, obovate to elliptic-oblong, some sometimes nearly orbicular, their bases usually cuneate, occasional ones rounded, apices mostly short-acuminate, margins entire or with 1–several dentate teeth, not uncommonly with at least some toothed leaves on a given tree; blades variable in size, even on an individual branch, from about 3–10 or even 15 cm long; surfaces usually glabrous above at maturity and lustrous or sublustrous, varying from glabrous to densely pubescent beneath.

Staminate flowers in short, moderately loose racemes, the pistillate or bisexual ones (2–) 3–4 in a bracted cluster at the end of a stalk, occasionally 5 or 6 interruptedly spicate. Drupes ellipsoid or oblong-ellipsoid, about 1 cm long, blue-black.

Chiefly in well drained woodlands of various mixtures; throughout our area. (S. Maine to s. Mich., Ind., s. Ill., southward in the east to pen. Fla., southwesterly from s. Mo. to e. Okla. and e. Tex.)

Co-national champion *N. sylvatica* var. *sylvatica* (1969), 16'7" circumf., 117' in height, 69' spread, Harrison Co., Tex.; (1971), 15'1" circumf., 139' in height, 83' spread, Easterly, Tex.

3b. Nyssa sylvatica var. biflora (Walt.) Sarg. BLACK GUM.   Fig. 243

A notably gregarious tree particularly in those places where surface water stands much of the time, less gregarious in bottomland places where flooding is intermittent. In the former situations the bases of the trunks are usually conspicuously swollen; in wet pinelands subject to periodic burning having large subterranean bases and shrubby, multistemmed tops (*N. ursina* Small). Bark of larger trunks light gray and roughened with interlacing ridges and fissures.

Leaf blades relatively thick and stiff, narrowly elliptic to lance-elliptic, lanceolate, or oblanceolate, cuneate basally, apically acute to obtuse, less frequently rounded, variable in size, 3–8 (10) cm long, margins entire (some on seedlings, saplings, or sprouts with 1-few dentate teeth on a side).

Staminate flowers as in var. *sylvatica*; pistillate flowers commonly 2, sometimes 1 or 3, in a bracted cluster at the end of a stalk. Drupes ellipsoid, oblong-ellipsoid, or subglobose, about 1 cm long.

Swamps, ponds, pineland ponds or depressions, wet pinelands, shrub-tree bogs or bays, bogs; throughout our area. (Chiefly but not exclusively coastal plain, Del. and Md. southward to pen. Fla., westward to e. Tex., northward to w. and s. Tenn.)

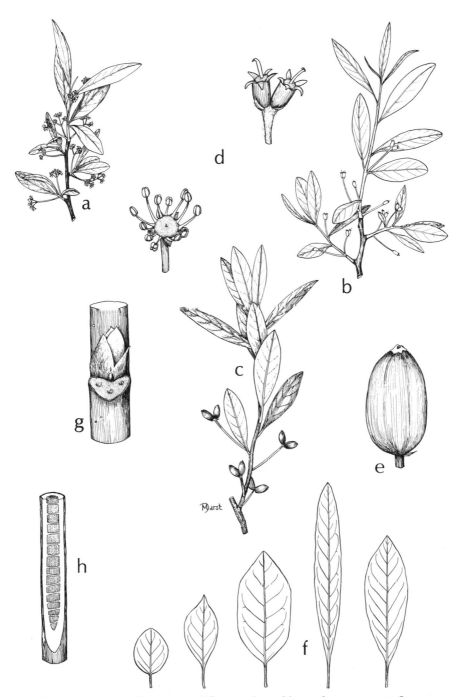

Fig. 243. **Nyssa sylvatica** var. **biflora:** a, branchlet with staminate inflorescences; b, branchlet with pistillate flowers or inflorescences; c, fruiting branch; d, flowers, staminate to left, pair of pistillate to right; e, fruit; f, variable leaves; g, node with leaf scar and bud; h, longitudinal section of stem showing diaphragmed pith, homogeneous between the diaphragms.

495

Co-national champion *N. sylvatica* var. *biflora* (1976), 13'1" circumf., 112' in height, 52' spread, Richland Co., S.C.; (1981), 13' circumf., 120' in height, 39' spread. Congaree Swamp National Monument, S.C.

Sterile specimens of *Nyssa sylvatica* and *Diospyros virginiana* are sometimes difficult to distinguish, one from the other; see *Diospyros virginiana* for features which may be of help.

Olacaceae (XIMENIA FAMILY)

Ximenia americana L. TALLOW-WOOD, HOG-PLUM.   Fig. 244

A glabrous, hemi-root-parasitic shrub or small tree to about 6 m tall, sometimes a small, sprawling shrub. Bark of main stem reddish brown and smoothish. Twigs brown or grayish brown and somewhat roughened by small raised lenticels. Leaf scars crescent-shaped to triangular, each with 3 vascular bundle scars. Leader shoots with longish internodes and armed with thorns 1–2 cm long; eventually the thorns elongate somewhat, often bear a leaf or two, ultimately become thorn-tipped short-shoots that elongate periodically and bear alternating series of scale and foliage leaves.

Leaves alternate, simple, petioles short and shallowly grooved abaxially, stipules none. Blades pinnately veined, lateral veins not prominent, oblong, elliptic, oblanceolate, or nearly orbicular, the latter usually on scrubby, sprawling individuals, varying from about 3 to 7 cm long and to 4 cm broad, extremities rounded to obtuse, apices sometimes shallowly emarginate.

Flowers for the most part in stalked clusters of 4–6 axillary to bracts or leaves. Flowers small, fragrant, stalked, bisexual, radially symmetrical. Calyx very small, about 1 mm high, lobes 4, deltoid. Corolla cream-colored, petals 6–10 mm long, linear-oblong, erect for a little over half their length, apices recurved at anthesis, each bearing a knob adaxially near the tip and each bearing adaxially two vertical bands of dense, stiff hairs from near their bases to a bit back of their tips. Stamens 8 in two series of 4, reaching the throat of the corolla. Pistil 1, ovary somewhat immersed in a disc, 4-locular, style slender, stigma capitate. Fruit a fleshy 1-stoned drupe 2–3 cm long, ripening yellow.

Inhabiting hammocks, well-drained pinelands, and scrub; n.e. Fla. (N.e. and pen. Fla., Fla. Keys; circumtropical and subtropical.)

National champion tallow-wood (1975), 1'4" circumf., 25' in height, 21' spread, Homestead, Fla.

Oleaceae (OLIVE FAMILY)

1. Leaf blades 1-pinnately compound; fruit a samara.   3. *Fraxinus*
1. Leaf blades simple; fruit drupaceous.
  2. The leaf blades with at least part of their margins serrate.   2. *Forestiera*
  2. The leaf blades with entire margins.
    3. Leaves sessile or subsessile, their blades mostly oblanceolate or spatulate, strongly tapered toward their bases from their middles or somewhat above, 2–4 (5.5) cm long and 1–1.5 cm broad. (*F. segregata* in genus)   2. *Forestiera*
    3. Leaves neither oblanceolate nor spatulate, some of them much larger than

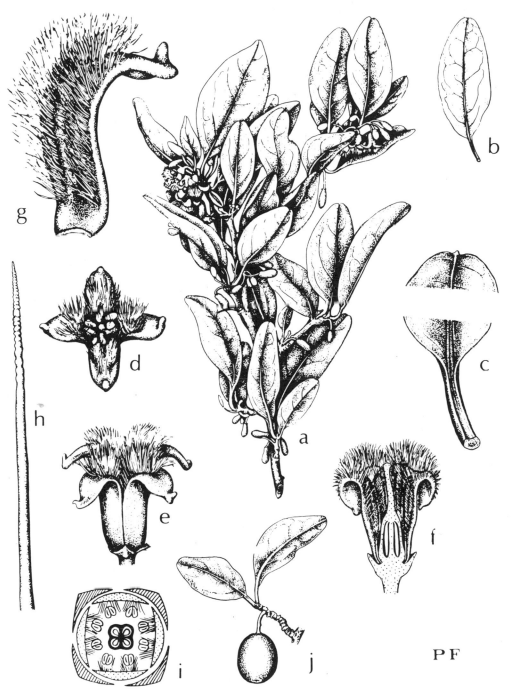

Fig. 244. **Ximenia americana:** a, flowering branchlet; b, leaf; c, base and apex of leaf; d, flower, from above; e, flower, side view; f, flower longitudinally dissected; g, petal, inner surface; h, single hair of petal; i, floral diagram; j, fruit. (Courtesy of Fairchild Tropical Garden.)

497

described above save for those of *Ligustrum sinense* which are elliptic, oval, or subrotund.

  4. Plants evergreen, the leaf blades leathery or subleathery; twigs glabrous.
    5. Leaf blades mostly broadest at or above their middles, margins revolute; flowers or fruits borne in short, jointed, axillary panicles with small scaly bracts, mostly on twigs of the previous season.   4. Osmanthus
    5. Leaf blades mostly broadest below their middles, margins not revolute; flowers or fruits borne in many-flowered or -fruited panicles terminating branchlets of the season.   5. *Ligustrum*
  4. Plants deciduous and leaf blades membranous, or, if tardily deciduous, then the twigs pubescent.
    6. Twigs glabrous; larger leaf blades up to 15 cm long and 5–8 cm broad.
                                                   1. *Chionanthus*
    6. Twigs densely short-pubescent; larger leaf blades 2–3 cm long and usually not over 2 cm broad. (*L. sinense* in genus)   4. *Ligustrum*

1. Chionanthus

1. Chionanthus virginicus L. FRINGE-TREE, OLD-MAN'S-BEARD, GRANDSIE-GRAY-BEARD.   Fig. 245

A shrub or small tree to about 10 m tall; in spring before or as the leaves unfold with abundant, dangling, loose, showy panicles of white or creamy white flowers. Leafy twigs reddish brown, glabrous, older twigs gray, with numerous, minute, dark dots and widely scattered, oval to circular, slightly raised, tan lenticels, the leaf scars about U-shaped in outline, concave centrally, distinctly raised-roughened on the twigs, the vascular bundle scars indistinct, in a semicircular line or U-shaped; pith white, continuous.

Leaves opposite, simple, deciduous, tapered basally to short, narrowly winged petioles, these more or less suffused with purplish red pigment; blades oblong, lanceolate, or oval, variable in size, to 20 cm long and 10 cm broad; margins entire, upper surfaces dark green and glabrous, the lower paler, often, but not always, softly short-pubescent when unfolding, glabrous or short-pubescent on the principal veins in age.

Inflorescence a pendent panicle bearing some leaflike bracts, borne from twigs of the previous season, axes of the panicle usually glabrous. Flowers mostly functionally unisexual and plants functionally dioecious, but some flowers sometimes functionally bisexual as well. Calyx minute, united below and with 4 short, deltoid lobes above. Petals 4, white or creamy white, linear or nearly so, separate nearly to the base, very much longer than the calyx. Stamens 2, short, inserted on the short corolla tube. Ovary superior, 2-locular, 2 ovules in each locule, style short, stigma 2-lobed.

Fruit an oval-ellipsoid drupe, dark blue at maturity, 1–1.7 cm long.

In diverse habitats, upland hardwood or pine forests, pine-oak scrub, rock outcrops, savannas, pine flatwoods, shrub bogs, throughout our area. (N.J. to Ohio, generally southward to cen. pen. Fla., Ark., s.e. Okla., and e. Tex.)

National champion fringe-tree (1977), 3'6" circumf., 24' in height, 26' spread, Mount Vernon Estate, Va.

2. Forestiera

Deciduous or tardily deciduous, straggly shrubs or small trees, the main stems tending to be strongly leaning and smaller branches rigidly divaricate. Leaf scars approximately round, each with 1 vascular bundle scar. Pith continuous.

Fig. 245. **Chionanthus virginicus:** a, fruiting twig; b, flowering branchlet; c, flower.

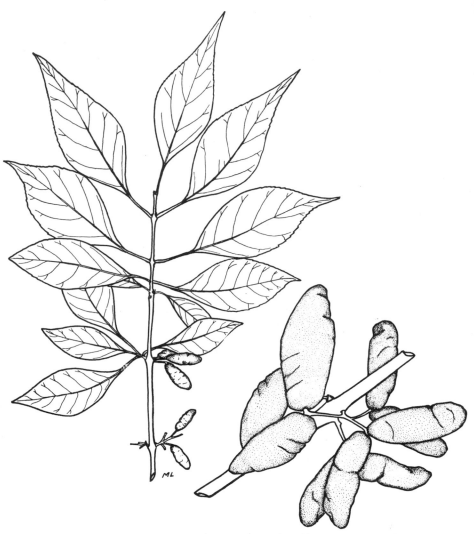

Fig. 246. **Forestiera acuminata.** (From Kurz and Godfrey, 1962. *Trees of Northern Florida.*)

Leaves simple, opposite, petiolate, without stipules, blades pinnately veined. Inflorescences short axillary clusters or racemes borne on twigs of the preceding season before new shoot emergence or in axils of leaves on twigs of the season (in *F. segregata* either condition may obtain). Flowers individually small, with 4 minute, quickly deciduous sepals or none and without petals; unisexual or structurally bisexual and functionally pistillate. Staminate flowers sessile and borne in tightly fastigiate clusters, *or* single, sessile flowers of 1 stamen each intermixed with umbels each with 2–4 stalked flowers with 3 or 4 stamens each. Pistillate flowers in short, few-flowered racemes or subumbellate, each flower with up to 4 abortive stamens or none. Ovary superior, 2-locular, 2 ovules in each locule. Fruit a 1–2-stoned drupe.

1. Leaf blades distinctly acute or acuminate apically, the tips sharply pointed.
1. F. acuminata

1. Leaf blades obtuse to rounded apically, or if abruptly short-tapered apically then the tips of the taper blunt.
    2. Upper and lower surfaces of leaf blades glabrous, the lower punctate; margins entire.
3. F. segregata

    2. Upper surfaces of leaf blades with at least some hairs on their midribs proximally, the lower with pubescence on the larger veins or uniformly over the surfaces, not punctate.
        3. Pubescence on internodes of stems of the season in 2 longitudinal (sometimes interrupted) bands, one on either side; petioles with few hairs, if any; plant flowering in mid to late summer, the clusters in leaf axils.
2. F. ligustrina

        3. Pubescence on internodes of stems of the season evenly distributed, petioles at least moderately pubescent; plant flowering in early spring from buds on twigs of the previous season before new shoot growth commences.
4. F. godfreyi

1. Forestiera acuminata (Michx.) Poir. in Lam. SWAMP-PRIVET.   Fig. 246

Deciduous shrub with few to many stems to about 6 m long, less frequently with the form and stature of a small tree. Young stems glabrous, woody twigs grayish brown, with prominent lenticels.

Leaves glabrous (in ours), petioles slender, 0.5–2 cm long; larger leaf blades mostly 4–8 cm long and 2–3.5 cm broad, not punctate beneath, lanceolate to subrhombic, usually broadest just below their middles, bases acute and somewhat decurrent on the petioles, apices acuminate or acute, tips strongly pointed for the most part, margins inconspicuously serrate with low teeth on their middle portions. Flowering occurs in early spring before new shoot emergence; staminate in tightly fastigiate, sessile clusters, the pistillate in sessile clusters terminating slender stalks or in umbels or subumbels at the ends of slender stalks (as are the drupes). Drupes essentially lanceolate and firm until just prior to final ripening which occurs quickly and after which oblong-ellipsoid (tending to dry ovoid), 10–15 mm long and 7–10 mm thick, surfaces irregularly wrinkled, dull reddish purple and with a slight bloom.

River banks, sloughs, alluvial swamps, margins of fluctuating ponds; Fla. Panhandle from about the Suwannee River westward, adjacent Ga. and Ala. (Chiefly coastal plain, S.C. to cen. n. Fla., westward to Tex., northward in the interior to Ind., Ill., and Kan.)

National champion swamp privet (1971), 2'7" circumf., 42' in height, 25' spread, Richland Co., S.C.

2. Forestiera ligustrina (Michx.) Poir. in Lam.   Fig. 247

Deciduous shrub, usually with several, often arching or leaning stems 3–4 m long. Woody twigs gray, with widely scattered, moderately prominent lenticels.

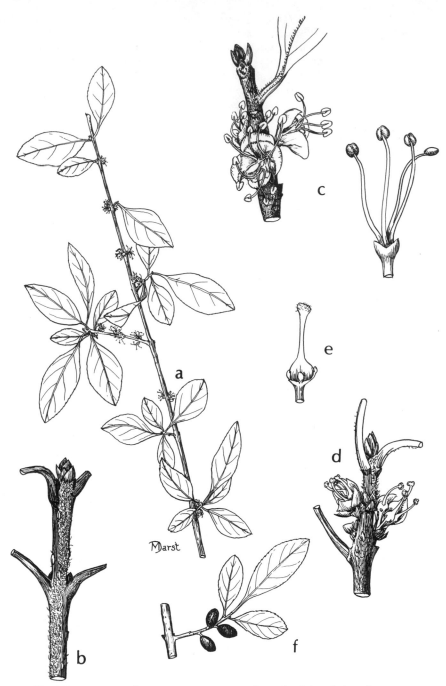

Fig. 247. **Forestiera ligustrina:** a, staminate flowering branch; b, enlargement of tip of stem of the season (leaves cut off); c, staminate flower clusters to left, staminate flower to right; d, pistillate flower clusters; e, pistillate flower; f, fruiting branchlet.

Larger leaf blades mostly 2–5 cm long, ovate to lance-ovate or elliptic, abruptly narrowed basally and slightly decurrent on slender petioles 1–1.5 cm long, the petioles in turn rounded-decurrent the length of the internodes, the decurrencies themselves glabrous but the internode pubescent as a longitudinal band between those of the leaf pairs; apices of blades obtuse, tipmost portions blunt or rounded, margins inconspicuously serrate from just above their bases nearly or quite to their tips, upper surfaces commonly sparsely short-pubescent on the midribs, lower varyingly short-pubescent on the midribs, midribs and major lateral veins, or sparsely to densely pubescent throughout, not punctate. Flowers in sessile or very shortly stalked clusters in leaf axils on branchlets of the season in mid- to late summer. Drupes single and essentially sessile or 2–several in a sessile cluster terminating a very short stalk, broadly ellipsoid, 7–8 mm long, blue-black.

In our area infrequent, in upland mixed woodlands, where or near where limestone outcrops, shell middens; n.w. pen. Fla. westward through the Panhandle of Fla., adjacent s.w. Ga., s. Ala. (Coastal plain and adjacent piedmont, Ga., n. Fla., westward to e. and s.e. Tex., northward in the interior to Tenn. and Ky.)

3. Forestiera segregata (Jacq.) Krug & Urban. INKBERRY, FLORIDA-PRIVET.

Fig. 248

Evergreen or tardily deciduous, glabrous shrub to about 3 m tall. Twigs gray or brownish gray, with scattered, prominent lenticels.

Leaves sessile or very shortly petiolate, blades mostly oblanceolate or spatulate, strongly tapered basally, some sometimes elliptic or oval, even obovate, 1.5–5 cm long, finely and sparsely to abundantly punctate beneath, apices blunt or narrowly rounded, margins entire.

Sometimes deciduous plants flowering in spring before new shoot emergence, or if leaves overwintering flowering in the leaf axils on branchlets of the previous season, sometimes flowering later in the season on branchlets of the season. Staminate flowers in short-panicled umbels, the pistillate in short racemes or stalked umbels or subumbels. Drupes in short racemes, stalked umbels or subumbels, oblong-oblate to globose, 5–7 mm in diameter, often a little broader than long, blue-black and thinly glaucous.

For the most part coastal or near-coastal, in scrub or scrub barrens, coastal thickets, shell middens, coastal hammocks; in our area, n.e. Fla., s.e. Ga., Dixie Co., Fla., near the Gulf Coast. (S.e. Ga., Atlantic Coast of Fla., Gulf Coast northward to Dixie Co.; Bermuda, W.I.)

4. Forestiera godfreyi L. C. Anderson

Fig. 249

Deciduous shrub or small tree, its habit very much like that of *F. acuminata*, the main stem arching or leaning, the branches rigid and divaricate. Young stems, petioles, and lower leaf surfaces (of half-grown leaves) with copious, soft, gray pubescence, upper blade surfaces very sparsely short-pubescent.

Leaves with slender petioles 2–10 mm long; larger blades of mature leaves ovate, lance-ovate, or elliptic, abruptly tapered basally, usually a little decurrent on the petioles, if the petioles very short then winging the petiole to their bases, sometimes bases scarcely tapered at all; upper surfaces glabrous or with minute hairs along the midribs, the lower uniformly pubescent, the hairs shorter and less dense than on the young blades; apices mostly obtuse, if abruptly short-tapered then the tips blunt; margins inconspicuously finely serrate from about their middles nearly to the tips.

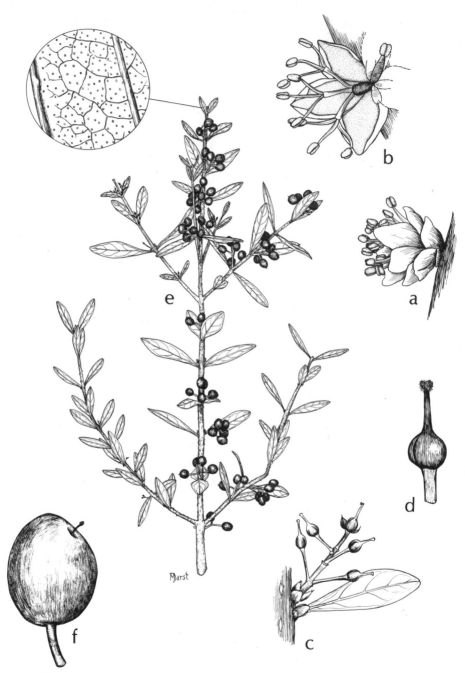

Fig. 248. **Forestiera segregata:** a, staminate inflorescence; b, staminate inflorescence (some bud scales and two inflorescence branches removed); c, pistillate inflorescence; d, pistil; e, fruiting branch and (upper left) small portion of lower leaf surface enlarged; f, fruit.

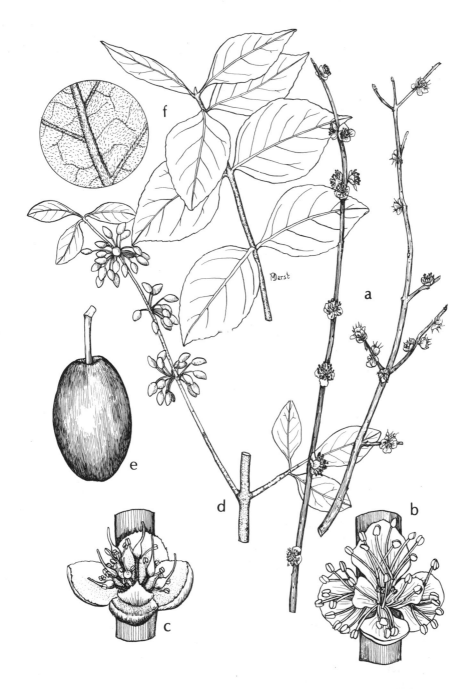

Fig. 249. **Forestiera godfreyi:** a, flowering branches, staminate to left, functionally pistillate to right; b, cluster of staminate flowers; c, cluster of functionally pistillate flowers; d, fruiting branch (fruits immature); e, mature fruit; f, branch with mature leaves and enlargement of a small portion of lower surface of leaf to the left.

505

Flowering occurring in very early spring on twigs of the previous season before any new shoot growth commences; plants functionally dioecious. Functionally staminate inflorescences with 6 basal bracts, the outer 4 largest, obovate-reniform, fringed apically, flowers 12–15, mostly in 3-flowered umbellate fascicles, but with a few stalked flowers borne singly; stalks of the inflorescence 0.25 mm long, flower stalks 0.3–0.6 mm long, sepals 4–5, 0.2–0.7 (0.9) mm long, unequal in length, sometimes one sepal much longer than the others and petaloid; stamens (2) 3–5 per flower, the petaloid sepals, if present, usually associated with tiny rudimentary pistils. Functionally pistillate inflorescences with 6 basal bracts, with (5) 7–10 flowers, some borne singly, others in 3-flowered umbels; sepals 0.3–0.5 mm long; abortive stamens 2–4 per flower. Fruiting stalks 2.5–7 mm long, flower stalks 5–7 mm long, drupes lance-ovoid for a considerable time in preripening, finally ripening very quickly, then slightly ovoid, or ovoid-ellipsoid, (8) 10–12 mm long and 6–9 mm in diameter, surfaces smooth, dark blue beneath a grayish bloom.

Mesic woodlands, usually where calcareous; presently known to me from the following counties of Fla.: Alachua, Gadsden, Gilchrist, Jackson, Levy, Liberty, and Jefferson.

The plant described above may be representative of what Small (1933) treated as *F. pubescens* Nutt., the range for which he gave as "Coastal Plain, Fla. to Tex. and Ark."

3. Fraxinus (ASHES)

Trees or shrubs. Twigs stoutish and stiffish, with buff, ashy, or reddish bark, thick, continuous pith; bark of older, larger trunks usually roughly ridged and furrowed; buds somewhat compressed, with 1–3 pairs of densely brown-pubescent outer scales, the terminal buds larger than the lateral ones. Leaves opposite, 1-odd-pinnately compound, without stipules. Flowers small, greenish yellow, unisexual (in ours, and plants dioecious), borne in dense fascicles, short compact racemes, or panicles axillary to leaf scars on twigs of the previous season before or as shoots of the season commence development. Calyx small, tubular, 4-lobed or irregularly and minutely cut or toothed at the summit, present and usually persisting at the base of the fruits in ours (sometimes deciduous by the time the fruits are fully matured but usually some of the fruits will have them then), absent in some kinds beyond our range. Corollas none in ours. Stamens usually 2 in the staminate flowers. Pistillate flowers with 0–2 abortive stamens, pistil 1, ovary superior, 2-locular, 2 ovules in each locule, style 1, stigma 2-lobed. Fruit a 1–2-seeded samara, the wing usually in 1 plane, the samara rarely 3-winged in one of ours.

In identification of our ashes, it is in many, if not most, instances necessary to have samaras. Since our trees are dioecious, probably considerably less than half the trees in any given location will bear them; thus one may have to search diligently. The fact that samaras are present through much of the growing season on those individuals that do bear them is an advantage.

1. Plants inhabiting well-drained woodlands; wing of the samara essentially terminal, i.e., not or very little decurrent along its plump (terete) seed-bearing portion.

1. *F. americana*

1. Plants inhabiting poorly drained places, places where surface water stands much of the time or where subject to periodic flooding; wing of the samara decurrent along at least the distal half of the seed-bearing portion, the latter sometimes not clearly distinguishable from the wing.

2. Samara essentially flat throughout or essentially flat but with a concavity over the seed-bearing portion, the wing extending to below, often well below, the seed-bearing portion and commonly stipelike below the wing. *2. F. caroliniana*

2. Samaras subterete or at least surficially rounded in the seed-bearing portion, the latter extending to the very base; wing decurrent along the seed-bearing portion to about its middle and the latter usually clearly evident in contrast to the wing *or* the wing merging imperceptibly with the seed-bearing portion.

   3. Wing of the samara decurrent along the seed-bearing portion to about its middle and the seed-bearing portion usually clearly evident; pubescence, if any, to either side of the midrib and along the lateral veins of the lower surface of leaflets usually short, in narrow bands and not notably tangled. *3. F. pennsylvanica*

   3. Wing of the samara imperceptibly merging with the seed-bearing portion; pubescence present at either side of the midrib and proximally along the lateral veins of the lower surfaces of leaflets conspicuously banded, the edges of the bands usually irregular, the hairs longish and notably tangled. *4. F. profunda*

1. Fraxinus americana L. WHITE ASH.   Fig. 250

Large tree, to 25 (35) m tall. Young twigs glabrous or on some individuals densely short-pubescent (in which case leaf axis and stalks of leaflets are similarly pubescent). Leaf scars roughly horseshoe-shaped, the arms extending to a little above the axillary bud on either side.

Leaves 20–30 cm long. Leaflets 5–9, commonly 7, stalks of the lateral ones 0.5–1.5 cm long; blades 6–15 cm long and 3.5–7.5 cm broad, ovate, lance-ovate, or oblongish, bases rounded, nearly truncate, or shortly and broadly tapered, sometimes slightly and obliquely inequilateral, apices acuminate, less frequently acute; margins entire or more rarely wavy or bluntly toothed, upper surfaces green and glabrous, lower surfaces whitish, pubescent along the veins, or uniformly softly short-pubescent along the veins, or uniformly softly short-pubescent throughout.

Samaras 2.5–6.5 cm long, with the wings essentially terminal above the seed-bearing portion or only slightly decurrent on its summit, the seed-bearing portion plump, the wing linear-oblong, narrowly elliptic, or spatulate, less frequently lanceolate, often minutely notched apically. [Incl. *F. biltmoreana* Beadle]

Well-drained woodlands or on floodplains where only occasionally and temporarily inundated; in our area, n. cen. Fla., Alachua to Jackson Cos., s.w. Ga., s. Ala. (N.S. to Ont. and Minn., generally southward to n. Fla. and n.e. Tex.)

Co-national champion white ashes (1976), 20′5″ circumf., 114′ in height, 126′ spread, Lenawee Co, Mich.; (1976), 24′ circumf., 80′ in height, 82′ spread, Glen Mills, Pa.

2. Fraxinus caroliniana Mill. POP ASH, WATER ASH, CAROLINA ASH.   Fig. 251

Relatively small tree, commonly with several trunks. Unfolding branchlets often heavily pubescent, much or all of the pubescence soon sloughing. Leaf scars mostly shield-shaped, apical auricles, if any, short.

Leaves 16–30 cm long. Leaflets 5–7 (9), rarely 3, stalks of the lateral leaflets mostly 0.5–2 cm long; blades very variable in size and shape, lanceolate, ovate, oval, elliptic, or obovate, 2.5–10 (15) cm long and 2–6 cm broad, bases acute to rounded or nearly truncate, often slightly and obliquely inequilateral, apices rounded to acute, infrequently acuminate; margins entire or less frequently irregularly serrate, upper surfaces glabrous or less frequently irregularly serrate, upper surfaces glabrous and green, glabrous or short pubescent to either side of the midrib and proximally for a short distance along the lateral veins.

Fig. 250. **Fraxinus americana:** a, fruiting branch; b, fruit, enlarged.

Fig. 251. **Fraxinus caroliniana:** a, fruiting branch; b, variable fruits; c, fruit, enlarged.

Fig. 252. **Fraxinus pennsylvanica:** a, fruiting branch; b, variable fruits; c, fruit, enlarged.

Samaras (infrequently 3-winged) usually winged to below, often to well below, the seed-bearing portion (which is usually indistinct) and commonly stipelike below the wing, flat throughout or concave over the seed-bearing portion; overall shape very variable from plant to plant, spatulate, oblanceolate, long-elliptic, lanceolate, broadly oblong, oval, ovate, or suborbicular, the narrower ones 2–5 cm long and 0.5–1.5 cm broad at their broadest places, the broader ones to 3 cm broad, apices rounded, bluntly pointed, obtuse, rarely acute, sometimes shallowly notched apically.

Swamps, flatwoods, depressions, wet shores, lagoons, sloughs, oxbows, pond margins, commonly where the tree bases are inundated for long intervals; throughout our area. (Chiefly coastal plain, Va. to s. pen. Fla., westward to Ark. and e. Tex.)

National champion Carolina ash (1981), 2'10" circumf., 58' in height, 35' spread, Congaree National Monument, S.C.

3. Fraxinus pennsylvanica Marsh. GREEN ASH, RED ASH. Fig. 252

A large tree, to about 30 m tall, the base of the trunk enlarged when growing where inundated for long intervals. Young twigs sometimes pubescent, soon becoming glabrous. Leaf scars shield-shaped.

Leaves 20–30 cm long. Leaflets 5–9, commonly 7, their stalks mostly 5–10 mm long, blade tissue narrowly decurrent-winged along most of their length; blades 4–10 (15) cm long and 2–6 cm broad, oblong-elliptic, lance-ovate, lanceolate, or rarely (the lowermost pair) suborbicular, bases of the lowermost pair often rounded or very shortly and broadly tapered, others obtuse to acute, commonly slightly and obliquely inequilateral, apices abruptly short-acuminate, the acumens blunt; margins usually entire, sometimes irregularly and remotely toothed, upper surfaces dark green and glabrous, lower paler and duller green, glabrous or pubescent in a narrow band to either side of the midribs and proximally on the lateral veins, the hairs short and not tangled, on some individuals uniformly pubescent throughout.

Samaras 3–6 cm long and 4–6 (8) mm broad at their broadest places, narrowly linear, oblanceolate, or spatulate, the wing decurrent to about the middle of the usually distinct seed-bearing portion which is subterete below the wing and 1–2 mm wide; apex of the wing truncate, bluntly tapered, or rounded, often minutely notched. [Incl. F. darlingtonii Britt., F. smallii Britt.]

River floodplains and swamps, moist to wet woodlands; throughout our area except perhaps northeasternmost Fla. (Que. to Man., generally southward to n. Fla. and e. Tex.)

National champion green ash (1981), 20'2" circumf., 131' in height, 121' spread, Cass Co., Mich.

4. Fraxinus profunda (Bush) Bush. PUMPKIN ASH. Fig. 253

A large tree, bases of the trunks enlarged if growing where inundated for long intervals. Young stems sometimes densely brown-velvety-pubescent, sometimes glabrous. Leaf scars shield-shaped to crescent- to half-moon-shaped.

Leaves from about 12–30 cm long. Leaflets 5–9, usually 7, stalks of the lateral ones variable in length from about 0.5–4 cm, in most cases less than 4 cm, not decurrent-winged; blades varying greatly in shape and size, ovate, oblong-ovate, oblong, or lanceolate, infrequently obovate, 4–15 cm long and 2–7 (10) cm broad, bases rounded, broadly and shortly tapered, infrequently acute, equilateral or slightly to markedly obliquely inequilateral, apices acuminate to long-acute, infrequently emarginate, margins entire, upper surfaces glabrous

511

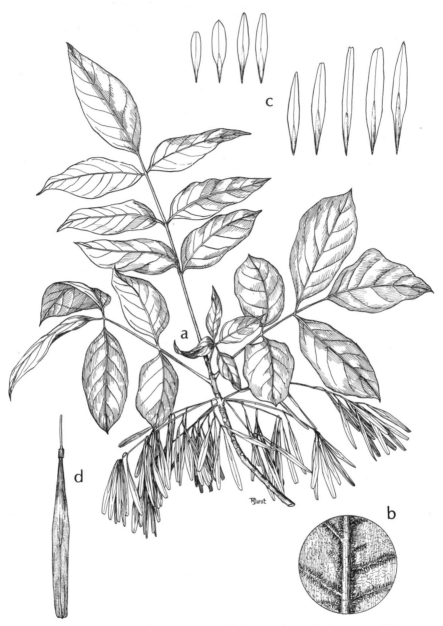

Fig. 253. **Fraxinus profunda:** a, fruiting branch; b, medial portion of lower surface of leaflet, much enlarged; c, variable fruits, somewhat enlarged; d, fruit, much enlarged.

and dark green, lower paler and duller green, generally with a relatively broad, sometimes narrow, irregularly edged, band of tangled pubescence to either side of the midribs and proximally for a short distance along the lateral veins; on some plants the lower surfaces softly, sparsely to copiously short-pubescent throughout but with denser and more tangled hairs along the midribs and lateral veins.

Samaras 4–8 cm long and 0.7–1.2 cm broad at their broadest places on the wings; wing downwardly imperceptibly merging with the surficially rounded seed-bearing portion which is strongly tapered to the very base of the samara; outline of the samaras spatulate, oblanceolate, oblong-spatulate or -oblanceolate, often minutely notched apically. [*F. tomentosa* Michx. f.; incl. *F. michauxii* Britt.]

River swamps and floodplains, low, wet woodlands generally; in our area, n.e. Fla. (southward to Marion Co.), s.e. Ga., perhaps elsewhere. (Overall distribution apparently very much interrupted and spotty, N.Y. to s. Ill. and s.e. Mo., generally southward to n. cen. pen. Fla. and La.; as mapped by Elias (1980), not in Ala.)

National champion pumpkin ash (1977), 18'3" circumf., 86' in height, 84' spread, Warrenton, Va.

4. Osmanthus

Osmanthus americanus (L.) A. Gray. **WILD-OLIVE, DEVILWOOD.**   Fig. 254

Shrub or small tree. Twigs gray or brownish gray, with scattered, moderately prominent lenticels; leaf scars broadly U-shaped, with a ring of small indistinct vascular bundle scars; pith white, continuous.

Leaves simple, opposite, evergreen, petioles 1–2 cm long; blades leathery, glabrous, pinnately veined, variable in shape and size, the larger ones about 14 cm long, 5.5 cm broad, elliptic, oblong-elliptic, oblanceolate, or obovate, margins entire and revolute, bases cuneate, apices generally acute, more rarely short-acuminate, obtuse, rounded, or notched.

Flowers small, creamy white, in short, scaly-bracteate axillary panicles, mostly on twigs of the previous season, the developing flower clusters usually evident during autumn and over winter, reaching full anthesis in early spring. Calyx tubular, with 4 minute lobes. Corolla tubular, with 4 short-ovate, spreading lobes. Stamens 2. Ovary superior, 2-locular, 4-ovulate.

Fruit an oval, 1-stoned, dark bluish purple drupe 1–1.5 cm long.

In a wide range of wooded habitats; throughout our range. (Coastal plain, s.e. Va. to Fla., westward to La.)

Distinguishing characteristics which may be seen on most plants at any time are the opposite leaves and the axillary, short-jointed, thickish flowering or fruiting panicles with small scaly bracts. The leaves of *Symplocos tinctoria*, the sweetleaf, are similar to those of *O. americanus*, although not so leathery, and they are alternately arranged.

National champion devilwood (1972), 5'6" circumf., 37' in height, 41' spread, Mayo, Fla.

5. Ligustrum

Evergreen, tardily deciduous, or deciduous shrubs or small trees. Leaves opposite, short-petiolate, their blades having entire, nonrevolute margins. Inflorescences many-flowered panicles terminating branchlets of the season

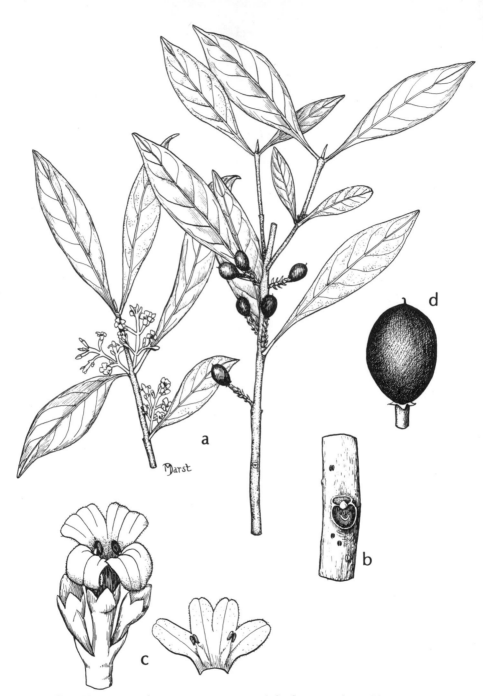

Fig. 254. **Osmanthus americanus:** a, at left, flowering branchlet, at right, fruiting branch; b, piece of two-year-old woody stem with leaf scar, axillary bud, and above that a scar left by dehiscence of an inflorescence branch; c, flower with flower bud to either side of it, at left, and at right, corolla, opened out; d, fruit.

514

after the leaves are fully matured. Flowers small, radially symmetrical, bisexual. Calyx united and campanulate, with 4 minute teeth on its truncate edge. Corolla white or off-white, cylindric- or funnelform-tubular below, with 4 spreading lobes above. Stamens 2, inserted on the corolla tube, included or exserted. Pistil 1, ovary superior, 2-locular, 2 ovules in each locule, style 1, stigma 2-lobed. Fruit a blue-black berrylike drupe.

The ligustrums of our area are of Asiatic origin, are cultivated as ornamentals or for hedges and screens. Insofar as we are aware, three species are naturalized here, one of them common and abundant, the other two much less frequent although sometimes abundant locally.

1. Leaf blades leathery or subleathery; twigs glabrous.
  2. The leaf blades with blunt tips; corolla tubes well exserted from the calyces, longer than the corolla lobes.   1. *L. japonicum*
  2. The leaf blades (many of them on a given plant, at least) with acuminate or acute tips; corolla tubes very little exserted from the calyces, equaling or shorter than the corolla lobes.   2. *L. lucidum*
1. Leaves membranous, twigs densely short-pubescent.   3. *L. sinense*

1. Ligustrum japonicum Thunb. JAPANESE LIGUSTRUM.   Fig. 255

An evergreen shrub to about 3 m tall. Twigs glabrous, eventually much roughened by raised, corky lenticels.

Leaf blades leathery, ovate to elliptic, smaller ones sometimes orbicular or nearly so, 3–8 cm long and 2.5–4 cm broad, rounded or little tapered basally, blunt apically, surfaces finely punctate, sparsely so above, abundantly so beneath.

Flowering panicles roughly pyramidal in outline, 8–10 cm across their bases and about as long. Corolla tubes well exserted from the calyces, longer than the corolla lobes; stamens exserted.

Mature drupes ellipsoid, not curvate, 8–12 mm long.

Sporadically naturalized in the area of our coverage, sometimes locally abundant, chiefly in mesic woodlands within or near towns or cities.

2. Ligustrum lucidum Ait. WAX-LEAF LIGUSTRUM.   Fig. 256

An evergreen shrub or small tree, to 5 m tall, perhaps taller, similar in general appearance to *L. japonicum*. Leaf blades averaging somewhat larger, to 12 cm long and 5 cm broad, their apices mostly acuminate or acute rather than blunt. Flowering panicles similar but averaging larger. Corolla tubes little exserted from the calyces, equaling or shorter than the corolla lobes; stamens exserted. Mature drupes globose to short-ellipsoid, 4–8 mm long.

Sporadically naturalized throughout our area, chiefly in upland woodlands, sometimes in low-lying ones.

3. Ligustrum sinense Lour. CHINESE PRIVET.   Fig. 257

A shrub or slender tree to at least 10 m tall, the leaves tardily deciduous. Young twigs densely short-pubescent; woody twigs buffish, irregularly low-ridged, with abundant pale lenticels; leaf scars raised, half-moon-shaped, each with 1 vascular bundle scar.

Characteristically, leafy branchlets are numerous and their leaves are disposed at right angles to the stems, at a glance giving the appearance of compound leaves. In flower the many such short, leafy branchlets bearing relatively small, many-flowered panicles produce the effect of a great profusion of bloom.

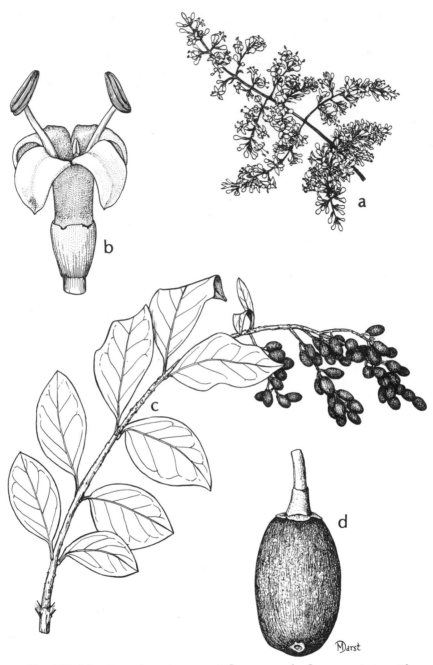

Fig. 255. **Ligustrum japonicum:** a, inflorescence; b, flower; c, twig with infructescence; d, fruit.

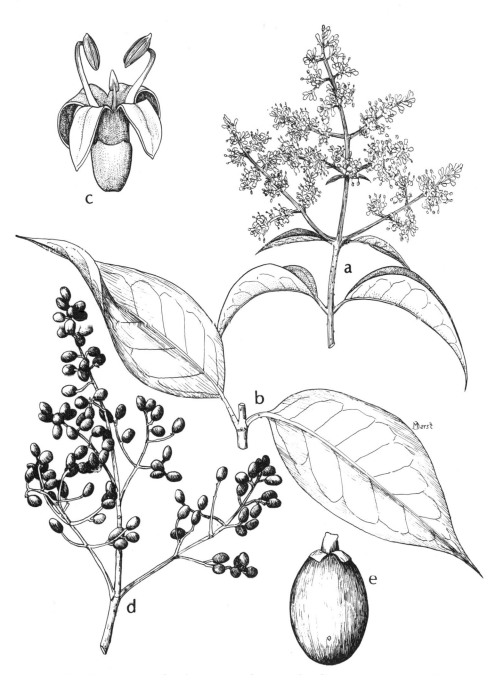

Fig. 256. **Ligustrum lucidum:** a, tip of twig with inflorescence; b, node with pair of leaves, from medial portion of twig; c, flower; d, infructescence; e, fruit.

Fig. 257. **Ligustrum sinense:** a, branch with branchlets bearing terminal inflorescences; b, flower; c, infructescence; d, fruit.

Leaf blades relatively small, 1.5–3 cm long and 1–2 cm broad, mostly elliptic or oval varying to subrotund, occasional ones larger and varying to ovate, broadly short-tapered basally, apices little tapered and blunt; petioles markedly short-pubescent, upper blade surfaces often glabrous at maturity, the lower usually persistently pubescent on the midribs.

Panicles usually narrowly conical in outline, 3 cm broad or a little more basally and 5–8 cm long, their axes copiously short-pubescent. Flowers ill-scented, their usual profusion making the ambient air malodorous. Corolla tube about equaling or shorter than the lobes, stamens exserted.

Mature drupes subglobose or short-ellipsoid, about 4–5 mm long.

The chinese privet flourishes around abandoned homesites, particularly in rural areas, along fence and hedge rows, is sometimes very abundant along courses of small streams, and is occasional in woodlands generally.

Onagraceae (EVENING-PRIMROSE FAMILY)

Ludwigia peruviana (L.) Hara. PRIMROSE-WILLOW.

Plant very coarse, woody below, herbaceous above, commonly with numerous stems from near the base, to 3 m tall or more, lower stems often attaining a diameter of 3–4 cm in a single season, the wood becoming very hard; in areas where freezing occurs, depending upon the severity of the cold, tops of plant dying back to near the base, or in extreme cold being killed altogether. Her-

518

baceous portions of stem copiously shaggy-pubescent, hairs tawny, older stems becoming glabrous, eventually the woody portions with thin brown bark that exfoliates in longish, thin strips.

Leaves alternate, lanceolate, lance-elliptic, or lance-ovate, cuneate basally to subpetiolate bases, acute to acuminate apically, 5–15 cm long and to 3 cm wide medially, both surfaces tawny shaggy-pubescent, the lower more copiously so than the upper. Stipules none.

Flowers solitary in axils of reduced leaves near the tips of branchlets, bisexual, radially symmetrical, ovary inferior; flower stalks 1–2 (–3.5) cm long, shaggy-hairy, minute scalelike bractlets at the base of the floral tube or distally on the flower stalks. Calyx segments 4 (rarely 5), triangular and with acuminate tips, 10–15 mm long and 5–7 mm wide basally, shaggy-pubescent. Petals 4 (rarely 5), spreading radially, bright yellow and showy, obovate with broadly rounded, truncate or broadly and shallowly concave apices, mostly about 2.5 cm long and as broad distally, very easily detached. Stamens 8 (rarely 10). Fruit a capsule, broadly obconic in outline, quadrangular, 1–3 cm long, shaggy-pubescent, sepals persistent at its summit, dehiscence irregularly loculicidal. Seeds in several series in each locule, free of endocarp, buff to light brown, plump, elliptic to oblongish and slightly falcate, usually varying in size in a given capsule from about 0.3 to 0.8 mm long. [*Jussiaea peruviana* L.]

Weedy, commonly in shallow water, ditches and drainage canals, swales, sloughs, marshy shores, wet clearings, introduced and local in our area. (Common and abundant in cen. pen. and s. Fla.; trop. Am.)

The primrose-willow, occurring sporadically in our area, is notable for its fast-growing coarse stems which become woody below and remain for the most part herbaceous above, the herbaceous branches bearing flowers with 4 showy, bright yellow petals that are very easily detached. The wood of the lower stems is very hard but is usually winter-killed in our areas.

Platanaceae (PLANE-TREE FAMILY)

Platanus occidentalis L. SYCAMORE, PLANE-TREE, BUTTONWOOD.   Fig. 258

A large deciduous tree, to 35 (50) m tall, picturesque particularly because of its bark which on young stems or branches of older trees is light gray but becomes mottled or dappled in several colors, tan, green, or chalky white, as portions peel off in thin, irregular plates. Bark of large, old trunks brown, sloughing in thinnish, relatively small, oblongish scales or plates. Twigs sometimes zigzagging, brown, becoming gray by developing a thin, waxy covering; terminal bud none; lateral buds forming in summer within the bases of petioles, each covered by 3 scales, one within the other, these dehiscing circumscissilely at their bases, the shoot-bud within very densely covered with long, silky hairs; leaf scar angularly somewhat raised, narrowly nearly encircling the base of the bud, vascular bundle scars 5–9; stipule scar narrow, obliquely encircling the stem just above the bud.

Young developing stems, petioles, and both surfaces of unfurling leaf blades loosely but copiously gray-pubescent with stellate hairs whose branches intermesh; pubescence of upper leaf surfaces becoming thin as the leaf expands, that of lower surfaces gradually loosening and gathering into cotton or woolly flocs or tufts.

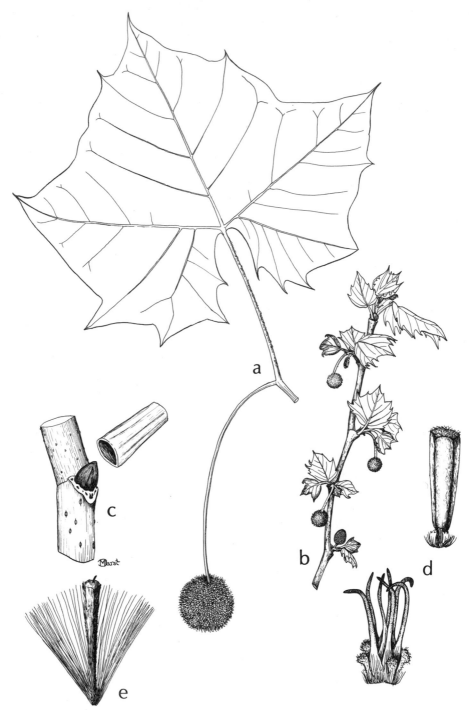

Fig. 258. **Platanus occidentalis:** a, mature leaf with an infructescence; b, twig with emerging branchlets and inflorescences (only the lowermost on the drawing is staminate, others all pistillate); c, node with piece of petiole (diagrammatic) at the side (to indicate that the petiole base covered the axillary bud); d, staminate flower, above, and pistillate flower, below; e, fruit.

Leaves simple, alternate, stipulate, and petiolate; stipules foliaceous, completely encircling the twig, their edges toothed or lobed, usually falling early but sometimes their remains persisting, even on leafless twigs; petioles with hollow, dilated bases, overall of variable lengths, 1–5 cm, surfaces patchily floccose. Mature blades variable in size, the lowermost on a given branch usually smallest, distal ones largest; smaller blades about as long as broad, larger ones mostly broader than long, 8–15 cm broad (to 25 cm broad on vigorous sprouts), in outline, ovate, suborbicular, or reniform; venation palmate or subpalmate, with 2 major lateral veins and the midrib from the base of the blade or the 2 major laterals arising from the midrib somewhat above the base of the blade; bases shortly tapered, truncate, or with a cordation well to either side of a shortly tapered base; edges of blades variably with irregular, conspicuous, salient teeth or both lobed and toothed; upper surfaces glabrous, lower surfaces sometimes more or less floccose well into the season, eventually usually pubescent only along the major veins.

Flowers minute, a very large number of them borne in stalked, ball-like heads, one at the end of each lateral branchlet of the season as it commences to develop from a bud on a twig of the previous season, each inflorescence having only staminate or pistillate flowers (plants monoecious); at anthesis, stalks of the inflorescences 1–2 cm long, those of the staminate heads stiffish and erect, those of the pistillate more lax and generally recurved or becoming so, after anthesis elongating considerably. Each flower head has a circular bract basally and small bractlets interspersed among the flowers. Pistillate flowers with several separate pistils per flower. Fruiting heads ("buttonballs") spherical, 2.5 cm or a little more in diameter, dangling on stalks 8–15 cm long. Fruit an achene or nutlet about 8 mm long, subtended by a ring of long, tawny hairs.

Chiefly in bottomlands along rivers and streams; in our area, occurring naturally in s. cen. and s.w. Ga., s. Ala., and drainages of the Apalachicola, Choctawhatchee, and Escambia Rivers in Fla. Panhandle; commonly planted as a shade or ornamental tree and to some extent naturalizing in various habitats in and beyond its natural range. (S. Maine to s. Mich., s. Minn., and cen. and s. Iowa, generally southward to Ga. (excepting southeasternmost Ga.), Fla. Panhandle, westward to e. and s. cen. Tex.

National champion sycamore (1973), 39′ circumf., 96′ in height, 100′ spread, Estill County, Ky.; Glendon Newton.

Polygonaceae (SMARTWEED OR KNOTWEED FAMILY)

1. Plant a high-climbing vine.   1. *Brunnichia*
1. Plant a woody, tap-rooted, suffrutescent perennial or subshrub.   2. *Polygonella*

1. Brunnichia

Brunnichia ovata (Walt.) Shinners. LADIES' EARDROPS, EARDROP VINE.

Fig. 259

High-climbing vine, stems slenderly woody below, growth of the season herbaceous, distal portions dying back over winter, climbing by means of branched tendrils terminating at least some of the short, lateral shoots of the season and terminating at least some of the inflorescences/infructescences, the branches of

Fig. 259. **Brunnichia ovata**: a, very small portion of vine; b, node; c, portion of leafless stem; d, flower; e, semidiagrammatic section of young fruit; f, mature fruit.

the tendrils very tightly twining. Bark of woody stems reddish with grayish stripes and pustulate lenticels; herbaceous portions of the stem angled-ribbed, glabrous, summits of the internodes flaring to one side into the petiole bases, a band of hairs often present at the edge of the summit of the internode.

Leaves simple, alternate, short-petiolate, petioles glabrous or sparsely pubescent; blades pinnately-veined, ovate, variable in size, to 10 cm long, mostly shorter, and 3–5 cm broad basally; bases mostly truncate to slightly cordate, apices short-acuminate; margins entire; upper surfaces glabrous, lower usually sparsely short-pubescent.

Flowers borne in bracteate fascicles arranged spicately, the fascicles sessile, the spikes simple from leaf axils or often in panicles leafy-bracted proximally. Stalk of the flower articulated to an unbranched, shorter stalk, the flower stalk itself winged on one side, gradually dilated upwardly to the floral tube (the wing becoming pronounced as the fruit matures). Flowers bisexual, radially symmetrical. Floral tube surmounted by 5 oblongish, subequal sepals spreading at full anthesis, later erect, the flower stalk, the floral tube, and the sepals increasing greatly in size as the fruit matures, the floral tube closely investing but not fused with the base of the fruit, the sepals erect and obscuring the remainder of it.

Fruit a narrowly ovoid, shortly beaked achene about 8 mm long, its surface brown, subglossy. [*Brunnichia cirrhosa* Banks ex Gaertn.]

River banks, borders of and clearings of floodplain forests, bayous, thickets bordering ponds near rivers; s.w. Ga., Panhandle of Fla., chiefly from the Ochlockonee River westward, s. Ala. (Coastal plain, S.C. to Fla. Panhandle, westward to e. Tex., northward in the interior to s.e. Mo., w. Ky., s. Ill.)

2. Polygonella (JOINTWEEDS)

Annuals or suffrutescent perennials with woody taproots. Stems little- to much-branched, the branches appearing internodal owing to some coalescence of the stem and branch internodes. Leaves alternate, articulated near the summits of the ocreae and appearing sessile. Inflorescences racemelike and these in panicles, the flowers borne singly on unbranched stalks articulated with unbranched stalks that arise from axils of imbricated ocreolae. Flowers bisexual or functionally unisexual, radially symmetrical. Calyx petaloid, persistent, consisting of 2 outer and 3 inner sepals or 2 outer, 2 inner, and sometimes 1 transitional, the inner sepals enlarging as the fruits mature. Corolla none. Stamens 8, in 2 series, 5 outer and 3 inner. Pistil 1, ovary superior or nearly so, 1-locular, 1-ovuled, styles 3. Fruit an achene.

1. Larger leaves cuneate-obovate, mostly 30–40 (–60) mm long and 10–20 mm broad just below their apices; older woody portions of stem usually dark reddish brown and with "stubble" of the ocreae present, or with ring-scars marking the insertions of the shed ocreae.   1. *P. macrophylla*
1. Larger leaves cuneate-spatulate or clavate, 4.5–18 mm long and 2–4 mm broad just below their apices; older woody stems tan or grayish, the bark sloughing in narrow shreds, neither "stubble" of the ocreae nor line scars indicating the former presence of ocreae evident.   2. *P. polygama*

1. Polygonella macrophylla Small.   Fig. 260

Suffrutescent perennial, woody portions of the stems to about 1 m tall, usually unbranched below the inflorescence; older portions of the woody stem markedly roughened by the "stubble" of shattered ocreae, or merely with line scars indicating the insertion of the wholly shed ocreae.

Leaves glabrous, cuneate-obovate, apices relatively broadly rounded, larger ones 30–40 (–60) mm long and 10–20 mm broad just back of their apices, some leaves usually a little smaller or a little larger; margins entire; venation obscure, or sometimes 5–8 veins evident beneath and ascending more or less parallel to each other; leaves all deciduous in autumn or some persisting through a second season. Ocreae with entire margins.

Overall panicle size varying from 1.5–4 dm long, compactly to loosely, few to moderately branched. Flowers all bisexual on some plants or on occasional plants all staminate or all pistillate. Outer sepals more or less cupped or folded, somewhat subrhombic in outline, the inner flat, obovate to rotund, both varyingly nearly wholly white, more or less flushed with pink or red, often wholly red, the latter especially lending widely scattered, brilliant flashes of autumn color in the sand pine–oak scrub which the plant inhabits. Racemes mostly 1.5–5.5 cm long. Flowers loosely spreading at anthesis, soon becoming reflexed and remaining so as the fruits mature.

Achene brown, smooth, and glossy, the body trigonous, narrowed to a stipe-like base and abruptly narrowed to a slightly longer apical point, about 3.5 mm long overall.

Widely scattered in pine-oak scrub of Gulf coastal sand ridges and stabilized dunes from Franklin Co., Fla. to Baldwin Co., Ala.

2. Polygonella polygama (Vent.) Engelm. & Gray.   Fig. 261

Suffrutescent perennial, woody portions of stems usually reaching only 1–2 dm above ground level, herbaceous flowering stems of the season mostly to about 4 dm long, sometimes 1 or few in number on a given plant, more often numerous;

Fig. 260. **Polygonella macrophylla:** a, habit; b, branchlet of inflorescence; c, flowers, staminate below, bisexual one above (probably functionally pistillate); d, fruiting calyx (nutlet within).

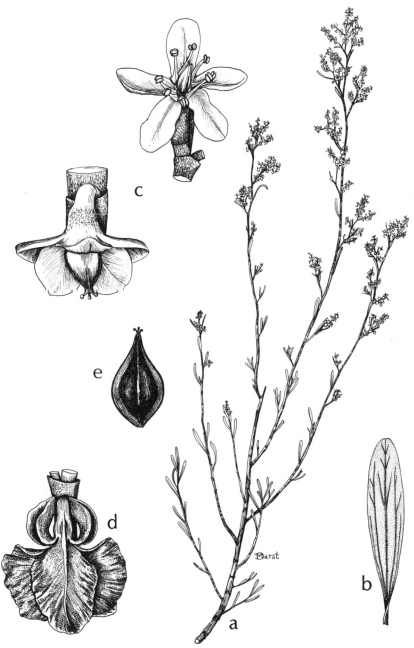

Fig. 261. **Polygonella polygama:** a, flowering branch; b, leaf, enlarged; c, flowers, functionally staminate above, pistillate below; d, fruiting calyx, nutlet within; e, nutlet.

older woody portions with tan or grayish bark sloughing in narrow shreds, no remains of the ocreae evident.

Leaves glabrous, cuneate-spatulate or clavate, apices relatively narrowly rounded, larger ones 4.5–18 mm long and 2–4 mm broad just below their apices, rarely as narrow as 1 mm; margins entire; venation obscure; leaves deciduous in autumn; ocreae with entire margins.

Inflorescence branches usually making up two-thirds the length of a given branch, overall abundantly floriferous, racemes loosely spreading, mostly 0.4–2 cm long. Some plants with functionally staminate flowers (pistils aborting), others with pistillate flowers, thus plants functionally dioecious. Staminate flowers: sepals white, irregularly spreading at anthesis, outer ones oblanceolate to elliptic, margins entire, inner broadly oblong to obovate, margins entire or wrinkled-wavy. Pistillate flowers: a not inconsiderable variation in coloration of sepals from plant to plant; sometimes all sepals white at anthesis, sometimes the outer ones more or less suffused with pink or red, the inner white, sometimes both very much suffused with pink or red, usually one or the other of these in local populations; anthers pale pink; in cases where all the sepals are white at anthesis, they change to greenish yellow as they age (occasionally becoming pink as they age), the plants especially attractive during the period of maturation of the fruits; in cases where the sepals have significant pink or red coloration, the fruiting plants are very showy indeed during the period of fruit maturation; outer sepals elliptic, spreading at anthesis, often eventually folded and becoming reflexed, margins entire; inner sepals erect, ovate, broadly elliptic, or rotund, margins entire or slightly wavy, the edges sometimes crinkled; stalks of the fruits reflexed.

Achene brown, smooth, glossy, ovate, strongly wing-angled, sessile or with a very short stipelike base and a stubby beak, 1.5–2 mm long.

On well-drained, sandy soils, longleaf pine–scrub oak sandhills and ridges, sand pine–oak scrub, adjacent roadsides, stabilized dunes, old, stabilized spoil mounds or banks (sandy dredged material); in our area, s. Ga., n. Fla., s.w. Ala. (Coastal plain, s.e. N.C. to s. pen. Fla., westward along the Gulf Coast to s.e. Miss.; e. and s.e. Tex.)

Ranunculaceae (CROWFOOT FAMILY)

1. Plant a vine, old portions of the stem woody, branches of the season herbaceous; wood of the stem and roots not yellow.   1. *Clematis*
1. Plant a low, upright shrub; wood of the stem and roots yellow.   2. *Xanthorhiza*

1. Clematis

Vines (those treated here), climbing by means of bending or twining of petioles, leaf rachi, or stalks of the leaflets. Older portions of stems woody, branches of the season herbaceous. Bark of woody stems buff or tan in color, longitudinally ridged and grooved, eventually the outer bark sloughing in narrow, elongate shreds. Leaves opposite, petiolate, the principal ones ternately, biternately, or pinnately compound, some of those distally on branchlets with fewer leaflets than major leaves or even simple, leaflets palmately or subpalmately veined. Inflorescences sessile or stalked, axillary cymes, panicled cymes, or flowers solitary. Flowers radially symmetrical, bisexual, or unisexual and plants dioecious. Sepals 4, valvate in the bud, petaloid. Petals none, somewhat petaloid staminodes present in some. Stamens numerous. Pistils separate, few to numerous, ovaries superior, styles plumose, persistent, greatly elongating and becoming more markedly plumose as the fruits mature. Fruit an achene.

Reference: Keener (1975).

1. Flowers bisexual; leaflets with entire margins.
  2. Flowers relatively numerous in axillary, sessile or stalked, panicled cymes; sepals off-white, not thick-leathery, spreading at anthesis.   3. *C. terniflora*
  2. Flowers axillary, solitary and with a pair of ovate to subrotund bracteal leaves more or less medially on the floral branch, or 2–3 flowers above the bracteal leaves, the stalks of each with or without a pair of smaller bracteal leaves; sepals rose-purple, thickish-spongy, essentially erect at anthesis.   2. *C. glaucophylla*
1. Flowers unisexual, plants dioecious; leaflets usually coarsely toothed or toothed-lobed.
  3. Leaves biternately compound or pinnately compound and with 5 leaflets.
                      1. *C. catesbyana*
  3. Leaves, the principal ones, ternately compound.   4. *C. virginiana*

1. Clematis catesbyana Pursh. VIRGIN'S-BOWER, WOODBINE.   Fig. 262

Vine, in our area at least, the older portions of woody stems to at least 3 cm in diameter, herbaceous growth of the season commonly luxuriantly growing through and over shrubs and well into the crowns of small trees; herbaceous stems prominently ribbed, moderately short-pubescent at first, usually becoming glabrous save about the nodes.

Principal leaves (of medial portions of the stem) biternately compound or pinnately compound and with 5 leaflets; leaflets varying a great deal in size on a given plant, to about 6 cm long and 4.5 cm broad basally, mostly ovate in overall outline, coarsely few-toothed varying to shallowly to deeply, regularly or irregularly, 3-lobed, sometimes with only 1 lateral lobe, the lobes sometimes without teeth, sometimes with a tooth on 1 or both sides, bases truncate to subcordate, sometimes slightly oblique, apices or apices of lobes acute or acuminate, lower surfaces of very young leaflets relatively densely pubescent, much of the pubescence sloughing, both surfaces of mature leaflets sparsely pubescent throughout or only along the veins. Leaves (and leaflets) of medial portions of herbaceous stems much larger than those of more slender, longish distal portions of the branches.

Cymes vary from simple, short ones to panicled and to 2 dm long, larger ones usually irregularly leafy-bracted proximally. Axes of cymes moderately pubescent, flower stalks densely so as are both surfaces of the sepals. Sepals spreading or reflexed, off-white, linear-oblong, blunt apically, 6–12 mm long. Stamens off-white, filaments flat, somewhat shorter than the sepals, anthers 1–1.5 mm long. Achenes tightly congested, short-pubescent, hairs short white-silky, lanceolate to elliptic, at maturity dark reddish or purplish brown to dark blackish purple, about 4 mm long, the long, white-plumose styles curling-spreading in all directions, about 4–5 cm across the spread.

In and on the borders of upland and lowland woodlands, woodland clearings, river banks, moist to wet thickets, fence and hedge rows; s.w. Ga., Fla. Panhandle, s. Ala. (All provinces, Va. to Ky. and Mo., southward to Fla. and Ark.)

2. Clematis glaucophylla Small. LEATHER-FLOWER.

Vine, in our area, the older, lower portions of stems woody, perhaps not exceeding 1 cm in diameter, herbaceous growth of the season sometimes relatively

Fig. 262. **Clematis catesbyana:** a, habit; b, woody stem (piece drawn was 2.5 cm in diameter); c, staminate flower; d, functionally pistillate flower; e, fruiting branch; f, "head" of fruits.

thinly branched, sometimes luxuriantly so; herbaceous stems ribbed, glabrous, commonly reddish or purplish brown.

Principal leaves with 2, 3, 4, or 5 leaflets, varying to biternately compound; leaflets palmately veined, mostly unlobed, sometimes lobed more or less mittenlike, margins entire; varying in size to 8–10 cm long and to 6–8 cm broad basally, broadly to narrowly ovate, occasionally elliptic, bases cordate to truncate, sometimes oblique, apices broadly obtuse to short-acuminate, glabrous, grayish-glaucous beneath at first, eventually becoming shiny green.

Floral branches axillary, glabrous, often with a single flower and with a pair of ovate to subrotund, rather conspicuous, bracteal leaves medially, or with 2 or 3 flowers above the bracteal leaves, the stalks of each with or without a pair of smaller bracteal leaves. Flowers bisexual. Sepals thickish-spongy, essentially erect at anthesis, 1.5–2.5 cm long, ovate, acuminate-attenuate apically, exteriorly a rich rose-purple, glabrous, the inner margins whitish-tomentose. Stamens somewhat shorter than the sepals, filaments and anthers pubescent, both flat or flattish, the anthers longer than the filaments. Numerous tightly congested achenes produced from each flower, their bodies ovate-attenuate, about 8 mm long, surfaces brown but with whitish or grayish pubescence, the long, tawny-plumose styles mostly arching-reflexed and forming a roundish ball-like mass of plumes about 8 cm across. [*Viorna glaucophylla* (Small) Small]

River banks, floodplain woodlands and their clearings, adjacent richly wooded slopes, thickets; s.w. Ga., w. half of Fla. Panhandle, s. Ala. (Various provinces, roughly W.Va. to (Mo.?) Ark., s.e. Okla., southward to Fla. Panhandle and Miss.)

3. Clematis terniflora DC.

Vine, the older woody portions of stem to at least 1.5 cm in diameter, the growth of the season generally luxuriantly climbing into and through shrubs or sapling trees; herbaceous portions of stems ribbed, glabrous or very sparsely pubescent internodally, moderately short-pubescent about the nodes.

Principal midstem leaves pinnately compound and mostly with 5 leaflets, some leaves on a given plant with 3 leaflets, some simple; occasionally at a given node the leaf on one side may have 5 leaflets, that on the other side 3, or that on one side 3, the opposite one simple, etc.; leaflet blades palmately veined, varying considerably in size to about 10 cm long and 6 cm broad basally, thickish-membranous, short-ovate to long-ovate, unlobed, bases truncate or subcordate, apices rounded, obtuse, or acute, surfaces glabrous or sparsely pubescent along the veins, margins entire.

Inflorescences axillary, stalked or sessile panicled cymes 5–25 cm long, the larger ones sometimes leafy-bracted proximally, the axes and flower stalks sparsely pubescent.

Flowers bisexual. Sepals spreading, off-white, woolly-pubescent marginally beneath, narrowly oblong-oblanceolate, blunt apically, 1–2 cm long. Stamens off-white, shorter than the sepals, filaments flat, anthers 2–3 mm long. Few achenes produced from each flower, commonly 3–6, loosely spreading, elliptic to fusiform, flattened, 8 mm long or a little more, at maturity dull brown, the surfaces with relatively sparse, brown, appressed pubescence, the long, white-plumose styles arched-spreading. [*Clematis dioscoreifolia* Levl. and Vaniot., *C. maximowicziana* Franch. and Savat., *C. paniculata* Thunb.]

Native to e. Asia, cultivated as an ornamental and sporadically naturalized in our area and beyond in thickets, fence and hedge rows, clearings, vacant lots, and margins of lakes and streams.

Fig. 263. **Clematis virginiana:** a, portion of flowering stem (staminate); b, infructescence; c, fruit.

4. Clematis virginiana L. VIRGIN'S-BOWER, WOODBINE.   Fig. 263

Similar in most general features to *C. catesbyana*, the woody portions of the stems to about 1 cm in diameter (perhaps more). Principal leaves with 3 leaflets, their blades narrowly to broadly ovate, usually unlobed, the margins regularly or irregularly with few large teeth, less frequently entire or with 1–few, small teeth on a side, or only on one side, 3-lobed-toothed on an occasional plant. Achene light to dark brown or greenish brown, otherwise as in *C. catesbyana*.

Lowland woodlands, stream banks, thickets, waste places; apparently rare in w. parts of our area of coverage. (All provinces, N.S. and e. Que. to Man., generally southward to Fla. Panhandle, cen. pen. Fla., and La.)

2. Xanthorhiza

Xanthorhiza simplicissima Marsh. YELLOW-ROOT, BROOK-FEATHER.   Fig. 264

A deciduous, colonial shrub with the wood of both stems and roots yellow. Stems slender, brittle, usually unbranched, to about 8 dm tall, the alternate leaves closely spiraled on a short increment of growth of the current season, more widely spaced on vigorous shoots or sprouts. Twigs tan or buff-colored, for the most part very irregularly angled and ridged, the ridges rounded and with rounded furrows between; lenticels scattered, knobby at first, scars of closely set leaves nearly linear or vaguely long crescent-shaped, with a line of numerous vascular bundle scars, scars of widely spaced leaves triangular, the longest side of the triangle flat below the axillary bud.

Leaves odd-pinnately compound, petiolate, the petioles dilated at the base and clasping much of the stem, otherwise slender and usually as long as or longer than the blades; blades with 3–5, usually 5, pinnately veined leaflets, bases cuneate, apices of leaflets or their lobes acute; lowermost pair of leaflets usually deeply cleft only on the abaxial side, the terminal one usually symmetrical and deeply 3-lobed, all margins strongly toothed except on their cuneate basal portions; blades on a given stem usually very unequal in size, the larger about 10–12 cm long overall and as wide across the basal pair of leaflets; petioles and axes of the leaves pubescent with strongly curved or both curved and straight hairs, upper surfaces of leaflets with very short, strongly curved hairs on the midribs and larger veins, longer hairs variously distributed or absent on the margins (even of a single leaflet), lower surfaces relatively sparsely pubescent with mostly straight and some slightly curved hairs both on the veins and to a lesser extent between them.

Flowers in flexuous, drooping racemes or narrow panicles 5–15 cm long, one inflorescence from the axil of each of several of the lowermost leaves, full anthesis occurring well before the leaves are fully grown; axes of the inflorescence and the flowers commonly maroon or brownish maroon, sometimes greenish yellow, the axes and flower stalks pubescent with short, curly hairs. Sepals 5, brown-purple, somewhat unequal, more or less clawed, varying in shape, 3–5 mm long. Petals none. Staminodia 5, stalked, their summits with 2 diverging, rounded, nectariferous maroon lobes. Stamens usually 5, sometimes 10. Pistils usually 5–10, each ripening into an obliquely oblong, 1-seeded follicle.

River and stream banks, moist thickets, springy places, usually where shaded; s.w. Ga., s.e. Ala., cen. and w. Panhandle of Fla. (S.w. N.Y. to w. Va. and Ky., generally southward to Fla. Panhandle, Miss., e. Tex.)

Distinctive identifying features of this shrub include the yellow wood of stems

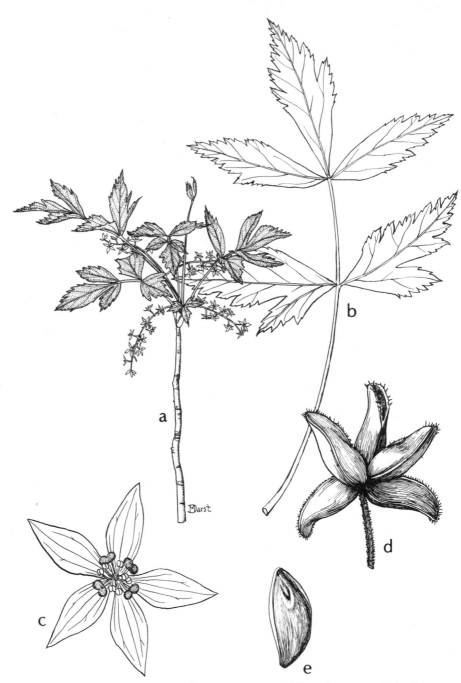

Fig. 264. **Xanthorhiza simplicissima:** a, top of flowering stem; b, leaf; c, flower, face view; d, follicles (from one flower); e, seed.

and roots, slender, brittle, low, usually unbranched stems terminated by closely set, odd-pinnately compound leaves whose leaflets are strongly toothed or toothed-cleft.

Rhamnaceae (BUCKTHORN FAMILY)

1. Lateral veins of the leaf blades equal in prominence and angling-ascending approximately parallel to each other from the midvein toward the leaf edges.
   2. Plant a twining-scrambling vine.   1. *Berchemia*
   2. Plant an erect shrub or small tree.   3. *Rhamnus*
1. Lateral veins of the leaf blades not as described above.
   3. Leaf blades with 2 principal, lateral, strongly ascending veins from the base.
      4. Plant a low shrub, seldom, if ever, exceeding 1 m tall; stems slender, not zigzagging, without spur-shoots bearing herbaceous branchlets resembling compound leaves, never armed; flowers borne in stalked, terminal and/or axillary corymblike thyrses composed of few-flowered umbellike cymes.   2. *Ceanothus*
      4. Plant a large shrub or small tree; woody stems zigzagging, having short spur-shoots at the nodes that seasonally produce more or less herbaceous leafy shoots resembling compound leaves, the nodes of the woody stems *sometimes* bearing a pair of unequal stipular (?) spines; flowers borne in short, irregular cymose clusters in the leaf axils.
                                                                      5. *Ziziphus*
   3. Leaf blades with alternating, arching, pinnate lateral veins.   4. *Sageretia*

1. Berchemia

Berchemia scandens (Hill) K. Koch. SUPPLE-JACK, RATTAN-VINE.   Fig. 265

Deciduous, glabrous, flexible, tough, unarmed, woody vine with short to widely divergent, reddish brown, smooth branches, to some extent twining but mostly scandent-scrambling on and over shrubs and low trees, in taller trees scrambling in the crowns.

Leaves alternate, simple, short-petiolate; blades elliptic, oval, or ovate, their bases broadly short-tapered to rounded, apices obtuse or abruptly short-acuminate, usually tipped by a mucro, 3–8 cm long and to 4 cm broad, pinnately veined, the lateral veins about 10 to a side, conspicuous and angling-ascending parallel to each other, distally curvate and ending on the leaf margin, veinlets reticulate between them; margins slightly wavy to entire; upper surfaces bright green and shiny, the lower pale and dull; stipules lance-acute, eventually deciduous.

Flowers small, about 2 mm across, greenish, functionally unisexual (plants functionally dioecious), borne on the branchlets in axillary and terminal panicles. Floral tube very short, a nectariferous disc within, this barely embracing the base of the ovary in the pistillate flowers, the floral tube surmounted by 5 deltoid sepals and 5 somewhat obovate petals, petals in the staminate flowers each somewhat hooded about a stamen. Fruit an oblong-ellipsoid drupe 5–7 mm long, blue-black and glaucous when ripe, the thin disc below persistent.

Bottomlands, swamps, wettish pine flatwoods, pineland bogs, wet thickets, also in well drained woodlands throughout our area. (Chiefly [not exclusively] coastal plain, s.e. Va. to s. Fla., westward to Tex., northward in the interior, Ark., s. Mo., Tenn.)

The twining-scrambling viny habit of the supple-jack, together with its alter-

Fig. 265. **Berchemia scandens:** a, fruiting branch; b, leaf from vigorous vegetative branch; c, flowering branchlet; d, flower; e, portion of perianth of staminate flower, opened out, petals more or less cupping the stamens; f, pistil, its base embedded in a disc; g, fruit.

nately disposed leaves whose blades have conspicuously subequal, angling-ascending, parallel lateral veins, serves to distinguish it from others of our woody plants.

2. Ceanothus

Low shrubs with alternate simple, stipulate leaves and small, white flowers borne in stalked, terminal and/or axillary corymblike thyrses composed of few-flowered umbellike cymes. Flowers with a short, cuplike floral tube. Sepals 5, triangular, not persistent. Petals 5, white or whitish (in ours), prominently clawed, claws longer than the calyx, their blades hooded, in bud each clasping an anther, at anthesis spreading or deflexed between the sepals by the curved claws. Stamens exserted at anthesis. Pistil 1, the upper portion of the semi-inferior, 3-locular ovary surrounded by the nectariferous disc, style 3-forked distally, the 3 stigmas capitate. Fruit a capsulelike, 3-lobed drupe subtended by the persistent floral tube, containing 3 coherent nutlets, the 3 parts eventually separating and dehiscing explosively, ejecting the seeds, the floral tube persisting.

1. Leaves very much larger than 6 mm long, venation evident on both surfaces, 3-nerved from the base but the lateral nerves arching-ascending and reaching the leaf edge some-what back from the tip.   1. *C. americanus*
1. Leaves minute, 2–6 mm long, venation not evident on their upper surfaces, 3-nerved on the lower, the lateral nerves subparallel to the midvein and extending nearly to the leaf tips.   2. *C. microphyllus*

1. Ceanothus americanus L. NEW JERSEY TEA.   Fig. 266

A shrub or subshrub, 5–10 dm tall, with few to numerous slender, ascending branches; commonly with a stout, gnarled or burllike, hard, woody rootstock, the branches more or less herbaceous distally, sometimes new each year from the base or sometimes dying back little overwinter. Younger parts of the stems tan, pubescent with long, shaggy-spreading and short curly hairs, older parts developing a brown, thin, slightly roughened bark retaining some of the short pubescence.

Leaves short-petiolate, petioles pubescent like the young stems, blades variable, 2–8 cm long and 1–4 cm broad, ovate to lanceolate (the former more characteristic of plants growing in fertile soils, the latter of plants growing in relatively sterile, often sandy, soils), bases slightly cordate, rounded, or short-tapered, apices acute, subacuminate, or obtuse; venation evident on both surfaces, more so beneath than above, venation palmate, the two lateral veins subequal to the midvein, these arching-ascending and ending at or near the margins above mid-leaf; upper surfaces essentially glabrous, pubescent only along the principal veins, or relatively evenly sparsely short-pubescent, lower surfaces usually (but not always) pubescent at least on the major veins, sometimes on most of the more pronounced veins, sometimes velvety to the touch (not in our range); margins finely serrate.

Thyrses ovoid to short-cylindric, long-stalked, axillary to upper leaves, sometimes branched, stalks naked or with few bracteal leaves at the bases of the thyrses, those lower on the branches much longer than those near the branch tips.

Drupes dark brown to nearly black, 3–5 mm broad, a little broader than long, 3-lobed but not prominently so, each lobe with a crest or stubby protuberance medially on its upper part, surface otherwise roughish. Seeds obovate in out-

Fig. 266. a–f, **Ceanothus americanus:** a, flowering branch; b, small portion of lower leaf surface; c, flower; d, fruiting branchlet; e, fruit; f, seed. g–k, **Ceanothus microphyllus:** g, flowering branch; h, leaves, lower surface above, upper surface below; i, fruiting branchlet; j, fruit; k, seed.

line, one side broadly rounded, the other slightly flattened, surface dull, brown and with very minute whitish papillae. [Incl. *Ceanothus intermedius* Pursh]

Well-drained mixed woodlands, woodland borders, glades, prairies, well-drained pinelands, local in our area. (Que., s. Ont. to Minn. and Neb., generally southward to cen. pen. Fla. and cen. Tex.)

2. Ceanothus microphyllus Michx.   Fig. 266

A low, bushy-branched evergreen shrub, the principal branches from near the base and usually not over 6 dm long, infrequently to 8 dm. Branches slender, yellow and sparsely dusty-pubescent at first, becoming irregularly striped with narrow patches of brown bark in age.

Leaves minute (commonly with very short leafy branches in their axils thus appearing fasciclelike), very short petiolate, blades elliptic or elliptic-obovate, 2–6 mm long, tapered basally and blunt to rounded apically, upper surfaces glabrous, shiny and without evident venation, the lower with 3 palmate veins, the 2 laterals subparallel to the midvein and extending nearly to the leaf tips, with stiff, antrorsely appressed hairs on the midvein or on all 3 veins, margins entire or with 1–2 obscure teeth to a side.

Thyrses terminal on most branchlets, with few leafy bracts proximally, 0.5–2.5 cm long, many of the thyrses as broad as long.

Drupes smooth, black or dark purplish black, mostly 4–5 mm broad, a little broader than long, with 3 broadly rounded lobes, styles persisting during maturation. Seeds suborbicular in outline, about 2 mm across, with 1 broadly rounded face and 2 flat or concave faces, surfaces smooth and sublustrous, dark reddish brown when fully mature.

Well-drained pinelands. (Coastal plain, s. Ga., southward in Fla. to about Lake Okeechobee, Fla. Panhandle, s. Ala.)

C. microphyllus is our only low shrub with yellow stems and minute (2–6 mm long), elliptic to elliptic-obovate leaf blades the upper surfaces of which have no evident venation, the lower surfaces 3-nerved, the lateral nerves subparallel to the midvein and extending nearly to the leaf tips.

3. Rhamnus

Rhamnus caroliniana Walt. CAROLINA BUCKTHORN.   Fig. 267

An unarmed, deciduous shrub or small tree to about 10 m tall. Young twigs reddish brown, pubescent, later becoming gray and glabrous; vascular bundle scars 3 in each leaf scar; pith continuous, white; buds all axillary, without covering scales, densely short-pubescent, only the uppermost bud on a given branchlet prominent, other lateral buds minute if visible at all.

Leaves simple, alternate, short-petiolate, stipulate; stipules subulate, very quickly deciduous as the leaves commence to grow; petioles and lower leaf surfaces notably soft-pubescent as they unfold, much, usually not all, of the pubescence sloughing as they age. Leaf blades broadly oblong, oblong-elliptic, or slightly obovate, usually the 1–3 proximal leaves on a given branchlet considerably smaller than the others, larger blades 5–12 cm long and 3–5 cm broad, bases usually rounded, some sometimes broadly short-tapered, apices varyingly obtuse, acute, or obscurely acuminate; lateral veins prominently ascending parallel or nearly parallel to each other to near the leaf edges where they become much fainter, strongly curve and run close to the leaf edges for some distance; margins obscurely and irregularly serrate.

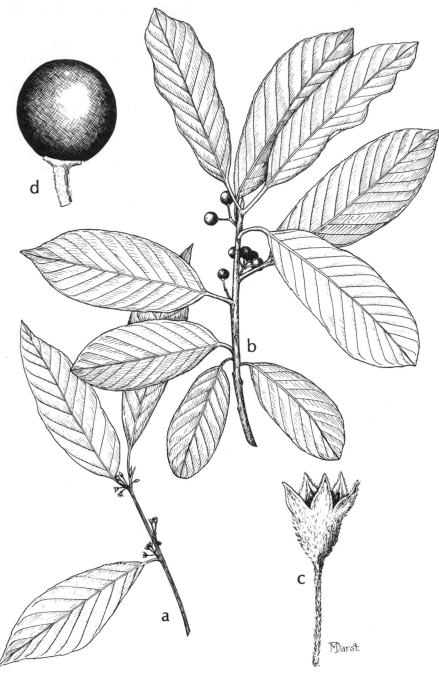

Fig. 267. **Rhamnus caroliniana:** a, branchlet with flowers; b, branchlet with fruits; c, flower; d, fruit.

Flowers bisexual, small, with a cuplike floral tube free from the ovary, a nectariferous disc beneath the ovary. Sepals 5, erect, triangular or ovate-triangular, greenish or greenish yellow. Petals 5, shorter than the sepals, subclawed basally, obovate above, notched apically, yellowish or whitish, each usually clasping a filament. Stamens 5, as long as or slightly longer than the petals. Ovary 3-locular, style 3-forked.

Fruit a subglobose drupe, black and juicy when fully ripe, containing 3 nutlets, the dried disc persisting beneath the drupe.

Usually in woodlands or copses, often in calcareous rocky woodlands, shell middens, or glades, local, n. Fla., s.w. Ga., s. Ala. (Western Va. to Ill. and Neb., southward to pen. Fla. and w. Tex.).

Co-national champion Carolina buckthorns (1974), 3'5" circumf., 27' in height, 23' spread, Middleburg, Va.; (1974), 1'9" circumf., 46' in height, 21' spread, Norris Dam State Park, Tenn.

4. Sageretia

Sageretia minutiflora (Michx.) Mohr. BUCKTHORN.   Fig. 268

Often a somewhat straggly, tardily deciduous, slender shrub to about 3 m tall, the branches divaricate or only moderately ascending. Potentially a llana: vigorous leader shoots usually bear alternate (owing to one of a pair of buds, in alternating sequence, remaining dormant) relatively short, stiff, sharply thorn-tipped, divaricate branches, these serving to hold or anchor the stems as they grow into and through other vegetation. Such stems may eventually self-prune as growth proceeds terminally resulting in ropelike lianas to at least 3 cm in diameter and with diffuse branch systems scrambling in the crowns of tallish trees.

Leaves simple, opposite, sometimes subopposite, very short-petiolate; blades somewhat leathery, varying not a little in size, 0.5–6 cm long, ovate, very slightly cordate to rounded basally, obtuse, acute, or vaguely acuminate apically, smaller ones lowest on short branches often suborbicular; margins shallowly crenate-serrate, teeth tipped by mucros; venation pinnate, major lateral veins few, arching-ascending and mostly anastomosing somewhat back of the leaf edges, upper surfaces finely reticulate-veined, dark green and glossy in age, somewhat duller beneath. Young branches, petioles, and leaf surfaces cottony-pubescent at least when young, glabrous or nearly so later.

Inflorescences narrowly interrupted-spicate, spikes axillary or terminal on the branches, sometimes panicled-spikes at branch tips, axes shaggy pubescent. Flowers fragrant, minute, bisexual, each subtended by 2 bractlets, with a saucerlike floral tube surmounted by 5 deltoid sepals keeled on their inner surfaces. Petals white, alternate with and shorter than the sepals, very short-clawed, blades suborbicular to ovate, minutely notched apically, somewhat concave and each somewhat enfolding a stamen. Stamens adnate to the bases of the petals inserted below the margin of a collarlike nectariferous disc. Pistil 1, ovary surrounded by the disc, style short, stigmas 3, very small. Fruit a subglobose to somewhat oblate, purplish red to purplish black drupe about 5–8 mm in diameter, the floral tube, calyx, and usually shriveled stamens persistent below, persistent over winter, eventually splitting into 3 leathery nutlets.

Calcareous rocky bluffs, woodlands, hammocks, shell mounds, local throughout our area. (Coastal plain, s.e. S.C., southward to Fla., Brevard Co. on the east and Lee Co. on the west, westward to Miss.)

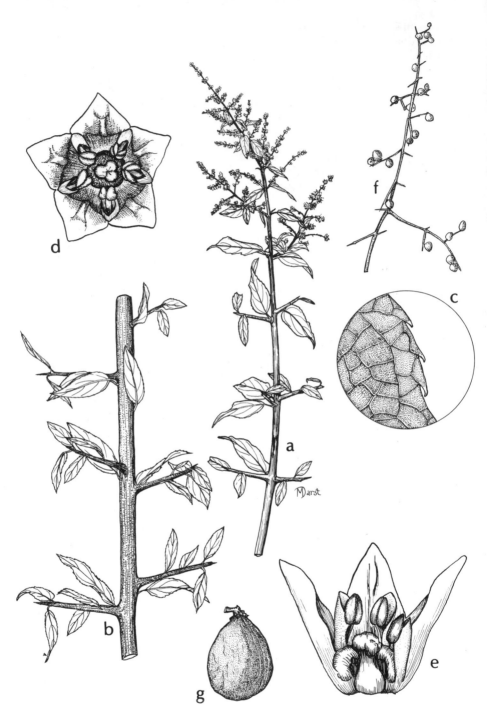

Fig. 268. **Sageretia minutiflora:** a, lateral flowering branch; b, piece of vigorous, elongation branch with its leafy branch thorns (each of which may eventually become a flowering branch as in (a); c, greatly enlarged portion of leaf to show character of marginal teeth; d, flower, face view; e, flower with part of perianth removed (note petals are smaller than the sepals and are more or less cupped about the stamens); f, part of an infructescence; g, fruit.

5. Ziziphus

Ziziphus jujuba Mill. JUJUBE.

Glabrous, deciduous tree to 10 m tall. Leafless twigs in winter stoutish, geniculate, bearing knobby to conical "spurs" at the nodes; pith homogenous. In spring most of the "spurs" produce 1–several, usually several, slender, determinate, subherbaceous shoots 1–2.5 dm long, their leaves essentially 2-ranked; by the time leaves of such shoots are fully grown, short, irregular cymes of small yellowish green flowers appear in the axils of all but the proximal 2–3 leaves. These determinate shoots very much resemble compound leaves, and they are deciduous in their entirety in autumn. The "spurs" produce determinate shoots annually for several years, increasing in size as they do so, and close examination shows them to have an ever-increasing number of roundish "branch-scars" over their entire surfaces. Coincidental with growth as described above in a given season, from each of a few "spurs" scattered about the crown of the tree, besides the determinate shoots, a single extension shoot is produced, the branches of which are determinate shoots that are deciduous in autumn and they, too, bear a few flowers. Some nodes of the extension shoots bear a pair of short, sharp nonpersistent spines.

Leaves simple, alternate, unlobed, petioles to about 5 mm long, blades lance-ovate to ovate, 2–5 cm long and 1–2.5 cm broad near their bases, apices mostly narrowly rounded, bases rounded or shortly and broadly tapered, upper surfaces dark green and moderately lustrous, the lower paler and dull green, margins crenate-serrate, venation subpalmate (best observed from beneath), with two lateral angling-ascending veins arising basally with the midrib.

Flowers small, 5–6 mm across at full anthesis, greenish yellow; floral tube cuplike, a prominent, thick nectariferous disc surrounding but not adherent to the ovary; calyx segments 5, triangular, tips slightly incurved, adaxial surfaces thickened medially; petals 5, shorter than the calyx segments, each strongly cupped and embracing a stamen as the flower opens, each stamen as long as or slightly longer than the petal. Pistil 1, ovary 2-locular, style short, with 2 recurved stigmatic lobes. Fruit an ellipsoidal or oblong-cylindric, reddish drupe about 3 cm long and 2.5 cm in diameter.

Native to warmer and drier parts of the Old World. In our area, infrequently planted, sparingly naturalized, sometimes at a considerable distance from the parent tree.

Rhizophoraceae (RED MANGROVE FAMILY)

Rhizophora mangle L. RED MANGROVE.   Figs. 269, 270

Evergreen shrubs (in our area), or trees to about 20 m tall, bowed or arching prop- or stilt-roots arising from the trunk and branches, eventually plants gregarious and forming inpenetrable thickets. Terminal buds narrowly elongated, pointed, covered by 2 green, scalelike stipules which when shed leave ring scars. Bark gray or grayish brown, smooth on small trunks, irregularly furrowed and thick on large ones, inner bark reddish or pinkish; bark of twigs reddish brown; leaf scars horizontally short-elliptical, vascular bundle scars apparently several (at area of leaf dehiscence, presumably 3 traces in the petioles).

Leaves opposite (some rarely appearing alternate), short-petiolate. Blades

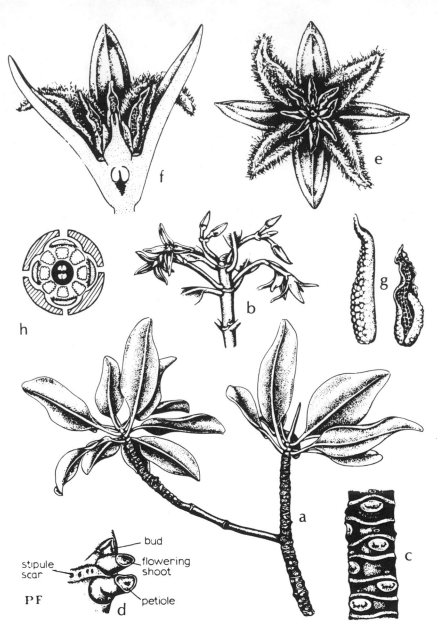

bud

stipule
scar

flowering
shoot

PF

petiole

d

a

b

c

e

f

g

h

Fig. 269. **Rhizophora mangle:** a, distal part of horizontal branch; b, part of
flowering shoot with the stipular bud removed to show 3–flowered dichasia
of various ages; c, detail of scar pattern on short-shoot; d, leaf insertion with
petiole and flowering shoot removed to show stipule scar and suppressed bud
above inflorescence axis; e, flower, from above; f, flower longitudinally dissected; g, stamen, undehisced (left) and dehisced (right); h, floral diagram.
(Courtesy of Fairchild Tropical Garden.)

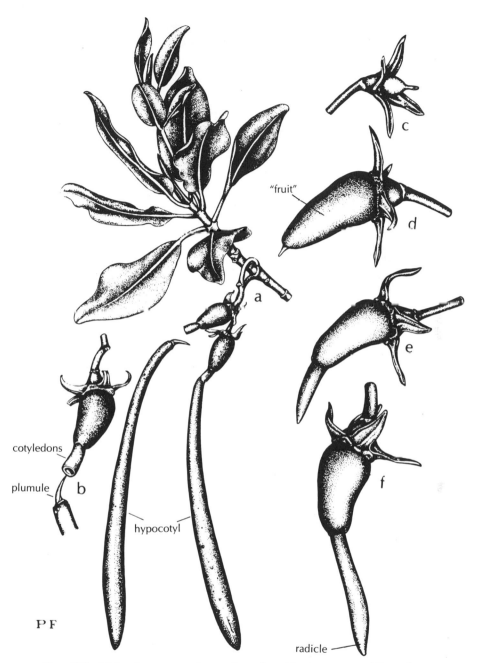

Fig. 270. **Rhizophora mangle:** a, branch with mature seedlings (one detached); b, detail of plumule and cotyledonary collar; c, young fruit soon after the petals have fallen; d, e, and f, stages in the development of fruits and viviparous seedlings. (Courtesy of Fairchild Tropical Garden.)

dark green and shiny above, paler beneath, leathery, elliptic to oblong, or elliptic-obovate, 4–12 (15) cm long and 1.5–5 cm broad, apices obtuse to rounded, margins entire; upper surfaces glabrous, sometimes with a few widely scattered, elevated punctae, lower surfaces glabrous, or on some plants moderately pubescent with straight, subappressed, grayish hairs, usually with at least some elevated punctae, these often numerous; punctae purplish red, or purplish red peripherally and with gray moundlike to crateriform centers.

Inflorescence axillary, the stalk articulated and bearing (1) 2–4 stalked flowers from the articulation which has 2 low, broad, fleshy bracts more or less united basally, a pair of similar bracts just beneath each flower. Flowers bisexual, radially symmetrical, floral tube fleshy, obconic, ovary inferior, solid above the 2-locular base, a disc around its summit. Sepals 4, inserted laterally on the rim of the disc, fleshy, elongate triangular, involute, 1 cm long or a little more, persistent and eventually reflexed. Petals 4, similar in shape to the sepals, pale yellow, alternating with the sepals inwardly on the disc and a little shorter than the sepals, involute distally, copiously silky-hairy adaxially. Stamens 8, filaments very short, anthers surrounding the style. Ovary producing 1 seed, soon protruding from the floral tube within the disc, leathery, becoming 2–3 cm long, its seed germinating while the fruit still on the plant and producing a long, fleshy, clavate-fusiform, pointed hypocotyl to 2.5–3 dm long. When the torpedolike seedling is released, it may lodge in the mud and commence forming shoots, or it may be carried elsewhere in water currents.

Shallow water of coastal areas, shores of lagoons, creeks, and rivers mostly where the water is saline or brackish, in fresh water somewhat inland, especially in solution holes in limerock. Coastal or near coastal areas of Fla., northward to about Flagler Co. on the Atlantic Coast, to Levy Co. on the Gulf Coast; W.I., trop. Am., W. Africa, Pacific Islands.

National champion red mangrove (1975), 6'5" circumf., 75' in height, 41' spread, Everglades National Park, Fla.

Rosaceae (ROSE FAMILY)

1. Leaves compound.
   2. The leaves with conspicuous, persistent stipules at least partially attached to proximal portions of the petioles; floral tube urceolate, wholly enclosing numerous separate pistils.   9. *Rosa*
   2. The leaves with inconspicuous, narrow stipules wholly free from the petioles; floral tube flat to hemispheric, the numerous separate pistils borne on a convex or conic receptacle elevating them above the floral tube.   10. *Rubus*
1. Leaves simple.
   3. Leaf blades subpalmately veined (midrib and 2 lowest lateral veins arising together near the base of the blade and lowest laterals more pronounced than others); pubescence of parts stellate.   5. *Physocarpus*
   3. Leaf blades pinnately veined; pubescence of parts not stellate.
      4. Upper surfaces of leaf blades bearing dark, purplish red glands irregularly distributed along their midribs, some blades with a few glands along the proximal portions of the principal lateral veins as well (the glands not visible to the unaided eye).
                                                                                      2. *Aronia*
      4. Upper surfaces of leaf blades without glands as described above.
         5. Principal lateral veins of leaf blades appressed to the midribs for a little distance (visible with some magnification).   1. *Amelanchier*

544

5. Principal lateral veins of the leaf blades not at all appressed to midribs.

    6. Ovary superior, pistil seated within a floral tube and wholly free from it; fruit usually having the shriveled floral tube persisting beneath its base.   6. *Prunus*

    6. Ovary or ovaries inferior, wholly concealed by the floral tube; fruit usually having the shriveled calyx segments persistent at its summit.

        7. Thorns (usually present on at least some of the branches) naked, axillary to leaves.   3. *Crataegus*

        7. Thorns (usually present on at least some of the branches) tipping leafy branches or tipping branches bearing leaf scars.

            8. Leaf blades, most of them, cuneate-oblanceolate or cuneate-spatulate.

                        7. *Pyracantha*

            8. Leaf blades not as above, oblong, elliptic, or ovate.

                9. Blades of the leaves blunt apically, even rounded.   4. *Malus*

                9. Blades of leaves abruptly cuspidate-acuminate apically.   8. *Pyrus*

1. Amelanchier

Amelanchier arborea (Michx.) Fern. DOWNY SERVICEBERRY.   Fig. 271

A deciduous shrub or small tree, the latter probably not exceeding 10 m tall (in our range). Young shoots sparsely cottony-pubescent, much of the pubescence soon sloughing, twigs brown, leaf scars narrowly crescent shaped, vascular bundle scars 3 per leaf scar, usually indistinct; old stems ashy gray, pith tan, continuous.

Leaves simple, alternate, slenderly petiolate, petioles much shorter than the blades, stipulate, stipules linear-subulate, membranous, pinkish, silky-pubescent, mostly 8–10 mm long, quickly deciduous. Petioles and both surfaces of blades densely woolly-pubescent as the leaves unfold, quickly becoming sparsely pubescent, usually glabrous at full maturity. Blades oblong-oval, oval, slightly ovate or slightly obovate, pinnately veined, 3–7 cm long and 2–4 cm broad, bases rounded to very slightly cordate, occasional ones short-tapered, apices acute or short-acuminate, margins finely serrate, lateral veins of leaf blades appressed to the midribs for a short distance.

Inflorescence an erect or flexuously arching raceme terminating a short branchlet of the season before or as the leaves unfold; flower stalks subtended by a bract and with a pair of bracts suboppositely disposed more or less medially, both similar to the sepals; inflorescence axis and flower stalks woolly-pubescent. Flowers bisexual, radially symmetrical, ovary inferior; floral tube usually glabrous, surmounted by 5 lance-attenuate sepals whose inner surfaces are densely woolly-pubescent, sepals shorter than the petals, usually becoming reflexed by the time the petals are fully spread radially. Petals 5, white or pinkish, linear-oblong, 15–20 mm long, erect at first, then spreading. Stamens 10–20, shriveled filaments more or less persistent at the summit of the fruit. Pistil 1, ovary 5-locular, ovules 2 in each locule but in development of the fruit ingrowing partitions form between the ovules of each locule; styles 5, essentially free. Fruit a reddish purple, berrylike, subglobose pome 5–8 mm in diameter, the sepals persistent at its summit.

Occasional in mesic woodlands, well-drained, relatively open woodlands of various mixtures and banks of streams; s. Ga., Fla. Panhandle, s. Ala. (S.w. N.B., Maine, s.w. Que., s. Ont. to Mich. and n.e. Minn., generally southward to Fla. Panhandle and n.e. Tex.)

National champion downy serviceberry (1959), 8′6″ circumf., 50′ in height, 50′ spread, New Philadelphia, Ohio.

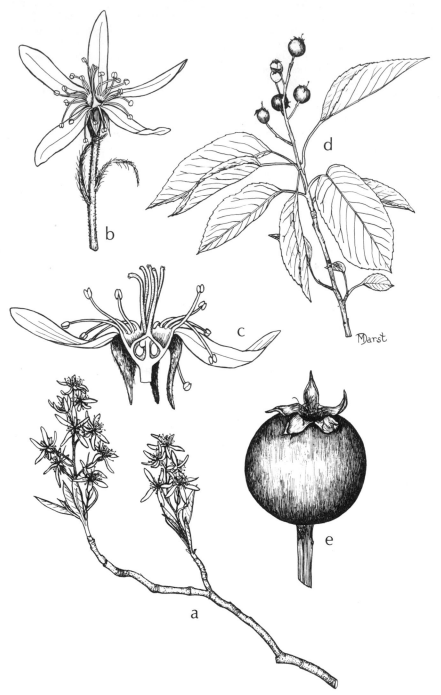

Fig. 271. **Amelanchier arborea:** a, inflorescences; b, flower; c, longitudinal
section of flower; d, fruiting branch; e, fruit.

2. Aronia

Aronia arbutifolia (L.) Ell. RED CHOKEBERRY. Fig. 272

A shrub, to 2–3 m tall, rarely taller, commonly spreading by subterranean runners and colonial; stems wandlike or wandlike below and with relatively few ascending branches. Leaf scars narrow and vaguely crescent-shaped, each with 3 vascular bundle scars. Young branchlets, petioles, stipules, lower leaf surfaces, inflorescence axes, flower stalks, floral tubes, and calyces densely pubescent. Leaves simple, alternate, deciduous, very short-petiolate, with narrow, soft stipules basally adnate to the bases of the petiole, margins of the stipules commonly with purplish red glands; blades elliptic, oval, oblanceolate, or obovate, at maturity variable in size on a given plant, 4–10 cm long and 1.5–4 cm broad, tapered basally, apically obtuse to rounded and usually short-cuspidate, margins finely toothed, the teeth tipped with purplish red glands, upper surfaces with purplish red glands irregularly distributed along their midribs, occasionally with a few glands along the proximal portions of the principal lateral veins as well.

Flowering in early spring, the flowers in corymbose clusters of up to 25 at least, terminating developing branchlets of the season. Expanding flower buds usually pinkish, the outer surfaces of the petals then pink, usually white on both surfaces when fully expanded. Flowers about 1 cm across, the ovary inferior; sepals 5, short-deltoid, borne around the rim of the floral tube (becoming thickish and stiffish and persisting on the fruits); petals 5, short-clawed, the claws inserted on the inner rim of the floral tube and spreading outward between the sepals, their blades rotund to obovate; stamens 15–20, erect, filaments white, anthers lavender-pink, drying brown (shriveled filaments commonly persistent on the summits of the fruits); styles 5, united basally, stigmas capitate. Fruit a small subglobose berrylike, bright red pome 6–9 mm long; fruits commonly persisting into or through the winter when the stems are bare of leaves. [*Pyrus arbutifolia* (L.) L.f., *Sorbus arbutifolia* (L.) Heynhold]

Bogs, wet pine savannas and flatwoods, creek banks; throughout our area. (Nfld,. N.S. to N.Y., e. Pa., W.Va., Ky., to s. and e. Ark., e. Tex., generally eastward to cen. pen. Fla.)

Characteristic features distinctive for the red chokeberry: alternate leaves with stipules or stipule scars, leaf margins finely toothed, the tips of the teeth with purplish red glands, the upper leaf surfaces with purplish red glands irregularly distributed along their midribs, and flowers or fruits in corymbs terminating branchlets of the season.

3. Crataegus (HAWTHORNS, HAWS)

Small trees or shrubs, usually (not always) armed with thorns on at least some of the branches. Branchlets rigid, often more or less zigzagging.

Leaves petiolate or basally tapered to subpetioles; blades simple, deciduous, pinnately veined, margins serrate, dentate, or lobed, in most species those on sterile shoots that are produced after flowering larger and tending to be more lobed or incised than on the flowering or fruiting branchlets; stipules small, often quickly deciduous on the short shoots, often persistent on the sterile shoots.

Flowers produced early in the season, borne in simple or branched corymbs terminal on short branchlets of the season, sometimes borne singly or 2 or 3 together. Flowers bisexual, radially symmetrical, with a cuplike, campanulate,

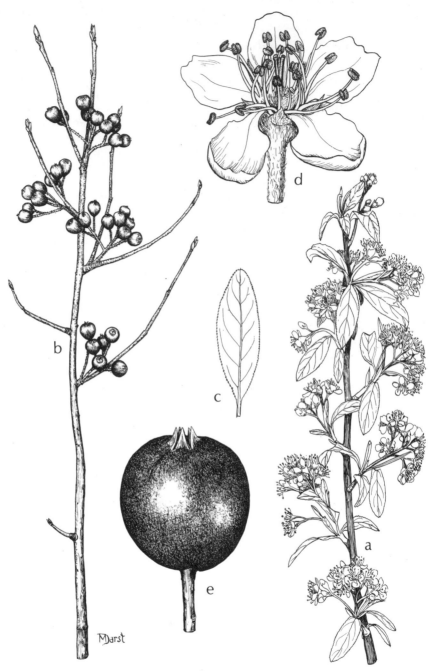

Fig. 272. **Aronia arbutifolia:** a, flowering branch; b, fruiting branch; c, mature leaf; d, flower; e, fruit.

obconic, or urceolate floral tube wholly or almost wholly adnate to the ovary. Sepals 5, persistent. Petals 5, often pink in the bud, white to pink at full anthesis. Stamens 5–20 (25), borne on the rim of the floral tube, the shriveled filaments often persistent at the summit of the fruit. Styles 1–5, free. Fruit a berrylike pome containing 1–5 one-seeded nutlets.

Most of ours are of potential small tree stature, but flower and fruit when of small shrublike form.

Little (1959) states: "*Crataegus*, with perhaps between 100 and 200 species of small trees and shrubs in the United States (nearly all in the eastern half), remains the largest and taxonomically most difficult genus of native trees. More than 1,100 specific names have been published for the native plants of this genus, nearly all in the quarter century beginning about 1899 by three investigators working independently, C. S. Sargent, W. W. Ashe, and C. D. Beadle. About 700 were proposed by Sargent alone.

Crataegus is regarded as an unstable genus characteristic of openings and exposed areas, which expanded and evolved rapidly following clearings of the forests, and the origin of vast new areas suitable for colonization. The variable, expanding populations probably produced numerous hybrids. Progeny tests have shown that many variations are perpetuated, or true breeding. However, cytological evidence indicates that a large number, perhaps a majority, of the supposed species are "asexual apomictic triploids"; that is, they are clonal populations of hybrid origin with one and one-half the normal number of chromosomes but form viable seeds vegetatively without benefit of pollination and thus perpetuate their characters the same as if they were propagated by grafting."

Although the problem for our small geographic area of coverage at the southeastern edge of the range is doubtlessly minuscule relative to the whole, it is nonetheless perplexing. In the following treatment, I have sought to generalize concepts of those occurring here and have reduced the number to nine. This is a reduction from eighteen in the treatment by Kurz and Godfrey (1962) for northern Florida. For the nine recognized, synonymy is given only as it relates to the treatment in Kurz and Godfrey; neither are general or overall ranges for the taxa treated here stated, for to do either is far beyond the intended scope of this book.

1. Lateral veins or veinlets of the leaf blades ending in both the marginal teeth or lobes, if any, and in the sinuses between them.
    2. Leaf blades, at least those of the flowering/fruiting short-shoots, broadest above their middles and longish-cuneate basally.   7. *C. spathulata*
    2. Leaf blades definitely broadest at or just above their bases.
        3. Blades of most of the leaves pinnately dissected-lobed, main lobes irregularly sharply toothed or with smaller toothed sublobes.   4. *C. marshallii*
        3. Blades of most of the leaves of short-shoots, at least, with 3 major lobes, many of them distinctly maplelike, the marginal teeth irregular but short and blunt.
                                        5. *C. phaenopyrum*
1. Lateral veins or veinlets of the leaf blades ending only in the teeth or points of the lobes, if any.
    4. The leaf blades of short-shoots mostly ovate, ovate-oblong, or broadly elliptic, their bases not longish-tapered.
        5. Lower surfaces of mature leaf blades with tufts of hairs in the principal vein axils; petioles and marginal teeth or lobes of blades without glands.   9. *C. viridis*
        5. Lower surfaces of mature leaf blades glabrous to sparsely pubescent throughout but not with tufts of hairs in the principal vein axils; small, dark purplish red glands on most petioles and tipping marginal teeth of leaf blades.   6. *C. pulcherrima*

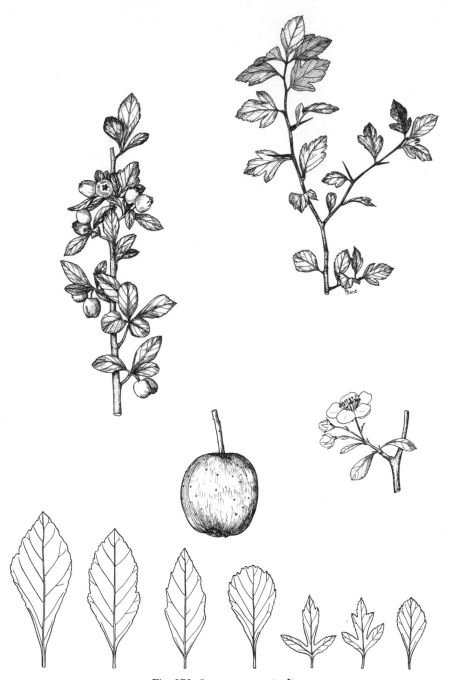

Fig. 273. **Crataegus aestivalis.**

4. The leaf blades of short-shoots broadest above their middles, tapered toward their bases.

6. Glands on the leaf margins pinheadlike and relatively conspicuous, *or* if minute and inconspicuous then many branchlets on the plant "weeping." 3. *C. flava*

6. Glands on the leaf margins, if any, tiny and inconspicuous, branchlets not "weeping."

7. Leaf blades with tufts of hairs at least in the proximal vein axils beneath, *or* if uniformly pubescent beneath then the pubescence copious, pale grayish on very young leaves, rusty-brown on mature leaves. 1. *C. aestivalis*

7. Leaf blades glabrous or sparsely pubescent beneath, not with tufts of hairs in vein axils.

8. Leaves of short-shoots mostly not exceeding 3 cm long. 8. *C. uniflora*

8. Leaves of short-shoots mostly 3–5 cm long or a little more. 2. *C. crus-galli*

1. **Crataegus aestivalis** (Walt.) T. & G. MAY HAW, APPLE HAW. Fig. 273

Shrub or tree to about 8 m tall, outer bark gray, exposed inner bark reddish, branches, if armed, with thorns 1.5–4 cm long.

Blades of leaves on flowering or fruiting branches mostly oblanceolate, spatulate, or obovate, 2–5 cm long and 1.5–2 cm broad, subleathery at maturity, cuneate basally, the petioles usually distinct although short, 0.5–1.5 cm long, apices commonly rounded or obtuse, infrequently acute; blades on proximal portions of later formed sterile shoots usually similar but larger, those on their distal portions or on especially vigorous shoots obovate or ovate, 3–8 cm long and to 3 cm broad or a little more, unlobed or slightly lobed to conspicuously 3–5-lobed; leaf margins crenate-serrate upwardly from about their middles or a little below, the teeth tipped by small red glands that eventually dry and shrivel and are then scarcely noticeable; upper surfaces of leaves usually dark glossy green and glabrous, the lower paler and duller green and with tufts of gray hairs in at least the proximal vein axils; in some populations, however, upper surfaces, at least at first, sparsely gray-pubescent, the lower densely gray-pubescent throughout, later the upper surfaces more or less rusty-brown and with sparse pubescence of the same color, the lower sparsely to rather heavily rusty-brown-pubescent throughout, petioles and young twigs similarly pubescent in either case.

Flowers borne singly or 2–4 subumbellately, their stalks usually very short but varying to 1.5 cm long, usually glabrous or nearly so. Floral tube glabrous to sparsely cottony-pubescent. Sepals triangular-acute to -attenuate, 2–4 mm long, usually sparsely pubescent at least on their inner surfaces, their margins sometimes minutely glandular. Petals white or white flushed with pink, broadly obovate, about 1 cm long. Fruit subglobose, about 1 cm in diameter, red, the flesh slightly acid, prized for making jelly.

For the most part inhabiting places where water stands much of the time, in and about small pools and ponds, floodplain pools, sloughs and oxbows, wet or swampy woodlands; throughout our area.

National champion may haw (1979), 2′6″ circumf., 38′ in height, 33′ spread, San Felasco Hammock, Fla.

2. **Crataegus crus-galli** L. COCKSPUR HAW, HOG-APPLE. Fig. 274

Tree up to about 10 m tall, generally with a broad crown of widely spreading branches (the lower ones often declining but not "weeping"), thorny flexuous branchlets, and dark, scaly bark.

Blades of leaves on flowering or fruiting branchlets subleathery, mostly oblan-

Fig. 274. **Crataegus crus-galli.** (From Kurz and Godfrey, 1962. *Trees of Northern Florida.*)

ceolate to spatulate, mostly 3–5 cm long and 0.8–3 cm broad, bases cuneate and partially decurrent on the short petioles, apices rounded to obtuse, upper surfaces lustrous dark green and glabrous, the lower paler and dull green, usually glabrous, infrequently pubescent along the midveins, without tufts of hairs in the vein axils, margins finely crenate-serrate or serrate above their middles, sometimes from a little below their middles; blades of leaves on later-formed sterile shoots similar but larger, those on their distal portions often largest and obovate, oval, or nearly rotund, the marginal teeth sometimes a little more pronounced.

Flowers several to numerous in branched clusters; inflorescence axis, flower stalks, and floral tube glabrous or shaggy-pubescent. Sepals deltoid to subulate, 3–5 mm long, sparsely pubescent, their margins entire or minutely toothed. Petals often pink in the bud, usually white at full anthesis, obovate, 5–7 mm long. Fruit subglobose or obovoid, 8–12 mm in diameter, dull red, rusty-orange, or green and dark-mottled.

Upland, open woodlands, thickets, pastures, fence and hedge rows, less frequently in bottomland woodlands, pine flatwoods and flatwoods depressions; throughout our area.

Co-national champion cockspur haws (1981), 4'10" circumf., 38' in height, 44' spread, Manassas, Va.; (1965), 6'2" circumf., 27' in height, 37' spread, Wayne Co., Ohio.

3. Crataegus flava Ait. *sens. lat.* YELLOW HAW. Fig. 275

Plants of *C. flava*, as treated here, are the most common and abundant of the several hawthorns occurring in our geographic range. In general they inhabit well-drained sandy or gravely soils of open or semiopen places and as a whole have such a "look-alike" appearance as to contrast effectively with those representative of other species save, perhaps, for *C. uniflora* with which it is sometimes associated. (See remarks under *C. uniflora*.)

Plants of *C. flava* exhibit a not inconsiderable variation as regards particular characteristics; for example: habit of branchlets, whether showing varying degrees of "weepiness," or a lack thereof; amount and nature of pubescence, if any, on early-developing shoots and their unfurling leaves; amount and kind of pubescence, if any, on inflorescence axes, bracts, flower stalks, floral tubes, calyx segments, and this variability in connection with glandularness or lack thereof; leaf size and shape, both leaves of flowering/fruiting short-shoots and sterile, later long-shoots; color, texture, lustrousness or its lack of leaf blades; number of flowers per shoot and their size; fruit size, shape, color at maturity, and nature of the flesh. If these variables and others correlate in any way such that individual specimens can be arranged into any satisfactorily distinctive groups, I have been unable to perceive it: however, specimens at hand do suggest the possibility of some poorly defined geographic trends.

Shrubs, relatively infrequently multistemmed, arborescent, or small trees, usually with rounded crowns, branchlets of which may be spreading, ascending, or commonly with some degree of "weepiness." Young shoots with their unfurling leaves varyingly glabrous to densely and compactly pubescent. Bark of trunks thickish, small-blocky, or irregularly ridged and furrowed, usually partially gray, partially almost black. Branches, usually at least some of them, with slender, sharp axillary thorns, occasional plants with none.

Leaves of flowering/fruiting short-shoots distinctly, slenderly petiolate, or bases of blades decurrently tapering to various degrees, some sometimes sessile; petioles, if any, glabrous to shaggy-pubescent, stipules linear-oblanceolate to setaceous, sometimes toothed, lacerate, or fimbriate on distal margins; blades mostly spatulate-obovate to rounded-obovate, their apices mostly rounded, truncate, or obtuse, less frequently acute, marginal toothing absent or, usually, obscure proximally, more distinct on the broader distal portions, the teeth there often irregular in size, infrequently toothed-lobed; at maturity surfaces of blades, *one or the other or both*, sparsely pubescent to glabrous, principal vein axils beneath sometimes with tufts of pubescence, color of upper surfaces dull, light green or olive-green to dark green and sublustrous, texture membranous to subcoriaceous; overall leaf length (petiole plus blade) 1–6 cm long, commonly not over 4 cm; pinheadlike glands dark purplish red, those of petioles few to numerous or none, always present on leaf margins, either at ends of veinlets or tipping teeth, varying from relatively tiny to relatively conspicuous, always present on stipules. Sterile, later, long-shoots of the season with obliquely rounded, inequilateral, sometimes lobed stipules with glandular mar-

Fig. 275. **Crataegus flava.**

gins. Leaves generally petiolate although blades often shortly decurrent on them, blades considerably larger than on short-shoots, very shortly and broadly tapered basally to broadly cuneate basally, broadly rounded distally, or shortly tapered above broadest places, obovate to suborbicular, often broader than long, unlobed and margins irregularly dentate above their broadest places, obscurely crenate proximally, varying to shallowly toothed-lobed or incised-lobed, glandularness as on leaves of short-shoots.

Flowers sometimes borne singly in axils of leaves of short-shoots of the season; inflorescence axes terminating short-shoots of the season, more often in several-flowered corymbs; inflorescence axes and flower stalks irregularly bracteate (bracts like the stipules), axes, flower stalks, floral tubes, and calyx segments densely and compactly pubescent, varying to sparsely shaggy-pubescent, or glabrous, the same parts variously glandular or without glands, infrequently with stipitate glands other than on the calyx segments. Calyx segments broadest basally then abruptly narrowed to narrowly triangular to subulate tips 2–3 mm long. Petals usually white, often pink in the swollen, unopened bud, rounded-obovate to nearly orbicular, 5–8 mm long or a little more, often as broad as or broader than long. Fruit subglobose, oblongish, greenish, yellow, yellow with reddish patches, russet, orange, orange-red, or purplish, variable in size to about 15 mm in diameter, perhaps a little more [Incl. *C. audens* Beadle; *C. egregia* Beadle; *C. floridana* Sarg.; *C. lacrimata* Small; *C. leonensis* Palmer; *C. ravenelii* Sarg.; *C. visenda* Beadle.]

Inhabiting open, well-drained, sandy or gravely pinelands, upland mixed woodlands and their borders, old fields, clearings, fence and hedge rows, highways and railway rights-of-way; throughout our area.

National champion yellow haw (1974), 3'3" circumf., 22' in height, 30' spread, Gainesville, Fla.

4. Crataegus marshallii Eggl. PARSLEY HAW.   Fig. 276

A small understory tree, to 6–8 m tall, with thin, scaly bark. Branches thornless or with slender, sharp thorns 1–3 cm long; young branchlets woolly-pubescent, later finely roughened by persistent, enlarged bases of the hairs.

Leaves alike on both flowering or fruiting and later sterile shoots; petioles slender, 0.5–3 cm long, more or less woolly-pubescent; blades membranous, not lustrous, triangular in overall outline, essentially as broad at the base as long, the smaller ones often broader than long, 1–3 (5) cm long, most of them deeply incised-lobed, the main lobes irregularly sharply toothed or with smaller toothed sublobes, bases truncate or broadly short-tapered, surfaces rather copiously pubescent when young, less so above than beneath, some, sometimes all, of the pubescence sloughing as the blades mature; stipules narrowly linear, soft, pubescent.

Flowers few to numerous in branched clusters, the axes, flower stalks, and floral tubes woolly-pubescent. Sepals triangular-subulate, 3–4 (5) mm long, margins usually with gland-tipped teeth or stipitate glands distally. Petals white or pale pink, sometimes white with a tinge of pink, oblong, oblanceolate, or narrowly obovate, corollas 1.5–2 cm across. Stamens 10–20. Fruit bright red, glabrous, ellipsoid, 5–7 mm long and about 5 mm in diameter, the sepals often deciduous by the time the fruit is fully ripe, nutlets usually 3.

Floodplain forests, adjacent wooded slopes, wet woodlands astride small streams, wooded ravine slopes, river banks; throughout our area.

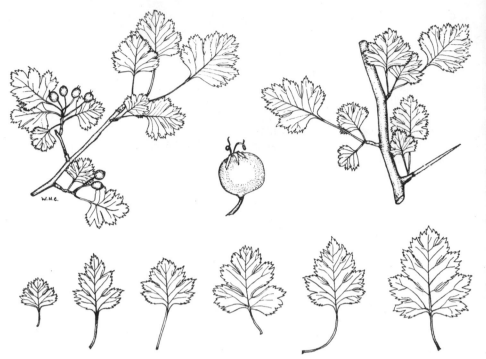

Fig. 276. **Crataegus marshallii.** (From Kurz and Godfrey, 1962. *Trees of Northern Florida.*)

National champion parsley haw (1974), 1′4″ circumf., 33′ in height, 23′ spread, Gainesville, Fla.

5. **Crataegus phaenopyrum** (L. f.) Medic. WASHINGTON THORN. Fig. 277

Shrub or small tree to about 10 m tall. Young stems slender, reddish purple, loosely gray-hairy only proximally, those of the long-shoots of the season bearing some slender, reddish purple, sharp thorns 1–2 cm long; older woody twigs gray, their thorns having become stouter and more rigid. Bark of trunks gray, relatively smooth.

Petioles slender, reddish purple, 1–2 cm long on the short-shoots, longer on some particularly vigorous long-shoots. Leaf blades subtly variable, those of the flowering/fruiting short-shoots for the most part, at a glance, giving the impression of small red maple leaves, their blades ovate in outline, 1.5–2.5 cm long, mostly with 3 main lobes, the lateral lobes usually shorter and often more bluntly tipped than the terminal lobe, margins variably and irregularly dentate to crenate-serrate, teeth tipped with purple glands; bases truncate, rounded, or subcordate, some broadly short-tapered; upper surfaces bright green, glabrous or sparsely short-pubescent on the veins, lower surfaces paler and duller green, usually glabrous. Blades of long-shoot leaves more variable, both in size and shape, those of some shoots much like those of short-shoot blades, those of other shoots to about 3 times as large; those on some very slender shoots prominently 3-lobed, the lateral lobes shortly divaricate and blunt- or rounded-tipped, the terminal lobe ovate and merely toothed or with a pair of sublobes; the blades on

556

Fig. 277. **Crataegus phaenopyrum.**

some relatively vigorous, stouter shoots relatively large, 5–7 cm long and as broad basally, irregularly wavy-lobed or merely irregularly toothed.

Flowers small, relatively numerous in more or less flat-topped, branched cymes. Branches of the cyme subtended by small, reddish purple, glandular-toothed bracts deciduous well prior to anthesis, similar bracts subtending the 3–8 (10) mm long, glabrous flower stalks and usually 1–2 such bracts variably positioned on the flower stalks, these usually deciduous by anthesis. Calyx glabrous exteriorly, pubescent interiorly, lobes short-deltoid, entire. Petals white, suborbicular. Stamens about 20. Fruits subglobose, 4–6 mm in diameter, shiny red. [Incl. *C. youngii* Sarg.]

Chiefly inhabiting low woodlands and thickets; s. Ala., apparently absent in s. Ga., known to me only from Wakulla and Washington Cos. in Fla. Panhandle. (Del. and Pa., westward to s. Ill., Ky., Tenn., s. Mo., generally southward to piedmont of Ga., locally in Ala., and Fla. Panhandle).

National champion Washington thorn (1965), 3'3" circumf., 24' in height, 20' spread, Secrest Arboretum, Wooster, Ohio.

6. Crataegus pulcherrima Ashe. Fig. 278

Small tree, branches commonly with slender, sharp thorns 2–4 cm long; bark of trunk grayish to blackish, relatively thick, irregularly blocky; young stems of short-shoots glabrous or with sparsely distributed long hairs, those of sterile, long-shoots glabrous.

Leaves slenderly petiolate, on some plants bases of blades decurrent on the petioles for a short distance, on others not at all decurrent. Blades of mature leaves of short-shoots variable in shape and size and in marginal toothing, ovate, obovate, or subrotund, infrequently some elliptic or lanceolate, often about as broad as long, some even broader than long, in general 2–6 cm long and 1.5–4 cm broad, irregularly serrate or dentate, sometimes doubly serrate, or shortly several-lobulate and the lobes serrate, bases sometimes truncate, more frequently shortly tapered, apices acute to broadly rounded, principal lateral veins relatively prominent and angling-ascending approximately parallel to each other; upper surfaces glabrous, lower sometimes so, often sparsely pubescent throughout but not with tufts of hairs in the vein axils; blades of leaves of sterile long-shoots, many of them, similar in shape and marginal toothing to those of short-shoots but averaging considerably larger, others, on some plants, irregularly incised-lobed. Sessile, small, dark purplish red glands, variable in number, on most petioles, on margins of stipules (where some shortly stipitate), and tipping the marginal teeth of leaf blades.

Inflorescences simple or few-branched, bracteate corymbs terminal on leafy short-shoots of the season, bracts lanceolate to linear-oblanceolate, with shortly stipitate, purplish red glands; axes of corymbs, flower stalks, and floral tubes with few sessile glands, sometimes none. Calyx segments triangular, or linear-oblong to linear-attenuate, 3–5 mm long, minutely or saliently toothed, teeth tipped with glands. Petals white or white tinged with pink, broadly rounded-obovate, 5–10 mm long, their distal edges irregularly minutely lacerated. [Incl. *C. opima* Beadle; *C. robur* Beadle]

Upland open woodlands and their borders, hedge rows, open pine–scrub oak ridges; s.w. Ga., s.e. Ala.(?), n. Fla., mostly from Jefferson Co. westward to Walton Co., Columbia and Alachua Cos. in n.e. Fla., perhaps elsewhere.

National champion *Crataegus pulcherrima* (1968), 1'11" circumf., 44' in height, 30' spread, Tallahassee, Fla.

Fig. 278. **Crataegus pulcherrima.**

Fig. 279. **Crataegus spathulata.**

7. Crataegus spathulata Michx. LITTLEHIP OR SMALL-FRUITED HAWTHORN, RED HAW.
Fig. 279

Arborescent shrub or small tree to about 8 m tall. Bark of trunks brownish gray, exfoliating in large, thin plates and exposing orange-brown inner bark. Branches thornless or armed with sharp thorns 2.5–6 cm long.

Leaves of flowering or fruiting branches mostly narrowly spatulate or oblanceolate, strongly tapered basally to subpetioles, 1–4 cm long overall, mostly crenately few-toothed distally on the expanded portions of the blades, sometimes only at their tips; leaves of sterile shoots with winged subpetioles, expanded portions of the blades of the larger leaves irregularly crenate-serrate to incised-lobed (commonly 3-lobed), ovate in overall outline, 2–3 cm broad at their broadest places; for the most part, veinlets ending both in the teeth and in the sinuses between them; upper surfaces of blades variably glabrous or sparsely pilose along the midrib, sometimes along the lateral veins and margins, lower surfaces glabrous, stipules obliquely arching, sometimes 2-lobed with few salient serrations.

Flowers few to relatively numerous in branched, glabrous corymbs, their stalks slender. Sepals broadly triangular, obtuse apically, margins entire. Petals white to pale pink, 7–10 mm across the corollas. Mature fruits glabrous, bright red, ellipsoid, 4–7 mm in diameter. Fruits tending to persist for a short while after leaf fall, some of them even over winter, and, if very numerous as they sometimes are, the plant is then very decorative.

Wooded slopes and moist bottomland woodlands, stream banks; cen. Panhandle of Fla., s.w. Ga., s. Ala.

8. Crataegus uniflora Muenchh. DWARF HAW, ONE-FLOWERED HAW.
Fig. 280

Arborescent shrub, usually little if any exceeding 3 m tall. Young branchlets of the season and petioles copiously shaggy-pubescent with translucent hairs having slightly enlarged bases, hairs eventually sloughed leaving the stem surface slightly roughened by the persistent bases and raised lenticels; branches thornless or armed with slender, sharp thorns 1–3 cm long.

Leaves of flowering or fruiting stems short-petiolate, oblanceolate to broadly short-tapered basally, mostly obtuse to rounded apically, 1–3 cm long, margins unevenly serrate or crenate-serrate distally from somewhat below their middles; leaves of sterile shoots with narrowly winged subpetioles, expanded portions irregularly toothed or shallowly toothed-lobed, ovate to suborbicular; leaf surfaces variously pubescent, usually more so beneath than above, upper surfaces eventually becoming sublustrous, usually becoming glabrous, sometimes finely roughened by persistent bases of hairs; principal veins well elevated beneath on mature leaves.

Flowers usually solitary, sometimes 2 or 3, at the tips of developing branchlets of the season, their stalks short and stoutish, shaggy-pubescent. Sepals lanceolate, oblong, or oblanceolate, acute apically, 4–7 mm long, pubescent, their margins serrate to incised-serrate, the teeth tipped by red glands. Petals white, obovate, corolla 1–1.6 cm across. Fruits glabrous or sparsely shaggy-pubescent, dull red or brownish, subglobose, mostly about 1 cm in diameter.

Well-drained open woodlands and their borders, pine-oak sand ridges; n.e. Fla., Fla. Panhandle, adjacent s. Ga. and Ala.

In general appearance *C. uniflora* often closely resembles some plants of *C. flava* with which it may grow intermixed or in close proximity. Young stems, petioles, stipules, inflorescence axes, and bracts of *C. uniflora* are without glands

Fig. 280. **Crataegus uniflora.** (From Kurz and Godfrey, 1962. *Trees of Northern Florida*.)

while those of *C. flava* are notably glandular on some parts, always on the leaf margins or tips of marginal teeth.

National champion dwarf or one-flowered haw (1974), 2'1" circumf., 30' in height, 19' spread, San Felasco Hammock, Fla.

9. Crataegus viridis L. GREEN HAW. Fig. 281

Shrub or small tree, unarmed or with slender, sharp thorns. Bark of trunk grayish, thin, sloughing in irregular, thin plates exposing cinnamon-brown inner bark; young twigs sparsely pubescent only at first.

Leaves slenderly petioled, petioles 1–3 cm long; blades of short-shoots mostly ovate, lance-ovate, or oblong-elliptic, infrequently lanceolate, 3–7 cm long and 2–5 cm broad, shortly and broadly tapered basally, mostly acute apically, some sometimes obtuse, upper surfaces sparsely pubescent only as leaves unfurl, quickly becoming glabrous, lower surfaces with tufts of hairs in the vein axils, the tufts partially sloughing as the blades mature, margins serrate or slightly lobulate-serrate, incised, or often conspicuously 3–5-lobed.

Flowers 5–20 in branched corymbs, axes of corymbs, floral tubes and calyx segments glabrous or with few, scattered longish hairs. Calyx segments broadest basally, then straplike above, about 3 mm long, tips blunt. Petals white or pinkish, broadly rounded obovate, about 4–8 mm long. Fruit subglobose or oblongish, 5–8 mm in diameter, red or orange-red at maturity. [Incl. *C. arborescens* Ell., *C. paludosa* Sarg.]

Swamps, lowland woodlands, small depressions in woodlands, margins of ponds, throughout our area.

National champion green haw (1981), 5'1" circumf., 40' in height, 45' spread, Pocahontas Co. Historical Museum, W.Va.

562

Fig. 281. **Crataegus viridis.**

Fig. 282. **Malus angustifolia:** a, stem with flowers on short-shoots; and b, stem with fruits on short-shoots; c, twig with branch-thorns; d, tip of vigorous elongation shoot; e, face view of flower; f, enlargement of fruit.

4. Malus

Malus angustifolia (Ait.) Michx. SOUTHERN CRABAPPLE. Fig. 282

A deciduous shrub or small tree to about 8–10 m tall, commonly forming thickets from root-sprouts, branches usually widely spreading and forming a rounded crown. Twigs stiff, reddish brown, leaf scars linear, with 3 vascular bundle scars; pith continuous. Outer bark of trunks breaking up into grayish plates exposing reddish brown inner bark. Vigorous branches elongate, upon them are eventually produced relatively short, slow-growing spur shoots, some spur shoots sometimes form leafy thorns (these remaining as naked thorns with leaf scars after leaves are shed). Young twigs, very young petioles, and lower surfaces of leaves as they unfurl densely woolly-pubescent, much of the pubescence quickly sloughing, mature petioles sparsely to moderately pubescent, surfaces of mature leaf blades glabrous.

Leaves simple, alternate, slenderly short-petiolate, stipulate, stipules membranous, reddish, narrowly linear-subulate, 1–3 mm long, sometimes 1 or more similar bracts on the petiole, both quickly deciduous; mature blades of firm texture, those of vigorous shoots generally larger than those of spur shoots; blades of spur shoots mostly oblong, elliptic, or ovate, 2.5–5 cm long and up to about 2.5 cm broad, bases shortly and broadly tapered to rounded, sometimes cuneate, apices mostly obtuse or rounded, occasional ones emarginate, less frequently acute, margins crenate-serrate, serrate, or essentially entire; blades of leaves of vigorous long shoots larger, ovate, saliently toothed or toothed-lobed.

Flowers in umbels of 3–5 from the ends of spur shoots, bisexual, radially symmetrical, ovary inferior. Flower stalks sparsely pubescent, purplish red, 1.5–2 cm long, usually with several, alternately disposed, membranous, reddish, stipulelike bracts on their proximal halves; floral tube glabrous. Sepals 5, triangular-acuminate, about 2 mm long, woolly-pubescent interiorly. Corolla about 2.5 cm across the 5 spreading, deep pink to nearly white petals each of which has a short basal claw and a broadly rounded obovate blade. Stamens numerous, shorter than the petals, some or all of their shriveled filaments persisting at the summits of fruits. Pistil 1, ovary 5–locular, styles 5, short, united basally. Fruit a nearly globular, yellowish green pome about 2.5 cm in diameter, the sepals persistent at its summit.

This plant, in flower in early spring, makes a beautiful show, the more so when it happens to occur in rather extensive thickets. [*Pyrus angustifolia* Ait.]

Relatively open, well-drained, woodlands of various mixtures, thickets, fence and hedge rows; s. cen. and s.w. Ga., Fla. Panhandle, and s. Ala. (Va. to Ky., and Mo., generally southward to Fla. Panhandle and La.)

National champion southern crabapple (1981), 77.5" circumf. [at 18"], 35.5' in height, 48.5' spread, Swannanoa, N.C.

5. Physocarpus

Physocarpus opulifolius (L.) Maxim. NINEBARK. Fig. 283

The ninebark has a wide geographic range and apparently varies considerably, especially with reference to quantity and distribution of pubescence (or lack of it) on its parts. Several segregate taxa have been described, none of them sufficiently distinctive that recent authors have recognized them. The description below is drawn from plants occurring in our local area of coverage and is not meant to encompass variability obtaining elsewhere.

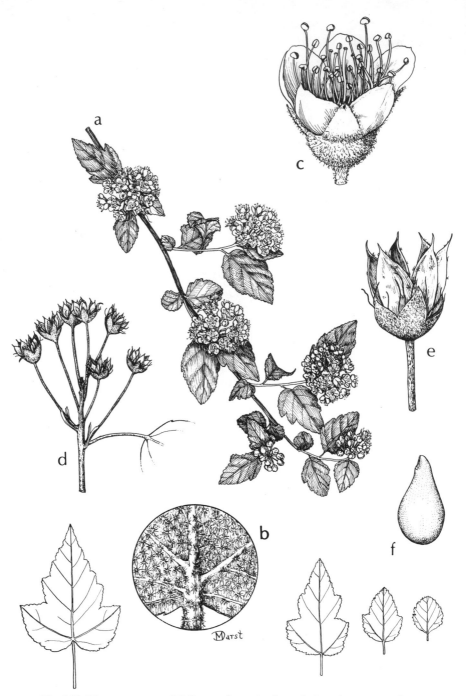

Fig. 283. **Physocarpus opulifolius:** a, flowering branch; b, enlargement of portion of lower surface of leaf, and, to either side, variable leaves; c, flower; d, infructescence; e, fruit, dehisced; f, seed.

A multistemmed, deciduous shrub to about 3 m tall. Vigorous leader-shoots for the most part stiffly erect, subsequent branches usually arching. Bark of older woody stems loose and peeling and shredding into long, thin, buff-colored layers or strips. Young stems densely stellate-pubescent.

Leaves simple, alternate, petioles 0.5–2 cm long, at first densely stellate-pubescent, much of the pubescence eventually sloughing. Blades subpalmately veined (3 major veins from near the base), those of vigorous leader shoots usually much larger and more uniform in size and shape than those of subsequent branches and branchlets. Blades of leader shoots ovate, with a pair of short, blunt lobes somewhat below midleaf, several wavy sublobes on either side distally, margins irregularly, finely crenate-serrate, in size mostly 4–7 cm long and 4–5 cm across the major lobes, bases slightly cordate, apices obtusish. Blades of secondary and other branches or branchlets much more variable, the larger ones closely similar to leader-shoot blades but about half as large, varying to ovate, elliptic, or suborbicular, the smallest ones about 1.5 cm long and broad; bases mostly truncate, some shortly and broadly tapered, apices obtuse excepting those of the suborbicular ones, margins finely crenate-serrate. All leaves having the upper surfaces sparsely stellate-pubescent at first, little of it retained, or eventually becoming glabrous; lower surfaces moderately pubescent throughout or somewhat more densely so along the major veins. Stipules narrowly lanceolate to linear-oblong, quickly deciduous.

Inflorescences short and broad umbellike racemes terminating the numerous short branchlets of the season, thus the floriferous arching branches spraylike, very much resembling some spiraeas. Inflorescence axes, flower stalks, floral tubes, and calyces densely stellate pubescent. Floral tubes cuplike, glistening-nectariferous within, surmounted by 5 deltoid sepals about 2 mm long, densely pubescent interiorly as well as exteriorly, the sepals persistent. Petals 5, white or white tinged with pink, ovate to suborbicular, about 4 mm broad. Stamens numerous, borne on the rim of the floral tube, filaments white, anthers red. Pistils 3–5, the ovaries of each cohering at least basally at first, later separating, styles slender, stigmas capitate. Ovaries of each flower maturing into 2-valved, beaked pods about 7 mm long, their surfaces with little persisting stellate pubescence or smooth and shiny-brown. Seeds 2–4 in each pod, obpyriform, tan or buff colored, lustrous, 1.5–2 mm long. [*Opulaster opulifolius* (L.) Kuntze]

Open wooded banks of small, often intermittent, streams, Jackson and Calhoun Cos., Panhandle of Fla., s.w. Ga., perhaps s.e. Ala. (Que. to Minn., S.D., Colo., generally southward to cen. Panhandle of Fla., Ala., Ark.)

6. Prunus (CHERRIES, PLUMS, PEACH)

Shrubs or small to medium-sized trees, the bitter astringent bark exuding a gummy substance. Twigs smooth and somewhat shiny and having lenticels eventually conspicuous and transversely elongate; leaf scars angularly somewhat raised, each with 2–3 vascular bundle scars, stipule scars thinly linelike to either side of the summit of the leaf scar. Pith continuous. Leaves alternate, simple, short-petiolate, petioles glandular in most, stipules attached to the twigs, quickly deciduous, blades pinnately veined. Flowers borne singly, in racemes, umbels, or in fascicles, bisexual, radially symmetrical, with a floral cup bearing upon its rim 5 small sepals, 5 spreading petals, and 10 or more stamens. Pistil 1, ovary superior, style 1. Fruit a 1-stoned drupe.

1. Flowers and fruits borne in racemes.

2. Leaf blades relatively membranous and pliable, deciduous, evenly finely serrate, the tips of the teeth incurved; floral branches usually bearing 1–4 reduced leaves on the axis, these below the raceme proper or sometimes one or more of them with an axillary flower.

    3. Young branchlets, petioles and inflorescence axes glabrous; leaf blades oval-oblong, elliptic, rarely ovate, acute or acuminate apically.   5. *P. serotina*

    3. Young branchlets, petioles, and inflorescence axes pubescent, at least at first; leaf blades, most of them at least, obovate, blunt apically.   6. *P. alabamensis*

2. Leaf blades stiff-leathery, evergreen, entire or with a few small remote straight, sometimes bristle-tipped teeth; floral branches without reduced leaves.   3. *P. caroliniana*

1. Flowers or fruits borne singly, in fascicles, or umbels of 2–5.

    4. Petals deep pink, flowers sessile or very nearly so, their stalks, if any, stoutish and not evident beyond the bud scales; ovary and fruit densely pubescent.   4. *P. persica*

    4. Petals white or white with a tinge of pink, flower stalks slender, clearly evident beyond the bud scales; ovary and fruit glabrous.

        5. Bark of trunks shaggy, the plates tan or buff colored; mature leaf blades mostly conspicuously acuminate apically.   1. *P. americana*

        5. Bark of trunks not shaggy, dark; leaf blades acute apically.

            6. Teeth of the leaf blades blunt and tipped with red or yellow glands; some at least of the leaf blades folded upwardly from the midribs and troughlike.   2. *P. angustifolia*

            6. Teeth of the leaf blades not glandular; leaf blades usually flat.   7. *P. umbellata*

1. Prunus americana Marsh. AMERICAN PLUM.   Fig. 284

A small deciduous tree, flowering/fruiting when of shrublike stature, usually with a single main stem (allegedly often thicket-forming). Twigs infrequently sharply thorn-tipped; bark of trunks tan or buff-colored, exfoliating in conspicuous curled plates.

Emerging leaf blades sometimes uniformly pubescent beneath, much of the pubescence quickly sloughing (in ours); mature leaf blades 4–12 cm long, 2.5–5 cm broad, oblong-oval, elliptic, some broadest above the middle, bases rounded or shortly and broadly tapered, occasional ones oblique, apices acuminate; upper surfaces glabrous, lower sometimes glabrous, often pubescent along the midrib and along proximal portions of the lateral veins; margins serrate to irregularly doubly serrate, the teeth broad basally and abruptly narrowed to finely pointed tips; petioles glabrous or sparsely pubescent, occasional ones with a prominent gland just below the base of the blade; stipules linear-attenuate, 5–7 mm long, glandular-toothed marginally, quickly deciduous.

Flowers pleasantly fragrant, borne in fascicles of 2–5 on wood of a previous season before or as new shoot growth starts; flower stalks slender, 1.5–2.5 cm long; floral tubes and exterior of sepals partially to wholly suffused with brownish red pigment which contrasts sharply with the white of the petals; sepals spreading, very sparsely short-pubescent exteriorly and with more and longer hairs interiorly, margins finely glandular-toothed. Petals white (often becoming pale pink before they fall), very short-clawed, blades mostly elliptic, varying to ovate, slightly obovate, or rotund, most of them 7–8 mm long, corolla about 2 cm across the spreading petals. Stamens 20–30, filaments white, anthers yellow. Drupe globose or nearly so, glaucous, red (ours) at maturity, 2–2.5 cm long, flesh flavorful.

Rich loamy, often calcareous, woodlands, slightly elevated places in floodplain woodlands, near-coastal hammocks, hedge rows; s.w. Ga., e. and cen. Panhandle of Fla., perhaps s.e. Ala. (widely distributed northward and northwestward of our range).

P. americana, P. angustifolia, and *P. umbellata,* at time of full flowering, may,

Fig. 284. **Prunus americana:** a, piece of fruiting branch; b, leaf, enlarged, to accentuate marginal toothing, at left, and, at right, marginal toothing more exaggerated; c, piece of woody twig with branch-thorns.

for many persons, be difficult to distinguish one from the other. Note that the bark of trunks of *P. americana* is buffish or tan in color and that the outer bark peels in curled plates unlike in either of the other two. In *P. americana*, moreover, the corollas of its flowers are sensibly larger (compare measurements across corollas) and the floral tubes and calyces of the flowers, at full anthesis, are sufficiently suffused with red pigment that they contrast sharply with the white petals. Later, in full leaf, the leaf blades of *P. americana* are acuminate apically, acute in the other two. (See also discussion under *P. angustifolia*.)

National champion American plum (1972), 3' circumf., 35' in height, 35' spread, Oakland Co., Mich.

2. Prunus angustifolia Marsh. CHICKASAW PLUM. Fig. 285

Deciduous shrub or small tree, noted for forming thickets, some of them dense and extensive; in open places of old fields or waste places, the dense thickets often approximately circular, the largest (oldest) stems centrally located, outwardly the stems gradually diminished in size, even the smallest peripheral ones flowering/fruiting. Small lateral twigs, many of them rigid and their extremities sharply thorn-tipped. Bark of trunks dark, not exfoliating in curled plates.

Leaf blades lanceolate, lance-oblong, or elliptic-oblong, 3–8 cm long and 1–2.5 cm broad, many of them, sometimes all of them, somewhat folded upwardly to either side of the midrib, the blade then appearing troughlike; bases mostly cuneate, apices acute, margins finely crenately toothed, the teeth tipped by yellowish or red glands, these more pronounced when the leaves are young and before the glands shrink or are detached; upper surfaces glabrous, bright green, the lower usually sparsely pubescent, at least along the veins, sometimes glabrous.

Flowers with an insipid aroma, borne singly or 2–several in fascicles from buds on wood of a previous season in spring before or as new shoot growth commences, usually the former. Flower stalks slender, 2–10 mm long. Sepals, at full anthesis, somewhat ascending, 1–1.5 mm long, triangular, blunt apically. Petals white, spreading, shortly clawed basally, blades ovate to rotund, often crinkled marginally, 2–3 mm long, corolla 7–10 mm across the spreading petals. Stamens 10–20, filaments white, anthers yellow. (After the petals have fallen and the anthers shriveled, the filaments, floral tubes, and calyces become brownish, the flowers being numerous, the plant has a muddy-brown appearance for a short while.) Drupes oval to rotund, red to yellow, glaucous, 1.5–2.5 cm long, the flesh somewhat acid-flavored, rather pleasant to taste.

Old fields, about homesites, especially rural ones, waste places, fence and hedge rows, throughout our area of coverage. (Widespread to the northward and westward of our range.)

When plants of *P. angustifolia* and *P. umbellata* are at full anthesis and in case no leaves are present at that time, they are, for me, hardly distinguishable apart from the notably thicket-forming habit of the former and the usually single-stemmed habit of the latter. (At full anthesis of the flower, the sepals of *P. angustifolia* tend to be ascending, those of *P. umbellata* horizontally spreading.) After the petals have fallen and the anthers shriveled, plants of *P. angustifolia* have for a short while a dirty-brownish aspect. *P. umbellata*, at the same stage (postanthesis) has a pinkish aspect owing to its filaments, floral tubes, and calyces having changed color. (See also comments under *P. americana*.)

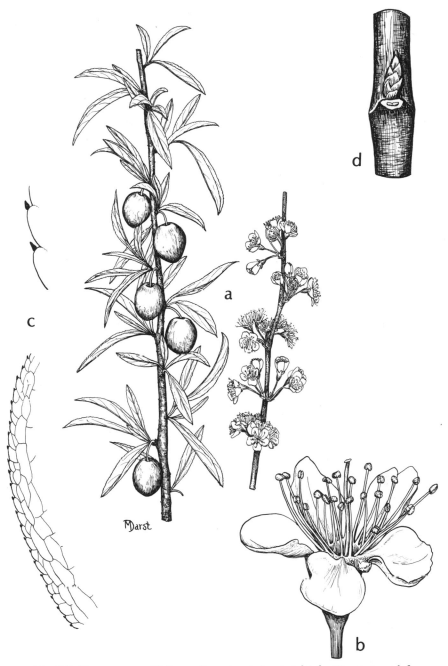

Fig. 285. **Prunus angustifolia:** a, flowering twig at right, fruiting twig at left; b, flower; c, enlargement of edge of leaf to accentuate gland-tipped teeth, and, above, a couple of teeth more exaggerated; d, piece of woody twig with leaf scar, stipule scars to either side, and axillary bud.

571

Fig. 286. **Prunus caroliniana:** a, flowering branch (previous year's increment of growth); b, tip of emerging branch of the season; c, serrated leaf characteristic for some individual plants; d, flower, face view; e, flower with part of floral tube cut away, side view; f, cluster of fruits.

National champion chickasaw plum (1978), 3'6" circumf., 29' in height, 28' spread, Gainesville, Fla.

3. Prunus caroliniana (Mill.) Ait. CAROLINA LAUREL CHERRY. Fig. 286

Evergreen tree to about 12 m tall, commonly flowering/fruiting when of low, shrublike stature. Bark of older stems dark gray, with transverse lenticels, eventually coarsely granular and almost black.

Leaves short-petiolate, blades glabrous, leathery, dark green and sublustrous above, paler and duller green beneath, oblong-elliptic, elliptic, oval, or oblanceolate, mostly 5–12 cm long and 1.5–4 cm broad, bases tapering, apices acute or short-acuminate; margins entire, or with few, remote spiculelike teeth, oftentimes some of each on the same branchlet.

Flowers in very short-stalked racemes 1–4 cm long borne singly in the axils of leaves or leaf scars distally on branchlets of the previous season very early in spring (occasionally leaves below the ones subtending racemes have but a single flower in their axils); racemes much shorter than the subtending leaves and no reduced leaves on the raceme axes. Sepals little more than blunt knobs on the outer rim of the floral tube and commonly pink just before the flower opens. Petals white, spreading, rotund to obovate, entire or minutely toothed distally, 1–2 mm long. Stamens mostly 10–15. Drupes 1–1.5 cm long, dull black, ovate, oval, or subglobose, the style bases persisting as an apiculation at the summit, many of the drupes persisting overwinter, some sometimes into the second season. [*Laurocerasus caroliniana* (Mill.) Roem.]

The Carolina laurel cherry has been extensively used as an ornamental or for hedges and screens. It naturalizes so freely that it is scarcely, if at all, possible to know what its original range or habitat may have been, but perhaps maritime communities. It now occurs in our range in various mixed upland woodlands, on wooded slopes and bluffs, old, well-formed levees along rivers and streams, maritime hammocks and scrub, in fences and hedge rows, vacant lots, and it volunteers readily in and about towns and cities. (Coastal plain, s.e. N.C. to cen. pen. Fla., westward to e. Tex.)

National champion Carolina laurel cherry (1972), 10'5" circumf., 44' in height, 48' spread, Dellwood, Fla.

4. Prunus persica (L.) Batsch. PEACH. Fig. 287

A small tree, native to China, its horticultural varieties widely cultivated for their luscious fruit, the feral form sporadically naturalized, commonly shrubby, flowering when of small stature.

Woody twigs relatively rigid, unarmed, glabrous, buds copiously short-pubescent.

Leaves glabrous, short-petiolate, petioles troughed, a pair of glands opposite or subopposite on the distal, adaxial rims of the trough, or, in many cases, the glands absent on the petioles and present at either side of the base of the blade; stipules subulate, 1–1.4 cm long, their margins subpectinately glandular-toothed. Blades lanceolate, oblanceolate, elliptic, or sometimes lance-ovate, 4–15 cm long, bases rounded or shortly tapered, apices long-acute or -acuminate, both surfaces bright green, margins finely serrate, the teeth gland-tipped.

Flowers appearing before new shoots emerge or as they emerge, usually solitary from winter (flower) buds, the latter alternately arranged or in pairs, each pair with a shoot-bud between them. Floral tube purplish red exteriorly, green within. Sepals ovate-oblong, 4–5 mm long, truncate basally and rounded api-

Fig. 287. **Prunus persica:** a, leafless flowering branch; b, fruiting branch; c, longitudinal section of flower.

cally, purplish and white-pubescent exteriorly, glabrous interiorly. Petals pink, with a very short claw, the blades broadly oval, obovate, or rotund, mostly 1.5 cm long or a little more, corolla 2.5–3.5 cm across the spreading petals. Stamens numerous, filaments pale pink at early anthesis, soon becoming deep rose, anthers (individually) partially yellow, partially orange. Drupes more or less ellipsoid, 4–6 cm in diameter; stones large, their surfaces deeply ridged and furrowed. [*Amygdalus persica* L.]

Thickets, fence and hedge rows, waste places, near dumps, edges of upland woodlands, sporadic in our area of coverage.

5. **Prunus serotina** Ehrh. BLACK CHERRY.   Fig. 288

Deciduous tree, potentially to 30 m tall, commonly flowering and fruiting when of small shrublike stature. Emerging shoots with prominent, purplish red scales below the first leaves, these quickly deciduous; stems of young branchlets glabrous, green or partially suffused with red; stipules subulate, their margins variably glandular or pectinately toothed, quickly deciduous. Sap bitter and with the odor of almond. Bark of woody twigs at first having a thin grayish waxy bloom, dark reddish brown after that is sloughed, lenticels inconspicuous and dotlike the first year, eventually prominent and transversely elongate; bark of older stems dark reddish brown to nearly black, varying to gray, on old trunks forming an irregular patchwork of plates separated by a network of narrow fissures.

Petioles slender, grooved, mostly 5–10 mm long, sometimes to 20 mm, with 1 or more variously disposed glands, similar glands often at or near the bases of the blades. Mature blades generally drooping, mostly 5–15 cm long and 2.5–4 cm broad, oblong-oval, elliptic, less frequently lanceolate or lance-ovate, bases rounded or shortly tapered, apices short-acuminate, the acumens pointed; upper surfaces glabrous, bright green, lustrous at first, lower paler and duller, glabrous or with continuous or intermittent woolly pubescence running along one or both sides of about the proximal one-third of the midrib, pubescence sometimes covering the major lateral veins of the proximal portion of the leaf for a short distance, pubescence pale or buff-colored at first, eventually becoming chestnut-brown; margins of blades finely appressed-serrate, the teeth with minute, red glandular tips.

Inflorescences from buds all along twigs of the previous season, stalked, each an oblong-cylindric raceme, stalk of the raceme leafless or bearing 1–3 somewhat reduced leaves; inflorescence axes glabrous, often curvate, mostly 3–8 cm long, some sometimes longer. Flowers small, about 7 mm across the spreading petals. Sepals short-deltoid, they and the floral tubes glabrous. Petals white, barely clawed, obovate, 2–3 mm long. Stamens about 20, disposed on the inner rim of the floral tube in 3 series outermost longest and more or less erect, others about equal in length, the filaments incurved. Drupes nearly globular but flattish apically, mostly 7–10 mm in diameter, the floral tube and filaments persisting at their bases, surfaces lustrous, purplish black to black, the flesh purplish and juicy. (*Padus virginiana* sensu Small.)

In mixed forests having varying moisture regimes; throughout much of our area very common in fence and hedge rows, often abundantly colonizing old fields and cut-over places. (N.S. to Minn., generally southward to cen. pen. Fla. and e. Tex.)

National champion black cherry (1980), 14′4″ circumf., 132′ in height, 126′ spread, Washtenaw Co., Mich.

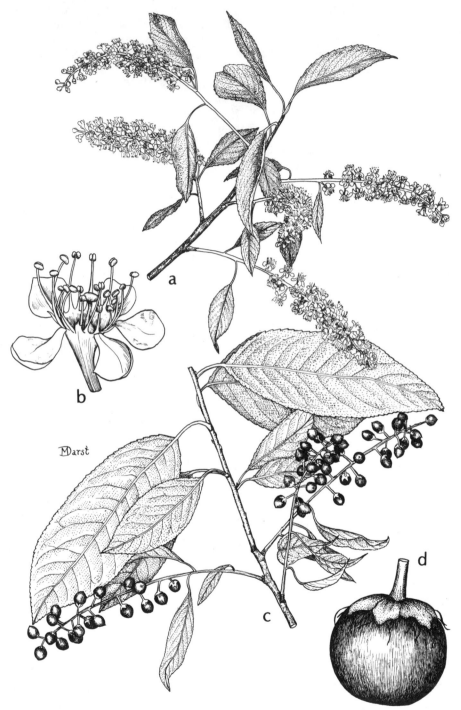

Fig. 288. **Prunus serotina** subsp. **serotina:** a, flowering branch; b, flower; c, piece of fruiting branch; d, fruit.

6. Prunus alabamensis Mohr. ALABAMA CHERRY.   Fig. 289

Young stems, petioles, inflorescence axes, and flower stalks pubescent. Emerging leaf blades often with lower surfaces uniformly pubescent, the pubescence sometimes persistent, often quickly sloughing save along the proximal portion of the midrib and proximal portions of the lower lateral veins. Mature blades stiffish, not drooping, 5–8 cm long and 3–5 cm broad, some, sometimes most of them, distinctly obovate, varying to oval-oblong but at least slightly broadest above their middles; bases obtuse to rounded, apices rounded to broadly obtuse, or abruptly contracted to a short, blunt acumen or cusp. Mature drupes red in preripening, eventually black, not flattened apically.

Excepting saplings, shoot growth of the season commences later than is the case for *P. serotina*, the latter having fully grown shoots and leaves by the time the buds of *P. alabamensis* burst. Flowering and fruit ripening is also later for the latter. [*Padus alabamensis* (Mohr) Small; *P. serotina* var. *alabamensis* (Mohr) Little; *P. serotina* subsp. *hirsuta* (Ell.) McVaugh.]

Infrequent in longleaf pine–scrub oak forests on sand ridges, or in thin stands of planted slash pine on sand ridges; also in mixed woodlands on slopes of steepheads and ravines; not "weedy" as in *P. serotina*. In our area, in w. half of the Panhandle of Fla., adjacent s.w. Ga., s. Ala. (N.C., S.C., Ga., Ala., Fla. Panhandle)

7. Prunus umbellata Ell. FLATWOODS OR HOG PLUM.   Fig. 290

Deciduous shrub or small tree, usually occurring singly and not thicket-forming. Small lateral twigs thinnish, infrequently thorn-tipped.

Leaf blades usually flat, elliptic, oval, or oblong, 2–6 cm long, 1–3 cm broad, broadly short-tapered basally, mostly acute apically, margins serrate, tips of the teeth pointed, not gland-tipped; upper surfaces glabrous or sparsely short-pubescent, the lower usually pubescent at least along the veins; commonly a gland is present on one side or both sides of the base of the blade or on one side or both sides of the summit of the petiole.

Flowers borne singly or more frequently in fascicles of 2–5 from buds on wood of a previous season, the flower buds often clustered at the ends of short spur-shoots; flowering occurring mostly before emergence of new shoots of the season. Flower stalks slender, 4–10 mm long. Sepals horizontally spreading at full anthesis, 2–3 mm long, oblong with obtuse to rounded apices varying to triangular-acute, short-pubescent marginally varying to pubescent on both surfaces. Petals spreading, white, shortly clawed basally, blades obovate, subrotund, or elliptic, 3 mm long or a little more, corolla mostly 12–15 mm across the spreading petals. Stamens 10–20 or more, filaments white, markedly unequal in length, anthers yellow. Usually before the petals fall they become at least partially pink, the filaments, sepals, floral tubes, and flower stalks become dull rose-pink and remain so for a short period after the petals fall, the flowers being very numerous, the plant then has a pinkish aspect as seen from a little distance.

Drupes red or yellow in preripening, becoming dark purplish or greenish blotched with purple, rarely red, strongly glaucous, globose or nearly so, about 1.5–2 cm in diameter, the flesh bitter-sour and astringent to the taste.

Open pine forests, mixed pine-hardwood forests, hammocks, coastal scrub, throughout our area. (Coastal plain and piedmont, s. N.C. to cen. pen. Fla., Ala., s.e. La.; n.w. La., s.w. Ark., e. Tex.)

For comparative features, see comments under *P. americana* and *P. angustifolia*.

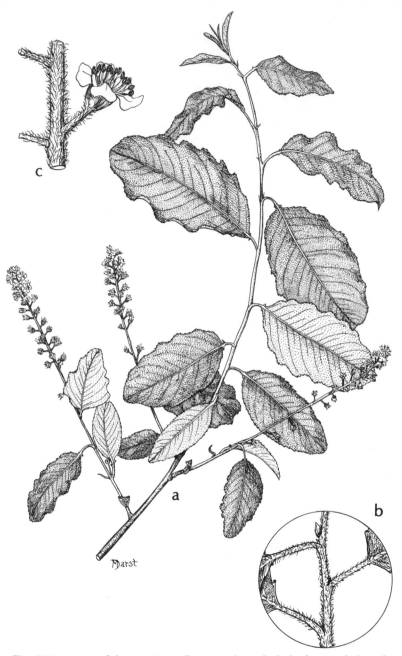

Fig. 289. **Prunus alabamensis:** a, flowering branch; b, leafy stem below the inflorescence; c, piece of axis of inflorescence with a flower.

Fig. 290. **Prunus umbellata.**

National champion flatwoods plum (1974), 3'4" circumf., 33' in height, 28' spread, Colclough Pond Wildlife Sanctuary, Fla.

7. Pyracantha (PYRACANTHAS, FIRETHORNS)

Evergreen shrubs, loosely to compactly branched, stems rigid, lateral branchlets sharply thorn-tipped; leaf scars crescent-shaped or 3-lobed, each with 3 vascular bundle scars; first-year stems glabrous or pubescent.

Leaves alternate, simple, petioles short; stipules quickly deciduous; blades pinnately veined, elliptic, oblong, cuneate-oblanceolate to cuneate-spatulate, a few sometimes orbicular, 5–40 mm long and about 3–10 mm broad, apices rounded, emarginate, obtuse, or acute, margins obscurely crenate or serrate, teeth often gland-tipped, sometimes entire, surfaces glabrous or one or the other sparsely pubescent.

Inflorescences several-flowered corymbs or cymes terminating very short branchlets of the season, usually present in profusion; inflorescence axes, flower stalks and calyces glabrous or pubescent. Flowers bisexual, radially symmetrical, with a floral tube having a nectariferous ring at its mouth. Calyx segments 5, broadly triangular. Petals 5, small, white or creamy white, spreading, shortly clawed basally, blades obovate, inserted on the rim of the floral tube. Stamens 20, inserted in 2 series on the rim of the floral tube between the perianth and the nectar ring, filaments free. Pistils 5, adaxially free, abaxially half adnate to the floral tube, styles terminal on the adaxial margins, stigmas discoid; ovaries 2-ovulate, each maturing into a 2- or 1-seeded nutlet, these surrounded by the eventually fleshy floral cup with the calyx segments persistent on its summit, the floral cup usually red at maturity, orange or orange-red in some.

Pyracanthas are native to Europe and w. and e. Asia. The species are said to be reasonably distinct in nature. In cultivation, there appear to be a number of garden varieties which intergrade a great deal so that it is, for me, not practicable to distinguish named variants. Plants become naturalized in some places, especially, it appears, where the soil is chalky, and they persist about abandoned homesites.

8. Pyrus (PEARS)

Deciduous, small trees. Twigs stoutish, short- or spur-shoots sometimes thorn-tipped. Leaves alternate, simple, slenderly petiolate, petioles one-third to one-half as long as the blades or a little more; stipules subulate, pubescent, about 1 cm long, very quickly deciduous; blades pinnately veined, roundish to ovate or ovate-oblong, bases rounded, truncate, or very little tapered, apices cuspidate-acuminate; margins minutely serrate, teeth gland-tipped at first, surfaces irregularly patched with woolly pubescence when unfurling, soon becoming glabrous.

Inflorescences of several- to numerous-flowered corymbs borne terminally from winter buds on short spur-shoots, mostly before, less frequently as, leaves are unfolding; flower stalks slender, unequal, 1–5 cm long or a little more. Flowers bisexual, radially symmetrical, ovary inferior. Calyx segments small, 5, red-glandular marginally. Petals 5, shortly clawed, blades round-obovate, white or sometimes white tinged with pink, inserted on the rim of the floral tube. Stamens numerous, alternating in 2 series, outer series on the rim of the floral tube within the perianth, inner series just below; anthers red. Pistil 1, ovary with as many 2-ovuled locules as there are styles. Fruit a pome, its flesh containing large quantities of grit-cells until fully ripe.

1. Calyx segments densely and compactly woolly-pubescent adaxially, not persistent on the fruit; styles (2) 3; flowers mostly 2.5–3 cm across the corollas; fruit globular, surface bronzy-brown, roughish-granular. 1. *P. calleryana*
1. Calyx segments loosely woolly-pubescent proximally adaxially, persistent on the fruit; styles 5; flowers mostly about 4 cm across the corollas; fruit usually broadest distally, tapered toward the base (pyriform), surface greenish, or yellow (fully ripe), smooth.
2. *P. communis*

1. Pyrus calleryana Decne. RED-SPIRE.   Fig. 291

Native of China, used as a pear-stock in the nursery trade. Naturalized in a few places in our area or persisting on abandoned homesites. A form, "Bradford," which flowers profusely in spring and has handsome red foliage in autumn, is now in the local trade.

2. Pyrus communis L. COMMON PEAR.

This "species" is presumed to be of hybrid origin, several species having contributed to its genealogy, known only in cultivation or naturalized from cultivation; sporadic in our area along fence and hedge rows, thickets, clearings, usually near human habitations; also persisting about abandoned homesites.

9. Rosa (ROSES)

Upright, arching, trailing, or clambering shrubs with prickly stems. Leaves alternate, odd-pinnately once compound, with stipules at least partially attached to proximal portions of the petioles. Flowers solitary in leaf axils or in terminal corymbs or panicles; bisexual, radially symmetrical, with a conspicuous urceolate to globose floral tube bearing the sepals, petals, and stamens on the rim of its constricted orifice. Ovaries ripening into achenes within the eventually fleshy floral tube, the aggregate referred to as a hip.

1. Stipules notably pectinately toothed on their margins or free distal portions.
  2. Corolla of a single whorl of white petals.
    3. Stems of branchlets glabrous; flowers in many-flowered panicles.   4. *R. multiflora*
    3. Stems with loosely matted, grayish hairs and stoutish, stiffish, gland-tipped ones; flowers solitary or few in a cluster. 1. *R. bracteata*
  2. Corolla "double," of numerous pink petals. 6. *R. wichuraiana*
1. Stipules entire or glandular on their margins, not pectinate.
  4. Stipules attached to the petioles for less than half their length; corolla white.
3. *R. laevigata*
  4. Stipules attached to the petioles for almost their entire length; corolla pink.
    5. Major prickles on the stem straight and diverging at right angles to the stem; plants of upland, well-drained places. 2. *R. carolina*
    5. Major prickles on the stem recurved-hooked; plants of wet places.   5. *R. palustris*

1. Rosa bracteata Wendl. MCCARTNEY ROSE.   Fig. 292

Shrub with arching or clambering branches. Stems with loosely matted, grayish hairs and numerous, stoutish, stiffish, purplish red, stipitate-glandular ones well exserted from the mat, a pair of broad-based, straight or curved, sharp prickles at the nodes, the prickles terete above the bases.

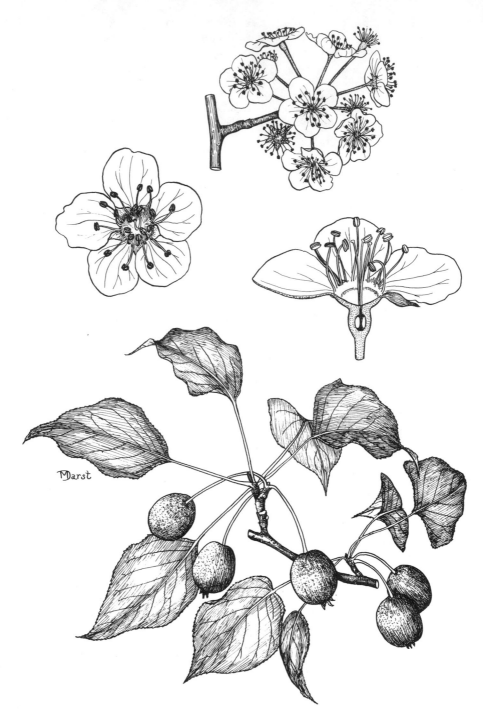

Fig. 291. **Pyrus calleryana.**

Fig. 292. **Rosa bracteata.**

Leaves semievergreen. Leaflets 5–9, commonly 7, short-stalked; blades oblong-elliptic to obovate, those on the flowering branchlets mostly 1–2 cm long and 0.8–1.2 cm broad, those on vigorous shoots larger; bases rounded to broadly tapered, apices mostly rounded; upper surfaces glabrous and glossy green, the lower paler and dull green, sparsely pubescent along the midrib, sometimes with a few stipitate glands on the midrib, occasionally with a sharply hooked prickle on the midrib proximally; margins serrate, the teeth tipped with purplish red glands; leaf axes usually with a few hooked prickles, some at the nodes, some internodally, sometimes having stipitate glands as well; stipules narrowly adnate to petioles for a short distance, variably pectinate-toothed and/or pectinate-branched distally.

Flower usually solitary at the end of a branchlet, subtended by an involucrelike cluster of broad, toothed bracts. Floral tube and both surfaces of the sepals densely matted-pubescent, or the interior of the sepals sparsely pubescent to glabrous. Sepals ovate or oblong-ovate, bases truncate, apices attenuated, 1.5–2 cm long. Petals white, obovate, commonly asymmetric, apices broadly rounded or emarginate, about 6–8 cm across the spreading petals. Copiously pubescent stigmas moundlike at the orifice of the floral tube. Hip dark brown, nearly globose, 1.5–2 cm in diameter. Achenes numerous, about 5 mm long, oblongish to obovate in outline, many of them irregularly angled owing to compaction, portions of their surfaces with straight, fragile, lustrous tan, easily detached hairs, otherwise the surfaces tan and smooth.

Native to s. China, cultivated, in our area locally persisting in fence or hedge rows, about abandoned homesites, infrequently naturalizing.

2. Rosa carolina L. CAROLINA ROSE. Fig. 293

Deciduous, upright shrub, stems to about 1 or 1.5 m tall, relatively sparingly branched, usually arising singly from subterranean runners; prickles few to numerous, usually broadly-based, essentially terete, straight and horizontally divergent, 5–10 mm long, often a given stem with more numerous, shorter but unequal, sometimes gland-tipped prickles between the major ones.

Leaflets 3, 5, or 7 (or 9), elliptic to oblanceolate or narrowly obovate, smaller ones sometimes rotund, margins serrate from a little above the base, bases tapered, apices acute to rounded, surfaces glabrous or pubescent; stipules narrow, pubescent, entire or with gland-tipped teeth; petioles and leaf axes short-shaggy-pubescent, stipitate-glandular, or both.

Flowers usually borne singly in the axils of the uppermost leaves of branchlets, shortly stalked, the stalk and floral tube stipitate-glandular. Sepals lance-acute to oblongish-acute below, with a longish, taillike tip, this sometimes dilated distally, the dilated portion entire or toothed; margins of basal portion of some of the sepals of a given flower often with 2–4 long, narrow, lobulate enations, sometimes the taillike extremities tripartite; inner surface of bases of sepals densely woolly-pubescent, the outer surfaces much less so and stipitate-glandular as are the taillike extremities. Petals pink, broadly obovate, 2–3 cm long, their extremities broadly rounded. Copiously pubescent stigmas compacted-moundlike just without the orifice of the floral tube. Hip red, subglobose, about 1 cm long. Achenes brown, plump, obliquely obovoid, surfaces smooth but slightly uneven, about 5 mm long.

Upland woodlands, shaded roadside banks, glades, thickets; infrequent in our range. (N.S. to Minn., generally southward to n. Fla. and e. Tex.)

Fig. 293. **Rosa carolina.**

Fig. 294. **Rosa laevigata:** a, flowering branch; b, node, enlarged to show stipules; c, fruit.

3. Rosa laevigata Michx. CHEROKEE ROSE. Fig. 294

Evergreen shrub, its glabrous leader stems robust and elongate; when suitable other vegetation or support occurs in its surroundings, the vigorous, stout, leader shoots and their branches clamber to considerable heights, the recurved prickles serving to hold or anchor in the process. Prickles reddish brown, with broad, flattened bases.

Leaflets 3, elliptic or lance-elliptic, glabrous, dark green and lustrous above, paler and dull beneath; bases broadly and shortly tapered, occasionally somewhat inequilateral and oblique; margins finely and sharply serrate from a little above their bases, usually with a few recurved prickles on the petioles and often with a few widely spaced, straight ones on the midribs of the leaflets beneath; stipules adnate to the petioles only for a very short distance, the free tips commonly subulate, sometimes more foliose, their margins glandular-toothed.

Flowers borne singly in the axils of uppermost leaflets of lateral branchlets, their stalks and floral tubes with straight, horizontally divergent, stiff trichomes tipped by minute, eventually deciduous, red glands. Sepals ovatish-acute basally, variable distally, even on a single flower, some gradually narrowed to an attenuate tip, others narrowed then expanded to a slightly foliose tip, or narrowed to a stalklike portion, then expanded to a more conspicuously foliose tip, in either of the latter two cases, the expanded tip usually glabrous or only sparsely soft-pubescent, varyingly entire to toothed marginally; inner surfaces of basal portions of sepals copiously and compactly woolly-pubescent, outer surfaces varying from nearly glabrous to woolly-pubescent near and at the edges, sometimes with a few trichomes like those on the floral tubes. Petals white, 3–4 cm long, broadly obovate, the edges irregularly wavy or wavy and dentate. Pubescent stigmas compactly moundlike, elevated somewhat above the orifice of the floral tube. Hip red, body ellipsoid, about 3 cm long, below which abruptly tapered to a stalklike base, thus, overall, somewhat pyriform. Achenes stramineous, angular, with 2 flat faces and 1 rounded, narrowly and obliquely obovoid in outline, about 3 mm long.

Native to China, widely cultivated and frequently naturalized in our area, chiefly near the edges of upland and lowland woodlands or in fence and hedge rows.

4. Rosa multiflora Thunb. MULTIFLORA ROSE, JAPANESE ROSE. Fig. 295

Shrub with erect or arching, glabrous branches. Prickles subnodal, conical-recurved, larger ones, especially, with a notably longitudinally spreading base, upward part of the spread often relatively short and rounded, that of the downward part relatively longer and tapering.

Leaflets usually 7, occasionally 5 or 9, elliptic or obovate, margins finely and sharply serrate from near the base, teeth sometimes gland-tipped; bases rounded or obtuse, apices rounded, acute, or short-acuminate, upper surfaces glabrous, lower glabrous or more commonly sparsely short-pubescent, principally along the veins; stipules pectinately toothed marginally, the free tips similar but larger; surfaces of the stipules, their teeth, petioles, leaf axes, and stalks of the leaflets variably softly pubescent or softly pubescent and glandular, the glands sometimes both sessile and stipitate, if stipitate, the stipes of unequal lengths.

Flowers borne in relatively floriferous panicles on lateral branchlets, the lowermost panicle branches usually subtended by leaves, otherwise bracts and stipules very variable in form and pubescence throughout the inflorescence;

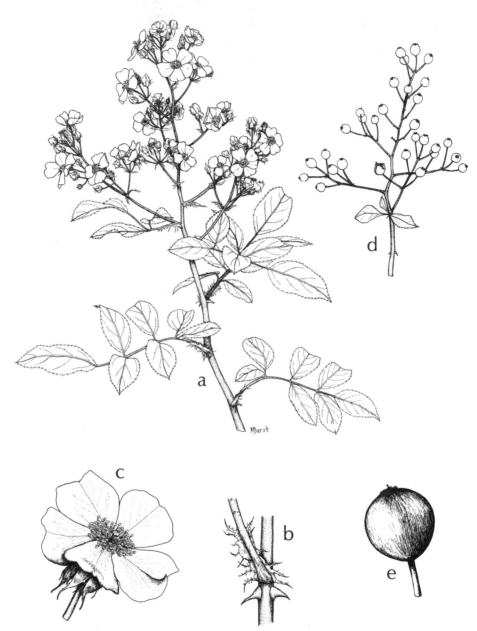

Fig. 295. **Rosa multiflora:** a, flowering branch; b, node with stipule and infrastipular prickles; c, flower; d, fruiting branch; e, "fruit" (hip).

flower stalks, floral tubes and exterior of the sepals vary from glabrous to cottony-pubescent to stipitate-glandular, although the floral tubes appear to be singularly free of stipitate glands even in those specimens otherwise having them. Sepals ovate-attenuate or triangular-attenuate, the lower margins usually with at least 1 longish subulate lobulate-enation on a side; interior of sepal copiously woolly-pubescent, the exterior variable as indicated above. Petals white or pale pink, 1.5–2 cm long, broadly obovate, equally or unequally notched at the summit. Styles elevated considerably above the orifice of the floral tube. Hip red, subglobose, about 7 mm long. Achene light brown, irregularly angular-obovoid, surface slightly papillate, 3–5 mm long.

An Asian species, cultivated for ornament or sometimes planted for screening, infrequently naturalized in our area in the vicinity of plantings.

5. **Rosa palustris** Marsh. SWAMP ROSE. Fig. 296

Upright, deciduous shrub, commonly with numerous, bushy-branched stems to 2 m tall; prickles broad-based, bases usually flattened, sometimes straight and horizontally divergent, more frequently somewhat recurved or hooked, most of them subnodal, some sometimes internodal, especially on vigorous shoots.

Leaflets 5, 7, or 9, elliptic or lanceolate, margins finely serrate from somewhat above their bases, bases shortly and broadly tapered to rounded, apices mostly acute, sometimes rounded, notably on smaller leaflets; upper surfaces glabrous, lower copiously softly pubescent as they unfold, later moderately softly pubescent, or pubescent only along the midribs; midrib beneath sometimes with a few small prickles; stipules narrow, usually pubescent, sometimes becoming glabrous; petioles and axes of mature leaves usually pubescent, sometimes glabrous, either or both often armed with prickles.

Flowers borne singly in the axils of 1 or more of the uppermost leaves of a branchlet, or in few-flowered corymbs in the axil of the uppermost leaf, flower stalks and floral tubes stipitate-glandular, the stipitate glands more numerous on the tubes than on the stalks. Sepals ovate, triangular, or oblongish below, abruptly contracted to longish, taillike extremities that are somewhat dilated distally; inner surfaces softly short-pubescent, outer stipitate-glandular and usually softly short-pubescent at least marginally, the taillike extremities usually sparsely pubescent and sparsely stipitate-glandular. Petals pink, 2–3 cm long, broadly obovate, their broad extremities emarginate, irregularly wavy, or crinkled. Copiously pubescent stigmas just exserted from the orifice of the floral tube. Hip red, somewhat oblate, about 1 cm broad, broader than long. Achenes light brown, irregularly angular-obovoid, surfaces semicorrugated, 7–8 mm long. [*R. floridana* Rydb.]

Marshy shores of streams, ponds, lakes, swamps, and their borders, marshes, wet thickets, drainage ditches and canals; throughout our range. (N.S. to Minn., generally southward to cen. pen. Fla., Miss., Ark.)

6. **Rosa wichuraiana** Crepin.

The form here described is the "Dorothy Perkins" cultivar.

Stems glabrous, usually long, trailing or clambering; prickles scattered internodally, with broad, conical bases, straight to slightly or distinctly recurved; similar but smaller prickles and dark purplish red glands on petioles, leaf axes, and sometimes on midribs of leaflet blades beneath.

Leaves semievergreen, blades, overall, 2.5–4 cm long, leaflets 5–9, mostly 7, laterals very short-stalked, terminal one with a stalk about 1 cm long; blades of

Fig. 296. **Rosa palustris.**

laterals mostly obovate, some elliptic, some nearly rotund, 0.8–2 cm long, terminal ones largest, bases little if at all tapered, apices rounded, terminal leaflet elliptic, margins of all leaflets finely serrate all around from a little above their bases; stipules prominent, adnate to petioles for half their length or a little more, their margins irregularly lacerate-toothed, teeth gland-tipped.

Flowers sometimes solitary terminating branchlets, more frequently in terminal more or less paniculate clusters. Floral tube and interior of ovate-attenuate sepals felty-pubescent. Corolla pink, about 3–4 cm across, petals numerous.

Infrequently and locally naturalized in our range, road and railroad banks, waste places.

10. Rubus (BRAMBLES)

Perennial shrubs (those treated here), many bearing bristles, bristles and prickles, or prickles; sending up from the base biennial stems, these in the first year usually unbranched and not bearing flowers or fruits (primocanes), in the second year not elongating but forming relatively short, lateral branches most of which are flower/fruit-bearing (floricanes), the flowers solitary terminating branchlets, or in racemes or panicles. Leaves mostly compound, slenderly petiolate, blades of those of the primocanes may differ in their complexity, in size, and in the form of the leaflets from those of the floricanes. Stipules narrow, inconspicuous, persistent or tardily deciduous. Floral tube small, flat to hemispheric. Sepals 5, spreading to reflexed. Petals 5, ascending or spreading. Stamens numerous. Pistils numerous, borne on a convex or conic receptacle which in some elongates as the ovaries mature into a cluster of drupelets, these falling separately or together.

In general, brambles (those comprising the blackberries), have been shown to hybridize, to exhibit various degrees of polyploidy, to have developed means of producing seeds without sexual reproduction; thus they exhibit an extraordinarily complex diversity of variable characteristics such that delimitation of species in the ordinary sense is apparently not possible. The number of "species" attributed to eastern North America by "specialists" is as high as 400 or more, the number varying greatly depending upon the particular "specialist," and "microspecies" perhaps as many as 10,000. Oversimplification appears to be the only way to achieve a practicable solution to the dilemma. Any two or even all of those treated here may grow intermixed.

1. Stems erect, arching-erect, or clambering.
  2. Leaflets glabrous to relatively densely pubescent beneath, the pubescence, if any, not pale grayish and not felty.   1. *R. betulifolius*
  2. Leaflets densely pale grayish and felty-pubescent beneath.   2. *R. cuneifolius*
1. Stems prostrate, trailing, or sprawling, those of primocanes commonly eventually rooting at their tips.
  3. Primocane stems essentially glabrous other than having relatively few, minute, sessile or very short-stipitate purplish red glands; leaves of primocanes mostly with 3 leaflets, one or both of the laterals often (not always) with 1 or a pair of subbasal lobes, or occasionally a few or all leaves on a primocane with 5 leaflets.   3. *R. flagellaris*
  3. Primocane stems with a few to numerous, straight, purplish red, gland-tipped bristles of variable length, some of them, at least, to 2 mm long; leaves of primocanes with 5, 3, or 3, 4, and 5 unlobed leaflets.   4. *R. trivialis*

1. Rubus betulifolius Small. HIGHBUSH BLACKBERRY.   Fig. 297

Stems erect, arching erect, or clambering, to 3 m tall or a little more, coarse, ridged and grooved, sparsely pilose on young portions, older portions glabrous.

Fig. 297. **Rubus betulifolius:** a, portion of primocane; b, flowering branchlet; c, distal portion of floricane (fruiting stage).

592

Armature of stems, petioles, stalks of leaflets, midveins of lower surfaces of leaflets, inflorescence axes, and flower stalks, consisting of hard, large-based, sharply pointed, straight or usually hooked prickles. Leaves of primocanes 5-foliolate, occasionally 4-foliolate, petioles 6–10 cm long, stalk of terminal leaflet longest, 2.5–4 cm long, those of intermediate laterals mostly 1–2 cm long, those of the lower laterals varying from sessile to 5 or 6 mm long; blades elliptic, elliptic-oblong, lance-ovate, or lance-obovate, bases rounded to cuneate, apices acuminate, margins sharply serrate; blade of terminal leaflet largest, 6–10 cm long or a little more, lowest pair smallest, 4–6 cm long. Leaves of floricanes ternate, varying considerably in size but all much smaller in all respects than primocane leaves; leaflet shape similar, glabrous above, varying from glabrous to densely pubescent beneath but not pale grayish felty-tomentose.

Floricane branchlets bearing loose, open racemes of flowers, the racemes often leafy-bracted proximally; axis of raceme and flower stalks bearing long, soft hairs, often with short, gland-tipped hairs intermixed. Sepals 4–6 mm long, ovate-triangular, becoming reflexed, densely tomentose interiorly, less so without, apex of each with a glandular mucro. Petals often pink to rose in bud, white or white suffused with pink at full anthesis, obovate or rhombic-obovate, 2–2.5 cm long. Aggregate fruit 1–2.5 cm long, 1–1.5 cm broad, very juicy and flavorful.

Wet swampy woodlands and especially their borders, often where surface water stands much of the time, open stream banks, moist to wet clearings, shoreline thickets, margins of cypress-gum ponds and depressions, shrub-tree bogs; also in old fields and pastures, fence and hedge rows, borders of rich woodlands; throughout our area. (Va. southward to cen. pen. Fla., westward to La.)

2. Rubus cuneifolius Pursh. SAND BLACKBERRY. Fig. 298

Plants suckering from subterranean parts and commonly forming relatively dense stands. Stems erect or arching-erect, 3–15 dm tall, densely short-pubescent, the pubescence mostly sloughing on older portions, armed with stiff and sharp, large-based, straight, recurved, or hooked prickles. Primocanes frequently branching, their leaves sometimes all 3–foliolate, sometimes 3– and 4– and 5–foliolate; floricane leaves ternate; primocane leaves and their leaflets much larger than those of floricanes; petioles and stalks of the leaflets armed like the stems. Leaflets with a grayish green cast, glabrous or pubescent above, if the latter then not felty, lower surfaces wholly clothed with a pale grayish, compact, felty tomentum; terminal leaflets with stalks 5–20 mm long, blades oblanceolate, spatulate, rhombic, or narrowly obovate, marginally serrate from near the base distally varying to serrate only somewhat above the middle; lateral leaflets sessile or subsessile, blades commonly inequilateral, oblanceolate, spatulate, obovate, or elliptic, varying in size but smaller than the terminal one; in the case of ternate leaflets, blades of the laterals sometimes with 1 or a pair of subbasal lobes.

Floricane branchlets terminated by single flowers or in loose, open racemes with up to 10 flowers, those with few flowers leafy-bracted, those with the larger number of flowers leafy-bracted proximally. Sepals oblong, elliptic-oblong, or lance-ovate, 5–7 mm long, both surfaces densely tomentose, with a thickened glandular mucro at their tips. Petals white, 1–1.5 cm long, clawed basally, blades obovate to oval. Aggregate fruit 1–2.5 cm long and about 0.8–1.5 cm broad, juicy and flavorful if well developed.

In both well-drained and wet places, abundantly colonizing old fields,

Fig. 298. **Rubus cuneifolius:** a, distal portion of primocane; b, node with leaf from primocane; c, distal portion of floricane; d, fruiting branchlet.

burned-over places, and places where the soil has been mechanically disturbed as in pine plantations, also shrub-tree bogs, edges of cypress-gum ponds, or depressions, pine–scrub oak sand ridges; throughout our area. (Conn., L.I., southward to s. pen Fla., westward to Miss.)

3. Rubus flagellaris Willd. NORTHERN DEWBERRY. Fig. 299

Primocanes erect at first, eventually arching and then trailing, usually rooting at the tips, relatively slender, at first little-armed, eventually with few to numerous, relatively short, broad-based, sharply pointed, recurved or hooked prickles; varyingly glabrous or with minute, sessile or very short-stipitate, purplish glands, the glands, like the prickles, apparently produced tardily. Floricane branchlets, petioles, and flower stalks similarly armed and glandular, often both branchlets and flower stalks softly pubescent, the flower stalks sometimes densely and compactly pubescent distally.

Primocane leaves usually 3-foliolate, frequently the lateral leaflets inequilateral, one or both often with 1 or a pair of prominent subbasal lobes, *or* occasionally a few or all leaves with 5 leaflets; primocane leaflets all ovate in outline, to 6 cm long and to 5 cm broad, infrequently the terminal leaflet lanceolate, stalk of the terminal one 1–2.5 cm long, laterals subsessile, margins irregularly serrate or irregularly serrate *and* serrate-lobed (apart from the prominent subbasal lobes, if any); bases broadly short-tapered to rounded, apices usually acuminate or acute, surfaces of all leaves glabrous, midrib beneath sometimes prickly. Floricane leaves 3-foliolate, lateral leaflets commonly inequilateral and sometimes with one or a pair of subbasal lobes, lanceolate, lance-ovate, or the terminal one ovate, sometimes all lance-ovate.

Flowers sometimes borne singly but varying to a raceme with 5 or 6 (10) flowers, leafy-bracted at least proximally, flower stalks slender and relatively long if only 1–3 flowers per branchlet, relatively short-stalked if more. Sepals triangular, 4–5 mm long or a little more, becoming reflexed, pubescent on both surfaces, usually densely and compactly so interiorly, tipped by a glandular apicule. Petals white, 1–1.5 cm long, subclawed basally, their blades narrowly obovate.

I have, as yet, not seen other than small, dryish aggregate fruits on plants in our range.

Open or wooded stream banks, lowland woodlands, borders of second-growth, upland woods, clearings, moist old fields, thickets; s.w. Ga., Fla. Panhandle, s. Ala. (Maine, Que., s. Ont. to Minn., generally southward to Fla. Panhandle and e. third of Tex.)

4. Rubus trivialis Michx. SOUTHERN DEWBERRY. Fig. 300

Primocanes trailing, reclining, prostrate, sprawling, or sometimes clambering, armed with large-based, straight, recurved, or hooked prickles, in addition sparsely to copiously pubescent with straight, stiff, purplish red, minutely gland-tipped hairs of variable length, to at least 2 mm long; similar armament and pubescence may occur on petioles, stalks of leaflets, flower stalks, and sometimes on midribs of leaflets beneath. Primocane leaves mostly 5-foliolate, sometimes 3-foliolate, usually overwintering and some of them, at least, present until well after the new floricane branchlets and flowers develop; leaflets elliptic or lanceolate, sometimes lance-ovate, 2–6 cm long, surfaces glabrous, margins serrate or doubly serrate, bases rounded or shortly tapered, apices usually acute. Floricane leaves ternate, mostly lanceolate or elliptic, less frequently oblanceo-

Fig. 299. **Rubus flagellaris:** portions of a primocane having 3 leaflets per leaf (by far the more common number), a, from near the base, b, from medial portion, and c, from the tip; d, a portion of primocane having 5 leaflets per leaf; e, portion of floricane (fruits unripe).

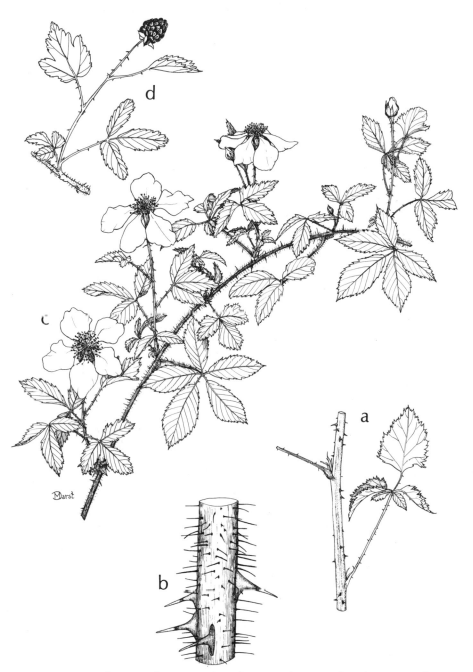

Fig. 300. **Rubus trivialis:** a, segment of young primocane; b, enlargement of stem of primocane to show vestiture; c, distal portion of floricane; d, fruiting branchlet.

597

late, lateral ones sometimes with one or a pair of subbasal lobes, all smaller than the leaflets of the primocanes.

Flowers solitary, less frequently 2–3 on floricane branchlets, their stalks mostly 2.5–4 cm long. Sepals elongate-triangular, becoming reflexed, attenuate apically, both surfaces pubescent, the upper densely so. Petals commonly pink to rose in bud, white or white tinted with pink at full anthesis, shortly sub-clawed basally, blades broadly elliptic to obovate, 1–2.5 cm long. Aggregate fruit 1.5–3 cm long and 1–1.5 cm broad, juicy and flavorful.

In both well-drained and poorly drained places, fields, waste places, road and railroad banks, fence and hedge rows, old fields, marshy swales, pineland depressions and clearings, stabilized dunes and interdune swales, banks and borders of marshes, bottomland clearings; throughout our area. (Coastal plain and outer piedmont, s.e. Va. to s. Fla., westward to Tex., northward in the interior to Mo. and s. Ill.)

Rubiaceae (MADDER FAMILY)

1. Stems prostrate-creeping, rooting at the nodes, or trailing; leaf blades 5–20 (25) mm long; flowers in axillary pairs, usually a pair axillary to one of the pair of leaves at the tip of a branchlet, their floral tubes joined laterally.   1. *Mitchella*
1. Stems of plant erect; leaf blades very much larger; flowers and fruits in dense, globose heads, *or* in racemes, panicles, or loose, open cymes, never paired and with floral tubes of the pair joined laterally.
  2. Flowers and fruits in dense, globose heads, the heads slenderly long-stalked, solitary or more frequently in few-headed cymes.   2. *Cephalanthus*
  2. Flowers and fruits in axillary racemes or panicles *or* in loose, open cymes terminal on branchlets.
    3. Flowers and fruits in axillary racemes or panicles.   3. *Chiococca*
    3. Flowers and fruits in loose, open cymes terminating branchlets.
      4. Calyx segments 5, in some flowers all of them lance-subulate and 1–1.5 cm long; in some flowers one of the segments becomes much enlarged, leaflike in form and petaloid, pink or yellowish, mostly ovate, 6–7 cm long; young twigs of the season, petioles, and inflorescence axes densely soft-pubescent; lateral veins, on upper leaf surfaces, raised.   4. *Pinckneya*
      4. Calyx of 5 very minute teeth at the summit of the floral tube; young twigs of the season, petioles, and inflorescence axes glabrous or sparsely pubescent; lateral veins, on upper leaf surfaces, impressed.   5. *Psychotria*

1. Mitchella

Mitchella repens L. PARTRIDGE-BERRY, TWIN-FLOWER, TWO-EYED-BERRY.

Leafy stems very slender, glabrous or with minute patchy pubescence, prostrate-creeping and rooting at the nodes or trailing; subwoody and perennial, often loosely mat-forming.

Leaves evergreen, opposite, pairs relatively remote from each other, slenderly petiolate, petioles shorter than the blades, often sparsely and minutely pubescent; stipules minute, deltoid, connecting the petioles of the leaf pair on either side of the stem. Blades ovate, to suborbicular, pinnately veined, 5–20 (25) mm long, a little longer than broad for the most part, smaller ones sometimes as broad as long; bases truncate or vaguely subcordate, apices rounded to bluntly obtuse; lower surfaces sometimes minutely pubescent about the base of the midrib; margins entire.

Flowers borne in axillary, singly-stalked, pairs, usually a pair axillary to one of the pair of leaves at the tip of a branchlet, their floral tubes joined laterally, rarely a flower solitary. Floral tube tubular-campanulate, with several, unequal, small, erect calyx segments at its summit. Corolla white, sometimes pink-tinged, tube narrowly cylindric-funnelform, about 10–15 mm long, limb composed of 4 (6) spreading to recurved, ovate to ovate-oblong lobes less than half as long as the tube and densely shaggy-pubescent interiorly. (Rarely the 2 flowers coherent into 1.) Stamens 4. Ovary inferior, style 1, stigmas 4. In some plants stamens included and styles exserted, in other plants stigmas included, stamens exserted. Ovary ripening into a single, rotund, bright red berrylike drupe 7–10 mm in diameter, the minute, persistent calyces of the flowers evident subapically (hence fruit two-eyed), the (combined) fruit having 8 seedlike stony nutlets.

Mesic or submesic to nearly xeric woodlands, banks of streams; throughout our area. (N.S. to Ont. and Minn., generally southward to cen. pen. Fla. and e. Tex.)

2. Cephalanthus

Cephalanthus occidentalis L. BUTTONBUSH, HONEY-BALLS, GLOBE-FLOWERS.

Fig. 301

A scrubby, deciduous shrub, rarely with the form of a small tree, to about 3 m tall. Twigs reddish brown, glabrous or short-pilose at first and becoming glabrous, with raised, corky lenticels; bark of older stems rough, ridged and furrowed, or bumpy; leaf scars U-shaped to nearly round, each with a relatively broad, crescent-shaped vascular bundle scar, axillary buds sunken and obscure, pith tan, continuous.

Leaves deciduous, opposite or in whorls of 3 or 4 (varying on a given plant), petiolate, petioles varying from 0.5–3 cm long, pubescent or glabrous; blades pinnately veined, oval, oblong-oval, elliptic, or ovate, very variable in size, 7–15 cm long, 3–10 cm broad, their bases broadly rounded to cuneate, apices acute or acuminate, upper surfaces glossy green, lower dull, sometimes glabrous, ours mostly short-pilose at least on the principle veins, sometimes uniformly softly pubescent; stipules short-deltoid, leaving stipular lines between the petioles after being sloughed.

Flowers small, very numerous in dense globose heads 3–4 cm in diameter, the heads long-stalked, solitary or in few-headed cymes borne both terminally and axillary on twigs of the season. Flowers bisexual, radially symmetrical, sessile, ovary inferior, each flower subtended by several hairy bractlets dilated distally. When the flowers of a given head are at anthesis, the short, pilose floral tubes, each with 5 short-ovate calyx segments at the summit, are so compacted beneath the more loosely spreading corolla tubes as to be obscured. Corolla white, with a slenderly funnelform tube 6–10 mm long, pubescent interiorly, with 4 very short, rounded, spreading lobes, in bud a black gland present at the base of some or all of the sinuses, these usually persisting at anthesis. Stamens 4, filaments short, inserted just below the sinuses of the corolla limb, anthers barely or only partially exserted. Style long and slender, very much exserted from the corolla tube, stigma capitate, slightly 4-lobed. Fruit hard, narrowly obconical, about 5 mm long, eventually splitting from the base upwardly into indehiscent nutlets.

The buttonbush has a disheveled, scrubby appearance owing to the die-back of leader shoots leaving dead and dying stumps.

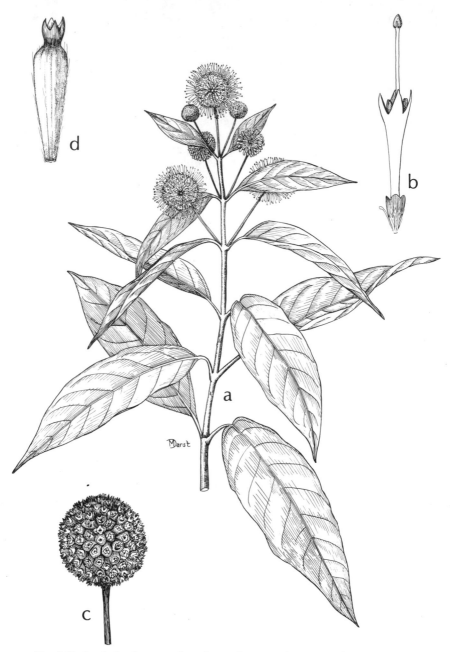

Fig. 301. **Cephalanthus occidentalis:** a, flowering branch; b, flower; c, "ball" of fruits; d, fruit.

Swamps, sloughs, shallow ponds, in and on the banks of small streams, marshes, ponds and lakes; throughout the area of our coverage. (E. Can. to Minn., generally southward to s. pen. Fla. and Tex., western states [as var. *californicus* Benth.]; Mex.; W.I.)

The buttonbush, a plant of wet places, commonly where surface water stands much of the time, may be recognized by its opposite or whorled, simple leaves, and, through much of the season, either white, pincushionlike, globose heads of flowers or globose, rough heads of fruits.

National champion buttonbush (1977), 4′1″ circumf., 23′ in height, 22′ spread, near High Springs, Fla.

3. Chiococca

Chiococca alba (L.) Hitchc. SNOWBERRY. Fig. 302

Glabrous shrub with diffusely branching, erect, reclining, or clambering stems. Young stems green, older woody stems pale brown to buff-colored, a raised band of stipular tissue around the nodes, this eventually sloughed.

Leaves opposite, simple, evergreen, broad-based cuspidate stipules 2–2.5 mm long connecting between the petiole bases, each stipule broad-based and very abruptly narrowed to a subulate tip; petioles slender, short, not exceeding 1 cm long; blades pinnately veined, lateral venation not pronounced, firmly textured, ovate to oblong, lanceolate, or slightly obovate, varying a great deal in size, in our range seldom exceeding 6 cm long and 4 cm broad, bases shortly tapered, apices obtuse, short-acuminate, or less frequently acute, margins entire.

Inflorescences axillary, stalked, lax racemes or panicles shorter than to somewhat exceeding the leaves. Flower stalks slender, to 5 mm long; floral tube campanulate, about 2 mm long, bearing 5 minute lobes on its rim, these persistent on the fruit. Corolla greenish white, or pale yellow to golden yellow, tube funnelform, about 5 mm long, the 5 lobes bluntly triangular, shorter than the tube. Stamens 5, not exserted, filaments united basally around the style. Pistil 1, ovary inferior, 2-locular, 1 ovule in each locule, style slender, stigma slightly clavate. Fruit a compressed-globose, leathery, white drupe 4–5 mm in diameter.

In our range in hammocks, particularly on near-coastal shell mounds, n.e. Fla. and along the Gulf Coast northward to Dixie Co. (Fla., s.e. Tex., trop. and subtrop. Am.)

4. Pinckneya

Pinckneya bracteata (Bartr.) Raf. PINCKNEYA, MAIDEN'S BLUSHES, FEVER-TREE, GEORGIA-BARK. Fig. 303

A shrub or small tree, very handsome during its flowering season. Young twigs of the season, petioles, inflorescence axes, floral tubes, and perianth parts densely soft-pubescent. Twigs tawny in color at first, becoming reddish brown as some or all of the pubescence is sloughed and with pale, raised, corky or warty lenticels; leaf scars more or less heart-shaped, each with a narrow, crescent-shaped vascular bundle scar; pith white, continuous.

Leaves simple, deciduous, opposite, petiolate, petioles mostly 1–3 cm long; stipules triangular, very quickly sloughed leaving a stipular line-scar between the petioles, blades pinnately veined, oval, elliptic, or ovate, varying in size on a given plant from 4–20 cm long and 2.5–12 cm broad, their bases broadly cuneate and somewhat decurrent on the petioles, apices obtuse to acute, upper sur-

PF

Fig. 302. **Chiococca alba:** a, twig with flowers; b, details of node and branch attachment; c, flower, front-side view; d, flower, side view, with developing fruit at right; e, flower longitudinally dissected; f, floral diagram; g, fruiting branch. (Courtesy of Fairchild Tropical Garden.)

Fig. 303. **Pinckneya bracteata:** a, flowering branch; b, c, portions of winter twig; d, flower; e, fruit; f, seed.

faces with scattered short hairs, lower usually uniformly but moderately soft-pubescent.

Flowers in loose, few-flowered cymes terminally on branches of the season and often from the 1 or 2 nodes below. Calyx segments 5, at least 1 segment of some flowers of each cyme becoming greatly enlarged as the flower develops, these variable in size on individual flowers of a given cyme, leaflike in form but petaloid, mostly pink, sometimes yellowish, largest ones ovate, 6–7 cm long and 4–5 cm broad, obtuse to rounded apically, lasting several weeks and rendering the cymes showy from a distance; other calyx segments lance-subulate, 1–1.5 cm long. Corolla greenish yellow, mottled with brown or purple, tube narrowly funnelform, 1.5–2.5 cm long, lobes long-triangular, shorter than the tube, usually somewhat curved-reflexed. Stamens 5, inserted somewhat above the bases of the corolla tube, exserted considerably beyond the throat. Pistil 1, ovary inferior, 2-locular, style longer than the stamens, stigma capitate. Fruit a sub-globose to ovoid, 2-valved, hard, brown capsule, its summit flattish and ringed by a scar left by the deciduous perianth. Seeds tan, flat, numerous, in a vertical stack in each locule; seed body elliptic in outline, with a thin, membranous, markedly reticulate covering extending outwardly from it, sometimes spreading equilaterally, more frequently variously obliquely, thus the overall outline very variable, 5–8 mm in length. [*Pinckneya pubens* Michx.]

Bays, branch bays, seepage swamps, hillside bogs in pinelands, often associated with poison-sumac. Local, coastal plain, s.e. S.C., Ga., n. Fla. westward to about Bay Co.

National champion pickneya (1972), 10″ circumf., 21′ in height, 14′ spread, Gadsden Co., Fla.

5. Psychotria

Psychotria nervosa Sw. WILD-COFFEE.   Fig. 304

A shrub to about 3 m tall, in the area of our coverage sometimes low and sub-shrubby with its stem in the leaf litter. Twigs at first smooth and glabrous, sometimes pubescent with brown hairs; later the twigs roughened between the nodes by irregular corkiness and at the nodes by somewhat raised, opposite, corky leaf scars with stipular remains connecting between their summits. Growth is periodic during a season; terminal buds on first growth of the season composed of 2 enlarged stipules enclosing the initials of a terminal inflorescence; later growth in length is from one or both uppermost axillary buds (renewal growth) and may be contemporaneous with or after first flowering on a given twig and such shoots may produce terminal inflorescences and more renewal growth.

Leaves simple, opposite, short-petiolate, with a prominent stipule between the leaves on each side of the stem and their bases meeting over the bases of the opposite petioles, the stipules relatively soon breaking away and leaving more or less of a fringe of stipular remains between the petioles and over their bases. Blades long-elliptic to narrowly obovate, the larger ones to about 15 cm long and 5 cm broad, acute to acuminate basally and apically; venation pinnate, the lateral veins equal and prominently angling-arching forward toward the leaf edges parallel with each other, impressed on the upper blade surfaces, sometimes the interveinal tissue slightly bulging upward, thus the impressed veins in shallow troughs, all the major veins raised beneath; in axils formed by the mid-rib and lateral veins are minute pockets (domatia) each marked by a small tuft of hairs extruding from it; upper surfaces glossy green, glabrous or with minute

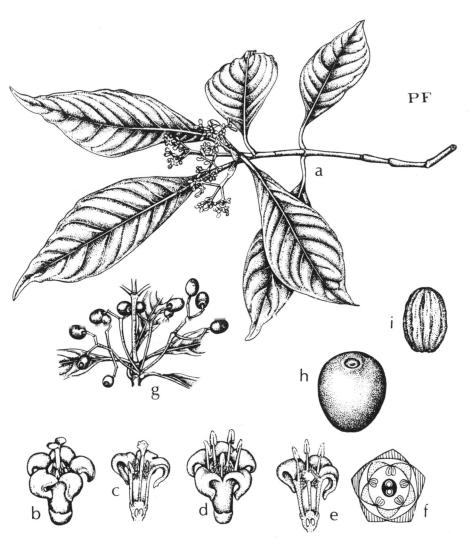

Fig. 304. **Psychotria nervosa:** a, flowering branch; b, long-styled flower, side view; c, long-styled flower longitudinally dissected; d, short-styled flower, side view; e, short-styled flower longitudinally dissected; f, floral diagram; g, fruits; h, single fruit; i, seed. (Courtesy of Fairchild Tropical Garden.)

pubescence along the veins, lower surfaces much paler green and dull, glabrous (apart from the domatia), or pubescent, the pubescence denser along the veins than between them; margins entire or irregularly very slightly wavy.

Inflorescences terminal on branchlets, each a sessile or stalked, open relatively short cyme. Flowers small, short-stalked, radially symmetrical, heterostylous, ovary inferior. Floral tube campanulate, about 1 mm long, with 5 minute calyx segments or teeth at its summit. Corolla white, its tube 2.3–4 mm long, nearly cylindric, a circle of intruding hairs more or less closing the throat about the bases of the filaments, lobes 5, more or less triangular, reflexed. Stamens 5. Flowers "long-styled" on some plants, the style exserted, the short stamens not exserted; on other plants, flowers "short-styled," the stamens exserted, the styles not exserted. Pistil 1, style slender and with a 2-lobed stigma, a nectariferous disc about the base of the style; ovary 2-locular, a single ovule in each locule. Fruit an ellipsoid or subglobose drupe, yellow preripened, red at maturity, 5–8 mm long, the narrow, brownish, differentiated ringlike apex of the floral tube with its minute calyx teeth persistent beyond the fruit body. Stones of the fruit with low, longitudinal ridges. [*Psychotria undata* Jacq.]

Inhabiting near-coastal hammocks, in our area of coverage only in n.e. Fla. (N.e. Fla. southward generally in the peninsula; W.I., trop. Am.)

Rutaceae (CITRUS FAMILY)

1. Leaves, most of them on a given plant, with 3 leaflets, or all of them on a given plant simple.
   2. Twigs bright green, some of them bearing stout, sharp, axillary thorns; petioles of at least some leaves winged for a part or all of their length; crushed foliage without a musky or fetid odor.
      3. Leaves, most of them on a given plant, with 3 leaflets.   1. *Poncirus*
      3. Leaves all simple.   2. *Citrus*
   2. Twigs not bright green, thornless, petioles not winged; crushed fresh foliage with a distinctly musky or fetid odor.   3. *Ptelea*
1. Leaves with 7–19 leaflets, commonly 7–9.   4. *Zanthoxylum*

1. Poncirus

Poncirus trifoliata (L.) Raf. MOCK ORANGE, HARDY ORANGE, TRIFOLIOLATE ORANGE.   Fig. 305

A deciduous shrub or small tree with green twigs bearing stout, sharp, green, basally flattened, axillary thorns. Bark of older stems with longish, irregular green and buff-colored stripes.

Leaves alternate, petiolate, petioles 0.5–3 cm long, winged their full length or only on their distal halves or a little more; most blades with 3 sessile, pinnately veined leaflets, the base of each articulated at its juncture with the petiole; lateral leaflets smaller than the central one and more broadly tapered basally than the latter which is abruptly or acuminately tapered; fully developed leaflets to 6 cm long, mostly obovate, varying to suborbicular, the lateral usually inequilateral or nearly so; apices broadly obtuse or rounded, margins minutely crenate distally from about their middles, upper surfaces minutely pubescent along the midvein, sometimes on the lateral veins as well. (Occasionally, aberration during development causes the central and a lateral leaflet to be fused giving a leaflet with 2 principal veins from the base and a bifoliolate leaf; another

Fig. 305. a–d, **Poncirus trifoliata:** a, leafy twig; b, leafless twig with flowers; c, flower; d, twig with fruits. e, **Citrus aurantium:** leaf and proximal portions of two other leaves to show variation of winging of the petiole.

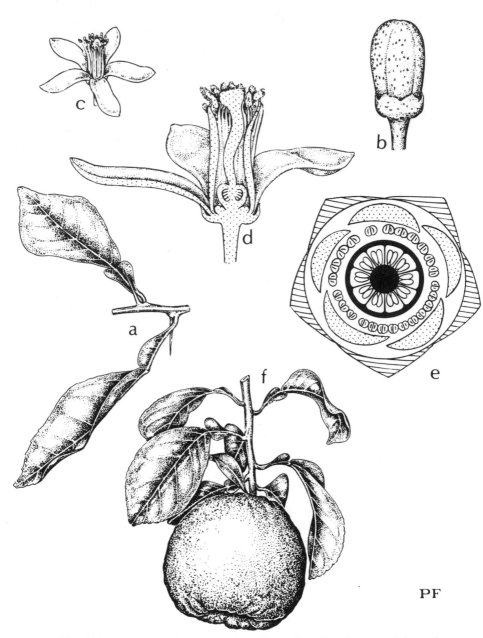

Fig. 306. **Citrus aurantium:** a, section of branch to show leaves and thorn; b, flower bud; c, flower, front-side view; d, flower longitudinally dissected; e, floral diagram; f, branch with fruit. (Courtesy of Fairchild Tropical Garden.)

instance of aberrant development results in the central leaflet having a deep division on one or both sides near the base.) Leaves borne relatively distant from each other on vigorous long shoots of the season; in addition, very short spur-shoots bearing closely set leaves may develop from buds in the axils of thorns on twigs of the previous season.

Conspicuous sessile flowers about 3–6 cm across the corollas occur singly or in pairs from buds axillary to thorns on the previous season's growth, usually in advance of the appearance of new leaves. Flowers bisexual, radially symmetrical. Sepals 4–7, small, obovate, with somewhat crinkled membranous margins, apices blunt to broadly rounded. Petals 4–7, white, somewhat clawed basally, dilated distally to spatulate or obovate, usually with upturned margins. Stamens numerous, filaments unequal in length. Pistil 1, ovary superior, subglobular, usually 7-locular, surface densely short-pubescent; style short and stout, stigma capitate. Fruit orangelike, 4–5 cm in diameter, dull yellow and fragrant at maturity, densely short-pubescent.

Native of China, cultivated for ornament, for hedges, and used as stock on which to graft commercial citrus. Sporadically naturalized in our area and in other parts of the s.e. U.S., chiefly coastal plain, in woodlands, woodland borders, along fences and in hedge rows.

Recognition features for this plant include stout, sharp, basally flattened axillary thorns on bright green stems and twigs, mostly trifoliolate leaves, their petioles wholly winged or winged on their distal halves.

2. Citrus

Citrus aurantium L. SOUR ORANGE, SEVILLE ORANGE. Figs. 305, 306

Evergreen, glabrous shrub or small tree with green twigs some of which bear green, terete, sharp, axillary thorns.

Leaves alternate, simple, petiolate, petioles 1–3 cm long, some on a given plant sometimes unwinged, usually broadly winged distally; blades punctate, elliptic to ovate, 6–12 cm long and 3–6 (10) cm broad, bases broadly and shortly tapered, apices acute or short-acuminate, tips of acuminations often blunt, margins entire but sometimes irregularly slightly wavy.

Flowers axillary, solitary or in small clusters, bisexual, radially symmetrical, sweet-scented. Calyx short-cupulate, with 5 small, low-triangular-obtuse lobes. Petals 4–8, spreading, white, linear-oblong. Stamens numerous, filaments nearly equal in length. Pistil 1, ovary superior, subglobose, 10–12-locular, style stout, stigma terminal. Fruit orangelike, glabrous, 7–9 cm in diameter, with a hollow core when fully ripe, rind thick and roughened on the surface, pulp acidic and bitter.

Native of s.e. Asia; used as a stock for sweet orange; occasional specimens naturalized in various parts of our range, in woodlands of various mixtures, not exclusively but perhaps especially on coastal shell mounds.

3. Ptelea

Ptelea trifoliata L. WAFER-ASH, STINKING-ASH, COMMON HOP-TREE, SKUNK-BUSH.
Fig. 307

A deciduous shrub or small tree, the bark bitter, the crushed foliage with a disagreeable, musky or somewhat fetid odor. Twigs of the season terete, soon becoming woody, the bark light brown to dark reddish brown, pustular-dotted,

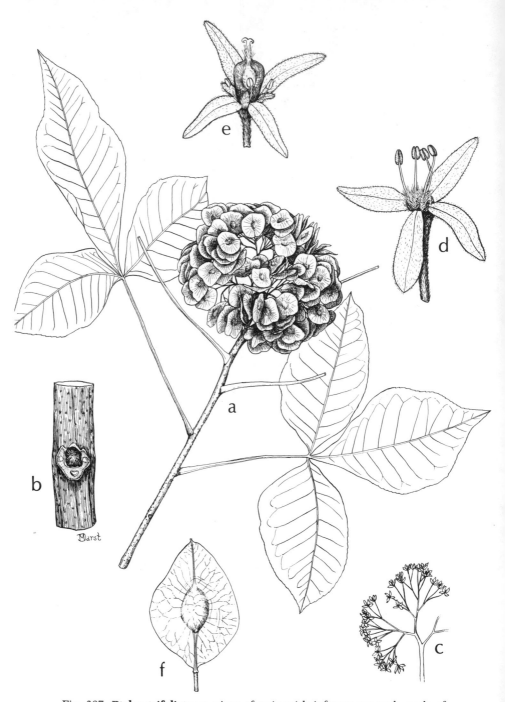

Fig. 307. **Ptelea trifoliata:** a, piece of twig with infructescence; b, node of woody twig with leaf scar and bud; c, diagrammatic view of portion of inflorescence; d, staminate flower; e, functionally pistillate flower; f, fruit.

glabrous or pubescent; leaf scars obcordate to shield-shaped, nearly enclosing the tomentose, depressed, axillary buds, vascular bundle scars 3.

Leaves alternate, long-petiolate, compound, with 3, rarely 5, leaflets. Leaflet blades pinnately veined, variable in size, shape, amount of pubescence if any, size and relative conspicuousness of glandular dots on the surfaces, and character of the margins; lateral pair of leaflets essentially sessile, usually inequilateral, ovate to lanceolate, margins entire to irregularly crenate-serrate or irregularly vaguely wavy; terminal leaflet cuneately narrowed to a sessile base or to a short, winged petiole, usually more nearly equilateral, elliptic, oval, or obovate; apices very variable, acute to broadly obtuse or rounded, sometimes acuminate; upper surfaces glabrous or sparsely pubescent, often only along the midrib or along that and the principal lateral veins, lower surfaces glabrous, sparsely pubescent, or often evenly pubescent with soft hairs and soft to the touch; in color the lower surfaces varyingly paler than the upper, on some plants almost chalky white.

Inflorescence a panicle subtended by the uppermost leaf on a shoot of the season, the subtending leaf and others below it usually extending well beyond the inflorescence. Flowers radially symmetrical, functionally unisexual or bisexual (plants bearing functionally unisexual and bisexual ones on the same plant probably less frequent than those bearing only functionally staminate or pistillate ones); panicle axes, flower stalks, calyces, and corollas varying from glabrous to densely pubescent. Sepals 4 or 5, minute. Petals 4 or 5, spreading, greatly exceeding the sepals, 4–6 mm long, narrowly oblong, elliptic-oblong, or oblanceolate, greenish white. Stamens usually 4 or 5, alternating with the petals, the functionally pistillate flowers with imperfect sterile anthers. Ovary superior, inserted on a low disc in bisexual and functionally pistillate flowers, raised on a conspicuous disc in the functionally staminate ones; style short, stigmas capitate, usually 2-lobed.

Fruit a flat, suborbicular, glandular-punctate samara 1.5–2 cm across, the two conspicuous wings united around the indehiscent 1–2-seeded body.

Rich woodland slopes, bluffs, and calcareous woodlands in scattered localities of our area. (Conn. to s. Ont., Mich., Iowa, and Neb., southward to cen. pen. Fla. and Tex.; Mex.)

P. trifoliata, broadly conceived, has a wide geographic range; the description above is meant to be limited to plants as I perceive them in our local area. For the range as a whole, segregate taxa have been recognized on the specific, subspecific, and varietal levels. I am unable, with relatively cursory study, to delimit segregate taxa locally.

The 3- (rarely 5-) foliate leaves which, when crushed, have a decidedly musky odor, and the waferlike, winged fruits aid in identification of this plant.

4. Zanthoxylum

Deciduous (ours) shrubs or trees, partially to wholly armed with prickles. Leaf scars shield-shaped to obcordate or crescent-shaped and with 3 vascular bundle scars, the axillary bud above low-knobby and seated within an elevated rim; pith white, continous. Leaves alternate, in ours odd-pinnately once-compound, petiolate; leaflets pinnately veined, the lateral ones opposite, subsessile or shortly stalked; leaflet margins with very low, longish crenulations, a sessile gland in each of the sinuses between the crenulations, or glands distributed along the edge of the blade if the blade is essentially entire; crushed leaves aromatic, the odor citruslike.

611

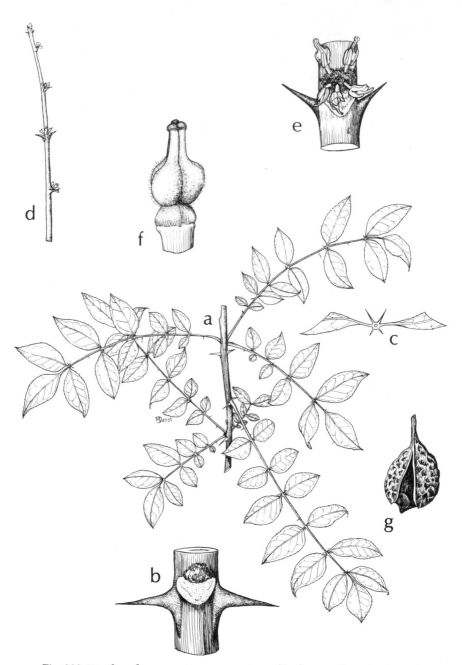

Fig. 308. **Zanthoxylum americanum:** a, piece of leafy stem; b, node of woody twig with prickles, leaf scar, and axillary bud; c, node of compound leaf showing prickles; d, piece of woody twig with clusters of shortly-stalked pistillate flowers; e, node from preceding with cluster of flowers, enlarged; f, pistil; g, fruit, dehisced.

Flowers unisexual or bisexual (plants dioecious or polygamous), radially symmetrical. Sepals 4 or 5, minute, or none. Petals 4 or 5. In the functionally staminate flowers, stamens 4 or 5, alternate with the petals, pistil rudimentary, a nectariferous intrastaminal disc present. Functionally pistillate flowers with 1–5 pistils, rudimentary stamens sometimes present, ovary 1-carpellate and 1-locular, styles lateral or sublateral, stigmas capitate; rarely the pistil with carpels united, then 2–5-carpellate and -locular. Fruit a 2-valved, usually 1-seeded follicle, its surface glandular-punctate. Seeds obovoid to subglobular, sometimes more or less lenticular, their surfaces more or less wrinkled-reticulate, lustrous black.

1. Plant a shrub (rarely with form and stature of a very slender, very small tree) usually forming clones by subterranean runners; prickles in pairs at the nodes of the stem, few if any on the leaves; flowers or fruits solitary or in axillary umbellate clusters on wood of the previous season before new shoot growth commences; leaflets membranous, equilateral or nearly so, their apices mostly blunt, dull green above, usually sparsely pubescent on the veins beneath.   1. *Z. americanum*
1. Plant potentially a tree (to at least 10 m tall), not clonal; prickles, where present on the stems, mostly scattered internodally, on many plants also on petioles and axes of the leaves; flowers or fruits borne in panicles at the ends of twigs of the season; leaflets firm, mostly markedly inequilateral and more or less falcate, their apices acuminate or acute, shiny green above, both surfaces glabrous.   2. *Z. clava-herculis*

1. Zanthoxylum americanum Mill. NORTHERN PRICKLY-ASH, TOOTHACHE-TREE.
Fig. 308

Commonly a clonal shrub, infrequently of slender, small-tree stature. Stem armed with pairs of short, broadly flat-based prickles at the nodes, the prickles sloughed on older, larger stems.

Leaves with petioles 3–6 cm long, petioles, leaf axes, leaf stalks, and lower leaflet surfaces pubescent, often some of the pubescence stipitate-glandular; upper leaf surfaces sparsely pubescent at least when young; surfaces of leaflets, especially the lower, more or less glandular-punctate; usually few, if any, prickles on the petioles or leaf axes. Leaflets 5–11 per leaf, commonly 5–7, blades membranous, ovate, lance-ovate, or oval, 1.5–6 cm long, only slightly inequilateral, bases mostly rounded, apices blunt, sometimes bluntly acuminate; margins with pale, sessile glands in the weak sinuses of the crenations or, if margin entire, the glands along the edge of the blade, in either case usually visible only from above; lateral leaflets subsessile, the terminal one usually distinctly petiolate.

Inflorescences axillary clusters on wood of the previous season. Sepals none. Petals sepallike, 4–5, oblongish, fringed apically. Pistillate flowers with 2–5 pistils, ovary slightly oblique, styles approximately terminal but usually bent laterally. Seed about 5 mm long.

In our area *Z. americanum* (pistillate plants only), known to me to occur only in two localities, one each in Jackson and Gadsden Cos., Fla. (Que. to Minn. and S.D., southward to Ga., cen. Fla. Panhandle, cen. Ala., Okla.)

2. Zanthoxylum clava-herculis L. SOUTHERN PRICKLY-ASH, HERCULES-CLUB, TICKLE-TONGUE.

Fig. 309

A shrub or tree to at least 10 m tall, the latter usually with a short trunk and rounded crown, armed with short, sharp, broadly flat-based prickles, these always present on the lower stem, present or absent on larger branches, twigs,

Fig. 309. **Zanthoxylum clava-herculis:** a, tip of branch with terminal inflorescence (several leaves removed); b, flowers, staminate, to right, pistillate to left; c, node of woody twig; d, piece of bark from trunk showing how, eventually, prickles are elevated on more or less conical corky mounds; e, fruit, dehisced; f, node of compound leaf.

and on petioles and axes of the leaves; prickles usually numerous on the lower trunks and eventually becoming elevated on broadly conical, domelike, or pyramidal corky excrescences, some, at least, of the prickles themselves eventually sloughed from the excrescences; on a given plant, prickles often absent from some branches, twigs, or leaves, if present, sometimes evenly, sometimes irregularly distributed, but not in pairs at nodes. Young twigs minutely pustular-punctate, glabrous; year-old twigs with brown bark that splits irregularly into longitudinal lines.

Leaves with petioles mostly 4–5 cm long, green or varying to purple throughout as are the leaf axes, both glandular-punctate; leaflets 5–19, commonly 7–9, per leaf, their stalks short, somewhat fleshily and irregularly knobby-winged; blades firm, lanceolate to ovate, mostly markedly inequilateral and falcate, 2.5–7 cm long, dark green and lustrous above, paler and duller beneath, glabrous, margins flat or somewhat crinkled, with low and longish crenations, conspicuous yellowish, buttonlike glands in the notches formed by the teeth, irregularly glandular-punctate on the surfaces, bases rounded, obliquely rounded, or strongly oblique, apices acute, long-acute, or those of the basal leaflets sometimes obtuse.

Inflorescences diffuse panicled cymes at the ends of branchlets of the season, the uppermost leaves greatly exceeding the inflorescences. Sepals 4 or 5 or none, minute. Petals 4 or 5, greenish yellow, often unequal, 3–4 mm long, oblong-ovate or ovate. Pistillate flowers with 1–5 pistils, ovary oblique, style lateral. Follicles 5–6 mm across, barely stipitate. Seeds 5–6 mm long.

Sand dunes, shell middens, sandy hammocks, along fences and in hedge rows, in widely scattered localities in the area of our coverage although often common locally. (Coastal plain, s.e. Va. to s. pen. Fla., westward to e., s.e. and n. cen. Tex., s. Ark., Okla.)

Distinctive features aiding in recognition of this plant are the 1-pinnate leaves whose leaflets are mostly markedly inequilateral and falcate, crenate marginally, and with buttonlike glands in the notches formed by the teeth; larger lower trunks have prickles elevated on broadly conical, domelike, or pyramidal corky excrescences; prickles may be present or absent on branches, petioles, and leaf axes.

National champion Hercules-club (1961), 7'6" [at 2'] circumf., 38' in height, 59' spread, Little Rock, Ark.

Salicaceae (WILLOW FAMILY)

1. Leaf blades broadly ovate, triangular-ovate, or cordate-ovate, nearly as broad as long to broader than long; buds covered by several imbricated scales.   1. *Populus*
1. Leaf blades lanceolate to oval, oblong, or ovate-oblong, larger ones much longer than broad; buds covered by a single scale.   2. *Salix*

1. Populus (POPLARS, COTTONWOODS)

Fast-growing trees with bitter-astringent bark and light wood. Buds with several imbricated, usually gummy scales. Leaf scars prominent, with 3 vascular bundle scars. Leaves simple, long-petiolate, alternate, deciduous; blades (in ours) often as broad as long or broader than long, if longer then not usually exceeding 2 times as long as broad, pinnately veined. Stipules present but quickly deciduous. Flowering occurring prior to new shoot emergence in spring. Flowers individually small, borne in pendulous catkins, unisexual, the plants dioecious, perianth none. Staminate flowers with 4–12 (60) stamens borne on a flat to shallowly cuplike disc, each flower subtended by a bract. Pistillate flowers with 1 pistil, ovary sessile within a disc, 1-locular, with 2–4 parietal placentae, style single or divided distally, stigmas same in number as the placentae. Fruit an ovate 2–4-valved capsule. Seeds abundant, minute, with a basal tuft of long, silky hairs that extend upward surrounding the seeds.

1. Leaf blades acuminate apically, the tips distinctly pointed; petiole eventually flattened and somewhat dilated distally, 2–5 glands at the very base of the blade or on the summit of the petiole, these often becoming erect and somewhat foliar; petioles and blades glabrous or nearly so. 1. *P. deltoides*
1. Leaf blades bluntly obtuse apically; petioles essentially terete throughout and not dilated distally, glands, if any, on the edges of the blade at either side of the summit of the petiole; petioles, young leaf blades, and twigs cottony-pubescent and grayish, most of the pubescence sloughing during maturation usually leaving a patch persisting around the veins at the base. 2. *P. heterophylla*

1. **Populus deltoides** Bartr. ex Marsh. var. deltoides. EASTERN COTTONWOOD.
Fig. 310

Tree to 30 m tall or more, bark of trunk grayish, roughly and irregularly fissured and ridged. Twigs yellowish green, glabrous, lustrous, with low, narrow ridges running longitudinally, older branches much roughened by raised leaf scars.

Leaf blades, many of them, copper-colored when first unfolding, quickly turning lustrous green, flaccid, very sparsely pubescent. Blades at maturity leathery and stiffish, prominently pinnately veined, mostly 5–15 cm long (longer on sprouts and saplings), deltoid, deltoid-ovate, to suborbicular-ovate, mostly truncate basally, sometimes broadly cordate, infrequently with small, cordate lobing near the summit of the petiole, acuminate apically, the tips distinctly pointed; margins crenate-dentate or crenate-serrate, surfaces glabrous or less frequently with sparse, short pubescence near the base, on the lower margins, or distally on the petioles; petioles flattened and usually dilated distally, the flattening at right angles to the surface of the blade, commonly a pair, sometimes 3, 4, or 5, often but not always somewhat foliose glands at the base of the blade or some or all of the glands on the petiole just below the blade.

Staminate catkins variable in length, to as much as 15 cm fully expanded, the flowers congested and red before anthesis, becoming loosely spreading and yellow by full anthesis, their slender stalks 5–8 mm long; each flower subtended by a broad, fimbriate-margined bract, the bracts falling before or by full anthesis; flower composed of numerous to many stamens borne over the upper surface of a flattish, asymmetric disc, the filaments finely threadlike. Pistillate flowers somewhat crowded on the catkins at first, rather quickly becoming loosely spreading-ascending from the axis, the stoutish flower stalks about 5 mm long; pistil seated in a shallow cuplike disc with a slightly undulating rim, the ovary ovoid, capped by a sessile, fleshy, spreading, prominently lobed and crinkled stigma. Staminate catkins shrivel and drop soon after the pollen is shed; fruiting catkins maturing after the leaves are fully developed. Capsules ovoid, 6–8 mm long, dehiscing by 2–3 recurved valves.

Lowland woodlands and wet disturbed places, spottily distributed; Fla. Panhandle (and n.w. pen. Fla.), s. cen. and s.w. Ga., s. Ala. (Ranging widely to the n.e., n., and w. of our area).

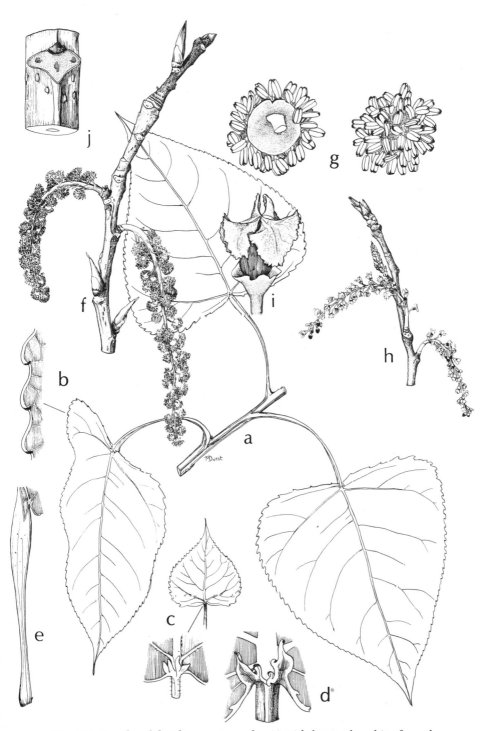

Fig. 310. **Populus deltoides:** a, piece of twig with leaves; b, a bit of much enlarged margin of leaf; c, leaf blade with enlargement of base to show glands (when leaf is young); d, enlargement of base of mature blade to show nature of enlarged and dessicated area of glandularness; e, enlargement of petiole to show dilation of its distal portion; f, leafless twig with staminate catkins; g, staminate flower, two views; h, leafless twig with pistillate flowers; i, pistillate flower; j, node with leaf scar and bud.

2. Populus heterophylla L. SWAMP COTTONWOOD. Fig. 311

Tree to about 30 m tall. Year-old twigs stout, reddish brown, glabrous, longitudinally irregularly fluted; leaf scars 3-lobed. Bark of old trunks thick, dirty brown tinged with red, irregularly ridged and furrowed, breaking into long, narrow plates.

Emerging shoots, petioles, and leaf surfaces felty with flocs or tufts of shaggy grayish or tawny-grayish hairs giving the shoots and their young leaves a decidedly grayish cast seen from a distance; much of the pubescence of upper leaf surfaces quickly sloughed, that of the lower, the petioles, and shoots sloughing more gradually, eventual exposed surface of young twigs dark reddish brown. Petioles not dilated distally; mature leaf blades thickish and stiffish, 10–20 cm long, ovate, from about as broad as long to twice as long as broad, bases truncate or more commonly broadly rounded on either side, often with cordate lobing just next to the petiole and the lobes overlapping the summit of the petiole, apices bluntly obtuse to rounded, margins shallowly crenate or crenate-serrate; upper surfaces becoming dark green and glabrous or nearly so save for a patch of pubescence around the veins at the base, lower surfaces grayish green, sparsely and evenly pubescent to glabrous; edges of blades close to the petiole usually glandular.

Inflorescences essentially as in *P. deltoides*. Discs bearing the numerous to many stamens in the staminate flowers varying from entire-margined to distinctly lacerate. Pistillate flowers loosely spreading-ascending from the axis of the catkin, their stalks slender, 0.5–1.5 cm long, the disc united only basally and with several erect, oblong lobes surrounding the base of the ovary; style slender, with 3 distal branches each bearing a 2- or 3-lobed, spreading, somewhat fleshy stigma. Capsule ovoid, 11 mm long or a little more, dehiscing by 2–3 recurved valves, maturing well after the shoots of the season and their leaves are fully developed.

Swamps along rivers and smaller streams; in our area, Panhandle of Fla. from Leon Co. westward, s. Ala. (Spottily distributed, coastal plain, s.e. Conn. to s.e. S.C. and just into e. Ga., Fla. Panhandle to s.e. and n.e. La., northward in the interior to s. Ill., s. Mich., Ohio.)

2. Salix (WILLOWS)

Trees or shrubs, often with bitter, aromatic, astringent bark. Buds enclosed by a single, usually nonresinous scale fused into a cap or with free overlapping margins; terminal bud none. Leaf scars essentially linear and horizontal, with 3 vascular bundle scars. Uppermost bud on a twig usually aborting, often few to several buds below it aborting and that portion of the twig aborting. In some kinds, in spring, some buds produce only catkins before others produce leafy extension shoots. In other kinds first buds to open produce short-shoots with small or bracteal leaves and catkins terminally; often, not always, the uppermost leaf or bracteal leaf of these short-shoots produces a bud that develops into an extension shoot of the season and if so the catkins would seem to be lateral at that stage (if the catkin-bearing short-shoot does not produce an extension shoot, it usually deteriorates after the catkin falls or the branchlet falls with the catkin); other buds may produce only extension shoots. Leaves proximally on the extension shoots generally are considerably smaller than the more numerous distal ones and it is the latter which are usually described as characteristic for the species. Leaves simple, alternate, deciduous, short-petiolate,

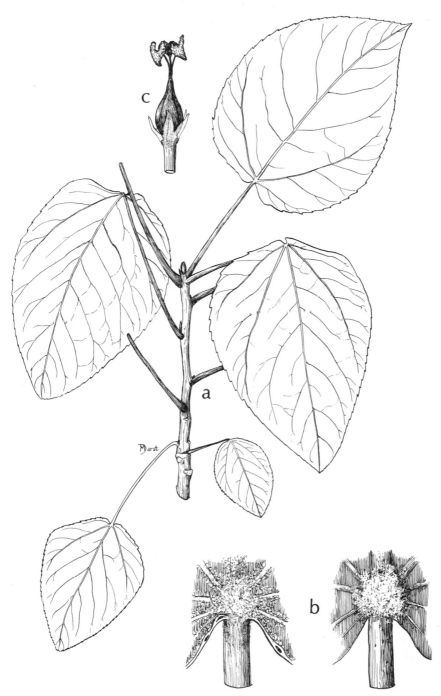

Fig. 311. **Populus heterophylla:** a, branchlet (some leaf blades removed); b, lower surface at left, upper surface at right, of basal portion of mature leaf to show pubescence; c, pistillate flower.

with or without stipules, blades pinnately veined, the larger ones several times longer than broad. Flowers individually small, unisexual (plants dioecious), without perianths, each subtended by a bract. Staminate flowers consisting of (1) 2 or 3–12 stamens, their filaments distinct or partially united, sometimes of varying lengths, 1 or more nectariferous glands basally. Pistillate flowers consisting of a single bicarpellate, 1-locular, stipitate (ours) pistil, 1–2 nectariferous glands basally; stigmas with 2 recurved lobes. Fruit a 2-valved capsule, the valves recurved-gaping at dehiscence. Seeds few to numerous, minute, bearing a coma of silky hairs basally that extend around and beyond the seed body.

1. Mature leaf blades persistently and densely gray-tomentose beneath.   1. *S. humilis*
1. Mature leaf blades glabrous beneath, or, if pubescent, the pubescence relatively sparse, definitely not densely and compactly covering the surfaces.
  2. Larger, mature leaf blades elliptic or oblong-elliptic, 8–15 cm long and 3–5 cm broad, bases rounded, apices moderately but not attenuately tapered, gray-glaucous beneath and with relatively prominent brownish midrib and lateral veins, the main veins relatively heavily pubescent, pubescence otherwise shaggy and more sparse but evenly distributed, soft to the touch.   2. *S. floridana*
  2. Larger, mature leaf blades narrowly to broadly lanceolate, bases cuneate to rounded, long-tapered distally, gray-glaucous or green beneath, *or* lance-oblong and mostly with very small auriculate lobing at either side basally, short tapered distally.
    3. Mature leaf blades on extension shoots lance-oblong, bases mostly with very small auricles on either side, apices acute to subacuminate; all leaves with prominent stipules.   3. *S. eriocephala*
    3. Mature leaf blades on extension shoots narrowly to broadly lanceolate, bases tapered to rounded, attenuately tapered distally; stipules commonly absent.
      4. Blades of mature leaves of extension shoots gray-glaucous beneath, often with a bluish hue; stipes of mature capsules mostly 4–5 mm long, diameter of mature fruiting catkins 15–17 mm.   4. *S. caroliniana*
      4. Blades of mature leaves of extension shoots green beneath; stipes of mature capsules about 1.5 mm long, diameter of mature fruiting catkins about 1 cm. 5. *S. nigra*

1. Salix humilis Marsh. SMALL PUSSY WILLOW, PRAIRIE WILLOW.   Fig. 312

Deciduous, slender shrub to about 3 m tall, often forming clones by layering. Branchlets not brittle at the joints. Buds reddish brown, oblongish, about 3 mm long, humped abaxially, blunt apically, sparsely short-pubescent. Young developing stems and petioles usually tomentose with gray stellate hairs, twigs as they age and become woody reddish brown and patchily ashy-pubescent, larger older woody stems grayish brown, relatively smooth. Upper surfaces of unfurling leaf blades green, with sparse, gray, short-stellate pubescence, margins revolute, lower surfaces densely pale gray–tomentose.

Mature leaves acutely, rarely obtusely, narrowed basally to short petioles, blades of vigorous, young shoots prevailingly oblanceolate or spatulate (ours), 4–10 cm long and 1.5–2 cm broad, apices acute to rounded, dark green above and essentially glabrous, somewhat revolute, lower surfaces densely gray-tomentose, margins obscurely undulate to obscurely toothed, undulations or teeth often gland-tipped, stipules oblique, broader than long to narrow and longer than broad; blades of diffuse branches generally averaging notably shorter and narrower (often all or almost all on a given plant thus), prevailingly narrowly oblanceolate but some narrowly oblong or elliptic, mostly 1.5–4 cm long and 0.5–1 cm broad, rarely with undulate margins, with or without stipules.

Catkins with 2–3 bracts basally, borne on wood of the previous season in early

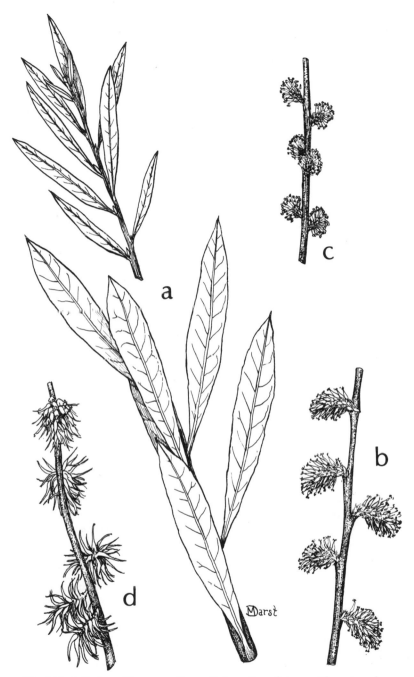

Fig. 312. **Salix humilis:** a, small, medial portion of stem with mature leaves, below, and small tipmost portion with partially developed leaves, above; b, piece of stem with staminate catkins at anthesis; c, piece of stem with pistillate catkins at anthesis; d, piece of stem with approximately mature pistillate catkins.

spring before any vegetative growth of the season commences. Staminate catkins sessile, 0.5–1 cm long, subtending bracts densely silvery-silky-pubescent as are the bracts subtending flowers, stamens 2 per flower, nectary 1, adaxial. Pistillate catkins, fully developed, 1–3.5 cm long or a little more, short-stalked, stalks and axes bearing long, silky hairs, subtending bracts and bracts subtending the flowers as in the staminate; pistils green, flask-shaped, surfaces with short, grayish pubescence, stipes 0.3–0.5 mm long, with longish, spreading, silky hairs, stigmas 2–4, recurved, nectary 1, adaxial; capsules 7–12 mm long, becoming tawny as they ripen.

In our area, in somewhat boggy openings, in wet pine flatwoods, low, seasonally wet areas in pastures, borders of low woodlands, ditches, highway, railway, and powerline rights-of-way where seasonally wet; local, s. cen. and s.w. Ga., in Fla. from about the Suwannee River to the cen. Panhandle. (Wide ranging northward and westward in N.Am.)

2. Salix floridana Chapm. FLORIDA WILLOW.   Fig. 313

Plant usually a few-branched shrub to about 4 m tall,* the branchlets notably brittle at the joints. Buds reddish brown, short-pubescent to glabrous, about 5 mm long, bluntly pointed apically, bud scale with free overlapping margins. Young developing stems and petioles copiously to moderately short-pubescent, much of the pubescence quickly sloughing, the stem soon becoming dark reddish brown and with minute, rather sparse pubescence. Unfurling leaf blades with upper surfaces copiously short-pubescent on their midribs, sparsely so elsewhere, most of the pubescence quickly sloughing, lower surfaces densely short-gray-pubescent throughout but the pubescence usually becoming relatively sparse. Rapidly growing shoots sometimes with conspicuously stipulate leaves, the stipules half-round to oblique and broader than long, glandular along their margins, often, however, no stipules present.

Mature leaves with definite petioles 1–2 cm long. Blades, the larger ones, 8–16 cm long and 3–5 cm broad, broadly lanceolate, elliptic, or oblong; bases predominantly rounded, occasionally subcordate, apices acute to obtuse; margins glandular-serrate, glandular-denticulate, or if not toothed then having small, sessile glands protruding from minute veinlets ending at the margins; upper surfaces dark green, short-pubescent along the midribs, more sparsely so along the lateral veins, and with few scattered hairs between them, lower surfaces having prominent, brownish veins, otherwise gray-glaucous, midrib more heavily clothed with hairs than the lateral veins, islets formed by the anastomosing veinlets sparsely hairy; lower surfaces usually moderately soft to the touch.

Catkins borne singly terminating very short, small-leaved, lateral branchlets early in the season. Staminate catkins short-stalked, 3–8 cm long, usually 1–3, green, membranous, oblongish bracts about 2–2.5 mm long borne distally on the stalks and a similar roundish to spatulate bract subtending each flower, abaxial surfaces of bracts and axis of the catkin more or less woolly-pubescent; stamens 3–7 per flower, filaments woolly-pubescent proximally; nectaries 2, adaxial and abaxial. Pistillate catkins loosely flowered, 3–8 cm long, short-stalked, usually with a subtending, early-deciduous bract; bracts subtending

---

*Parenthetical statement in Godfrey and Wooten (1981), "trees known to us about 20 m tall and 35 cm d.b.h.," has proved to be erroneous; specimens were misidentifications of particularly large-leaved *S. caroliniana.*

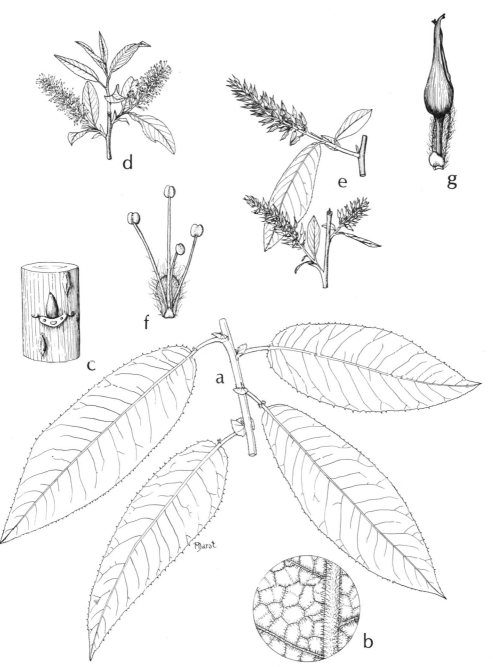

Fig. 313. **Salix floridana**: a, small portion of stem with mature leaves; b, enlargement of a small area of lower surface of leaf; c, node of twig with leaf scar and axillary bud; d, branchlets with staminate catkins; e, branchlets with pistillate catkins; f, staminate flower; g, pistillate flower.

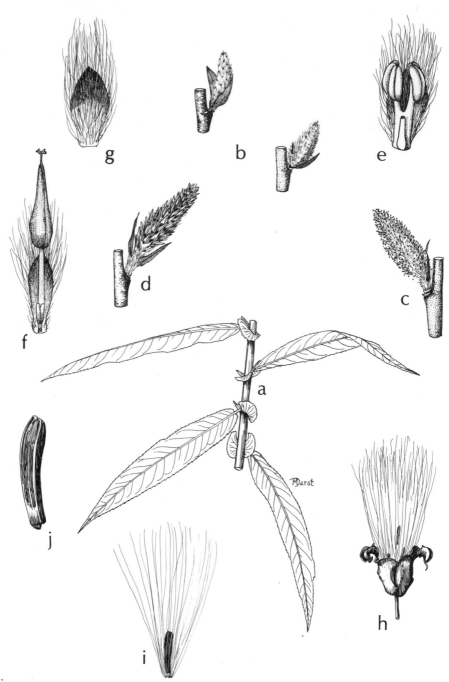

Fig. 314. **Salix eriocephala:** a, small portion of stem with mature leaves; b, nodes bearing emerging catkins; c, staminate catkin at full anthesis; d, pistillate catkin at full anthesis; e, staminate flower; f, pistillate flower; g, abaxial view of bract that subtends pistillate flower; h, dehisced capsule; i, seed; j, seed, enlarged, basal hairs removed.

the flowers persistent, abaxially woolly-pubescent as is the axis of the catkin; pistils green and glabrous, obconic in outline, their stipes 3.2–5.6 mm long, style short, terminated by a minute, 2–lobed stigma; nectary 1, adaxial; capsules mostly 6–7 mm long, yellowish brown, usually 4 seeds per capsule; diameter of mature fruiting catkins 2–2.3 cm. [Incl. *Salix chapmanii* Small]

An understory shrub inhabiting very wet, swampy woodlands; local, known from Pulaski and Early Cos., Ga., in Fla., Jackson Co. (where apparently it has not been seen since Chapman's original collection), in Jefferson Co., and in several localities in n.e. and n. cen. Fla.

3. **Salix eriocephala** Michx.   Fig. 314

Plant a shrub to about 6 m tall. Buds about 8 mm long, yellowish brown or brown, pubescent, plumply humped abaxially, somewhat flattened adaxially, abruptly tapered to a blunt, sometimes retuse tip, bud scale caplike (margins fused). Stems pubescent at first, woody twigs becoming dark reddish brown, eventually glabrous. Unfurling leaf blades glabrous (on the few plants I have seen in the field).

Mature leaves with petioles 5–8 mm long; stipules conspicuous and persistent, for the most part broader than long, 8–15 mm broad, prevailingly oblique, half-ovate or -cordate, apices obtuse to broadly rounded, margins finely glandular serrate or -dentate. Leaf blades, the larger ones, mostly 8–10 cm long and 2–3 cm broad, rounded or slightly cordate basally (small ones sometimes cuneate), proximal one-half to two-thirds oblongish, distal portion tapering to acute or subacuminate apices; upper surfaces dark green and glabrous, lower surfaces grayish-glaucous; margins finely glandular-serrate.

Catkins borne laterally on very short shoots with 2–3 tiny bractlike leaves below them. Staminate catkins, as they emerge from the buds, densely gray-woolly (pussy willow–like), eventually, when the flowers are at full anthesis, 3–4 cm long and 1–1.3 cm across, the flowers so numerous and closely aggregated that the axis is not visible, the bracts scarcely so; bracts dark brown distally, pale proximally, oblongish, about 1.5 mm long, bearing silky hairs abaxially, these extending well beyond their tips; stamens 2 per flower, filaments slender, glabrous; nectary 1, adaxial. Pistillate catkins 2.5–4.5 cm long, flowers densely aggregated and obscuring the axis, bracts as in the staminate catkins; pistils green and glabrous, flasklike, their stipes 1–1.5 mm long, very slender, obscured because of the density of the pistils; style slender 0.3–0.6 mm long, stigmatic lobes 2, minute; nectary 1, adaxial; capsules yellowish brown, 3.5–4 (7) mm long, seeds several. [*Salix cordata* Muhl.; *S. rigida* Muhl.]

Alluvial woodlands, stream banks, marshy areas in fields; known to me only from three localities in our area of coverage, Lowndes Co., Ga., and Gadsden and Jackson Cos., Fla.; perhaps being overlooked or misidentified. G. W. Argus (pers. comm.) informs me that it occurs (but has rarely been recognized) in much of Ala. and in scattered localities in w. Ga. and that that part of its distribution is disjunct from its main widespread occurrence to the north.

4. **Salix caroliniana** Michx. CAROLINA WILLOW.   Fig. 315

Tree to about 10 m tall, flowering/fruiting when of small stature. Buds small, reddish brown, usually short-pubescent with grayish hairs, less frequently glabrous, sometimes falcate, bluntly pointed, bud scale with free overlapping edges. Stems of young developing branchlets variably rather densely short-

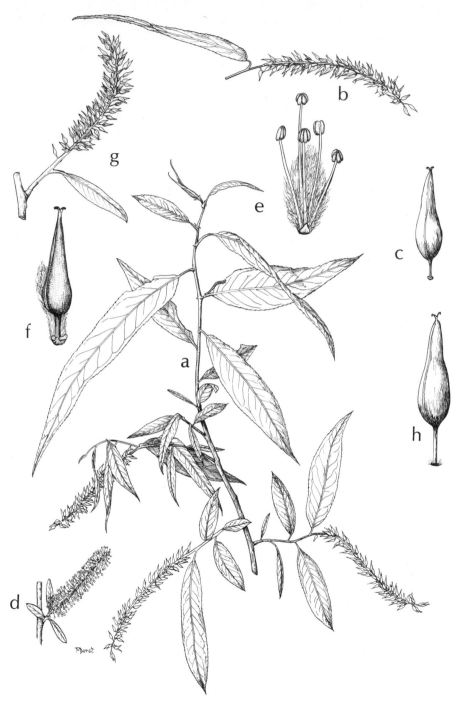

Fig. 315. a–c, **Salix nigra:** a, branch with pistillate catkins; b, fruiting catkin; c, fruit. d–h, **Salix caroliniana:** d, staminate catkin; e, staminate flower; f, pistillate flower; g, fruiting catkin; h, fruit. (b and g drawn to same scale and c and h drawn to same scale)

pubescent, pubescent with soft spreading hairs, or glabrous, if pubescent, the shriveled hairs persisting after the twigs become woody, even into the second year. Leaf blades when first starting to unfurl commonly pubescent on both surfaces, much of the pubescence, sometimes all of it, rather quickly sloughing as the blades expand. Woody twigs dull reddish brown, grayish, or a rather bright light brown. Bark of trunks grayish brown, divided into broad ridges whose surfaces are scaly.

Mature leaves of extension shoots with petioles (3) 4.5–14 (22) mm long; stipules, if present, sometimes prominent, sometimes inconspicuous, commonly oblique and broader than long but very variable in shape, margins usually irregularly glandular-toothed. Leaf blades showing much variation in average size from plant to plant, varying from linear-lanceolate, about 8 cm long and 0.8 cm broad, to broadly lanceolate, about 20 cm long and 3.5 cm broad near the base; bases of narrow blades slightly tapered, others predominantly rounded, upwardly predominantly long-tapered to acute tips, less frequently more abruptly short-tapered to acute tips; upper surfaces glossy green, glabrous throughout or pubescent along the midribs, lower surfaces glaucous, dull grayish and often with a bluish hue, glabrous or pubescent along the midribs, rarely persistently pubescent on both surfaces; margins very finely glandular-toothed.

Catkins borne terminally on small-leaved, lateral, short-shoots early in spring. Staminate catkins 2–9 cm long, 1–1.5 cm across at full anthesis, sometimes loosely flowered and the axis visible, sometimes compactly flowered and the axis obscured, axis gray-tomentose; bracts obovate, 1–3 mm long, usually sparsely to densely woolly-pubescent on both surfaces; stamens usually 6 per flower, less frequently 4, 5, or 7, filaments pubescent proximally; nectaries 1, 2, or several, one adaxial, the others abaxial or intermixed with stamens. Pistillate catkins loosely flowered, axis evident, sparsely or densely woolly-pubescent; bracts ovate, oblong, or obovate, 1–3 mm long, both surfaces copiously woolly-pubescent, quickly deciduous after anthesis; pistils flasklike, they and their stipes green and glabrous, stipes mostly about 2 mm long at anthesis, often elongating to 4–5 mm as the capsules mature; nectary 1, adaxial; capsules usually green when starting to dehisce, becoming tan as the seeds are shed; mature, unopened capsules 6–7 mm long, diameter of the catkins 15–17 mm when capsules are mature. [*Salix amphibia* Small; *S. longipes* Shuttlw. ex Anders.]

Swamps, marshes, river banks and bars, interdune and other swales, pond and lake shores, along ditches and canals, mucklands, commonly colonizing wet clearings, including wet areas at the edges of highway or railway rights-of-way after construction; throughout our area, often growing intermixed with *S. nigra*. (Md., D.C., westward to s. Ind., s. Ill., Mo. and e. Kan., generally southward to s. Fla. and s. cen. Tex.; Cuba.)

In our area, especially in the Panhandle of Florida, many plants having leaf blades thinly to conspicuously glaucous beneath are considered by me to be intermediate between *S. caroliniana* (as I know it in peninsular Florida where *S. nigra* does not occur) and *S. nigra*.

5. Salix nigra L. BLACK WILLOW. Fig. 315

Tree to about 20 m tall, flowering/fruiting when of small stature. Buds small, glabrous, reddish brown, often falcate, apices sharply pointed, bud scale with free overlapping margins. Stems of young branchlets generally pubescent at their growing tips where the leaves are just unfurling, often quickly becoming glabrous or nearly so. Petioles and surfaces of blades, as they commence to unfurl, especially the lower surfaces of the latter, thinly to densely pubescent, usually but not always much of the pubescence quickly sloughing. Woody twigs reddish to yellowish brown, glabrous or sometimes persistently short-pubescent. Bark of trunks brown, deeply divided into broad, flat, connecting ridges whose surfaces break into platelike scales, often shaggy on old trunks.

Mature leaves of extension shoots with petioles 2–10 mm long, glabrous or persistently pubescent; stipules, if present, small, usually oblique, glandular-toothed marginally. Leaf blades lanceolate, often falcate, mostly 5–18 cm long and 0.5–1.5 (2) cm broad, bases short-tapered to rounded, commonly narrowed distally to acute, acuminate, or finely attenuated tips; upper surfaces dark green, glabrous or less frequently sparsely pubescent, sometimes short-pubescent only along the midrib, lower surfaces green, not glaucous or only thinly glaucous, glabrous or sparsely pubescent, or short-pubescent only along the midrib; margins finely glandular-toothed.

Catkins borne terminally on small-leaved, lateral short-shoots in spring. Staminate catkins 1.5–7.5 cm long, 0.5–1.0 cm across at full anthesis, loosely flowered and the axis visible or densely flowered and the axis obscured, axis gray-pubescent with spreading hairs; bracts obovate or spatulate, 1–3 mm long, sparsely to densely pubescent on both surfaces or abaxial surfaces partially glabrous; stamens usually 6 per flower, infrequently 4 or 5, filaments pubescent proximally; nectaries usually 2, adaxial and abaxial, the latter usually 2–3-lobed. Pistillate catkins loosely flowered, axis evident, sparsely to densely woolly-pubescent; bracts mostly oblong, pubescent as in staminate catkins, very quickly deciduous; pistils flasklike, they and their stipes green and glabrous, stipes usually less than 1 mm long at anthesis, elongating to 1.5 mm as the capsules mature; nectary 1, adaxial; capsules green when starting to dehisce, becoming tan as the seeds are shed; mature, unopened capsules 3–4 mm long, diameter of the catkins about 10 mm.

Habitats as given for *S. caroliniana*; throughout our area but less frequent east of the Okefenokee Swamp and Suwannee River than west of them. (N.B. and Que., s. Ont. to s.e. Minn., generally southward to n. Fla. and cen. Tex.)

Note: See statement at end of account of *S. caroliniana*.

Sapindaceae (SOAPBERRY FAMILY)

Sapindus marginatus Willd. FLORIDA SOAPBERRY.   Fig. 316

Small tree with a rounded crown, bark of trunk pale brown or grayish brown, sloughing in small, loose plates; twigs of the season glabrous, green and dotted with longish, narrowly lenticular, buff-colored lenticels; older woody twigs brown, roughened by corky, somewhat raised brown lenticels; leaf scars more or less shield-shaped, tending to be 3-lobed, vascular bundle scars 3, often indistinct; axillary buds superposed (at least after the uppermost "bursts" through the bark); terminal buds none.

Leaves glabrous, alternate, deciduous, 1-pinnately compound; stipules none; petioles 5–12 cm long, overall length of petiole and axis about 15–32 cm; leaflets shortly stalked, pinnately veined, sometimes opposite, more frequently subopposite, in either case leaf usually evenly pinnate, sometimes more distinctly alternate in which case there is usually an obviously terminal leaflet, the leaf

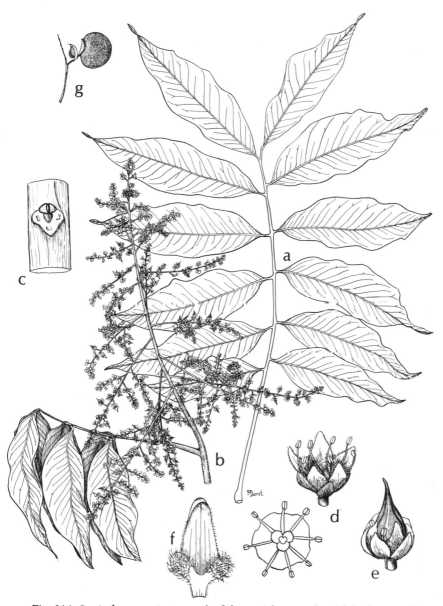

Fig. 316. **Sapindus marginatus:** a, leaf; b, panicle; c, node with leaf scar and superposed buds; d, staminate flower and, below left, diagrammatic sketch of disc and stamens; e, functionally pistillate flower; f, petal, enlarged; g, fruit.

then odd-pinnate, leaflet number varying from 6 or 7 to 12 or 13; leaves proximal on branchlets often much smaller than most of those higher on branchlets; leaflets mostly lanceolate, some lance-ovate, bases rounded to shortly tapered, sometimes oblique and blades falcate, apices usually acuminate and the tip of the acumination blunt; lowermost leaflets of a given leaf usually not much smaller than the others; in general leaflet size varies from about 5 to 15 cm long and 2–7 cm broad; margins of leaflets not toothed, surfaces dark green, the lower little if any paler and only slightly duller than the upper. Leaf rachis not winged.

Inflorescence an abundantly floriferous, thyrsoid panicle, roughly triangular in outline, borne above the uppermost leaf on the shoot of the season, panicles varying greatly in size; inflorescence axes, tiny bracts, and very short, stoutish flower stalks short-pubescent; stalks of the ultimate flower-bearing branches mostly 2–5 mm long and each bearing 1–several flower clusters. Flowers radially symmetrical, unisexual and staminate, or structurally bisexual and functionally pistillate (anthers not producing pollen). Perianth campanulate. Sepals 5, short, united only basally, the outer 2 smaller than the inner 3, more or less triangular. Petals 5, equal, longer than the sepals, shortly clawed basally, blades infolded, each bearing a copiously hairy, bifid scale proximally on its abaxial surface, a flattish nectariferous disc seated within the perianth. Stamens 8, borne on the disc, filaments broadest basally and tapering gradually distally, copiously pilose from the base to midway their length or a little more, erect in the staminate flowers and well exserted from the perianth, a yellow glandlike pistillode centrally on the disc; in the structurally bisexual flowers stamens little exserted, the anthers sterile, pistil 1, seated centrally on the disc, glabrous, ovary 3-lobed, 1-locular, style short, stigma 3-lobed. Both kinds of flowers in each cluster, the earliest to reach anthesis in each cluster staminate, later ones opening together over a relatively short period of time. Petals cream-yellow, anthers bright yellow, inflorescence appearing greenish yellow from a little distance. Two of the locules of the ovary usually abort, the resultant fruit a peculiar, lopsided drupelike structure with a globose fertile portion to one side and 2 dried-up ovary remains on the other, the whole hard and heavy, about 2 cm across.

Local, chiefly but not exclusively near-coastal, in woodlands, often where calcareous, wooded coastal shell mounds, cen. pen. Fla. northward and westward in Fla. Infrequently planted as an ornamental or shade tree and sometimes naturalized in the vicinity, thus may be expected anywhere in our range.

Sapotaceae (SAPODILLA FAMILY)

Bumelia (BUCKTHORNS)

Shrubs or small trees, sap milky (usually observable on severed fresh petioles and abundant in unripe fruits). Growth in length is by long-shoots with relatively distant leaves some of which subtend axillary thorns or thorn-tipped branches both of which become spur-shoots that may or may not remain thorn-tipped. Well-established plants have alternating and irregular periods of slow-growing spur-shoots and accelerated long-shoot production yielding a crooked branch system. Leaves simple, alternate (obviously alternate on long-shoots, approximate and appearing fascicled on spur-shoots), tardily deciduous, sessile or short-petiolate; stipules none. Flowers small, borne in axillary umbels, sometimes singly, often mostly on spur-shoots, or on both spur-shoots and long-shoots; radially symmetrical, bisexual. Calyx united only basally, 5-lobed, persistent below the fruit. Corolla united basally, 5-lobed, each lobe with a lateral appendage on each side near the base, the calyx opening at anthesis only sufficiently for the erect (later spreading) corolla lobes to emerge. Stamens 5, opposite the corolla lobes, 5 petaloid staminodia alternating with the corolla lobes, both inserted on the corolla tube. Pistil 1, ovary superior, 5-locular, a single ovule in each locule. Fruit a drupelike berry, commonly having the style base persisting apically.

1. Stems of shoots of the season glabrous save in the very early stage of development, the pubescence then blond or whitish.
   2. Mature leaf blades having upper surfaces green and glabrous, the lower surfaces with dense, silvery (rarely slightly tawny) pubescence, the hairs as though brushed anteriorly; stems of shoots of the current season with a definite pale gray or silvery hue.
   1. B. anomala
   2. Mature leaf blades having both surfaces green and glabrous (or sparsely pubescent beneath and never with a silvery hue); stems of shoots of the season green to brown.
      3. Upper surfaces of mature leaf blades (observed with suitable magnification) relatively faintly and coarsely reticulate-veined, the veins of the reticulum not at all raised, usually somewhat impressed, and, although pale, not bony-cartilaginous. 7. B. thornei
      3. Upper surfaces of mature leaf blades (as observed with suitable magnification) notably finely reticulate-veined, the veins of the reticulum usually raised above the enclosed islets and bony-cartilaginous in contrast to the green of the islets.
         4. Larger leaf blades mostly not exceeding 5 cm long (rarely some to 7 cm), sometimes all blades on a given plant shorter than 5 cm, most of them broadest distally; shrub, usually with several to numerous stems from the base, these to 4 m tall; berries 5–8 mm long and about 5 mm in diameter.   4. B. reclinata
         4. Larger leaf blades 8–12 (14) cm long, most of them broadest at or below their middles; stem usually solitary, sometimes flowering/fruiting when of shrub stature, potentially a tree to at least 10 m tall; berries mostly 10–15 mm long and 10–12 mm in diameter.   3. B. lycioides
1. Stems of the season copiously to moderately pubescent with rusty reddish brown, or tawny to dark brown hairs.
   5. Pubescence of lower surfaces of mature leaf blades dense and matted, usually with a sheen.   6. B. tenax
   5. Pubescence of lower surfaces of mature leaf blades copious, but not matted, to sparse, without a sheen.
      6. Lower surfaces of leaf blades at first with copious pubescence, this mostly quickly sloughing and the lower surfaces then green and glabrous or nearly so; plants commonly extensively cloning from subterranean runners, stems 1–3 dm tall in some clones, to 6–8 dm in other clones.   5. B. rufotomentosa
      6. Lower surfaces of leaf blades persistently pubescent with tawny to brown hairs; plant with a single main stem or several from the base but not producing extensive clones.   2. B. lanuginosa

1. Bumelia anomala (Sarg.) Clark.   Fig. 317

An apparently rare and little known arborescent plant to at least 3 m tall. Leader, long shoots bright green at first, with very few, scattered, pale hairs, soon becoming grayish, glabrous, and dotted with small, pale lenticels; long shoots bearing sharp, green thorns axillary to most leaves, these eventually becoming short spur-shoots.

Leaves very short-petiolate. Blades on elongation shoots ovate-elliptic, larger ones 5–6 cm long and 3–4 cm broad, those of spur-shoots generally smaller, oblanceolate to elliptic; all blades lustrous, deep green above, metallic-silvery

Fig. 317. **Bumelia anomala:** a, habit; b, portion of a leader-shoot; c, enlargement of medial portion of lower surface of leaf.

beneath, the metallic-silvery hue deriving from a dense, relatively compact, silvery pubescence; upper surfaces of mature leaves finely reticulate-veined, the veinlets of the reticulum raised and bony-cartilaginous.

Flowers and fruits not seen by me.

The only place at which I have seen this plant was at the Alachua Sink area, Gainesville, Fla., where it occurs as scattered individuals in a mesic hardwood forest on ridges that separate a rather large number of impressive sinkholes. I was taken there (summer of 1984), together with Angus Gholson, by Robert W. Simons of Gainesville, expressly for the purpose of seeing it. Mr. Simons tells me it occurs in hardwood hammocks elsewhere in Fla.: by the Silver River in Marion Co., near St. Augustine, and at Blue Sink by the Withlacoochee River, Madison Co. Clark (1942) records its occurrence (besides at Gainesville) near Orlando, as well.

2. **Bumelia lanuginosa** (Michx.) Pers., subsp. lanuginosa. GUM BUMELIA.

Fig. 318

An irregularly shaped shrub or small tree, rarely to 12 m tall, flowering/fruiting when only of shrub stature. Bark of trunks furrowed and divided into narrow ridges with flaky reddish brown scales.

Young developing stems, petioles, and both surfaces of unfurling leaves densely, dull gray-, tan-, brown-, or rusty- pubescent, pubescence not appressed and never lustrous; most of the pubescence of upper leaf surfaces, or all of it, quickly sloughing, the surfaces then appearing sublustrous; pubescence of lower leaf surfaces becoming much less dense as the blades expand, but usually persistent, that of petioles and stems usually persisting through much or all of the first year. Some individual plants unarmed, others with some branches armed with branch thorns, lateral branches and spur-shoots often remaining thorn-tipped as they elongate. Blades oblanceolate, sometimes some obovate or elliptic, rounded apically, infrequently emarginate, variable in size on a given plant and from plant to plant, varying in average size, generally larger on elongation-shoots than on spur-shoots, 1–10 cm long and 0.8–4 cm broad; margins entire; upper surfaces of mature leaf blades notably finely reticulate-veined, the veins of the reticulum usually raised above the enclosed islets and bony-cartilaginous.

Flowers borne in sessile, axillary umbels of up to 40 or more per umbel, their stalks 3–4 mm long on some plants, varying to 8–20 mm long on others, the stalks and calyces sparsely or densely grayish- or rusty-pubescent, more commonly dense and rusty than otherwise. Calyx lobes erect, oblongish to suborbicular, about 2–3 mm long, apices broadly rounded. Corolla white, lobes just exserted from the calyx. Berries slightly obovoid or ellipsoid, shiny black at maturity, 6–8 mm long, usually beaked by the persistent style base, short-pubescent mainly at the summit.

Inhabiting upland, mostly well-drained, sandy woodlands of various mixtures, sandy old fields, coastal sandy woodlands and sand pine–oak scrub, longleaf pine–oak sand ridges; mainly s. cen. and s.w. Ga., n. cen. Fla. and across n. Fla., s. half of Ala., westward to La.

3. **Bumelia lycioides** (L.) Pers. BUCKTHORN BUMELIA.

Fig. 319

A shrub or small tree, shrubs sometimes bushy-branched and flowering when no more than 1–2 m tall, varying to trees to at least 20 m tall and 3.5 dm d.b.h., crowns often somewhat columnar. Bark of trunks grayish brown, eventually

Fig. 318. **Bumelia lanuginosa:** flowering branch, center; at upper right, small portion of twig, winter condition; lower right, flower; center right, part of corolla opened-out; lower left, fruit.

Fig. 319. **Bumelia lycioides:** flowering branch, upper right, and below a short-shoot with inflorescence; part of current season's elongation shoot at left; flower at lower right; cluster of fruits at lower left.

breaking into irregularly scaly strips or small plates which slough and expose reddish brown inner bark beneath, innermost bark with a somewhat rancid odor. Very young developing stems, petioles, and both surfaces of unfurling leaf blades with rather silky, blond pubescence, that of lower surfaces more dense and more appressed than that of upper surfaces, both surfaces and petioles becoming glabrous as the leaves expand. Stems very quickly becoming glabrous, commonly bearing supra-axillary thorns or thorn-tipped branches.

Mature leaves shortly cuneately narrowed to petioles 0.5–1.5 cm long. Blades long-elliptic, long-oval, elliptic, or less frequently oblanceolate, larger ones mostly 8–12 cm long and 2–4 cm broad; apices acute, obtuse, or rounded; margins entire.

Inflorescences and flowers like those of *B. lanuginosa* but without pubescence. Berries black at maturity, sublustrous, subglobose, slightly ovoid, or obovoid, mostly 10–15 mm long and 10–12 mm in diameter.

Inhabiting wooded natural levees, bottomland woodlands where flooding is of brief duration, adjacent wooded slopes or bluffs, usually where the soil is circumneutral; in our area, chiefly in s.w. Ga., s.e. Ala., cen. Panhandle of Fla. from Jefferson and Wakulla Cos. to Jackson Co. (Irregularly distributed, s.e. Va. to s. Ill., s.e. Mo., generally southward to n.w. Fla. and s.e. Tex.)

4. Bumelia reclinata (Michx.) Vent. var. reclinata. SMOOTH BUMELIA.

Fig. 320

Shrub to about 5 m tall (infrequently arborescent), frequently with several to numerous stems from the base; plants occasionally numerous on open, gladelike calcareous areas where, when stressed, forming very crooked, scrubby branch systems abounding with lichens. Stem glabrous, usually the leader,

Fig. 320. **Bumelia reclinata:** a, six pieces of flowering or fruiting stems, each one from a different plant to show some of the leaf variation; b, flower; c, fruit (drawn to same scale as fruit of B. rufotomentosa).

long-shoots bearing naked or leafy thorns which eventually become spur-shoots.

Leaves subsessile or cuneately narrowed to short petioles. Blades variable in size on a given plant, and from plant to plant showing much variation in average size and shape; mostly spatulate or oblanceolate, sometimes elliptic, obovate, or suborbicular, 1–5 cm long (rarely a few leaves on a given plant to 7 cm) and 0.4–2 cm broad distally, apices rounded to obtuse, infrequently emarginate; upper surfaces glabrous, notably finely reticulate-veined, the veinlets of the reticulum raised and bony-cartilaginous, lower surfaces and petioles pubescent with relatively sparse, whitish-cottony pubescence when young, all of the pubescence usually quickly sloughing on most individuals.

Flowers usually borne in sessile, axillary umbels of 2–3 to 20, occasionally solitary; flower stalks 10–15 mm long on some individuals, much shorter on others. Calyx about 2 mm high, lobes oval, broadly rounded apically. Corolla white, 2–3 mm high, lobes 1.5 mm long, obovate, cupped, slightly spreading, exserted from the calyx. Berries lustrous-black when fully ripe, subglobose or slightly obovate, 5–8 mm long, 4–6 mm in diameter.

Mesic wooded bluffs or ravine slopes, wooded river or stream banks, wooded areas about lime-sinks, hammocks with shallow soil over limestone; s.w. Ga., n. and pen. Fla.

5. Bumelia rufotomentosa Small.   Fig. 321

Shrub, frequently forming clones from subterranean runners, clones sometimes several meters long and broad and the stems closely disposed and not over 2–3 dm high, sometimes cloning less extensively with stems less densely disposed and to about 10 dm tall. Stems and petioles of elongation shoots densely short-shaggy pubescent with darkish rusty-red hairs, much of the pubescence retained through the first year, some of it longer; both surfaces of unfurling leaf blades heavily but loosely clothed with a mixture of whitish and dark rusty-red pubescence (the former masked by the latter as seen with the unaided eye), the pubescence becoming thinner as the blades expand to full size, most of it soon sloughing from the upper surfaces which then appear green to the unaided eye, with magnification the upper surfaces of mature blades are seen to be reticulate-veined, the veinlets of the reticulum raised and bony-cartilaginous; pubescence of lower blade surfaces sloughing much more gradually, thus the surfaces gradually becoming more green but still with more or less of a rusty hue; often some or nearly all lower blade surfaces becoming essentially glabrous and green; branches of leader long-shoots variously bearing axillary thorns or thorns in all or nearly all leaf axils, the thorns mostly eventually becoming thorn-tipped spur-shoots. The main axis of elongation shoots, their branches, or thorn-tipped spur-shoots sometimes have irregularly swollen, woody galls (the causal agent unknown to me). Such galls may occur on plants of other species of *Bumelia*, but not, insofar as I have observed, so regularly or to the same extent as on plants of this species.

Leaf blades variable in shape and size (even on a single branch), spatulate, obovate, oval, elliptic, or subrotund, 1.5–3 (6) cm long and 0.5–2 (3) cm broad, bases cuneate to rounded, apices rounded, occasional ones emarginate.

Flowers mostly in sessile umbels, occasionally solitary, axillary to leaves or leaf scars, up to at least 24 per umbel; flower stalks 2–10 mm long, shaggy rusty-pubescent throughout or only distally. Calyx 2–3 mm long, lobes ovate, oval, or obovate, rounded apically, rusty-pubescent basally or throughout. Corolla

Fig. 321. **Bumelia rufotomentosa:** a, flowering branch; b, tip of elongation shoot of the season; c, enlargement of a small piece of an elongation shoot to show pubescence of stem and lower surface of leaf; d, flower; e, fruit (drawn to same scale as fruit of B. reclinata).

white, lobes cupped, oval to suborbicular. Berries shiny black, subglobose, oval, or slightly ovoid or obovoid, (8) 10–13 mm long and 8–12 mm in diameter. [*B. reclinata* (Michx.) Vent. var. *rufotomentosa* (Small) Cronq.]

Inhabiting longleaf pine–scrub oak or sand pine–scrub oak sandy ridges and slopes, and pine plantations on such ridges and slopes; known to occur in the following counties in Fla.: Suwannee, Columbia, Alachua, Putnam, Levy, Citrus, Pasco, Hillsboro, and Orange.

Plants of *B. rufotomentosa* apparently occur very locally in the kinds of habitats in which they do occur and they are not easily perceived from even a short distance. This may account for the paucity of records for it. Cronquist (1945) considers it only varietally distinct from *B. reclinata*. Wunderlin (1982) places the name in the synonymy of *B. reclinata* without comment.

6. Bumelia tenax (L.) Willd. TOUGH BUMELIA, IRONWOOD.   Fig. 322

A shrub or small tree, as a shrub often irregularly branched and scrubby, as a tree presumably to 8 m tall. Bark of trunks reddish brown, eventually fissured and divided into flat ridges that become scaly. Young stems, petioles, and lower surfaces of leaf blades with dense, more or less appressed, silky and lustrous or sublustrous pubescence varying in color from silvery to goldish, coppery, rusty or chocolate-brown, most or all of the pubescence of petioles and lower leaf surfaces persisting, that on the stems persisting through the first year; upper surfaces of unfurling leaf blades sparsely and irregularly pubescent, the hairs quickly sloughing.

Mature leaves cuneately narrowed to subsessile bases or to petioles to about 1 cm long. Blades mostly spatulate or oblanceolate, some elliptic or narrowly obovate, 2–5 (7) cm long and 0.5–2 (3) cm broad, bases cuneate, apices rounded to obtuse, veinlets of upper surfaces finely raised-reticulate, those of lower surfaces obscured by pubescence; margins entire.

Flowers borne in sessile, axillary umbels, usually numerous, 10 to about 50 per umbel; flower stalks mostly 6–12 mm long, they and the calyces usually densely pubescent, rarely sparsely so, hairs more or less silky, in color matching that on the lower leaf surfaces. Calyces 2–2.5 mm high, lobes oblongish, rounded apically. Corolla white, lobes oblong-elliptic to subrotund, distal margins sometimes slightly jagged. Berries obovoid, 10–14 mm long, black. [Incl. *B. lacuum* Small]

Occurring in near-coastal scrub, maritime woodlands, and sand pine–oak scrub; n.e. S.C. to s. pen. Fla.

7. Bumelia thornei Cronq.

A relatively spindly shrub to about 2.5 m tall with few principal branches, vigorous specimens bushy-branched, to 6 m tall; thorns when present of variable lengths up to about 8 cm the first year, in the second year usually bearing several essentially sessile spur-shoots each of which may produce 1–several flowers/fruits. Thorns or stems of elongation shoots tawny to rusty pubescent at first, the hairs usually sloughing leaving the thorns or twigs dark brown.

Leaves mostly tapering basally to persistently pubescent petioles 2–8 mm long; blades variable in size on a given plant and varying in average size from plant to plant, 1–7 cm long and 0.5–2.5 cm broad, in shape predominantly oblanceolate but varying to narrowly to broadly elliptic, obovate, or nearly round, apices bluntly tapered to rounded; upper surfaces of mature blades glabrous or with very sparse pale pubescence along the proximal portions of the

639

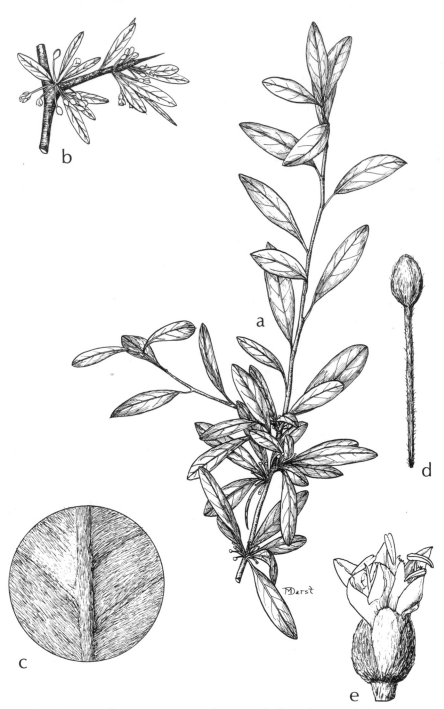

Fig. 322. **Bumelia tenax:** a, piece of stem with short-shoots and elongation shoot; b, short-shoot with inflorescences; c, enlargement of a portion of lower surface of leaf; d, enlargement of flower bud with its stalk; e, flower.

midribs, to either side of the midrib faintly and relatively coarsely reticulate-veined, the veinlets of the reticulum not raised, sometimes impressed, pale but not bony cartilaginous; lower blade surfaces persistently and uniformly moderately woolly-pubescent, the hairs grayish to tawny or rust-colored; margins entire.

Flowers not seen by me.

Berries dull black, short-obovoid, 8–10 mm long, broadly rounded apically, 8–10 mm long, their stalks slender, 6–8 mm long, lobes of the subtending calyx ovoid, about 2 mm long, rounded apically.

In woods bordering ponds (and creeks?), apparently where some surface water stands during wet seasons; known to occur in s.w. Ga. (Early, Baker, Calhoun Cos.) and well to the n.e. of there in Tatnall Co., Ga., and known to me at one locality, Jackson Co., Fla.

Saxifragaceae (SAXIFRAGE FAMILY)

1. Leaves opposite.
  2. Plant sometimes running along the ground, usually a vine climbing by means of aerial roots on the stem; petals 7 or more   1. *Decumaria*
  2. Plant an erect shrub; petals 4 or 5 (in fertile flowers).
    3. Flowers both sterile (with showy calyx and no sexual parts) and fertile, the latter small, with 5 petals only 1–2 mm long and 8–10 stamens; capsule ribbed, dehiscing by a pore between the persistent styles.   2. *Hydrangea*
    3. Flowers all fertile, relatively large, with 4 petals 15–25 mm long; stamens 20–40; capsule not ribbed, dehiscing longitudinally.   4. *Philadelphus*
1. Leaves alternate.
  4. Stem unarmed; leaf blades pinnately veined, unlobed; flowers in cylindrical racemes terminating branchlets of the season.   3. *Itea*
  4. Stem with awllike, terete, sharp-pointed, simple or forked spines at the nodes; leaf blades palmately veined and palmately 3-lobed; flowers, in early spring, borne singly and pendently from the axil of one of the 2–several leaves on short, lateral shoots.
                                                  5. *Ribes*

1. Decumaria

Decumaria barbara L. WOOD-VAMP, CLIMBING HYDRANGEA.   Fig. 323

Deciduous woody vine with abundant, short, adventitious roots from the stem by which climbing is achieved. Vines sometimes running along the ground where forming loose mats, or on rocks, but not observed flowering in such instances; more commonly climbing on the trunks of trees and with branches spreading laterally or declining; older stems to at least 4 cm in diameter, very old ones eventually become free from tree trunks, their leafy branches all high in the crowns of the trees; leaf scars more or less crescent-shaped, vascular bundle scars 3.

Leaves opposite, simple, without stipules, petioles 1–3 cm long, much shorter than the blades, usually pubescent; blades fully developed, ovate, oval, or subrotund, less frequently obovate (the several shapes often on a single branch and varying considerably in size); longest blades 10–12 cm long and about 6 cm broad, bases broadly tapered or rounded, gradually tapered from the broadest places to their tips; subrotund blades usually mucronate apically; margins

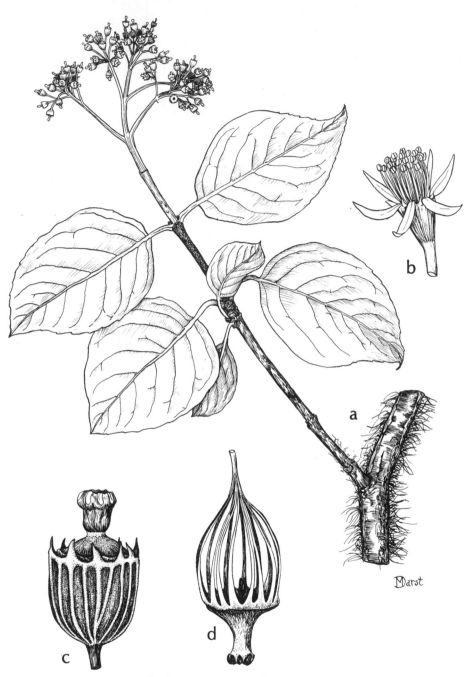

Fig. 323. **Decumaria barbara:** small piece of vine with its aerial, adventitious roots, and branch with young infructescence; b, flower; c, fruit; d, persistent remains of fruit after having shed its seeds (at which time it is pendulous).

entire, vaguely wavy, or with short, blunt teeth from about their middles upwardly; upper surfaces dark green and lustrous, glabrous, lower paler and a little duller, sparsely short-pubescent, sometimes mainly on the veins; venation pinnate, not prominent, lateral veins with branches anastomosing and only a veinlet extending into the teeth (if any). Internodes of vigorous shoots 3–6 cm long, but leaves on short, spurlike shoots or spurlike tips of longer shoots closely set.

Inflorescence terminal on a shoot of the season, a convex, compound cyme 3–10 cm across with numerous small flowers; cymes nakedly stalked above a pair of leaves or the stalks with a pair of foliar bracts subtending the first branches. Flowers bisexual, radially symmetrical, with a turbinate 7–10 (–12) -ribbed floral tube (adnate to the ovary) bearing 7–10 (–12) short-triangular, erect, creamy white sepals on its outer rim. Petals of the same number as the sepals, creamy white, lanceolate, about 3 mm long, alternating with and within the sepals on the rim of the floral cup. Stamens 20–30, unequal in length, the longer ones about the length of the petals. Style 1, stout, broadest at base, the stigma knoblike, capitate, and with 7–12 radiating stigmatic lines.

Fruit a cuplike capsule, prominently ribbed, ribs pale, greenish brown between the ribs, greenish sepals persistent on the rim at its summit, the broadly based persistent style whose narrowed distal portion bears the prominently radially lobed persistent stigma capping the capsule. Seed small, with a yellowish brown, sublustrous body whose surface is coarsely roughened-reticulate, a short flattish flaplike projection at one extremity, attenuate taillike at the other, seeds gradually released as capsular tissue between the ribs decays, eventually tiny jack-o-lanternlike skeletal remains of all the capsules dangle on their frayed stalks, usually still attached overwinter and into the next season.

Rich woodlands or, more commonly, in very moist to wet woodlands and swamps, throughout the area of our coverage. (Coastal plain, s.e. Va. and N.C., throughout S.C., in s.w. N.C. and Tenn., generally southward to cen. pen. Fla. and s.e. La.)

2. Hydrangea

1. Leaves all unlobed.   1. *H. arborescens*
1. Leaves (excepting sometimes those just below inflorescences) prominently lobed.
                                                                        2. *H. quercifolia*

1. Hydrangea arborescens L. subsp. **arborescens.** WILD HYDRANGEA, SEVEN-BARK.

Shrub to 2 m tall, young stems sparsely short-pubescent, the bark of older stems brown, exfoliating in irregular papery patches; leaf scars approximately crescent-shaped, vascular bundle scars 3.

Leaves opposite, simple, slenderly petioled, petioles usually shorter than the blades; blades mostly ovate or oval, conspicuously serrate with many teeth, variable in size, largest ones to 20 cm long and 12 cm broad but commonly 10–12 cm by 6–8 cm, pinnately veined, the lateral veins ascending, their few branches anastomosing back of the margins and then branchlets ending in the serrations; both surfaces green, the lower paler than the upper, commonly sparsely short-pubescent on the principal veins above and beneath, mostly rounded basally, occasionally truncate or subcordate, short-acuminate apically.

Inflorescence terminal on a shoot of the season, a convex to flat-topped, compound cyme up to 15 cm across, usually with a pair of bracteal leaves at or a little below its base, sometimes distinctly stalked above a smallish pair of foliage leaves; inflorescence axes pubescent with short, curvate hairs. Some plants having cymes with all fertile flowers, others having cymes with many fertile flowers and few to numerous neuter ones, rarely an individual having cymes with all neuter flowers. In cymes with a mixture of fertile and neuter flowers, the neuter ones with 3 or 4 broadly oval to obovate, white, pale greenish, or cream-white, reticulately veined, radially spreading, petaloid sepals; neuter flowers borne terminally on a cymule and having stalks much longer than those of fertile flowers extending them well outward marginally on the cymes. Fertile flowers bisexual, radially symmetrical, with a cuplike, strongly ribbed floral tube (adnate to the ovary) surmounted by 5 minutely triangular sepals. Petals 5, white, oblongish, 1–2 mm long, erect or ascending alternately with the sepals on the rim of the floral tube. Stamens 8–10, filaments straight, unequal in length, long-exserted. Styles 2, each with a broad, short stigma.

Fruit a brown, campanulate capsule with 8–10 conspicuous, longitudinal, pale-bonelike ribs connected to a similarly textured band encircling the summit; dehiscence by a terminal pore between the persistent styles. Seeds small, obliquely fusiform or obliquely ellipsoid, somewhat translucent, longitudinally ribbed, sometimes with a few connecting cross-ribs, about 1 mm long.

Rich woodlands, wooded bluffs, shady ledges, and stream banks. In our area, s.w. Ga., s. Ala., cen. Fla. Panhandle. (S. N.Y. to Ohio, Ill., Mo., generally southward to s.w. Ga., Fla. Panhandle, Okla., La.)

2. Hydrangea quercifolia Bartr. OAK-LEAF HYDRANGEA, SEVEN-BARK.

Fig. 324

Deciduous shrub, to about 3 m tall, the branching arching or straggly. Young twigs with a covering of copious, rust-colored, cottony pubescence, that of year-old twigs usually gray, the bark eventually splitting irregularly longitudinally, the gaps formed rust-colored; bark on older, larger stems peeling in papery patches yielding an attractive, varicolored mottling effect; leaf scars varying from crescent- to half-moon shaped, vascular bundle scars 3–11, the number sometimes thus variable on single plants; pith relatively large, continuous, somewhat spongy, buff-colored.

Leaves opposite, simple, petioled, blades pinnately veined, those on flowering shoots smaller and less deeply lobed (even unlobed) than those on vigorous sterile shoots; larger leaves with petioles 10–15 cm long, blades 25–30 cm long and about two-thirds as broad (including the lobes); blades with 2 pairs of broad, angling-ascending lateral primary lobes and a broad terminal lobe, each of the lobes sometimes sublobed; margins mostly with small dentations or serrations, petioles thinly rusty-cottony-pubescent; upper surfaces deep green at maturity, with sparse, short, white hairs and only at first with a thin cottony pubescence, lower surfaces similarly pubescent but the short white hairs much more copious, the thin cottony pubescence mostly persisting along the midvein or along both it and the principal laterals.

Inflorescence a stalked, relatively large, pyramidal panicle, the branches bearing lateral, umbellike cymes of fertile flowers and each branch terminated by a single neuter flower having 4 much enlarged, broadly ovate to suborbicular, radially spreading, reticulately veined, white-petaloid sepals and only rudiments of sexual parts in the center, the neuter flowers giving the inflorescence much or most of its attractive showiness; neuter flowers persisting, turning pink or pinkish, then purple or purplish. Fertile flowers bisexual, radially symmetri-

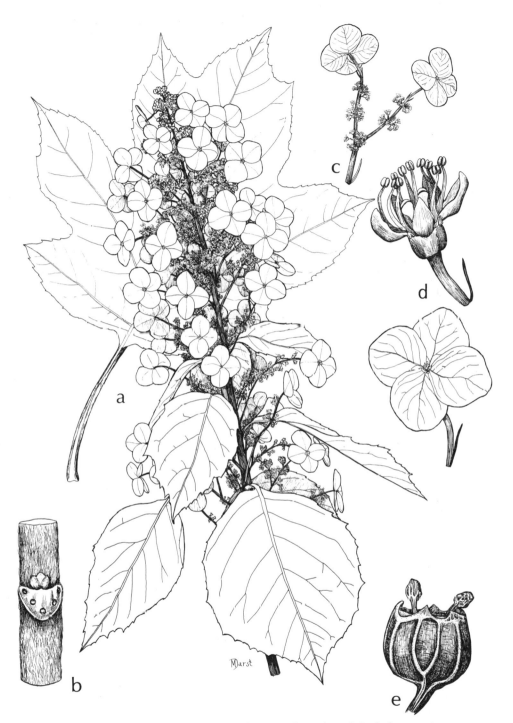

Fig. 324. **Hydrangea quercifolia:** a, flowering branch with leaf of vegetative shoot in background; b, node with leaf scar; c, flowering branchlet; d, fertile flower above, sterile flower below; e, fruit.

cal, with a cuplike, white or nearly white, floral tube (adnate to the ovary) surmounted by 5 white, triangular sepals 1 mm long. Petals 5, white, inserted on the rim of the floral tube, alternating with the sepals and spreading between them, oblong or obovate and cupped, twice as long as the sepals or a little more. Stamens 8–10, inserted on the rim of the floral cup, filaments usually somewhat arching outward then inward, much exceeding the corolla. Styles 2, each with a 2-lipped stigma.

Fruit a campanulate, strongly 8–10-ribbed dull brown capsule dehiscing by a pore between the persistent styles. Seeds small, dull brown, fusiform-falcate, strongly ribbed, 0.5–1 mm long.

In rich, mixed woodlands, stream banks, ravines, usually in calcareous areas; in the area of our coverage, interior of Fla. Panhandle from about the Ochlockonee River westward, s.w. Ga., s. Ala. (Cen. and s.w. Tenn., southward to the Fla. Panhandle, and La. eastward of the Mississippi River.)

H. quercifolia is in some favor as an ornamental, but would probably be much more appreciated for this purpose were it not deciduous. Its large leaves (which become reddish before dropping in autumn), large showy inflorescences, which are handsome both in flower and in fruit in summer, are features which principally earn it plaudits; however, the stem architecture and the patina of its interesting bark are considered by some as lending it notable attractiveness in winter.

3. Itea

Itea virginica L. VIRGINIA-WILLOW, TASSEL-WHITE, SWEETSPIRE. Fig. 325

Openly, slenderly branched, tardily deciduous shrub to about 2 m tall. Axillary buds superposed, the upper one much larger than the lower, sometimes both developing shoots so that two branches, one above the other, appear at a node. Leaf scars crescent-shaped with 3 vascular bundle scars.

Leaves simple, alternate, without stipules, very short-petioled, blades pinnately veined, very variable in shape and size even on a single branch, oblong, oval, oblanceolate, or infrequently somewhat obovate, cuneate basally, apices varyingly short-acuminate, acute, or obtuse, larger ones 5–10 cm long and 3–4 cm broad at maturity; margins minutely serrate, the teeth pointed; petioles and lower leaf surfaces with few short hairs.

Flowers small, borne on stalked, usually arching, cylindrical racemes 3–10 cm long and about 2 cm across, the short flower stalks diverging at right angles to the axis, racemes terminating branchlets of the current season; axis of raceme and flower stalks varying from sparsely to copiously short-pubescent, floral tubes similarly throughout or only basally. Floral tube hemispherical, free from the ovary. Sepals 5, erect, subulate, 1 mm long or a little more, deciduous. Petals 5, white, linear-subulate, arching-ascending, pubescent at the base on the upper side, 6–7 mm long. Stamens 5, about half as long as the petals. Ovary superior, pubescent, 2-locular, styles 2, confluent, persistent, stigmas capitate.

Fruit capsular, pubescent, conic-cylindric, hard, grooved along the two sutures, tardily splitting downward through the styles and along the sutures about halfway, sometimes wholly, to the base; capsules more or less reflexed on the axis, shriveled floral tubes persistent. Seeds several, irregularly oblongish in outline, a little over 1 mm long, a reticulum of isodiametric surface cells evident, surface gold-colored, lustrous.

Swamps, wet woodlands, along wooded streams; throughout our area. (s. N.J.

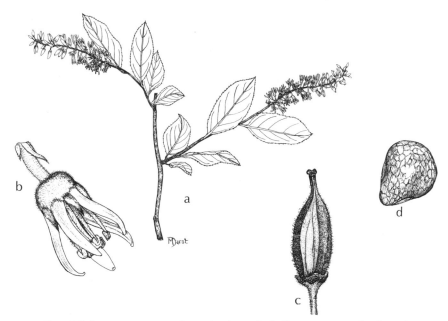

Fig. 325. **Itea virginica:** a, flowering branch; b, flower; c, capsule; d, seed.

and e. Pa., southward through e. and cen. Va., most of N.C. thence to pen. Fla. and westward to e. Tex., northward in the interior, s.e. Okla., Ark., w. Tenn., s.w. Ky., s. Ill., and s.e. Mo.)

4. Philadelphus

Philadelphus inodorus L. MOCK-ORANGE, SUMMER-DOGWOOD, SYRINGA.

Fig. 326

Shrub to about 4 m tall. Young twigs smooth, brown, outer bark breaking up and exfoliating on year-old and older twigs. Leaf scars narrow, vascular bundle scars 3.

Leaves deciduous, simple, opposite, short-petiolate, without stipules; blades pinnately veined, those of lowermost leaves on a given branchlet usually much smaller than others, larger ones mostly oval, ovate, or elliptic, 4–8 cm long and 3–4 cm broad, rounded to broadly and shortly tapered basally, apices varyingly sharply acuminate, abruptly and bluntly short-acuminate, or occasionally rounded; margins entire, with few small dentate teeth, or sometimes barely a suggestion of 1 or 2 teeth on a side; upper surfaces glabrous or sparsely strigose, lower glabrous or sparsely pubescent.

Inflorescence a 3-flowered cyme terminating a slender stalk above the upper-most pair of leaves of a branchlet, or flower solitary and shortly stalked above an articulation of the "inflorescence stalk", i.e., a 1-flowered cyme by reduction. Flowers bisexual, radially symmetrical. Floral tube cuplike, almost wholly adnate to the ovary, sepals 4, lance-ovate to triangular-ovate, the floral tube and sepals glabrous exteriorly, the sepals felty-pubescent on their inner surfaces. Petals 4, white, spreading, broadly obovoid, inserted on the rim of the floral tube, open corolla 3.5–5.5 cm across. Stamens 20 or more, borne on the rim of

647

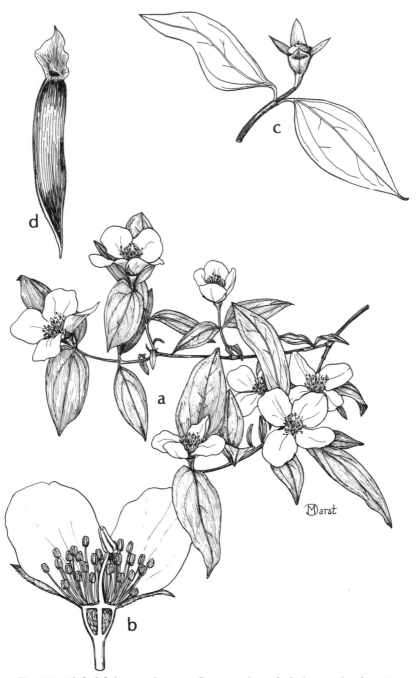

Fig. 326. **Philadelphus inodorus:** a, flowering branch; b, longitudinal section of flower; c, fruiting branchlet; d, seed.

the floral tube. Pistil 1, ovary inferior, 4-locular, style united below, stigmas 4, distinct.

Fruit a loculicidal capsule, its summit protruding above the floral tube, sepals persistent, erect. Seeds numerous, red, smooth, narrowly elliptic, 2–3 mm long.

Mesic woodlands, s. cen. and s.w. Ga., s. Ala., cen. Panhandle of Fla. (Va. and Tenn. southward to Ga., Ala., and the Fla. Panhandle.)

This and other species of *Philadelphus* are old-time favorites as ornamentals, principally because of their sprays of showy white flowers in early summer. As is the case with so many plants, contemporaneously their deciduousness mitigates against their more frequent use.

5. Ribes

Ribes echinellum (Coville) Rehd. MICCOSUKEE GOOSEBERRY.   Fig. 327

Bushy-branched, low shrub, branches erect-ascending to spreading-recurved, the latter sometimes rooting at their tips. Twigs with 2 low ridges, outer bark buff-colored, on older stems splitting and exposing very dark purplish brown inner bark; at the nodes are reddish brown, awllike, terete, sharp-pointed spines, simple or 2–3-forked from the base, each to about 1.5 cm long, many of them shorter especially on younger stems; main woody stems produce very short branches with 2 several leaves appearing fascicled; young leader shoots pubescent, their leaves relatively widely spaced.

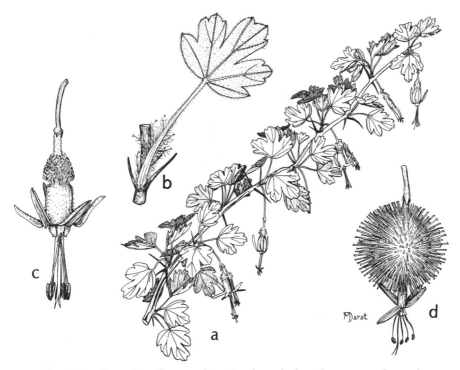

Fig. 327. **Ribes echinellum:** a, flowering branch; b, enlargement of a node with leaf (lower surface) and stipules; c, flower; d, maturing fruit with some-what shriveled perianth and stamens intact.

649

Leaves simple, alternate (sometimes most of them deciduous in midsummer, new leaves appearing on short-shoots in autumn and overwintering), petiolate, petioles slender, 1–2 cm long, dilated at base and with a narrow, membranous wing on each side proximally, each wing with several divaricately spreading, longish, ciliate, minutely gland-tipped hairs; blades palmately 3-lobed, the lobes themselves bluntly toothed or toothed-lobed; larger blades of branches 1.5–2 cm across the basal lobes, larger blades of vigorous elongation shoots 5–6 cm across the basal lobes; both surfaces of blades pubescent, some of the pubescence or all of it gradually sloughing from the upper surfaces.

Flowers, in early spring, borne singly, usually only one from one of the axils of the 2–several leaves of short-shoots, pendent, bisexual, radially symmetrical, ovary inferior; flower stalks 5–12 mm long, jointed and with a pair of unequal, broad, short, clasping bracts somewhat above their middles, the stalks copiously shaggy-pubescent and with spreading gland-tipped hairs as well. Floral tube distinctly into 2 parts: the lower part, more or less globular and adnate to the ovary, its surface completely covered with elongate, soft and membranous, greenish, gland-tipped hairs that are broadest basally and flat, taper gradually upwardly and gradually become terete (these harden and stiffen, tend to become terete throughout and lustrous-brown as the fruit matures); the upper half of the floral tube, all of which extends beyond the ovary, is cylindrical, short-pubescent, bears on its rim 5 reflexed, yellowish green, linear-oblong, obtuse, exteriorly pubescent, calyx segments 4–6 mm long, 5 erect, inconspicuous petals 1–1.5 mm long, each rolled in such a way as to appear tubular, petals alternating with the sepals, and 5 erect stamens with long, slender filaments extending for about 8–10 mm beyond the perianth. Style slender, bifid into 2 linear parts at its tip which reaches to the bases of the anthers.

Fruit a many-seeded, rotund, spiculate berry 1.5–2 cm across (including the spicules), the shriveled upper portion of the floral tube and other parts of the flower persisting at least until the fruit is fully ripe.

Mesic woodlands. Known to occur only in Jefferson Co., Fla., and McCormick Co., S.C.

Schisandraceae (STAR-VINE FAMILY)

Schisandra glabra (Brickell) Rehd. STAR-VINE, WILD SARSAPARILLA.   Fig. 328

A twining woody vine forming bowers in the crowns of trees or sometimes trailing along the ground, glabrous throughout. Twigs brown, leaf scars raised on year-old stems, nearly circular, with 3 vascular bundle scars; pith tan at first, eventually becoming dark brown, continuous but consisting of loose strips and flakes; old stems to at least 3 cm in diameter.

Leaves without stipules, simple, deciduous, alternate (remote from each other on long shoots, closely set on short shoots), somewhat aromatic, slenderly petioled, blades pinnately veined, ovate, oval, or elliptic, mostly 4–12 cm long and 2–6 cm broad (much larger on vigorous shoots), wedgelike basally, tips short-acuminate, margins usually with a few almost imperceptible dentations.

Flowers inconspicuous, about 1 cm across, borne singly on dangling slender stalks from leaf axils, unisexual, staminate and pistillate on the same plant, the former somewhat earlier in the season than the latter on a given plant. Perianth

Fig. 328. **Schisandra glabra:** a, piece of vine; b, flowering short-shoot; c, staminate flower, face view; d, 5 stamens fused into a pentagonal shield, above, and same with 5 holes formed by collapse of anther sacs, below; e, pistillate flower, face view, with several spirally arranged pistils; f, mature elongated receptacle with fruits.

651

of 9–12 separate parts similar in form, the outer greenish white, the inner red to strawberry-pink. Staminate flowers with 5 stamens fused into a pentagonal shield. Pistillate flowers with numerous pistils closely set in the flower but the receptacle greatly elongating as the ovaries mature; each ovary ripening into a 2-seeded, subglobose or ellipsoid, red berry. [*Schisandra coccinea* Michx.]

In mesic woodlands, in our area of coverage known to us from cen. Fla. Panhandle, perhaps overlooked in some other parts of our range. (Coastal plain, N.C. to Fla. Panhandle, westward to La., e. Ark., w. Tenn.)

At a glance, the star-vine is easily mistaken for the wood-vamp, *Decumaria barbara*. The latter has opposite leaves, climbs by means of aerial roots on the stem, the star-vine has alternate leaves and climbs by twining.

Simaroubaceae (QUASSIA FAMILY)

Ailanthus altissima (Mill.) Swingle. AILANTHUS, TREE-OF-HEAVEN.   Fig. 329

Native to northern China, this irregularly branched, rapidly growing, short-lived tree was introduced into North America as an ornamental during the latter part of the nineteenth century and has become naturalized, even weedy, in much of eastern North America and to some extent in the southwestern United States.

Twigs stout, pubescent at first, quickly becoming glabrous and smooth, eventually with squarish, slightly raised lenticels; leaf scars large, obcordate, vascular bundle scars about 8, usually indistinct; axillary buds low-moundlike, terminal bud none; pith large, continuous, dark brown.

Leaves alternate, deciduous, most of them large, but varying from 1.5 to 9 dm long or a little more, the smaller ones proximally on branchlets; petioles from 2 to about 15 cm long; blades 1-pinnately compound, leaflets pinnately veined, opposite, subopposite, or less frequently distinctly alternate on the rachis, predominantly odd-pinnate, but many even-pinnate (by abortion of the terminal leaflet in development); all leaflets shortly stalked, blades of lower lateral leaflets usually smallest, others increasing in size to the medial ones then becoming somewhat smaller distally, or the size gradually increasing from the lowest to the uppermost; terminal leaflets, if present, very variable, from threadlike to roughly similar to some of the larger laterals, occasional ones 1–2-incised-lobed basally; leaflets 2–15 cm long, bases rounded, truncate, or oblique and inequilateral, lance-ovate, ovatish-oblong, lance-oblong, or oblong, sometimes falcate, apices mostly acute or acuminate, margins usually with 1–2 blunt teeth on either edge of the base of the blade, the lower side of each tooth bearing a gland. As the leaves unfurl, they are commonly purplish red, the petioles, leaf axes, stalks of the leaflets and their lower surfaces copiously softly pubescent, much of the pubescence quickly sloughing but all of the leaf parts usually remaining sparsely pubescent at maturity. Crushed herbage with a mildly disagreeable musky odor.

Inflorescences panicles borne in the axils of approximate uppermost leaves on twigs of the season, sometimes only one in the axil of the uppermost leaf. Flowers radially symmetrical, unisexual or bisexual, plants polygamodioecious. Calyx small, united basally and with 5 minute lobes. Corolla of 5 separate, infolded (boatlike), greenish yellow petals 2–3 mm long, copiously hairy proximally on their abaxial sides. Stamens 10, 5 in each of 2 series (in flowers exam-

Fig. 329. **Ailanthus altissima:** a, leaf with overlay of a portion of functionally pistillate inflorescence; b, node with leaf scar; c, a small portion of staminate inflorescence; d, staminate flower; e, functionally pistillate flower; f, a flower stalk with developing fruits showing that pistils separate after anthesis; g, fruit.

ined by me), fertile in the staminate flowers, apparently not producing pollen in many of the structurally bisexual ones; outer stamens opposite the petals and at early anthesis lying flat within their infolded sides, later angling-ascending, the inner 5 alternate with them and essentially erect. Petals and stamens arising from beneath the edge of a moundlike, nectariferous disc upon which the pistils, if any, are borne. Pistils 5, connivent at anthesis and appearing as 1, ovary of each winged on the outer side, collective styles short and stoutish, collective stigmas umbrellalike; each ovary 1-ovulate, maturing into a samara, the latter with a membranous wing completely surrounding the centrally located seed-bearing portion, the wing elongate-tapering to its extremities, there somewhat twisted, the whole commonly somewhat falcate, 4–5 cm long.

In our area of coverage, infrequently planted and sparingly naturalized on woodland borders, in fence and hedge rows, in weedy places such as vacant lots.

Characteristics aiding in the identification of this tree are the musky odor of its crushed herbage, the large, alternate, 1-pinnately compound leaves, leaflets having 1-few teeth on each side near their bases, teeth glandular beneath.

National champion ailanthus (1972), 19'8" circumf., 60' in height, 80' spread, Head of Harbor, Long Island, N.Y.

Solanaceae (POTATO OR NIGHTSHADE FAMILY)

Lycium carolinianum Walt. CHRISTMAS-BERRY.   Fig. 330

Sparingly to densely bushy-branched, glabrous shrub to about 3 m tall, the main branches bearing shorter, more or less divaricate, rigid, buff-colored branchlets many of which are sharply thorn-tipped.

Leaves simple, sessile, alternate (commonly with very short, leafy branchlets in the axils giving a fascicled appearance), tardily deciduous, succulent, linear-oblanceolate or narrowly clavate, variable in length to 2.5 cm long.

Flowers bisexual, radially symmetrical, solitary in leaf axils, often in axils of the small leaves of the short axillary branchlets, the flower stalks slender, a little shorter than to equaling the subtending leaves. Calyx about 4 mm long, tubular below and with 4 short-deltoid, erect lobes, persistent. Corolla blue or lavender, sometimes nearly white, usually purple-streaked within, tubular below, the limb with 4 (5) rotate lobes about 5 mm long, slightly longer than the tube, pubescent in the throat and around the bases of the filaments which are inserted below the sinuses between the corolla lobes; anthers exserted. Fruit a bright, lustrous red, ovoid to ellipsoid berry 8–15 cm long.

Coastal sand spits, shell beaches and mounds, borders of, or in, salt and brackish marshes and mangrove swamps, coastal spoil areas; in places where high tides may inundate bases of the plants and extending above highest tides as well. (Ga. and Fla., westward to n. Tex.; W.I.)

Features which, taken together, help to distinguish this shrub: stiff, commonly divaricate, buff-colored branches some of which end in sharp thorns, succulent, alternate, linear-oblanceolate or narrowly clavate leaves, flowers solitary in leaf axils and having blue, lavender, or nearly white, 4 (5) -lobed corollas, the lobes rotate, and ovoid to elliptic, lustrous red berries.

Flowering is most abundant in autumn, the fruits ripening in late autumn or early winter. Oftentimes, flocks of birds thrash about devouring the ripe berries and in the process detach most or all of the leaves.

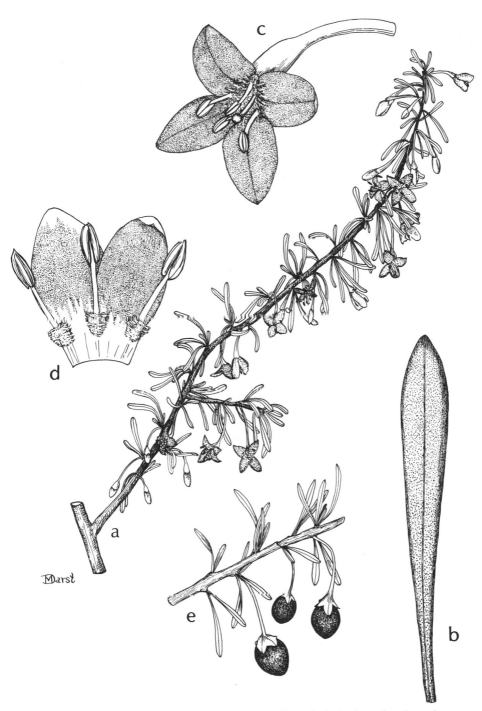

Fig. 330. **Lycium carolinianum:** a, flowering branch; b, leaf, much enlarged; c, flower; d, portion of corolla, detached showing insertion of stamens within; e, piece of stem with fruits.

655

Staphyleaceae (BLADDERNUT FAMILY)

Staphylea trifolia L. BLADDERNUT. Fig. 331

Deciduous shrub, rarely becoming of small treelike stature, often colonial. Bark of twigs brown, dotted with pale, elongate to rounded lenticels, bark of older stems becoming gray to nearly black, somewhat mottled.

Leaves opposite, long-petiolate, with linear-subulate, quickly deciduous stipules, compound, usually with 3, rarely 5, leaflets; lower pair of leaflets sessile or nearly so, the terminal one usually with a stalk 1.5–3 cm long; leaflets glabrous above, or short-pubescent when young and becoming glabrous, sparsely pubescent beneath, broadly elliptic to ovate-elliptic, 3–10 cm long and 2–5 cm broad, bases rounded or broadly tapering, sometimes oblique, apices short-acuminate, margins finely serrate, lateral veins adnate to the midrib for a short distance then abruptly divergent.

Inflorescence a loosely few-flowered panicle axillary to the uppermost leaf of a short, lateral branchlet of the season as the new growth of the season is emerging; flower stalks slender, flexuous, 1–2 cm long, usually pendent, each jointed at or just above the middle, each subtended by a pair of long, linear-subulate, pubescent, membranous bracts. Flower bisexual, corolla campanulate. Calyx united only at base, lobes 5, erect, somewhat shorter than the petals, greenish white, linear-oblong, 6–8 mm long. Petals 5, erect, cream-colored striped with green. Stamens 5, inserted outside and below a fleshy disc, alternate with the petals and about equaling them in length. Ovary partially embedded in the disc, 3-locular, ovules 4–12 in each locule, styles 3, free below, united at the blunt stigmatic apex. Fruit elliptic to obovate, inflated, bladderlike, thin-walled, 3–5 cm long and to 3 cm broad, the carpels separating at the summit as the fruit matures thus becoming 3-lobed and weakly 3-beaked by the persistent styles, fruits persisting for a time after leaf-fall. Seeds gray to brown, 5–7 mm long, subspherical, plump or somewhat flattened, seed coat hard. Often no ovules maturing into seeds, if seeds are produced then mostly 1 or 2 per locule.

Infrequent in our area, floodplain woodlands where flooding is of brief duration, stream banks, wooded bluffs and ravine slopes, along the Chattahoochee and upper Apalachicola Rivers. (Que. to Minn., generally southward to s.w. Ga., cen. Panhandle of Fla., cen Ala., n. Miss., Okla.)

Sterculiaceae (STERCULIA FAMILY)

Firmiana simplex (L.) W. F. Wight. CHINESE PARASOL-TREE, VARNISH-TREE.

Fig. 332

The Chinese parasol-tree, native to tropical and subtropical China, is grown as an ornamental and has become naturalized very locally, generally in the near vicinity of plantings.

Twigs green, very stout, smooth, older stems grayish green, the grayness owing to a coating of waxy bloom; trunks eventually gray and somewhat roughened; leaf scars round or broadly oval, small narrow stipule scars to either side of their apices; buds brown-velvety, the axillary ones relatively small, moundlike, the terminal one broad, as broad as or a little broader than the stout tip of the twig; pith large, white, continuous.

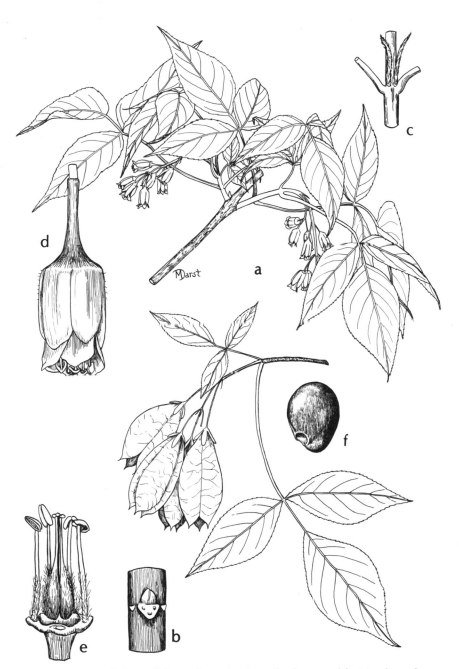

Fig. 331. **Staphylea trifolia:** a, flowering branch, above, and fruiting branch-let, below; b, piece of twig with leaf scar, stipule scars to either side of it, and axillary bud; c, node of developing branchlet (leaves cut off) showing stip-ules; d, flower; e, portion of flower, perianth removed; f, seed.

Fig. 332. **Firmiana simplex:** a, piece of twig with annual node and leaf scars;
b, leaf in background with overlay of portion of an inflorescence centrally; c,
functionally staminate flower; d, functionally pistillate flower; e, stalked ovaries (half mature) of one flower, each an opened follicle.

Leaves simple, alternate, deciduous, stipulate, stipules quickly deciduous, petiolate; varying not a little in length of petioles and size of blades even on a single branch; larger leaves with petioles 2–5 dm long, blades 1–3 dm long, mostly as broad as or a little broader than long, palmately veined, few of them broadly ovate and unlobed, usually 3–5-lobed, lobes sometimes somewhat fingerlike, sometimes broad, bases of the blades varying from subtruncate to deeply cordate, the cordations forming an inverted U-shaped sinus, sometimes their large bases touching or overlapping; other than the lobing, margins entire or vaguely and shallowly wavy; upper surfaces bright green and glabrous, the lower softly and compactly stellate-pubescent. Leaf blades turn golden yellow before falling in autumn.

Inflorescences produced in late spring or early summer, composed of many-flowered, open panicles or thyrses closely clustered at and approximate to the tips of very stout twigs of the previous season, leaves of lateral branches of the present season fully developed below them; panicles up to at least 12 in a cluster, the separate ones 1.5–6 dm long; main axis of the panicle green, the branches paler green proximally, cream-colored distally, the axes, flower stalks, and flower buds densely short-stellate-pubescent, some of the pubescence eventually sloughing, especially on the main axis. Flowers slightly fragrant, unisexual and staminate, or structurally bisexual but functionally pistillate, a given cluster of panicles with predominantly staminate flowers, an occasional functionally pistillate one, another cluster with the opposite combination. Staminate flower shortly stalked, the calyx tubular only basally and more or less campanulate, the calyx lobes 5, 7–8 mm long, narrowly oblanceolate and somewhat straplike, as the flower opens the lobes quickly becoming curled-reflexed; proximal portions of the calyx lobes yellow interiorly, the remainder cream-colored, the yellow part becoming deep pink at late anthesis; throat of the calyx tube and interior bases of calyx lobes clothed with straight hairs; within the base of the calyx tube is a fleshy nectariferous disc surrounding the base of the androphore, a stoutish cylindrical, white column extending beyond the calyx 7–8 mm and bearing at its expanded summit the fused stamens in the form of a yellow domelike hood, an insignificant mere rudiment of a pistil enclosed within it. Functionally pistillate flowers like the staminate respecting the perianth, the nectariferous disc within it, and the column emanating from that, the column, in this case, somewhat shorter, bearing at its summit a ring of more clearly discernible but indehiscent (sterile) anthers, above that ring the column supporting terminally a densely hairy gynoecium, at anthesis its parts appearing fused into one, soon appearing as 5 separate ovaries whose styles are fused at least above and terminally with 5 stigmas so closely appressed as to appear as one 5-lobed, disclike stigma. Flowers of the ultimate divisions of any given cluster of panicles open one by one over a relatively short period of time, in the case of the staminate, each falling shortly after full anthesis.

After pollination and fertilization ovaries rather quickly become stalked and expanding into 5 radiating, stalked follicles, each opening long before maturity into an ovate, leaflike body with 1–4 exposed, subglobular seeds on each of its margins. There being numerous of these on a given cluster of infructescences, they are notably conspicuous through the early summer. [*Firmiana platanifolia* (L. f.) Schott & Endl.]

Of sporadic occurrence in our area of coverage, usually in mixed, upland woodlands or on their borders usually in and about towns and cities. (Chiefly coastal plain, N.C. to n. Fla., westward to La.)

The Chinese parasol-tree is notable for its large, palmately, mostly 3–5-lobed leaf blades, by its smoothish greenish gray trunk bark, green twigs, and conspicuous panicles of open spoonlike fruits bearing exposed seeds on their margins during summer.

Styracaceae (STORAX FAMILY)

1. Pith diaphragmed, chambered between the diaphragms; uppermost axillary winter bud ovoid; flowers and fruits in fascicles or short racemes axillary to leaf scars on wood of the previous season; corolla lobes 4; ovary virtually wholly inferior; fruit 2- or 4-winged.
1. *Halesia*
1. Pith homogeneous and continuous; uppermost axillary winter bud thumblike in shape; flowers and fruits on short shoots of the current season; corolla lobes 5, rarely 6 or 7; ovary about one-third inferior; fruit not winged.
2. *Styrax*

Halesia (SILVERBELLS)

Deciduous shrubs or trees. Bark of trunks finely furrowed and ridged, sloughing in thin, narrow plates. Axillary winter buds 2 and superposed, the uppermost evident, ovoid, the lower about wholly hidden by the leaf scar. Leaf scars crescent- to horseshoe-shaped, each with a single crescentlike vascular bundle scar. Pith diaphragmed, chambered between the diaphragms. Pubescence, wherever present, stellate. Leaves alternate, simple, slenderly short-petiolate, without stipules; blades pinnately veined. Young stems, petioles, inflorescence axes, if any, flower stalks, floral tubes, and calyces sparsely to densely pubescent. Flowers pendent, bisexual, radially symmetrical, ovary virtually wholly inferior; borne in fascicles or short racemes 2 to 6 or 7 from axils of leaf scars on wood of the previous season before and as new shoots emerge. Floral tube with 4 very small, triangular sepals at its summit. Corolla white, showy, 4-cleft or -lobed, open long before reaching full anthesis (thus size measurements sometimes conflicting). Stamens 8–16, filaments united for about one-third of their length and adnate to the base of the corolla. Pistil 1, ovary 2–4-locular, each locule 4-ovuled, sterile tip of ovary tapering into the slender style, stigma minutely 4-lobed. Ovary greatly increasing in size after anthesis, eventually producing a dry, indehiscent, 2- or 4-winged, hard fruit bearing 1–3 seeds, at least a portion of the persistent style beaking the fruit apically.

The silverbells are deserving subjects for ornamental use, their dangling white flowers rather showy before and as new shoots develop in early spring.

1. Blades of larger mature leaves broadly oval, broadly obovate, or suborbicular, for the most part varying from half as broad as long or slightly more to as broad as long; petals united only basally; fruit broadly 2-winged.
1. *H. diptera*
1. Blades of larger mature leaves elliptic-oblong, elliptic, or slightly obovate, for the most part twice as long as broad or a little more; petals fused for more than half their length; fruits narrowly 4-winged.
2. *H. carolina*

1. Halesia diptera Ellis. TWO-WING SILVERBELL.
Fig. 333

A tree to 10 (15) m tall, flowering/fruiting when of small shrub stature. Mature blades of larger leaves broadly oval, broadly obovate, or suborbicular, 9–16 cm long and 4–10 cm broad, for the most part half as broad as long or slightly more to as broad as long; bases broadly rounded to shortly and broadly tapered,

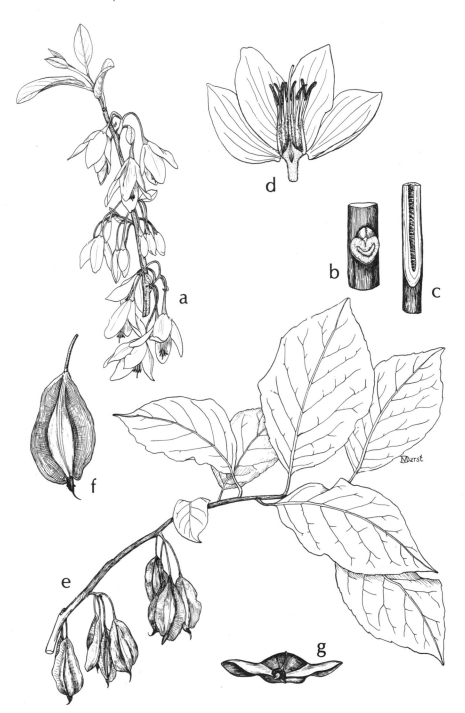

Fig. 333. **Halesia diptera:** a, flowers from buds on previous year's wood and emerging shoot at tip; b, node with leaf scar and superposed buds; c, longitudinal section of twig showing chambered pith; d, flower, opened out; e, fruiting branch; f, fruit; g, fruit, view toward apex.

661

apices abruptly short-acuminate or mucronate; margins irregularly and inconspicuously dentate-serrate; teeth glandular-tipped, both surfaces sparsely pubescent, the lower paler green than the upper (surfaces of emerging blades usually moderately to densely pubescent).

Corollas united only basally, petals at full anthesis 1–1.5 cm long in var. *diptera*, 2–3 cm long in var. *magniflora* Godfrey. Fruits broadly 2-winged, overall mostly 2.5–5 cm long and 1.5–2.5 cm broad, elliptic-oblong, broadly oval, or oblanceolate, shorter ones sometimes nearly orbicular.

The vars. *diptera* and *magniflora*, insofar as I know them, appear to be distinguishable morphologically only in respect to flower size, corollas of the former being, for the most part, on the order of half as large as those of the latter; it appears, in addition, that the var. *diptera* grows in generally wetter places, slight rises in river and stream floodplains, and on adjacent lower slopes, the var. *magniflora* in mesic woodlands of bluffs, ravines, and uplands (and is geographically much more restricted). The ranges roughly as follows: var. *diptera*, s.e. S.C. to w. Panhandle of Fla. (Choctawhatchee River to Escambia River), chiefly s. half of Ala., s. Miss. to e. Tex.; var. *magniflora*, s.w. Ga., Panhandle of Fla. (Tallahassee Red Hills, Ochlockonee, Apalachicola, and Chipola River bluffs, ravines, and uplands), and s.e. Ala.

National champion two-wing silverbell (1971), 4′1″ circumf., 55′ in height, 38′ spread, Maclay Gardens State Park, Fla.

2. Halesia carolina L. LITTLE SILVERBELL. Fig. 334

A small tree with rather tightly striated bark. Blades of larger mature leaves elliptic-oblong, elliptic, or slightly obovate, 7–12 (18) cm long and 3–7 cm broad, many of them twice as long as broad or a little more; bases rounded or little tapered, apices abruptly short-acuminate to gradually tapered to a point; margins irregularly and inconspicuously serrate, entire, or occasionally regularly finely serrate, teeth, if any, glandular-tipped; upper surfaces sparsely pubescent to glabrous (usually sparsely pubescent when emerging), lower surfaces sparsely pubescent throughout or only on the major veins (sparsely to very densely pubescent when emerging).

Corollas fused to well above their middles and bell-like, lobes broadly rounded, overall 1.2–1.5 cm long at full anthesis. Fruits narrowly 4-winged, oblanceolate or obovate in outline, 2–4 cm long and to 1.5 cm broad distally. [*H. tetraptera* sensu Kurz & Godfrey (1962); *H. parviflora* Michx.]

Wooded slopes, wooded river banks, wooded rises in floodplains; s. Ga., s. Ala., n. Fla. (Coastal plain, s.e. S.C. to n. cen. pen. Fla., westward to Ala., Miss.)

Use of the name, *H. carolina*, herein is in accordance with the nomenclatural analysis of Reveal and Seldin (1976) and according to them the name for the plant designated *H. carolina* in recent manuals is *H. tetraptera* Ellis. The latter occurs to the north and west of our range.

2. Styrax (STORAXES)

Deciduous (ours) shrubs or small trees. Axillary winter buds superimposed, the upper two relatively distinct, uppermost largest and thumblike in shape, lowermost smallest, appressed against the base of the bud above and sometimes obscure, usually all three evident prior to leaf fall on vigorous shoots of the season, densely pubescent. Leaf scars more or less shield-shaped but with a lobulate projection extending to either side of the lowermost bud and the base of

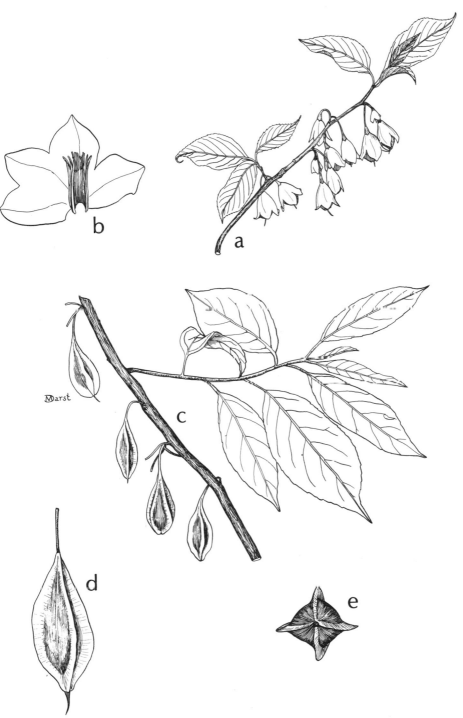

Fig. 334. **Halesia carolina:** a, flowering branch; b, corolla, opened out; c, fruiting branch; d, fruit; e, fruit, view toward the apex.

663

the one above, vascular bundle scar 1, crescent-shaped. Pith homogeneous and continuous. Pubescence stellate wherever present. Leaves alternate, simple, short-petiolate, without stipules; blades pinnately veined. Inflorescences racemes terminating short branchlets of the season, or flowers borne singly, or in few-flowered clusters from axils of small leaves of short-shoots of the season giving the appearance of foliose racemes. Flowers bisexual, radially symmetrical, ovary partially inferior. Floral tube surmounted by 5 very short-triangular sepals or narrow teeth. Corolla white, tubular below, 5 (rarely 6 or 7) -lobed above. Stamens usually 10, adnate to the corolla tube in a ring, filaments united below, free above, anthers dehiscing longitudinally. Pistil 1, ovary adnate to the basal part of the floral tube, 3-locular basally, 1-locular distally, each locule usually with 4–6 ovules; style linear throughout, stigmatic at the tip, somewhat exceeding the anthers, often, not always, all or part of it persistent at the tip of the developing fruit. Fruit globose or nearly so, densely, often compactly pubescent, crustaceous, loculicidally 3-valved, or irregularly dehiscent, or indehiscent, the floral tube adnate about the base or to half its length, each fruit 1–2-seeded.

1. Blades of larger mature leaves on a given plant not exceeding 8 cm long and 4 cm broad, many of them, even most of them, smaller, elliptic, oval, or narrowly obovate, lower surfaces, if softly pubescent, having a rusty color at least along the principal veins; flowers, for the most part, in leaf axils on short shoots appearing like foliose racemes.

1. *S. americanum*

1. Blades of larger mature leaves on a given plant 8–12 (15) cm long and 4–10 (15) cm broad, broadly oval, broadly obovate, to suborbicular, lower surfaces softly pubescent, the pubescence pale and imparting a grayish green color to the entire surface; flowers in drooping racemes terminating short branchlets of the season, only the lowermost 1 or 2 flowers subtended by a leaf or foliose bract, the others having only minute bracts subtending them.

2. *S. grandifolium*

1. **Styrax americanum** Lam. AMERICAN SNOWBELL. Fig. 335

Commonly a shrub, occasional specimens reaching small tree stature, usually abundantly floriferous and attractive when in bloom. Young stems varyingly glabrous, sparingly or densely pubescent. Petioles varying from densely pubescent to glabrous. Blades of larger mature leaves usually not exceeding 8 cm long and 4 cm broad, those of short-shoots generally smaller, elliptic, oblong-elliptic, oval, or narrowly obovate; bases broadly to acutely tapered, apices mostly broadly obtuse to rounded, sometimes cuspidate; margins entire, wavy, or obscurely and irregularly dentate or dentate-serrate; upper surfaces dark green, on some plants glabrous, on others sparsely pubescent, lower surfaces paler green, on some plants glabrous, on others sparsely to densely pubescent.

Flowers, for the most part, solitary in leaf axils on short-shoots, sometimes 2–4 terminating the shoots and subtended only by minute bracts; flower stalks, if as much as 10–14 mm long, usually glabrous or sparsely pubescent, the floral tube and calyx glabrous, sparsely pubescent, or densely pubescent, *or* if flower stalks shorter, then they and the floral tube generally densely pubescent; corollas lightly soft-pubescent, lobes oblongish, recurved, 10–12 mm long; ovary always densely pubescent, sometimes very compactly so. Fruits mostly 6–8 mm in diameter. [Incl. *S. pulverulentum* Michx. (*S. americanum* var. *pulverulentum* (Michx.) Rehd. in Bailey; *S. americanum* f. *pulverulentum* (Michx.) Perkins)]

Inhabiting moist to wet places, commonly where water stands much of the time, marshy shores, swamps and wet woodlands, cypress-gum ponds and depressions, ditches; throughout our area. (Chiefly but not exclusively coastal

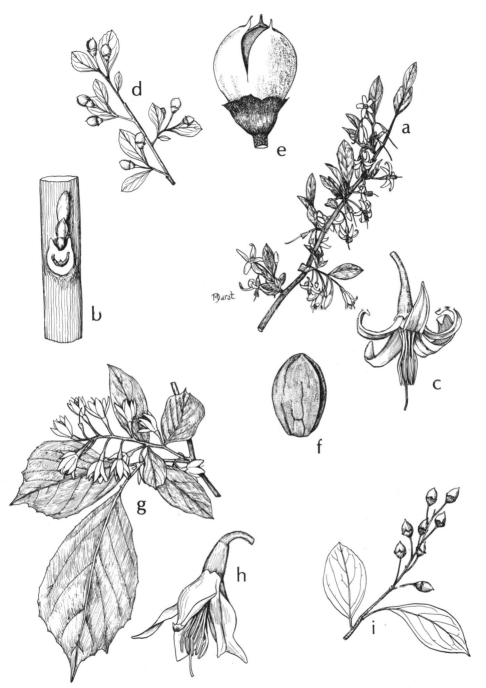

Fig. 335. a–f, **Styrax americana:** a, flowering branch; b, node with leaf scar and superposed buds; c, flower; d, fruiting branch; e, capsule, partially dehisced; f, seed. g–i, **Styrax grandifolia:** g, flowering branchlet; h, flower; i, fruiting raceme.

plain from Va. to cen. pen. Fla., thence westward to e. Tex., northward in the interior, s.e. Okla., Ark., Tenn., s.e. Mo., Ill., Ind., Ohio.)

Gonsoulin (1974) treats *S. americanum* as composed of two varieties, var. *americanum* and var. *pulverulentum* on the basis of amount and distribution of pubescence on various parts, although he states: "Intermediates are often more numerous than either of the two well-defined varieties, and positive identification may be impossible."

National champion American snowbell (1981), 8" circumf., 15' in height, 11' spread, Clemson University Forest, S.C.

2. Styrax grandifolium Ait. BIG-LEAF SNOWBELL. Fig. 335

Most commonly a shrub, sometimes with the stature of a small tree, conspicuous and pretty when in bloom. Buds, young stems, petioles, inflorescence axes, flower stalks, floral tubes, corollas, and ovaries pubescent, pubescence varying on some parts from sparse to dense. Larger mature leaf blades 8–12 (15) cm long and 4–10 (15) cm broad, broadly oval, broadly obovate, to suborbicular; bases broadly tapered to rounded, apices rounded, obtuse, abruptly short-acuminate, or merely cuspidate; margins entire or less frequently irregularly and obscurely dentate; upper surfaces dark green, glabrous or sparsely pubescent, especially along the major veins, lower surfaces softly pubescent throughout, the pubescence pale and imparting a grayish color to the entire surface.

Flowers borne in drooping racemes to about 15 cm long terminating short-shoots of the season when the leaves are essentially fully grown, the lowermost 1 or 2 flowers subtended by a leaf or foliose bract, the others by minute bracts, up to about 12 flowers per raceme. Corolla lobes elliptic to oblong-elliptic, 15–22 mm long, spreading to recurved. Fruits subglobose to ellipsoid, 7–9 mm in diameter.

In widely scattered localities, inhabiting well-drained, mesic woodlands of bluffs and ravines, on rises in floodplain woodlands; s. cen. and s.w. Ga., Panhandle of Florida from about the Ochlockonee R. westward, s. Ala. (Chiefly coastal plain and piedmont, s. Va. to Fla. Panhandle, westward to e. Tex., northward in the interior to s.e. Mo., Ky., Tenn., s.e. Ohio.)

Symplocaceae (SWEETLEAF FAMILY)

Symplocos tinctoria (L.) L'Her. SWEETLEAF, HORSE-SUGAR. Fig. 336

A shrub or small tree, the leaves sometimes deciduous in autumn, commonly most of them persisting through the winter and shed in spring just before or during the flowering period. Younger twigs sparsely pubescent, grayish, older twigs brown but more or less covered with a waxy ash-colored bloom, the bloom eventually sloughed and the twigs brown, somewhat furrowed; leaf scars half-round to more or less crescent-shaped, each with a single transverse bundle scar; pith chambered; twigs with a terminal winter bud.

Leaves simple, alternate, short-petioled; stipules none; blades pinnately veined, mostly elliptic, sometimes oblong or oblanceolate, cuneate at base, short-acuminate or acute apically, 5–15 cm long, margins entire or shallowly toothed, glabrous or pubescent above, moderately pubescent beneath, dull green, becoming yellowish green late in the season. Tissues of old leaves, espe-

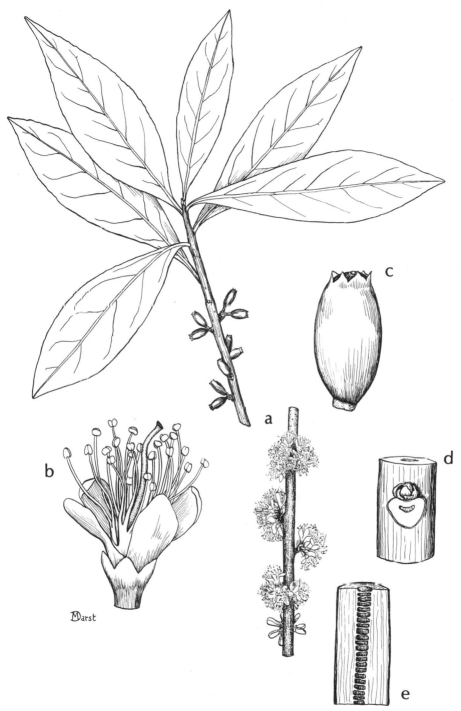

Fig. 336. **Symplocos tinctoria:** a, flowering branch, at right below, and fruiting branch, above at left; b, flower; c, fruit; d, piece of stem with leaf scar and axillary bud; e, piece of stem, longitudinal section showing diaphragmed pith.

cially near their midribs, slightly sweet to the taste; leaves relished by browsing animals.

Flowers bisexual, yellow or creamy yellow, fragrant, closely set in sessile, almost ball-like clusters on twigs of the previous season before new growth commences; ovary half-inferior at anthesis, wholly-inferior in the fruit, the floral tube surmounted by 5 very small, deltoid-obtuse, persistent calyx segments; petals 5, united at base, the lobes obovate, oblanceolate or spatulate, 6–8 mm long; stamens numerous, in fascicles, each fascicle inserted at the base of a corolla lobe; filaments free above and variously united basally. Ovary, surrounded at the summit by a nectariferous orange disc, usually 3-locular with 2 ovules in each locule; style slender, stigma capitate. Fruit an oblong, green drupe about 1 cm long, with small, erect, deltoid calyx segments at its summit, usually containing only one seed.

Upland woodlands, moist ravines, bottomland woodlands, floodplain forests and palmetto-marsh areas where seasonally wet, stream banks; throughout our area. (Del. to Tenn., Ark., and s.e. Okla., generally southward to n. Fla. and e. Tex.)

National champion sweetleaf (1972), 3'11" circumf., 55' in height, 27' spread, Tallahassee, Fla.

Tamaricaceae (TAMARISK FAMILY)

Tamarix

Tamarix gallica L. TAMARISK, SALT-CEDAR.   Fig. 337

Deciduous shrub or small tree with irregularly spreading-ascending, elongate branches, the leafy branchlets very slenderly flexuous.

Leaves alternate, minutely scalelike, sessile, thickish, broadest basally and more or less clasping, their abaxial surfaces punctate, margins entire; leaves of main axis of shoots of the season relatively remote from each other, those of secondary branchlets from the main axis more closely set, both abruptly attenuately tapered toward their tips, those of tertiary branchlets imbricated, more gradually tapered from their bases to acute tips.

Flowers minute, bisexual, radially symmetrical, borne in few to numerous, spreading, shortly stalked, spikelike racemes on inconspicuously leafy terminal portions of branchlets of the season. Calyx of 5 tiny, green, appressed sepals, triangular in outline and with entire margins. Petals 5, pink, small but much larger than the sepals, inserted beneath a staminal disc, narrowly obovate, their tips rounded, sometimes with 1–few very small notches. Staminal disc tiny, 5-lobed, the broadened base of each of the 5 filaments confluent with each of the lobes of the disc; filaments white, anthers pink, exserted from the corolla. Pistil 1, ovary superior, 1-locular, tapered distally and with 3 short styles stigmatic terminally. Fruit a capsule, lance-attenuate in outline, becoming longitudinally 3-angled at maturity, 2–2.5 mm long, styles persistent, dehiscing by 3 longitudinal valves. Seeds minute, each with a terminal, relatively conspicuous tuft of hairs.

Native to s. Eur., weedy in coastal areas of s.e. Ga. where locally abundant, perhaps sporadic elsewhere along the coasts of our area.

Fig. 337. **Tamarix gallica:** a, flowering branch; b, enlargement of small portion of vegetative branch; c, raceme; d, flower; e, staminal disc; f, pistil; g, fruiting raceme; h, fruit; i, seed.

Theaceae (TEA FAMILY)

1. Plant evergreen, the leaf blades leathery; flowers solitary in the axils of closely set leaves on twigs of the season, usually several on a given twig but blooming one at a time, thus a given tree with scattered flowers on its crown over several weeks during summer; stamens yellow.   1. *Gordonia*

1. Plant deciduous, leaves membranous; flowers solitary on twigs of the season, axillary to well-spaced leaves, all blooming nearly or quite simultaneously after the leaves are fully developed, the short flower stalks turned so that all flowers extend in one plane above the leafy branchlets; stamens with purple filaments and bluish anthers.   2. *Stewartia*

1. Gordonia

Gordonia lasianthus (L.) Ellis. LOBLOLLY BAY.   Fig. 338

Evergreen, potentially a tree to about 25 m tall with a relatively narrow, conical to columnar crown; however, producing flowers and fruits when of low stature, sometimes when as little as 5–10 dm tall.

Bark of older trunks dark gray, very thick, roughened by coarse interlacing, flat-topped ridges separated by rough, narrow furrows; twigs dark brown and often somewhat glaucous when fresh, those of the season glabrous to copiously short-pubescent, the hairs single or commonly in clusters of 2–4, the hairs usually eventually sloughed.

Leaves alternate, simple, very short-petiolate, petioles pubescent proximally; blades leathery, pinnately veined, mostly long-elliptic, 8–16 cm long and 3–5 cm broad, cuneate basally, apices obtuse to acute, sometimes notched, margins very shallowly appressed-crenate-serrate, upper surfaces glabrous, dark green, the lower paler, more or less olive-green, sparsely pubescent with hairs as on the twigs, the hairs sometimes sloughed as the leaves age.

Flowers handsome, solitary, axillary to close-set leaves on twigs of the season, their stiff stalks 5–8 cm long, usually several on a given twig but blooming one at a time, usually a relative few at one time on the tree as a whole, the flowering period extended over a number of weeks in summer. Flowers about 8 cm across; sepals 5, short-clawed at base, their blades suborbicular, silky-pubescent exteriorly, deciduous; petals 5, white, united basally, margins crinkly-fringed, their broadly rounded tips turned up, silky pubescent exteriorly; stamens numerous, yellow, the filaments united basally into a 5-lobed cup, each lobe flush against the base of the petal; pistil one, ovary superior, ovoid, 5-locular. Fruit a hard, woody, ovate-oblong capsule about 1.5 cm long, its surface appressed-silky-pubescent, dehiscing loculicidally, each locule with 4–8 flat, winged seeds about 1 cm long.

Evergreen shrub-tree bogs and bays, pond cypress depressions, acid swamps, scattered localities in our area. (Coastal plain, N.C. to about n. of Lake Okeechobee, pen. Fla., westward to La.)

National champion loblolly bay (1972), 12′9″ circumf., 84′ in height, 58′ spread, Ocala National Forest, Fla.

2. Stewartia

Stewartia malacodendron L. STEWARTIA, SILKY-CAMELLIA.   Fig. 339

Shrub or small tree, probably not exceeding 6 m tall, relatively inconspicuous save during its flowering period, then very attractive. The branches and leaves

Fig. 338. **Gordonia lasianthus;** a, flowering branch; b, small portion of lower leaf surface very much enlarged to show pubescence, and c, smaller portion of that surface much more enlarged; d, face view of flower; e, petal with one lobe of staminal cup appressed to its base; f, capsule, dehisced; g, seed.

671

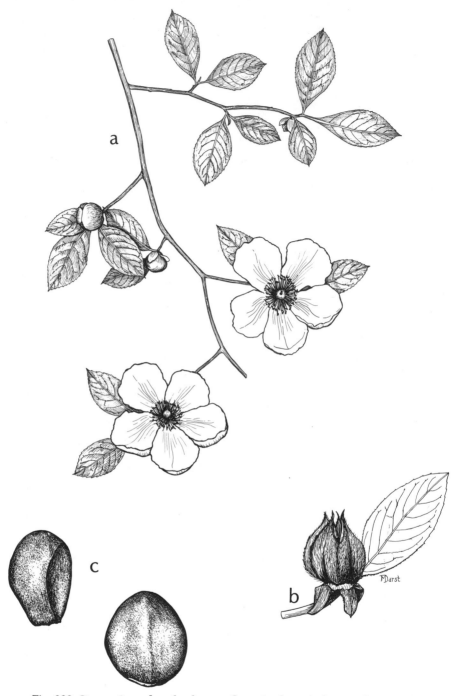

Fig. 339. **Stewartia malacodendron:** a, flowering branch; b, capsule; c, seed, two views.

(flowers when present) tend to be oriented more or less in one plane, thus spraylike, somewhat reminiscent of those of the flowering dogwood.

Leaves 2-ranked, deciduous, simple, alternate, pinnately veined; blades membranous, oval, broadly elliptic, some sometimes ovate or obovate, larger ones 5–10 cm long and 3–5 cm broad, major lateral veins arching-ascending and branching well before reaching the margin; margins obscurely serrate and somewhat ciliate, bases rounded to broadly and shortly tapered, apices mostly short-acuminate. Very young twigs and lower surfaces of developing leaves copiously silky-pubescent, some but not all of the pubescence eventually sloughed.

Flowers borne singly from leaf axils, radially symmetrical, bisexual, about 7–8 cm across the corollas, their short stalks turned so that all flowers extend in one plane above the leafy branchlets, thus giving the effect of floral sprays since the flowers bloom nearly or quite simultaneously. Sepals 5, broadly ovate with short tapered tips, silky-pubescent exteriorly. Petals 5, creamy white, obovate, spreading but with thin, crinkled, broadly rounded apices turned up, outer surfaces copiously silky pubescent on their proximal halves. Stamens numerous, filaments purple, flat and dilated basally, gradually narrowed to filiform distal portions, anthers bluish. Pistil 1, ovary superior, very densely silky-pubescent, 5-locular, style 1 (or styles 5 and coherent), stigma shortly 5-lobed. Fruit a woody, short-ovoid capsule 1–1.5 cm long, pubescent, loculicidally dehiscent (the more or less shriveled calyx persistent and sublustrous), lenticular or asymmetrically somewhat angled, about 6–7 mm long, a little longer than broad.

Understory plant of rich wooded bluffs and ravine slopes, creek banks; scattered localities, s.w. Ga., Fla. Panhandle, from the Ochlockonee River westward, s. Ala. (Chiefly, but not exclusively, coastal plain, Va. to Fla. Panhandle, westward to La., s. cen. Ark.)

Thymelaeaceae (MEZERON FAMILY)

Dirca palustris L. LEATHERWOOD.   Fig. 340

A rather straggly-branched, deciduous shrub 5–20 dm tall, older main stems to at least 2 cm in diameter, the bark very pliable, its surface smooth. Growth slow, the annual increments of growth commonly very short. The distal portions of the annual increments of growth becoming dilated-funnelform, in the winter condition capped by a short-conical, cottony-cushionlike, pseudo-terminal bud; lateral buds similar and borne angularly diverging on raised leaf scars, the leaf scars proper being narrow, nearly encircling the buds. Very early in spring flowers emerging from buds in short stalked clusters of 2–4, the 3 or 4 bud scales very hairy exteriorly, expanding considerably and subtending the flower cluster like an involucre during anthesis; new leafy shoots emerging quickly during anthesis and quickly reaching full size.

Leaves simple, alternate, pinnately veined, very short-petiolate; blades varying considerably in size, the larger ones 5–9 cm long and 3–6 cm broad, in shape oval, oblanceolate, broadly elliptic, or obovate, bases rounded to broadly short-tapered, apices blunt, upper surfaces green and glabrous, lower pale green, usually glabrous, sometimes sparsely short-pubescent.

Flowers radially symmetrical, bisexual, the superior ovary seated in a short, narrowly funnelform floral tube surmounted by a somewhat broader, funnelform, yellowish calyx 5–6 mm long and shallowly 5-notched at the summit.

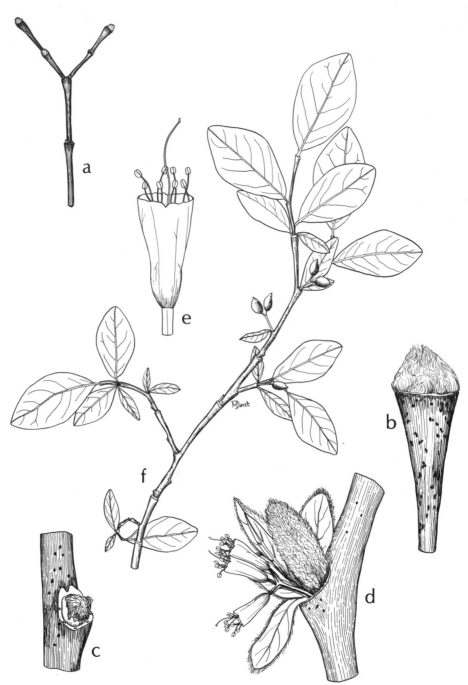

Fig. 340. **Dirca palustris:** a, winter twig; b, tip of winter twig with terminal bud; c, node of winter twig with leaf scar and lateral bud; d, node with emerging branchlet and its flowers; e, flower; f, branch with fruits.

Petals very minute and inconspicuous at the base of the calyx tube and between the insertion of the filaments, or none. Stamens 8 in 2 closely adjacent whorls of unequal length, the anthers exserted. Pistil 1, ovary with a minute ring surrounding its base; style long and slender, exserted, stigma minutely capitate. Fruit a 1-seeded, slightly asymmetrical, subellipsoid berry 5–6 mm long, yellow or greenish yellow approximately at maturity, ripening and falling about the time the leaves are fully grown, often whitening, then becoming purplish in after-ripening.

Of sporadic occurrence on rich wooded ravine slopes and bluffs, usually frequent where it does occur, s.w. Ga., Fla. Panhandle, s. Ala. (N.B., Que., Ont., westward to Minn., generally southward to s.w. Ga., Fla. Panhandle, Ark.)

A relatively low shrub of rich wooded bluffs and ravine slopes, the very pliable twigs with leathery bark are distinctive as are the somewhat funnelform dilations on the twigs marking the terminus of a year's increment of growth.

Tiliaceae (LINDEN FAMILY)

Tilia americana L. BASSWOOD, LINDEN.   Fig. 341

For eastern North America, authors have differed greatly in their interpretations of *Tilia*, the number of named species varying from (1) 3 to about 16, the delimitations being based primarily on differences in amount, kind, and distribution of pubescence, chiefly on leaves and twigs. Little (1953) and Ashby (1964) have stated that the leaf characters vary with season, environment, and position on the tree. Hickok and Anway (1972) indicate that populations studied by them exhibit both random character variations and continuous latitudinal variations, and conclude that the genus in eastern North America should be considered as composing a single highly variable species, *T. americana*. I choose to follow their interpretation.

Small to large deciduous tree, wood light in color and relatively soft. Bark of trunks gray to brown or reddish brown, relatively thick and tough, its surface sometimes broken up into small scales or furrowed, the inner bark with long, tough fibers. Twigs terete, sap mucilaginous, terminal buds none, axillary buds 2-ranked, leaf scars somewhat raised-angled-ascending, each with several, scattered vascular bundle scars. Pubescence of all parts, if any, stellate. Young twigs varyingly pubescent to glabrous, if pubescent most or all of the hairs sloughing.

Leaves simple, alternate, 2-ranked, slenderly petiolate, petioles mostly one-fourth to one-half as long as the blades. Stipules 0.5–1.5 cm long and generally quickly deciduous, occasionally a petiole of a young leaf bearing bracts like the stipules. Blades ovate to ovate-oblong, infrequently suborbicular, venation palmate or subpalmate (lower lateral veins arising with the midrib at the base of the blade), very variable in size, 5–20 cm long and 5–12 cm broad, bases commonly inequilateral and oblique, sometimes equilateral, truncate, subcordate to cordate, seldom tapered, apices usually abruptly short-acuminate; margins saliently serrate to crenate-serrate; pubescence of lower surfaces, if present, whitish, grayish, tan, or brown, sparse to dense (often soft and densely felty on unfurling leaves), varyingly partially to wholly persistent, or sloughing entirely, sometimes without pubescence and when young green or glaucous.

Inflorescences axillary, with few to numerous flowers in long-stalked cymes, the stalks of the cyme united part of the way to a distinctive and conspicuous,

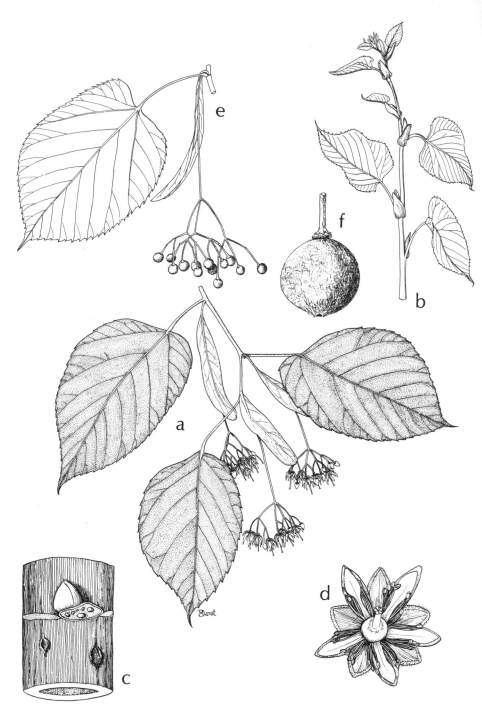

Fig. 341. **Tilia americana:** a, flowering branchlet; b, emerging branch showing prominent stipules; c, node with leaf scar, stipule scars to either side, and axillary bud; d, flower, face view; e, node with infructescence; f, fruit.

straplike, oblong, or spatulate, glabrous or pubescent bract, the distal portion of the bract extending beyond the point of departure of the stalk of the cyme and about equal in length to that of the cyme. The weight of the flowers and later the fruits tend to twist the bract so that flower and fruit clusters dangle from their lower sides. Flowers small, bisexual, radially symmetrical, pleasantly fragrant. Sepals 5, thickish, glabrous to densely pubescent exteriorly, usually pubescent interiorly, eventually deciduous. Petals 5, alternate with the sepals and a little longer than the sepals, cream-colored or yellow, oblong-spatulate, a petallike staminode opposite each one. Stamens numerous, cohering basally into 5 groups (in ours), each group opposite a petal, filaments sometimes forked distally, each fork bearing a half-anther. Pistil 1, ovary superior, densely pubescent, 5-locular (eventually 1-locular), 2 ovules in each locule, style slender to stout, partially to wholly pubescent or glabrous, stigma capitate, 5-lobed. Each flower and each cyme branch subtended by a small bract that is sloughed much before anthesis. Fruit dry and hard, indehiscent, subglobose to short-oblong or ovoid, mostly 5–7 mm in diameter, usually bearing 1–2 seeds. [Incl. (for our area of coverage) the following: *T. caroliniana* Mill.; *T. creno-serrata* Sarg.; *T. eburnea* Ashe; *T. floridana* Small; *T. georgiana* Sarg.; *T. heterophylla* Vent.; *T. lasioclada* Sarg.; *T. poracea* Ashe.]

Inhabiting mesic to submesic mixed woodlands; throughout our area. (Widespread in temperate e. N.Am.)

Ulmaceae (ELM FAMILY)

1. Leaf blades with 4–8 principal, ascending or arched-ascending, lateral veins, the lowest pair arising at the base of the blade and forming there a 3-nerved or a V-nerved part with the midrib (palmate), the principal lateral veins not reaching the blade margin but forming an anastomosis short of the margin; pith of the twigs chambered in the nodal region; fruit a smooth drupe.   1. *Celtis*
1. Leaf blades with 6–20 principal lateral veins, these angling-ascending approximately straight and parallel to each other, ending in marginal teeth or forking and the forks ending in marginal teeth; pith at the nodes of the twigs continuous; fruit a burlike drupe or a samara.
  2. Bark scaly and flaky, exposing reddish brown inner bark as it sloughs; flowers at least partially unisexual, appearing in fascicles on emerging shoots of the season and usually from buds on wood of the previous season as other buds are producing developing shoots; fruit a soft, wingless, burlike nut; most leaf blades broadest at their bases.
                                                                                    2. *Planera*
  2. Bark ridged and furrowed or scaly, not exposing reddish brown inner bark as it sloughs; flowers bisexual, appearing in fascicles or racemes either from buds on twigs of the previous season before new shoot growth commences or from axillary buds in autumn; fruit a samara; leaf blades broadest somewhat above their bases.   3. *Ulmus*

1. Celtis (HACKBERRY, SUGARBERRY)

Shrubs or small to moderately large trees. Bark gray, smoothish, or often roughened by corky outgrowths in the form of warty bumps or irregularly ridged masses. Pith of twigs diaphragmed, eventually chambered between the diaphragms. Leaf scars oval to crescent-shaped, vascular bundle scars 3. Leaves simple, alternate, 2-ranked, short-petiolate, stipulate, the stipules very quickly deciduous. Leaf blades commonly inequilateral and oblique basally, the principal lateral veins ascending or arched-ascending, the lowest pair arising at the base of the blade and with the midrib forming a 3–nerved part, the principal lateral veins not reaching the blade margins but forming an anastamosis short of the margins. (Note that the blades of leaves occurring on first, relatively short, usually flowering/fruiting shoots of the season may differ in size, sometimes in marginal toothing, if any, from those of sterile leader-shoots produced later or on sprouts.) Flowers small, functionally unisexual, the staminate with a vestigial pistil, the pistillate with sterile stamens; staminate flowers solitary or few in clusters, usually from buds on leafless proximal portions of the developing shoots, less frequently in fascicles on leafless shoots that are deciduous after flowering; structurally bisexual (functionally pistillate) flowers usually solitary in axils of leaves of developing shoots. Corolla none. Calyx deeply 5-lobed, stamens as many as the calyx lobes and opposite them, arched inwardly as the flower bud opens, then exserted; pistillate flowers similar structurally, the stamens infertile, not exserted, the pistil with a flasklike ovary, its short style bearing two prominent, outward-curved stigmas papillate on their inner faces. Fruits 1-stoned, small subglobose or globose drupes.

1. Leaf blades lanceolate or lance-ovate, less frequently ovate, most of them twice as long as broad or a little longer, long-tapered from the broadest places to acute, or more frequently to attenuate-acuminate tips; blades, many of them, notably inequilateral and falcate, bases distinctly oblique; upper surfaces of blades on older larger trees smooth to the touch, those of seedlings, saplings, and sprouts may be scabrid and sandpapery to the touch; plants potentially moderately large trees. 1. *C. laevigata*
1. Leaf blades distinctly ovate, less than twice as long as broad for the most part, varyingly equilateral or moderately inequilateral, the distal taper relatively short, varyingly acute, obtuse, or abruptly short-acuminate, the tip of the acumination often blunt; upper surfaces generally scabrid; plants shrubs or small trees. 2. *C. tenuifolia*

1. **Celtis laevigata** Nutt. Fig. 342

Potentially a moderately large deciduous tree, often flowering/fruiting when of small stature.

Petioles slender, mostly 5–15 mm long, somewhat longer on vigorous shoots, pubescent. Leaf blades mostly lanceolate or lance-ovate, few if any ovate, predominantly twice as long as broad or a little more, size varying considerably depending upon the position on the branch-system, from about 3 to 10 (15) cm long and 1.5–6 cm (8) broad; blades varying from equilateral (with rounded to truncate bases) to varyingly inequilateral (with moderately to markedly oblique bases), inequilateral blades commonly falcate; blade apices commonly attenuate-acuminate, less frequently acute, the taper of the former generally relatively long; upper surfaces of blades of mature leaves on relatively mature individuals usually essentially glabrous and smooth to the touch, those on seedlings, saplings, and sprouts usually scabrid; lower surfaces often, not always, shaggy-pubescent basally, sparingly pubescent along the veins, sometimes glabrous; margins variably entire, slightly wavy, irregularly few-toothed, or with regular toothing except around the bases.

Drupes subglobose, 5–8 mm in diameter, style usually sloughed when mature, surfaces orangish to brown or red, glaucous. [*C. mississipiensis* Spach, incl., *C. smallii* Beadle in Small]

Occurring in a wide variety of habitats, bottomland woodlands subject to some periodic flooding, mixed woodlands of river bluffs, ravines, upland mixed woodlands, old fields, fence and hedge rows; throughout our area. (S.e. Va., N.C., Tenn., w. Ky., s. Ind. and Ill., s. Mo., generally southward to s. pen. Fla. and e. half of Okla. and e. two-thirds of Tex.; n.e. Mex.)

Fig. 342. **Celtis laevigata:** a, flowering branchlet (staminate flowers below and pistillate flowers on leafy portion); b, staminate flower at right, pistillate flower at left; c, fruiting branchlet; d, elongation shoot.

2. Celtis tenuifolia Nutt. DWARF HACKBERRY.

A shrub, often scraggly, or a small irregular tree, often with a long-persistent shrub stage which may or may not be followed by a tree stage in which one or few trunks take over. Crown often compact, with rigid intergrown branches, the branch tips sometimes spinose.

Petioles slender, 2–8 mm long, short-pubescent; leaf blades predominantly ovate, equilateral and their bases barely subcordate, rounded, or truncate, or inequilateral and their bases oblique, the tips obtuse or bluntly short-acuminate, predominantly less than twice as long as broad, in general 3.7–6.3 cm long and 2.4–4.5 cm broad; margins entire or variably serrate; upper surfaces usually retrorsely scabrid, sometimes smooth, the lower with stiff straight or hooked trichomes along the veins or more or less throughout, sometimes glabrous.

Drupes globular, 5–8 mm in diameter, the style often persisting, surfaces orange to reddish brown. [Incl. *C. georgiana* Small (*C. occidentalis* L. var. *georgiana* (Small) Ahles)]

Occurring in xeric (less frequently submesic or mesic) habitats (perhaps mostly where limestone outcrops or where it is not far below the soil surface), glades, fields, openly wooded bluffs, ridges, and slopes, rocky woods, hedge and fence rows; local in our area. (S. N.J., w. W.Va., Ky., southernmost Ohio, Ind., and Ill., southeasternmost Kan., n.e. and s. two-thirds of Mo., generally southward to n. Fla. and e. and s.e. Tex., outlying n. stations in n. Ohio, n. Ind., s. Mich., s. Ont.)

2. Planera

Planera aquatica J. F. Gmel. PLANER-TREE, WATER-ELM.   Fig. 343

A shrub or small tree, rarely more than 18 m tall, often with a short trunk and a low, spreading crown, older specimens commonly with considerable sprouting from the base of the trunks; very old specimens may have had the main trunks die after which several trunks will have developed from basal sprouts. Bark of trunks scaly or flaky, sloughing in long, grayish brown plates and exposing reddish brown inner bark.

Leaves simple, alternate, 2–ranked, deciduous, short-petiolate, stipulate, stipules ovate or ovate-oblong, about twice as long as the petioles, quickly deciduous. Leaf blades 2–3.5 cm long and 2–4 cm broad, pinnately veined, lateral veins angling-ascending approximately straight and parallel with each other, forking before reaching the leaf margins, the forks ending in marginal teeth; in shape, blades mostly ovate, deltoid-ovate, or subrhombic-ovate, tapered to rounded basally, occasionally slightly inequilateral basally and with one side rounded and the other tapered, acute apically; surfaces glabrous or less frequently sparsely pubescent mostly on the veins beneath, upper surfaces dark green, the lower paler and the veins becoming brownish, margins somewhat irregularly serrate or obscurely doubly serrate.

Flowers small, bisexual or unisexual, both on the same plant, with a calyx of 2–5 separate or basally united sepals and no corollas. Staminate flowers from overwintering buds, borne in clusters below one or a few bisexual ones; on most specimens, some of the overwintering buds (before new shoot growth commences) are infested with insect larvae causing the buds to develop into ball-like, usually reddish, velvety-pubescent galls from which irregularly "erupt"

Fig. 343. **Planera aquatica:** a, fruiting branch; b, gall (many of which are usually present on trees as flowering season commences and from which "erupt" numerous staminate and 1–few bisexual flowers); c, developing branchlet with bisexual flowers; d, flowers, bisexual at left, staminate at right; e, fruit; f, seed.

clusters of staminate flowers and one or more bisexual flowers; in addition, bisexual flowers are borne singly or 2–3 from the axils of leaves of developing shoots of the season. Staminate flowers with a campanulate, 4–5-lobed calyx, as many stamens as calyx lobes, or (apparently) sometimes only one stamen (or occasionally none and the flowers pistillate). Pistil 1, ovary superior, with 2 prominent, densely pubescent stigmas.

Fruit a soft, stipitate, 1-seeded drupe about 1 cm long, its surface with irregular, fleshy projections and somewhat burlike, maturing in spring.

The water-elm inhabits river shores, banks of bayous and backwaters, oxbow lakes, sand and gravel bars, locally throughout our range. (Coastal plain, s.e. N.C. to n. Fla., westward to e. Tex., northward in the interior to s.e. Mo., and southernmost Ill.)

Distinguishing features helpful in its identification are the scaly-flaky bark which as it sloughs exposes reddish brown inner bark, 2-ranked leaves whose blades are ovate, deltoid-ovate, or subrhombic-ovate, their margins serrate, bases slightly, if at all, oblique.

National champion planer-tree (1971), 8′4″ circumf., 77′ in height, 47′ spread, Gadsden Co., Fla.

3. Ulmus (ELMS)

Small to medium-sized deciduous trees having ridged and furrowed or scaly bark, in some species some twigs with corky wings. Leaves simple, alternate, short-petiolate, 2-ranked, stipules thin-membranous, subulate, quickly deciduous. Leaf blades pinnately veined, the lateral veins angling-ascending straight and parallel to each other and ending in marginal teeth or distally forked and the forks ending in teeth, in some ending in the sinuses between the teeth as well, margins conspicuously to obscurely doubly serrate. Flowers borne in fascicles or short racemes from buds on twigs of a previous season before new shoot emergence, or (in one of ours) appearing in autumn on twigs of the season from buds in leaf axils or from buds on leafless nodes. Flowers small, usually bisexual. Calyx campanulate and cuplike, often oblique, with 3–9 lobes, or united only basally and 3–9-parted. Corolla none. Stamens 3–9. Pistil 1, ovary superior, sessile or stipitate, 1-locular, styles 2. Fruit a 1-seeded samara, the wing membranous, surrounding the seed-bearing portion, the withered calyx persistent at the base, in some kinds the style persistent and not at all shriveled.

1. Leaf blades with rounded to obtuse tips; flowers and fruits appearing in autumn.
                                                3. *U. crassifolia*
1. Leaf blades with pointed tips; flowers and fruits appearing in late winter or early spring before new shoot growth commences.
  2. The leaf blades lanceolate to narrowly elliptic, their apices acute; surfaces of samaras pubescent throughout and their margins copiously ciliate-fringed.   1. *U. alata*
  2. The leaf blades broadly elliptic, oblong, oval, or obovate, their apices acuminate; surfaces of the samaras glabrous, their margins copiously ciliate-fringed, *or* their surfaces pubescent only over the seed-bearing portion and otherwise glabrous, margins not ciliate.
    3. Inner bark mucilaginous; scales of the buds, at least the upper (inner) ones, woolly-pubescent with reddish brown hairs; flowers and fruits with very short stalks or sessile; faces of the samaras pubescent only over the seed-bearing portion, margins not ciliate; upper surfaces of leaf blades always harsh-scabrid to the touch.   4. *U. rubra*
    3. Inner bark not mucilaginous; scales of buds glabrous or minutely pubescent; flowers and fruits with longish slender stalks; faces of the samaras glabrous, their mar-

gins copiously ciliate-fringed; upper surfaces of leaf blades smooth to the touch (except those on stump sprouts, which are scabrid). 2. *U. americana*

1. Ulmus alata Michx. WINGED ELM. Fig. 344

Medium-sized to large tree, the bark of trunks with irregular, flat ridges composed of closely pressed scales separating shallow furrows. Corky, flat wings may form on either side of the twigs during the second year and on older but still small twigs may attain a breadth of 1.5–2 cm on each side; on older twigs, the corkiness more irregular, having been broken up and partially sloughed. Not all twigs of a given tree winged in many instances, and some trees have no winged twigs. In a given restricted locality where the tree occurs, usually one may see some individuals with some of the twigs winged.

Leaves very short-petiolate. Blades lanceolate to narrowly elliptic, 1.5–10 cm long and 1–4 cm broad, bases equilateral or infrequently slightly inequilateral and rounded on one side and slightly oblique on the other, apices acute; margins doubly serrate, upper surfaces dark green and smooth, the lower paler and duller, softly short-pubescent mainly on the principal veins.

Samaras notably stipitate, their bodies narrowly ovate to narrowly elliptic, about 5 mm long, the curvate (incurved) styles persistent; faces of the samara relatively sparsely pubescent, the margins, stipes, and styles copiously ciliate-fringed.

Upland well-drained woodlands, wooded slopes, moist to seasonally wet hammocks, and floodplains subject to short-term flooding; throughout our range. (Va., Ky., s. Ind., s. Ill., Mo., generally southward to n. cen. Fla. and e. Tex.)

Features serving to aid in identification of the winged elm are the 2-ranked, lanceolate or narrowly elliptic, apically acute, marginally doubly serrate leaf blades, and in many cases the presence of corky-winged twigs. It is most nearly like the cedar elm; it too may have corky-winged twigs, but its leaf blades are smaller and are for the most part rounded or obtuse apically.

National champion winged elm (1977) 11'1" circumf., 116' in height, 56' spread, Torreya State Park, Fla.

2. Ulmus americana L. AMERICAN ELM, WHITE ELM. Fig. 345

Medium-sized to large tree with a vaselike crown, the spraylike branches commonly with pendulous branchlets. Bark of trunk grayish brown, fissured and ridged, sloughing in flakes or scales from the ridges. Young emerging branchlets pubescent, most or all of the hairs quickly sloughing or sometimes to some extent persistent on year-old wood; woody twigs slender, brown or grayish brown, not winged. Scales of winter buds glabrous or minutely pubescent.

Leaves short-petiolate (sometimes sessile or subsessile on sprouts). Blades oblong-elliptic, oval, or ovate, less frequently lanceolate, variable in size, those lowest on branchlets smallest, 2–15 cm long and 1–8 (10) cm broad, moderately to significantly inequilateral for the most part, the bases correspondingly moderately to significantly oblique, if the former then both sides of the base more or less rounded, if the latter then one side of the base usually rounded, the other strongly slanted upwardly from the midrib, apices mostly acuminate, infrequently acute; margins occasionally singly serrate, usually doubly serrate, the latter varying considerably in the size of the primary teeth and number of secondary teeth, smaller teeth usually merely pointed and with 1–2 secondary teeth, grading to deeply cut primary teeth whose tips are acuminate, often curved upward and inward and hooklike and with 2–4 secondary teeth; upper

Fig. 344. **Ulmus alata:** a, portion of elongation shoot; b, woody twig with short branches of the season; c, clusters of staminate flowers; d, staminate flower, side view, at left above and staminate flower, face view, at right below; e, raceme of fruits; f, fruit.

Fig. 345. **Ulmus americana:** a, branchlet; b, node of winter twig with leaf scar and axillary bud; c, twig with clusters of staminate flowers; d, staminate flower, enlarged; e, staminate flower, face view; f, cluster of fruits; g, fruit, enlarged.

685

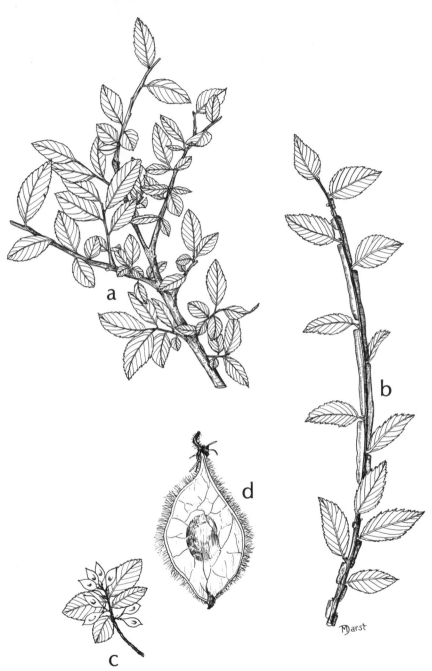

Fig. 346. **Ulmus crassifolia:** a, twig from a tree not having winged twigs; b, twig from a tree having winged twigs; c, branchlet with fruits; d, fruit, much enlarged.

surfaces dark green and sublustrous, usually smooth to the touch (except on leaves of stump sprouts and small saplings, which are notably scabrid), lower surfaces paler and duller green, with inconspicuous tufts of hairs in the principal vein axils, those of some trees sparsely soft-short-pubescent throughout.

Samaras stipitate, body ovate to oval, symmetrical or asymmetrical, 1 cm long or a little more, the wing reticulate-veined, faces glabrous, margins notably ciliate, apices notched by the 2 prominent, straight or more commonly incurved styles. [Incl. *U. floridana* Chapm. (*U. americana* var. *floridana* (Chapm) Little)]

Chiefly inhabiting bottomland woodlands or wet hammocks, occasionally on wooded slopes; throughout our area. (Nfld. to Man., generally southward to s. cen. pen. Fla. and e. third of Tex.)

National champion American elm (1978), 23'2" circumf., 99' in height, 133' spread, east of Louisville, Kan.

3. Ulmus crassifolia Nutt. CEDAR ELM. Fig. 346

Medium-sized tree, crown with crooked limbs and interlacing branches. Bark of trunk irregularly and moderately furrowed and with more or less interlacing flat ridges or simply broken into flat scales, sloughing in scalelike plates. Young twigs copiously very short-pubescent, most or all of the hairs sloughing during the first year or persisting; woody twigs brown, stiffish; none of the twigs on some trees winged, on others some with opposite, lateral, brown, corky wings, some not, the wings usually interrupted at the nodes, often sporadically produced along the twigs.

Leaves subsessile, tardily deciduous. Blades elliptic, oblong-elliptic, ovate, or the smaller ones proximally on branchlets often rotund or nearly so, mostly 1–3 (4) cm long and the larger ones to 3 cm broad, equilateral and bases rounded, moderately inequilateral and bases slightly oblique, or large ones often notably inequilateral, falcate, and one side of the base rounded, the other strongly slanting upwardly from the midrib, apices rounded to obtuse, sometimes acute on vigorous growth; margins crenate-serrate, sometimes irregularly doubly so, varying to singly serrate, teeth relatively small; upper surfaces darkish green, somewhat scabrid, sublustrous, petioles pubescent, lower surfaces shaggy-pubescent along the major veins or throughout.

Flowering and fruiting on twigs of the season in autumn. Samaras stipitate, body more or less oval and usually asymmetric, about 1 cm long, apices notched by the 2 short-triangular, incurved styles, faces sparsely short-pubescent, margins copiously ciliate.

In our area, inhabiting floodplain woodlands over limestone where only periodically and briefly flooded, local in Fla. along the Suwannee River, and in coastal or near-coastal hammocks over limestone from Levy to Hernando Co.; also near the Silver River, Marion Co. (Otherwise, n.w. Miss., s. Ark., cen. and n. La., cen. and s. Tex.)

National champion cedar elm (1969), 15'11" circumf., 94' in height, 70' spread, Bryan, Tex.

4. Ulmus rubra Muhl. SLIPPERY ELM, RED ELM. Fig. 347

Medium-sized tree with an open crown, its branches and twigs relatively few and far apart. Bark of trunk reddish brown to grayish, with flat, nearly parallel or interlacing ridges, shallow fissures between the ridges, inner bark mucilaginous when first exposed. Young twigs clothed with short, spiculelike, broad-

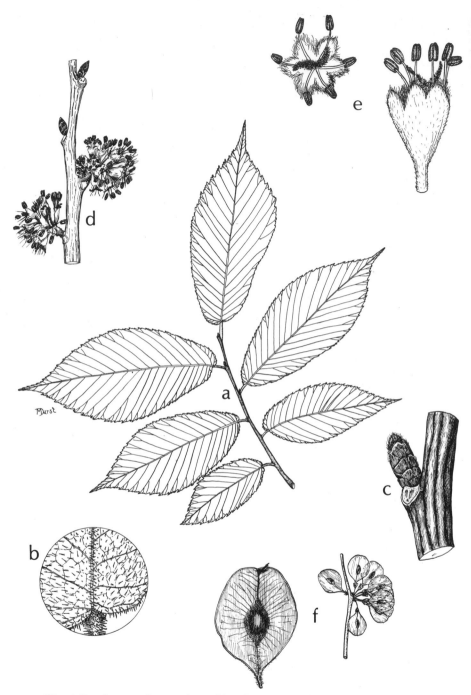

Fig. 347. **Ulmus rubra:** a, branchlet; b, portion of upper surface of leaf, enlarged; c, node of woody twig with leaf scar and axillary bud; d, twig with flower clusters; e, bisexual flower, face view to left and side view to right; f, fruiting cluster to right, and to left fruit, enlarged.

based hairs and notably scabrid, year-old twigs light brown and roughened by the persistent bases of hairs, tips of which have been sloughed, and by raised lenticels. Scales of winter buds clothed with rusty brown hairs, the hairs on inner scales especially prominent and abundant.

Petioles short, clothed with divergent, finely spiculelike hairs. Blades obovate, oval, or oblongish, proximal ones on branchlets much the smallest and often ovate, 3–18 cm long and the larger ones to 4–8 cm broad, occasional ones equilateral and bases rounded, mostly moderately inequilateral and bases moderately oblique with one side rounded and the other curvate or slanting; apices mostly acuminate; margins doubly and saliently serrate; upper surfaces very rough to the touch, the hairs erect, enlarged-based above which spiculelike, the latter often mostly sloughing but the surfaces still rough to the touch, hairs of the lower surfaces not enlarged-based, much softer to the touch, commonly tufted in the principal vein axils.

Samaras barely if at all stipitate, broadly oval or suborbicular, mostly 0.8–2 cm broad, wing not pubescent marginally, entire or slightly wavy, apically rounded, very slightly and narrowly notched, or sometimes with a concave outline, pubescent over the faces of seed-bearing portion. [*U. fulva* Michx.]

Mesic wooded bluffs and slopes, clay and calcareous woodlands, s.w. Ga., s. Ala., cen. and w. Fla. Panhandle. (S. Maine and s. Que. to e. N.D. and S.D., generally southward to Fla. Panhandle and e. third of Tex.)

National champion slippery elm (1972), 19'11" circumf., 90' in height, 80' spread, Perry Co., Pa.

Verbenaceae (VERVAIN FAMILY)

1. Leaves simple.
  2. Inflorescences and infructescences sessile, compound, short cymes borne in the axils of successive pairs of leaves.   1. *Callicarpa*
  2. Inflorescences and infrutescences congested, headlike spikes terminating 4-angled stalks solitary from the leaf axils.   2. *Lantana*
1. Leaves palmately compound.   3. *Vitex*

1. Callicarpa

Callicarpa americana L. BEAUTYBERRY, FRENCH-MULBERRY.   Fig. 348

Shrub to about 2 m tall. Young stems, young leaves, inflorescences axes, and calyces abundantly but loosely clothed with whitish stellate pubescence. Shoot growth occurs throughout much of the season and the terete stems remain essentially green and herbaceous until late in the season; superposed buds axillary to new leaves 3, the middle one potentially a branch bud, the upper one potentially an inflorescence bud, the lowermost and smallest one perhaps potentially with long dormancy; older woody stems light brown, leaf scars roundish-obcordate, vascular bundle scar 1 per leaf scar.

Leaves deciduous, simple, opposite or subopposite, pinnately veined, petiolate, petioles shorter than the blades; stipules none; blades ovate to lance-ovate, variable in size, from about 7 to 15 cm long and 3–10 cm broad, bases tapered, apices acute or acuminate, margins crenate to serrate distally above the tapered bases; lower surfaces retaining much of the pubescence, much of it, but not all, eventually sloughed from the upper surfaces.

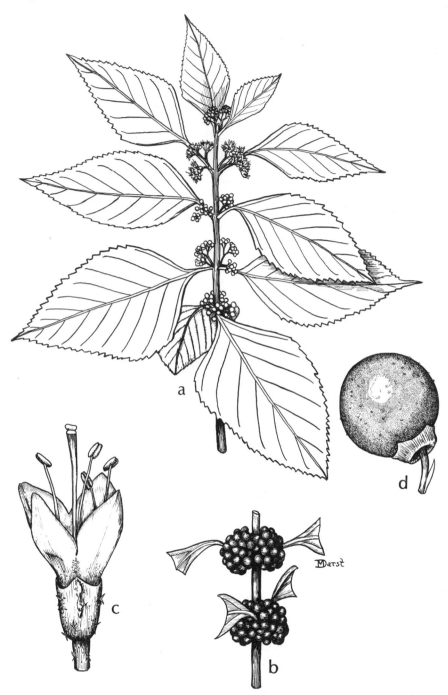

Fig. 348. **Callicarpa americana:** a, flowering branch; b, small piece of stem with clusters of fruits; c, flower; d, fruit.

Inflorescences essentially sessile, compound cymes axillary to a number of the successive pairs of leaves. Flowers very small, relatively numerous on each cyme, bisexual, radially symmetrical. Calyx pale yellowish green, campanulate with 5 low, triangular, erect lobes, the whole closely investing much of the corolla. Corolla usually lavender-pink (rarely all white on an individual plant), the tube funnelform, the limb with 5 ovate, apically rounded, somewhat incurved lobes. Stamens 5, inserted near the base of the corolla. Pistil 1, ovary superior, style slender, with a slightly dilated, slightly 2-lobed stigmatic tip, well exserted from the corolla.

Fruit a 4-stoned, lavender-pink, magenta, or violet, globose drupe 4–5 mm in diameter, the shriveled calyx loosely subtending it. (An occasional individual plant may have white fruits.)

Late in the season, the numerous clusters of fruits which compactly surround the nodes are very ornamental for some time before and after leaf-fall.

In diverse relatively open or closely canopied, usually well-drained woodlands and their borders, thickets, fence and hedge rows; throughout our area. (Md. southward to s. Fla., from Md. southwesterly to Tenn., Ark., and Tex.; n.e. Mex., Berm., Bah.Is., w. Cuba.)

2. Lantana

Deciduous subshrubs, herbage with a rank odor, the young branches and inflorescence stalks 4-angled, tardily becoming woody; leaf scars divaricately raised, circular, with 1 vascular bundle scar; pith white, continuous. Leaves simple, opposite, petiolate; stipules none; blades ovate to ovate-lanceolate, pinnately veined. Flowers in stalked, compacted, headlike spikes, stalks shorter than to much exceeding the leaves, axis of the spike thickened-domelike, elongating as the fruits develop, each flower in the axil of a bract, usually no empty (involucral) bracts at the base of the spike (in those treated here). Flowers bisexual, radiating from the axis in a very close spiral, overall the spike nearly flat-topped, the peripheral flowers reaching anthesis first and forming a ring of open flowers, gradually inwardly the remainder reaching anthesis, none withering until all have reached full anthesis. Calyx short-cylindric, irregularly notched at the summit, very closely investing only the base of the corolla tube. Corolla tube slender, nearly cylindric, slightly arching, the 4-lobed, asymmetrical limb abruptly spreading at the summit of the tube, about 7 mm across. Stamens in 2 pairs, filaments short, 1 pair inserted above the other more or less medially within the corolla tube, anthers not exserted. Pistil 1, ovary superior, 2-locular, deeply seated within the corolla tube; style slender, stigma obliquely capitate, reaching a little less than half the length of the tube. Fruit a subglobose or slightly oblate, 2-stoned drupe 5–6 mm in diameter; bracts of the spike persistent and remaining green, exserted from between the developing fruits until just before they ripen then shriveling and falling.

1. Woody stems prickly, erect or spreading.   1. *L. camara*
1. Woody stems not prickly, prostrate-decumbent, arching, or clambering, sometimes rooting at the nodes.   2. *L. montevidensis*

1. Lantana camara L. SHRUB-VERBENA, LANTANA.   Fig. 349

Erect shrub, in our range distal portions of branches commonly winter-killed and plants not usually exceeding 2 m tall; herbaceous portions of stems and inflorescence stalks hispid and atomiferous-glandular, shedding much of the

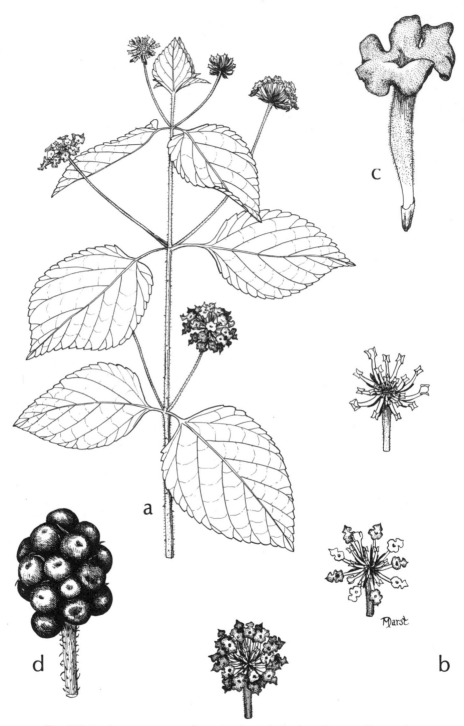

Fig. 349. **Lantana camara:** a, flowering branch; b, three flower spikes to show sequential anthesis of flowers, earliest at top; c, flower; d, spike of fruits.

pubescence and glandularness as they become woody and then having scattered, flattened, short, broad-based, recurved prickles on the tan bark.

Petioles 0.5–2 cm long; blades ovate, truncate to abruptly short-tapered basally, obtuse to acute or very short acuminate apically, margins crenate to bluntly serrate, surfaces hispid and scabrous; principal blades of flowering stems mostly 3–8 cm long and 2–5 cm broad, those of vigorous shoots often much larger, to at least 15 cm long and 6 cm broad.

Flowers usually 20–30 per spike; bracts bright green, oblongish or oblong-oblanceolate, acute, sparsely pubescent, reaching to about the middle of the fully developed corolla tubes. Corolla pubescent in the throat, 2 lobes of the limb approximately reniform, the other two narrower, essentially obovate but with very broadly rounded apices; on some plants lobes of the limb at first yellow, the center golden yellow, the lobes soon changing to salmon-pink or lavender-pink, the center of the limb to orange-red; owing to the way the flowers open gradually from the outside inwardly, the nearly flat-topped inflorescence has at first a ring of yellow flowers, finally all of them are salmon-pink to lavender-pink with reddish orange centers; on other plants limb of the corolla golden yellow at first with a tinge of orange at the throat, later the limb turning completely orange. The foregoing color combinations appear to be the prevailing ones in our area of coverage but perhaps others occur. Drupes a metallic blue at maturity.

Native of the Am. tropics, widely cultivated, sporadically to commonly naturalized in our area in waste places, vacant lots, roadside banks, banks of marshes, fence and hedge rows, thickets.

2. Lantana montevidensis (Spreng.) Briq. WEEPING OR TRAILING LANTANA.

Stems slender, pliable, prostrate, decumbent, arching, or clambering, herbaceous portions and inflorescence stalks hispid and minutely atomiferous-glandular, woody portions not prickly, bark tan.

Petioles 0.5–1 cm long; blades ovate, mostly 2–3 cm long, abruptly narrowed centrally at their bases and shortly decurrent on the petioles, truncate or very slightly cordate at either side of the petioles, apices obtuse, margins evenly crenate; upper surfaces moderately shaggy-pubescent, lower copiously and softly so.

Lowermost inflorescence bracts ovate, those above obovate, scarcely half as long as the corolla tubes, obtuse to rounded apically, upper surfaces hispid, hairs of the lower surfaces not enlarged-based, both surfaces atomiferous glandular; in most spikes all bracts subtend flowers, occasional ones have 1–3 empty bracts basally. Tube and limb of the corolla lilac, limb at first with a yellow throat bordered irregularly by white and whitish around the throat exteriorly. [*Lantana sellowiana* Link & Otto]

Native to S.Am., cultivated and locally naturalized in our area in vacant lots, on roadside banks, various weedy places about habitations.

3. Vitex

Vitex agnus-castus L. CHASTE-TREE, SAGE-TREE, HEMP-TREE.

Native to s. Europe, infrequently cultivated as an ornamental and infrequently naturalized.

In our area with a bushy-branched shrubby form; strongly aromatic. Young twigs densely clothed with short, brown, or grayish pubescence and resinous; woody stems smooth, brown, with unraised, paler, inconspicuous lenticels.

Leaves opposite, deciduous, palmately compound, petiolate, petioles 1.5–7.5 cm long, densely clothed with very short, grayish pubescence; stipules none; leaflet number variable, even on an individual plant, commonly 5, varying from 3 to 9; blades in general varying in size, unequal in size on each leaf, the larger central ones with subpetioles, pinnately veined, all lanceolate or lance-elliptic, to 10 cm long, rarely longer, 1.2–3 cm broad, relatively short-attenuate basally, long-acute distally; margins generally entire, some saliently serrate on an occasional individual; upper surfaces green and glabrous or nearly so (drying to dark brown), the lower surfaces tightly and densely gray-pubescent.

Inflorescence a thyrse, terminal on the branchlets or terminal and with branches from the upper leaves, overall somewhat pyramidal, the cymules often spaced apart on the axes; all axes of the inflorescence densely and compactly grayish-pubescent as are the minute bracts and calyces. Flowers sessile or subsessile. Calyx tubular-campanulate, 2–3 mm long, with 5 minute, triangular lobes at its summit. Corolla lavender or lilac, blue or white on occasional plants, tube funnelform, densely gray-pubescent exteriorly on the portion beyond the calyx, the limb obliquely 2-lipped and 5-lobed. Stamens 4, in pairs of unequal length, exserted. Pistil 1, ovary superior, becoming 4-locular, 1 ovule in each locule. Fruit a globose, dry hard drupe (about half enclosed by the persistent calyx), with a 4-celled stone.

Only occasionally naturalized in our area, in vacant lots and fence or hedge rows.

Vitaceae (GRAPE FAMILY)

1. Plant a semiwoody vine with succulent or subsucculent branches and leaves. 2. *Cissus*
1. Plant a woody vine with nonsucculent branches and leaves.
  2. Leaves compound.
    3. Leaf blades palmately compound.   3. *Parthenocissus*
    3. Leaf blades unevenly 2-pinnately, 2-ternately, or partially tripinnately compound. (*A. arborea* in genus)   1. *Ampelopsis*
  2. Leaves simple (in some sometimes deeply lobed or divided).
    4. Pith white; inflorescence a repeatedly bifurcate cyme; petals free, falling separately. (*A. cordata* in genus)   1. *Ampelopsis*
    4. Pith brown; inflorescence thrysoid-paniculate; petals united apically, connivent below where, at anthesis, their edges separate, dehiscing basally, the whole corolla then falling exposing the sexual parts.   4. *Vitis*

1. Ampelopsis

Woody vines (growth of the season herbaceous until late in the season), old woody stems to at least 6–7 cm in diameter, somewhat swollen at the older nodes, climbing by means of tendrils opposite leaves on vigorous leader shoots, each tendril bifurcate 4–5 cm above the base. Leaves deciduous, alternate, simple or compound, petiolate, stipulate, stipules small, soon deciduous. Inflorescence a bifurcately compound cyme borne oppositely to the leaves on branchlets of the season, their stalks longer than the leaves. Flowers small, mostly bisexual, radially symmetrical, a few of any given cyme at anthesis simultaneously. Calyx united, not distinguishable from the dilated summit of the flower stalk, with 5 low, merely undulate lobes at the upper edge. Petals 5, oblong, yellowish green, each with a stamen opposite it, both arising from below a conspicuous, cuplike, nectariferous disc with a crenate margin, the disc adnate to the base of the ovary. Pistil 1, with a short, conical style stigmatic at its tip. Inflorescence axes, flower stalks, calyx, exterior of the petals, and ovary very short-pubescent, pubescence of the ovary soon deciduous, its surface during rapid development into a fruit becoming glossy green with scattered purplish papillae, these later more or less tuberculate. Fruit a berry.

1. Leaves compound. 　　　　　　　　　　　　　　　　　1. *A. arborea*
1. Leaves simple. 　　　　　　　　　　　　　　　　　　2. *A. cordata*

1. **Ampelopsis arborea** (L.) Koehne. PEPPER-VINE. 　　　Fig. 350

Leaves compound, blades unevenly 2-pinnate, ternate, or partially tripinnate, overall more or less triangular in outline. Leaflets short-stalked (terminal ones longer stalked than the lateral ones), blades pinnately veined, 1–3 (7) cm long, ovate, bases rounded, truncate, or shortly tapered, margins saliently few-toothed, apices pointed and cusplike above the uppermost teeth; petioles, leaf axes, leaflet stalks and major veins of the lower surfaces of leaflets sparsely shaggy-pubescent.

Flowers very small. Petals arched-reflexed at full anthesis. Berry lustrous-black, oblate to subglobose (or asymmetric by abortion of 1 or 2 carpellate portions), mostly 8–14 mm broad fully developed.

Stream banks, floodplain forests, in and on the banks of marshes, sand and gravel bars, moist to wet hammocks, fence and hedge rows, thickets; throughout our area. (Md. to s. Ill. and Mo., generally southward to s. Fla. and e. Tex.)

2. **Ampelopsis cordata** Michx. RACCOON-GRAPE. 　　　Fig. 350

Leaves simple, pinnately or subpalmately veined, petioles slender, often about as long as the blades; blades of larger leaves to about 9 cm long and about as broad at the base, ovate, truncate to subcordate basally, apices short-acuminate, margins coarsely and often irregularly serrate, occasionally with 1 or 2 small lateral lobes; petioles at first sparsely pubescent proximally, usually densely pubescent at juncture with the blades, blades at first sparsely pubescent, upper surfaces soon becoming essentially glabrous, the lower usually remaining somewhat pubescent.

Petals apparently erect at full anthesis. Berries 7–10 mm broad, somewhat broader than long, usually 3-lobed, ripening from green to orange, rose, purple, finally blue, somewhat iridescent, with pale brownish lenticels or lenticellike, low excrescences.

Stream banks, floodplain forests, sand and gravel bars, borders of wet woodlands, lowland thickets, glades; s.w. Ga., s.e. Ala., Apalachicola River, Fla., possibly elsewhere in our area. (Va. to Fla. Panhandle, westward to Tex., northward in the interior to s. Ohio, s. Ind., s. Ill., Mo., and Nebr.; Mex.)

2. Cissus

Cissus trifoliata L. MARINE-IVY, MARINE-VINE.

A stoutish vine to 10 m long, older stems woody, the bark warty, young stems herbaceous and succulent, roots tuberous.

Leaves alternate, slenderly petiolate, deciduous or tardily deciduous. Blades variable: simple and broadly ovate or oblong, 3–divided, or 3–foliolate, to about 8 cm long overall and about as broad; blade divisions or leaflets ovate to

Fig. 350. a, **Ampelopsis cordata,** small portion of fruiting branch. b–g,
Ampelopsis arborea: b, portion of flowering branch; c, node with leaf and
basal portion of fruiting cyme; d, tip of growing shoot to show tendrils; e,
flower; f, flower with perianth and stamens cut away from the base of nectariferous disc; g, fruit.

oblong, cuneate basally, margins coarsely and irregularly toothed; long, spiraling, unbranched tendrils opposite some of the leaves.

Inflorescences axillary, long-stalked cymes, stalks exceeding the leaves, the ultimate divisions with the small flowers in subumbellate clusters. Flowers bisexual or unisexual (plants polygamomonoecious), greenish, creamy yellow, whitish, or purplish. Calyx cuplike, indistinctly 4-lobed at the summit. Petals 4, spreading, about 2 mm long. Nectariferous disc cuplike and adnate to the ovary nearly to its summit. Ovary incompletely 2-locular, style slender, longish, stigmatic at the tip.

Fruit an ovoid or obovoid, blue-black, 1–4-seeded berry 6–8 mm long, its stalk usually curvate or recurved. [*Cissus incisa* (Nutt.) Desmoul.]

Maritime woodlands, shell mounds in salt marshes, rocky ledges and rocky woodlands inland; apparently sporadic and coastal in our area. (Gulf Coast of Fla. to Ariz., northward in the interior to Ark., Mo., Kan.; n. Mex.)

3. Parthenocissus

Parthenocissus quinquefolia (L.) Planch. VIRGINIA-CREEPER, WOODBINE.

Fig. 351

Woody vine (growth of the season herbaceous), old stems to at least 6 cm in diameter, climbing by means of very slenderly pinnately branched tendrils opposite the leaves of vigorously growing leader shoots; branch tips of tendrils, if in contact with a suitable substrate, dilate somewhat and the lower surfaces of the dilations become markedly adhesive; without such contact, the tendrils quickly shrivel. Old stems high in trees or in other high places commonly form leaders with short leafy branches which grow freely downward as festoons. Vines from seedlings in places where there is nothing upon which to climb run along the ground and attach to it by adventitious roots.

Leaves palmately compound, petiolate, stipulate, stipules membranous, linear-oblong, about 5 mm long, quickly deciduous. Leaves of short branches from old stems much larger than those of vigorously growing leader shoots, however, leaves of the short branches vary greatly in size. Leaflets 5, 4, or 3 on the leader shoots, 5 on branches from older stems. Leaflets tapered proximally to sessile or subsessile bases, pinnately veined, mostly acuminate apically, margins serrate from somewhat above their bases, the teeth of large leaves coarse, sometimes some of them with small subteeth; leaflets of small leaves of leader shoots mostly lanceolate or lance-ovate; lower leaflets of short branches from older stems usually broadest a little below their middles, varying from lanceolate to ovate and usually inequilateral, other three leaflets generally broadest above their middles and tending to be mostly obovate or oblanceolate; largest leaves have petioles 15–20 cm long and their largest leaflets about 18 cm long and 5 cm broad. Stems, petioles, leaflet surfaces, and inflorescences glabrous on some plants, softly pubescent on some, the latter referable to forma *hirsuta* (Donn) Fern.

Flowers very small, borne in paniclelike compound cymes of very variable size, smaller ones few-branched and with few flowers, larger ones much branched and with many flowers, in either case few flowers at anthesis at any one time; inflorescence branches and flower buds usually a dull brick red. Flowers mostly bisexual, radially symmetrical. Calyx red, bowllike or saucerlike, with 5 low lobes at the summit; petals 5, more or less boatlike from a little above their bases, reddish with greenish margins, strongly reflexed at anthesis.

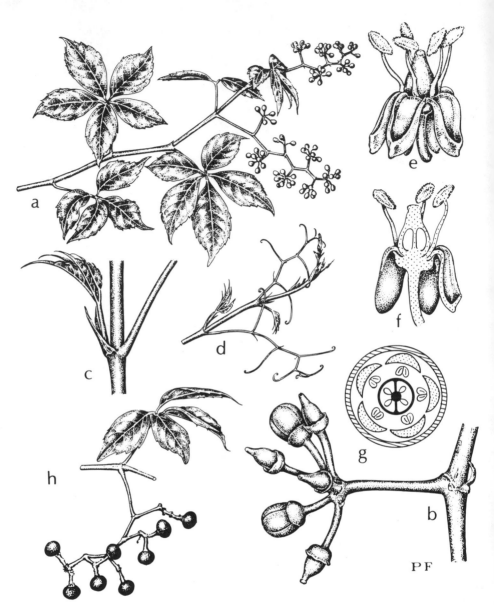

Fig. 351. **Parthenocissus quinquefolia:** a, branchlet with inflorescences (flowers at bud stage); b, one unit of inflorescence; c, node to show stipule; d, tip of growing branch to show stipules; e, flower, side view; f, flower longitudinally dissected; g, floral diagram. (Courtesy of Fairchild Tropical Garden.)

Stamens 5, opposite the petals, anthers attached at their middles. Pistil 1, ovary ovoid, 2-locular, style thickish conical, stigmatic at the tip; nectariferous disc none. Fruit a slightly obovoid, slightly 2-lobed berry essentially truncate at the apex.

In diverse mixtures of upland and bottomland woodlands, thickets, hedge rows; throughout our area. (S.e. Maine, s. N.H., Vt., s.w. Que. to Minn., generally southward to s. Fla. and Tex.)

4. Vitis (GRAPES)

Contributed by Michael O. Moore, Botany Department, University of Georgia.

Deciduous woody vines or viny shrubs climbing by tendrils. Stems generally regarded as sympodial; bark adherent with prominent lenticels (*V. rotundifolia*) or shredding in strips and without evident lenticels, then exfoliating with age. Pith brown, interrupted by nodal diaphragms (except *V. rotundifolia*). Tendrils unbranched (*V. rotundifolia*) or bifurcate to rarely trifurcate, without adhesive discs at their tips, borne at only two consecutive nodes in our species. Leaves petiolate, blades simple, unlobed to lobed, occasionally deeply so, palmately veined, cordate to orbicular, toothed to nearly entire, frequently mucronate, bases mostly cordate, occasionally truncate. Stipules caducous, 1–7 mm long, promptly deciduous. Inflorescence thrysoid-paniculate, in our species borne opposite only two consecutive nodes on branchlets of the season. Flowers pedicellate, perigynous, functionally unisexual, occasionally bisexual; plants polygamodioecious. Calyx minute and fused to form a collar at the base of the flower, essentially absent. Corolla of 5 (3–9) petals united apically, 1.5 to 3.0 mm long, lobes connivent below in bud, separating at anthesis and dehiscing at the base, the whole corolla then falling exposing the sexual parts, the fallen corolla usually with the lobes spreading radially in the fashion of a 5-bladed propeller. Stamens 5 (3–9), filaments erect in staminate and bisexual flowers, 2–7 mm long, reflexed (rarely absent) in functionally pistillate flowers; anthers dorsifixed, valvate, introrse, ca. 0.5 mm long. Nectariferous intrastaminal disc of 5 more or less separate glands alternating with the stamens. Pistil 1, 0.7–2.0 mm long; ovary 2 (3–4) -locular, each locule with 2 ovules; style very short; stigma capitate. Fruit a pulpy 1–4-seeded globose or subglobose berry. Seeds obovoid to pyriform, 3–8 mm long, the ventral surface with two longitudinal grooves on either side of the attached funiculus (raphe).

1. Tendrils unbranched; bark closely adherent, with prominent lenticels; pith continuous through nodes.   5. *V. rotundifolia*
1. Tendrils bifurcate to trifurcate; older bark shredding, with inconspicuous lenticels or lenticels absent; pith interrupted by nodal partitions (diaphragms).
  2. Leaves glaucous beneath, the glaucescence occasionally somewhat obscured by arachnoid pubescence; nodes frequently glaucous; berries glaucous.   1. *V. aestivalis*
  2. Leaves not glaucous beneath, various types of lower leaf surface pubescence present; nodes not glaucous; berries with little or no glaucescence.
    3. Leaves densely and evenly white to rusty tomentose beneath, concealing the leaf undersurface but not always the veins; nodal diaphragms greater than 8 mm wide; berries greater than 9 mm in diameter; infructescences with less than 20 berries.
                                              6. *V. shuttleworthii*
    3. Leaves variously pubescent beneath but never so dense as to conceal the leaf undersurface; nodal diaphragms less than 6 mm wide; berries usually less than 9 mm in diameter; infructescences usually with more than 20 berries.
      4. Branchlets of the season slightly to distinctly angled, pubescent, often densely so,

with either short, straight trichomes which are perpendicular to the branchlet surface or with appressed arachnoid trichomes, or both; nodes frequently banded with red pigmentation; leaves arachnoid-pubescent beneath.

    5. Branchlets distinctly angled and covered with short, straight trichomes which are perpendicular to the branchlet surface.   2. *V. cinerea* var. *cinerea*

    5. Branchlets slightly angled and covered with appressed arachnoid pubescence.
                                    3. *V. cinerea* var. *floridana*

4. Branchlets of the season terete, glabrous or only very sparsely arachnoid-pubescent; nodes not banded with red pigmentation; leaves more or less glabrous beneath or with tufts of short, straight trichomes along the veins and in their axils, occasionally with very sparse arachnoid pubescence as well.

    6. Branchlets of the season entirely purplish red; nodal diaphragms 2–5 mm thick; leaves generally 3-lobed, the terminal lobe acuminate.   4. *V. palmata*

    6. Branchlets generally green to gray or brown, *or* with red pigmentation only on the upper surface; nodal diaphragms less than 2 mm thick; leaves unlobed or with 2 shoulderlike lateral lobes, the terminal lobe more or less acute.   7. *V. vulpina*

1. Vitis aestivalis Michx. SUMMER GRAPE.   Fig. 352

High-climbing vine, branchlets of the season terete, arachnoid floccose to glabrous. Bark exfoliating in shreds on mature stems, lenticels absent or inconspicuous, pith brown, interrupted by diaphragms at the nodes, diaphragms 2–4 mm thick. Tendrils bifurcate, a tendril or inflorescence present at only 2 consecutive nodes, nodes frequently glaucous, not banded with red pigmentation.

Petioles about as long as the blades, glabrate to pubescent. Blades cordiform to orbicular, unlobed to more frequently 3-shouldered, often 3–5-lobed, often deeply so, when lobed the lobes mostly acute, the sinuses rounded to acute; margins crenate to dentate; upper surface of mature leaves glabrous to puberulent, lower surface glaucous with varying degrees of arachnoid, floccose pubescence, when heavy the glaucescence somewhat obscured, the pubescence whitish to more commonly rusty, hirtellous trichomes also frequently present along the veins and as tufts in the vein axils; stipules 2–4 mm long.

Panicles 5–16 cm long, usually narrowly triangular in outline, infructescences usually with more than 20 berries (or pedicels). Berries 5–12 mm in diameter, black-glaucous, without lenticels. Seeds tan to brown, pyriform, 3–6 mm long. [*V. rufotomentosa* Small]

Generally found on well drained sites, woodlands of various mixtures, woodland borders, thickets, fence and hedge rows, scrub, stabilized dunes, less often along stream or river banks, rarely in floodplains or lowland woods; throughout our area. (Mass. to Ont. and s. Minn, southward to s. pen. Fla. and e. third of Tex.) Flowers April to mid-May; fruits June to September in our area.

This species is frequently confused with both varieties of *V. cinerea* that are found in our area. However, the glaucous leaf undersurfaces, more heavily glaucous and larger berries, terete branchlets which are less evenly pubescent, preference for more well drained, drier habitats and earlier blooming period distinguishes *V. aestivalis* from *V. cinerea*.

2. Vitis cinerea (Engelm. ex Gray) Millardet var. cinerea. DOWNY OR SWEET WINTER GRAPE.

High-climbing vine in floodplains and lowland woods, branchlets angled (the angling difficult to see with the unaided eye), branchlets of the season covered with dense short, straight (hirtellous) trichomes, occasionally thinly arachnoid also. Bark exfoliating in shreds on mature stems, lenticels absent or inconspic-

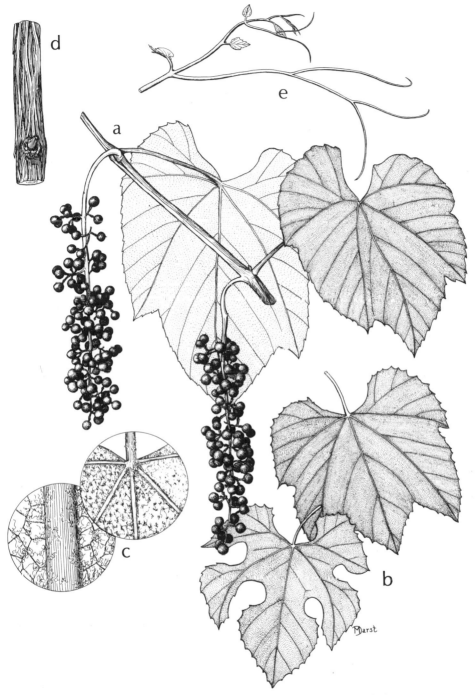

Fig. 352. **Vitis aestivalis:** a, piece of fruiting branch; b, variable leaves; c, portions of lower leaf surface, enlarged at right and more enlarged at left; d, piece of woody stem; e, elongation tip of branch with tendrils.

uous, pith brown, interrupted by diaphragms at nodes, diaphragms 3–5 mm thick. Tendrils bifurcate, a tendril or inflorescence present at only 2 consecutive nodes, nodes of branchlets of the season commonly banded with red pigmentation, nodes not glaucous.

Petioles about as long as the blades, puberulent to pubescent with hirtellous trichomes, thin arachnoid pubescence commonly present as well. Blades cordiform, unlobed to 3-shouldered, rarely 3-lobed, the apex acute to acuminate; margins crenate; upper surface of mature leaves glabrous to pubescent, lower surface not glaucous, more or less evenly arachnoid pubescent, the pubescence mostly whitish, hirtellous trichomes also commonly present along the veins and as small tufts in the vein axils; stipules 3–5 mm long.

Panicles 10–20 cm long, usually broadly triangular in outline, infructescences usually with more than 20 berries (or pedicels). Berries 4–9 mm in diameter, black, with little or no glaucescence, lenticels absent. Seeds brown, obovoid, 2–4 mm long.

Found in floodplains, lowland woods, pond and stream margins, rare in our area, recorded only from Jackson, Wakulla and Washington Cos. Native to the rich bottomlands of the Mississippi Basin, this taxon reaches its easternmost distribution in our area and intergrades into the more common *V. cinerea* var. *floridana* here. (W. Tenn. and Ill. southward to Tex. and eastward to a few scattered localities in Ala. and Panhandle Fla.) Flowers late May to June; fruits July to October in our area. This species is frequently confused with *V. aestivalis*. See the discussion provided under *V. aestivalis*.

3. **Vitis cinerea** var. **floridana** Munson. SIMPSON'S GRAPE.

Similar in appearance to *V. cinerea* var. *cinerea* but differs from this latter variety by having branchlets of the season which are only slightly angled, arachnoid pubescent, often densely so, and generally lacking the dense hirtellous pubescence characteristic of *V. cinerea* var. *cinerea*. The leaf undersurfaces of *V. cinerea* var. *floridana* also tend to be more densely arachnoid pubescent than is common in *V. cinerea* var. *cinerea*. [*V. simpsonii* Munson (1887, not 1890)]

Common in floodplains, lowland woods, stream and pond margins, throughout our area. (Coastal plain of Va., S.C., N.C., Ga., Fla., Ala., and Miss.) Flowers late May to June; fruits July to October in our area. This species is frequently confused with *V. aestivalis*. See the discussion provided under *V. aestivalis*.

4. **Vitis palmata** Vahl. RED, CAT OR CATBIRD GRAPE. Fig. 353

Relatively slender, high-climbing vine, the branchlets of the season subterete and entirely dark crimson or purplish red until mature, upon maturity the branches then of a reddish brown to chestnut color, glabrous to very thinly arachnoid. Bark exfoliating in shreds on mature stems, pith brown, interrupted by nodal diaphragms, diaphragms 2–5 mm thick. Tendrils bifurcate, red-pigmented when young, a tendril or inflorescence present at only 2 consecutive nodes, nodes not glaucous.

Petioles slender, somewhat shorter than the blades, glabrous to puberulent. Blades generally cordiform, commonly deeply 3 (5) -lobed, the lobes attenuate acuminate, sinuses acute (V-shaped) to rounded (U-shaped); margins dentate-serrate; upper surface of mature leaves glabrous, lower surface not glaucous, glabrous or pubescent with only hirtellous trichomes along the veins and in their axils; stipules 3–4 mm long.

Panicles 6–15 cm long, usually narrowly triangular in outline, infructescences usually with more than 20 berries. Berries 5–8 mm in diameter, bluish

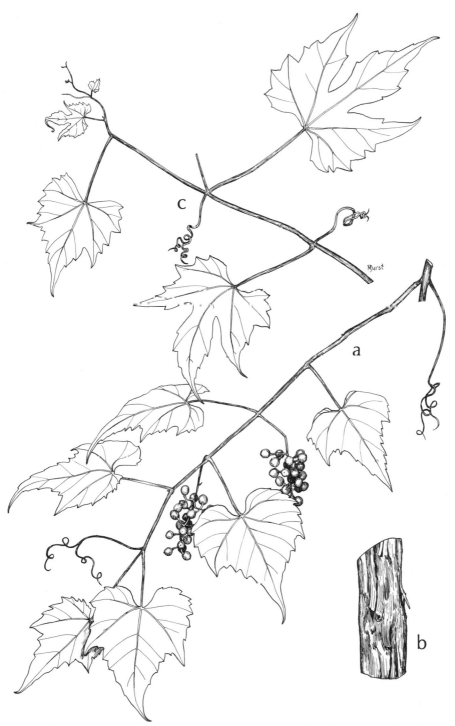

Fig. 353. **Vitis palmata:** a, fruiting branch; b, piece of woody stem; c, piece of
leafy stem from basal sprout.

black to black, with very little or no glaucescence, lenticels absent. Seeds dark brown, globose, 4–7 mm long, nearly filling the berry.

River banks and alluvial floodplain woodlands, in our area along the Apalachicola River in Fla., possibly along the Chattahoochee River in s.w. Ga. and s.e. Ala. (Ill. and Ind. southward to Mo., Tex., w. cen. Ala., cen. Panhandle of Fla.) Flowers the latest of all native species, mid-June; fruits late July to October in our area.

5. Vitis rotundifolia Michx. MUSCADINE, BULLACE. Fig. 354

High-climbing vine, branchlets of the season terete to slightly angled. Bark of younger woody stems with evident lenticels, that of older stems tight, not exfoliating, that of still older stems exfoliating in plates, pith brown, continuous through nodes, diaphragm absent. Tendrils unbranched, a tendril or inflorescence present at only 2 consecutive nodes, nodes not glaucous but often banded with red pigmentation. Very young, rapidly growing stems and leaf surfaces usually with thin, loose, grayish arachnoid pubescence or with dense, rusty, arachnoid pubescence at the nodes of the stems and pinkish on leaf surfaces, the pubescence eventually deciduous.

Petioles mostly as long as the blades, glabrous to glabrate. Blades cordiform to nearly reniform, very rarely lobed; margins crenate to dentate, apices very short-acuminate; upper surface of mature leaves glabrous and lustrous, lower surface not glaucous, glabrous or pubescent with few to many hirtellous trichomes along the veins and in their axils; stipules 1–2 mm long.

Panicles 3–8 cm long, rarely longer, usually more or less globose in outline, infructescences usually with less than 12 berries (or pedicels). Berries 10–25 cm in diameter, generally black or purplish, occasionally bronze when ripe, glaucescent, with tan, circular lenticels present on the skin. Seeds brown, oval to ellipsoidal, 5–8 mm long. [*Muscadinia rotundifolia* (Michx.) Small]

Inhabiting a very wide variety of sites, both upland and well-drained and lowlands and poorly drained, including intermittently flooded bottomlands; throughout our area. (Del. to Ky., s. Ind., Mo., generally southward to Fla. and e. Tex.) Flowers late April to May; fruits late July to September in our area.

Individuals of this species with smaller leaves, fruits only 5–9 mm in diameter and infructescences with more than 12 berries (or pedicels) have been treated as *V. munsoniana* Simpson ex Munson [*Muscadinia munsoniana* (Simpson ex Munson) Small] by several authors in the past. In our area, *V. munsoniana* is listed as occurring in northeast Fla. and the peninsular region, while *V. rotundifolia* is restricted to the Panhandle and n. cen. Fla. The aforementioned characters used to distinguish *V. munsoniana* from *V. rotundifolia*, particularly leaf size, appear to intergrade abundantly, making assignment of intermediate individuals to one species or the other most difficult. Studies are needed to determine the most practical means of treating *V. munsoniana*, but it seems that this taxon, at the very best, is deserving of only varietal status under *V. rotundifolia*.

6. Vitis shuttleworthii House. CALUSA GRAPE.

Moderately high-climbing vigorous vine, branchlets of the season oval to terete, densely tomentose when young, becoming more thinly tomentose with age. Bark exfoliating in shreds on 2 year old stems, lenticels absent or inconspicuous, pith brown, interrupted by diaphragms at nodes, diaphragms typically 10 mm thick but frequently continuing halfway into the internode. Tendrils bifurcate to

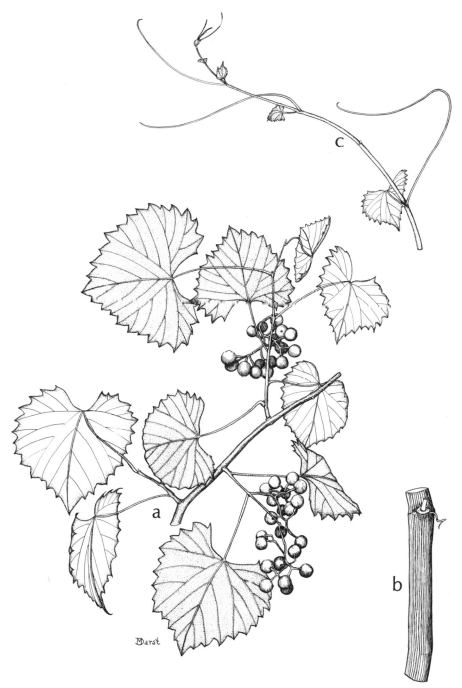

Fig. 354. **Vitis rotundifolia:** a, fruiting branch (from form with relatively small fruits); b, piece of woody stem; c, elongating branch with tendrils.

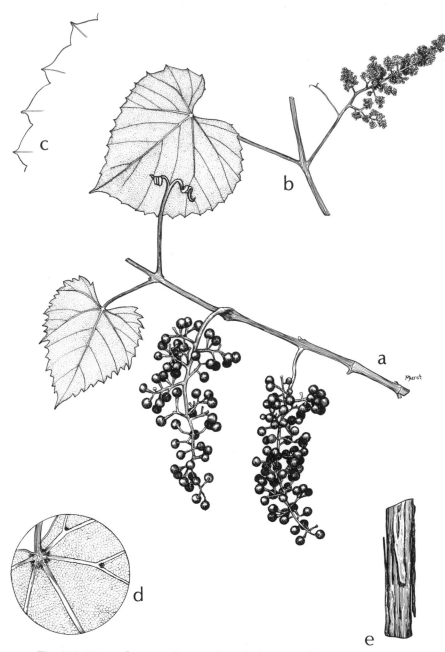

Fig. 355. **Vitis vulpina:** a, fruiting branch; b, piece of branch with inflorescence; c, leaf margin enlarged to accentuate toothing; d, portion of lower surface of leaf, enlarged; e, piece of woody stem.

trifurcate, a tendril or inflorescence present opposite only 2 consecutive nodes, nodes not glaucous, not banded with red pigmentation.

Petioles about half to three-quarters the length of the blade, densely tomentose. Blades broadly cordate to nearly reniform, typically unlobed but varying to 3-shouldered or, less often, deeply 3–5-lobed, when lobed the lobes acute and the sinuses rounded; margins with shallow, broadly scalloped, obtuse teeth, typically nearly entire, leaf bases cordate to truncate; upper surface of mature leaves floccose to glabrous, lower surface not glaucous but densely and evenly covered with white to rusty tomentum, typically concealing the leaf undersurface but not always the veins; stipules minute, 1 mm or less long, promptly deciduous.

Panicles 4–10 cm long, the rachis arachnoid floccose, usually broadly short triangular in outline, infructescences with fewer than 20 berries (or pedicels). Berries large, 9–18 mm in diameter, dark red to purple-black, with little or no glaucescence, lenticels absent. Seeds dark brown, ovoid to rounded, 5–6 mm long.

Generally found in woodlands of various mixtures, woodland borders, thickets and lowland woods in peninsular Fla. This species is recorded from Volusia, Marion and Citrus Cos. and south, just peripheral to our area, but could possibly occur further north, within our area. (Endemic to pen. Fla.) Flowers early April to early May, fruits June to August.

The mustang grape, *V. mustangensis* Buckley, is closely related and similar to *V. shuttleworthii*, but is a more western species that also occurs in Lowndes and Wilcox counties in Alabama and could possibly occur within our range. It can be distinguished from *V. shuttleworthii* by larger, more prominent stipules, measuring 2–3 mm in length.

7. Vitis vulpina L. WINTER, FROST, OR CHICKEN GRAPE.   Fig. 355

High-climbing vine, branchlets of the season slightly angled when very young but becoming terete, very young stems and emerging leaves glabrous to sparsely arachnoid pubescent. Bark exfoliating in shreds on mature stems, lenticels absent or inconspicuous, pith brown, interrupted by nodal diaphragms, diaphragms 1–2 mm thick. Tendrils bifurcate, a tendril or inflorescence present at 2 consecutive nodes only, nodes not glaucous, not banded with red pigmentation.

Petioles about as long as the blades, sparsely to moderately pubescent with hirtellous trichomes or glabrous. Blades cordiform, often 3-shouldered to shallowly 3-lobed, deeply lobed only on ground shoots; margins irregularly dentate-serrate, bases typically cordate, apices acute to short acuminate; upper surface of mature leaves typically glabrous to very sparse hirtellous pubescent, often lustrous, lower surface not glaucous, typically green, with short, straight hirtellous pubescence along the veins and in their axils, varying to more or less glabrous or also with very sparse arachnoid pubescence; stipules 2–4 mm long.

Panicles 10–17 cm long, usually narrowly triangular in general outline, infructescences typically with more than 20 berries (or pedicels). Berries 5–10 mm in diameter, black, very slightly, or more typically, not at all glaucous, lenticels absent. Seeds dark brown, ovoid, 3–5 mm long. [*V. cordifolia* Michx.]

In upland, well-drained woodlands of various mixtures, woodland borders, fence and hedge rows, thickets, less commonly in floodplains or lowland woods; more or less throughout our area. (S.e. N.Y. to Mo. and e. Kan., generally southward to pen. Fla. and n. cen. Tex.) Flowers May; fruits July to August in our area.

References

Adams, R. P. 1986. "Geographic Variation in *Juniperus silicicola* and *J. virginiana* of the Southeastern United States: Multivariant Analyses of Morphology and Terpenoids." *Taxon* 35:61–75.

Anderson, L. C. 1985. "*Forestiera godfreyi*, a New Species from Florida and South Carolina." *Sida* 11:1–5.

Argus, G. W. 1980. "The Typification and Identity of *Salix eriocephala* Michx. (Salicaceae)." *Brittonia* 32:170–177.

_____. 1984. "*Salix occidentalis* Walter, the Correct Name for *S. tristis* Aiton (Salicaceae)." *Brittonia* 36:328–329.

_____. 1986. "The Genus *Salix* (Salicaceae) in the Southeastern United States." *Systematic Botany Monographs* 9:1–170.

Ashby, W. C. 1964. "A Note on Basswood Nomenclature." *Castanea* 29:109–116.

Ashe, W. W. 1931. "Polycodium." *Jour. Elisha Mitch. Sci. Soc.* 46:196–213.

Bailey, L. H. 1944. "Revision of the American Palmettoes." *Gentes Herbarum* 6:365–459.

Baum, Bernard R. 1967. "Introduced and Naturalized Tamarisks in the United States and Canada (Tamaricaceae)." *Baileya* 15:19–25.

Boothroyd, Lucy E. 1930. "The Morphology and Anatomy of the Inflorescence and Flower of the Platanaceae." *Am. Jour. Bot.* 17:678–693.

Braun, E. L. 1950. *The Deciduous Forests of Eastern North America*. Philadelphia: Blakiston.

Brizicky, G. K. 1962. "The Genera of Rutaceae in the Southeastern United States." *Jour. Arn. Arb.* 43:1–22.

_____. 1962. "The Genera of Anacardiaceae in the Southeastern United States." *Jour. Arn. Arb.* 43:359–375.

_____. 1963. "The Genera of Sapindales in the Southeastern United States." *Jour. Arn. Arb.* 44:462–501.

_____. 1964. "The Genera of Celastrales in the Southeastern United States." *Jour. Arn. Arb.* 45:206–234.

_____. 1964. "The Genera of Rhamnaceae in the Southeastern United States." *Jour. Arn. Arb.* 45:439–463.

_____. 1965. "The Genera of Vitaceae in the Southeastern United States." *Jour. Arn. Arb.* 46:48–67.

_____. 1966. "The Genera of the Sterculiaceae in the Southeastern United States. *Jour. Arn. Arb.* 47:60–74.

Brown, K. E. 1976. "Ecological Studies of the Cabbage Palm, *Sabal palmetto*." *Principes* 20:3–10, 49–56, 98–115, 148–157.

Channell, R. B., and C. E. Wood, Jr. 1962. "The Leitneriaceae in the Southeastern United States." *Jour. Arn. Arb.* 43:435–438.

Clark, Robert B. 1942. "A Revision of the Genus *Bumelia* in the United States." *Ann. Mo. Bot. Gard.* 29:155–182.

Clark, Ross C. 1971. "The Woody Plants of Alabama." *Ann. Mo. Bot. Gard.* 58:99–242.

Clewell, A. F. 1966. "I. Identification of the Lespedezas in North America; II. A Selected Bibliography on *Lespedeza*." *Bull. Tall Timbers Res. Sta.* No. 7, Tallahassee, Fla.

————. 1977. "Geobotany of the Apalachicola River Region." *Fla. Marine Res. Bull.* 26:6–15.

————. 1981. "Natural Setting and Vegetation of the Florida Panhandle." Report submitted under Contract No. DACW 01-77-C-0104, U.S. Army Corps of Engineers, Mobile, Ala.

————. 1985. *Guide to the Vascular Plants of the Florida Panhandle.* Tallahassee: University Presses of Florida.

Coker, W. C., and H. R. Totten. 1934. *Trees of the Southeastern States.* Chapel Hill: University of North Carolina Press.

————. 1944. "The Woody Smilaxes of the United States." *Jour. Elisha Mitch. Sci. Soc.* 60:27–62.

Correll, D. S., and M. C. Johnston. 1970. *Manual of the Vascular Plants of Texas.* Renner: Texas Research Foundation.

————, and Helen B. Correll. 1972. *Aquatic and Wetland Plants of Southwestern United States.* Environmental Protection Agency, Washington. Reprint (in 2 vols.) Stanford, Calif.: Stanford University Press. 1975.

Cronquist, Arthur. 1945. "Studies in the Sapotaceae. III. *Dipholis and Bumelia.*" *Jour. Arn. Arb.* 25:435–471.

————. 1980. *Vascular Flora of the Southeastern United States.* Vol. I. *Asteraceae.* Chapel Hill: University of North Carolina Press.

Davis, D. E., and N. D. Davis. 1965. *Guide and Key to Alabama Trees.* Dubuque, Iowa: Kendall/Hunt.

Duncan, W. H., and Nell E. Brittain. 1966. "The Genus *Gaylussacia* (Ericaceae) in Georgia." *Bull. Georgia Acad. of Sci.* 24:13–26.

————. 1974. *Woody Vines of the Southeastern United States.* Athens: University of Georgia Press.

Eckenwalder, J. E. 1980. "The Taxonomy of the West Indian Cycads." *Jour. Arn. Arb.* 61:701–722.

Edmisten, J. A. 1963. "The Ecology of Florida Pine Flatwoods." Ph.D. dissertation. University of Florida, Gainesville.

Elias, T. S. 1970. "The Genera of Ulmaceae in the Southeastern United States." *Jour. Arn. Arb.* 51:18–40.

————. 1971. "The Genera of Fagaceae in the Southeastern United States." *Jour. Arn. Arb.* 52:159–195.

————. 1972. "The Genera of the Juglandaceae in the Southeastern United States." *Jour. Arn. Arb.* 53:26–51.

————. 1980. *The Complete Trees of North America. Field Guide and Natural History.* New York: Van Nostrand Reinhold.

Ernst, W. R. 1964. "The Genera of Berberidaceae, Lardizabalanaceae, and Menispernaceae in the Southeastern United States." *Jour. Arn. Arb.* 45:1–35.

Ewel, K. C., and H. T. Odum, eds. 1984. *Cypress Swamps.* Gainesville: University Presses of Florida.

Eyde, R. H. 1966. "The Nyssaceae in the Southeastern United States." *Jour. Arn. Arb.* 47:117–125.

Faircloth, W. R. 1971. "The Vascular Flora of Central South Georgia." Ph.D. dissertation. University of Georgia, Athens.

Fernald, E. A., ed. 1981. *Atlas of Florida.* Tallahassee: Florida State University Foundation.

Fernald, M. L. 1950. *Gray's Manual of Botany*, 8th ed. New York: American Book Co.

Ferguson, I. K. 1966. "The Cornaceae in the Southeastern United States." *Jour. Arn. Arb.* 47:106–116.

———. 1966. "The Genera of Caprifoliaceae in the Southeastern United States." *Jour. Arn. Arb.* 47:33–59.

Garren, K. H. 1943. "Effects of Fire on Vegetation of the Southeastern United States." *Bot. Rev.* 9:617–654.

Gillis, William T. 1971. "The Systematics and Ecology of Poison-ivy and the Poison-oaks." *Rhodora* 73:72–159; 161–237; 370–443; 465–540.

Gonsoulin, G. J. 1974. "A Revision of *Styrax* (Styracaceae) in North America, Central America, and the Caribbean." *Sida* 5:191–258.

Gleason, H. A. 1952. *The New Britton and Brown Illustrated Flora of the Northeastern United States and Adjacent Canada.* 3 vols. New York: New York Botanical Garden.

Graham, Shirley A. 1964. "The Genera of Lythraceae in the Southeastern United States." *Jour. Arn. Arb.* 45:235–250.

———. 1964. "The Genera of Rhizophoraceae and Combretaceae in the Southeastern United States." *Jour. Arn. Arb.* 45:285–301.

———. 1966. "The Genera of Araliaceae in the Southeastern United States." *Jour. Arn. Arb.* 47:126–136.

———, and C. E. Wood, Jr. 1965. "The Genera of Polygonaceae in the Southeastern United States." *Jour. Arn. Arb.* 46:91–121.

Hardin, James W. 1971. "Studies of the Southeastern United States Flora. I. Betulaceae." *Jour. Elisha Mitch. Sci. Soc.* 87:39–47.

———. 1971. "Studies of Southeastern United States Flora. II. The Gymnosperms." *Jour. Elisha Mitch. Sci. Soc.* 87:43–50.

———. 1972. "Studies in the Southeastern United States Flora. III. Magnoliaceae and Illiciaceae." *Jour. Elisha Mitch. Sci. Soc.* 88:30–32.

———. 1974. "Studies of the Southeastern United States Flora. IV. Oleaceae." *Sida* 5:274–285.

———. 1979. "*Quercus prinus* L.—*nomen ambiguum.*" *Taxon* 28:355–357.

Harper, R. M. 1914. "Geography and Vegetation of Northern Florida." *Ann. Re. Fla. Geol. Surv.* 6: 163–416.

Harrar, E. S., and J. G. Harrar. 1962. *Guide to Southern Trees.* 2d ed. New York: Dover Publications.

Hickok, L. G., and J. C. Anway. 1972. "A Morphological and Chemical Analysis of Geographical Variation in *Tilia* L. of Eastern North America." *Brittonia* 24: 2–8.

Isely, D. 1969. "Legumes of the United States. I. Native *Acacia.*" *Sida* 3:365–386.

Johnston, M. C. 1957. "Synopsis of the United States Species of *Forestiera* (Oleaceae)." *Southwestern Naturalist* 4:140–151.

Jones, G. N. 1968. "Taxonomy of American Species of Linden (*Tilia*)." *Illinois Biol. Monogr.* 39.

Judd, Walter S. 1981. "A Monograph of *Lyonia* (Ericaceae)." *Jour. Arn. Arb.* 62:1–209; 315–436.

Keener, Carl S. 1975. "Studies in the Ranunculaceae of the Southeastern United States. III. Clematis L." *Sida* 6:33–47.

Kral, Robert. 1960. "A Revision of *Asimina* and *Deeringothamnus* (Annonaceae)." *Brittonia* 12:233–278.

Kurz, H. 1942. "Florida Dunes and Scrub, Vegetation and Geology." *Fla. Geol. Surv. Bull.* 23:1–154.

———, and R. K. Godfrey. 1962. *Trees of Northern Florida*. Gainesville: University of Florida Press.

Lakela, Olga. 1963. "The Identity of *Bumelia lacuum* Small." *Rhodora* 65:280–282.

———, and R. P. Wunderlin. 1980. *Trees of Central Florida*. Miami: Banyan Books.

Laessle, A. M. 1958. "The Origin and Successional Relationship of Sandhill Vegetation and Sand-pine Scrub." *Ecol. Monogr.* 28:361–387.

Li, Hui-Lin. 1962. "A New Species of *Chamaecyparis.*" *Morris Arb. Bull.* 13:43–46.

Little, E. L., Jr., and K. W. Dorman. 1952. "Slash Pine (*Pinus elliottii*): Its Nomenclature and Varieties." *Jour. of Forestry* 50:918–923.

———. 1953. *Check List of Native and Naturalized Trees of the United States (Including Alaska)*. U.S.D.A., Forest Service, Agric. Handbook no. 41.

———. 1971. *Atlas of United States Trees*. Vol. 1, *Conifers and Important Hardwoods*. U.S.D.A., Forest Service, Misc. Pub. no. 1146.

———. 1977. *Atlas of United States Trees*. Vol. 4, *Minor Eastern Hardwoods*. U.S.D.A., Forest Service, Misc. Pub. no. 1342.

McDaniel, J. C. 1966. "Variations in the Sweetbay Magnolia." *Morris Arb. Bull.* 17:7–12.

Miller, R. F. 1975. "The Deciduous Magnolias of West Florida." *Rhodora* 77:64–75.

Monk, C. D. 1965. "Southern Mixed Hardwood Forest of Northcentral Florida." *Ecol. Monogr.* 35:335–354.

Nicely, Kenneth A. 1965. "A Monographic Study of the Calycanthaceae." *Castanea* 30:38–81.

Peattie, D. C. 1950. *A Natural History of Trees of Eastern and Central North America*. Boston: Houghton Mifflin.

Prance, G. T. 1970. "The Genera of Chrysobalanaceae in the Southeastern United States." *Jour. Arn. Arb.* 51:521–528.

Quaterman, E., and C. Keever. 1962. "Southern Mixed Hardwood Forest: Climax in the Southeastern Coastal Plain." *Ecology* 31:234–254.

Radford, A. E., Harry E. Ahles, and C. R. Bell. 1964. *Manual of the Vascular Flora of the Carolinas*. Chapel Hill: University of North Carolina Press.

Reed, Clyde. 1970. *Selected Weeds of the United States*. U.S.D.A., Agr. Res. Ser., Agr. Handbook no. 336.

Robertson, K. R. 1974. "The Genera of Rosaceae in the Southeastern United States." *Jour. Arn. Arb.* 55:303–332; 344–401; 611–662.

Rogers, D. J., and J. A. Mortensen. 1979. "The Native Grape Species of Florida." *Proc. Fla. State Hort. Soc.* 92:286–289.

Sargent, C. S. 1918. "Notes on North American Trees. III. *Tilia.*" *Bot. Gaz.* 66:421–438; 494–511.

———. 1933. *Manual of the Trees of North America (Exclusive of Mexico)*. 2d ed. Boston and New York: Houghton Mifflin Co.

Shinners, L. H. 1962. "Synopsis of *Conradina* (Labiatae)." *Sida* 1:84–88.

Shuey, A. G., and R. P. Wunderlin. 1977. "The Needle Palm: *Rhapidophyllum hystrix.*" *Principes* 21:47–59.

Small, J. K. 1933. *Manual of the Southeastern Flora*. Reprint. New York: Hafner, 1972.

Spongberg, S. A. 1971. "The Genera of Staphyleaceae in the Southeastern United States." *Jour. Arn. Arb.* 52:196–203.

_____. 1972. "The Genera of Saxifragaceae in the Southeastern United States." *Jour. Arn. Arb.* 53:409–498.

_____. 1976. "Magnoliaceae Hardy in Temperate North America." *Jour. Arn. Arb.* 57:250–312.

Steyermark, Julian A. 1949. "*Lindera melissaefolia*." *Rhodora* 51:153–162.

Stone, D. E. 1968. "Cytological and Morphological Notes on the Southeastern Endemic *Schisandra glabra* (Schisandraceae)." *Jour. Elisha Mitch. Sci. Soc.* 84:351–356.

Thomas, J. L. 1960. "A Monographic Study of the Cyrillaceae." *Contr. Gray Herb.* 186:1–114.

_____. 1961. "The Genera of the Cyrillaceae and Clethraceae of the Southeastern United States." *Jour. Arn. Arb.* 42:96–106.

Thorne, R. F. 1954. "The Vascular Plants of Southwestern Georgia." *Am. Midl. Naturalist* 52:257–327.

Tomlinson, P. B. 1960. "Seedling Leaves in Palms and Their Morphological Significance." *Jour. Arn. Arb.* 41:414–428.

_____. 1980. *The Biology of Trees Native to Tropical Florida*. Pub. by the author.

Vander Kloet, S. P. 1980. "The Taxonomy of the Highbush Blueberry, *Vaccinium corymbosum*." *Can. Jour. Bot.* 58:1187–1201.

Wagner, W. H., Jr. 1975. "Notes on the Floral Biology of Box-elder (*Acer negundo*)." *Michigan Botanist* 14:73–82.

Ward, D. B. 1974. "Contributions to the Flora of Florida—6, *Vaccinium* (Ericaceae)." *Castanea* 39:191–205.

_____, and J. B. Burkhalter. 1977. "Rediscovery of Small's *Acacia* in Florida." *Fla. Scientist* 40:267–270.

_____, ed. 1979. *Rare and Endangered Biota of Florida*. Vol. 5, *Plants*. Gainesville: University of Florida Presses.

Watson, F. D. 1983. "A Taxonomic Study of Pondcypress and Baldcypress." Ph.D. dissertation. North Carolina State University, Raleigh.

Webster, G. L. 1967. "The Genera of Euphorbiaceae in the Southeastern United States." *Jour. Arn. Arb.* 48:303–430.

West, Erdman, and L. E. Arnold. 1956. *The Native Trees of Florida*. Rev. ed. Gainesville: University of Florida Press.

Wharton, C. H. 1978. *The Natural Environments of Georgia*. Atlanta: Geologic and Water Resources Division and Resource Planning Section, Office of Planning and Research, Georgia Department of Natural Resources.

Wilbur, R. L. 1963. *The Leguminous Plants of North Carolina*. N.C. Agricultural Exp. Sta. Bull. no. 151, Raleigh, N.C.

Wilson, K. A., and C. E. Wood, Jr. 1959. "The Genera of Oleaceae in the Southeastern United States." *Jour. Arn. Arb.* 40:369–384.

_____. 1965. "Variation of Three Taxonomic Complexes of the Genus *Cornus* in Eastern United States." *Trans. Kan. Acad. Sci.* 67:747–817.

Wood, C. E., Jr. 1959. "The Genera of Woody Ranales in the Southeastern United States." *Jour. Arn. Arb.* 39:296–346.

_____. 1959. "The Genera of Theaceae of the Southeastern United States." *Jour. Arn. Arb.* 40:413–419.

_____, and R. B. Channell. 1960. "The Genera of Ebenales in the Southeastern United States." *Jour. Arn. Arb.* 41:1–35.

_____. 1961. "The Genera of Ericaceae in the Southeastern United States." *Jour. Arn. Arb.* 42:10–80.

713

————, and Preston Adams. 1976. "The Genera of Guttiferae (Clusiaceae) in the Southeastern United States." *Jour. Arn. Arb.* 57:74–90.

Wunderlin, R. P. 1977. "The Florida Species of *Ilex* (Aquifoliaceae)." *Florida Scientist* 40:7–21.

————. 1982. *Guide to the Vascular Flora of Central Florida.* Tampa: University Presses of Florida.

Zona, Scott. 1983. "A Taxonomic Study of the *Sabal Palmetto* Complex (Palmae) in Florida." M.S. thesis. University of Florida, Gainesville.

Index to Common Names

Running oak, 317
Rusty black-haw, 191

S

Sage-tree, 693
St. John's-wort[s], 346
  family, 346
St. Peter's-wort[s], 346,
  349
Salt-cedar, 668
Saltwort, 159
  family, 159
Sand
  blackberry, 493
  hickory, 397
  holly, 153
  live oak, 313
  pine, 45
  post oak, 306
Sapodilla family, 630
Sarsaparilla
  -vine, 91
  wild, 87
Sarvis holly, 149
Sassafras, 417
Saw palmetto, 76
Saxifrage family, 641
Scalybark hickory, 395
Scarlet
  basil, 401
  maple, 101
  oak, 344
Scrub
  hickory, 390
  holly, 136
  palmetto, 83
Sea
  -myrtle, 203
  -oxeye, 206
Sebastian-bush, 285
Serviceberry
  downy, 545
Sesban[s], 439
  purple, 441
Seven-bark, 643, 644
Seville, orange, 609
Shagbark hickory, 395
Shellbark hickory, 385
Shining sumac, 110
Shiny blueberry, 275
Shortleaf pine, 47
Shrub
  highwater-, 213

strawberry-, 177
sweet-, 177
-verbena, 691
Shumard oak, 342
Silk
  bay, 415
  -tree, 421
Silkgrass, 72
Silky-camellia, 670
Silverbell[s], 660
  little, 662
  two-wing, 660
Silverling, 203
Silver maple, 103
Simpson's grape, 702
Skunk-bush, 609
Slash pine, 49
Slippery elm, 687
Small
  -fruited hawthorn, 561
  pussy willow, 620
  viburnum, 191
Small's
  acacia, 421
  lupine, 433
Smartweed family, 521
Smooth
  bumelia, 636
  sumac, 112
Snailseed, 475
Snowbell
  American, 664
  big-leaf, 666
Snowberry, 601
Soapberry
  family, 628
  Florida, 628
Soft maple, 103
Sour
  gum, 492, 494
    family, 490
  orange, 609
Sourwood, 256
Southern
  bayberry, 485
  black-haw, 191
  catalpa, 171
  crabapple, 565
  dewberry, 595
  -gooseberry, 270
  magnolia, 469
  pinxterbloom, 265
  prickly-ash, 613
  red oak, 337
  sugar maple, 103

Spanish
  bayonet, 71
  dagger, 69
  oak, 337
Sparkleberry, 272
Spice
  pond-, 413
Spicebush, 411
Spire
  red-, 581
Spruce pine, 49
Spurge family, 278
Squaw-huckleberry, 270
Staff-tree family, 194
Stagger-bush, 251, 253,
  255
Star
  -anise, 383
    family, 381
  -vine, 650
    family, 650
Stem
  blue-, 80
Sterculia family, 656
Stewartia, 670
Stiff cornell or dogwood,
  217
Stink-bush, 381
Stinking
  -ash, 609
  cedar, 59
  laurel, 381
Storax family, 660
Storaxes, 662
Strawberry
  -bush, 194
  -shrub, 177
    family, 177
Sugar
  horse-, 666
Sugarberry, 677
Sumac
  family, 108
  fragrant, 108
  poison, 118
  shining, 110
  smooth, 112
  winged, 110
Summer
  -dogwood, 647
  grape, 700
Sump-weed, 213
  dune, 213
Sunflower family, 200
Supple-jack, 533

721

Index to Scientific Names

725

Vitis (continued)
  simpsonii Munson [1887, not 1890], 702
  vulpina L., 707

W

Wallia nigra (L.) Alef., 384
Wisteria, 444
  frutescens (L.) Poir. in Lam., 444
  sinensis (Sims) Sweet, 447

X

Xanthorhiza simplicissima Marsh., 531
Ximenia americana L., 496
Xolisma
  ferruginea (Walt.) Heller, 251
  fruticosa (Michx.) Nash, 253

Y

Yucca, 69
  aloifolia L., 69

filamentosa L. var. smalliana (Fern.)
    Ahles, 74
  flaccida Haw., 72
  gloriosa L., 71
  smalliana Fern., 74

Z

Zamia
  angustifolia Jacq., 68
  floridana A. DC., 68
  integrifolia Ait., 68
  pumila L., 66
  silvicola Small, 68
  umbrosa Small, 68
Zamiaceae, 66
Zanthoxylum, 611
  americanum Mill., 613
  clava-herculis L., 613
Ziziphus jujuba Mill., 541